The Principles of
Semiconductor Laser Diodes and Amplifiers

The Principles of
Semiconductor Laser Diodes and Amplifiers

Analysis and Transmission Line Laser Modeling

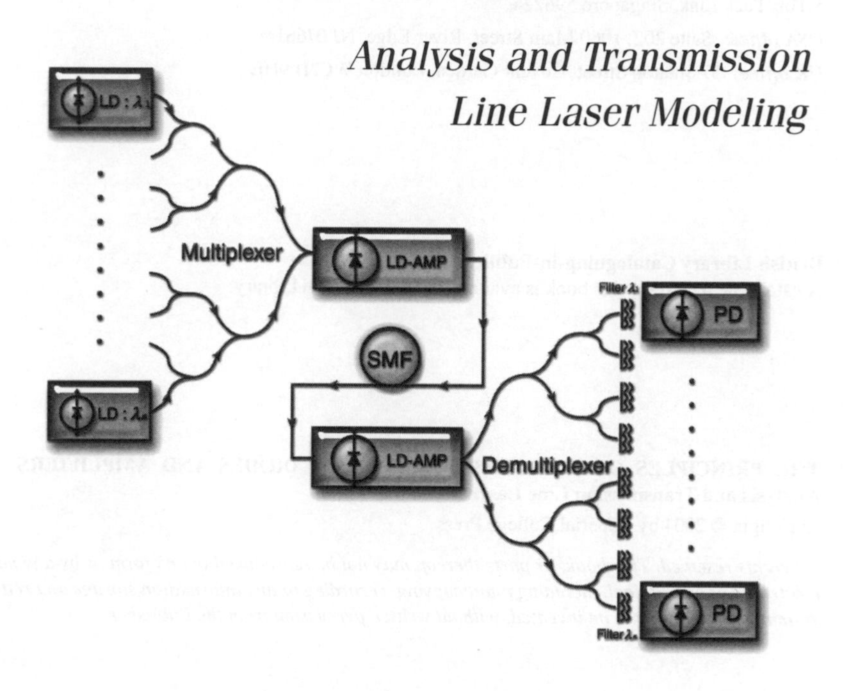

H Ghafouri-Shiraz
The University of Birmingham, UK
and
Nanyang Technological University, Singapore

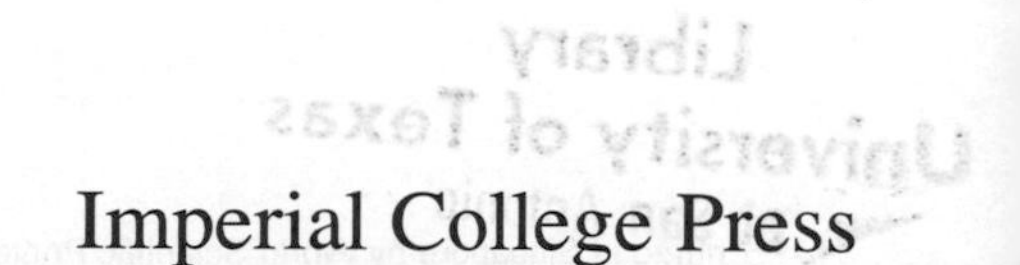

ICP
Imperial College Press

Published by

Imperial College Press
57 Shelton Street
Covent Garden
London WC2H 9HE

Distributed by

World Scientific Publishing Co. Pte. Ltd.
5 Toh Tuck Link, Singapore 596224
USA office: Suite 202, 1060 Main Street, River Edge, NJ 07661
UK office: 57 Shelton Street, Covent Garden, London WC2H 9HE

British Library Cataloguing-in-Publication Data
A catalogue record for this book is available from the British Library.

THE PRINCIPLES OF SEMICONDUCTOR LASER DIODES AND AMPLIFIERS
Analysis and Transmission Line Laser Modeling

ISBN 1-86094-339-X
ISBN 1-86094-341-1 (pbk)

Printed in Singapore by World Scientific Printers (S) Pte Ltd

This book is dedicated to

My Father, The Late Haji Mansour, for the uncompromising principles that guided his life.

My Mother, Rahmat, for leading her children into intellectual pursuits.

My Supervisor, The Late Professor Takanori Okoshi, for his continuous guidance, encouragement, inspiring discussion and moral support. A distinguished scientist and a great teacher who made me aware of the immense potential of optical fibre communications.

My Wife, Maryam, for her magnificent devotion to her family. My constant companion and best friend, she has demonstrated incredible patience and understanding during the rather painful process of writing this book while maintaining a most pleasant, cheerful and comforting home.

My Children, Elham, Ahmad-Reza and Iman, for making everything worthwhile.

To all of my Research and Undergraduate Students, for their excellent and fruitful research works, and for many stimulating discussions, which encouraged and motivated me to write this book.

Preface

It was in April 1976 that I published my first book entitled *Fundamentals of Laser Diode Amplifiers.* Since then we have witnessed rapid and dramatic advances in optical fiber communication technology. To provide a comprehensive and up-to-date account of laser diodes and laser diode amplifiers I decided to publish this new book, which in fact is an extensive extension to my above book. The main objective of this new book is also to serve both as a textbook and as a reference monograph. As a result, each chapter is designed to cover both physical understanding and engineering aspects of laser diodes and amplifiers.

With the rapid growth and sophistication of digital technology and computers, communication systems have become more versatile and powerful. This has given a modern communication engineer two key problems to solve: (i) how to handle the ever-increasing demand for capacity and speed in communication systems and (ii) how to tackle the need to integrate a wide range of computers and data sources, so as to form a highly integrated communication network with global coverage.

The foundations of communication theory show that by increasing the frequency of the carrier used in the system, both the speed and capacity of the system can be enhanced. This is especially true for modern digital communication systems. As the speeds of computers have increased dramatically over recent years, digital communication systems operating at a speed which can match these computers have become increasingly important. Rather than the electronic circuitry, it is now apparent that the upper bound on the speed of a communication system is limited by the transmission medium. An example which illustrates fast development in recent communication is that today's PC generally uses PCI bus as the electrical interconnect, which can provide data transfer rate up to 8.8 Gb/s. However, the speed of the modem normally connected to such a PC has just recently reached 8 Mb/s over copper lines using ADSL technology in commercial broadband access networks. This is at least 1100 times slower than the current electrical interconnect in the PC. One of the reasons for such mismatch is that modems use telephone lines (which are typically twisted-pair transmission lines) and these cannot operate at very high frequencies. To improve the speed and hence capacity of the system, one does not only need to switch to a carrier with a higher frequency, but to switch to an alternative transmission medium.

Given the preceding argument, one will not be surprised by the rapid development of optical communications during the past 30 years. Ever since Kao and his co-workers discovered the possibility of transmitting signals using light in circular dielectric waveguides, research in optical communication systems has developed at an unprecedented pace and scale. Optical communications offer two distinct advantages over conventional cable or wireless systems. Firstly, because the carrier frequency of light is in the region of THz (i.e. 10^{14} Hz), it is possible to carry many more channels than radio wave or even microwave systems. Secondly, the former advantage can be realised because of the development of a matching transmission medium, namely optical waveguides (including fibres and planar structures). Optical waveguides not only provide the necessary frequency bandwidth to accommodate a potentially large number of channels (and hence a huge capacity), but also offer an immunity from the electromagnetic interference from which the traditional transmission medium often suffers.

In addition to optical waveguides, another key area of technological development which plays a crucial role in the success of optical communication systems is optical devices. The rapid growth of semiconductor laser diodes has allowed optical transmitters to be miniaturised and become more powerful and efficient. Both the fabrication and theoretical research in semiconductor lasers have given rise to a wide range of components for optical communication systems. For example, from conventional buried heterostructure laser diodes to the recent development of multiple quantum-well lasers and from simple Fabry-Perot structures to (i) distributed feedback (DFB) structures, (ii) single cavity laser diodes and (iii) multiple cavity laser diodes. Laser diodes are not only important in compact disc players, but they also provide coherent light sources which are crucial in enhancing the speed and range of transmission of optical communication systems.

The technological forces which gave us optical waveguides and semiconductor laser diodes have recently explored theoretical research and manufacturing technology to develop further innovative devices that are crucial in optical communications, for example, optical amplifiers, optical switches and optical modulators. Previously *optical/electronic* conversion devices had to be used for performing these functions, but the bandwidth of these was limited. The integration of semiconductor laser diodes with optical waveguide technology allows such components to be developed specifically for optical communications. This force of integration does not stop here. The advent of photonic integrated circuits (PIC), which are ICs built entirely with optical components, such as laser diodes, waveguides and modulators, will further enhance the power and future prospects of optical communication networks.

In view of the increasing pace of development and growing importance of optical communication technology, I believe students, researchers and practicing engineers should be well equipped with the necessary theoretical foundations for this technology, as well as acquiring the necessary skills in applying this basic theory to a wide range of applications in optical communications. There are of course many good books about optical communication systems, but they seldom direct their readers to concentrate on the two key aspects behind the success in optical communications which we have discussed above. I am attempting to fill this gap with this book. I will be concentrating on the basic theory of optical waveguides and semiconductor laser technology, and I will illustrate how these two aspects are closely related to each other. In particular, I will examine how semiconductor laser amplifiers have been developed based on applications of the basic theory of these two areas.

Throughout this book, it is intended that the reader gains both a basic understanding of optical amplification and a factual knowledge of the subject based on device analysis and application examples. I hope that this book will be beneficial to students aiming to study optical amplification, and to the active researchers at the cutting edge of this technology. This book is organised as follows:

Chapter 1 explores the state of the art of optical fibre communication systems in this rapidly evolving field. A short introduction includes the historical development, the principles and applications of semiconductor laser amplifiers in optical fibre communications, the general optical system and the major advantages provided by this technology. In Chapter 2, the fundamentals and important performance characteristics of optical amplifiers will be outlined. Chapter 3 gives an introduction to optical amplification in semiconductor laser diodes. Chapters 4 to 6 deal with the analysis of semiconductor laser amplifiers (SLAs). In these chapters the waveguiding properties and the basic performance characteristics of SLAs (i.e. amplifier gain, gain saturation and noise) will be studied. Also a new technique, which is based on an equivalent circuit model, will be introduced for the analysis of SLAs. Implications of SLAs on optical fibre communication system performance will also be discussed. In Chapter 7 the accuracy and limitations of the equivalent circuit model will be investigated by comparing both theoretical and experimental results for actual devices. In Chapter 8 we introduce a new semiconductor laser diode amplifier structure. Chapter 9 deals with amplification characteristics of pico-second Gaussian pulses in various amplifier structures. Chapter 10 studies the sub-pico-second gain dynamic in a highly index-guided tapered-waveguide laser diode amplifier. In Chapter 11 we introduce a novel approximate analytical expression for saturation intensity of tapered travelling-wave semiconductor laser amplifier structures. Wavelength conversion using cross-gain modulation in linear tapered-waveguide semiconductor laser amplifiers is studied in Chapter 12.

The main theme of the work presented in Chapters 13 to 17 is microwave circuit principles applied to semiconductor laser modelling. The advantages and additional insight provided by circuit models that have been used for *analytical* analysis of laser diodes have long been acknowledged. In these chapters, we concentrate on the derivation, implementation, and application of *numerical* circuit-based models of semiconductor laser devices.

Design automation tools are playing an increasingly important role in today's advanced photonic systems and networks. A good photonic computer aided design (PCAD) package must include a model of the semiconductor laser, one of the key optoelectronic devices in fibre-optic communications. In this part of the book, the feasibility and advantages of applying microwave circuit techniques to semiconductor laser modelling for PCAD packages are investigated.

Microwave circuit models allow us to explore fundamental properties of electromagnetic waves without the need to invoke rigorous mathematical formulations. These equivalent circuit models are easy to visualise, providing a simple and clear physical understanding of the device. Two types of circuit models for semiconductor laser devices have been investigated: (i) lumped-element model, and (ii) distributed-element model based on transmission-line laser modelling (TLLM). The main differences between the lumped circuit and distributed circuit models have been compared in this book.

Most other dynamic models of laser diodes have failed to consider the high-frequency parasitics effect and impedance matching. These microwave aspects of the laser diode can be conveniently included in microwave circuit models. The matching network has been, for the first time, included in the integrated TLLM model, based on monolithically integrated lumped elements. The parasitics effect and matching considerations have been included in both small-signal and large-signal RF modulation of the laser transmitter module. The carrier dependence of the laser impedance within the TLM network has also been investigated.

Computational intensive two-dimensional (2-D) models of tapered laser devices are unattractive for PCAD packages. An efficient 1-D dynamic model of tapered structure semiconductor lasers has been developed based on TLLM, in which a semi-analytical approach was introduced to further enhance the computational efficiency. The tapered structure transmission-line laser model (TS-TLLM) includes inhomogeneous effects in both lateral and longitudinal directions, and is used to study picosecond pulse amplification. Previous models of tapered semiconductor amplifier structures failed to consider residual reflectivity but in TS-TLLM, reflections have been taken into account. Furthermore, the stochastic nature of TS-TLLM allows the influence of noise to be studied.

The TS-TLLM developed in this book has been combined with other existing TLLM models to form a multisegment mode-locked laser incorporating distributed Bragg reflectors, and a tapered semiconductor amplifier. This novel design can be used to generate high-power mode-locked optical pulses for various applications in fibre-optic systems. Important design considerations and optimum operating conditions of the novel device have been identified in conjunction with the RF detuning characteristics. A new parameter to define stable active mode-locking, or locking range, is discovered. Microwave circuit models of semiconductor laser devices provide a useful aid for microwave engineers, who wish to embark on the emerging research area of microwave photonics, and bring on a fresh new perspective for those already in the field of optoelectronics.

In Chapter 13, first, a short historical background and the relevant physics behind the semiconductor laser will be given. Chapter 14 introduces the transmission-line matrix (TLM) method that provides the basic microwave circuit concepts used to construct the time-domain semiconductor laser model known as the transmission-line laser model (TLLM). We then proceed to compare two categories of equivalent circuit models, i.e. lumped-element and distributed-element, of the semiconductor laser in Chapter 15. In the same chapter, a comprehensive laser diode transmitter model is developed for microwave optoelectronic simulation. The microwave optoelectronic model is based on the transmission-line modelling technique, which allows propagation of optical waves, as well as lumped electrical circuit elements, to be simulated. In Chapter 16, the transmission-line modelling technique is applied to a new time-domain model of the tapered waveguide semiconductor laser amplifier, useful for investigating short pulse generation and amplification when finite internal reflectivity is present. The new dynamic model is based on the strongly index-guided laser structure, and quasi-adiabatic propagation is assumed. Chapter 17 demonstrates the usefulness of the microwave circuit modelling techniques that have been presented in this thesis through a design study of a novel mode-locked laser device. The novel device is a multisegment monolithically integrated laser employing distributed Bragg gratings and a tapered waveguide amplifier for high power ultrashort pulse generation. Finally, Chapter 18 is devoted to some concluding remarks and comments. The book is referenced throughout by extensive end-of-chapter references which provide a guide for further reading and indicate a source for those equations and/or expressions which have been quoted without derivation.

The principal readers of this book are expected to be undergraduate and postgraduate students who would like to consolidate their knowledge in lightwave technology, and also researchers and practicing engineers who need to equip themselves with the foundations for understanding and using the continuing innovations in optical communication technologies. Readers are expected to be equipped with a basic knowledge of communication theory, electromagnetism and semiconductor physics.

Finally, I must emphasize that optical communication is still a rapidly growing technology with very active research. After reading the book, I hope that the reader will be equipped with the necessary skills to apply the most up-to-date technology in optical communications.

A/Prof. Dr. H. Ghafouri-Shiraz
June 2003, Birmingham, UK

Acknowledgements

I owe particular debts of gratitude to my former research students, Dr C. Y. J. Chu and Dr W. M. Wong, for their excellent research works on semiconductor laser diode and amplifiers. I am also very grateful indeed for the many useful comments and suggestions provided by colleagues and reviewers which have resulted in significant improvements to this book. Thanks also must be given to the authors of numerous papers, articles and books which I have referenced while preparing this book, and especially to those authors and publishers who have kindly granted permission for the reproduction of some diagrams. I am also very grateful to both my many undergraduate and postgraduate students who have helped me in my investigations.

A/Prof. Dr H. Ghafouri-Shiraz
June 2003, Birmingham, UK

Contents

Chapter 1

The Evolution of Optical Fibre Communication Systems

1.1 Introduction

The demand for high-capacity long-haul telecommunication systems is increasing at a steady rate, and is expected to accelerate in the next decade [1]. At the same time, communication networks which cover long distances and serve large areas with a large information capacity are also in increasing demand [2]. To satisfy the requirements on long distances, the communication channel must have a very low loss. On the other hand, a large information capacity can only be achieved with a wide system bandwidth which can support a high data bit rate (> Gbit/s) [3]. Reducing the loss whilst increasing the bandwidth of the communication channels is therefore essential for future telecommunications systems.

Of the many different types of communication channels available, optical fibres have proved to be the most promising [4, 5]. The first advantage of an optical fibre is its low attenuation. Typical values of attenuation factor in Modified Chemical Vapour Deposition (MCVD) optical fibres are plotted against wavelength of the electromagnetic carrier in Fig. 1.1 [6]. At present, optical fibres with loss coefficients of less than 0.25 dB/km around emission wavelengths of 1.55 μm are available [7]. This remarkable progress in fibre manufacturing technology has led to wide applications of long distance optical fibre communications in recent years. Furthermore, optical fibres can also transmit signals over a wide bandwidth because the electromagnetic carrier in optical fibres has a frequency in the optical frequency region ($\approx 10^{14}$ Hz). Hence, optical fibres can also carry many baseband channels, each with a bandwidth of the order of GHz using wavelength division

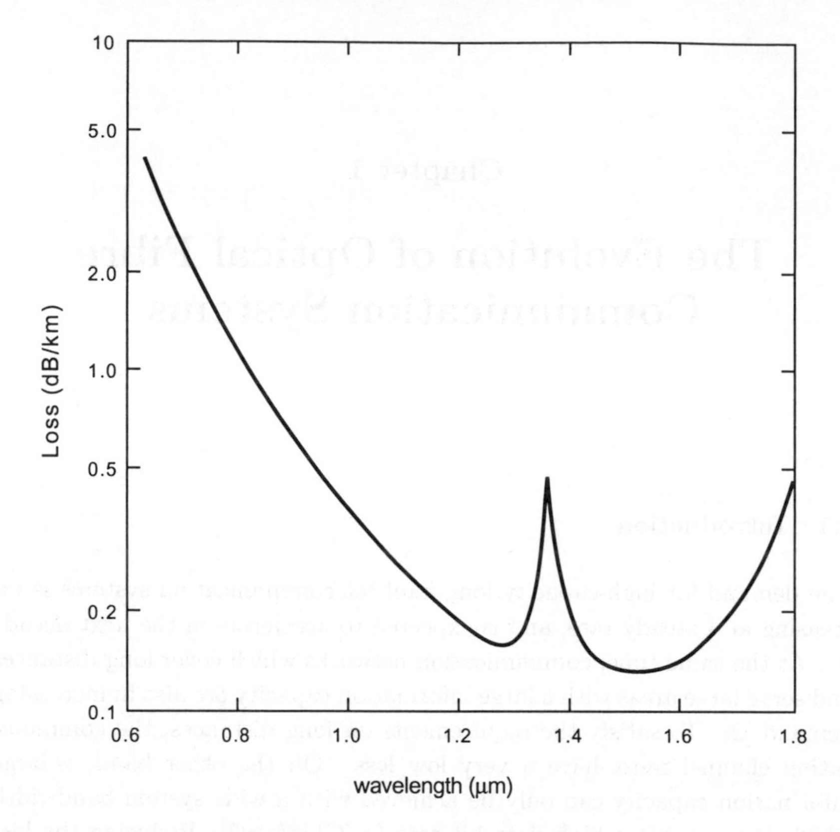

Fig. 1.1. Attenuation coeffiecient of MCVD optical fibres as a function of emission wavelength.

multiplexing (WDM) [8, 9]. For these reasons, optical fibre communication systems have attracted a lot of attention in recent years, and much research has been carried out to optimise their performance.

Figures 1.2(a) to (d), respectively, show the properties of various elements used in optical fibre communication systems, namely, the main materials and wavelengths used for different light sources, optical detectors, and optical amplifiers where there have been rapid recent advances. With semiconductor optical amplifiers, by changing the crystal composition the wavelength band (i.e. amplifiable waveband) can be selected as required from short to long wavelengths (see Fig. 1.2(c)). Furthermore, if a travelling wave device is used, broad band operation over 10 THz or so is possible. Rare-earth-doped optical fibre amplifiers, on the other hand, have an amplifiable waveband which is essentially determined by the dopant material,

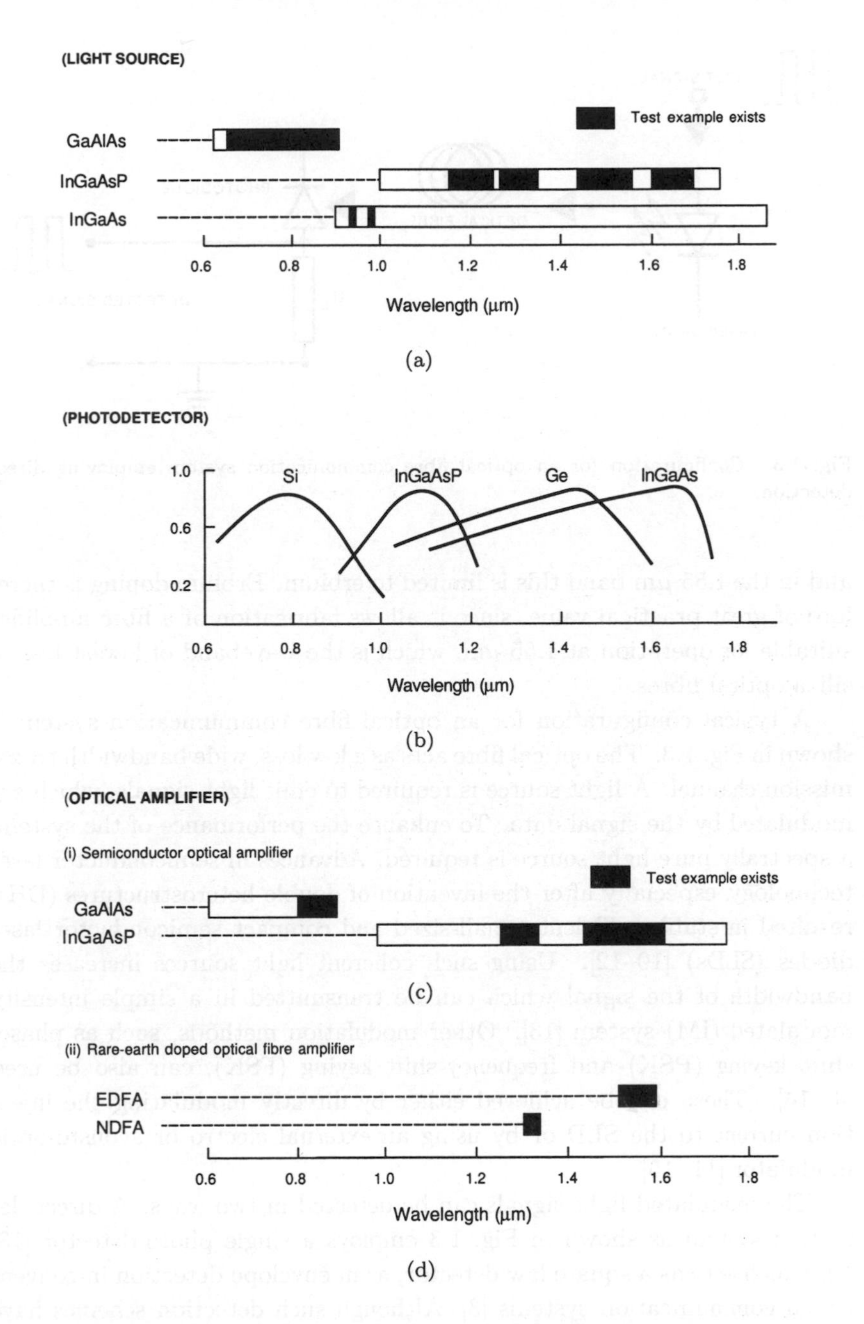

Fig. 1.2. Wavebands of components used in optical fibre communication systems. (after [28]).

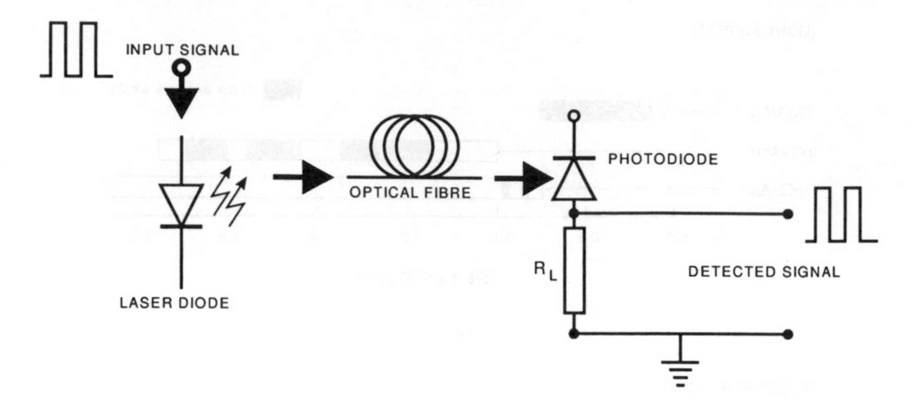

Fig. 1.3. Configuration for an optical fibre communication system employing direct detection.

and in the 1.55 μm band this is limited to erbium. Erbium doping is therefore of great practical value, since it allows fabrication of a fibre amplifier suitable for operation at 1.55 μm, which is the waveband of lowest loss in silica optical fibres.

A typical configuration for an optical fibre communication system is shown in Fig. 1.3. The optical fibre acts as a low loss, wide bandwidth transmission channel. A light source is required to emit light signals, which are modulated by the signal data. To enhance the performance of the system, a spectrally pure light source is required. Advances in semiconductor laser technology, especially after the invention of double heterostructures (DH), resulted in stable, efficient, small-sized and compact semiconductor laser diodes (SLDs) [10–12]. Using such coherent light sources increases the bandwidth of the signal which can be transmitted in a simple intensity modulated (IM) system [13]. Other modulation methods, such as phase-shift keying (PSK) and frequency-shift keying (FSK), can also be used [4, 14]. These can be achieved either by directly modulating the injection current to the SLD or by using an external electro or acousto-optic modulator [11, 15].

The modulated light signals can be detected in two ways. A direct detection system as shown in Fig. 1.3 employs a single photo-detector [13, 16] which acts as a square law detector, as in envelope detection in conventional communication systems [3]. Although such detection schemes have the inherent advantage of simplicity, the sensitivity of the receiver is limited [17]. In order to detect data transmitted across the optical fibre with a

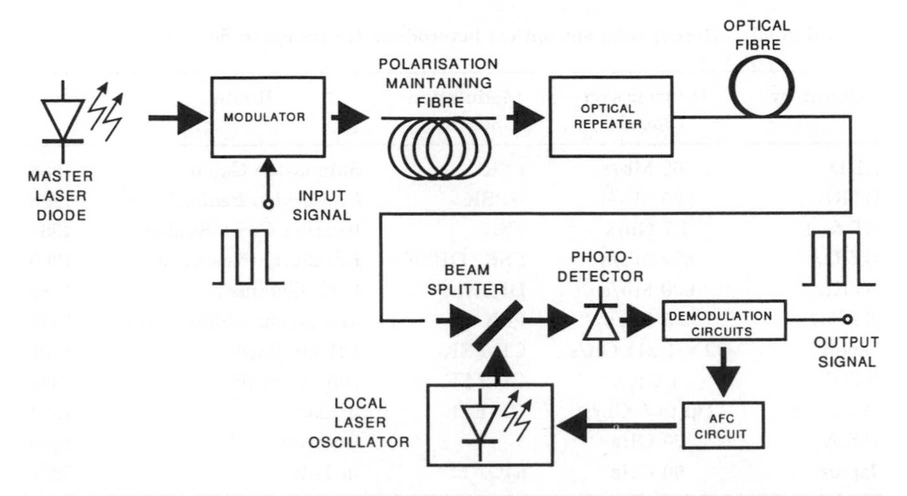

Fig. 1.4. Configuration for a coherent heterodyne optical fibre communication system.

higher bit-rate, the signal-to-noise ratio at the input to the receiver must be made as high as possible. In a system without repeaters, this will limit the maximum transmission span of the system [1]. An alternative detection method is to use coherent detection [4, 18] as shown in Fig. 1.4. By mixing the signal with a local oscillator at the input to the detector, it can be shown that a higher sensitivity can be achieved if the receiver is designed properly [5]. The principle is similar to that in a heterodyne radio [3]. In this system, one can easily, detect WDM transmission by tuning the local oscillator wavelength, as in a heterodyne radio system. In practice, however, because of the finite spectral width of the master and/or local oscillators which are usually SLDs, the limited tunability in SLDs and the extreme sensitivity of the receiver to the states of polarisation of the light signal will severely limit the performance of such complicated receivers [4]. Some of the recent field trials employing coherent detection are shown in Table 1.1 [5, 19–23].

Although coherent detection theoretically seems to offer a better performance for optical communications over direct detection, receivers employing this technique are very much at the research stage and their performance has yet to be improved [5]. On the other hand, many existing practical optical communication systems employ direct detection with intensity modulation. In order to use them for transmission of data with a higher data rate in the future, it is more economical if one can simply improve the input signal-to-noise ratio of the optical receiver instead of replacing or upgrading

Table 1.1. Recent coherent optical heterodyne transmission field experiments.

Laboratory	Transmission Speed	Modulation Scheme	Route	Year
KDD	565 Mb/s	FSK	Submarine Cable	1988
BTRL	565 Mb/s	DPSK	Cambridge-Bedford	1988
AT & T	1.7 Gb/s	FSK	Roaring Creek-Sunbury	1989
BTRL	622 Mb/s	FSK, DPSK	Edinburgh-Newcastle	1989
BTRL	620 Mb/s	DPSK	U.K.-Guernsey	1989
NTT	2.5 Gb/s	FSK	Matsuyama-Ohita-Kure	1990
Japan	32 × 1.244 Gb/s	CP-FSK	121 km SMF	1991
BTRL	4 Gb/s	CP-FSK	100 km SMF	1991
Denmark	Up to 5 Gb/s	CP-FSK	In Lab	1993
U.S.A.	26 GHz		Receiver	1995
Japan	60 GHz	64QAM	In Lab	2001

existing components in the systems like using new optical fibres or replacing the entire receiver using coherent detection with a new modulation scheme. In addition, the problem of retrieving WDM signals using direct detection has been overcome by using tunable optical filters, which are cheaper than tunable SLDs at the input of the receiver [1]. Hence, it appears that, if the input signal-to-noise ratio of the receivers can be improved, existing direct detection systems with intensity modulation can be used for transmissions with an even higher data rate.

The weak signal at the receiver in many optical communication systems arises because of the accumulation of losses along the optical fibres [1]. Although the loss can be as low as 0.2 dB/km for optical fibres operating around 1.55 μm, for a long transmission span this can build up to a significant loss, which will degrade signal power and hence the overall system performance [24]. Two ways of improving the signal-to-noise ratio of an optical receiver are possible. One can either boost the optical signal power along the transmission path using in-line repeaters [25], or boost the optical signal power at the input of the receiver by a pre-amplifier [26]. For many applications, both methods must be used to improve the system performance. In-line repeaters can be constructed using electronic circuits, which consist of photodetectors, electronic circuits for demodulation of the signals, amplification circuits for loss compensation, and laser diode driving circuits for regeneration. These conventional electronic repeaters are known commonly as regenerative repeaters. With them, the signal-to-noise ratio at the input of the receiver can indeed be improved. However, since the specification and subsequent design and configuration of this type of regenerator depends heavily on the modulation format, data bit-rate, multiplexing scheme and, in the case of optical networks, the number of branches emerging from a node, they are uneconomical because of their poor flexibility [27].

To solve the flexibility problem for in-line repeaters and to provide a pre-amplifier for optical receivers, one must be able to amplify light signals directly. Direct optical amplification avoids regeneration circuits in the in-line repeaters, so they can be used for any modulation format of the signal [28] and provides a maximum flexibility for applications in systems [27]. Repeaters employing such techniques are commonly known as non-regenerative repeaters, and the devices which perform such tasks are called optical amplifiers, or quantum amplifiers [29]. These optical amplifiers are usually called laser amplifiers because stimulated emissions are involved in the amplification process, which is also responsible for oscillations in lasers. These optical amplifiers can also be used as pre-amplifiers to receivers to enhance their sensitivities further [30]. Improvement in system performance by using optical fiber and laser diode amplifiers as in-line repeaters and/or pre-amplifiers to optical receivers has been reported in numerous experiments, some of which are tabulated in Tables 1.2 and 1.3 [1, 22–49].

The future prospects of long distance optical communication systems thus depend heavily on the availability of low-cost optical amplifiers which can compensate for the build-up of losses in optical fibre cables over long distances [2, 4]. Two types of optical amplifier exist: semiconductor laser amplifiers (SLAs) and fibre amplifiers (FAs). SLAs are essentially laser diodes operating in the linear amplification region below oscillation threshold [28, 5–51], whereas FAs are optical fibres doped with Erbium ions (Er^{+3}) to provide optical gain [24]. SLAs have the inherent advantage of compactness and the possibility of integration with other opto-electronic components, whereas FAs have the advantages of easy and efficient coupling with optical fibres. The design and analysis of both these types of optical amplifiers are therefore crucial for future development in optical fibre communication systems.

In this book, the principles and applications of semiconductor laser amplifiers in optical communications will be explored. In Chapter 2, the fundamentals and important performance characteristics of optical amplifiers will be outlined. An introduction to optical amplification in semiconductor lasers will be described in Chapter 3. A formal treatment of the analysis of semiconductor laser amplifiers will be given in Chapters 4

Table 1.2. Recent transmission experiments with erbium doped fibre amplifiers (EDFAs).

Year	Laboratory	Bit Rate (Gbit/s)	Distance (km)	Comments
1989	NTT	1.8	212	Booster + pre-amplifier used
1989	BTRL	0.565	—	DPSK system
1989	KDD	1.2	267	Two amplifiers used
1989	NTT	20	—	Soliton transmission
1989	Bell Core	11	260	Two amplifiers used
1989	Fujitsu	12	100	—
1989	KDD	1.2	904	12 amplifiers used
1990	NTT	2.5	2223	25 amplifiers used
1991	BTRL	2.5	10^4	Recirculating loop
1991	NTT	10	10^6	Soliton transmission, 12 amplifiers used
1991	NTT	20	500	Soliton transmission
1992	NTT	2.4	309	4 Repeaters + pre-amplifier
1992	NTT	10	309	4 repeaters + pre-amplifier
1993	BELL	10	309	EDFA + Dispersion Compensation
1994	BELL	16 × 2.5	1420	14 amplifiers
1995	BELL	2.5	374	1 local EDFA + 1 remotely-pumped EDFA + pre-amplifier
1997	BELL	32 × 10	640	9 Gain-flattened broadband EDFA with 35 nm Bandwidth (Total Gain 140 dB and total gain ununiformity 4.9 dB between 32 channels spaced by 100 GHz)
1998	Alcatel	32 × 10	500	4 EDFA + pre-amplifier (with 125 km amplifier spacing)
2000	KDD	50 × 10.66	4000	EDFA + low-dispersion slope fiber (40 km span)
2000	BELL	100 × 10 (25 GHz spacing)	400	4 EDFA + 4 Raman Amplifier

to 6, where the waveguiding properties, and the basic performance characteristics such as gain, gain saturation and noise will be studied. A new technique for analysing SLAs using an equivalent circuit model will also be introduced. Implications for system performance will also be discussed. In Chapter 7, the accuracy and limitations of this model will be investigated by comparing theoretical predictions with the results of experimental measurements on actual devices. In Chapter 8 we introduce a new

Table 1.3. Recent transmission experiments with semiconductor laser amplifiers.

Year	Laboratory	Bit Rate (Gbit/s)	Distance (km)	Comments
1986	BTRL	0.14	206	2 amplifiers used
1988	AT & T	1	313	4 amplifiers used
1988	AT & T	0.4	372	4 amplifiers + FSK
1988	Bell Core	—	—	20 Channels transmission
1989	BTRL	0.565	400	5 amplifiers + DPSK
1989	KDD	2.4	516	10 amplifiers used
1991	KDD	0.14	546	10 amplifiers used
1993	Japan	4 × 10	40	2 SOA preamplified receiver receiver with bandwidth of 40 nm
1994	PPT	10	89	2 SOA preamplified receiver
1995	PTT	2 × 10	63.5	2 SOA preamplified receiver
1997	BT	40	1406	2 mm-long SOA for dispersion compensation
1996		10	420	38 km interval
1997	Germany	10	550	34 km interval
1998	Germany	40	434	SOA for dispersion compensation + 3 EDFA
1998	Germany	10	1500	38 km interval
1999	Germany	80	106	SOA for dispersion compensation
2000	Germany	10	5000	25 km interval
2000	USA	32 × 10	160	4 SOA (40 km interval) + 1 EDFA
2000	AT & T	8 × 20	160	4 SOA (40 km interval)
2001	BELL	10 (WDM)	500	50 km interval

semiconductor laser diode amplifier structure. Chapter 9 deals with amplification characteristics of pico-second Gaussian pulses in various amplifier structures. Chapter 10 studies the sub-pico-second gain dynamic in a highly index-guided tapered-waveguide laser diode amplifier. In Chapter 11 we introduce a novel approximate analytical expression for saturation intensity of tapered travelling-wave semiconductor laser amplifier structures. Wavelength conversion using cross-gain modulation in linear tapered-waveguide semiconductor laser amplifiers is studied in Chapter 12. The main theme of the work presented in Chapters 13 to 17 is microwave circuit principles applied to semiconductor laser modelling. The advantages and additional insight provided by circuit models that have been used for analytical analysis of laser diodes have long been acknowledged.

In these chapters, we concentrate on the derivation, implementation, and application of numerical circuit-based models of semiconductor laser devices.

In Chapter 13 first, a short historical background and the relevant physics behind the semiconductor laser will be given. Chapter 14 introduces the transmission-line matrix (TLM) method that provides the basic microwave circuit concepts used to construct the time-domain semiconductor laser model known as the transmission-line laser model (TLLM). We then proceed to compare two categories of equivalent circuit models, i.e. lumped-element and distributed-element, of the semiconductor laser in Chapter 15. In the same chapter, a comprehensive laser diode transmitter model is developed for microwave optoelectronic simulation. The microwave optoelectronic model is based on the transmission-line modelling technique, which allows propagation of optical waves as well as lumped electrical circuit elements to be simulated. In Chapter 16, the transmission-line modelling technique is applied to a new time-domain model of the tapered waveguide semiconductor laser amplifier, useful for investigating short pulse generation and amplification when finite internal reflectivity is present. The new dynamic model is based on the strongly index-guided laser structure, and quasi-adiabatic propagation is assumed. Chapter 17 demonstrates the usefulness of the microwave circuit modelling techniques that have been presented in this thesis through a design study of a novel mode-locked laser device. The novel device is a multisegment monolithically integrated laser employing distributed Bragg gratings and a tapered waveguide amplifier for high power ultrashort pulse generation. Finally, Chapter 18 is devoted to some concluding remarks suggestions and comments.

References

[1] C. Rolland, L. E. Tarof and A. Somani, "Multigigabit networks: The challenge," *IEEE LTS Magazine*, Vol. 3, pp. 16–26, May 1992.

[2] M. G. Davis and R. F. O'Dowd, "A new large-signal dynamic model for multielectrode DFB lasers based on the transfer matrix method," *IEEE Photo. Tech. Lett.*, Vol 4, No. 8, pp. 838–840, 1992.

[3] H. Taub and D. L. Schilling, *Principles of Communication Systems*, 2nd Edition (MacGraw-Hill, 1986).

[4] T. Kimura, "Coherent optical fibre transmission," *IEEE J. Lightwave Technology*, Vol LT-5, No. 4, pp. 414–428, 1987.

[5] R. E. Wagner and R. A. Linke, "Heterodyne lightwave systems: Moving towards commercial use," *IEEE LCS Magazine*, Vol. 1, No. 4, pp. 28–35, 1990.

[6] M. J. Adams, *An Introduction to Optical Waveguides* (John Wiley and Sons, 1981).
[7] T. Miya, Y. Teramuna, Y. Hosuka and T. Miyashita, "Ultimate low-loss single mode fibre at 1.55 μm," *Electronics Letters*, Vol. 15, No. 4, pp. 106–108, 1979.
[8] E. Dietrich *et al.*, "Semiconductor laser optical amplifiers for multichannel coherent optical transmission," *J. Lightwave Technology*, Vol. 7, No. 12, pp. 1941–1955, 1989.
[9] W. I. Way, C. Zah and T. P. Lee, "Application of travelling-wave laser amplifiers in subcarrier multiplexed lightwave systems," *IEEE Trans. Microwave Theory Tech.*, Vol. 38, No. 5, pp. 534–545, 1989.
[10] G. H. B. Thompson, *Physics of Semiconductor Laser Devices* (New York, John Wiley and Sons, 1980).
[11] A. Yariv, *Optical Electronics*, 3rd Edition (Holt-Saunders, 1985).
[12] G. P. Agrawal and N. K. Dutta, *Long-wavelength Semiconductor Lasers* (New York, Van-Nostrad Reinhold, 1986).
[13] J. M. Senior, *Optical Fibre Communications: Principle and Practice*, 2nd Edition (Prentice-Hall, 1992).
[14] D. J. Maylon and W. A. Stallard, "565Mbit/s FSK direct detection system operating with four cascaded photonic amplifiers," *Electronics Letters*, Vol. 25, No. 8, pp. 495–497, 1989.
[15] A. Yariv and P. Yeh, *Optical Waves in Crystals* (John Wiley and Sons, 1984).
[16] P. E. Green and R. Ramaswami, "Direct detection lightwave systems: Why pay more?," *IEEE LCS Magazine*, Vol 1, No. 4, pp. 36–49, 1990.
[17] H. Kressel, *Semiconductor Devices for Optical Communications*, 2nd Edition (Springer-Verlag, 1982).
[18] T. Okoshi, "Ultimate performance of heterodyne/coherent optical fibre communication," *IEEE J. Lightwave Technology*, Vol. LT-4, No. 10, pp. 1556–1562, 1986.
[19] H. Tsushima *et al.*, "1.244 Gbit/s 32 channel 121 km transmission experiment using shelf-mounted continuous-phase FSK optical heterodyne system," *Electron. Lett.*, Vol. 27, No. 25, pp. 2336–2337, Dec. 1991.
[20] M. A. R. Violas, D. J. T. Heatley, X. Y. Gu, D. A. Cleland and W. A. Stallard, "Heterodyne detection at 4 Gbit/s using a simple pin-HEMT receiver," *Electron. Lett.*, Vol. 27, No. 1, pp. 59–61, Jan. 1991.
[21] R. J. S. Pedersen *et al.*, "Heterodyne detection of CPFSK signals with and without wavelength conversion up to 5Gb/s," *IEEE Photonic Technol. Lett.*, Vol. 5, No. 8, pp. 944–946, August 1993.
[22] H. C. Liu *et al.*, "Optical heterodyne detection and microwave rectification up to 26 GHz using quantum well infrared photodetectors," *IEEE Electron Device Lett.*, Vol. 16, No. 6, pp. 253–255, June 1995.
[23] Y. Shoji, M. Nagatsuka, K. Hamaguchi and H. Ogawa, "60 GHz band 64QAM/OFDM terrestrial digital broadcasting signal transmission by using millimeter-wave self-heterodyne system," *IEEE Transactions on Broadcasting*, Vol. 47, No. 3, pp. 218–227, Sept. 2001.
[24] N. Nakagawa and S. Shimada, "Optical amplifiers in future optical communication systems," *IEEE LCS Magazine*, Vol. 1, No. 4, pp. 57–62, 1990.

[25] S. D. Personick, "Applications for quantum amplifiers in simple digital optical communication systems," *Bell Sys. Tech. J.*, Vol. 52, No. 1, pp. 117–133, 1973.
[26] A. J. Arnaud, "Enhancement of optical receiver sensitivity by amplification of the carrier," *IEEE J. Quantum Electron.*, Vol. QE-4, No. 11, pp. 893–899, 1968.
[27] H. Nakagawa, K. Aida, K. Aoyama and K. Hohkawa, "Optical amplification in trunk transmission networks," *IEEE LTS Magazine*, Vol. 3, pp. 19–26, Feb. 1992.
[28] G. Eisenstein, "Semiconductor optical amplifiers," *IEEE Circuits and Devices Magazine*, pp. 25–30, July 1989.
[29] Y. Yamamoto and T. Mukai, "Fundamental of optical amplifiers," *Opt. Quantum Electronics*, Vol. QE-21, pp. S1–S14, 1989.
[30] Y. Yamamoto and H. Tsuchiya, "Optical receiver sensitivity improvement by a semiconductor laser amplifier," *Electronics Letters*, Vol. 16, No. 6, pp. 233–235, 1980.
[31] H. Nakano, S. Tsuji, S. Sasaki, K. Uomi and K. Yamashita, "10 Gb/s, 4-channel wavelength division multiplexing fiber transmission using semiconductor optical amplifier modules," *IEEE/OSA Journal of Lightwave Technology*, Vol. 11, No. 4, pp. 612, April 1993.
[32] H. Gnauck, R. M. Jopson and R. A Derosier, "10 Gb/s 360-km transmission over dispersive fiber using midsystem spectral inversion," *IEEE Photonics Technology Letters*, Vol. 5, No. 6, pp. 663–666, June 1993.
[33] H. de Waardt, L. F. Tlemeijer, and B. H. Verbeek, "89 km 10 Gbit/s 1310 nm repeaterless transmission experiments using direct laser modulation and two SL-MQW laser preamplifiers with low polarization sensitivity," *IEEE Photonics Technology Letters*, Vol. 6, No. 5, pp. 645–647, May 1994.
[34] A. R. Chraplyvy *et al.*, "1420-km transmission of sixteen 2.5 Gb/s channels using silica-fiber-based EDFA repeaters," *IEEE Photonics Technology Letters*, Vol. 6, No. 11, pp. 1371–1373, Nov. 1994.
[35] H. de Waardt, L. F. Tierneijer and B. H. Verbeek, "2 × 10 Gbit/s WDM 1310-nm optical transmission over 63.5-km standard single-mode fiber using optical preamplifiers," *IEEE Photonics Technology Letters*, Vol. 7, No. 1, pp. 104–107, Jan. 1995.
[36] P. B. Hansen *et al.*, "374 km transmission in a 2.5 Gb/s repeaterless system employing a remotely pumped erbium-doped fiber amplifier," *IEEE Photonics Technology Letters*, Vol. 7, No. 5, pp. 588–590, May 1995.
[37] A. Shipulin *et al.*, "10 Gbit/s signal transmission over 550 km in standard fibre at 1300 nm using semiconductor optical amplifiers," *Electronics Letters*, Vol. 33, No. 6, pp. 507–509, March 1997.
[38] D. D. Marcenac *et al.*, "40 Gbit/s transmission over 406 km of NDSF using mid-span spectral inversion by fourwave-mixing in a 2 mm long semiconductor optical amplifier," *Electronics Letters*, Vol. 33, No. 10, pp. 879–880, May 1997.
[39] Y. Sun *et al.*, "Transnussion of 32 WDM 10 Gb/s channels over 640 km using broad-band, gain-flattened erbium-doped silica fiber amplifiers," *IEEE Photonics Technology Letters*, Vol. 9, No. 12, pp. 1652–1654, Dec. 1997.

[40] G. Onishchukov, V. Lokhnygin, A. ShiPulin and P. Riedel, "10 Gbit/s transmission over 1500 km with semiconductor optical amplifiers," *Electronics Letters*, Vol. 34, No. 16, pp. 1597–1598, August 1998.
[41] S. Bigo *et al.*, "320-Gb/s (32 × 10 Gb/s WDM) Transmission over 500 km of conventional single-mode fiber with 125 km amplifier spacing," *IEEE Photonics Technology Letters*, Vol. 10, No. 7, pp. 1045–1097, July 1998.
[42] U. Feiste *et al.*, "40 Gbit/s transmission over 434 km standard fibre using polarisation independent mid-span spectral inversion," *Electronics Letters*, Vol. 34, No. 21, pp. 2044–2045, Oct. 1998.
[43] U. Feiste *et al.*, "80-uu/s Transmission over 106-km standard-fiber using optical-phase conjugation in a sagnac-interferometer," *IEEE Photonics Technology Letters*, Vol. 11, No. 8, pp. 1063–1065, August 1999.
[44] L. H. Spiekman *et al.*, "Transmission of 8 DWDM channels at 20 over 160 km of standard fiber using a cascade of semiconductor optical amplifiers," *IEEE Photonics Technology Letters*, Vol. 12, No. 6, pp. 717–719, June 2000.
[45] L. H. Spiekman, A. H. GnaUCk, J. M. Wiescnfeld and L. D. Gaffett, "DWDM transmission of thirty two 10 Gbit/s channels through 160 krn link using semiconductor optical amplifiers," *Electronics Letters*, Vol. 36, No. 12, pp. 1046, June 2000.
[46] Kaoru Imai *et al.*, "500 Gb/s (50 × 10.66 Gb/s) WDM transmission over 4000 km using broad-band EDFA's and low dispersion slope fiber," *IEEE Photonics Technology Letters*, Vol. 12, No. 7, pp. 909–911, July 2000.
[47] Z. Bakonyi, G. Onishchukov, C. Kn, A Ges and F. Lederer, "10 Gbit/s RZ transmission over 5000 krn with gain-clamped semiconductor optical amplifiers and saturable absorbers," *Electronics Letters*, Vol. 36, No. 21, pp. 1790–1791, Oct. 2000.
[48] A. K. Srivastava *et al.*, "Ultradense WDM transmission in L-band," *IEEE Photonics Technology Letters*, Vol. 12, No. 11, pp. 1570–1572, Nov. 2000.
[49] H. K. Kim, S. Chandrasekhar, A. Srivastava, C. A. Burrus and L. Buhl, "10 Gbit/s based WDM signal trans over 500 km of NZDSF using semiconductor optical amplifier as the in-line amplifier," *Electronics Letters*, Vol. 37, No. 3, pp. 185–187, Feb. 2001.
[50] Y. Mukai, Y. Yamamoto and T. Kimura, "Optical amplification by semiconductor lasers," *Semiconductor and semimetals*, Vol. 22, Part E, pp. 265–319, Academic Press, 1985.
[51] S. Shimada and H. Ishio, *Optical Amplifiers and their Applications* (John Wiley and Sons, 1994), p. 3.

Chapter 2

Basic Principles of Optical Amplifiers

2.1 Introduction

The future prospects of high-speed long distance optical fibre communication systems depend heavily on the availability of low-cost optical amplifiers which can compensate for the build up of losses in optical fibre cables over long distances. Two types of optical amplifier exist: (i) semiconductor laser diode amplifiers and (ii) fibre amplifiers. Semiconductor laser diode amplifiers are essentially laser diodes operating in the linear amplification region below oscillation threshold, whereas fibre amplifiers are optical fibres doped with Erbium ions (Er^{3+}) to provide optical gain. Semiconductor laser diode amplifiers have the inherent advantages of compactness and the possibility of integration with other optoelectronic components, whereas fibre amplifiers have the advantages of easy and efficient coupling to optical fibres. The design and analysis of both types of optical amplifier are therefore crucial for the future development of coherent optical communications.

To understand fully how optical amplification can be achieved, the interaction of electromagnetic radiation with matter must first be understood. Therefore, in this chapter, we will first explore the interaction of radiation with a simple two-level atomic system. This simple model provides the basis for studies of more complex quantum mechanical systems, including those of semiconductors [1].

An understanding of the interaction of radiation with a two-level system enables one to understand the operation of optical amplifiers and from this their fundamental performance characteristics can be derived. In general, such characteristics can be used to describe both fibre amplifiers and semiconductor laser amplifiers [2], so that any optical communication system incorporating either type of optical amplifier can be analysed in a formal and consistent way [3–5]. The performance characteristics of an ideal optical amplifier will be derived after the above discussion. The ideal optical amplifier can be used as a reference to assess the ultimate performance of real semiconductor laser amplifiers [6]. Finally, the performance limitations of optical amplifiers, which will determine the ultimate performance of an optical system, will be analysed.

2.2 Interaction of Radiation with a Two-Level System

One way to understand the physics behind optical amplification processes in any optical amplifier is by considering a simple two-level system as shown in Figs. 2.1(a)–(c). This description is sufficient to give a fairly accurate qualitative picture of the physical processes that take place inside gas, or solid state, semiconductor lasers or optical amplifiers [1, 7]. There are three fundamental radiative processes that may take place when an electromagnetic wave interacts with a lasing material. These are spontaneous emission, stimulated emission and absorption. The spontaneous emission, because of its very nature, is distributed over a wide range of frequencies. The dynamic behavior of a laser or an optical amplifier is often described with reasonable precision by a set of coupled rate equations involving the three radiative processes. In their simplest form, these are a pair of simultaneous differential equations describing the population inversion and the laser radiation field as functions of time. A more accurate picture for

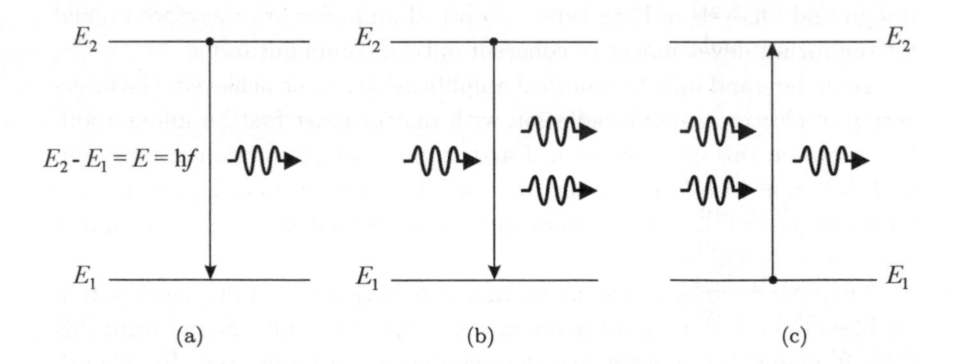

Fig. 2.1. Radiative processes in a two level system; (a) spontaneous emission, (b) stimulated emission and (c) absorption.

semiconductor laser amplifiers will be treated in Chapter 3. Meanwhile, we will first examine how electromagnetic radiation interacts with a two-level system, and then we will see how these interactions give rise to optical gain in optical amplifiers.

2.2.1 *Radiative processes*

In a two-level optical amplification system, the atoms can occupy one of two discrete energy levels, which are separated by a finite energy E. On excitation, the atoms gain energy and transit from the lower energy level 1 to the higher energy level 2. Conversely, the atoms may lose energy and transit from the higher energy level 2 to the lower energy level 1. The transitions between the energy levels of the system can be described accurately by quantum mechanics [1, 7, 8], although a semi-classical treatment can also be used [9, 10]. In the following discussion, the latter approach will be used to clarify the analysis. This problem was first analysed by Einstein based on some physically reasonable postulates concerning radiative transitions [9].

Consider an atomic system that has two nondegenerate energy states E_1 and E_2 such that $E = E_2 - E_1$. When an atom undergoes transitions between these two states, it either emits or absorbs a photon of frequency $f = E/h$, where $h = 6.626 \times 10^{-34}$ Js is Planck's constant [1]. Under normal circumstances, the atom will be in the lower level because physical systems tend to be more stable at the lowest energy state. Given the presence of an incoming photon, there is a high probability that an atom will configure itself at the excited energy state, but it returns to the ground state again by giving up a photon of energy E. In the following we examine the details of the three radiative processes for a proper understanding of the process of achieving population inversion.

2.2.2 *Spontaneous emission*

This is a random radiative process whereby an electron in the excited state E_2 decays to state E_1 and, in so doing, gives out energy E in the form of a photon, as shown in Fig. 2.1(a). If the population density of the excited energy state is N_2 and that of the lower energy state is N_1, then the rate of spontaneous decay of these atoms is found to be proportional to N_2. Hence, we have

$$\left[\frac{dN_2}{dt}\right]_{\text{sp}} = -\left[\frac{dN_1}{dt}\right]_{\text{sp}} = -A_{21}N_2 \tag{2.1}$$

where A_{21} is called the spontaneous emission probability. The spontaneous emission lifetime $\tau_{21,\mathrm{sp}}$, defined as the average time during which the electron survives in the excited state before contributing to a spontaneous emission, is equivalent to $1/A_{21}$. The probability that a particular atom will undergo a spontaneous transition in time dt is $A_{21}dt$. The population density of the excited state is given by

$$N_2 = N_2^0 \exp\left(-\frac{t}{\tau_{21,\mathrm{sp}}}\right) \tag{2.2}$$

where N_2^0 is the value of N_2 at $t = 0$. Equation (2.2) justifies the fact that if an atomic system is subjected to spontaneous emission only, N_2 will deplete rapidly, thus causing a rise in the population density N_1 of the lower energy state.

2.2.3 *Stimulated emission*

It is possible that an input photon can interact with an excited system and trigger emission of excess energy. A photon passing sufficiently close to an excited atom may cause the atom to undergo a radiative transfer before it would otherwise do so spontaneously. The excess photon energy output has the same frequency, phase, direction and polarisation as the stimulating photons. Hence, the process involves an amplification of coherent light [8]. Stimulated emission is associated with a specific energy level difference in a material, and the emitted photons show a high degree of spatial and temporal coherence with the incident photon (see Fig. 2.1(b)). The fact that the stimulated and stimulating photons are in phase results in a high degree of coherence in lasers. In practice, a less than perfect temporal coherence is achieved because of the randomness present in radiative transfers due to thermal vibrations among other things. In comparison, the spontaneous emission has no definite phase and the resulting photon emission can be in any random direction. The stimulated emission rate is dependent on the incoming field as well as on the number of atoms to be stimulated and is given by

$$\left[\frac{dN_2}{dt}\right]_{\mathrm{st}} = -\left[\frac{dN_1}{dt}\right]_{\mathrm{st}} = -B_{21}N_2\rho(f) \tag{2.3}$$

where $\rho(f)$ is the average spectral energy density per unit frequency for blackbody radiation [9], and B_{21} is the stimulated emission probability per unit time per unit spectral energy density. Note that $\rho(f)df$ provides the total photon density within the frequency interval f and df. As will be shown in the next section, the stimulated emission is the inverse of stimulated absorption.

2.2.4 *Absorption*

An electron in a lower energy state, after having absorbed an energy equivalent to E, is raised to an excited energy state. In the presence of a quasi-monochromatic radiation field of frequency f, the population of the less excited state is depleted at a rate proportional to both the population of that state and radiation density. This de-amplification process, called absorption, is shown in Fig. 2.1(c). The absorption rate is given by

$$\left[\frac{dN_2}{dt}\right]_{\text{abs}} = -\left[\frac{dN_1}{dt}\right]_{\text{abs}} = B_{12}N_1\rho(f) \tag{2.4}$$

where B_{12} is the absorption probability per unit spectral energy density. The input energy is accordingly decreased by an amount that was absorbed by the atomic system. At thermal equilibrium, the photon absorption rate must equal the sum of the stimulated and spontaneous emission rates. An atomic system that is simultaneously subjected to all three radiative processes has an overall decay rate given by

$$\frac{dN_2}{dt} = -\frac{dN_1}{dt} = -A_{21}N_2 + B_{12}N_1\rho(f) - B_{21}N_2\rho(f)\,. \tag{2.5}$$

Since at equilibrium, $dN_2/dt = dN_1/dt = 0$, using Eq. (2.5) we can calculate the ratio of the population densities as

$$\frac{N_2}{N_1} = \frac{B_{12}\rho(f)}{A_{21} + B_{21}\rho(f)}\,. \tag{2.6}$$

There would be no absorption but only stimulated emission if N_1 were zero, but there would be only absorption if N_2 were zero. The ratio of the population densities at thermal equilibrium is also given by the Boltzmann distribution

$$\frac{N_2}{N_1} = \frac{g_2}{g_1}\exp\left(-\frac{E}{kT}\right) \tag{2.7}$$

where k is Boltzmann's constant (1.38×10^{-23} J/K), T is the absolute temperature and $g_i (i = 1, 2)$, referred to as the degeneracy, corresponds to the number of independent ways in which the atom can have the same energy E_i. Note that the ratio N_2/N_1 is always less than unity when

E_2/E_1 and $T > 0°$K. Usually, the energy level separation is of the order of 10^{-19} joules or more and, thus, at thermal equilibrium (i.e. $T = 300°$K) N_1 exceeds N_2 by a factor of the order of 10^8. This implies that optical amplification is not realizable at thermal equilibrium. The spectral energy density $\rho(f)$ can be found by equating Eqs. (2.5) and (2.7) as

$$\rho(f) = \frac{A_{21}}{B_{21}}\left[\frac{g_1 B_{12}}{g_2 B_{21}}\exp\left(\frac{E}{kT}\right) - 1\right]^{-1} . \tag{2.8}$$

It may be noted that the spectral energy density is also given by Planck's radiation law as

$$\rho(f) = \frac{8\pi h f^3 n^3}{c^3[\exp(E/kT) - 1]} \tag{2.9}$$

where n is the refractive index of the medium and c is the velocity of light in free space. The only way Eqs. (2.8) and (2.9) can then be equal is if and only if [9, 10]

$$\frac{g_1}{g_2} = \frac{B_{21}}{B_{12}} \tag{2.10}$$

and

$$A_{21} = \left[\frac{8\pi h f^3 n^3}{c^3}\right] B_{21} . \tag{2.11}$$

Equations (2.10) and (2.11) are referred to as Einstein's relations. These equations interrelate the coefficients A_{21}, B_{12} and B_{21}. The fact that $B_{12} = B_{21}$, when $g_1 = g_2$, for simplicity, implies that for a given radiation density, stimulated emission and absorption are equally probable. In addition, Eq. (2.11) confirms that for most applicable frequencies, the rate of spontaneous emission is insignificant when compared with that of stimulated emission. Hence, if favorable circumstances permit, the lasing within a material is expected to be dominated primarily by stimulated emission. Under conditions of thermal equilibrium, however, stimulated emission is not likely. The higher the frequency, the less likely the process. We may conclude that all three radiative processes compete against each other according to Eqs. (2.10) and (2.11). In building a laser or amplifier, we expect to find stimulated emission to be the most dominating radiative process. This scenario can be achieved only by increasing energy density and by making N_2 larger than $(g_1/g_2)N_1$. This condition is referred to as population inversion.

2.2.5 *Optical gain*

The previous analysis has neglected the spatial dependence of the radiation field in the system. In reality, light is often confined in an optical cavity and interacts along with the atoms which are distributed in the cavity. Hence the spatial dependence of radiation must be taken into account in analysing practical optical amplifiers. To tackle this problem, let N_1 and N_2 be the population density of atoms at point z along the propagation direction of radiation for levels 1 and 2, respectively, $A = A_{21}$ and $B = B_{12} = B_{21}$ (i.e. $g_1 = g_2$). Then Eq. (2.5) may be expressed as

$$\frac{dN_1}{dt} = -\frac{dN_2}{dt} = AN_2 + B(N_2 - N_1)\rho(f, z)\,. \tag{2.12}$$

Notice that the z-dependence of ρ has been introduced. In general, N_1 and N_2 are functions of both position z and time t. Since each transition of an atom between the two energy levels involves a generation (for spontaneous emission and stimulated emission) or loss (for absorption) of the photon, one can derive a rate equation in terms of photon density S_p at position z and time t as

$$\frac{dS_p}{dt} = AN_2 + B(N_2 - N_1)\rho(f, z)\,. \tag{2.13}$$

Because S_p is also a function of both z and t, one can write it as a total derivative [11], that is

$$\frac{dS_p}{dt} = \frac{\partial S_p}{\partial t} + \frac{\partial S_p}{\partial z} \cdot \frac{dz}{dt} = \frac{\partial S_p}{\partial t} + \frac{\partial S_p}{\partial z} \cdot \nu_g \tag{2.14}$$

where $\nu_g = dz/dt$, ν_g is the group velocity of the travelling photons. If the radiation has been interacting with the system for a sufficiently long time such that a steady state has been reached, $\partial S_p/\partial t$ becomes zero and Eq. (2.14) reduces to

$$\nu_g \frac{\partial S_p}{\partial z} = AN_2 + B(N_2 - N_1)\rho(f, z)\,. \tag{2.15}$$

Equations (2.12) and (2.15) are the fundamental equations for the analysis of optical amplifiers, which we will examine in more detail in later chapters. Meanwhile, we will concentrate on the solution of Eq. (2.15). To do this analytically, we assume that the spontaneously emitted photons are scattered and lost instead of guided along with the incident radiation (i.e. $A = 0$). Denoting the photon density of the guided modes as S_g, the following equation can be obtained:

$$\frac{\partial S_g}{\partial z} = \frac{B(N_2 - N_1)\rho(f,z)}{\nu_g} . \tag{2.16}$$

The photon density of the guided mode and the spectral energy density are not unrelated. In general, the transition frequencies of the atoms, even in a simple two-level system, possess a finite statistical spread due to finite temperature, and the proportion of atoms involved in transition with frequencies lying in a range df is given by $F(f)df$, where $F(f)$ is the statistical distribution function describing the spread of transition frequencies of the atoms in the cavity [9]. Hence, the photons generated due to radiative transitions will have their frequencies spread in the range df as well. Consider a small section of length dz along the cavity with a cross-section area σ. The spectral energy density $\rho(f,z)$ can be related to S_g by the following equation

$$\rho(f,z)df = S_g \cdot hf \cdot \sigma \cdot \nu_g \cdot F(f)df . \tag{2.17}$$

Substituting Eq. (2.17) into Eq. (2.16) gives

$$\frac{\partial S_g}{\partial z} = B(N_2 - N_1) \cdot hf \cdot \sigma \cdot S_g \cdot F(f)df . \tag{2.18}$$

Because it is optical power rather than photon density which is usually measured, it is more convenient to express Eq. (2.18) by the power P over the frequency range df of the guided modes, which is related to the photon density by

$$P(z) = S_g \cdot hf \cdot \sigma \cdot \nu_g \cdot F(f)df \tag{2.19}$$

then Eq. (2.18) can be re-written in the form of

$$\frac{dP}{dz} = hf \cdot \sigma \cdot \nu_g \cdot B(N_2 - N_1) \cdot P \cdot F(f)df . \tag{2.20}$$

For the frequency range df concerned, Eq. (2.20) can be solved by integration to give P at $z = z_0$ as:

$$P(z_0) = P(0)\exp(K(z_0)) \tag{2.21}$$

where $P(0)$ is the optical power at $z = 0$, and K is known as the *attenuation factor*, which is given by

$$K(z_0) = hf \cdot \sigma \cdot \nu_g \cdot F(f)df \cdot \int_0^{z_0} B(N_2 - N_1)dz \,. \tag{2.22}$$

It can be seen that the value of K is affected by (i) the population distribution of the atoms in the two energy levels, (ii) the statistical distribution of transition frequencies $F(f)$, (iii) the exact frequency range df of the measure optical power, and (iv) the structural parameters of the cavity (i.e. parameters σ and ν_g). These factors are very important and will be encountered again when discussing amplification in semiconductor lasers in later chapters. At the moment, we will examine (i) in more detail.

Under normal conditions with N_1 and N_2 independent of z (uniformly distributed along the cavity), the distribution of the population density between the two energy levels is described by the following relation [10]

$$\frac{N_2}{N_1} = \exp\left(-\frac{hf}{kT}\right) \tag{2.23}$$

where k is the Boltzmann constant and T is the absolute temperature of the system measured in Kelvin. When T is larger than absolute zero, it can be seen from Eq. (2.23) that $N_2 < N_1$, and hence K will be negative as depicted by Eq. (2.22). This implies that the optical power reduces as the photons travel along z and the radiation is absorbed within the cavity. To provide optical gain (i.e. optical power increasing with z instead of decreasing) the value of K *must be made positive.* As seen from Eq. (2.22), this can only be achieved if $N_2 > N_1$. Physically, this implies that the stimulated emission rate BN_2 is greater than that of induced absorption BN_1 along the cavity, generating sufficient coherent photons such that the numbers of photons are building up as they travel along the cavity. The condition of $N_2 > N_1$ is known as *population inversion*, and the resulting system becomes an optical amplifier [9]. From Eq. (2.23), it can be seen that this is possible only if T is maintained below absolute zero, and hence this condition is also known as *negative temperature.* According to thermodynamics, this means that energy has to be *pumped* continuously into the system to maintain such population inversion. Therefore, if optical gain has to be introduced into the system, an external energy source has to provide pumping to create a population inversion. In semiconductors, such an energy source can take the convenient form of an electrical current [12], whereas in gas and solid state material like fibre amplifiers, the external pumping can take the form of intense light pulses [1].

2.3 Characterisation of Optical Amplifiers

After identifying the physical origin of the optical gain in optical amplifiers, we need to investigate the essential characteristics of these devices. Mukai and Saito [13] identified four important parameters which could be used to describe the performance of any optical amplifier, including semiconductor laser amplifiers (SLAs) and fibre amplifiers (FAs). These parameters are important in defining the performance of optical amplifiers in an optical fibre communication system. They include: the *signal gain*, *frequency bandwidth*, *saturation output power* and the *noise figure*.

2.3.1 *Signal gain*

The most important property of an optical amplifier is its ability to amplify the power of light. As shown in Section 2.2.5, optical gain in an optical amplifier is created by external pumping. In SLAs, it is provided by the injection of carriers by an electrical current, whereas in FAs which are doped by rare-earth such as erbium (Er^{3+}), the external pumping is provided by a powerful light source [2]. In any case, the gain of the optical amplifier is of primary interest [14], as it determines many other essential factors, like signal-to-noise (S/N) ratio, when they are incorporated into systems [15]. The signal gain G of an optical amplifier is given, in decibels, by:

$$G = 10 \log[P_{\text{out}}/P_{\text{in}}] \tag{2.24}$$

where P_{out} is the light power measured at the output of the optical amplifier, and P_{in} is that measured at the input end [16].

One can further refine the definition of signal gain G by considering the light path. If Eq. (2.24) describes the input and output light power due to a single light path from input to output of the optical amplifier, the resulting gain is known as *single pass gain* G_s. If negative feedback is provided (i.e. by reflections from end-facets in semiconductor laser diode amplifiers) as shown by the feedback path in Fig. 2.2, the signal gain G becomes:

$$G = \frac{G_s}{1 + F_B G_s} \tag{2.25}$$

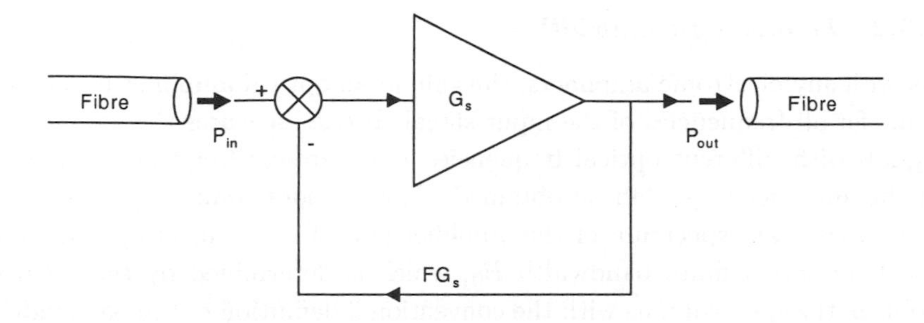

Fig. 2.2. An optical amplifier with feedback.

where F_B is the proportion of output signal which is fed back to the input. Eq. (2.25) is analogous to that of an electronic amplifier. When $G = G_s$, it corresponds to amplifiers without feedback ($F_B = 0$). This is true in measuring the gain of SLAs with zero or nearly zero facet reflectivities (known as travelling-wave or near-travelling-wave amplifiers, respectively [17]). If the reflectivities are finite so that some sort of optical feedback is provided to the optical amplifier, then F_B is finite and the resulting amplifier structure becomes a Fabry–Perot amplifier. The physical implications of Eq. (2.24) and F_B will be discussed again in more detail in Chapter 3 when different types of SLAs are explored. At this point, it is sufficient to know how to distinguish between the different meanings of G and G_s.

Ideally, an optical amplifier should have as high a gain as possible. Physically, optical amplifiers with infinite gain are impossible to achieve. An infinite amplifier gain requires an infinite pumping rate to maintain an infinite population inversion, and such an energy source does not exist in Nature. Furthermore, according to Eq. (2.16), an amplifier with infinite gain will actually generate an output without any input. This is equivalent to the condition for oscillation where the amplifier becomes a laser [10]. Therefore, the performance limit of the gain in any optical amplifier is that *the amplifier gain is not so large that self-sustained oscillations will be excited.* This is complicated by the fact that stray reflections in the system can provide additional feedback to the amplifier [18, 19], pushing the overall gain of the amplifier towards oscillation for moderate pumping (Eq. (2.25)). In addition, the maximum signal gain which can be obtained in an optical amplifier is further restricted by gain saturation mechanisms [20]. This subsequently affects the dynamic range of the amplifier, limiting the maximum optical power which can be input to and output from the amplifier [21].

2.3.2 *Frequency bandwidth*

As with any electronic amplifiers, the gain in an optical amplifier is not the same for all frequencies of the input signal. If one measures the gain G for signals with different optical frequencies f, an optical frequency response of the amplifier $G(f)$ can be obtained. This is more commonly known as the optical gain spectrum of the amplifier [22–24]. A typical optical gain spectrum has a finite bandwidth B_0, which is determined by the -3 dB point in the spectrum, as with the conventional definition of the bandwidth in linear systems. The value of B_0 therefore determines the bandwidth of signals which can be transmitted by an optical amplifier, hence affecting the performance of optical communication systems when using them as repeaters or preamplifiers [25].

In linear systems, finite bandwidth arises because of the presence of reactive components like capacitors and inductors. Different mechanisms, however, account for the finite bandwidth in optical amplifiers. Again, the exact mechanism varies in different types of optical amplifier. Nevertheless, two major reasons for finite bandwidth can be identified. The first reason is because of the waveguiding action of the amplifier (or dispersive effects) [26]. In order to amplify light signals, they must be confined within the region with optical gain. This can be achieved conveniently in most optical amplifiers by using optical waveguides (e.g. the active region in semiconductor laser amplifiers; see Chapters 3 and 4). Any optical waveguide, whether semiconductor laser diode amplifiers (SLAs) or circular dielectric types (as in fibre amplifiers), possesses cut-off frequencies and hence a finite bandwidth [27, 28]. The second reason is that the material gain itself has a finite bandwidth [29–31]. As we have discussed in Section 2.2.5, in real systems at finite temperature, each energy level of a system which an atom can occupy has a statistical spread according to the principles of thermodynamics and atomic bonding, and hence these two energy levels spread from two distinct levels to form two bands [1, 12]. This allows a range of frequencies to be amplified by the system, resulting a finite material gain bandwidth. The corresponding mechanisms for the material gain spectra in SLAs will be discussed in more detail in Chapter 3.

The above two reasons can account for the finite bandwidth in optical amplifiers. Usually, both dispersive effects and finite frequency range of the optical gain will occur in all types of optical amplifier and the resulting bandwidth of the amplifier is a summation of these two effects. Many proposed models of SLAs have neglected the dispersive effect in deriving the amplifier gain spectra $F(f)$ [11–32], because experimental work showed that material gain effects dominate in SLAs [33]. However, it can be shown that for many SLA structures, the full material gain bandwidth cannot be utilised because of the presence of resonant behaviour in the device, which reduces the actual bandwidth of the amplifier significantly [16]. This will be discussed in further detail in the next chapters.

2.3.3 *Saturation output power*

The signal gain of an optical amplifier is not only limited to a finite range of optical frequencies, but also by a finite range of input and hence output power, as discussed in Section 2.3.1. According to Eq. (2.24), whenever the input power to the optical amplifier is increased, the output power should be increased simultaneously by a scaling factor G. However, in real situations, the input power cannot be increased forever. Experimentally, it is observed that in all optical amplifiers, once the input power is increased to a certain level, P_s, the gain G starts to drop [13]. If the measured gain G is plotted against P_{out}, a curve similar to that in Fig. 2.3 can be obtained. The output

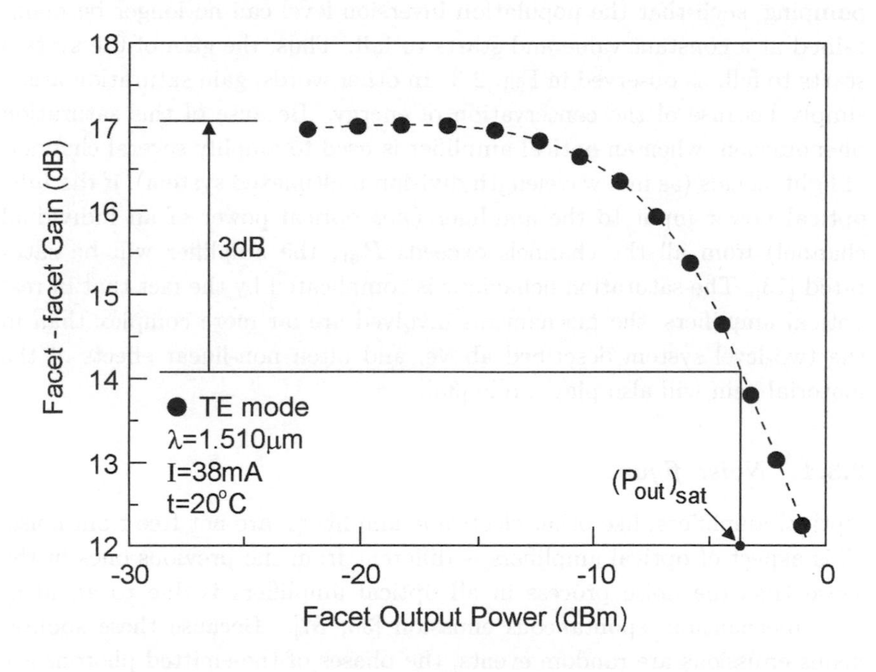

Fig. 2.3. A typical experimental result of amplifier gain against output optical power.

power at the −3 dB points, as shown in Fig. 2.3, is known as the *saturation output power* P_{sat}, and the corresponding *saturation output intensity* I_{sat} can be used to describe the gain saturation effect quantitatively. When the amplifier gain G is measured against the output light intensity I_{out}, a similar phenomenon of saturation can be observed, which can be described in an optical amplifier with gain given by [13].

$$G_{\text{s}} = \exp(g_{\text{m}}L) = \exp[g_{\text{o}}L/(1 + I_{\text{out}}/I_{\text{sat}})] \tag{2.26}$$

where g_m, measured in cm^{-1}, is the *material gain coefficient* which is determined by the population inversion level in the material [31], L is the length of the amplifier, and g_0 is the unsaturated value of g_{m}.

This behaviour of optical amplifiers is known as *gain saturation* [34]. A qualitative explanation of this phenomenon can be obtained from the two-level system model. The pumping source creates a fixed amount of population inversion at a particular rate, and on the other hand the amplification process is continuously draining the inverted population by creating stimulated emissions. As one increases the input power, a point comes where the rate of draining due to amplification is greater than the rate of pumping, such that the population inversion level can no longer be maintained at a constant value and starts to fall. Thus, the gain of the system starts to fall, as observed in Fig. 2.3. In other words, gain saturation arises simply because of the conservation of energy. Because of this saturation phenomenon, when an optical amplifier is used to amplify several channels of light signals (as in a wavelength division multiplexed system), if the *total* optical power input to the amplifier (*not* optical power of an individual channel) from all the channels exceeds P_{sat}, the amplifier will be saturated [13]. The saturation behaviour is complicated by the fact that in real optical amplifiers, the mechanisms involved are far more complex than in the two-level system described above, and often non-linear effects of the material gain will also play a role [35].

2.3.4 *Noise figure*

Optical amplifiers, like other electronic amplifiers, are not free from noise. This aspect of optical amplifiers is different from the previous ones in the sense that the noise process in all optical amplifiers is due to an identical mechanism: spontaneous emission [36, 37]. Because these spontaneous emissions are random events, the phases of the emitted photons are also random. This can be proved using quantum mechanics [8, 9]. If the spontaneous emission photons happen to be emitted close to the direction of travel of the signal photons, they will interact with the signal photons [38], causing both amplitude and phase fluctuations [39, 40]. In addition, these spontaneous emissions will be amplified as they travel across the optical amplifier towards the output [41]. Hence, at the output of the amplifier, the measured power consists of both the amplified signal power GP_{in} and the amplified spontaneous emissions (ASE) power P_n. This means that:

$$P_{out} = GP_{in} + P_n \,. \tag{2.27}$$

As with electronic amplifiers, a figure of merit can be attributed to an optical amplifier to describe its noise performance. In an electronic amplifier, its noise performance is measured by a *noise figure* F, which describes the degradation of signal-to-noise (S/N) ratio due to addition of amplifier noise [42]. Mathematically, it is given as:

$$F = \frac{(S/N)_{in}}{(S/N)_{out}} \tag{2.28}$$

where $(S/N)_{in}$ is the S/N ratio at the amplifier input, and $(S/N)_{out}$ is that at the amplifier output. This description can be applied equally to optical amplifiers [6]. It can be shown either by quantum mechanics [6] or semi-classical arguments [43] that, because spontaneous emission is unavoidable in any optical amplifier, the minimum noise figure of an ordinary optical amplifier is 3 dB, unless the optical amplification process is due to parametric amplification [10], for which a noise figure of 0 dB can be achieved [6, 43]. A more detailed analysis can be found in Section 2.5.4.

Noise in an optical amplifier is the most important parameter. It will not only limit the S/N ratio in systems incorporating optical amplifiers [3, 44, 45], but it also imposes other limitations on various applications of optical amplifiers in optical fibre communications. For example, consider several optical amplifiers cascaded in tandem along a transmission span as linear repeaters to compensate fibre losses [15, 46] (Fig. 2.4). The *amplified spontaneous emission* (ASE) noise power P_n contributes a part of the output power P_{out} of a particular amplifier in the chain, as described by Eq. (2.19), and becomes the input to the next amplifier. Therefore P_n can be further amplified by subsequent amplifiers. Because gain saturation depends on the total amount of power input to the amplifier (Section 2.3.3), the ASE noise from the output of the earlier stages in the optical amplifier chain can be so large that it will saturate the following ones. If the reflectivities on both input and output ends of the amplifier are low, backwardly

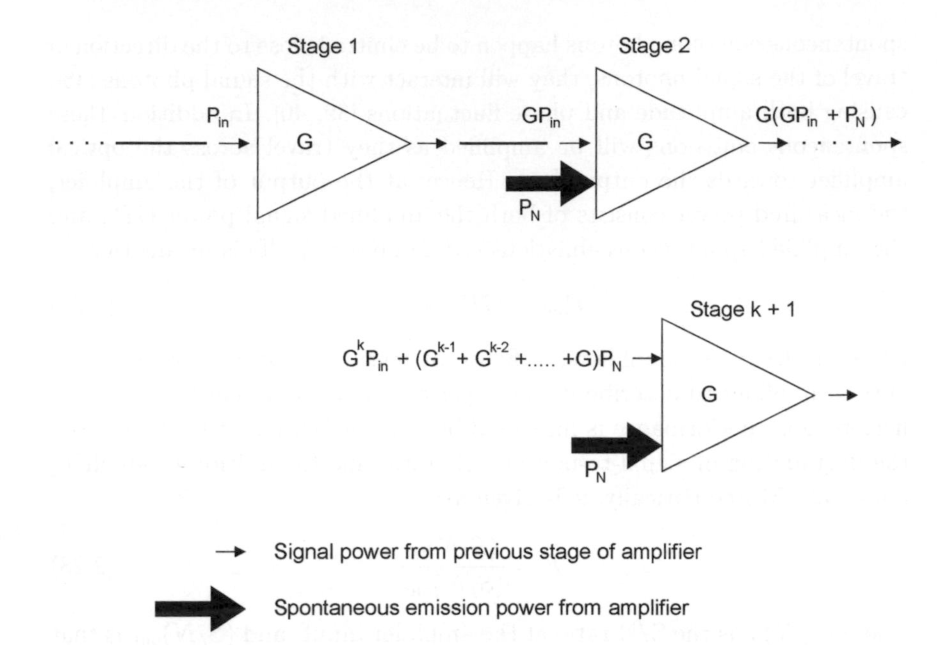

Fig. 2.4. A chain of cascaded optical amplifiers and the subsequent propagation of amplified spontaneous emissions.

emitted ASE from amplifiers of the later stages can also input to amplifiers of the earlier stages, enhancing saturation due to ASE. With FAs, this building up of ASE noise can actually result in self-sustained oscillations along the transmission span of the fibre if discontinuities and hence reflections are present along the path [18, 19]. Even though such back reflections are small, in a long transmission distance involving a relatively large number of optical amplifiers as linear repeaters, sufficient ASE power may build up along the amplifier chain as to trigger oscillation. To minimise this effect, optical isolators can be installed along the fibre link to cut off backward-emitted ASE [47], but this will prevent the system being used for bi-directional transmission [48].

In addition to the degradation of performance in terms of power, the phase contamination of the signal due to spontaneous emissions is also manifest as additional amplitude and frequency noise [49], especially due to stray reflections from optical interfaces [18]. As the input signal to optical amplifiers already has a finite amount of phase noise due to the finite spectral spread of the laser source [50, 51], further enhancement of noise from the amplifier is possible. This will further degrade the performance of optical communication systems using phase modulation and coherent detection (e.g. phase-shift keying PSK).

2.4 Ideal Optical Amplifiers

The four parameters discussed in the previous sections: gain, bandwidth, saturation output power and noise, are used by Mukai and Saito [13] to characterise an optical amplifier. Other factors such as linearity are important in some specific applications like multi-channel systems [52], but for general discussion these four parameters are sufficient. For instance, when amplifiers are used as linear repeaters, the gain-bandwidth determines how many optical amplifiers are needed along the transmission span of a particular distance [46]. On the other hand, the maximum number of optical amplifiers that can be used as repeaters is limited by the ASE power of each amplifier's output (i.e. their noise figures), as well as their saturation output power [17]. The exact performance requirements of optical amplifiers vary with different system applications. Therefore, for a particular system application, a specific set of *performance requirements* for the optical amplifiers must be specified [16]. This will be examined in further detail in Chapter 3. It should be noted, however, that all of these four parameters are important in determining the performance in many different applications of optical amplifiers. Therefore, it is possible to identify some basic performance requirements which are universal for all major applications. An optical amplifier which satisfies these performance requirements is described as an *ideal* optical amplifier. It is an ideal device in the sense that its performance characteristics are suitable for a very wide range of applications.

As the principal use of optical amplifiers is to amplify light signals, an ideal optical amplifier should have as high a value of gain G a possible. A wide optical bandwidth is also desirable, so that the amplifier can amplify a wide range of signal wavelengths. Saturation effects introduce undesireable distortion to the output, and hence an ideal optical amplifier should have a very high saturation output power to maximise its dynamic range with minimum distortion. Finally, an ideal amplifier should have a very low noise figure (minimum 3 dB), which will minimise ASE power at the output, hence maximising the number of optical amplifiers that can be cascaded.

The preceding discussion on the performance requirements for ideal optical amplifiers assumes that they are used as lightwave amplifiers. This type of application includes in-line repeaters and pre-amplifiers to optical receivers [16]. On the other hand, these performance requirements are equally valid for other applications. For instance, since an optical amplifier can be used as a gating optical switch ("on" when pumping is on, "off" when pumping is off), its response time (and hence its optical bandwidth) is a crucial factor (but note that for switching based on *optical bistability* instead of pumping this is not true; see next chapter for details), in addition to the requirement of low noise figures for the amplifiers (to avoid spurious switching). Similarly, when the amplifier is used as an optical modulator (by modulating the refractive index of the amplifier which can be achieved very easily in semiconductor laser amplifiers by simply modulating the electrical current [53]) and as a detector [54], its optical bandwidth should be wide, its optical gain should be high and the saturation output power should also be high (to achieve good linearity) with minimum noise. The detailed performance requirements for these various applications will be discussed in the following chapter, with special reference to semiconductor laser amplifiers.

Of course, real devices fail to meet the above criteria of an ideal optical amplifier. Nevertheless, these ideal characteristics can be seen as targets to which device engineers can aspire. To do so, a more detailed understanding of the operation of optical amplifiers is required.

2.5 Practical Optical Amplifiers

The preceding discussions on ideal performance requirements of optical amplifiers were purely qualitative. In practice, it is useful to have some quantitative criteria in order to assess how far actual performance of optical amplifiers falls short of the ideal optical amplifier. Furthermore, in real physical situations, optical amplifiers with infinite gain, bandwidth and saturation output power are impossible to obtain because of the physical limitations of various processes taking place inside the amplifier. Hence, it is more practical to study optical amplifiers which are operating within their performance limits in terms of the following four characteristics [6].

2.5.1 *Performance limits of the amplifier signal gain*

Physically, optical amplifiers with infinite gain are impossible to achieve, as an infinite pumping rate is required to maintain an infinite population inversion [1, 12]. Furthermore, according to Eq. (2.24), an amplifier with infinite gain will actually generate an output without having any input. This is equivalent to an oscillator and hence the amplifier becomes a laser [10]. This is actually the operational principle of a laser oscillator [1, 12]. Therefore, the performance limit of the gain of any optical amplifier is that the amplifier gain is not so large that self-sustained oscillations will be excited. In practice, the maximum signal gain which can be obtained in an optical amplifier is further restricted by gain saturation mechanisms [55]. This dynamic characteristic of the amplifier can severely reduce the maximum signal gain for high input and output powers [56]. Therefore, both gain and saturation output power are limited by gain saturation mechanisms, and are inter-related in this respect [57, 58]. The details of the effects of gain saturation will be discussed again in subsequent chapters with special reference to semiconductor laser diode amplifiers.

2.5.2 *Performance limits of the amplifier bandwidth*

Concerning bandwidth, a material which can amplify the whole electromagnetic spectrum (i.e. has infinite bandwidth) does not exist in nature. In other words, an optical amplifier can have a large but finite bandwidth. The size of bandwidth depends on the material gain spectra; hence, the performance limit in terms of bandwidth is one which can fully utilise the whole material gain spectrum. As mentioned in Section 2.3.2, the structure of an amplifier must be designed properly so that the amplifier bandwidth matches that of the material gain. One such semiconductor laser diode amplifier structure is the so called travelling-wave amplifier (TWA). The principle behind the TWA's capability to utilise the full possible bandwidth will be discussed in detail in subsequent chapters.

2.5.3 *Performance limits of saturation output power*

The limit on the maximum saturation output power which can be achieved in real devices depends strongly on the gain saturation mechanisms [55]. It also depends on the maximum amplifier gain which can be obtained, because, as we have seen, the gain saturation mechanisms also determine the maximum gain which can be achieved by the device. Usually, a high gain optical amplifier will have a high saturation output power because a high level of population inversion can be maintained for a wide range of input and output power in amplifiers [12]. However, there are also exceptions to this postulate (e.g. Fabry–Perot types of semiconductor laser diode amplifiers) [17, 24].

2.5.4 *Performance limits of the noise figure*

In contrast to the above performance limits of gain, bandwidth and saturation output power, which cannot reach thc ideal infinite limits, the noise figure of an optical amplifier can reach the theoretical 0 dB (i.e. noiseless) value even though spontaneous emissions are present [6]. This rather surprising characteristic will be discussed as follows. However, at this stage it should be noted that, as shown in reference [6], most of the optical amplifiers cannot achieve this noise figure unless they satisfy certain conditions, such as having a phase sensitive gain, as in parametric amplifiers [6, 10, 12]. When a semiconductor laser diode amplifier or a fibre amplifier is used in a linear application (i.e. as an in line repeater), the minimum noise figure which can be achieved is 3 dB. The amplifiers which satisfy this minimum noise figure requirement are said to be operating in the quantum limit. This quantum limit and the corresponding noise figure can be derived by quantum mechanics [6]. However, in what follows we present a more intuitive approach to obtain this limiting value of the noise figure [43]. Consider the input electric field E_{in} of the light signal into an optical amplifier. This field can be expressed in terms of quadrature components, as in [6,42];

$$E_{\text{in}} = A_1 \cos(\omega t) + A_2 \sin(\omega t) = R \sin(\omega t + \theta) \tag{2.29}$$

where

$$\omega = 2\pi f \tag{2.30a}$$

$$R = \sqrt{A_1^2 + A_2^2} \tag{2.30b}$$

$$\theta = \sin^{-1}\left[\frac{A_1}{\sqrt{A_1^2 + A_2^2}}\right]. \tag{2.30c}$$

At the output of the amplifier, both quadrature components are amplified. If the gain for the cosine component is G_1 and that for the sine component is G_2 we can express the output electric field of the amplifier E_{out} as

$$E_{\text{out}} = (\sqrt{G_1}A_1 + \delta_{1\text{i}}) \cos(\omega t) + (\sqrt{G_2}A_2 + \delta_{2\text{i}}) \sin(\omega t) \tag{2.31}$$

where $\delta_{1\text{i}}$ and $\delta_{2\text{i}}$ account for the amplitude and phase fluctuations due to the ith spontaneous emission event in the amplifier. Figures 2.5 and 2.6

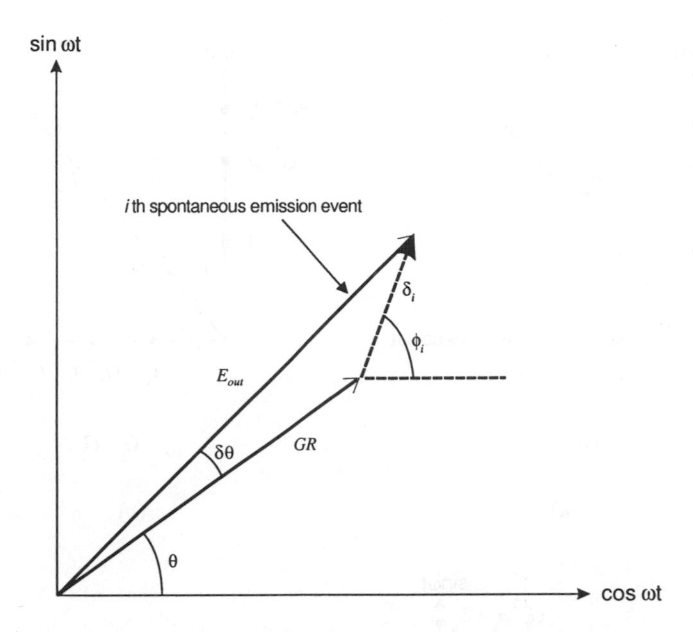

Fig. 2.5. Phasor representation of the electric field in an optical amplifier contaminated by spontaneous emissions.

illustrate the meaning of Eq. (2.31). In Fig. 2.5 a phasor representation of E_{out} is shown, illustrating the effect of the random phase of the ith spontaneous emission ϕ_i on the amplitude and phase θ of the electric field of the signal [51]. This fluctuation in amplitude and phase can be accounted for by the two random variables $\delta_{1\text{i}}$ and $\delta_{2\text{i}}$ for both quadrature components, as shown in Fig. 2.6. Figure 2.6(a) represents an amplification process with $G_1 = G_2$ and when $\delta_{1\text{i}}$ and $\delta_{2\text{i}}$ are both zero. An amplifier with $G_1 = G_2$ is known as a phase insensitive linear amplifier [6]. This relation is true for both semiconductor laser diode amplifiers and fibre amplifiers in most applications. It can be seen from Fig. 2.6(a) that the phase of E_{out} is the same as E_{in} and $G = G_1 + G_2$. In Fig. 2.6(b), the amplified field of a phase insensitive amplifier with non-zero $\delta_{1\text{i}}$ and $\delta_{2\text{i}}$ are shown, where it can be seen that the phase of E_{out} has been altered by these spontaneous emissions. Notice the similarities between Fig. 2.5 and Fig. 2.6(b), where they actually represent an identical situation, but in different mathematical representations (phasor form in Fig. 2.5 and quadrature components in Fig. 2.6(b)). Figure 2.6(c) shows amplification with phase sensitive gain (e.g. in parametric amplification [6, 12]), where $G_1 \neq G_2$. Again, the field

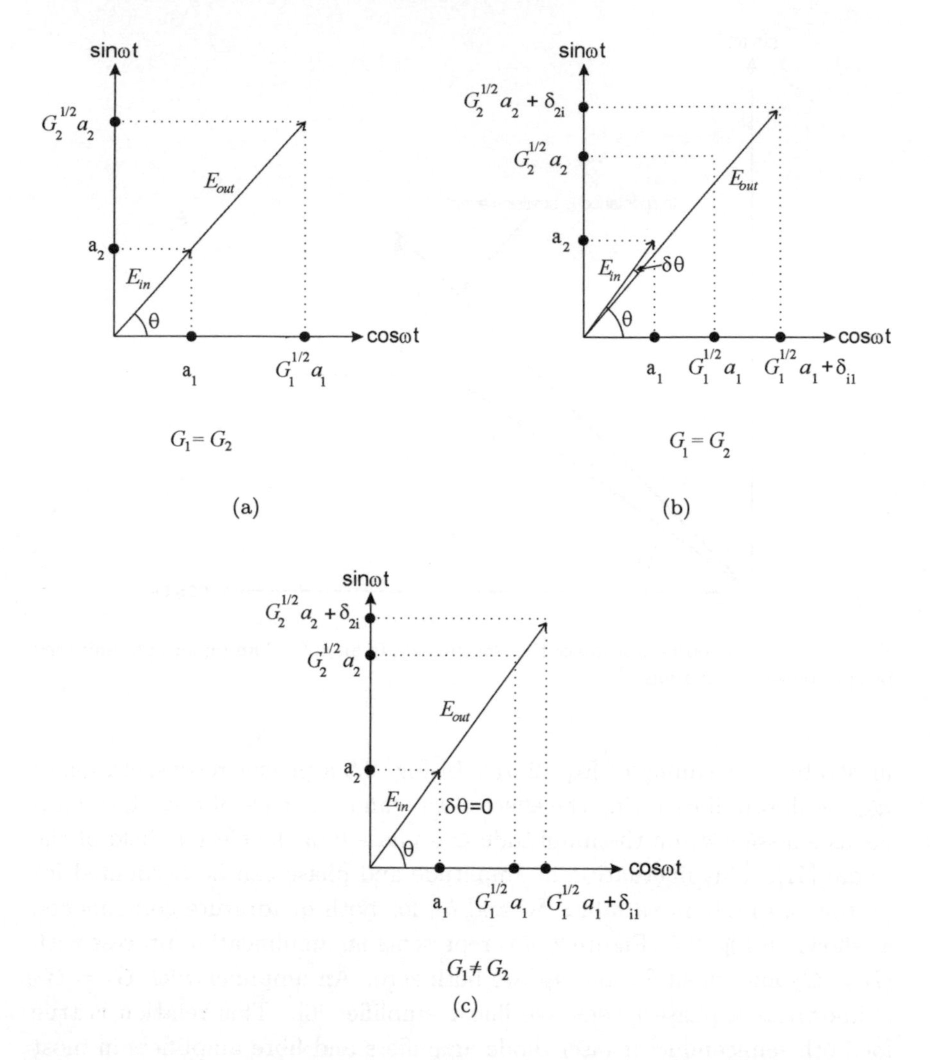

Fig. 2.6. Quadrature representation of the electric field in (a) a phase insensitive amplifier, (b) a phase insensitive amplifier with fluctuations due to spontaneous emissions and (c) a phase amplifier ($G_1 \neq G_2$) with fluctuations.

shows a fluctuation of amplitude and phase for the noiseless case. However, because of the phase sensitivity, G_1 and G_2 can be adjusted until the difference in amplification between the two phase quadrature components can counteract these fluctuations in amplitude and phase, resulting in an 0 dB noise figure. By knowing the physical meaning of Eq. (2.31) we can proceed to analyse the noise figure of an amplifier as follows. Because δ_{1i} and δ_{2i} are random quantities, and as the signal to noise ratio at the amplifier output has to be obtained, we can measure the output power of the amplifier to see the overall effects of these random spontaneous emission events. This involves an ensemble average of E_{out}^2 which is given by [6];

$$\langle E_{\text{out}}^2 \rangle = \left\langle \left(\sqrt{G_1} A_1 + \delta_{1\text{i}} \right)^2 \right\rangle + \left\langle \left(\sqrt{G_2} A_2 + \delta_{2\text{i}} \right)^2 \right\rangle \tag{2.32}$$

where $\langle \; \rangle$ represents a normalised ensemble average process, that is $\langle \cos^2(\omega t + \theta) \rangle = \langle \sin^2(\omega t + \theta) \rangle = 1$. Such an ensemble averaging includes a summation for all possible spontaneous emission events during the measuring period [38, 51]. Therefore, the ensemble average of the cross product of the sine and cosine quadrature terms goes to zero in the above equation. The above process usually occurs when the output quadrature components are detected by coherent methods [6, 42].

Suppose that A_1 and A_2 in the input signals are contaminated with noise. This can be represented by two fluctuating parameters ΔA_1 and ΔA_2 for both phase quadratures, respectively [39]. Equations (2.29) and (2.31) can be re-written as

$$E_{\text{in}} = (A_1 + \Delta A_1)\cos(\omega t) + (A_2 + \Delta A_2)\sin(\omega t) \tag{2.33}$$

$$\begin{aligned} E_{\text{out}} &= \left[\sqrt{G_1}(A_1 + \Delta A_1) + \delta_{1\text{i}} \right] \cos(\omega t) \\ &\quad + \left[\sqrt{G_2}(A_2 + \Delta A_2) + \delta_{2\text{i}} \right] \sin(\omega t) . \end{aligned} \tag{2.34}$$

In determining the noise figure of the amplifier, the power at the input has to be determined. This can be deduced from Eq. (2.33) as

$$\langle E_{\text{in}}^2 \rangle = (A_1^2 + A_2^2) + [\langle \Delta A_1^2 \rangle + \langle \Delta A_2^2 \rangle] \tag{2.35}$$

In the above equation the first term in brackets is the input signal power and the second term in brackets is the input noise power. Also we have assumed that the noise and signal of the amplifier are uncorrelated such that $\langle A_1 \Delta A_1 \rangle = \langle A_2 \Delta A_2 \rangle = 0$ [6]. In addition, A_1 and A_2 are considered to be stationary quantities [59]. By using the same assumption about A_i and ΔA_i, as well as a similar assumption between A_i and δ_{ji}, and ΔA_i and δ_{ji} $(j = 1, 2)$, Eq. (2.34) can be simplified as:

$$\langle E_{\text{out}}^2 \rangle = (G_1 A_1^2 + G_2 A_2^2) + [G_1 \langle \Delta A_1^2 \rangle + G_2 \langle \Delta A_2^2 \rangle] + [\langle \delta_{1\text{i}}^2 \rangle + \langle \delta_{2\text{i}}^2 \rangle] . \tag{2.36}$$

In Eq. (2.36) the first term in brackets is the amplified output signal power, the second term in brackets is the amplified input noise power and the third term in brackets is the noise power due to spontaneous emissions. From Eqs. (2.35) and (2.36), the signal to noise ratios at the input and output of the amplifier can be found, respectively, to be;

$$(S/N)_{\text{in}} = \frac{(A_1^2 + A_2^2)}{(\Delta A_1^2 + \Delta A_2^2)} \tag{2.37}$$

and

$$(S/N)_{\text{out}} = \frac{G_1 A_1^2 + G_2 A_2^2}{G_1\langle\Delta A_1^2\rangle + G_2\langle\Delta A_2^2\rangle + \langle\delta_{1i}^2\rangle + \langle\delta_{2i}^2\rangle}\,. \tag{2.38}$$

Consider a phase insensitive amplifier with $G = G_1 = G_2$. Then the noise figure F is given by;

$$F = \frac{(S/N)_{\text{in}}}{(S/N)_{\text{out}}} = \frac{G[\langle\Delta A_1^2\rangle + \langle\Delta A_2^2\rangle] + \langle\delta_{1i}^2\rangle + \langle\delta_{2i}^2\rangle}{G[\langle\Delta A_1^2\rangle + \langle\Delta A_2^2\rangle]}\,. \tag{2.39}$$

When an amplifier is operating at the quantum limit (i.e. a single noise photon can be detected) the uncertainty principle [8] tells us that we cannot distinguish between a photon due to spontaneous emission and one due to the amplified input noise. Under this circumstance, we can postulate that after ensemble averaging we have [6];

$$\langle\Delta A_1^2\rangle = \frac{\langle\delta_{1i}^2\rangle}{G} \tag{2.40a}$$

$$\langle\Delta A_2^2\rangle = \frac{\langle\delta_{2i}^2\rangle}{G}\,. \tag{2.40b}$$

Substituting Eq. (2.40) into Eq. (2.39) gives $F = 2$ (i.e. 3 dB). Hence, the basic cause of this limit of noise figure of an optical amplifier is the amplification of uncertainty (or fluctuation) at the input and output [6]. For most optical amplifier applications, this represents the minimum noise figure which any real device can ultimately achieve. F can reach 0 dB if gain is phase sensitive (i.e. $G_1 \neq G_2$). This corresponds to optical amplifiers operating as parametric amplifiers [10, 60, 61], that are beyond the scope of the present study.

2.6 Summary

In this chapter, the most important properties of an optical amplifier have been discussed. They are (i) amplifier gain, (ii) bandwidth, (iii) saturation output power, and (iv) noise figure. The particular importance of noise figure has been highlighted, and has been shown to be the parameter that is most significant in limiting the performance of any system incorporating optical amplifiers.

A two-level system has been used to discuss how these four properties arise in optical amplifiers. Although a more detailed treatment on optical amplifiers will be given in subsequent chapters, this simple model is sufficient to give a basic physical understanding of the fundamental principles of optical amplifiers.

The ideal performance characteristics of optical amplifiers have also been explored. To design real devices which can perform closely to this ideal, quantitative criteria are needed. These ideal characteristics are therefore taken further from qualitative terms to quantitative descriptions of the ultimate performance limits of an optical amplifier. It has been seen that the structure and material characteristics of the amplifier play an important role in determining its ultimate performance. Finally, the ultimate noise figure of an optical amplifier operating at the quantum limit has been analysed using a rather simplistic and intuitive approach. This quantum limit of optical amplifiers determines the physical limit at which a device can perform.

References

[1] A. Yariv, *Quantum Electronics*, 3rd Edition (John Wiley and Sons, 1989).

[2] N. Nakagawa and S. Shimada, "Optical amplifiers in future optical communication systems," *IEEE LCS Magazine*, Vol. 1, No. 4, pp. 57–62, 1990.

[3] C. H. Henry and B. H. Verbeck, "Solution of the scalar wave equation for arbitrarily shaped dielectric waveguides by two-dimensional Fourier analysis," *IEEE J. Lightwave Technology*, Vol. LT-7, No. 32, pp. 308–313, 1989.

[4] D. Marcuse, "Derivation of analytical expression for the bit-rate probability of lightwave systems with optical amplifiers," *IEEE J. Lightwave Technology*, Vol. LT-8, No. 12, pp. 1816–1823, 1990.

[5] A. Yariv, "Signal-to-noise considerations in fiber links with periodic or distributed optical amplification," *Opt. Lett.*, Vol. 15, No. 19, pp. 1064–1066, 1990.

[6] Y. Yamamoto and T. Mukai, "Fundamental of optical amplifiers," *Opt. Quantum Electronics*, Vol. QE-21, pp. S1–S14, 1989.

[7] M. I. Sargent, M. O Scully and W. E. Lamb Jr, *Laser physics* (Addison-Wesley, 1974).

[8] W. H. Louisell, *Quantum statistical properties of radiation* (New York; John Wiley and Sons, 1973).

[9] R. Loudon, *The quantum theory of light*, 2nd Edition (Oxford University Press, 1983).

[10] A. Yariv, *Optical Electronics*, 3rd Edition (Holt-Saunders, 1985).

[11] D. Marcuse, "Computer model of an injection laser amplifier," *IEEE J. Quantum Electronics*, Vol. QE-19, No. 1, pp. 63–73, 1983.

[12] G. H. B. Thompson, *Physics of semiconductor laser devices* (New York, John Wiley and Sons, 1980).

[13] Y. Mukai, Y. Yamamoto and T. Kimura, "Optical amplification by semiconductor lasers," *Semiconductor and semimetals*, Vol. 22, Part E, pp. 265–319, Academic Press, 1985.

[14] T. Mukai and Y. Yamamoto, "Gain, frequency bandwidth, and saturation output power of AlGaAs DH laser amplifiers," *IEEE J. Quantum Electronics*, Vol. QE-17, No. 6, pp. 1028–1034, 1981.

[15] T. Mukai and Y. Yamamoto, "Noise in an AlGaAs semiconductor laser amplifier," *IEEE J. Quantum Electronics*, Vol. QE-18, No. 4, pp. 564–575, 1982.

[16] G. Eisenstein, "Semiconductor optical amplifiers," *IEEE Circuits and Devices Magazine*, pp. 25–30, July 1989.

[17] M. J. O'Mahony, "Semiconductor laser optical amplifiers for use in future fibre systems," *IEEE J. Lightwave Technology*, Vol. LT-6, No. 4, pp. 531–544, 1988.

[18] J. L. Gimlett and N. K. Cheung, "Effects of phase-to-intensity noise conversion by multiple reflections on giga-bit-per-second DFB laser transmission system," *J. Lightwave Technology*, Vol. LT-7, No. 6, pp. 888–895, 1989.

[19] J. L. Gimlett *et al.*, "Impact of multiple reflection noise in Gbit/s lightwave system with optical fibre amplifiers," *Electronics Letters*, Vol. 25, No. 20, pp. 1393–1394, 1989.

[20] A. Elrefaire and C. Lin, "Performance degradations of multi gigabit-per-second NRZ/RZ lightwave system due to gain saturation in travelling-wave semiconductor optical amplifiers," *IEEE Photonics Technology Letters*, Vol. 1, No. 10, pp. 300–303, 1989.

[21] M. J. Adams, "Time dependent analysis of active and passive optical bistability in semiconductors," *IEE Proc.*, Part J, Vol. 132, No. 6, pp. 343–348, 1985.

[22] J. W. Crowe and R. M. Craig, "Small-signal amplification in GaAs lasers," *Appl. Phys. Lett.*, Vol. 4, No. 3, pp. 57–58, 1964.

[23] W. J. Crowe and W. E. Aheam, "Semiconductor laser amplifier," *IEEE J. Quantum Electronics*, Vol. QE-2, No. 8, pp. 283–289, 1966.

[24] J. Wang, H. Olesen and K. E. Stubkjaer, "Recombination, gain and bandwidth characteristics of 1.3 m semiconductor laser amplifiers," *IEEE J. Lightwave Technology*, Vol. LT-5, No. 1, pp. 184–189, 1987.

[25] T. Mukai, Y. Yamamoto and T. Kimura, "S/N and error rate performance in AlGaAs semiconductor laser preamplifier and linear repeater systems," *IEEE Trans. Microwave Theory Tech.*, Vol. MTT-30, No. 10, pp. 1548–1556, 1982.

[26] H. Ghafouri-Shiraz and C. Y. J. Chu, "Analysis of waveguiding properties of travelling-wave semiconductor laser amplifiers using perturbation technique," *Fiber & Integrated Optics*, Vol. 11, pp. 51–70, 1992.

[27] D. Marcuse, *Theory of dielectric optical waveguides* (Academic Press, 1974).

[28] M. J. Adams, *An Introduction to Optical Waveguides* (John Wiley and Sons, 1981).

[29] C. J. Hwang, "Properties of spontaneous and stimulated emissions in GaAs junction laser; I & II," *Phy. Rev. B*, Vol. 2, No. 10, pp. 4117–4134, 1970.

[30] F. Stern, "Calculated spectral dependence of gain in excited GaAs," *J. Appl. Phys.*, Vol. 47, No. 2, pp. 5382–5386.

[31] M. Osinski and M. J. Adam, "Gain spectra of quaternary semiconductors," *IEE Proc.*, Part J, Vol. 129, No. 6, pp. 229–236, 1982.

[32] I. D. Henning, M. J. Adams and J. V. Collins, "Performance prediction from a new optical amplifier model," *IEEE J. Quantum Electron.*, Vol. QE-21, No. 6, pp. 609–613, 1985.

[33] T. Saitoh and M. Mukai, "1.5 m GaInAs travelling-wave semiconductor laser amplifier," *IEEE J. Quantum Electronics*, Vol. QE-23, No. 6, pp. 1010–1020, 1987.

[34] T. Mukai, K. Inoue and T. Saitoh, "Homogeneous gain saturation in 1.5 m InGaAsP travelling-wave semiconductor laser amplifiers," *Appl. Phys. Lett.*, Vol. 51, No. 6, pp. 381–383, 1987.

[35] R. Frankenberger and R. Schimpe, "Origin of non-linear gain saturation in index-guided InGaAsP laser diodes," *Appl. Phys. Lett.*, Vol. 60, No. 22, pp. 2720–2722, 1992.

[36] G. Lasher and F. Stern, "Spontaneous and stimulated recombination radiation in semiconductors," *Phys. Rev.*, Vol. 133, No. 2A, pp. 553–563.

[37] T. Mukai and T. Yamamoto, "Noise characteristics of semiconductor laser amplifiers," *Electronics Letters*, Vol. 17, No. 1, pp. 31–33, 1981.

[38] C. H. Henry, "Theory of the phase noise and power spectrum of a single mode injection laser," *IEEE J. Quantum Electronics*, Vol. QE-19, No. 9, pp. 1391–1397, 1983.

[39] K. Hinton, "Optical carrier linewidth broadening in a travelling wave semiconductor laser amplifier," *IEEE J. Quantum Electronics*, Vol. QE-21, pp. 533–546, 1989.

[40] K. Kikuchi, C. E. Zah and T. P. Lee, "Measurement and analysis of phase noise generated from semiconductor optical amplifiers," *IEEE J. Quantum Electronics*, Vol. QE-27, No. 3, pp. 416–422, 1991.

[41] A. J. Lowery, "Amplified spontaneous emission in semiconductor laser amplifiers: validity of the transmission-line laser model," *IEE Proc.*, Part J, Vol. 137, No. 4, pp. 241–247, 1990.

[42] H. Taub and D. L. Schilling, "Principles of communication systems," 2nd Edition (MacGraw-Hill, 1986).

[43] C. Y. J. Chu, *Semiconductor laser optical amplifiers for optical communications*, M. Phil. (Eng.), qualifying thesis, University of Birmingham, UK.

[44] Y. Yamamoto, "Noise and error rate performance of semiconductor laser amplifiers in PCM-IM optical transmission systems," *IEEE J. Quantum Electronics*, Vol. QE-16, No. 10, pp. 1073–1081, 1980.

[45] D. Marcuse, "Calculation of bit-error probability for a lightwave system with optical amplifiers," *IEEE J. Lightwave Technology*, Vol. 9, No. 4, pp. 505–513, 1991.

[46] A. Lord and W. A. Stallard, "A laser amplifier model for system optimization," *Opt. Quantum Electron.*, Vol. QE-21, pp. 463–470, 1989.

[47] O. Lumholt *et al.*, "Optimum position of isolators within erbium-doped fibres," *IEEE Photonics Technology Letters*, Vol. 4, No. 6, pp.'568–569, 1992.

[48] D. J. Maylon and W. A. Stallard, "565 Mbit/s FSK direct detection syetem operating with four cascaded photonic amplifiers," *Electronics Letters*, Vol. 25, No. 8, pp. 495–497, 1989.

[49] K. Vahala and A. Yariv, "Semiclassical theory of noise in semiconductor lasers-Part I and II," *IEEE J. Quantum Electronics*, Vol. QE-19, No. 6, pp. 1096–1109, 1983.

[50] P. Spano, S. Piazzola and M. Tamburrini, "Phase noise in semiconductor lasers; a theoretical approach," *IEEE J. Quantum Electronics*, Vol. QE-19, No. 7, pp. 1195–1199. 1983.

[51] C. H. Henry, "Phase noise in semiconductor lasers," *IEEE J. Lightwave Technology*, Vol. LT-4, No. 3, pp. 298–311, 1986.

[52] B. Glance, G. Eisenstein, P.J. Fitzgerald, K. J. Pollack and G. Raybon, "Optical amplification in a multichannel FSK coherent system," *Electronics Letters*, Vol. 24, No. 18, pp. 1157–1159, 1988.

[53] M. A. Ali, A. F. Elrefaie and S. A. Ahmed, "Simulation of 12.5 Gb/s lightwave optical time-division multiplexing using semiconductor optical amplifiers as external modulators," *IEEE Photonics Technology Letters*, Vol. 4, No. 3, pp. 280–282, 1992.

[54] R. M. Fortenberry, A. J Lowery and R. S. Tucker, "Up to 16 dB improvement in detected voltage using two section semiconductor optical amplifier detector," *Electronics Letters*, Vol. 28, No. 5, pp. 474–476, 1992.

[55] G. P. Agrawal, "Effect of gain and index nonlinearities on single-mode dynamics in semiconductor lasers," *IEEE J. Quantum Electronics*, Vol. QE-26, No. 11, pp. 1901-1909, Nov. 1990.

[56] M. J. Adams, J. V. Collins and I. D. Henning, "Analysis of semiconductor laser optical amplifiers," *IEE Proc.*, Vol. 132, Part J, No. 1, pp. 58–63, Feb. 1985.

[57] A. J. Lowery, "Modelling spectral effects of dynamic saturation in semiconductor laser amplifiers using the transmission-line laser model," *IEE Proc.*, Vol. 136, Part J, No. 6, pp. 320–324, Dec. 1989.

[58] Z. Pan, H. Lin and M. Dagenais, "Switching power dependence on detuning and current in bistable diode laser amplifiers," *Appl. Phy. Lett.*, Vol. 58, No. 7, pp. 687–689, Feb. 1991.

[59] D. L. Snyder, *Random Point Processes*, John Wiley and sons, New York, 1975.

[60] Y. Yamamoto (Editor), *Coherence, Amplification, and Quantum Effects in Semiconductor Lasers*, John Wiley and sons, New York, 1991.
[61] K. Kikuchi, "Proposal and performance analysis of novel optical homodyne receiver having an optical preamplifier for achieving the receiver sensitivity beyond the shot-noise limit", *IEEE Photon. Technol. Lett.*, Vol. 4, No. 2, pp. 195–197, 1992.

Chapter 3

Optical Amplification in Semiconductor Laser Diodes

3.1 Introduction

In Chapter 2, the major characteristics of optical amplifiers were explored. These characteristics are important in determining the ultimate performance of any optical communication system using optical amplifiers [1]. However, as discussed in the previous chapter, the exact performance requirements of an optical amplifier vary with its type and the application for which it is used. In this chapter, we continue our discussion by considering how optical amplification can be achieved in semiconductor lasers. This will help us to understand the physical processes which determine the characteristics of semiconductor laser amplifiers (SLAs), and hence the performance requirements in different SLA applications. The mechanisms of optical gain in semiconductor materials will first be reviewed, and we will discuss how optical amplification can be realised in semiconductor laser structures. A summary of the historical development of SLAs from literature will then be reviewed, and different types of semiconductor laser amplifiers will also be examined. Finally, a review of current research trends in different applications, and the subsequent performance requirements of SLAs, will be presented.

3.2 Principles of Optical Amplification in Semiconductor Lasers

It is well known in linear systems theory that an oscillator is essentially an amplifier with positive feedback. The same is true for laser diode oscillators.

The gain of an amplifier with positive feedback as mentioned in Chapter 2 is given by:

$$G = \frac{G_s}{1 + FG_s} \tag{3.1}$$

where G_s is the forward path gain and F is the amount of feedbak gain. Equation (3.1) indicates that oscillation will occur when $FG_s = -1$. As long as $|FG_s| < 1$, the amplifier will not oscillate and will provide gain according to Eq. (3.1). This is true for semiconductor lasers as well. If one can reduce F, and/or G_s, such that the laser is operating below oscillation threshold, then the laser can be used as a simple optical amplifier for light signals [2]. Historically, the amplification using semiconductor lasers stemmed from this idea of using the laser below oscillation threshold. Therefore, the early development of SLAs was closely related to that of semiconductor laser diodes (SLDs), and in many early works, a SLA is identical to a SLD in terms of their structures. If one is to understand more fully the principles of optical amplification using semiconductor lasers, one must first understand (i) how optical gain is achieved in semiconductors, and (ii) the structural aspects of semiconductor lasers which will affect *both* the values of F and G_s. We will now review these two aspects briefly, to provide an introduction to the results of further investigations which will be presented later.

3.2.1 *Optical processes in semiconductors*

As we saw in Chapter 2, population inversion is the vital condition to provide optical amplification. In semiconductors, population inversion is formed by forward biasing a heavily doped p.n. junction [3, 4], for which the band diagram is shown in Fig. 3.1. For such a heavily doped *degenerate* semiconductor, the Fermi energy level lies above the conduction band edge in an n-type material, and below the valence band edge in a p-type material [5]. When these two types of material are joined to form a p.n. junction, diffusion of minority carriers occurs (i.e. electrons diffuse toward *p*-side, and holes toward *n*-side). The diffusion stops when the resulting build up of electric field can counteract the diffusion process (Fig. 3.1(a)). This field is known as the *depletion* electric field, and can be lowered by applying a forward bias across the junction as shown in Fig. 3.1(b). A drift of minority carriers across the junction occurs, in addition to the diffusion process. Because of the positions of the Fermi energy levels, these carrier movements will create a very narrow region in which there are both electrons in the

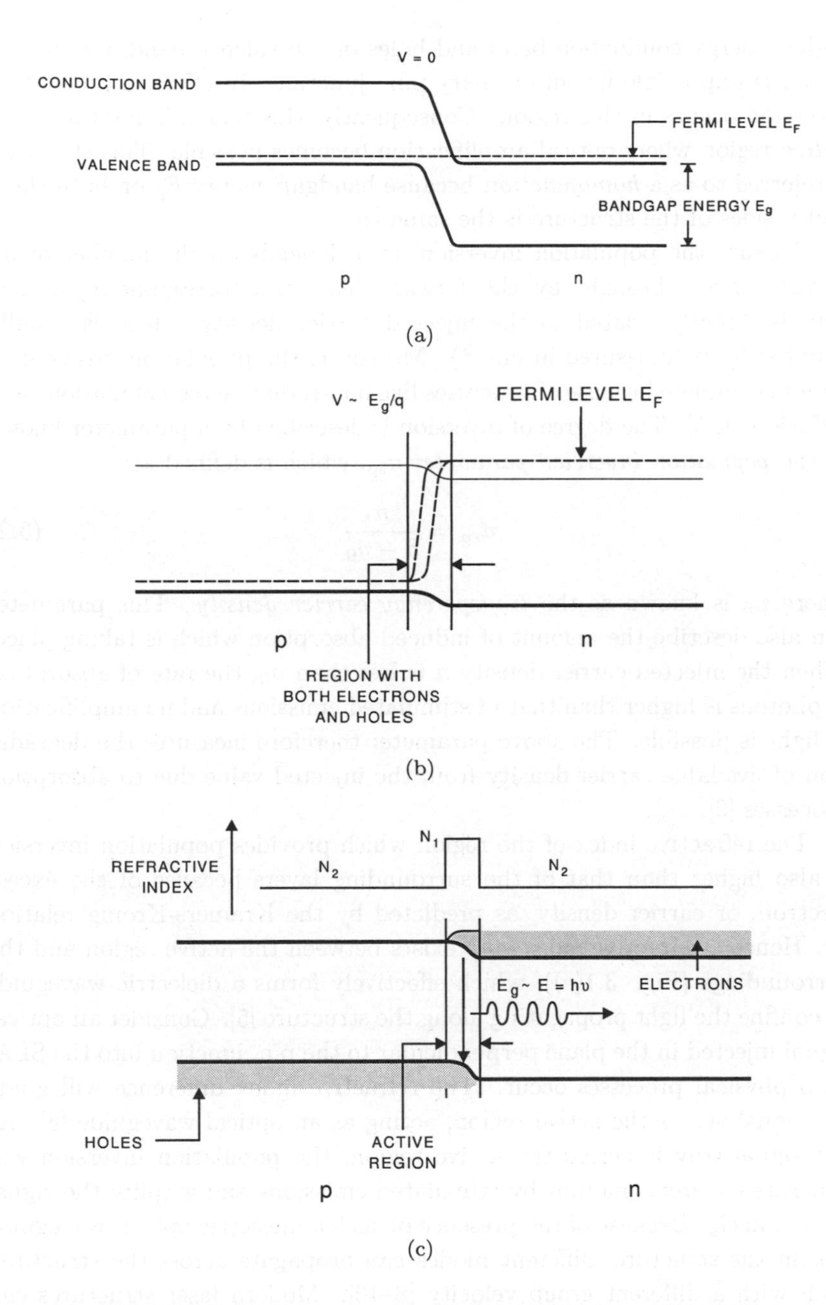

Fig. 3.1. (a) Unbiased p.n. junction under thermal equilibrium. (b) Forward biased heavily doped p.n. junction. (c) Formation of active region under forward bias.

higher energy conduction band and holes in the valence band, a situation which is impossible for an ordinary p.n. junction. In effect, a population inversion occurs in this region. Consequently, this region is known as the *active* region where optical amplification becomes possible. This structure is referred to as a *homojunction* because bandgap energy E_g on both the p and n sides of the structure is the same [4].

Because the population inversion level depends on the number of injected carriers brought by the forward bias, the corresponding optical gain is directly related to the injected carrier density, which is usually denoted by n (measured in cm^{-3}). Moreover, the population inversion is seldom complete because of processes like non-radiative recombinations and diffusion [4, 5]. The degree of inversion is described by a parameter known as the *population inversion parameter* n_{sp}, which is defined as:

$$n_{sp} = \frac{n}{n - n_0} \tag{3.2}$$

where n_0 is known as the *transparency carrier density.* This parameter can also describe the amount of induced absorption which is taking place. When the injected carrier density n is less than n_0, the rate of absorption of photons is higher than that of stimulated emissions and no amplification of light is possible. The above parameter therefore measures the degradation of available carrier density from the injected value due to absorption processes [6].

The refractive index of the region which provides population inversion is also higher than that of the surrounding layers because of the excess electron, or carrier density, as predicted by the Kramers-Kronig relation [7]. Hence a refractive index step exists between the active region and the surroundings (Fig. 3.1(c)), which effectively forms a dielectric waveguide to confine the light propagating along the structure [5]. Consider an optical signal injected in the plane perpendicular to the p.n. junction into the SLA. Two physical processes occur. The refractive index difference will guide the signal across the active region, acting as an optical waveguide [8]. As the signal travels across the active region, the population inversion will generate coherent photons by stimulated emissions and amplify the signal by a gain G_s. Because of the presence of such a dielectric optical waveguide within the structure, different modes can propagate across the structure, each with a different group velocity [8–10]. Modern laser structures can minimise the number of modes which can propagate to two by tailoring the active region dimensions [4, 5, 11]. It can be shown that the electric fields

of these modes are orthogonal to each other, and we regard each of these modes as having different *states of polarisation* [12]. We will discuss this in further detail in the following chapter. It should be noted, however, that because the active region can at least support propagation of signals with two different states of polarisation having different group velocities (hence confinement within the structure), one will also expect the gain for these two states of polarisation to be different. The resulting *polarisation sensitivity* of the optical gain in a laser amplifier is one of the crucial areas which has been addressed recently by many research workers [13–16], and will be discussed in depth in the next chapter.

In addition to losses due to induced absorption, photons can be lost via scattering [4]. The overall gain in the active region can be quantified by two material parameters: the *material gain coefficient* g_m and *loss coefficient* α. Both quantities are measured in cm^{-1}, and for a distance L, these two parameters are related to G_s by the following relation:

$$G_s = \exp(g_m - \alpha)L\,. \tag{3.3}$$

In the above equation, we have assumed a perfect guiding condition, that is, all signals are confined to the active region by the waveguiding action [8]. In practice, because the waveguide formed is only a weakly guiding one, with $(N_1 - N_2) \ll N_1$ (see Fig. 3.1(c)), power will "leak" out of the active region into the cladding (the bulk of dielectric material surrounding the active region) due to evanescent fields [9]. To take this into account, Eq. (3.3) is modified by a weighting factor Γ such that:

$$G_s = \exp[\Gamma(g_m - \alpha_a) - (1-\Gamma)\alpha_c]L \tag{3.4}$$

where α_a and α_c are the absorption coefficients in the active and claddings regions, respectively. The factor Γ is known as the optical confinement factor, and is defined as the ratio between the optical power confined in the active region to the total optical power flowing across the structure [8, 17]. The form of Eq. (3.4) suggests that the proportion of power in the active region is related to the magnitude of the material gain coefficient of the structure, and an alternative expression for more complex structures has been derived by the author using perturbation theory [18]. Alternatively, it can be shown that Eq. (3.4) can be obtained directly by analysing the power flow using Poynting vectors [8]. Furthermore, as we have discussed previously, the state of polarisation of the signals will affect the optical gain of the amplifier. In practice, it is the optical confinement factor Γ

which is actually affected by the polarisation of the signals. Hence, the value of G_s is polarisation-dependent according to Eq. (3.4). This is particularly important when the facet reflectivities of the amplifiers are zero (see Section 3.4.1), in which case the overall gain of the amplifier $G = G_s$.

3.2.2 *Analysis of optical gain in semiconductors*

The exact mathematical formulation of material gain coefficients in semiconductors had been investigated since the early developments in semiconductor lasers and laser amplifiers [19–21]. In practice, the population inversion in semiconductor lasers cannot be described quantitatively in a straightforward matter. Because of the presence of bands rather than sharp and distinct energy levels, the density of states and hence population densities of carriers in both conduction and valence bands are spread non-uniformly over a range of photon energies [5]. This statistical spread of carrier density results in different stimulated emission rates for different incident photon energy (hence signal frequencies and wavelengths). This spectral function of stimulated emission will ultimately determine the material gain coefficient, and has to be found by Fermi-Dirac statistics with some assumptions on the band structure as well as the nature of the radiative transitions [4, 22]. The detailed formulation is clearly beyond the scope of the present book, though interested readers can find a brief description in Section 3.6. In the following discussion, we will describe some of the most useful approximations which can be used conveniently in analysing the material gain coefficients in semiconductor laser devices without using complex statistical and quantum mechanical methods.

The results of the exact calculation [21] of the *peak* material gain coefficient g_{peak} can be approximated by the following expression [3, 23]:

$$g_{\text{peak}} = A(n - n_o) \tag{3.5}$$

where A is a constant. This linear relationship gives an accurate description of the relation between material gain and injected carrier density, except in the range close to the transparency region, that is, when g is very small. To overcome this problem, a more accurate parabolic approximation of the form:

$$g_{\text{peak}} = an^2 + bn + c \tag{3.6}$$

where a, b and c are constants, has been suggested [24]. The constants in Eq. (3.6) can be found by a least square fitting technique to the available

Table 3.1. Semiconductor laser parameters use to calculate Fig. 3.2.

Wavelength	λ	1.3 μm
Temperature	T	300 K
Coeffiecients	A	7.2543×10^{-16} cm^{-2}
	n_o	1.5034×10^{18} cm^{-3}
	a	3.9138×10^{-34} cm^5
	b	-8.2139×10^{-18} cm^2
	c	428.7016 cm

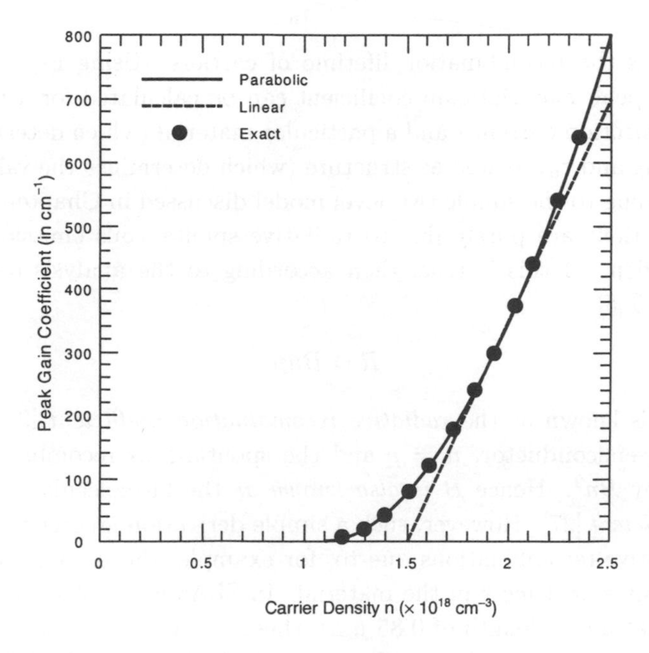

Fig. 3.2. Comparison between linear and parabolic models for peak gain coefficient calculated using the parameters listed in Table 3.1.

exact solution [19]. We have compared the difference between using Eq. (3.5) and (3.6) by calculating the peak gain coefficients of an SLA with the parameters listed in Table 3.1 [25], and the results are plotted on Fig. 3.2. Also shown are the exact solutions calculated by [19]. As shown in Fig. 3.2, a better fit can be obtained for a wider range of values of g_{peak} and n using Eq. (3.6), especially for small values of n.

In the absence of a significant degree of stimulated emission (below threshold in a semiconductor laser, or no signal input if the laser is used as an amplifier), the injected carrier density is related to the injection or bias current i by:

$$i = \frac{eVR}{\eta_\mathrm{i}} \tag{3.7}$$

where e is the electron charge, η_i is the internal quantum efficiency of the material [5], V is the volume of the active region, and R is the *carrier recombination rate* expressed as [4]:

$$R = \frac{n}{\tau_\mathrm{n}} \tag{3.8}$$

where τ_n is the recombination lifetime of carriers. Using Eqs. (3.7) and (3.8), the peak material gain coefficient can be calculated for a particular value of injection current i and a particular material (which determines the values of η_i and τ_n) as well as structure (which determines the value of V).

According to the simple two-level model discussed in Chapter 2, carrier recombinations are purely due to radiative spontaneous emissions in the active region. If this is true, then according to the analysis outlined in Section 3.5 [5]:

$$R = Bnp \tag{3.9}$$

where B is known as the *radiative recombination coefficient* [26]. For an undoped semiconductor, $n = p$ and the spontaneous recombination rate is given by Bn^2. Hence B *is also known as* the bi-molecular *recombination coefficient* [27]. However, such a simple derivation ignores the effect of non-radiative recombinations due to, for example, the presence of recombination sites or defects in the material. In SLAs made of AIGaAs/GaAs (working at a wavelength of 0.85 μm), these non-radiative recombinations of carriers are unimportant. However, for InGaAsP/InP SLAs operating at a longer wavelength of 1.3 and 1.55 μm, the contribution of these non-radiative recombinations is quite significant.

There are two major sources of non-radiative recombinations. We have mentioned recombination with defects above. A far more important recombination process is *Auger recombination* [4]. It involves four particle states (3 electrons and 1 hole, or 2 electrons and 2 holes, etc.). In this process, the energy released during the electron-hole recombination is transferred to another electron or hole, which gets excited and then relaxes back to the lower state by losing its energy to lattice vibration or *phonons*. Both of these recombination processes are similar to spontaneous emissions in that they are all random events. However, in Auger recombinations and recombinations with defects, no photons are emitted. If these non-radiative recombinations are included, the recombination rate R can then be expressed by [4, 26, 27, 33]:

$$R = A_{\mathrm{nr}}n + B_{\mathrm{rad}}n^2 + C_{\mathrm{Aug}}n^3\,. \tag{3.10}$$

The three terms represent recombination with defects or surface, spontaneous emission and Auger recombinations, respectively. The difference between using Eqs. (3.9) and (3.10) on calculating carrier recombination lifetime τ_{n} from Eqs. (3.7) and (3.8) is illustrated by the curves in Fig. 3.3. Here, the values of $1/\tau_{\mathrm{n}}^2$ calculated by the two different models of Eqs. (3.9)

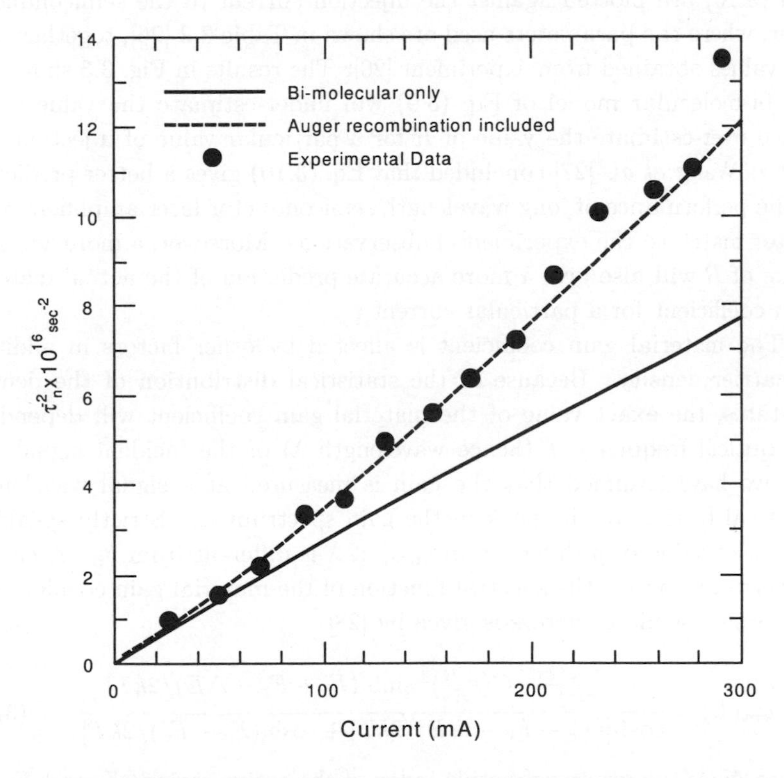

Fig. 3.3. Comparison of recombination lifetime calculated by bi-molecular term only and by including non-radiative recombinations with experimental measurements obtained by Olshansky *et al.* [27].

Table 3.2. Parameters of semiconductor laser used for comparison between different carrier recombination models.

Laser Length	200 μm
Active region width	25 μm
Active region thickness	0.2 μm
Wavelength	1.3 μm
Recombination Coefficients	
A_{nr}	2×10^8 s^{-1}
B_{rad}	3.8×10^{-17} m^{3}s^{-1}
C_{Aug}	5.1×10^{-42} m^{6}s^{-1}

and (3.10) are plotted against the injection current to the semiconductor laser, where the parameters used are shown in Table 3.2 [26], together with the values obtained from experiment [20]. The results in Fig. 3.3 show that the bi-molecular model of Eq. (3.9) will under-estimate the value of τ_{n}, hence over-estimate the value of R for a particular value of injection current i. Wang *et al.* [27] concluded that Eq. (3.10) gives a better prediction of the performance of long wavelength semiconductor laser amplifiers, with better match to the experimental observations. Moreover, a more accurate value of R will also give a more accurate prediction of the actual material gain coefficient for a particular current i.

The material gain coefficient is affected by other factors in addition to carrier density. Because of the statistical distribution of the density of states, the exact value of the material gain coefficient will depend on the optical frequency f (hence wavelength λ) of the incident signal. So far, we have assumed that the gain is measured at a signal wavelength identical to that of the peak in the gain spectrum λ_{o}. Strictly speaking, the exact value of g differs from g_{peak} if λ is different from λ_{o}. A closed-form expression for the spectral function of the material gain coefficient in quaternary semiconductors is given by [28]

$$g_m(E) = \frac{\frac{\pi^3 c^2 (h/2\pi)^3}{8N_{\mathrm{g}}^2} C(\frac{\Delta E}{E})^2 \sinh[(F_{\mathrm{c}} + F_{\mathrm{v}} - \Delta E)/2kT]}{\cosh[(F_{\mathrm{c}} + F_{\mathrm{v}} - \Delta E)/2kT] + \cosh[(F_{\mathrm{c}} - F_{\mathrm{v}})/2kT]} \tag{3.11}$$

where N_{g} is the group refractive index of the active region, F_{c} and F_{v} are the quasi-Fermi levels in the conduction and valence bands, respectively, ΔE is the energy difference between the signal photon and the band gap energy E_g (the energy difference between the conduction and valence band edges), T is the temperature, k is Boltzmann constant, and C is a constant which weights out the contribution of masses of electrons and holes in the spectral function of gain. This expression is true for semiconductors with parabolic bands and with no k-selection for radiative transitions [5, 7].

We have plotted the material gain spectrum obtained from Eq. (3.11) in Fig. 3.4 for undoped $\mathrm{In}_{1-x}\mathrm{Ga}_x\mathrm{As}_y\mathrm{P}_{1-y}$ having $x = 0.25$, $y = 0.55$ at room temperature ($T = 300°\mathrm{K}$) for six different values of carrier density n and two different wavelengths. It can be seen that the whole spectrum shifted upwards as n increases. There are two values of λ where g_m crosses over zero. First, there is a gradual decrease of material gain with longer wavelength. Note that this cut-off point for the longer wavelength end is almost identical for all values of injection carrier density (see Fig. 3.4). This can be explained as follows. As the wavelength increases, the energy of the photon reduces until it is less than the bandgap energy of the material (Fig. 3.1(c)). Emissions (and absorption) of photons with energies less than the bandgap energy are not allowed, and hence the optical gain reduces to zero for this particular value of photon energy. This explains why, for all levels of injection, the cut-offs at the longer wavelength end are almost identical. The slight differences between these zero-crossings for different injection carrier densities arise because of the slight modification of the shape of the band edge (known as *band tailing*) with injection current [22].

The cut-off of $g_\mathrm{m} = 0$ at the shorter wavelength end can be understood from the fact that optical gain can only be maintained if the signal photon frequency f satisfies the basic constraint of (see Section 3.6) [4, 7]

$$E = hf \leq (F_\mathrm{c} - F_\mathrm{v})\,. \tag{3.12}$$

The upper limit on frequency f in the above equation accounts for this shorter wavelength cut-off of the spectrum. As the carrier densities decrease for energy levels which are further away from the band edge in both conduction and valence bands, the stimulated emission rate also reduces for transitions involving larger photon energies. Hence the signal gain reduces with reducing wavelength until $g_\mathrm{m} < 0$, beyond which absorption predominates over stimulated emissions. Equation (3.11) can be approximated by the following parabolic form around the peak gain region [29]:

$$g_\mathrm{m}(n, \lambda) = g_\mathrm{peak}(n) - \gamma(\lambda - \lambda_0)^2 \tag{3.13}$$

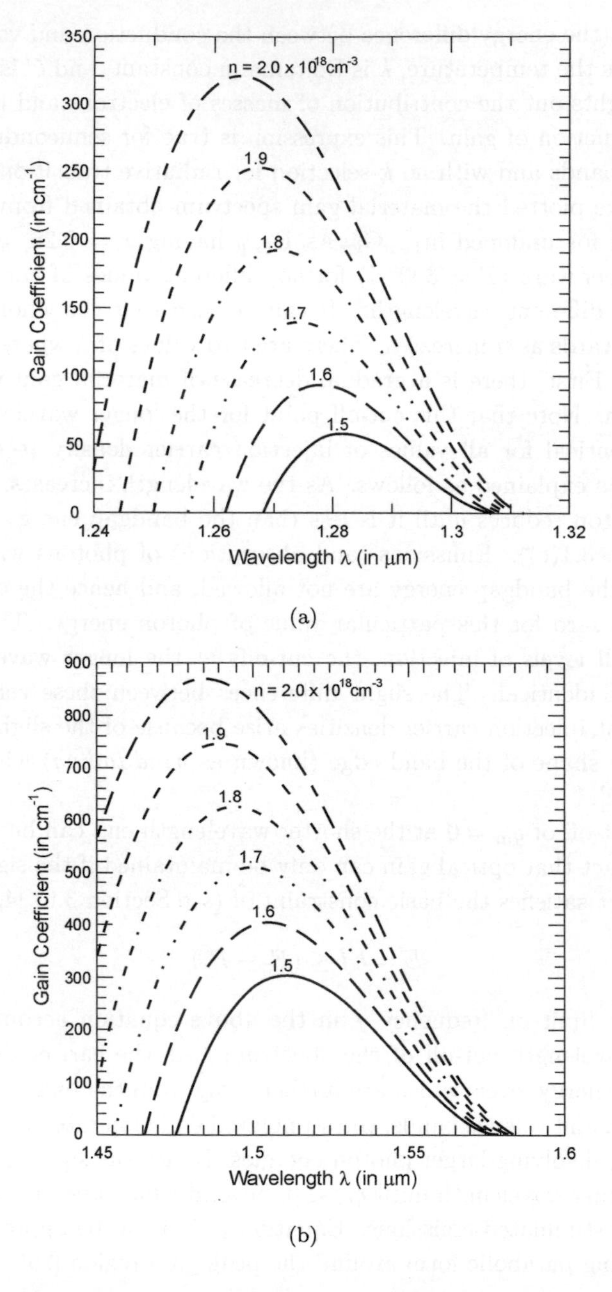

Fig. 3.4. Gain spectra for undoped InGaAsP at $T = 300$ K using Eq. (3.11) for injection carrier density ranging from 1.5×10^{18} to 2.5×10^{18} cm^{-3} (a) 1.3 μm and (b) 1.55 μm wavelengths.

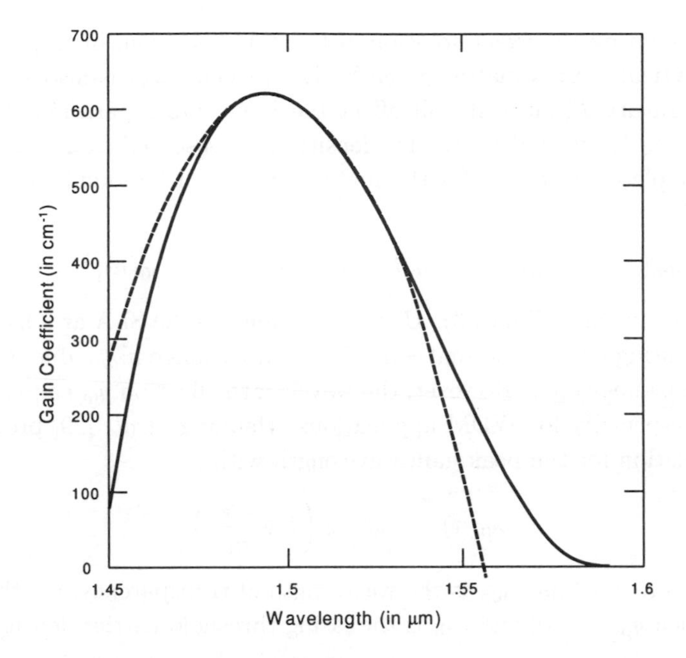

Fig. 3.5. An example of the parabolic fit to an exact gain spectrum of undoped InGaAsP. The solid line represents gain coefficients calculated by Eq. (3.11) with $n = 1.8 \times 10^{18}$ cm^{-3} for the emission wavelength of 1.5 μm, whilst dotted line represents the parabolic fit around gain peak region.

where γ is a constant related to the actual spectral width of the material gain spectrum. An example of such fitting to one of the curves in Fig. 3.4 is shown in Fig. 3.5. Notice that from Eq. (3.13), one can see that g_{m} will be increased by the same proportion for all wavelengths in the gain spectrum as n increases. The gain spectrum is said to be *homogeneously broadened.* By increasing the injection current, the peak gain value will increase until a point where saturation will occur (notice the difference between approaching saturation in this way with that discussed in Chapter 2). At this point, the peak material gain coefficient cannot be increased further by increasing pumping, and is effectively clamped [7]. A common situation to observe this phenomenon is when lasing occurs in semiconductor lasers. It is found that in semiconductor lasers, when the peak gain coefficient has reached the saturation value, the gain coefficients for other optical wavelengths across the whole gain spectrum are also clamped, although they are still far away from saturation value. This phenomenon is known as *homogeneous gain saturation.* This property was also shown to be true in SLAs [30].

As shown by the exact solution of the gain spectrum in [45] and from the analytical approximation given by Eq. (3.11), g_{peak} is also a function of temperature T, since it will affect the band gap separation [103]. In addition, λ_0 is affected by carrier density n as observed from Fig. 3.4. A more complete expression for the material gain can be found to be of the form [27].

$$g_{\mathrm{m}}(\lambda, n) = [a(T)n^2 + b(T)n + c(T)] - \gamma[\lambda - \lambda_0(n)]^2 \tag{3.14}$$

where we have used Eq. (3.6). Usually, we operate the SLA at a fixed temperature using electro-thermal control, and any change in g_{m} due to change in T can be neglected. However, the wavelength effect on g_{m} cannot be neglected, especially for WDM applications. Henning *et al.* [29] proposed a linear relation for the peak gain wavelength with n:

$$\lambda_0(n) = \lambda_{0\mathrm{p}} - \xi\left(1 + \frac{n}{n_{\mathrm{th}}}\right) \tag{3.15}$$

where ξ is a constant, $\lambda_{0\mathrm{p}}$ is the wavelength of transparency, i.e. the value of λ_0 when $g_{\mathrm{peak}} = 0$, and n_{th} is the lasing threshold carrier density.

3.3 Semiconductor Laser Diodes as Optical Amplifiers

The possibility of introducing optical gain in semiconductors has led to the birth of lasers fabricated by homojunctions, as described in the previous section, in the early 1960s. At the same time, it was realised that the active region in these devices can also be used as an active waveguide to provide optical amplification below the lasing threshold. This eventually gave birth to the first semiconductor laser amplifiers. In the following, we will briefly trace the development of semiconductor laser amplifiers, where we will find, not surprisingly, that their development was closely linked to advances in technologies related to semiconductor laser diodes.

3.3.1 *Optical amplification using homojunctions*

After the first successful operation of semiconductor laser diodes [31, 32], the development of laser diode amplifiers received considerable attention. In the early stages, SLDs were fabricated with homojunctions using GaAs as the amplifying material (see Fig. 3.6) [4, 5]. These homojunctions required a large threshold current to achieve laser oscillation, and could only be used in continuous-wave (cw) operation under very low sub-zero temperatures.

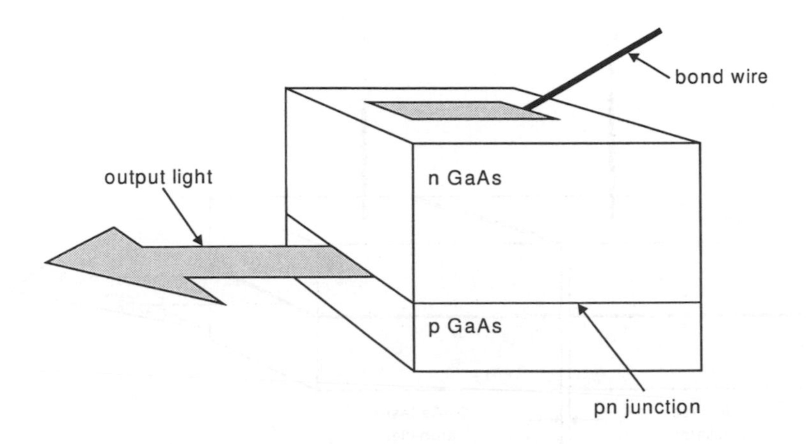

Fig. 3.6. A homojunction semiconductor laser.

At the same time, there were several successful attempts to use these homojunction SLDs as optical amplifiers by reducing the injection current, and hence the single pass gain G_s, below threshold [6]. Crowe and Craig [31] reported a small signal amplification with GaAs laser diodes at 77°K. The injection current required to create the active region was found to be in the range of several amperes. A maximum small signal gain of 1000 (= 30 dB) can be obtained with an injection current density of 6000 A/cm^2. The bandwidth of the amplifier was found to be 15 Å. An amplifier with a similar structure but higher gain was reported by [32] two years later. The maximum signal gain reported at 77°K was 2000, with an output power of 150 mW. The bandwidth of the amplifier was 30 Å, twice that in the former experiment. In both experiments, gain saturation occurred for high input light power with constant pumping (i.e. injection current). The gain-current characteristics were found to be exponential in both experiments, agreeing with the available theory at that time [32]. It was found that the gain measurement was very sensitive to coupling between the source and amplifier [31]. Side modes were observed at the output of the amplifier with a single mode input in the earlier experiment. Crowe and Ahearn [32] proposed the use of anti reflection (AR) coatings by deposition of multi-layer dielectric three-quarter-wavelength coatings using SiO_2 to suppress these side modes. Thus the application of AR coatings in SLAs can be traced back as early as 1966. They also discovered a shift in output wavelength of the amplifier towards the blue end of the optical spectrum as the pumping increases, which remained unexplained at that time.

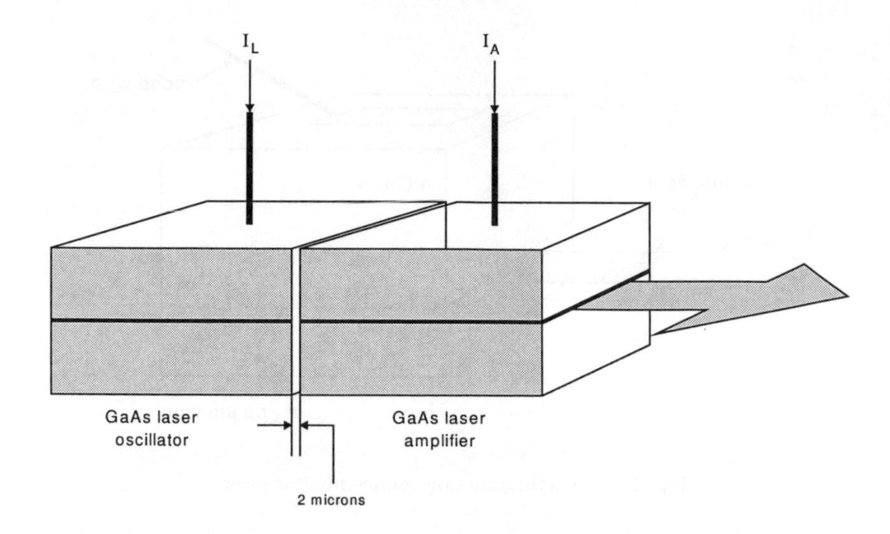

Fig. 3.7. A cleaved-substrate semiconductor laser amplifier.

Kosonocky and Cornley [33] reported a different device configuration as an optical amplifier. Their objective was to design an amplifier directly coupled to an oscillator. This was the first attempt to design a cleaved substrate amplifier [34]. A schematic of their device is shown in Fig. 3.7. Again, their experimental results showed a strong dependence of signal gain on input signal power. The maximum amplifier gain obtained was about 150 with an input signal power flux of 8 kW/cm^2. They also observed that spontaneous emission power was quenched as the input signal power increased, and the internal oscillation modes of the amplifier were suppressed as well. The amplifier gain was found to be inversely proportional to the separation between the oscillator and amplifier, since the coupling efficiency between the oscillator and the amplifier reduces with increasing separation. Their amplifier possessed a strong non-linear characteristic because of the non-uniform gain distribution along the longitudinal direction of the amplifier (see Chapter 4), hence they proposed a technique for maintaining a uniform gain by exciting oscillations along the *transverse* section of the amplifier. Although this technique is not used now, their effort highlighted the problem faced by research workers in the 1960s, of how to improve the gain uniformity along the longitudinal direction of the laser amplifier.

At the same time, much theoretical work was published, introducing new ideas on the applications of optical amplifiers. For example, Arnaud [35] proposed that optical amplifiers could be used as pre-amplifiers to optical receivers to enhance receiver sensitivities. Personick [36] showed that by using optical amplifiers as in-line repeaters, the performance of simple digital optical transmission using on-off keying could be improved. In addition, optical amplifiers could be used as font-end amplifiers for regenerative repeaters to improve their sensitivities. This early theoretical work laid down fundamentals for later work in system applications of optical amplifiers.

3.3.2 *Optical amplification using heterostructures*

The homojunction SLDs were soon found to be unsatisfactory, both as amplifiers and as oscillators, because of their operational requirements of low temperature, large threshold currents and difficulties in maintaining cw operation [5]. Both Kressel *et al.* in the United States and Alferov *et al.* in the U.S.S.R. discovered independently that the carrier recombinations, and hence threshold current, could be greatly reduced by introducing additional potential barriers around the active region [5]. Such structures proved to be a milestone in the development of modern semiconductor lasers and these are known as *heterojunctions* [4]. The most useful heterojunctions involve two potential barriers, one each on the p- and n-side respectively, and are hence called *double heterostructures* (DH) (Fig. 3.8). They were found to be extremely successful in avoiding excess carrier recombination, as well as providing waveguiding for the optical signals to travel across the active region, improving the gain uniformity along the longitudinal direction [5]. The device can then be used as a cw source at room temperature. Although complicating the band structure, the basic principles behind the formation of and amplification in an active region remain unchanged. The equations described in previous sections can be applied to heterostructures without any significant modifications, because the optical confinement factor Γ will take into account the effect of the structural effects on the gain of the amplifier [12].

Not surprisingly, experiments in optical amplification using DH semiconductor lasers biased below oscillation threshold were carried out soon after successful operations of DH SLDs. For example, Schicketanz and Zeidler [37] developed a GaAs DH SLA which can operate under both travelling-wave (i.e. non-resonant; see Section 3.4.1) and resonant conditions at room temperature. The non-resonant operations were achieved by immersing the amplifier in a fluid with refractive index matched to the amplifier. They measured the amplified spontaneous emissions (ASE)

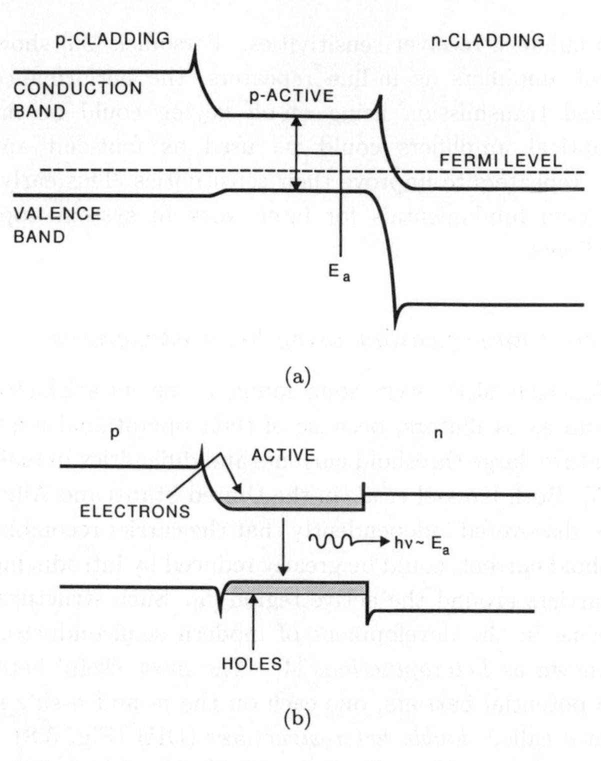

Fig. 3.8. Band diagram for a double heterostructure at (a) zero bias (b) forward bias.

at the amplifier output, and found that the amplifier gain spectrum was asymmetric, and its bandwidth would shrink as the gain increased. The shift in wavelength of maximum spontaneous emissions (i.e. the peak gain wavelength) to a shorter wavelength with pumping was also observed, as with Crowe and Ahearn's results [32]. They predicted from their results that noise should be reduced as the amplifier is driven close to saturation. For amplifiers operating under resonant conditions, a signal gain approaching 1000 (30 dB) can be obtained with an injection current of 500 mA (c.f. the range of injection current in homojunction SLAs). It was found that the signal gain was extremely sensitive to fluctuations of parameters like reflectivities under resonant operations. Again, gain saturation was observed in their experiments. Finally, although satisfactory amplification can be achieved at room temperature with a fairly low injection current, their results showed that their amplifier was extremely sensitive to temperature, an issue which, has received similar concern in recent years [38, 39].

Table 3.3. Recent transmission experiments with both SLA and fibre amplifiers.

Year	Laboratory	Bit Rate (Gbit/s)	Distance (km)	Comments
1989	NTT	1.8	212	Booster + pre-amplifier used
1989	BTRL	0.565	–	DPSK system
1989	KDD	1.2	267	Two amplifiers used
1989	NTT	20	–	Soliton transmission
1989	Bell Core	11	260	Two amplifier used
1989	Fujitsu	12	100	–
1989	KDD	1.2	904	12 amplifiers used
1990	NTT	2.5	2223	25 amplifiers used
1991	BTRL	2.5	10^4	Recirculating loop
1991	NTT	10	10^6	Soliton transmission, 12 amplifiers used
1991	NTT	20	500	Soliton transmission
1992	NTT	2.4	309	4 Repeaters + pre-amplifier
1992	NTT	10	309	4 repeaters + pre-amplifier
1993	BELL	10	360	EDFA + Dispersion Compensation
1994	BELL	16×2.5	1420	14 amplifiers used
1995	BELL	2.5	374	1 local EDFA + a remotely-pumped EDFA + pre-amplifier
1997	BELL	32×10	640	9 Gain-flattened broadband EDFA with 35 nm Bandwidth (Total Gain 140 dB and total gain non-uniformity 4.9 dB between 32 channels space by 100 GHz)
1998	Alcatel	32×10	500	4 EDFA + pre-amplifier (with 125 km amplifier spacing)
2000	KDD	50×10.66	4000	EDFA + low-dispersion slope fiber (40 km span)
2000	BELL	100×10 (25 GHz spacing)	400	4 EDFA + 4 Raman Amplifier

The development of SLAs gathered pace after the invention of heterostructures. Many experiments had been performed, e.g. by the NTT group in Japan led by Mukai and Yamamoto [40–42], by Simon in France [38] and by the BTRL group in England [43]. Theoretical modelling of SLAs became more sophisticated, e.g. analytical models proposed by Mukai and Yamamoto [41], Marcuse [44], Adams *et al.* [45] Buus and Plastow [46], and a recent one proposed by Lowery [47]. Experimental work in system applications with both SLAs and fibre amplifiers has also proceeded rapidly in recent years as shown in Table 3.3 [87]; also refer to references [27–49] in Chapter 1. SLAs have reached a high degree of sophistication and a systematic classification and discussion of the various types of SLAs is now possible as compared with the relative lack of diversity of types of SLAs in early 1970s. In the next section, we will assess the present state of the art for the different-types of SLAs that are now available.

3.4 Types of Semiconductor Laser Amplifiers

SLAs can be classified by their differences in operation, or by differences in their structures. The following discussion will cover both of these methods of classification.

3.4.1 *Operational classification*

Because any SLA is principally a structure derived from a SLD, one can generalise the structure of a SLA by an embedded active region with two end-facets for input and output of light signals (Fig. 3.9) [2]. The various structural details will be discussed in Section 3.4.2. With this simplified model, one can classify SLAs into two broad categories.

As we discussed in Section 3.2, the end-facets with power reflection coefficients R_1 and R_2 (Fig. 3.9) are responsible for providing optical feedback into the amplifier. These finite reflections arise because of the discontinuities of refractive index between the semiconductor and air [49, 50]. In a semiconductor laser, such reflections provide the necessary feedback F to excite oscillations when the single pass gain G_s satisfies the threshold condition of $FG_s = -1$, as depicted by Eq. (3.1). Because we are interested in linear amplification, the value of G_s can be reduced by reducing the injection current below threshold (as with the early experiments performed by Crowe and Craig [31] and Crowe and Ahearn [32], such that $|FG_s| < 1$).

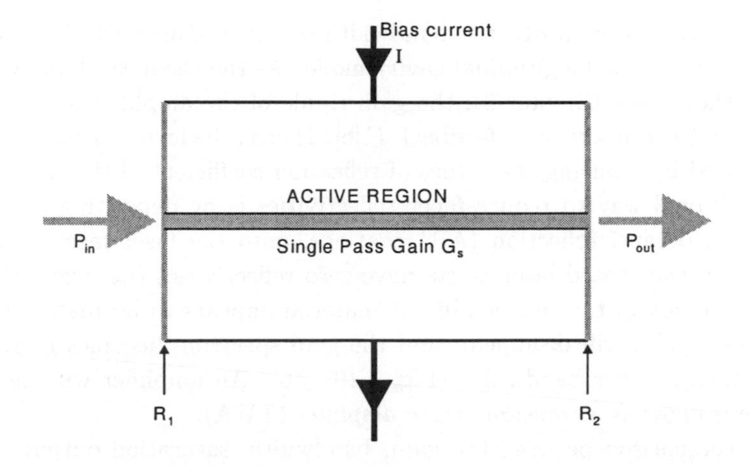

Fig. 3.9. A simplified schematic of a semiconductor laser amplifier.

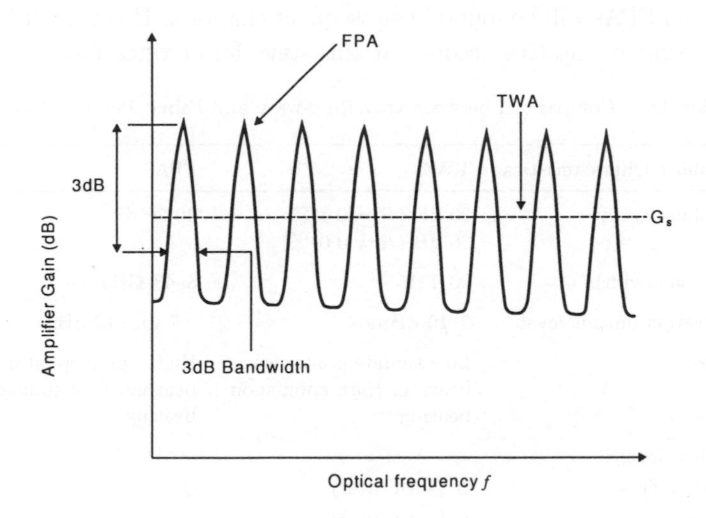

Fig. 3.10. Classification of semiconductorlaser amplifiers by their gain spectra.

In this case, the amplifier becomes a *Fabry–Perot amplifier* (FPA). It can be seen that because of the finite reflectivities of the end-facets, some resonant cavity properties will be observed. As shown in Chapter 4, the resonating characteristics of the cavity will modulate the material gain spectrum, introducing ripples along the amplifier gain spectrum [41] (see Fig. 3.10). The level of ripples will affect the bandwidth of the SLA [13–51]. If the ripple level exceeds 3 dB, the bandwidth of the amplifier will be reduced to that of a single longitudinal cavity mode. As the theoretical analysis on FPAs shows (see Chapter 8), the gain ripple of the amplifier actually depends on the proportion of feedback F [6]. Hence, the level of ripples can be minimised by reducing the values of reflection coefficient of the end-facets. A traditional way to reduce facet reflectivities is by depositing dielectric coatings, or anti reflection (AR) coatings, onto the laser facets [52–55]. Ideally, if the coated laser facets have zero reflectivities (i.e. when the refractive index of the semiconductor material appears to be matched with air, the ripples will disappear, and the gain spectrum becomes more uniform with a wider bandwidth (Fig. 3.10) [56]. An amplifier with no gain ripple is know as a *travelling-wave amplifier* (TWA).

A comparison between the gain, bandwidth, saturation output power and noise characteristics of these SLAs can be found in Table 3.4 [13, 38] and [51]. A detailed treatment on the differences in performance between TWAs and FPAs will be found in subsequent chapters. However, it is useful to have some qualitative account at this stage for clarification.

Table 3.4. Comparison between travelling-wave and Fabry–Perot amplifiers.

Amplifier Characteristics	TWA	FPA
Amplifier Gain	20 dB ($R = 0.1\%$) 30 dB ($R = 0.01\%$)	20-20 dB
Gain bandwidth	20 THz	8–43 GHz
Saturation output level	5–10 dBm	−7 to −12 dBm
Noise	Low signal-spontaneous beating; High spon-spon beating	High signal-spontaneous beating; Low spon-spon beating
Applications:		
Preamplifiers	◇ (with filter)	◇
Repeaters	◇ (with filter)	◇
Booster amplifiers	◇	◇
Post-amplifiers	◇	◇

FPAs have a very high gain when the wavelength of the input signals matches the peak transmission wavelength of (i.e. tuned to) the cavity resonance [37]. However, the amplifier gain is extremely sensitive to structural and operational parameters for FPAs [38]. The signal gain in TWAs is more stable than that of FPAs, and is less sensitive to structural parameters and operational conditions [23]. Because of the presence of cavity resonance in FPAs, the ripples along the amplifier gain spectrum reduce their bandwidth. Hence TWAs have wider bandwidth than FPAs, although the narrow bandwidth of FPAs is useful for active optical filters [2] and non-linear switching purposes [13]. Because of the presence of end-facet reflectivities, photons travel back and forth continuously across an FPA due to reflections. Some will be transmitted as output, but some will keep on travelling inside the cavity, generating more stimulated emissions from the population inversion. This enhances the pumping action and hence an FPA will saturate faster than a TWA without end reflections (where photons only travel across the cavity once). The output saturation power is therefore higher in TWAs. With respect to noise, the resonating action in FPAs can filter off some spontaneous emission power, especially that due to beating between spontaneous emissions [1]. However, the high gain due to cavity resonances in FPAs can enhance the beating between signal and spontaneous emission in the amplifier [42]. As we will see later, this can increase the overall noise figure in an FPA. Therefore the overall noise figure is lower in a TWA, compared with an FPA [104].

In general, it can be seen that TWAs appear to be more promising than FPAs, and perform closer to the ideal, and are found to be more useful in many applications [6]. Indeed, much research effort has been devoted recently to developing and fabricating TWAs [13, 23, 38] and [51]. In practice, TWAs with zero end-facet reflectivities are impossible to fabricate. This is because traditional AR coatings cannot reduce the facet reflectivities to absolute zero [57–60]. Residual reflectivities still exist, and ripples are still present in the amplifier gain spectrum [38]. O'Mahony [13] used the term *near travelling-wave amplifiers* (NTWAs) to describe these real devices, which have a characteristic lying between that of a FPA and an ideal TWA. The criteria for NTWA is that the ripple level in the amplifier gain spectrum should be less than 3 dB. Nearly all fabricated devices which are coated to reduce end-facet reflectivities are in fact NTWAs.

In addition to the residual reflectivities in AR coating, there is further difficulty in employing these coatings to eliminate facet reflectivities. No matter whether the AR coatings are single layered [59] or multi-layered [61, 62] they will have a minimum reflectivity for a finite range of optical wavelengths only. This so-called *window* of minimum reflectivity for the AR coatings is difficult to match with the peak gain region of the material gain spectrum [38, 63]. Even if the wavelength for minimum reflectivity of the AR coatings is *designed* to match with the peak gain wavelength of the material, it is difficult to *achieve* the design during the actual fabrication process [51]. A further complication is that in many applications of the TWAs a large input signal bandwidth is involved. The corresponding residual reflectivities of the end-facets due to the coatings are different for different wavelengths of the input signal. Hence, the result is a difference in signal gain for different wavelengths. Whilst some of the wavelengths of the input are amplified more, some will be amplified less, distorting the output signals. Such behaviour is far from that of ideal TWAs, because the amplifier gain spectrum should be uniform along the pass band in a TWA. Recently, it was discovered that the roughness which remains on the facets of the amplifier due to coatings (which is always finite because of limitations in the fabrication process) will also affect the values of their facet reflectivities [64].

In order to overcome these problems, other techniques are proposed to reduce facet reflectivities [2]. Recent techniques include using a window region [65], a buried facet structure [66] and an angled facet structure (i.e. tilted end-facets) [67–69]. Some work on minimising end-facet reflectivities in semiconductor waveguides for other opto-electronic devices may also be used in SLA fabrication [70, 71].

3.4.2 *Structural classification*

In the previous section, the structural details inside the SLA have been neglected by using the simplified model of Fig. 3.9. In real devices, there are several possible types of structural layout for SLAs. Some of them are shown in Fig. 3.11.

The simplest structure is a buried heterostructure (BH) (Fig. 3.11(a)), which is the most common structure found in SLDs as well as SLAs [37]. The active region is embedded in layers of semiconductors (i.e. in a *stratified* media) to improve the confinement of light. Further improvement in the performance of SLAs can be obtained by using separate confinement heterostructures (SCH) (Fig. 3.11(b)) [14], in which the carriers and light are separately confined (in contrast to BH, where both the light and carriers are confined within the active region [5]). Depending on the exact fabrication procedure, variations of this basic structure exist. An extensive review can be found in reference [4].

There are also SLA structures which are very different from this basic single waveguide, single active region structure. A twin-guided laser amplifier (TGLA) Fig. 3.11(c) is a SLA with two waveguides, which was proposed [72] for integration of a laser with an amplifier/detector for injection locked amplifier (ILA) applications [6]. This structure is found to be promising in active directional couplers and switches [73, 74]. A cleaved substrate laser amplifier (CSLA) (Fig. 3.11(d)) usually consists of an oscillator and an amplifier separated by a small cleaved gap [34], which is again useful for ILA applications (e.g. demodulation of FM signals [2]).

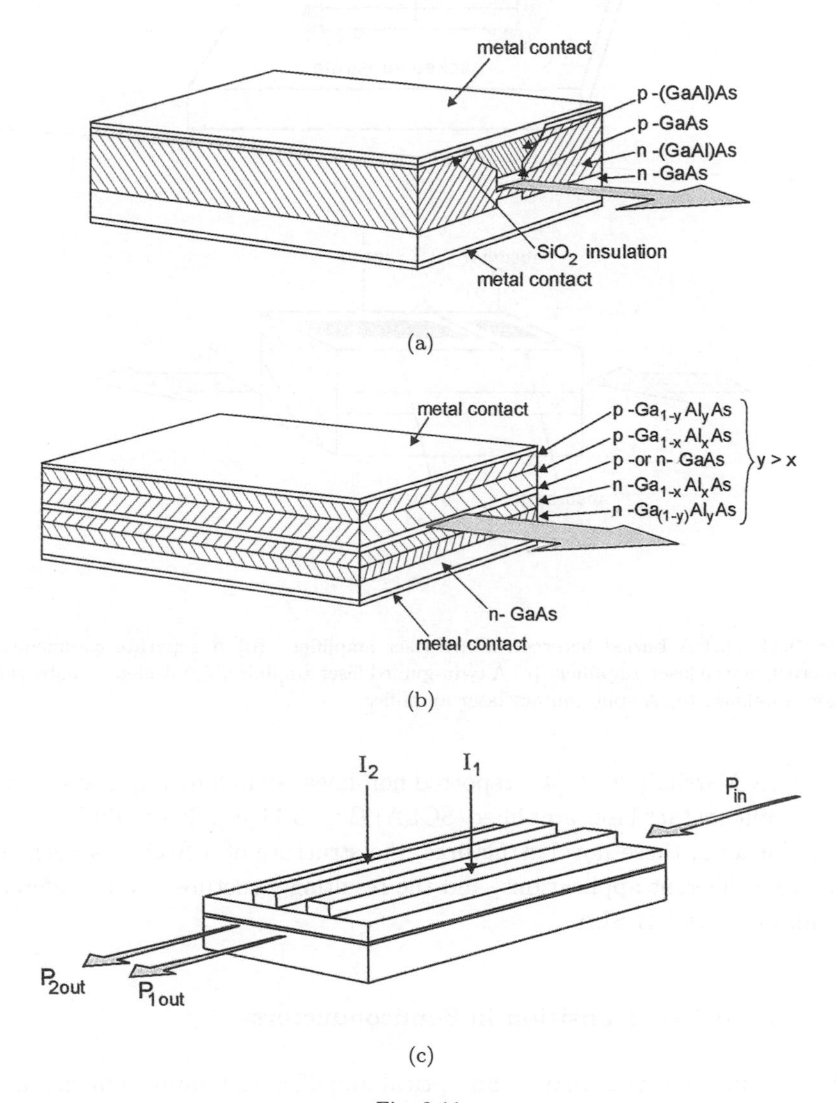

Fig. 3.11.

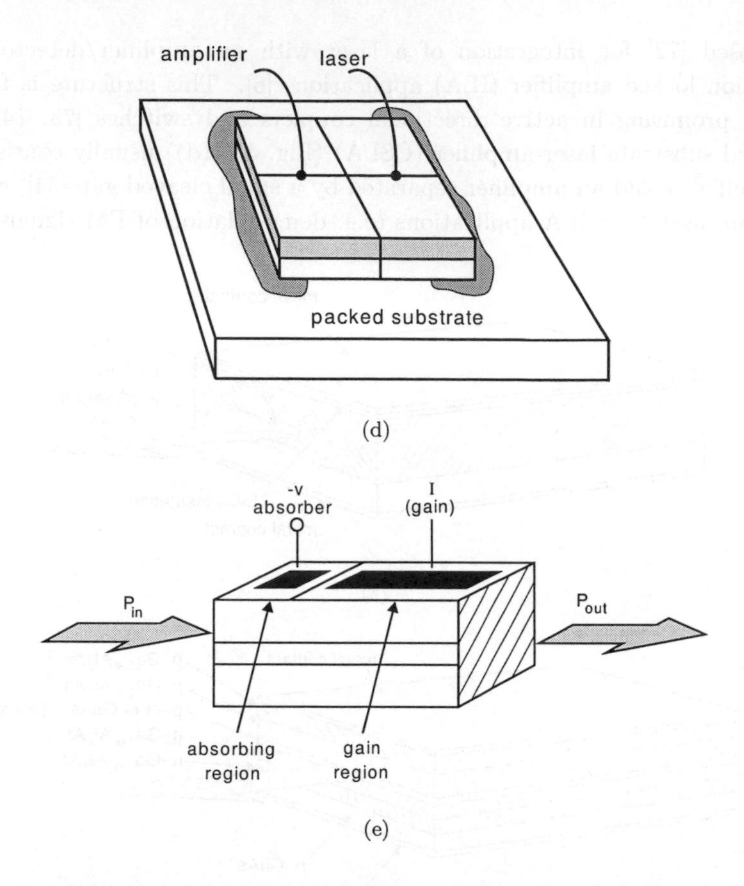

Fig. 3.11. (a) A buried heterostructure laser amplifier. (b) A separate confinement heterostructure laser amplifier. (c) A twin-guided laser amplifier. (d) A cleaved substrate laser amplifier. (e) A split-contact laser amplifier.

Recently Marshall *et al.* [43] reported non-linear switching applications using a split contact laser amplifier (SCLA) (Fig. 3.11(e)). It should be noted that for all of these novel structures, the structure of a SLA has been refined for different applications, and the resulting structure is very different to an ordinary BH SLA.

3.5 Radiative Transition in Semiconductors

We saw in Chapter 2 that in an optical amplifier at a finite temperature, random spontaneous emission events will occur. Einstein argued that to satisfy Planck's law of spectral distribution of electromagnetic radiation, stimulated emissions must also occur, which will manifest itself as optical gain if the amplifier is under negative temperature or population inversion [105]. This treatment was considered in some detail in Chapter 2 for a simple two-level system. In semiconductors, similar radiative transition of carriers occurs. However, instead of occurring between two discrete and distinct energy levels as postulated in Chapter 2, in semiconductors these radiative transitions will occur between two energy bands known as conduction and valence bands, respectively. The basics of semiconductor band theory are covered in many textbooks [5, 7]. Readers are recommended to refer to them for more fundamental questions about the definition of Fermi levels, the meaning of density of states, etc. In this section, we will give a very brief outline on the relevant equations for radiative transitions in a semiconductor where bands have exponential band tailing [5, 22]. Such band tailings occur because of the presence of randomly distributed impurities in the material due to heavy doping [4].

3.5.1 *Stimulated emissions*

It can be shown that by considering the density of states and population distribution of holes and electrons in a parabolic band as shown in Fig. 3.12, the net stimulated emissions will only occur if and only if [3, 4]

$$\exp\left(\frac{E_c - hf - F_v}{kT}\right) > \exp\left(\frac{E_c - F_c}{kT}\right) \tag{3.16}$$

where E_c is the energy of a state in the conduction band, F_c and F_v are the quasi-Fermi levels in the conduction and valence bands, respectively, h is the Planck's constant, f is the optical frequency, k is the Boltzmann constant, and T is the temperature of the semiconductor. The above equation can be simplified as

$$F_c - F_v > hf \tag{3.17}$$

which indicates the condition for a net positive value of stimulated emission; i.e. maintaining the population inversion. Under such conditions, the optical gain of the material can be found from the stimulated emission rate, and the corresponding spontaneous emission can be found from the spontaneous emission rate. The stimulated emission rate can be calculated if the transition probability B_{21} for the stimulated transition is known. To calculate B_{21} we must consider the mechanisms involved when the

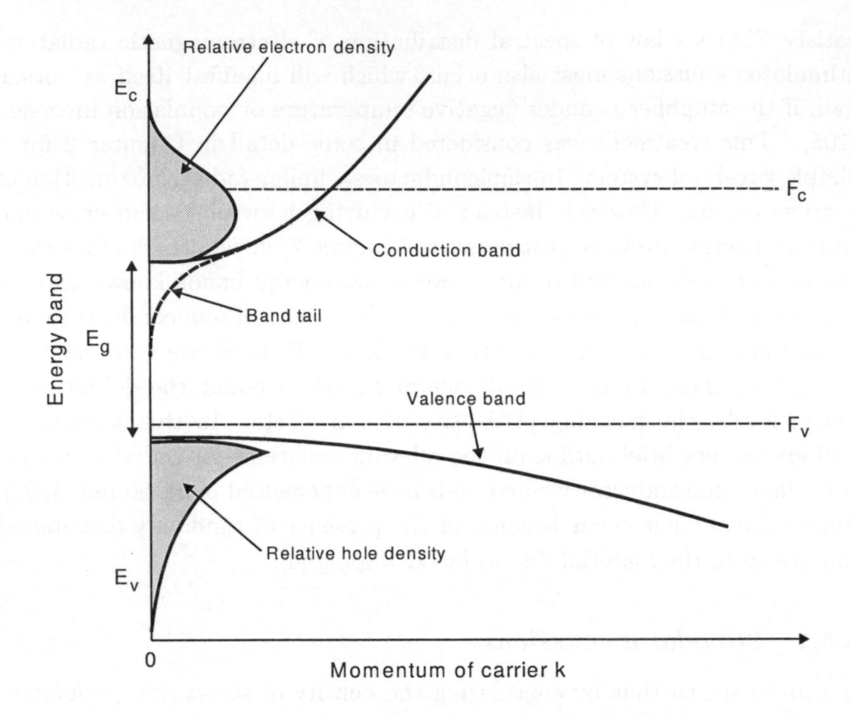

Fig. 3.12. Energy band versus momentum of carrier.

electrons in the semiconductor interact with the radiation field. This requires a knowledge of the quantum mechanics of the system, involving the wave functions of the electron. The appropriate approach is to use time dependent perturbation of the Schrodinger equation, such that the effects of the incident radiation on the system can be expressed in terms of the wave functions which describe the system in the absence of radiation [106]. We will simply quote the result here [5] as

$$B_{21} = \frac{e^2 h |M|^2}{2m^2 \varepsilon_o \mu^2} \tag{3.18}$$

where e and m are the charge and mass of an electron and M is the matrix element of the momentum operator connecting the wave function of the upper state in the conduction band to that of the lower state in the valence band [106]. The evaluation of the matrix element M involves finding the appropriate wave functions that describe the conduction and valence band states of the semiconductor. This in turn requires a knowledge of the band structure of the material.

For transitions under k selection the polarisation of the wave functions during transition is taken into account. Under these circumstances, it can be shown that the matrix element M requires the wave vector for wave functions during transitions to be identical [5]. In this situation, the value of M can be found approximately as

$$|M|^2 = \frac{m^2 E_\mathrm{g}[1+\Delta/E_\mathrm{g}](1-m_\mathrm{c}/m)}{3m_\mathrm{c}[1+2\Delta/3E_\mathrm{g}]}. \tag{3.19}$$

Where m is the free electron mass, m_c is the effective mass of the electron in the conduction band, E_g is the band gap energy and Δ is the spin-orbital splitting. Using this expression, the transition probability is given by

$$B_{21} = \frac{e^2 h[1+\Delta/E_\mathrm{g}](1-m_\mathrm{c}/m)}{6m_\mathrm{c}\varepsilon_\mathrm{o}\mu^2[1+2\Delta/3E_\mathrm{g}]} \tag{3.20}$$

with the value of B_{21} known, the material gain of the amplifying medium can be found following the procedure described in reference [28].

3.5.2 *Spontaneous emissions*

In general, the transition probability of spontaneous emission can be expressed in terms of B_{21} (see Section 2.2 of Chapter 2). Then the spontaneous emission recombination rate can be found from the following expression [5]

$$R_\mathrm{sp} = \int_{-\infty}^{-(b+a)} \int_{-b}^{\infty} Z(E_\mathrm{c}-E_\mathrm{v}) B(E_\mathrm{c}, E_\mathrm{v}) \times V\rho_\mathrm{c}(E_\mathrm{v})\rho_\mathrm{v}(E_\mathrm{c}) f_\mathrm{c}(1-f_\mathrm{c})\, dE_\mathrm{c}\, dE_\mathrm{v} \tag{3.21}$$

where $Z(E)$ indicates the average number of modes in the radiation field with photon energy E, V is the volume of the material, E_c and E_v are energy of states within the conduction and valence bands, respectively (see Fig. 3.12), f_c and f_v are the occupational probability of the state in the conduction and valence bands at energies of E_c and E_v, respectively, ρ_c and ρ_v are the density of states at the conduction and valence bands, respectively. The constants a and b are chosen such that the integration terminates within the forbidden bandgap. Notice that $B = B_{21}$. The above equation can be formulated by considering the number of transitions across the bands per second, similar to our analysis in the two level system in

Chapter 2. We can rewrite the above equation in the following form

$$R_{\text{sp}} = Z \int_0^n \int_0^p B(n_\rho, p_\rho) V dndp \tag{3.22}$$

where the variables n_ρ and p_ρ are alternatives to E_c and E_v, which represent the total electron and hole states (whether they are occupied or not) up to energy E_c and down to energy E_v. Z is taken outside the integral as being approximately constant over the small range of photon energy involved in the integration. Usually, the above form of B can be handled in two situations [5]: first the non-inverted Boltzmann condition with both quasi-Fermi levels lying inside the bandgap; and secondly when B can be approximated as independent of the carrier density n and p, and hence the spontaneous emission rate is given approximately by

$$R_{\text{sp}} \approx Bnp \tag{3.23}$$

which is the well-known expression used throughout this book and other references in considering the radiative recombination rate [4] and [27].

3.6 Applications of Semiconductor Laser Amplifiers

We will conclude this chapter by discussing various applications of SLAs. References [2] and [13] have given extensive reviews on various applications of SLAs. We saw in Chapter 2 that the exact performance requirements of SLAs vary with their applications. They are summarised in Table 3.5 [1] and will be discussed below. In addition, the problems in present designs of SLAs for different application requirements will also be introduced.

Table 3.5. Performance requirements in amplifier applications.

Performance requirements	Preamplifier	In-line Repeater	Booster Amplifier
1. Sufficient small signal gain	Required	Required	Required
2. Wide gain bandwidth	Required	Required	Required
3. Polarisation-insensitive signal gain	Required	Required	Required
4. High saturation output power	–	Indispensable	Indispensable
5. Small noise figure	Indispensable	Indispensable	Indispensable
6. Use of narrow-band optical filter	Indispensable	Required	Required

3.6.1 *Non-regenerative repeaters*

Conventionally, the in-line repeaters in optical fibre communications are *regenerative* (Fig. 3.13(a)). They are regenerative because the amplified signals at the output of the repeaters are actually regenerated signals. This is illustrated in Fig. 3.13(a). The whole process involves the detection of a weak incoming signal using a photodetector (PD), which converts the light signal into electrical current, and is then fed into an electronic amplifier for amplification. The amplified current is then used to drive a SLD and hence,

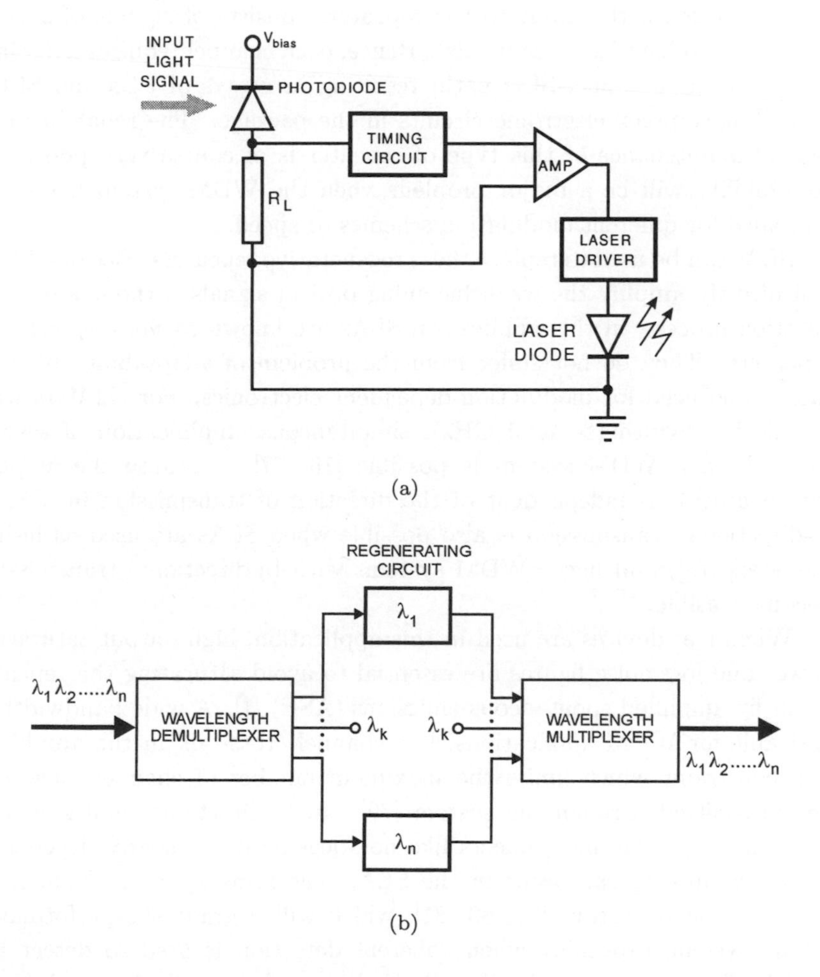

Fig. 3.13. (a) Schematic of a regenerative repeater. (b) Regenerative repeaters for WDM systems.

regenerating the required signal. Such repeaters are not only expensive but also suffer from poor sensitivity at the input to PDs, and noise is an inherent problem [2]. In addition, the laser driver electronics have to be designed for different schemes of modulation and signal shaping, and hence its versatility is very limited. It can also be expected that maintenance is difficult for these repeaters, especially when they are undersea [75]. Reliability is also a problem.

Additional problems arise when these regenerative repeaters are used in a wavelength division multiplexed (WDM) system (Fig. 3.13(b)). In a WDM system, the input to the repeaters consists of signals of several different wavelengths (or channels). Hence, each channel requires a distinct PD-electronic amplifier-SLD path, resulting in arrays of PDs and SLDs, as well as complex electronic circuits in the package. The reliability and ease of maintenance of this type of repeater is of course very poor, and adaptability will be a major problem when the WDM system has to be upgraded for different modulation schemes or speed.

SLAs can be used to replace these regenerative repeaters. Because SLAs will directly amplify the weak incoming optical signals without any regeneration process, in this application SLAs are known as *non-regenerative* repeaters. They do not suffer from the problem of adaptability because there is no need for modulation-dependent electronics. For NTWAs with a wide bandwidth ($>$ 4000 GHz), simultaneous amplification of several channels in a WDM system is possible [76, 77]. Because the amplification process is independent of the direction of transmission in a SLA, bi-directional transmission is also possible when SLAs are used as in-line repeaters [78], and hence WDM systems with bi-directional transmission become feasible.

When real devices are used in this application, high output saturation power and low noise figures are essential to avoid saturating the repeater chain by amplified spontaneous emissions (ASE) [1]. A wide bandwidth is desirable for WDM applications, but channel cross-talk in the amplifiers is a constraint which limits the maximum number of channels that can be transmitted through the system [79]. In bi-directional multi-channel transmissions, non-linear effects like non-degenerate or nearly degenerate four-wave mixing can occur in the SLA, generating spurious frequencies at the amplifier output [77, 80, 81], which will degrade the performance of the system, especially when coherent detection is used to detect the signals. The channel cross-talk and non-linear effects in SLAs also have to be investigated and minimised for improving the performance of SLAs as repeaters in WDM systems.

3.6.2 *Pre-amplifiers to optical receivers*

In optical communications, receivers are required to detect the optical signals transmitted along the fibre, and convert them back to electrical signals. A typical configuration of an optical receiver is shown in Fig. 3.14, which consists of a PD and a FET amplifier [82]. Using an avalanche photodetector (APD), the largest gain bandwidth which can be obtained from the receiver is $\sim$ 70 GHz [3]. For even wider bandwidth, e.g. sub-carrier multiplexed lightwave system or multi-channel video distribution [83], an optical pre-amplifier will be needed [35]. When a SLA is used as a front-end pre-amplifier to the receiver, it can amplify the incoming weak signal, hence improving the sensitivity of the receiver [84, 85]. Usually, when a SLA is incorporated into the receiver, a high speed p.i.n. photo-detector will be used to convert the light signals into electrical signals, and a receiver bandwidth as wide as the material gain bandwidth of the SLA can be achieved [86, 87]. For this particular application, SLAs with wide bandwidth and low noise figure are required [15]. TWAs are ideal candidates in this case, although optical filtering may be needed to eliminate the beating noise between spontaneous emissions [6]. In addition, high coupling efficiency between the SLA and the p.i.n. PD, as well as between the SLA and the optical fibre, is required, as the improvement in receiver sensitivity depends strongly on the coupling loss [2]. Therefore, reliable TWAs, as well as high coupling efficiency between the optical components, are the major

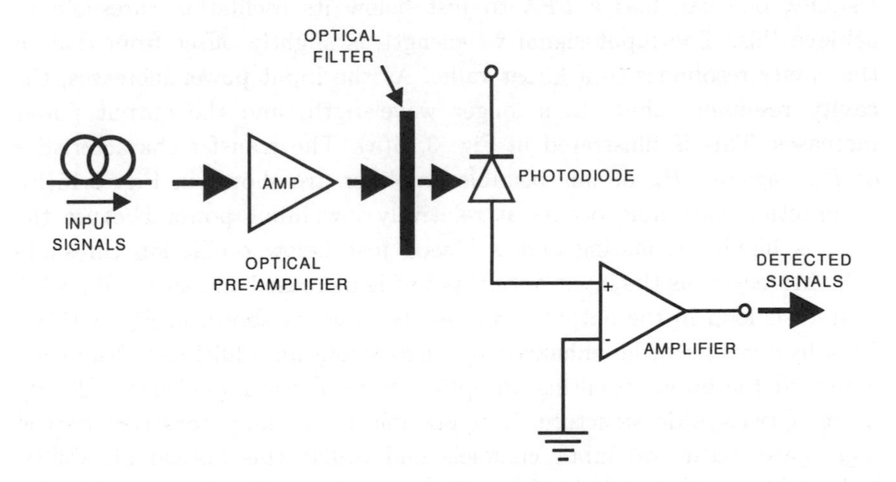

Fig. 3.14. Configuration of an optical receiver.

targets to be achieved if SLAs are to be used effectively as pre-amplifiers to optical receivers.

3.6.3 *Bistable and switching applications*

The phenomenon of optical bistability in semiconductor lasers has received a lot of attention in the development of semiconductor lasers [45]. There are two distinctive ways to operate SLAs as a bistable element [2, 3], i.e. electrically or optically. Because a SLA absorbs all the input light power when there is no injection or bias current (i.e. no pumping) and it provides gain to the input signals when the current is turned on fully, it can be used as a lossless gating switch [7]. In this application, the SLA will block all incoming signal when there is no electrical current flowing through it (i.e. the "off" state), whereas when the current flowing across it is turned on fully (i.e. the "on" state), the SLA can pass the signals with gain to compensate losses in the optical fibre network. Such gating switches are ideal for optical routing in an optical fibre network. A SLA can also be switched optically as well, due to the asymmetric gain spectrum of semiconductors. It was shown by many experiments that as the input power increases, the wavelength of maximum output power in a FPA shifts to the red side of the optical spectrum [13]. Such a detuning phenomenon can be used to provide optical bistability. This can be achieved by enclosing the amplifying medium in a highly resonating Fabry–Perot cavity [45, 89, 90]. Usually, one can bias a FPA to just below its oscillation threshold to achieve this. The input signal wavelength is slightly offset from that of the cavity resonance to a larger value. As the input power increases, the cavity resonance shifts to a longer wavelength, and the output power increases. This is illustrated in Fig. 3.15(a). The transfer characteristics of P_{out} against P_{in} of this bistable amplifier are shown in Fig. 3.15(b). In practice, switching occurs at relatively low input power because the SLA is highly resonating and is biased just below oscillation threshold [13]. Moreover, as the input optical power is increased and then reduced, a hysteresis loop in the output power is observed, as shown in Fig. 3.15(b). This hysteresis can be enhanced by introducing an additional absorption region at the input, resulting an optical bistable flip-flop element [2]. By using a twin-guide structure, it is possible to actually construct optical logic gates with two input channels and utilise this optical bistability. Indeed, this is the principle of an active optical directional coupler [74, 91]. Additional applications of optical bistability can be found in pulse shaping

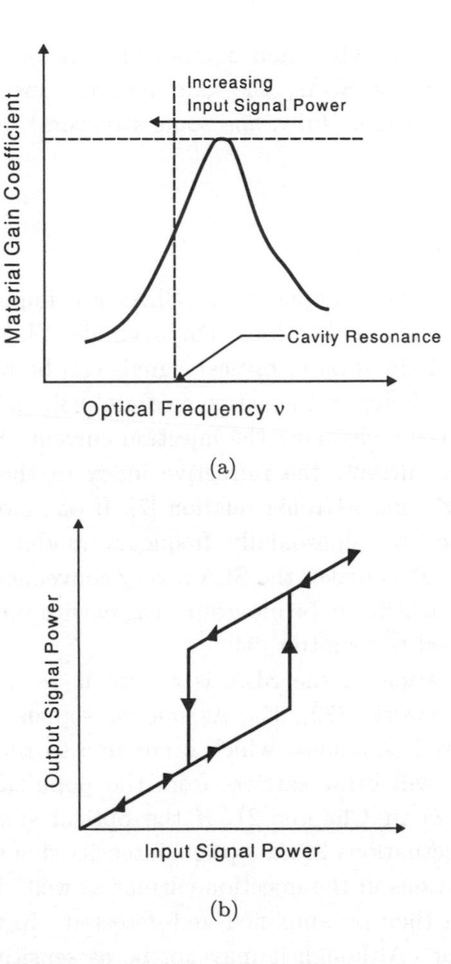

Fig. 3.15. (a) The detuning of a FPA with input signal power. (b) Resulting transfer characteristics of the bistable amplifier.

[13]. Generation and shaping of optical pulses with very short duration can be achieved by utilising the properties of optical bistability, which are extremely useful in high speed pico-and femto-second optics [14].

In contrast to the previous two applications, when SLAs have to be used as switches, they must be highly resonating with a narrow bandwidth [13]. Hence, FPAs become ideal candidates in this application. Although promising, at present the applications of SLAs as switches and bistable amplifiers are hindered by the relatively low operational speed and poor stability. These drawbacks arise because of the finite gain recovery time due to carrier recombination mechanisms. In order to improve the bistable performance of SLAs, the gain recovery mechanisms in SLAs should be understood more fully, and some structural refinements may be necessary [2].

3.6.4 *Other applications*

In addition to the above applications, SLAs are found to be potential devices in many other applications. For example, SLAs can be used as modulators [92]. Modulation of optical signals can be achieved by modulating the refractive index in the active medium [93]. In SLAs, this can be achieved very easily by changing the injection current. Since the material gain changes with current, the refractive index in the SLA changes as predicted by the Kramers-Kronig relation [7]. If one modulates the injection current of the SLA sinusoidally, frequency modulation will result in the optical signals. This makes the SLA a very convenient form of external optical modulator, which can be integrated easily with semiconductor lasers to form a high speed transmitter [94].

Another application of the SLA is to use it as a detector/monitor in optical fibre networks [72, 95]. As optical signals pass through the amplifier, stimulated emissions, which occur due to the material gain in the active region, will draw carriers from the population inversion (see Eq. (2.4) and (2.7) in Chapter 2). If the optical signals are intensity modulated, the fluctuations in the optical intensity due to the modulation will induce fluctuations in the injection current as well. These fluctuations in the current can then be amplified and detected. In this way the SLA becomes a detector. Although it may not be as sensitive as an ordinary photo-diode, the possibility of detecting signals using SLAs makes them suitable as monitors along the transmission span of an optical fibre communication system [96, 97]. These monitors will detect the intermediate optical power within the transmission span, and if this power gets too low, it indicates that there is probably a break in the preceding transmission path and appropriate maintenance steps should be undertaken. In addition, such monitors can be used to detect control protocols which are sent along with the optical signals in an optical network, which will be used to control the routing of optical signals.

Other recent proposed applications of SLAs include: use as a component in high-flexibility optical filters [98], as an injection-locked amplifier for demodulation of FM signals [12], as a component in active optical switching matrices [99], as a channel dropping node [100], in optical phase-locked-loops [101] and as a dual-function gate/amplifier in an optical fibre network [102]. One may ask why, in spite of the end less list of potential applications, the development of SLAs seemed to be much slower in comparison to that of SLDs. Eisenstein [2] considered this as the result of the slow speed of development in fabrication technology as well as the theoretical design of AR coatings. Indeed, even after the problem of reducing facet reflectivities has finally been overcome, there are many other problems left. The reasons behind the occurrence of these problems and the possible solutions will be examined in subsequent chapters.

References

[1] Y. Yamamoto and T. Mukai, "Fundamental of optical amplifiers," *Opt. Quantum Electronics*, Vol. QE-21, pp. S1–S14, 1989.
[2] G. Eisenstein, "Semiconductor optical amplifiers," *IEEE Circuits and Devices Magazine*, pp. 25–30, July 1989.
[3] A. Yariv, "Optical Electronics," 3rd Edition (Holt-Saunders, 1985).
[4] G. P. Agrawal and N. K. Dutta, *Long-wavelength semiconductor lasers* (New York, Van-Nostrad Reinhold, 1986).
[5] G. H. B. Thompson, *Physics of semiconductor laser devices* (New York, John Wiley and Sons, 1980).
[6] Y. Mukai, Y. Yamamoto and T. Kimura, "Optical amplification by semiconductor lasers," *Semiconductor and semimetals*, Vol. 22, Part E, pp. 265–319, Academic Press, 1985.
[7] A. Yariv, *Quantum Electronics*, 3rd Edition (John Wiley and Sons, 1989).
[8] M. J. Adams, *An Introduction to Optical Waveguides* (John Wiley and Sons, 1981).
[9] D. Marcuse, *Theory of dielectric optical waveguides* (Academic Press, 1974).
[10] A. W. Snyder and J. D. Love, *Optical waveguide theory* (Chapman & Hall, 1983).
[11] M. Cross and M. J. Adam, "Waveguiding properties of stripe-geometry double heterostructure injection lasers," *Solid State Electron.*, Vol. 15, pp. 919–921, 1972.
[12] C. Y. J. Chu and H. Ghafouri-Shiraz, "Structural effects on polarization sensitivity of travelling-wave semiconductor laser amplifiers," *3rd Bangor Communication Symposium*, United Kingdom, pp. 19–22, 1991.
[13] M. J. O'Mahony, "Semiconductor laser optical amplifiers for use in future fiber systems," *IEEE J. Lightwave Technology*, Vol. LT-6, No. 4, pp. 531–544, 1988.
[14] T. Saitoh and T. Mukai, "Structural design for polarization-insensitive travelling-wave semiconductor laser amplifiers," *Opt. Quantum Electron.*, Vol. 21, pp. S47–S48, 1989.

[15] A. G. Failla, G. P. Brava and I. Montrosset, "Structural design criteria for polarization insensitive semiconductor optical amplifiers," *J. Lightwave Technology*, Vol. 8, No. 3, pp. 302–308, 1990.
[16] H. Ghafouri-Shiraz and C. Y. J. Chu, "Refractive index control on the polarization sensitivity of semiconductor travelling-wave laser amplifiers," *Microwave and Optical Technology Letters*, Vol. 5, No. 3, pp. 152–154, 1992.
[17] H. Ghafouri-Shiraz and C. Y. J. Chu, "Analysis of waveguiding properties of travelling-wave semiconductor laser amplifiers using perturbation technique," *Fiber and Integrated Optics*, Vol. 11, pp. 51–70, 1992.
[18] H. Ghafouri-Shiraz and C. Y. J. Chu, "Analysis of polarization sensitivity in semiconductor laser amplifiers using the perturbation method," *Trans. IEICE Japan*, National Convention Record, Paper No. C-188, 1991.
[19] G. Lasher and F. Stern, "Spontaneous and stimulated recombination radiation in semiconductors," *Phys. Rev.*, Vol. 133, No. 2A, pp. 553–563, 1964.
[20] C. J. Hwang, "Properties of spontaneous and stimulated emissions in GaAs junction laser; I & II," *Phy. Rev. B*, Vol. 2, No. 10, pp. 4117–4134, 1970.
[21] F. Stern , "Calculated spectral dependence of gain in excited GaAs," *J. Appl. Phys.*, Vol. 47, No. 2, pp. 5382–5386, 1976.
[22] M. J. Adam, "Theoretical effects of exponential band tails on the properties of the injection laser," *Solid State Electron.*, Vol. 12, pp. 661–669, 1969.
[23] T. Saitoh and M. Mukai, "1.5 μm GaInAs travelling-wave semiconductor laser amplifier," *IEEE J. Quantum Electronics*, Vol. QE-23, No. 6, pp. 1010–1020, 1987.
[24] H. Ghafouri-Shiraz, "A model for peak-gain coefficient in InGaAsP/InP semiconductor laser diodes," *Opt. Quantum Electron.*, Vol. 20, pp. 153–163, 1988.
[25] H. Ghafouri-Shiraz, "Analysis of characteristics of long-wavelength semiconductor laser diode amplifiers," *Opt. Quantum Electron.*, Vo. 19, pp. 303–311, 1987. *Technology*, Vol. LT-5, No. 1, pp. 184–189, 1987.
[26] R. Olshansky, C. B. Su, J. Manning and W. Powazink, "Measurement of radiative and nonradiative recombination rates in InGaAsP and AlGaAs light sources," *IEEE J. Quantum Electron.*, Vol. QE-20, No. 8, pp. 838–854, 1984.
[27] J. Wang, H. Olesen and K. E. Stubkjaer, "Recombination, gain and bandwidth characteristics of 1.3 μm semiconductor laser amplifiers," *IEEE J. Lightwave.*
[28] M. Osinski and M. J. Adam, "Gain spectra of quaternary semiconductors," *IEE Proc.* Part J, Vol. 129, No. 6, pp. 229–236, 1982.
[29] I. D. Henning, M. J. Adams and J. V. Collins, "Performance prediction from a new optical amplifier model," *IEEE J. Quantum Electronics*, Vol. QE-21, No. 6, pp. 609–613.
[30] T. Mukai, K. Inoue and T. Saitoh, "Homogeneous gain saturation in 1.5 μm InGaAsP travelling-wave semiconductor laser amplifiers," *Appl. Phys. Lett.*, Vol. 51, No. 6, pp. 381–383, 1987.

[31] J. W. Crowe and R. M. Craig, "Small-signal amplification in GaAs lasers," *Appl. Phys. Lett.*, Vol. 4, No. 3, pp. 57–58, 1964.
[32] W. J. Crowe and W. E. Aheam, "Semiconductor laser amplifier," *IEEE J. Quantum Electronics*, Vol. QE-2, No. 8, pp. 283–289, 1966.
[33] W. F. Kosonocky and R. H. Cornely, "GaAs Laser amplifier," *IEEE J. Quantum Electronics*, Vol. QE-4, No. 4, pp. 125–131, 1968.
[34] M. B. Chang and E. Garmire, "Amplification in cleaved-substrate lasers," *IEEE J. Quantum Electron.*, Vol. QE-16, No. 9, pp. 997–1001, 1980.
[35] A. J. Arnaud, "Enhancement of optical receiver sensitivity by amplification of the carrier," *IEEE J. Quantum Electron.*, Vol. QE-4, No. 11, pp. 893–899, 1968.
[36] S. D. Personick, "Applications for quantum amplifiers in simple digital optical communication systems," *Bell Sys. Tech. J.*, Vol. 52, No. 1, pp. 117–133, 1973.
[37] D. Schicketanz and G. Zeidler, "GaAs-double-heterostructure lasers as optical amplifiers," *IEEE J. Quantum Electron.*, Vo. QE-11, No. 2, pp. 65–69, 1975.
[38] J. C. Simon, "GaInAsP semiconductor laser amplifiers for single-mode fiber communications," *J. Lightwave Technology*, Vol. LT-5, No. 9, pp. 1286–1295, 1987.
[39] M. S. Lin, A. B. Piccirilli, Y. Twu and N. K. Dutta, "Temperature dependence of polarization characteristics in buried facet semiconductor laser amplifiers," *IEEE J. Quantum Electron.*, Vol. 26, No. 10, pp. 1772–1778, 1990.
[40] T. Mukai and T. Yamamoto, "Noise characteristics of semiconductor laser amplifiers," *Electronics Letters*, Vol. 17, No. 1, pp. 31–33, 1981.
[41] T. Mukai and Y. Yamamoto, "Gain, frequency bandwidth, and saturation output power of AlGaAs DH laser amplifiers," *IEEE J. Quantum Electronics,* Vol. QE-17, No. 6, pp. 1028–1034, 1981.
[42] T. Mukai and Y. Yamamoto, "Noise in an AlGaAs semiconductor laser amplifier," *IEEE J. Quantum Electronics*, Vol. QE-18, No. 4, pp. 564–575, 1982.
[43] I. W. Marshall *et al.*, "Gain characteristics of a 1.5 μm non-linear split contact laser amplifier," *Appl. Phys. Lett.*, Vol. 53, No. 17, pp. 1577–1579, 1988.
[44] D. Marcuse, "Computer model of an injection laser amplifier," *IEEE J. Quantum Electronics*, Vol. QE-19, No. 1, pp. 63–73, 1983.
[45] M. J. Adam, J. V. Collins and I. D. Henning, "Analysis of semiconductor laser optical amplifiers," *IEE Proc.*, Part J, Vol. 132, pp. 58–63, 1985.
[46] J. Buus and R. Plastow, "A theoretical and experimental investigation of Fabry–Perot semiconductor laser amplifiers," *IEEE J. Quantum Electron.*, Vol. QE-21, No. 6, pp. 614–618, 1985.
[47] A. J. Lowery, "New in-line wideband dynamic semiconductor laser amplifier model," *IEE Proc.*, Part J. Vol. 135, No. 3, pp. 242–250, 1988.
[48] T. Mukai, Y. Yamamoto and T. Kimura, "S/N and error rate perfor-

mance in AlGaAs semiconductor laser preamplifier and linear repeater systems," *IEEE Trans. Microwave Theory Tech.*, Vol. MTT-30, No. 10, pp. 1548–1556, 1982.

[49] T. Ikegami, "Reflectivity of a mode at facet and oscillation made in double-heterostructure injection laser," *IEEE J. Quantum Electron.*, Vol. QE-8, No. 6, pp. 470–476, 1972.

[50] H. Ghafouri-Shiraz, "Facet reflectivity of InGaAsP BH laser diodes emitting at 1.3 μm and 1.55 μm," *Semiconductor Science Technology*, Vol. 5, No. 2, pp. 139–142, 1990.

[51] T. Saitoh and T. Mukai, "Recent progress in semiconductor laser amplifiers," *I. Lightwave Technology*, Vol. LT-6, No. 11, pp. 1156–1164, 1988.

[52] A. Hardy, "Formulation of two-dimensional reflectivity calculations based on the effective index method," *J. Opt. Soc. Am. A.*, Vol. 1, No. 5, pp. 550–555, 1984.

[53] D. R. Kaplan and P. P. Deimel, "Exact calculation of the reflection coefficient for coated optical waveguide devices," *AT&T Bell Lab. Tech. J.*, Vol. 63, No. 6, pp. 857–877, 1984.

[54] P. Kaczmarski, R. Baets, G. Franssens and P. E. Lagasse, "Extension of bi-directional beam propagation method to TM polarization and application to laser facet reflectivity," *Electronics Letters*, Vol. 25, No. 11, pp. 716–717, 1989.

[55] M. Yamada, Y. Ohmori, K. Takada and M. Kobayashi, "Evaluation of antireflection coatings for optical waveguides," *Appl. Opt.*, Vol. 30, No. 6, pp. 682–688, 1991.

[56] C. A. Balanis, *Advanced engineering electromagnetics* (John Wiley and Sons, 1989).

[57] R. H. Clark, "Theoretical performance of an anti-reflection coating for a diode laser amplifier," *Int. J. Electron.*, Vol. 53, No. 5, pp. 495–499, 1982.

[58] R. H. Clarke, "Theory of reflections from antireflection coatings," *Bell Sys. Tech. J.*, Vol. 62, No. 10, pp. 2885–2891, 1983.

[59] G. Eisenstein, "Theoretical design of single-layer antireflection coatings on laser facets," *AT&T Bell Lab., Tech. J.*, Vol. 63, No. 2, pp. 357–364, 1984.

[60] T. Saitoh, T. Mukai and O. Mikami, "Theoretical analysis and fabrication of antireflection coatings on laser-diode facets," *J. Lightwave Technology*, Vol. LT-3, No. 2, pp. 288–293, 1985.

[61] C. Vassallo, "Theory and practical calculation of antireflection coatings on semiconductor laser diode optical amplifiers," *IEE Proc.*, Part J. Vol. 137, No. 4, pp. 193–202, 1990.

[62] C. Vassallo, "Some numerical results on polarization insensitive 2-layer antireflection coatings for semiconductor optical amplifiers," *IEE Proc.*, Part J. Vol. 137, No. 4, pp. 203–204, 1990.

[63] J. Stone and L. W. Stulz, "Reflectance, transmission and loss spectra of multilayer Si/Sio2 thin film mirrors and antireflection coatings for 1.5 μm," *Appl. Opt.*, Vol. 29, No. 4, pp. 583–588, 1990.

[64] C. F. Lin, "The influence of facet roughness on the reflectivities of etched-angled facets for superluminescent diodes and optical amplifiers," *IEEE Photo Tech. Lett.*, Vol. 4, No. 2, pp. 127–129, 1992.

[65] I. Cha, M. Kitamura and I. Mito, "1.5 μm band travelling-wave semiconductor optical amplifiers with window facet structure," *Electronics Letters*, Vol. 25, No. 3, pp. 242–243, 1989.

[66] M. S. Lin, A. B. Piccirilli, Y. Twu and N. K. Dutta, "Fabrication and gain measurement for buried facet optical amplifier," *Electronics Letters*, Vol. 25, No. 20, pp. 1378–1380, 1989.

[67] C. E. Zah *et al.*, "1.3 μm GaInAsP near-travelling-wave laser amplifiers made by combination of angled facets and antireflection coatings," *Electronics Letters*, Vol. 24, No. 20, pp. 1275–1276, 1988.

[68] G. A. Alphonse, J. C. Connolly, N. A. Dinkel, S. L. Palfrey and D. B. Gilbert, "Low spectral modulation high-power output from a new AlGaAs superluminescent diode/optical amplifier structure," *Appl. Phys. Lett.*, Vol. 55, No. 22, pp. 2289–2291, 1989.

[69] D. Marcuse, "Reflection loss of laser mode from tilted end mirror," *IEEE J. Lightwave Technology*, Vol. LT-7, No. 2, pp. 336–339, 1989.

[70] T. Baba and Y. Kokubun, "New polarization-insensitive antiresonant reflecting optical waveguide (ARROW-B)," *IEEE Photo. Tech. Lett.*, Vol. 1, No. 8, pp. 232–234, 1989.

[71] K. W. Jelly and R. W. H Engelmann, "An etch tunable antireflection coating for the controlled elimination of Fabry–Perot oscillations in the optical spectra of transverse modulator structures," *IEEE Photo. Tech. Lett.*, Vol. 4, No. 6, pp. 550–553, 1989.

[72] K. Kishino, Y. Suematsu, K. Utaka and H. Kawanishi, "Monolithic integration of laser and amplifier/detector by twin-guide structure," *Japan J. Appl. Phys.*, Vol. 17, No. 3, pp. 589–590, 1978.

[73] M. J. Adam, "Theory of twin-guides Fabry–Perot laser amplifiers," *IEE Proc.*, Part J, Vol. 136, No. 5, pp. 287–292, 1989.

[74] M. J. Adam, D. A. H. Mace, J. Singh and M. A. Fisher, "Optical switching in the twin-guide Fabry–Perot laser amplifier," *IEEE J. Quantum Electron.*, Vol. 26, No. 10, pp. 1764–1771, 1990.

[75] T. Kimura, "Coherent optical fibre transmission," *IEEE J. Lightwave Technology,* Vol. LT-5, No. 4, pp. 414–428, 1987.

[76] B. Glance, G. Eisenstein, P. J. Fitzgerald, K. J. Pollack and G. Raybon, "Optical amplification in a multichannel FSK coherent system," *Electronics Letters*, Vol. 24, No. 18, pp. 1157–1159, 1988.

[77] E. Dietrich *et al.*, "Semiconductor laser optical amplifiers for multichannel coherent optical transmission," *J. Lightwave Technology*, Vol. 7, No. 12, pp. 1941–1955, 1989.

[78] D. J. Maylon and W. A. Stallard, "565 Mbit/s FSK direct detection syetem operating with four cascaded photonic amplifiers," *Electronics Letters,* Vol. 25, No. 8, pp. 495–497, 1989.

[79] G. Grosskopf, R. Ludwig and H. G. Weber, "Cross-talk in optical amplifiers for two-channel transmission," *Electronics Letters*, Vol. 22, No. 17, pp. 900–902, 1986.

[80] K. Inoue, T. Mukai and T. Saito, "Nearly degenerate four-wave mixing in a travelling-wave semiconductor laser amplifier," *Appl. Phys. Lett.*, Vol. 51, No. 14, pp. 1051–1053, 1987.

[81] S. Ryu, K. Mochizuki and H. Wakabayashi, "Influence of non-degenerate four-wave mixing on coherent transmission system using in-line semiconductor laser amplifiers," *J. Lightwave Technology*, Vol. L-T-7, No. 10, pp. 1525–1529, 1989.

[82] H. Kressel, "Semiconductor devices for optical communications," 2nd Edition (Springer-Verlag, 1982).

[83] W. I. Way, C. Zah and T. P. Lee, "Application of travelling-wave laser amplifiers in subcarrier multiplexed lightwave systems," *IEEE Trans. Microwave Theory Tech.*, Vol. 38, No. 5, pp. 534–545.

[84] K. Kannan, A. Bartos and P. S. Atherton, "High-sensitivity receiver optical pre-amplifier," *IEEE Photo. Tech. Lett.*, Vol. 4, No. 3, pp. 272–274, 1992.

[85] K. Kikuchi, "Proposal and performance analysis of novel optical homodyne receiver having an optical preamplifier for achieving the receiver sensitivity beyond the shot-noise limit," *IEEE Photo. Tech. Lett.*, Vol. 4, No. 2, pp. 195–197, 1992.

[86] Y. Yamamoto and H. Tsuchiya, "Optical receiver sensitivity improvement by a semiconductor laser amplifier," *Electronics Letters*, Vol. 16, No. 6, pp. 233–235, 1980.

[87] N. Nakagawa and S. Shimada, "Optical amplifiers in future optical communication systems," *IEEE LCS Magazine*, Vol. 1, No. 4, pp. 57–62, 1990.

[88] T. Nakai, R. Ito and N. Ogasawara, "Asymmetric frequency response of semiconductor laser amplifiers," *Japan J. Appl. Phys.*, Vol. 21, No. 11, pp. L680–L682, 1982.

[89] A. J. Lowery, "Modelling spectral effects of dynamic saturation in semiconductor laser amplifiers using the transmission-line laser model," *IEE Proc.*, Part J, Vol. 136, No. 6, pp. 320–324, 1989.

[90] C. T. Hultgren and E. P. Ippen, "Ultrafast refractive index dynamics in AlGaAs diode laser amplifiers," *Appl. Phys. Lett.*, Vol. 59, No. 6, pp. 635–637, 1991.

[91] J. Singh *et al.*, "A novel twin-ridge-waveguide optical amplifier switch," *IEEE Photo., Tech. Lett.*, Vol. 4, No. 2, pp. 173–176, 1992.

[92] D. Bakewell, "Amplitude and phase modulation of laser amplifier using travelling wave model," *Electronics Letters*, Vol. 27, No. 4, pp. 329–330.

[93] A. Yariv and P. Yeh, *Optical waves in crystals* (John Wiley and Sons, 1984).

[94] M. A. Ali, A. F. Elrefaie and S. A. Ahmed, "Simulation of 12.5 Gb/s lightwave optical time-division multiplexing using semiconductor optical amplifiers as external modulators," *IEEE Photonics Technology Letters*, Vol. 4, No. 3, pp. 280–282, 1992.

[95] M. Gustvasson, L. Thylen and D. Djupsjobacka, "System performance of semiconductor laser amplifier detectors," *Electronics Letters*, Vol. 25, No. 2, pp. 1375–1377, 1989.

[96] R. M. Fortenberry, A. J Lowery and R. S. Tucker, "Up to 16 dB improvement in detected voltage using two section semiconductor optical amplifier detector," *Electronics Letters*, Vol. 28, No. 5, pp. 474–476, 1992.

[97] K. T. Koai and R. Olshansky, "Simultaneous optical amplification, detection, and transmission using in-line semiconductor laser amplifiers," *IEEE photo. Tech. Lett.*, Vol. 4, No. 5, pp. 441–443, 1992.

[98] B. Moslehi, "Fibre-optic filters employing optical amplifiers to provide design flexibility," *Electronics Letters*, Vol. 28, No. 3, pp. 226–228, 1992.

[99] C. Burke, M. Fujiwara, M. Yamaguchi, H. Nishimoto and H. Honmou, "128 line photonic switching system using LiNbo3 switch matrices and semiconductor travelling wave amplifiers," *J. Lightwave Technology*, Vol. 10, No. 5, pp. 610–615, 1992.

[100] C. Jørgensen, N. Storkfelt, T. Durhuus, B. Mikkelsen, K. E. Stubkjaer, B. Fernier, G. Gelly and P. Doussiere, "Two-section semiconductor optical amplifier used as an efficient channel dropping node," *IEEE Photo. Tech. Lett.*, Vol. 4, No. 4, pp. 348–350, 1992.

[101] S. Kawamishi and M. Saruwatari, "10 GHz timing extraction from randomly modulated optical pulses using phase-locked loop with travelling-wave laser-diode optical amplifier using optical gain modulation," *Electronics Letters*, Vol. 28, No. 5, pp. 510–511, 1992.97.

[102] K. T. Koai and R. Olshansky, "Dual-function semiconductor laser amplifier in a broad-band subcarrier multiplexed system" *IEEE photo. Tech. Lett.*, Vol. 2, No. 12, pp. 926–928, 1990.

[103] H. Ghafouri-Shiraz, "Temperature, bandgap-wavelength, and doping dependence of peak-gain coefficient parabolic model parameters for InGaAsP/InP semiconductor laser diodes," *IEEE J. Lightwave Technology*, Vol. LT-8, No. 4, pp. 500–506, April 1988.

[104] J. C. Simon, J. L Favennec and J. Charil, "Comparison of noise characteristics of Fabry–Perot-type and travelling-wave-type semiconductor laser amplifiers," *Electronics Letters*, Vol. 19, No. 8, pp. 288–290, 1983.

[105] R. Loudon, *The quantum theory of light*, 2nd Edition (Oxford University Press, 1983).

[106] W. H. Louisell, *Quantum statistical properties of radiation* (New York; John Wiley and Sons, 1973).

Chapter 4

Analysis of Transverse Modal Fields in Semiconductor Laser Amplifiers

4.1 Introduction

In Chapter 3, we have examined how optical amplification can be achieved in semiconductor lasers. We have seen that the formation of an active region due to population inversion in a forward biased p.n. junction will introduce optical gain, and a simple dielectric waveguide will be formed due to the resulting refractive index step. The semiconductor laser amplifier amplifies signals which travel across the active region via the dielectric waveguide. In this chapter, we will apply some of the mathematical principles of electromagnetics to analyse how optical signals propagate across a semiconductor laser amplifier by treating them as electromagnetic waves [1, 2]. This investigation is important because the propagation characteristics of optical signals in the amplifier determine the electromagnetic field distribution inside the structure, which affects the photon density in the amplifier and is crucial in determining the amplifier gain, gain saturation and noise characteristics.

The electromagnetic analysis of a semiconductor laser amplifier (SLA) involves two steps. Firstly, one has to determine the *transverse field distributions in the amplifier*. This is affected by the dimensions of the active region on the $x-y$ plane of the dielectric optical waveguide formed in the active region [3, 4]. Secondly, one has to determine the *longitudinal field distributions*. This is affected by the amplifier length and the boundary conditions along the direction of propagation of signals. Both of these field distributions in the amplifier depends on the group velocity of signals, which is principally determined by the transverse field distributions.

Furthermore, as we shall see later, several discrete and distinct field patterns can exist in a dielectric optical waveguide for a particular signal frequency. We designate each of these patterns as different *modes*, and each field pattern is known as a *modal field distribution*. The number of these modes which can propagate across the SLA is determined by their group velocities and the dimensions of the waveguide (hence the transverse modal field distribution). Thus both the number and field patterns of the propagating modes of the signal in a particular SLA are determined ultimately by the transverse field distribution over its transverse $x - y$ plane. In this chapter, we will concentrate on the analysis of the transverse modal fields in SLAs, which can help us to determine the propagation characteristics, or waveguiding characteristics, of the amplifier. These characteristics are extremely important as they will also affect the signal gain of the amplifier, which we will also examine in some depth in this chapter.

This chapter is organised as follows: various methods of solving the transverse modal field distribution in a dielectric rectangular waveguide will be reviewed. We concentrate on rectangular structures, as in most SLAs their active regions have a rectangular cross-section [5]. Solution of this modal field distribution can determine two important parameters: the *effective index* [4, 6], and the *modal gain coefficient*, or the *effective gain*, of the amplifier [7]. The latter parameter can be used to study the polarisation sensitivity of the signal gain in an SLA [8, 9]. A method based on *perturbation analysis*, proposed by Ghafouri-Shiraz and Chu [10], will be used to calculate the modal gain coefficient of an SLA. This chapter is concluded by a discussion on the structural design implications for polarisation insensitive travelling-wave SLAs.

4.2 Solution of Transverse Modal Fields in Rectangular Optical Waveguides

The solution of transverse modal fields in a rectangular dielectric optical waveguide involves solving the wave equation for the structure. This topic has been covered in many textbooks on electro-optics, integrated optics or electromagnetics [1–4, 11, 12]. However, improvement in the technique for solving the wave equation in dielectric waveguides is still under active research. This is because to date no exact analytical solutions are available for this structure [13]. We will first briefly examine the simplest case of a three layer dielectric slab and the corresponding modal field solutions. We

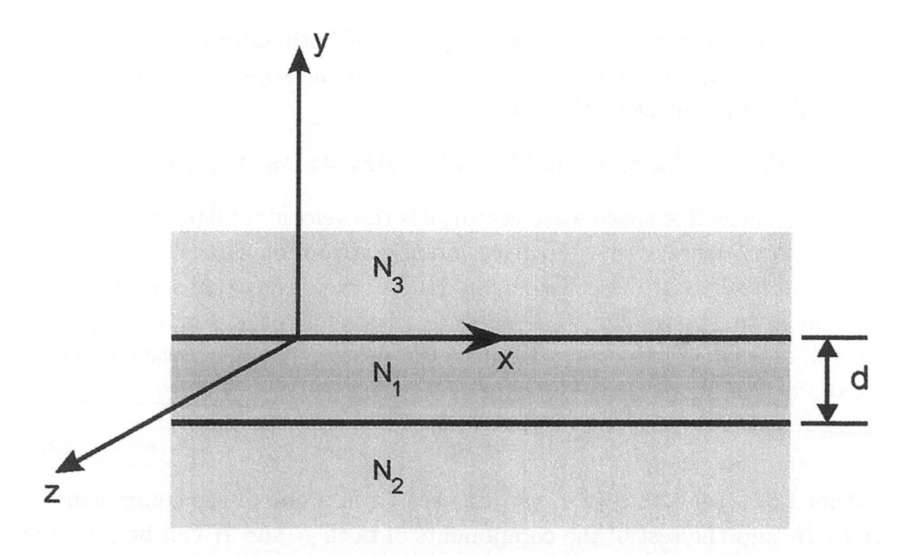

Fig. 4.1. A planar three-layer dielectric slab.

will then review possible methods of solving the problem for more complex rectangular dielectric waveguides.

4.2.1 *Solution for a three-layer slab (Planar optical waveguide)*

A schematic of the dielectric slab is shown in Fig. 4.1. It is well known in electromagnetics that the electric and magnetic field of a source free waveguide satisfies the homogeneous wave equation [2]. We can start to derive this equation from Maxwell's **curl** equations, which are:

$$\nabla x E = -j\omega\mu H \tag{4.1}$$

$$\nabla x H = j\omega\varepsilon E\,. \tag{4.2}$$

The bold symbols represent vector quantities. The above equations are expressed in time harmonic form without current sources. In these equations, ε and μ are the relative permittivity and relative permeability of the medium, respectively. They can be spatial dependent (in an inhomogeneous medium), or even depend on the magnitude and direction of the electric and magnetic field vectors themselves (in an anisotropic medium). In the latter case, tensor analysis has to be used [14–16], which is beyond the scope of this book. In the following discussion, we will consider a homogeneous medium only.

By using vector identities, taking **curl** of both sides of Eq. (4.1) and substituting Eq. (4.2) into it, one can arrive at the *scalar wave equation* for the scalar component of E as [4]:

$$\nabla^2 E_i = -\omega^2 \mu \varepsilon E_i = -(\omega^2/c^2) N^2(x,y) E_i = -{k_0}^2 N^2(x,y) E_i \tag{4.3}$$

where k_0 is the free space wave vector, c is the velocity of light in free space, and $N(x,y)$ denotes the refractive index distribution of the medium. We have assumed that μ is constant in the dielectric waveguides and is equal to that in free space [20]. The subscript i stands for x, y or z, depending on the direction of the component of the vector. A similar wave equation exists for the magnetic field vector H [4], i.e.:

$$\nabla^2 H_i = -k_0 N^2(x,y) H_i \,. \tag{4.4}$$

Either Eq. (4.3) or Eq. (4.4) can be used to find one of the components of E or H, and the rest of the components of both E and H can be found by Eq. (4.1). Usually one prefers to work with Eq. (4.3), as the electric field is responsible for most physical phenomena in optics [17], and is the principal field measured by photodetectors [12].

The solution of the wave Eq. (4.3) is quite complex for rectangular dielectric structures [3]. However, as we are considering a planar slab structure extending to infinity along both $+x$ and $-x$ directions (Fig. 4.1), a very simple form of analytical solution can be obtained [2, 4]. An inspection of the structure shows that either E_z or H_z can exist in this structure, but not both simultaneously (this is because of the absence of any boundary which is perpendicular to the x-direction), hence, by enforcing the continuity condition it can be seen that only the components E_z or H_z exist [11]. This leads to two possible solutions: a *transverse electric* (TE) mode with $E_z = 0$, and a *transverse magnetic* (TM) mode with $H_z = 0$. The notation used follows that which is commonly used in metallic waveguides [2]. Equations (4.1) and (4.2) can be expanded to give the following set of relations between the different components of the electric and magnetic field vectors:

$$\frac{\partial E_z}{\partial y} - \frac{\partial E_y}{\partial z} = -j\omega\mu H_x \tag{4.5}$$

$$\frac{\partial E_z}{\partial x} - \frac{\partial E_x}{\partial z} = j\omega\mu H_y \tag{4.6}$$

$$\frac{\partial E_y}{\partial x} - \frac{\partial E_x}{\partial y} = -j\omega\mu H_z \tag{4.7}$$

$$\frac{\partial H_z}{\partial y} - \frac{\partial H_y}{\partial z} = j\omega\varepsilon E_x \tag{4.8}$$

$$\frac{\partial H_z}{\partial x} - \frac{\partial H_x}{\partial z} = -j\omega\varepsilon E_y \tag{4.9}$$

$$\frac{\partial H_y}{\partial x} - \frac{\partial H_x}{\partial y} = j\omega\varepsilon E_z\,. \tag{4.10}$$

For the fields to propagate in the TE mode, it requires $H_z \neq 0$, $E_x \neq 0$, $E_y = 0$, $H_y \neq 0$ and $H_x = 0$, otherwise the above curl equations cannot be satisfied. Using similar arguments, it can be shown that for the TM mode, we must have $E_x = 0$, $E_y \neq 0$, $H_y = 0$ and $H_x \neq 0$ [3, 4]. It can be seen that the electric field on the transverse $x - y$ plane of the TE mode is orthogonal to that of the TM mode (in fact, one can prove this analytically from the solutions of TE and TM modes tabulated in Table 4.1 [18, 19]). The field solution for the TE mode can be found by substituting E_x into Eq. (4.3) with $E_x \propto \exp(-j\beta z)$ [2].

$$\frac{\partial^2 E_x}{\partial y^2} = -({k_0}^2 N_1^2 - {\beta_z}^2)E_x = -q^2 E_x \tag{4.11}$$

in a region with refractive index N_1 (hereafter denoted as region 1) (Fig. 4.1). We have assumed that E_x is dependent of x because of the infinite dimensions of the slab (see Eqs. (4.5) and (4.6)). The parameter β_z is known as the *longitudinal propagation constant.* It determines how the phase of the electric field varies with distance z in the waveguide. If the propagation constant is real, then we have travelling fields with phase varying sinusoidally with z [2]. If β_z is pure imaginary, the fields will vary exponentially with z and are commonly known as *evanescent waves* [11, 17]. In many cases, β_z is complex, and the resulting fields are both travelling with their phases varying sinusoidally with z and their amplitude changing exponentially with z. Let us proceed with the analysis in this dielectric slab (Fig. 4.1) by considering the fields outside region 1. In the cladding region (i.e. regions on top of and below region 1), if $N_2 = N_3$, the wave equation becomes:

$$\frac{\partial^2 E_x}{\partial y^2} = -({k_0}^2 N_2^2 - {\beta_z}^2)E_x = -r^2 E_x\,. \tag{4.12}$$

Both the constants q^2 and r^2 can be positive or negative, depending on the relative magnitude of N_1, N_2, k_0 and β_z. The possible solutions for different ranges of negative values of q^2 and r^2 are illustrated in Fig. 4.2 [12].

Table 4.1. Field expressions for both TE and TM modes in a slab waveguide.

Mode	Field solution	Eigen equation
TE		
$E_z = E_y = 0$; $H_y = H_z = 0$	$E_x = \begin{cases} A\cos(qy) + B\sin(qy)\,, & -d \le y \le 0 \\ A\exp(-ry)\,, & y \ge 0 \\ A\cos(qd) - B\sin(qd)\exp[r(y+d)]\,, & y \le -d \end{cases}$	$\tan(dq) = \dfrac{2rq}{q^2 - r^2}$
TM		
$H_z = H_y = 0$; $E_z = E_y = 0$	$H_x = \begin{cases} C\cos(qy) + D\sin(qy)\,, & -d \le y \le 0 \\ C\exp(-ry)\,, & y \ge 0 \\ C\cos(qd) - D\sin(qd)\exp[r(y+d)]\,, & y \le -d \end{cases}$	$\tan(dq) = \dfrac{2n_1^2 n_2^2 qr}{n_1^4 q^2 - n_1^4 r^2}$

Note: $\beta^2 = k^2 n_1^2 - q^2$

Only the solution of one transverse field component is given. The remaining field solutions can be found by using Maxwell's equations. The arbitrary constants A, B, C and D can be found by the output power condition of the amplifier.

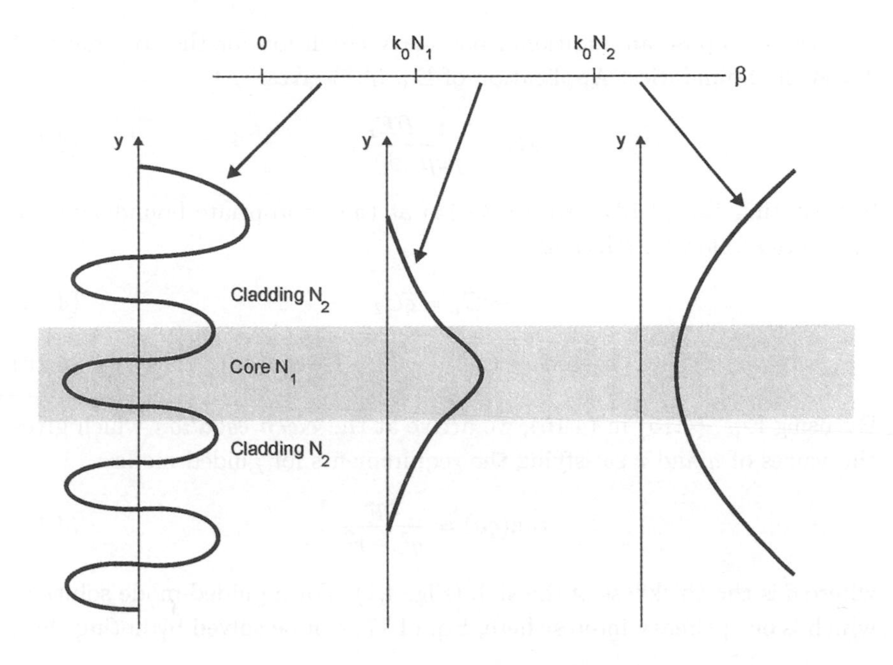

Fig. 4.2. Different types of modal solutions determined by the constants q and r.

If both q^2 and r^2 are positive (see Fig. 4.2), the corresponding values of the propagation constant β_z are in the *continuous radiating mode* regime [20]. The resulting waveguide does not guide any optical signals. For *guided modes* (case (b) in Fig. 4.2) to occur, q^2 has to be positive but r^2 must be negative, i.e. $k_0^2N_1^2 > \beta_z^2 > k_0^2N_2^2$ [21]. In this case, the solutions for the transverse fields can be found by imposing continuity requirements at the appropriate boundaries:

$$\begin{aligned} E_x &= C_1 \exp(-ry) \quad y \geq 0 \\ E_x &= C_1 \cos(qy) + C_2 \sin(qy) \quad -d \leq y \leq 0 \\ E_x &= [C_1 \cos(qd) - C_2 \sin(qd)] \exp[r(y+d)] \quad y \leq -d \end{aligned} \tag{4.13}$$

where we have dropped the term $\exp(-j\beta_z z)$, assuming that it is implicit. Equation (4.13) is a general solution. There are four unknowns yet to be determined: C_1 and C_2, which correspond to the amplitude of the transverse modal field, and q and r, which determine the exact "shape" of the field as well as the value of the propagation constant β_z. To find the values of q and r which satisfy Eq. (4.13) for a given value of k_0 and N_1 and N_2, one has to impose an additional boundary condition for the continuity of H_z at the boundaries. Application of Eq. (4.7) gives:

$$H_z = \frac{1}{j\omega\mu}\frac{\partial E_x}{\partial y}\,. \tag{4.14}$$

Substituting Eq. (4.13) into Eq. (4.14) at the appropriate boundaries, we obtain the following relations:

$$-rC_1 = qC_2 \tag{4.15}$$

$$\frac{q}{r}[C_1 \tan(qd) - C_2] = -C_1 - C_2 \tan(qd)\,. \tag{4.16}$$

By using Eqs. (4.15) in (4.16), we arrive at the *eigen equation* which gives the values of q and r satisfying the requirements for guided modes:

$$\tan(qd) = \frac{2qr}{q^2 - r^2} \tag{4.17}$$

where d is the thickness of the slab (Fig. 4.1). For a guided mode solution, which is our primary interest here, Eq. (4.17) can be solved by noting that:

$$q^2 = k_0^2 N_1^2 - \beta_z^2 \tag{4.18}$$

$$r^2 = \beta_z^2 - k_0^2 N_2^2 = k_0^2(N_1^2 - N_2^2) - q^2\,. \tag{4.19}$$

By solving Eqs. (4.17) and (4.19) iteratively, we can arrive at solutions of q and r for a particular structure (hence N_1 and N_2) and frequency (hence the wave number k_0). Once the values of q and r have been determined, the unknown C_1 and C_2 can be found easily if the total power of the propagating electromagnetic wave is known. It should be noted that, because of the periodicity of Eq. (4.17), each particular set of values of k_0, N_1 and N_2 will result in a number of possible values of q and r which will satisfy Eq. (4.17). We will discuss this implication later. Let us first examine the property of Eq. (4.17) at $y = -d/2$, i.e. at the plane situated at the centre in region 1. If we express $\tan(qd/2)$ as t, and by using trigonometric identities, we will arrive at the following relation.

$$\tan(qd) = \frac{2t}{1 - t^2} = \frac{2qr}{q^2 - r^2} = \frac{2(r/q)}{1 - (r/q)^2} \tag{4.20}$$

which, by solving in terms of t, gives the following two solutions [3]:

$$t_1 = \tan\left(\frac{qd}{2}\right) = \frac{r}{q} \tag{4.21}$$

$$t_2 = \tan\left(\frac{qd}{2}\right) = -\frac{q}{r}\,. \qquad (4.22)$$

Equation (4.21) represents *even* mode solutions, with a maximum in E_x at $y = -d/2$, whilst Eq. (4.22) represents *odd* mode solutions, with $E_x = 0$ at $y = -d/2$ [3]. The procedures outlined above are for solutions of TE mode only. The corresponding field solutions for TM modes and the eigen equations for determining the guided mode values of q and r can be found by following this approach, but with H_x replacing of E_x. Details of the solution procedure for the TM mode can be found in standard textbooks (e.g. [3, 4]). Here, for the sake of completeness, we will only quote the solutions and eigen equations for the TE and TM modes of the planar dielectric slab waveguide; these are tabulated in Table 4.1.

Before we examine dielectric waveguides with more complex structures, we must first discuss the concept of *modes*. The forms of eigen equations for q and r of TE and TM modes require both q and r to be a set of *discrete* values for a particular value of k_0 and a particular refractive index distribution (i.e. N_1 and N_2) (hence the term eigenequation — the values of q and r are the *eigenvalues* of the eigen equation). This is illustrated in Fig. 4.3, where the functional forms of Eqs. (4.21) and (4.22) are plotted with the constraints of Eqs. (4.18) and (4.19) [12]. A similar figure holds for the eigen equation of the TM mode. The circular arcs are derived from Eqs. (4.18) and (4.19), which have the functional forms of $(qd/2)^2 + (rd/2)^2 = k_0^2 d^2(N_1^2 + N_2^2)/4 = \text{constant}$. The intersections of Eq. (4.21) (solid lines) and Eq. (4.22) (dashed lines) with one of these arcs represent the possible solutions for the eigen equation with a particular value of $k_0^2 d^2(N_1^2 + N_2^2)/4$. It then follows from Eqs. (4.18) and (4.19) that β_z must also have a finite number of discrete values. Each of these discrete values of β_z is said to be the propagation constant of a *mode* with order m. Notice that the previous description of *even* and *odd* modes does not describe whether the value of m is even or odd. It should be emphasised that, on the contrary, the modal field of an *odd* mode (with a zero value of E_x at $y = -d/2$) has an *even* value of m, whereas the modal field of an *even* mode (with a maxima of E_x at $y = -d/2$) has an *odd* value of m. These different modes will also have distinct transverse modal field distributions, as they have different values of q and r in Eq. (4.13) [4]. It can also be seen that for some combinations of k_0 (i.e. frequency) and structural parameters d, N_1 and N_2, the number of possible solution modes which satisfy the eigen equation can be as few as one (e.g. the $k_0^2 d^2(N_1^2 + N_2^2)/4 = 1$ arc

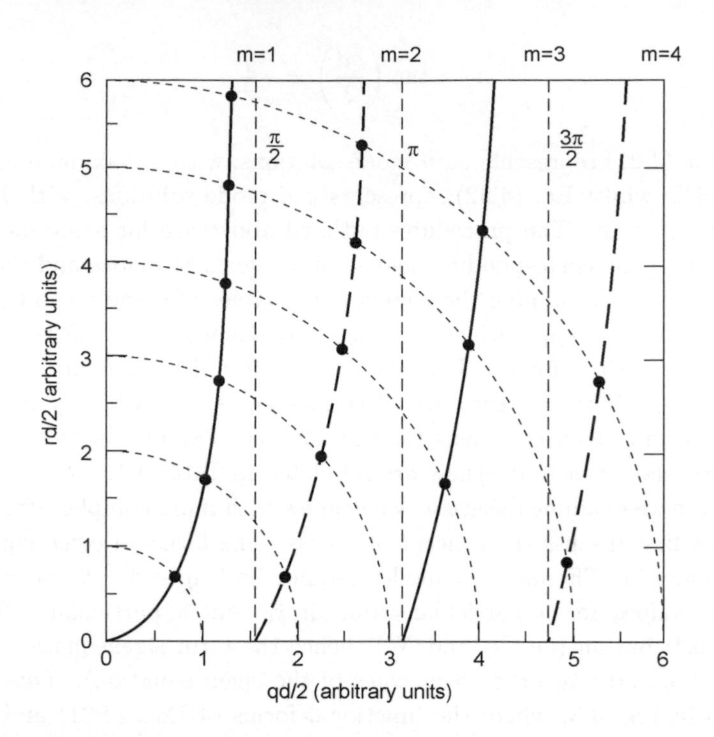

Fig. 4.3. Graphical method to solve the eigenequation for a symmetrical three-layer dielectric waveguide (after [12]).

has only one intersection with Eq. (4.21) at $m = 1$). Therefore, by using proper structural dimensions, it is possible to design a dielectric waveguide which only supports one single propagating mode *of a particular polarisation* (either TE or TM mode) for a particular range of signal frequencies. For simple three-layer structures, Fig. 4.3 proves to be an extremely useful graphical tool to aid such design. However, for more complex structures like buried heterostructures (BH) in SLAs, the waveguide structure involved is more complicated, and such simple techniques may not be applicable. We will next examine methods to find the transverse modal fields and the corresponding propagation constants in more complex waveguides.

4.2.2 *Solution for a rectangular dielectric waveguide using modal field approximations*

In theory, the above principles for analysing planar waveguides can be extended to rectangular dielectric structures. A typical rectangular dielectric

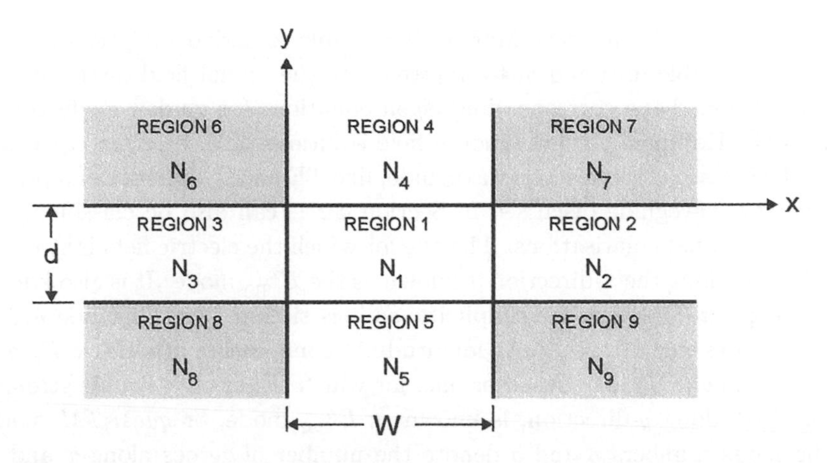

Fig. 4.4. The structure used to analyse the modal field distributions in rectangular waveguides.

waveguide is shown in Fig. 4.4, with width W and thickness d. To analyse the problem, one can follow the procedures applied to the planar waveguide in the previous section to find the field solutions for rectangular structures as well. We can first generate the wave equation for each region in Fig. 4.4, and solve them by matching proper boundary conditions to obtain the appropriate eigen equations and hence the necessary condition for guided modes. Because of the presence of boundaries (discontinuities in refractive indices) in both x and y directions, E_z and H_z will be finite for all propagating modes [3]. This will make the analysis more complex, although mathematically feasible.

However, there is an additional problem. The matching of boundary conditions can only be made by choosing either the continuous tangential fields at the four boundaries surrounding region 1 (the *core*) or the normal fields at the same boundaries [22]. This leaves other boundaries (e.g. between regions 3 and 5, 2 and 7, etc.) unmatched. In other words, the boundary conditions cannot be matched simultaneously at all boundaries [4]. Because of this, it is obvious that the field solutions are no longer separable functions of x and y [16, 23–25]. Marcatilli [26] suggested that if the waveguide was supporting a propagating mode which was well above cut-off, then the power flow in the corners (shaded regions) in Fig. 4.4 can be neglected. Then the matching of boundary conditions in these regions is unimportant. Hence one can simply match boundary conditions around the core to generate approximate modal field solutions for the unshaded regions 1 to 5. In this way, Marcatilli was able to derive analytical expressions, separable in x and y, to approximate the modal field distributions, and developed the corresponding eigen equations for guided mode conditions [3]. He found that the modal field solutions of these eigen equations for the rectangular dielectric waveguide, like TE and TM modes in a planar dielectric waveguide discussed in Section 4.2.1, can also be classified into two orthogonal polarisations. The one for which the electric field is strongly polarised along the x-direction is known as the $E^x{}_{\mathrm{pq}}$ mode. It is also known as the *quasi-TE* mode for simplicity, as it is similar to a TE mode *except* for the presence of *both* finite longitudinal components of electric E_z and magnetic field H_z [27]. Another one, for which the electric field is strongly polarised along y-direction, is known as $E^y{}_{\mathrm{pq}}$ mode, or *quasi-TM* mode. The mode numbers p and q denote the number of zeroes along x and y-directions, respectively [26]. Notice that the quantity q here is similar in meaning to that used in the previous section, which was used to denote the number of zeroes along y in a dielectric slab. The above modal fields in a rectangular dielectric waveguide are known as *hybrid modes* because they are mixtures of the TE and TM modes in the simple three-layer dielectric waveguide discussed in last section [2]. We have tabulated the analytical solutions for the approximate modal fields in regions 1 to 5 in Table 4.2 for the field components E_z and H_z, together with the eigen equations derived for guided modes. Other field components can be found from Eq. (4.5) to 4.10. The two parameters k_x and k_y in Table 4.2 are known as the *transverse wave numbers* of the modal fields, and are related to the propagation constant β_z by:

$$\beta_z{}^2 = k_0{}^2 N_1^2 - k_x^2 - k_y^2 \,. \tag{4.23}$$

Note the similarity of Eq. (4.23) with Eq. (4.18). These two parameters determine the sinusoidal variation form of the fields along x- and y-directions in the core. We can also relate the *decaying parameters* γ_i in the ith region of the cladding with β_z [4]:

$$\beta_z{}^2 = k_0{}^2 N_1^2 - k_y^2 + \gamma_i{}^2 \,; \quad i = 2, 3 \tag{4.24}$$

$$\beta_z^2 = k_0{}^2 N_1^2 - k_x^2 + \gamma_j^2 \,; \quad j = 4, 5 \,. \tag{4.25}$$

Again, γ_i, k_x and k_y are discrete, which can be found by the eigen equations in Table 4.2 for guided modes. Therefore, similar to the arguments outlined in the previous section with Fig. 4.3, each modal field solution has a particular value of β_z, and for a given value of frequency, there

Table 4.2. Field solutions of wave equations in a rectangular dielectric waveguide.

Field	Region	E_z	H_z
E^x_{mn}	1	$A\cos(k_x(x+\zeta))\cos(k_y(y+\eta))$	$A\sqrt{(\epsilon_0/\mu_0)}n_1^2(k_y/k_x)(k/\beta)\sin(k_x(x+\zeta))\sin(k_y(y+\eta))$
	2	$A\cos(k_x(W+\zeta))\cos(k_y(y+\eta))\exp[-\gamma_2(x-W)]$	$-A\sqrt{(\epsilon_0/\mu_0)}n_1^2(k_y/\gamma_2)(k/\beta)\cos(k_x(W+\zeta))\sin(k_y(y+\eta))\cdot \exp[-\gamma_2(x-W)]$
	3	$A\cos(k_x\zeta))\cos(k_y(y+\eta))\exp(\gamma_3 x)$	$A\sqrt{(\epsilon_0/\mu_0)}n_3^2(k_y/\gamma_3)(k/\beta)\cos(k_x\zeta)\sin(k_y(y+\eta))\exp(\gamma_3 x)$
	4	$A(n_1/n_4)^2\cos(k_x(x+\zeta))\cos(k_y\eta)\exp(-\gamma_4 y)$	$A\sqrt{(\epsilon_0/\mu_0)}n_1^2(\gamma_4/k_x)(k/\beta)\sin(k_x(x+\zeta))\cos(k_y\eta)\exp(-\gamma_4 y)$
	5	$A(n_1/n_5)^2\cos(k_x(x+\zeta))\cos(k_y(\eta-d))\cdot \exp[\gamma_5(y+d)]$	$-A\sqrt{(\epsilon_0/\mu_0)}n_1^2(\gamma_5/k_x)(k/\beta)\sin(k_x(x+\zeta))\cos(k_y(\eta-d))\cdot \exp[\gamma_5(y+d)]$
E^y_{mn}	1	$A\cos(k_x(x+\zeta))\cos(k_y(y+\eta))$	$-A\sqrt{(\epsilon_0/\mu_0)}n_1^2(k_x/k_y)(k/\beta)\sin(k_x(x+\zeta))\sin(k_y(y+\eta))$
	2	$A(n_1/n_2)^2\cos(k_x(W+\zeta))\cos(k_y(y+\eta))\cdot \exp[\gamma_2(x-W)]$	$-A\sqrt{(\epsilon_0/\mu_0)}n_1^2(\gamma_2/k_y)(k/\beta)\cos(k_x(W+\zeta))\sin(k_y(y+\eta))\cdot \exp[-\gamma_2(x-W)]$
	3	$A(n_1/n_3)^2\cos(k_x\zeta))\cos(k_y(y+\eta))\exp(\gamma_3 x)$	$A\sqrt{(\epsilon_0/\mu_0)}n_1^2(\gamma_3/k_y)(k/\beta)\cos(k_x\zeta)\sin(k_y(y+\eta))\exp(\gamma_3 x)$
	4	$A\cos(k_x(x+\zeta))\cos(k_y\eta)\exp(-\gamma_4 y)$	$A\sqrt{(\epsilon_0/\mu_0)}n_4^2(k_x/\gamma_4)(k/\beta)\sin(k_x(x+\zeta))\cos(k_y\eta)\exp(-\gamma_4 y)$
	5	$A\cos(k_x(x+\zeta))\cos(k_y(\eta-d))\exp[\gamma_5(y+d)]$	$-A\sqrt{(\epsilon_0/\mu_0)}n_5^2(k_x/\gamma_5)(k/\beta)\sin(k_x(x+\zeta))\cos(k_y(\eta-d))\cdot \exp[\gamma_5(y-d)]$

For the E^x_{mn} mode:

$$\tan(k_y d) = \frac{k_y(\gamma_4+\gamma_5)}{k_y^2-\gamma_4\gamma_5}, \quad \beta^2 = k_0^2 n_1^2 - k_x^2 - k_y^2$$

$$\tan(k_x W) = \frac{n_1^2 k_x(n_2^2\gamma_3 + n_3^2\gamma_2)}{n_2^2 n_3^2 k_x^2 - n_1^2\gamma_2\gamma_3}$$

$$\tan(k_x\zeta) = \left(\frac{n_3}{n_1}\right)^2\left(\frac{k_x}{\gamma_3}\right), \quad \tan(k_y\eta) = \frac{\gamma_4}{k_y}$$

For the E^y_{mn} mode:

$$\tan(k_x W) = \frac{k_x(\gamma_2+\gamma_3)}{k_x^2-\gamma_2\gamma_3}, \quad \beta^2 = k_0^2 n_1^2 - k_x^2 - k_y^2$$

$$\tan(k_y d) = \frac{n_1^2 k_y(n_4^2\gamma_5 + n_5^2\gamma_4)}{n_4^2 n_5^2 k_y^2 - n_1^2\gamma_4\gamma_5}$$

$$\tan(k_y\eta) = -\left(\frac{n_4}{n_1}\right)^2\left(\frac{k_y}{\gamma_4}\right), \quad \tan(k_x\zeta) = -\frac{\gamma_3}{k_x}$$

can be several possible modal field distributions as long as the structure satisfies the eigen equations for multiple guided modes. By manipulating the structural parameters of the waveguide, it is possible to reduce the number of possible modes which can propagate *for a particular polarisation* (i.e. quasi-TE or quasi-TM) to one. Furthermore, it is possible to obtain a structure such that the transverse field is cut off along the x-direction (i.e. $k_y = 0$). This condition is known as the *single transverse mode* (STM) condition [28]. Under such a condition, the rectangular waveguide structure can be approximated accurately by the three-layer slab discussed in the last section, [5] and [29]. Marcatilli's work provided a set of convenient analytical solutions to approximate the modal field distributions in a rectangular dielectric waveguide. However, the assumption of these functions being separable in x and y is not valid near cut-off, i.e. when the power flow in the corner regions cannot be neglected [13] and [30]. For a better range of solution, alternative numerical techniques have to be used. A very convenient technique known as the effective index method (EIM) provides a simple method to obtain more accurate values of β_z for a wider range of frequencies, and this will be discussed below [4, 31, 32].

4.2.3 *Application of Effective Index Method (EIM) for calculating propagation constants for transverse modal fields in rectangular dielectric waveguides*

In order to improve the solution range of β_z, an alternative method is to intercouple the modal field approximations in x- and y-directions using a concept known as the *effective index*. The accuracy of β_z can be improved by taking into account the inter-relation between the transverse wave numbers k_x and k_y, which were assumed to be independent of each other in Section 4.2.2. Physically, this accounts for the power flow in the active region more accurately than modal field approximations. The effective index for the mth propagating mode can be found by [1].

$$N_{em}^2 = \left(\frac{\beta_z}{k_0}\right)^2 = \frac{\iint N^2(x,y)|\Phi_m(x,y)|^2 dxdy}{\iint |\Phi_m(x,y)|^2 dxdy} \tag{4.26}$$

where Φ_m is the mth transverse modal field solution of the wave equation for the structure, and the integration is taken over the entire transverse $x-y$ plane. The index m indicates a particular combination of number of zeroes along x and y (i.e. p and q), respectively. This is particularly useful in actual programming to solve for β_z in computers (see, for example, the

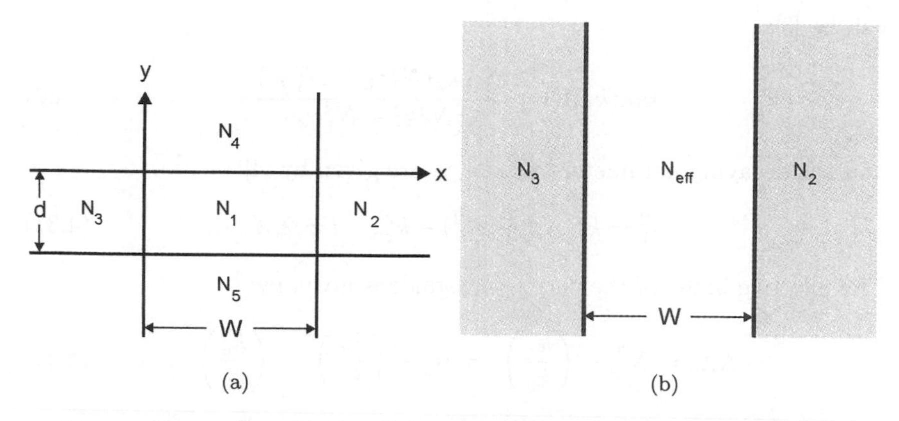

Fig. 4.5. Solution procedure for analysing a rectangular waveguide using EIM.

algorithms proposed by Henry *et al.* [33, 34]). It can be seen from the definition in Eq. (4.26) that N_{em} is a *group* index because it determines the group velocity of the propagating fields [2]. Suppose that $\Phi_m(x, y)$ is a separable function in x and y, such that $\Phi_m(x, y) = F(x)G(y)$. We can now first find the effective index N_{ey} of the dielectric slab of indices N_4, N_1 and N_5 with thickness d for the propagating mode (see Fig. 4.5(a)). For the quasi-TE mode, N_{ey} can be found by solving the eigen equation for the dielectric slab along y. This equation is given by [3]

$$\tan(k_y d) = \frac{k_y(\gamma_4 + \gamma_5)}{k_y^2 - \gamma_4\gamma_5} . \tag{4.27}$$

This formulation is possible because Φ_m is a separable function, hence eigen equations can still be used [2] and [25]. The parameters γ_4 and γ_5 are given by:

$$\gamma_j^2 = k_0{}^2(N_1^2 - N_j^2) - k_y^2 ; \quad j = 4, 5 . \tag{4.28}$$

Once k_y is found, N_{ey} can be found by [4]:

$$N_{ey}^2 = N_1^2 - \left(\frac{k_y}{k_0}\right)^2 . \tag{4.29}$$

We then solve for the dielectric slab in the x-direction with indices N_3, N_1 and N_2, as shown in Fig. 4.5(b). For the quasi-TE mode, the eigen equation will be [3]:

$$\tan(k_x W) = \frac{N_{ey}^2 k_x (N_2^2 \gamma_3 + N_3^2 \gamma_2)}{N_2^2 N_3^2 k_x^2 - N_1^4 \gamma_2 \gamma_3} \tag{4.30}$$

and the decaying parameters γ_2 and γ_3 are given by [4]:

$$\gamma_1^2 = k^2 (N_{ey}^2 - N_i^2) - k_x^2 \, ; \quad i = 2, 3 \, . \tag{4.31}$$

The effective index of the entire waveguide is given by:

$$N_{em}^2 = N_{ey}^2 - \left(\frac{k_x}{k_0}\right)^2 = N_1^2 - \left(\frac{k_x}{k_0}\right)^2 - \left(\frac{k_y}{k_0}\right)^2 . \tag{4.32}$$

Note that Eq. (4.32) is an equivalent of Eq. (4.23). From this equation, N_{em} can decalculated and β_z can subsequently be found from Eq. (4.26). The procedures outlined above are for solutions of quasi-TE modes only. A similar procedure is adopted for solutions of quasi-TM modes, except with different eigen equations. To find N_{ey}, the following eigenequation for quasi-TM mode (see Table 4.2) has first to be solved:

$$\tan(k_y d) = \frac{N_1^2 k_y (N_4^2 \gamma_5 + N_5^2 \gamma_4)}{N_4^2 N_5^2 k_y^2 - N_1^4 \gamma_4 \gamma_5} \tag{4.33}$$

followed by Eq. (4.29). The decaying parameters γ_4 and γ_5 can be found from Eq. (4.28). Then the effective index N_{em} can be found by first solving for k_x by:

$$\tan(k_x W) = \frac{k_4(\gamma_2 + \gamma_3)}{k_x^2 - \gamma_2 \gamma_3} \tag{4.34}$$

where γ_2 and γ_3 can be found from Eq. (4.31), and finally Eq. (4.32) can be used to calculate N_{em}. The major differences between Marcatilli's modal approximation and EIM can be seen by comparing Eqs. (4.24) and (4.25) with Eq. (4.31). Marcatilli assumed that the decaying parameters can be calculated from the same refractive index of the active region N_1, whereas with EIM, we have taken into account the energy leaked to the cladding of N_4 and N_5 before calculating γ_2 and γ_3 by using Eq. (4.31). Physically speaking, the inter-coupling effect between the decaying parameters and the transverse wave numbers has been accounted for, and hence a more accurate description of the power flow in the structure can be obtained. Consequently, the EIM generates a value of β_z which is more accurate than Marcatilli's modal field approximation technique when compared with more exact solutions obtained by rigorous numerical analysis, as well as with experimental measurements [4] and [35].

There are two major disadvantages in using EIM to solve the wave equations in rectangular dielectric waveguides. Firstly, it does not generate field solutions Φ_m automatically [31]. Additional numerical techniques have to be used if Eq. (4.26) has to be used to derive Φ_m from the effective index N_{em} (for example, see the complex field expressions used by Buss [6] in analysing gain guided semiconductor lasers). In addition, the value of β_z calculated from EIM, although more accurate than that obtained by the approximate modal field analyses proposed by Marcatilli, will not be the exact solution [30]. β_z is only exact when a variational analysis performed on Eq. (4.26) indicates that it is minimal [1, 14] and [18]. In EIM, we have assumed that Φ_m is a separable function in x and y, and we have used approximate eigen equations developed by Marcatilli. Hence, the same assumption on negligible power flow in the shaded regions in Fig. 4.4 still apply to EIM, and a variational analysis on Eq. (4.26) with the value of β_z obtained by EIM indicates that the expression is not yet minimal, i.e. the corresponding values of β_z do not give an exact solution. These two major disadvantages will create problems for more complex waveguide structures (e.g. rib waveguides [23]), but since many semiconductor laser amplifiers employ relatively simple rectangular buried heterostructures with index guiding (i.e. the fields are real [4–6, 29] and [30]), the value of β_z calculated by EIM is sufficiently accurate for most purposes [35], and the modal fields solutions listed in Table 4.2 can be used to a good approximation to describe the fields *well above cut-off*. Hence, in the foregoing analyses and discussions, the propagation constant β_z will be calculated using EIM, and the field solutions will be those proposed by Marcatilli, listed in Table 4.2.

Dispersion curves, i.e. the variation of propagation constant β_z against the frequency of the propagating fields, have been plotted in Fig. 4.6 for a rectangular dielectric waveguide with an aspect ratio $W/d = 2.0$ and $N_2 = N_3 = N_4 = N_5$ (i.e. a symmetrical waveguide). Because the value of β_z depends on the structure of the waveguide as well as the frequency of the propagating fields, these dispersion curves are conventionally plotted with *normalised* propagation constant b against *normalised frequency* ν_x [1, 35] and [38]. These parameters are defined by [4].

$$\nu_x^2 = W^2 k_0 (N_1^2 - N_2^2) \tag{4.35}$$

$$b^2 = \frac{(\beta_z^2 - k_0^2 N_2^2)}{(k_0^2 N_1^2 - k_0^2 N_2^2)} \,. \tag{4.36}$$

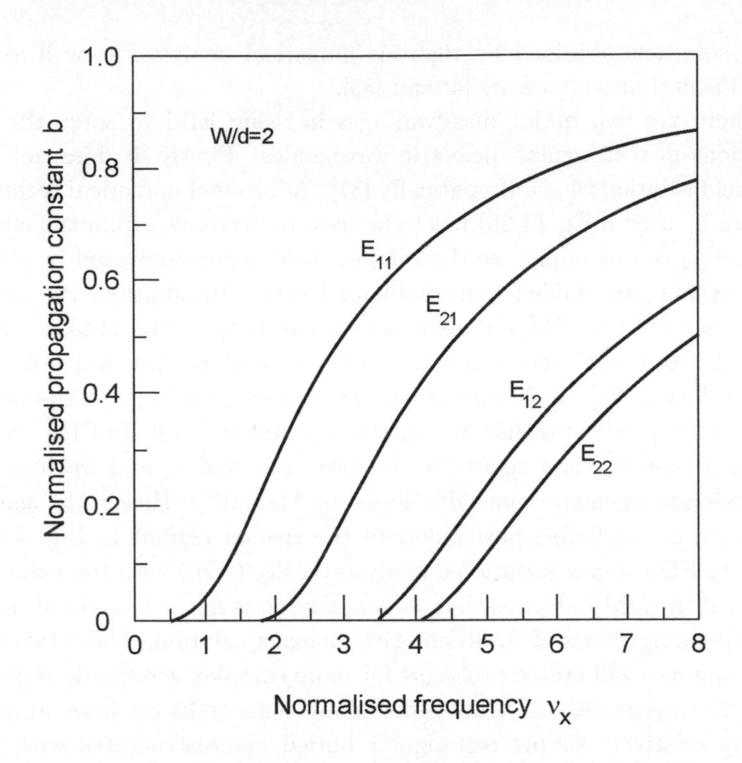

Fig. 4.6. Dispersion curves of b against ν_x for a weakly guiding rectangular waveguide with an aspect ratio $W/d = 2$ calculated using the effective index method.

The four lowest order modes are shown in Fig. 4.6. In the analysis, we have assumed that the waveguide is a *weakly guiding structure* (i.e. $N_1 - N_2 \ll N_1$). Under such circumstances, the quasi-TE and quasi-TM modes have virtually identical values of propagation constant [1] and [4]. It can be seen that for all modes, b increases with an increasing ν_x. This implies that the propagation constant β_z increases with increasing wave number k_0 and hence increasing signal frequency. Physically, as observed from Eq. (4.26), an increase in β_z means that more optical power will be confined within the core region (with refraction index N_1) of the rectangular waveguide (see Fig. 4.4).

In Fig. 4.7, we have compared the dispersion curves for the lowest order modes for several weakly guiding dielectric rectangular waveguides with the same refraction index distribution but different aspect ratios. It can be seen that the value of b changes with the value of the aspect ratio W/a for a particular value of ν_x. The above observation from these dispersion

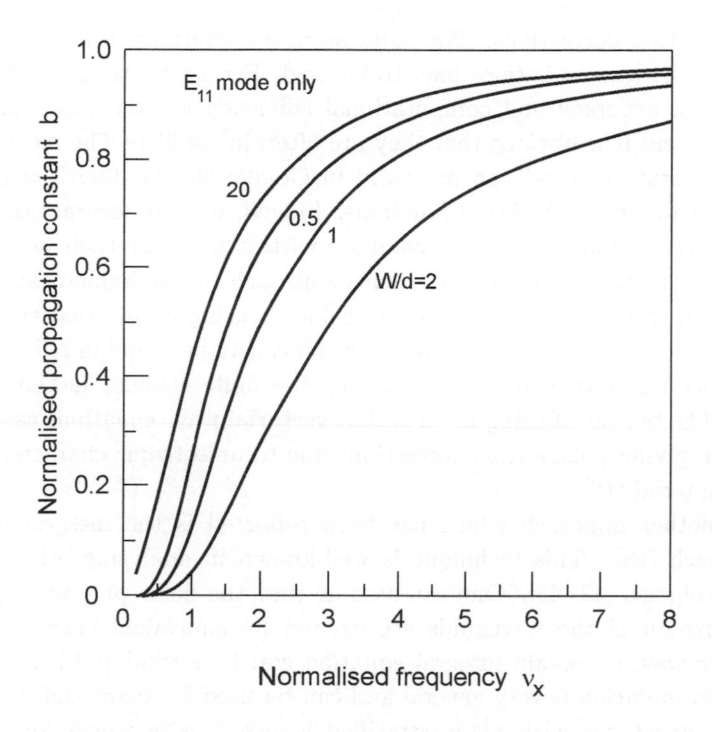

Fig. 4.7. Comparisons of dispersion curves for rectangular dielectric waveguides with different aspect rations.

curves is important in designing the dimensions of a rectangular dielectric waveguide. Moreover, these curves also provide important information for the analysis of semiconductor lasers and the optical gain in semiconductor laser amplifiers (which have structures that resemble that of a dielectric rectangular waveguide). An example of the former analysis was performed by Buss [6, 39] who used the EIM to evaluate the threshold current required for different semiconductor laser diode structures.

4.2.4 *Other methods to solve for transverse modal fields and the dispersion characteristics of rectangular dielectric waveguides*

Marcatilli's approximations and EIM form the basis for analysing rectangular dielectric waveguides. They also form the basis for a number of variant methods. A recent one can be found in Cheung and Lin's work [3] on the iterative equivalent index method and Chiang's [32] modified EIM with perturbation corrections. For more accurate solutions, other methods of solving the wave equations have to be used. For applications of all of these methods, accuracy and computational efficiency are the most important factors, and it is obvious that they are often in conflict. The most reliable and accurate method can be found in Goell's [38] landmark work using point-matching with Bessel functions. Indeed, it is so accurate that other numerical techniques have to compare with it to verify their accuracy [3, 4]. With powerful computers, Goell's approach can be implemented easily. A recent attempt to solve by exact solution using an alternative series of basic functions with the concepts of *residues* can be found in reference [40]. Another rigorous method is to employ the finite element method (FEM) [41]. The benefit of using FEM is that vectorial wave equations can also be solved, giving polarisation corrections due to anisotropic characteristics of the material [16].

Another approach which has been reported is the integral equation approach [18]. This technique is well-known in analysing other scattering problems [42, 43]. One can assume that the fields propagating in the *core* region of the waveguide are excited by equivalent current sources. In this way, a domain integral equation can be formulated [15] and [44]. This formulation is very general and can be used for many different waveguide structures with many stratified layers. It also works for vectorial wave equations. However, solving the integral equations requires the use of a very powerful computer, usually in the supercomputer regime, and is also very time consuming. Some faster methods based on modification of integral equations can be found in Hoekstra's work [45].

A more economical way is to use variational analysis on the expression of Eq. (4.26). By iteratively substituting approximations of $\phi_m(x, y)$ into it, we can obtain the exact solution of β_z as well as the modal field solutions ϕ_m by minimising the expression [4]. There are different ways to find the minimum of Eq. (4.26). It can be based upon expansion of a series of basic functions and the iterative deduction of coefficients for each term in the series until β_z is a minimum [4]. Alternatively, direct iteration by combining Eq. (4.26) with the perturbation expression (see next section) can also be used [30]. A third way is to start from a separable function for ϕ_m, and then proceed to find the solutions of $F(x)$ and $G(y)$ by weighting the refractive indices, i.e. using the method of moments [2]. This method, known as the weighted index method (WIM), is extremely fast, accurate as well as highly efficient, and can yield accurate solutions even for rib waveguides [23, 24]. Other methods of course exist: e.g. by the steepest descent approximation technique to solve the wave equation [46], by the vector beam propagation method [30], or by a two-dimensional Fourier analysis [33, 34].

4.3 Applications of Solutions of Transverse Modal Fields in SLAs

We have seen in Chapter 3 that in analysing the gain characteristics of SLAs, one of the crucial tasks is to relate the material gain coefficient g_m to the single pass gain G_s. In general, if the losses in the cladding can be neglected, it is given by:

$$G_s = \exp[\Gamma_m(g_m - \alpha_a)L] \tag{4.37}$$

where α_a is the loss in the active region of the amplifier, and Γ_m is the optical confinement factor of the mth propagating mode. We can replace the exponent in this equation by the *modal gain coefficient* g [47], such that:

$$G_s = \exp(gL)\,. \tag{4.38}$$

Buss [6] called this parameter the *effective gain*. For the cases of negligible cladding loss, g is simply given by:

$$g = \Gamma_m(g_m - \alpha_a)\,. \tag{4.39}$$

In the following sections, we will examine how to calculate g in more complicated cases with finite cladding losses, and how the structure of the semiconductor laser amplifier (SLA) will affect the value of g and hence G.

4.3.1 *Analysis of the modal gain coefficients*

We stated in Chapter 3 that when the loss in the cladding region was finite and uniform, the optical confinement factor Γ_m could be used to calculate the modal gain coefficient g. As shown by Adams [4], this is given by:

$$g = \Gamma_m(g_m - \alpha_a) - (1 - \Gamma_m)\alpha_c \tag{4.40}$$

where α_c is the losses in the cladding region. The above expression can be derived from the variational expression of Eq. (4.26) [47]. Assuming that, because of the presence of gain in the active region, there is amplitude variation along z as the field is being amplified as it propagates across the amplifier. This amplitude variation can be described by introducing a complex longitudinal propagation constant β_z (c.f. the complex propagation constant in transmission lines with losses [2]):

$$\beta_z = \beta_{zr} + j\beta_{zi} \,. \tag{4.41}$$

Since the total propagating field solution of the wave equation is given by the product of the transverse modal field solution $\phi_m(x, y)$ and $\exp(-j\beta_z z)$ (see Section 4.2), we can immediately see that the imaginary part β_{zi} takes into account the variation of the amplitude of the field due to optical gain, which is by a factor of $\exp(-\beta_{zi} L)$ over an amplifier of length L. In this case, the power amplification factor, i.e. the single pass gain is given by the square of the field amplitude, i.e.:

$$G_s = \exp(-2\beta_{zi} L) \,. \tag{4.42}$$

Equating Eq. (4.38) with Eq. (4.42) gives the following relation:

$$g = -2\beta_{zi} \,. \tag{4.43}$$

To proceed further, we consider the structure sketched Fig. 4.8, which represents the cross section of a buried heterostructure (BH) SLA [7]. We compare the wave equations for a passive waveguide shown in Fig. 4.4 and that with gain and losses (Fig. 4.8). They are listed below:

$$\lfloor \nabla_t^2 + k_0{}^2 \hat{N}^2(x, y) - \hat{\beta}_z{}^2 \rfloor \hat{\Phi} = 0 \tag{4.44}$$

$$\lfloor \nabla_t^2 + k_0{}^2 N^2(x, y) - \beta_z{}^2 \rfloor \Phi = 0 \tag{4.45}$$

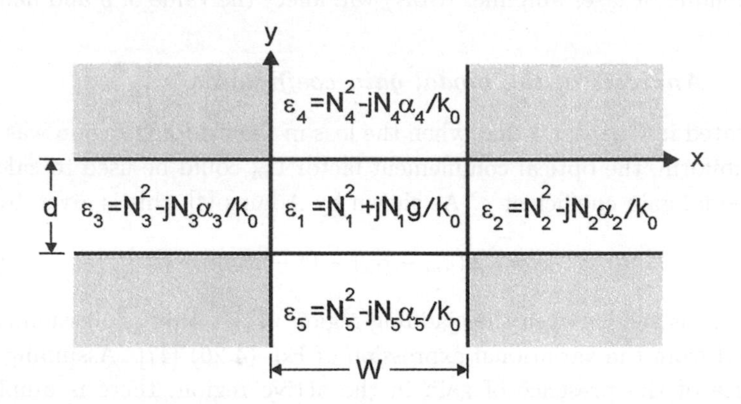

Fig. 4.8. The structure used in perturbation analysis for modal gain coefficients in BH SLAs.

where the quantities with $\wedge$ are those for the passive waveguide, assuming that the transverse field solutions are purely real (i.e. index guided structures) [4] and [5]. The subscript m for the transverse modal field solutions has been dropped for simplicity, assuming their dependence on m to be implicit. Multiplying Eqs. (4.44) and (4.45) by Φ and $\hat{\Phi}$ respectively, followed by subtraction from each other, one can obtain the following expression under a weakly guiding approximation after integrating over the entire cross section S [1]:

$$\beta_z{}^2 - \hat{\beta}_z{}^2 = \frac{\iint_s k_0{}^2[\varepsilon(x,y) - \hat{\varepsilon}(x,y)]\Phi \cdot \hat{\Phi} \cdot dS}{\iint_s \Phi \cdot \hat{\Phi} \cdot dS} . \tag{4.46}$$

The dielectric constants ε for the active structure of each region are shown in Fig. 4.8 [48]. It can be shown that the imaginary parts of the dielectric constants due to gain and losses have much smaller magnitudes compared with the real part N_j^2 [6]. Therefore Fig. 4.8 can be regarded as a slightly perturbed structure of that in Fig. 4.4. Under these conditions, the solutions Φ and $\hat{\Phi}$ are very close to each other, and it can be shown that [7]:

$$\beta_z{}^2 - \hat{\beta}_z{}^2 = \frac{\iint_s k_0{}^2[\varepsilon(x,y) - \hat{\varepsilon}(x,y)] \cdot \hat{\Phi}^2 dS}{\iint_s \hat{\Phi}^2 dS} . \tag{4.47}$$

The difference of the dielectric constants between the perturbed and unperturbed structures $[\varepsilon(x,y) - \hat{\varepsilon}(x,y)]$ are simply the imaginary parts $jN_i\chi_i/k$, where χ_i is either the net gain coefficient $g_m - \alpha_a$ in the active region or the losses in the cladding α_i ($i = 2, 3, 4$ and 5). Hence, if $\hat{\beta}_z = \beta_0$, then:

$$\beta_z^2 = \beta_0^2 + j\Lambda \tag{4.48}$$

and the quantity Λ can be written as [7]:

$$\Lambda = \frac{k_0{}^2 \iint_s (N_i\chi_i/k_0) \cdot \hat{\Phi}^2 \cdot dS}{\iint_s \hat{\Phi}^2 dS} . \tag{4.49}$$

Now since the imaginary parts of the dielectric constants of the perturbed structure of Fig. 4.8 are much smaller than their corresponding real parts, the quantity Λ will also be small compared with β_0^2. Applying the binomial theorem and neglecting terms with orders higher than 2, the difference between the propagation constants of a perturbed active waveguide with that of the unperturbed passive waveguide is found to be given by:

$$\Delta\beta_z = \beta_z - \beta_0 = \frac{j\Lambda}{2\beta_0} \cdot \tag{4.50}$$

Therefore the perturbation analysis shows that the introduction of gain and losses into a dielectric waveguide (e.g. in a SLA) actually introduces a small imaginary part to the original propagation constant of an identical unperturbed lossless waveguide [49]. Comparing Eq. (4.50) with Eq. (4.41), one can see that:

$$\beta_{zi} = \Delta\beta_z = \frac{j\Lambda}{2\beta_0} \tag{4.51}$$

$$\beta_{zr} = \beta_0 \tag{4.52}$$

and the modal gain coefficient is given by substituting the above equations into Eq. (4.42):

$$g = -2\Delta\beta_z = -\frac{j\Lambda}{\beta_0}. \tag{4.53}$$

Therefore if the modal field solutions of the passive (unperturbed) waveguide are known, then the quantity Λ can be calculated easily from Eq. (4.49), and the modal gain coefficients can then be found from Eq. (4.53). In general, the quantity for a general perturbed structure with dielectric constant distribution $\varepsilon(x, y)$ is given by the following expression:

$$\Lambda = \frac{k_0{}^2 \iint_s [\varepsilon(x,y) - \hat{\varepsilon}(x,y)]\hat{\Phi}^2 \cdot dS}{\iint_s \hat{\Phi}^2 \cdot dS} \tag{4.54}$$

where $\hat{\Phi}$ is the unperturbed modal field solutions found in the unperturbed structure. For a lossy structure, Λ is positive and g is negative. However, if sufficient material gain is introduced in region 1 of Fig. 4.8, such that Λ becomes negative, then g and hence G_s becomes positive, and the resulting structure is capable of optical amplification.

The advantages of using perturbation analysis can be seen when the loss distribution in the cladding is not uniform around regions 2–5. In this situation, the expression of Eq. (4.40) for modal gain coefficients is no longer valid [49]. This is because the power flow in the cladding cannot be represented by a simple factor of $(1-\Gamma_m)$ anymore. Instead, one has to resort to Poynting vectors to calculate the proportion of optical power flow in each region, and then evaluate the overall modal gain coefficient by weighting the losses and gains with these proportions [4] and [7]. Furthermore, if the loss and gain coefficients vary in a functional form of $x-y$ (e.g. in a medium with a parabolic gain profile [6]), then it is impossible to use Eq. (4.40). In this situation, the perturbation analysis offers a convenient way to solve the problem numerically using Eqs. (4.54) and (4.53). Because many numerical algorithms are available to solve for the modal field solutions of passive optical waveguides, the solutions can be directly input into Eq. (4.54) to compute the modal gain coefficient g rapidly. In the above derivation, we have assumed an index guided structure, such that the transverse field solutions are purely real [1]. However, there exist gain guided structures in which the gain and loss distribution accounts for the guiding action [4] and [29]. In these structures the modal field solutions are generally complex [36], and have expressions similar to the Gaussian beam description of the propagation of light [12]. In these circumstances, the above expressions are still applicable, but with the square of the field $\hat{\Phi}^2$ replaced by the product of the complex field with its conjugate $\hat{\Phi} \cdot \hat{\Phi}^*$. The study of these gain guided laser amplifiers is very challenging, but is beyond the scope of the present study.

4.3.2 *Design of a polarisation insensitive Travelling Wave Amplifier (TWA)*

In most of the relevant literature, the analyses of BH laser diodes and laser amplifiers have been performed under the assumption of single transverse mode (STM) operation, in which the fields are assumed to be cut-off along the x-direction (Fig. 4.9) [9] and [28]. In this case, the propagating fields are identical to those propagating in a planar dielectric slab waveguide, which we have analysed in Section 4.2.1. This is in fact necessary in designing a laser diode in order to avoid multi-mode emissions [5]. Saitoh and Mukai [9] analysed the polarisation dependence of signal gain in a travelling-wave amplifier (TWA) for TE and TM modes under STM operation. They discovered that the difference between the values of the optical confinement factors for these two polarisations, given by Eqs. (4.38) and (4.40), could be minimised by simply increasing the active layer thickness, d. From these studies, they derived design curves for polarisation insensitive TWAs.

However, when one increases the active layer thickness d, depending on the value of W, it is possible for the TWA to depart from STM operation by exciting higher order quasi-TE and quasi-TM modes, as shown by Ghafouri-Shiraz [28] in his studies on STM conditions for separate confinement heterostructure (SCH) semiconductor lasers. Therefore, for thick active layers, the polarisation sensitivities of the signal gain of hybrid modes must be considered as well to get a more complete picture before analysing other amplifier characteristics. This can be done easily by

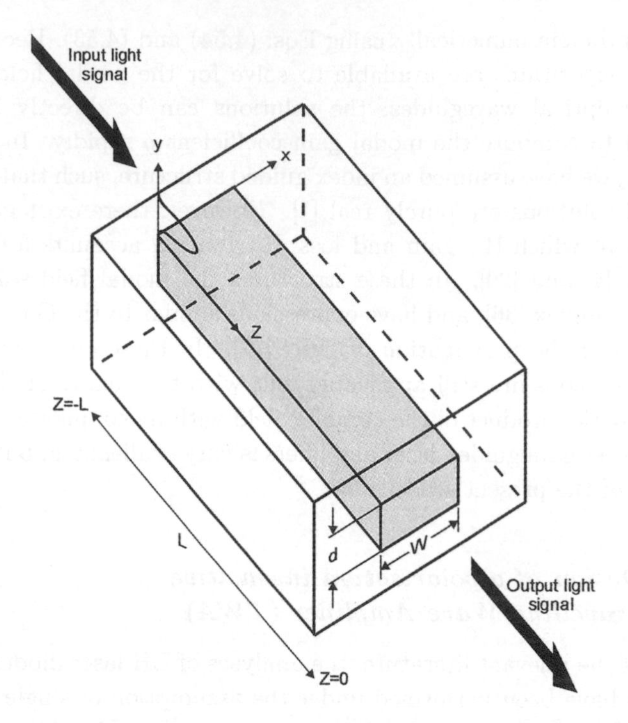

Fig. 4.9. Illustrating the form of electric field in a rectangular dielectric optical waveguide, with a single transverse mode condition (field variation is independent of x.)

analysing the modal gain coefficients for different hybrid modes in addition to those of TE and TM modes.

We have compared the accuracy of using the perturbation approximation of Eqs. (4.49)–(4.53) to derive the modal gain coefficients with the exact methods calculated using Poynting vectors [4]. In Fig. 4.10(a), we have calculated the optical confinement factor Γ_m from the modal gain coefficients derived by the above two methods using Eq. (4.40). The parameters used in the analysis are tabulated in Table 4.3. It can be seen that the perturbation approximation yields results in close agreement with those obtained by Poynting vector analysis. This proves the usefulness and accuracy of the perturbation approximation proposed by the author. The difference of taking into account the active layer width W (i.e. using hybrid modes and assuming that the fields are not cut-off along x [49] in the analysis of optical confinement factor Γ_m) is illustrated by the curves in Fig. 4.10(b). Here, the optical confinement factor of the fundamental TE

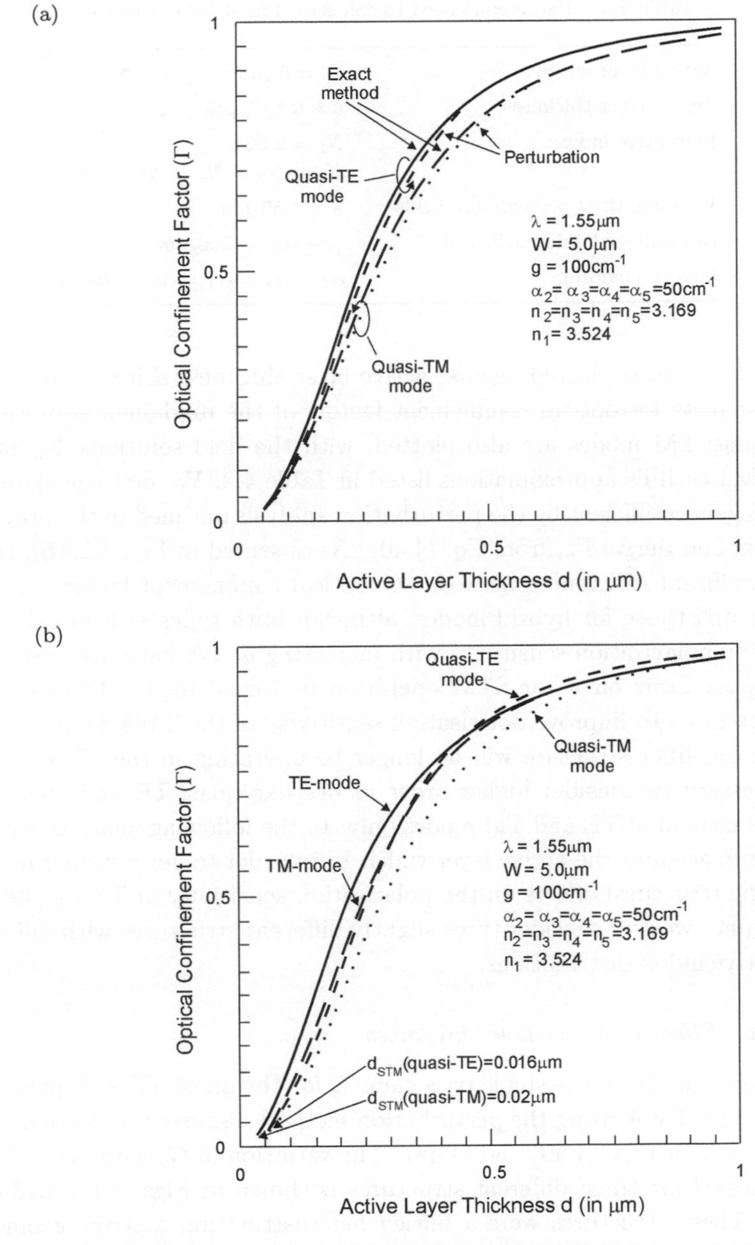

Fig. 4.10. (a) Comparison of calculated optical confinement factors for hybrid modes using Poynting vectors and perturbation approximation, respectively. (b) Variation of optical confinement factor with active layer thickness for TE and TM modes and quasi-TE quasi-TM modes.

Table 4.3. Parameters used to calculate Fig. 4.10(a) and (b).

Active layer width	$W = 5\ \mu m$
Active layer thickness	$d = 0$ to $1\ \mu m$
Refractive indices	$N_1 = 3.524$
	$N_2 = N_3 = N_4 = N_5 = 3.169$
Wavelength of propagating fields	$\lambda = 1.55\ \mu m$
Net material gain coefficient	$(g - \alpha_a) = 100$/cm
Loss in cladding	$\alpha_2 = \alpha_3 = \alpha_4 = \alpha_5 = 50$/cm

and TM mode is plotted against active layer thickness d from 0 to 1 μm. Similar plots for optical confinement factors of the fundamental quasi-TE and quasi-TM modes are also plotted, with the field solutions Φ_m taken from Marcatilli's approximations listed in Table 4.2. We first calculate the modal gain coefficient by the perturbation analysis outlined in the previous section, and derive Γ_m from Eq. (4.40). As observed in Fig. 4.10(b), there are significant differences between the optical confinement factors for slab modes and those for hybrid modes, although both types of field exhibit a reducing polarisation sensitivity with increasing d. We have also indicated the upper limit on d for STM operation in Fig. 4.10(b). Obviously, if one attempts to improve polarisation sensitivity of the TWA by increasing d, the amplifier structure will no longer be operating in the STM, and it is necessary to consider higher order modes like quasi-TE and quasi-TM modes instead of TE and TM modes only. In the following analysis, we will take into account the active layer width W in order to get a deeper insight into the structural effects on the polarisation sensitivity of TWAs [49]. In particular, we will consider three slightly different structures with different refractive index distributions.

4.3.2.1 *Effect of active layer thickness*

We have calculated the single pass gain G_s for the quasi-TE and quasi-TM modes of a TWA using the perturbation method discussed in Section 4.3.1 together with Eqs. (4.40) and (4.38). The variation of G_s with active layer thickness d for three different structures is shown in Figs. 4.11, 4.13 and 4.15. These structures were a buried heterostructure, a stripe geometry structure and a buried channel (BC) structure, respectively. We have illustrated their structural differences in Figs. 4.12, 4.14 and 4.16, and the parameters used in the analysis are tabulated in Table 4.4(a)–(c).

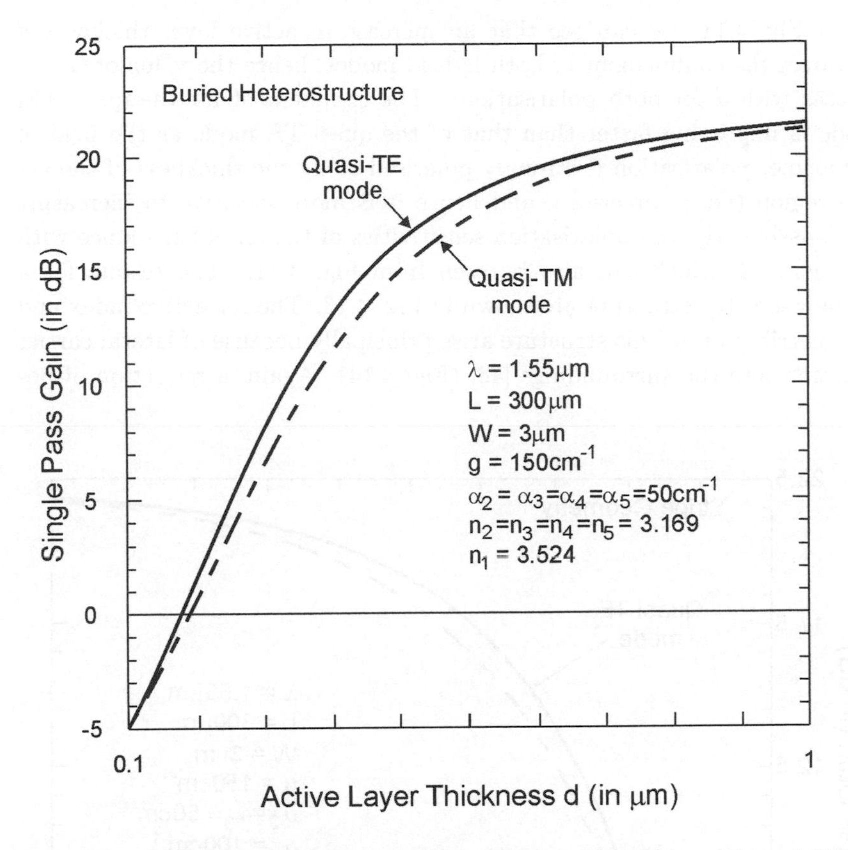

Fig. 4.11. Variation of gain with active layer thickness for fundamental quasi-TE and quasi-TM modes in a buried heterostructure TWA.

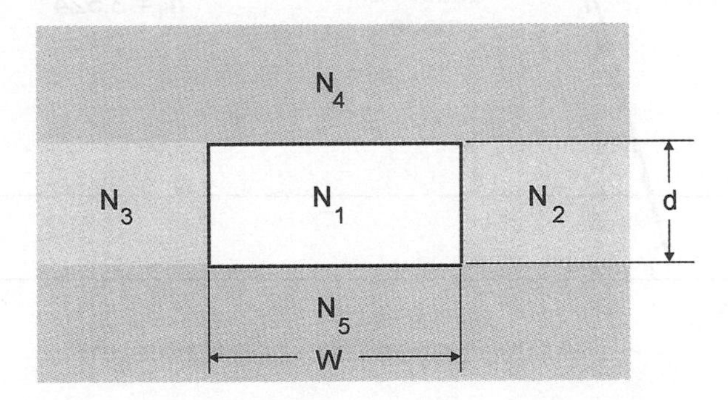

Fig. 4.12. Refractive index distribution for a buried heterostructure TWA.

In Fig. 4.11, we can see that an increase in active layer thickness d improves the confinement of both hybrid modes, hence the value of G_s increases with d for both polarisations. The confinement for the quasi-TM mode is improving faster than that of the quasi-TE mode as the field of the former polarisation is strongly polarised along the thickness of the active region (i.e. y-direction), and hence it is more sensitive to increasing d. Consequently, the polarisation sensitivities of the amplifier reduce with increasing d, which can also be seen from Fig. 4.11. The results for a stripe geometry structure are shown in Fig. 4.13. The refractive index and loss distribution in this structure arise principally because of lateral carrier diffusion into the surroundings [48] (Fig. 4.14). Again, a reduction of po-

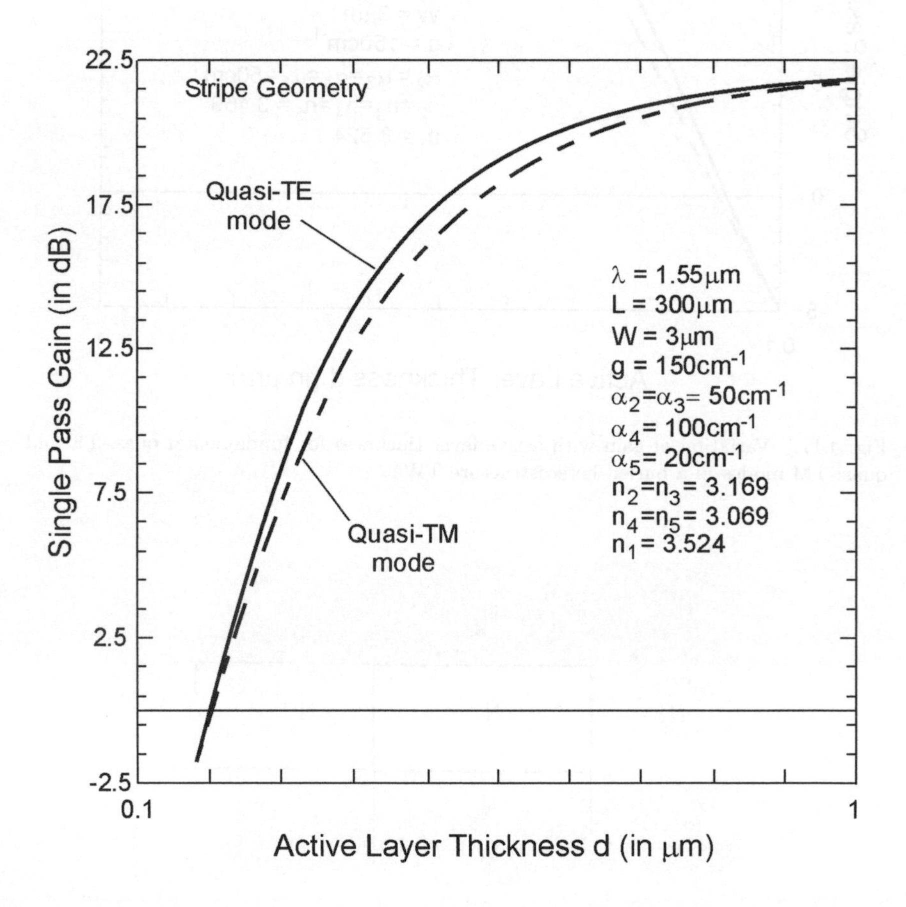

Fig. 4.13. The same as Fig. 4.11 but for a TWA with stripe geometry.

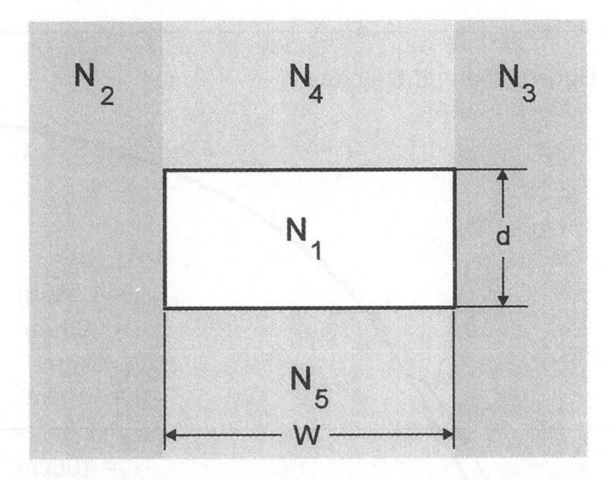

Fig. 4.14. Refractive index distribution for a TWA with stripe geometry (after [47]).

This can be attributed to the improvement in the confinement of the quasi-TM mode with a thick active layer, similar to the improvement of the confinement of the TM mode as discussed by Saitoh amd Mukai [9]. The results for a BC TWA in Fig. 4.15 illustrate the impact of refractive index distribution on the polarisation characteristics of the TWA. The active region is formed by diffusing a channel into the substrate [30]. The striking feature of Fig. 4.15 is the very small polarisation sensitivity of the structure as d approaches 1 μm. A virtually polarisation insensitive TWA with zero polarisation sensitivity is achieved for d larger than 0.55 μm. In addition, for small values of d, the signal gain for the quasi-TM mode is *larger* than that of quasi-TE mode. This can be explained by the asymmetric refractive index step along x and y. Such asymmetry actually reduces the initial differences between the confinement of these two modes, and for a small active layer thickness, the confinement for the quasi-TM mode is actually better than that for the quasi-TE mode. The subsequent improvement in confinement by increasing d further suppressed the gain dependence on the modal field distributions. Therefore, it appears that a BC structure is useful in minimising the polarisation sensitivities in TWAs, although STM operation may not hold for this refractive index distribution [4] and [16].

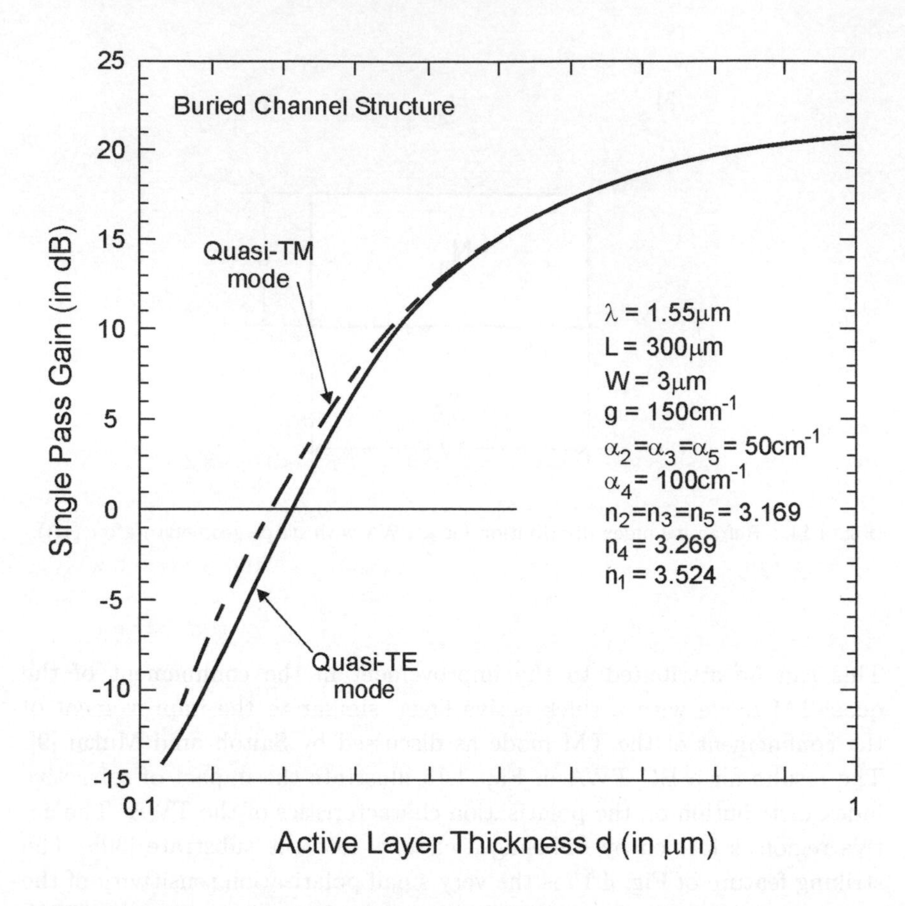

Fig. 4.15. The same as Fig. 4.11 but for a TWA with a buried stripe structure.

4.3.2.2 *Effect of refractive index distribution*

In order to investigate the effect of refractive index distribution on the polarisation characteristics of the hybrid modes, we introduce a parameter ΔN.

$$\Delta N = N_4 - N_2 . \tag{4.55}$$

We have considered two circumstances, both with $N_2 = N_3$: (a) a symmetrical case with $N_4 = N_5$; and (b) an asymmetric case with $N_4 \neq N_5$ but $N_2 = N_3 = N_5$. Notice the similarity of the refractive index distribution of

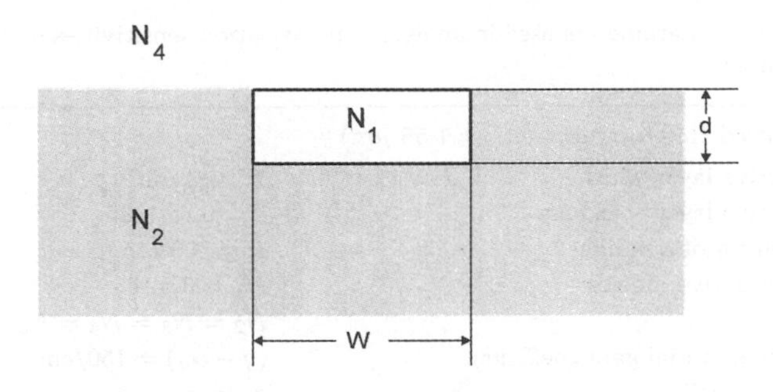

Fig. 4.16. Refractive index distribution for a TWA with a buried channel (BC).

the case (b) with that of a BC TWA (Fig. 4.16). In the following analysis, $N_2 = N_3 = 3.169$, $d = 1$ μm and other parameters are as listed in Table 4.4(a), except for the values of N_4 and N_5, which vary according to ΔN.

Case (a): Symmetrical Case

As shown in Fig. 4.17, the single pass gain G_s reduces as ΔN increases for both quasi-TE and quasi-TM modes. Increasing the value of ΔN will lead to a reduction of confinement for both modes because of the increasing refractive index differences between the active region 1 and regions 4 and 5 (i.e. a larger refractive index step along the y-direction), which results in a smaller value of propagation constant β_o in the perturbation analysis. The structure departs from the weakly guiding condition for bounded modes [1]. The reduction of confinement can be seen as roughly the same for negative values of ΔN, and as ΔN is increased, the gain of the quasi-TE mode begins to fall faster than that of the quasi-TM mode. For $\Delta N = 0.18$ the gain is the same for both modes: this corresponds to a polarisation insensitive structure. Beyond that, the gain of the quasi-TM mode is larger than that of the quasi-TE mode. The more rapid fall in gain for quasi-TE mode for positive values of ΔN can be explained by the more significant reduction of confinement for quasi-TE modes than that for the quasi-TM mode, which is due to the large refractive index step along the y-direction. As d is fixed to 1 μm, it can be seen that by using a combination of thick active layer with an appropriate refractive index profile, the polarisation sensitivity for the TWA can be reduced below that obtained by increasing the active layer thickness only.

Table 4.4. Parameters used in analysing polarisation sensitivities of different structures.

(a) Buried Heterostructure ($\lambda = 1.55\ \mu$m)	
Active layer width	$W = 3\ \mu$m
Active layer thickness	$d = 0$ to $1\ \mu$m
Length of amplifier	$L = 300\ \mu$m
Refractive indices	$N_1 = 3.524$
	$N_2 = N_3 = N_4 = N_5 = 3.169$
Net material gain coefficient	$(g - \alpha_a) = 150$/cm
Loss coefficients	$\alpha_2 = \alpha_3 = \alpha_4 = \alpha_5 = 50$/cm
(b) Stripe Geometry ($\lambda = 1.55\ \mu$m)	
Active layer width	$W = 3\ \mu$m
Active layer thickness	$d = 0$ to $1\ \mu$m
Length of amplifier	$L = 300\ \mu$m
Refractive indices	$N_1 = 3.524$
	$N_2 = N_3 = 3.169$
	$N_4 = N_5 = 3.069$
Net material gain coefficient	$(g - \alpha_a) = 150$/cm
Loss coefficients	$\alpha_2 = 50$/cm
	$\alpha_3 = \alpha_4 = 100$/cm
	$\alpha_5 = 20$/cm
(c) Buried Channel Structure ($\lambda = 1.55\ \mu$m)	
Active layer width	$W = 3\ \mu$m
Active layer thickness	$d = 0$ to $1\ \mu$m
Length of amplifier	$L = 300\ \mu$m
Refractive indices	$N_1 = 3.524$
	$N_2 = N_3 = N_5 = 3.169$
	$N_4 = 3.269$
Wavelength	$\lambda = 1.55\ \mu$m
Net material gain coefficient	$(g - \alpha_a) = 150$/cm
Loss coefficients	$\alpha_2 = \alpha_3 = \alpha_5 = 50$/cm
	$\alpha_4 = 100$/cm

Case (b): Asymmetrical case

In this case, only the refractive index in region 4 is varied. Therefore, we would expect that the reduction of confinement for both modes would not be as rapid as with the previous case. This is confirmed in Fig. 4.18, where the variation of G_s with ΔN is again calculated for this structure. Besides a slower rate of reduction in G_s with ΔN, observations similar to those made in the previous case can be made. Again, it can be seen

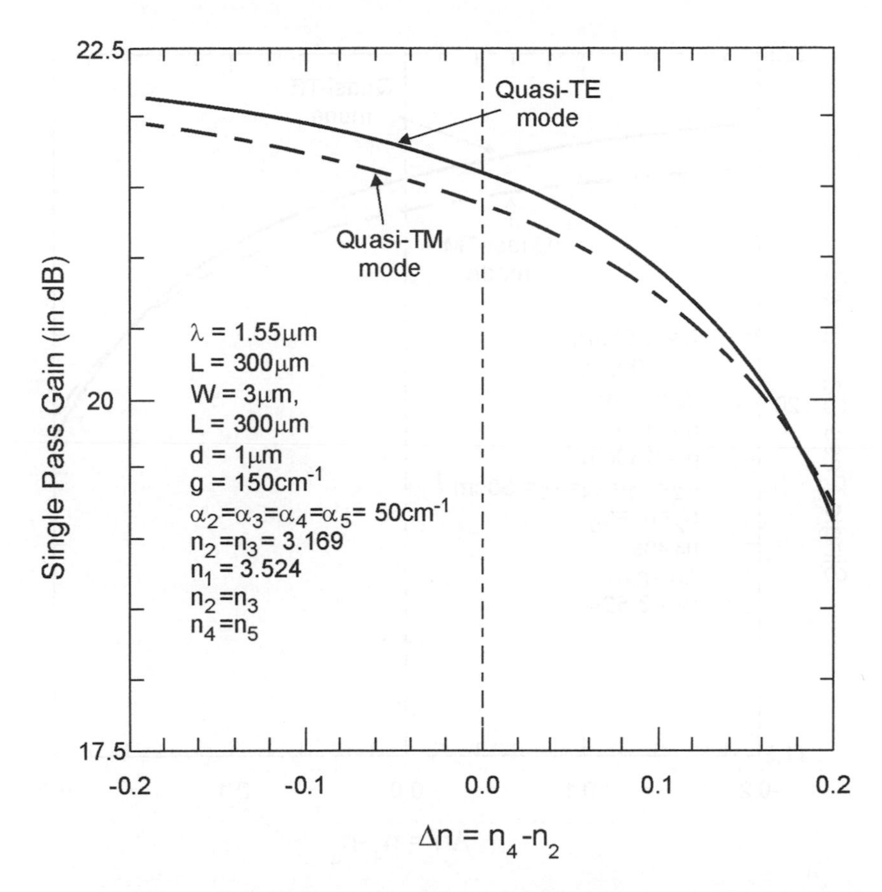

Fig. 4.17. Effect of refractive index profile on the polarisation sensitivities of the amplifier gain for the fundamental quasi-TE and quasi-TM modes.

that the confinement of the quasi-TE mode reduces more rapidly than that of the quasi-TM mode for positive values of ΔN, and again a cross over between the two curves is observed. This crossover point corresponds to the refractive index distribution for a polarisation insensitive TWA structure with the given active region dimensions. The observations in both cases prove that the refractive index profile can play a significant role in controlling the polarisation sensitivities of TWAs. This is exactly the reason why separate confinement structures were suggested by Saitoh and Mukai [9] to minimise polarisation sensitivities of TE and TM modes further. However, our results do not indicate whether scheme (a) or (b) is better. In addition, the refractive index profile control has to be matched with the

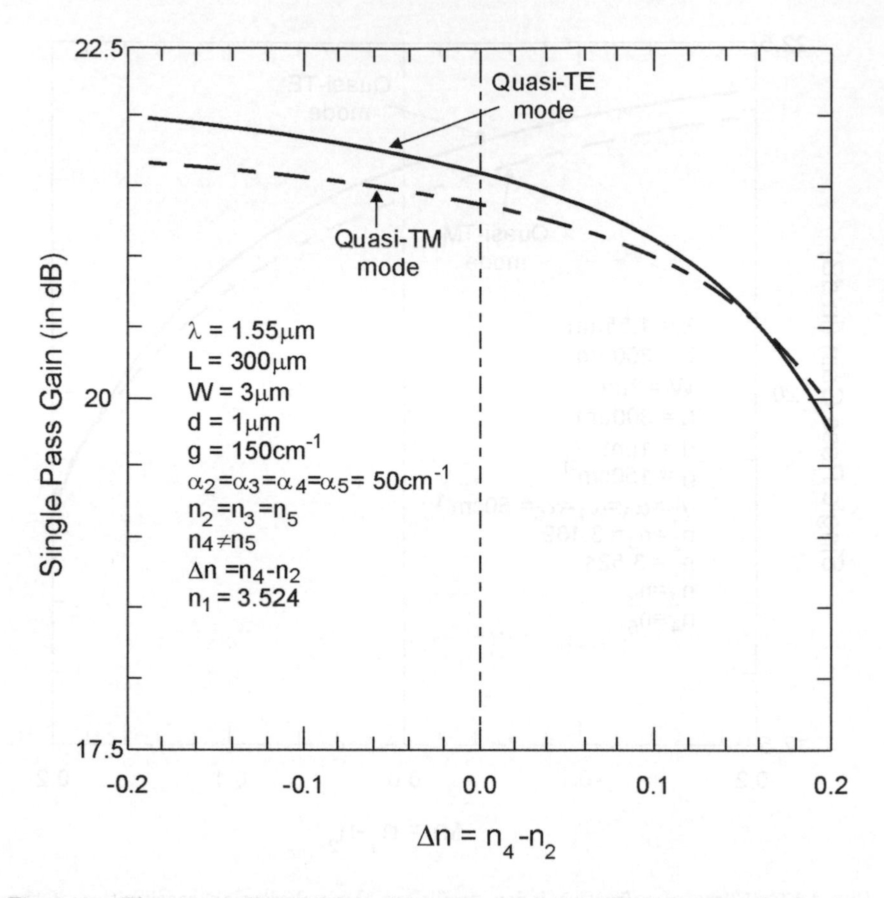

Fig. 4.18. The same as Fig. 4.17 but with an asymmetrical refractive index distribution.

material characteristics and the fabrication processes of the amplifier [5]. An optimum way of controlling the refractive index profile has yet to be found.

4.3.2.3 *Effect of active layer width*

The transverse modal field distributions are affected by the active layer width W, as discussed in Section 4.2 (see Table 4.2). Therefore, the possibility exists of minimising the polarisation sensitivity for hybrid modes in a TWA by altering the active layer width W. Figure 4.19 shows some of the results of our investigation into this possibility. We have calculated the variation of single pass gain G_s with active layer width W for fundamental

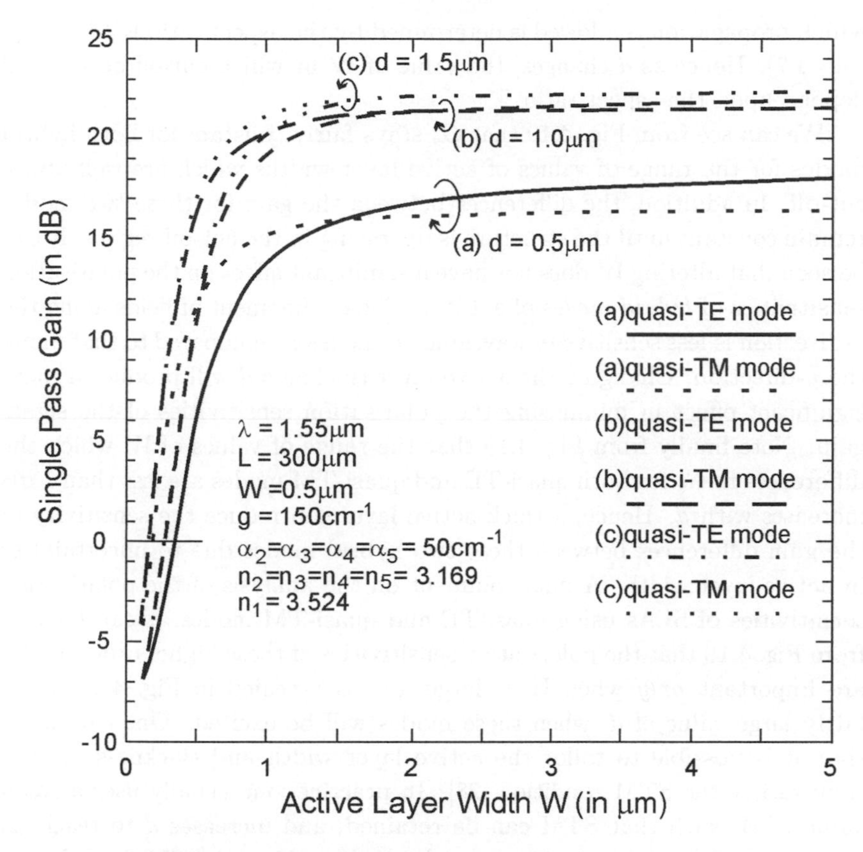

Fig. 4.19. Variation of gain for the fundamental quasi-TE and quasi-TM modes with active layer width for a buried heterostructure TWA.

quasi-TE and quasi-TM modes of a TWA with different values of active layer thickness d. The parameters used in the analyses are identical to those listed in Table 4.4(a), except that now the active layer width W is varied between 0 and 5 μm. Three values of active layer thickness d are used: 0.5, 1.0 and 1.5 μm respectively. It can be seen that for all values of d, as W is reduced from 5 μm, the values of G_s for both hybrid modes stay nearly constant until a certain critical value of W is reached, where the values of G_s fall off rapidly for both hybrid modes. In this region, the dimension of the structure is approaching the cut-off for the quasi-TE and quasi-TM modes for the wavelength of interest. The effect of varying d can be seen as altering this critical value of W only. This can be explained by the fact that for a particular wavelength, the waveguide dimensions for which propagation is allowed is determined by the aspect ratio W/d [4] (see Fig. 4.7). Hence as d changes, the value of W at which cut-off occurs will depend upon the aspect ratio.

We can see from Fig. 4.19 that G_s stays fairly constant for both hybrid modes for the range of values of active layer widths which are well above cut-off. In addition, the differences between the gain for these two modes remain constant until the structure is operating in the cut-off region. It can be seen that altering W does not have a significant effect on the polarisation sensitivities of *hybrid modes* of a TWA. The confinement of fields along the x-direction is less sensitive to waveguide dimensions compared to that along the y-direction. Changing the active layer thickness d will produce a more significant effect in minimising the polarisation sensitivities of the signal gain. Note finally from Fig. 4.19 that the range of values of W which the differences in G_s for both quasi-TE and quasi-TM modes are less than 1 dB increases with d. Hence, a thick active layer can reduce the sensitivity of the gain differences between these two hybrid modes due to uncertainties in active layer width. A final comment on the analysis of the polarisation sensitivities of SLAs using quasi-TE and quasi-TM modes, it can be seen from Fig. 4.19 that the polarisation sensitivities of these higher order modes are important *only* when W is large, or, as revealed in Fig. 4.11, for a fairly large value of d, when these modes will be excited. One can argue that it is possible to tailor the active layer width and thickness so that they satisfy the STM condition [28]. In practice, one usually uses a small value of W such that STM can be retained, and increases d to reach an optimum polarisation sensitivity without disturbing the STM condition. This has an additional advantage of generating a more circular far-field pattern, hence enhancing the coupling efficiency with single-mode fibres. However, there are also advantages from increasing both the active layer width and thickness. These include lower electrical resistance and hence thermal dissipation [50], and also increases in the saturation output power [51] (see next chapter). In these circumstances, analysis of the polarisation characteristics of these hybrid modes becomes important.

4.4 Importance of Transverse Modal Fields Properties in SLAs

We have seen in this chapter how to analyse the electric field distribution in a semiconductor laser amplifier by considering the rectangular dielectric waveguide formed in the active region. Also we have noted that for index guided structures, the presence of gain and losses simply perturbs, the propagation constant of an identical passive waveguide by adding an imaginary part, which accounts for the amplification of the field along the amplifier. We have also seen that these fields are, in general, $x - y$ dependent. Hence, the signal gain becomes dependent on the modal field distribution in the amplifier. Failla *et al.* [27] discovered that this can affect the design of anti-reflection coatings as well, as confirmed by Vassallo [52, 53]. An experimental study on the polarisation sensitivities of SLAs will be presented later in Chapter 7, together with the reported measurements on other SLA characteristics. Because the transverse modal field distribution ultimately determines the photon distribution within the amplifier, the modal fields should affect other amplifier characteristics like noise, output saturation power, etc. That is the pitfall in Saitoh and Mukai's analysis [9]. They have considered the polarisation sensitivity on gain, based only on the differences in optical confinement factors between the two polarisations, and utilise Eq. (4.38) in their subsequent analyses, probably for the sake of convenience. In practice, because the field distribution is different for different modes, not only will the optical confinement factor be affected, but also the photon distribution, hence recombination rate, gain profile along the amplifier etc. will be affected, and will ultimately affect the overall gain of the amplifier [54, 55]. Under such circumstances, calculating the gain dependence on polarisation with Eq. (4.38) based on the optical confinement factor only, will not give a sufficiently accurate model. We will proceed, using the knowledge and tools acquired in this chapter, to form a systematic framework for analysing the different characteristics of laser amplifiers, taking account of the above physical processes. In the following chapter, we will first examine how such a framework can be used to analyse the gain and saturation characteristics of SLAs.

References

[1] A. W. Snyder and J. D. Love, *Optical waveguide theory* (Chapman & Hall, 1983).

[2] C. A. Balanis, *Advanced engineering electromagnetic* (John Wiley and Sons, 1989).

[3] D. Marcuse, *Theory of dielectric optical waveguides* (Academic Press, 1974).

[4] M. J. Adams, *An Introduction to Optical Waveguides* (John Wiley and Sons, 1981).

[5] G. H. B. Thompson, *Physics of semiconductor laser devices* (New York, John Wiley and Sons, 1980).
[6] J. Buss, "The effective index method and its application to semiconductor lasers," *IEEE J. Quantum Electronics*, Vol. QE-18, No. 7, pp. 1083–1089, 1982.
[7] H. Ghafouri-Shiraz and C. Y. J. Chu, "Analysis of waveguiding properties of travelling-wave semiconductor laser amplifiers using perturbation technique," *Fiber and Integrated Optics*, Vol. 11, pp. 51–70, 1992.
[8] M. J. O'Mahony, "Semiconductor laser optical amplifiers for use in future fiber systems," *IEEE J. Lightwave Technology*, Vol. LT-6, No. 4, pp. 531–544, 1988.
[9] T. Saitoh and T. Mukai, "Structural design for polarization-insensitive travelling-wave semiconductor laser amplifiers," *Opt. Quantum Electron.*, Vol. 21, pp. S47–S48, 1989.
[10] H. Ghafouri-Shiraz and C. Y. J. Chu, "Analysis of polarization sensitivity in semiconductor laser amplifiers using the perturbation method," *Trans. IEICE Japan*, National Convention Record, Paper No. C-188, 1991.
[11] R. E. Collin, *Field theory of guided waves* (MacGraw-Hill, 1960).
[12] A. Yariv, *Optical Electronics*, 3rd Edition (Holt-Saunders, 1985).
[13] Sudbø, "Why are accurate computations of mode fields in rectangular dielectric waveguides difficult?," *J. Lighttwave Tech.*, Vol. 10, No. 4, pp. 418–419, 1992.
[14] P. M. Morse and H. Feshbach, *Methods of theoretical physics*: *Part I* (MacGraw-Hill, 1953).
[15] N. H. G. Baken, M. B. J., van Diemeer, J. M. Splunter and H. Blok, "Computational modelling of diffused channel waveguides using a domain integral equation," *IEEE J. Lightwave Technology*, Vol. 8, No. 4, pp. 576–586, 1990.
[16] W. P. Huang, "Polarization corrections of dispersion characteristics of optical waveguides," *Opt. Lett.*, Vol. 15, No. 19, pp. 1052–1054, 1990.
[17] M. Born and E. Wolf, *Principle of optics*, 6th Edition (Pergamon Press, 1983).
[18] S. G. Mikhlin and K. L. Smolitskiy, "Approximate methods for solution of differential and integral equations," American Elsevier Publishing Co., 1967.
[19] A. Yariv and P. Yeh, *Optical waves in crystals* (John Wiley and Sons, 1984).
[20] T. Ikegami, "Reflectivity of a mode at facet and oscillation made in double-heterostructure injection laser," *IEEE J. Quantum Electron.*, Vol. QE-8, No. 6, pp. 470–476, 1972.
[21] T. E. Rozzi and G. H. in'tVeld, "Variational treatment of the diffraction at the facet of d.h. lasers and of dielectric millimeter wave antennas," *IEEE Trans. Microwave Theory Tech.*, Vol. MTT-28, No. 2, pp. 61–73, 1980.
[22] L. Horn and C. A. Lee, "Choice of boundary conditions for rectangular dielectric waveguides using approximate eigenfunctions separable in x and y," *Opt. Lett.*, Vol. 15, No. 7, pp. 349–350, 1990.

[23] P. C. Kendall, M. J. Adam, S. Ritichie and M. J. Robertson, "Theory for calculating approximate values for the propagation constants of an optical rib waveguide by weighting the refractive indices," *IEE Proc.*, Part A, Vol. 134, No. 8, pp. 699–702, 1987.

[24] M. J. Robertson, P. C. Kendall, S. Ritchie, P. W. A. Mcllroy and M. J. Adam, "The weighted index method: a new technique for analysing planar optical waveguides," *IEEE J. Lightwave Technology*, Vol. LT-7, No. 12, pp. 2105–2111, 1989. "

[25] V. J. Menon, S. Bhattacharjee and K. K. Dey, "The rectangular dielectric waveguide revisited," *Opt. Commun.*, Vol. 85, Nos. 5 & 6, pp. 393-396, 1991.

[26] E. A. J. Marcatilli, "Dielectric rectangular waveguides and directional couplers for integrated optics," *Bell Syst. Tech. J.*, Vol. 48, pp. 2071–2102, 1969.

[27] A. G. Failla, G. P. Brava and I. Montrosset, "Structural design criteria for polarization insensitive semiconductor optical amplifiers," *J. Lightwave Technology*, Vol. 8, No. 3, pp. 302–308, 1990.

[28] H. Ghafouri-Shiraz, "Single transverse mode condition in long wavelength SCH semiconductor laser diodes," *Trans. IEICE*, Vol. E70, No. 2, pp. 130–134, 1987.

[29] G. P. Agrawal and N. K. Dutta, *Long-wavelength semiconductor lasers* (New York, Van-Nostrad Reinhold, 1986).

[30] W. Huang, H. A. Haus and H. N. Yoon, "Analysis of buried-channel waveguides and couplers: scalar solution and polarization correction," *IEEE J. Lightwave Technology*, Vol. 8, No. 5, pp. 642–648, 1990.

[31] Y. H. Cheung and W. G. Lin, "Investigation of rectangular dielectric waveguides: an iteratively equivalent index method," *IEE Proc.*, Part J, Vol. 137, No. 5, pp. 323–329, 1990.

[32] K. S. Chiang, "Analysis of rectangular dielectric waveguides: effective index with built-in perturbation correction," *Electron. Lett.*, Vol. 28, No. 4, pp. 388–390, 1992.

[33] C. H. Henry and B. H. Verbeck, "Solution of the scalar wave equation for arbitrarily shaped dielectric waveguides by two-dimensional Fourier analysis," *IEEE J. Lightwave Technology*, Vol. LT-7, No. 32, pp. 308–313, 1989.

[34] C. H. Henry and Y. Shani, "Analysis of mode propagation in optical waveguide devices by Fourier expansion," *IEEE J. Quantum Electron.*, Vol. QE-27, No. 3, pp. 523–530, 1991.

[35] G. B. Hocker and W. K. Burns, "Mode dispersion in diffused channel waveguides by the effective index method," *Appl. Opt.*, Vol. 16, No. 1, pp. 113–118, 1977.

[36] H. A. Haus and S. Kawakami, "On the excess spontaneous emission factor in gain-guided laser amplifiers," *IEEE J. Quantum Electronics*, Vol. QE-21, No. 1, pp. 63–69, 1985.

[37] H. A. Haus, *Wave and fields in optoelectronics* (Prentice-Hall, 1984).

[38] J. E. Goell, "A circular-harmonic computer analysis of rectangular dielectric waveguides," *Bell Sys. Tech. J.*, Vol. 48, No. 7, pp. 2133–2160, 1969.

[39] J. Buss, "Principles of semiconductor laser modelling," *IEE Proc.*, Part J, Vol. 132, No. 1, pp. 42–51, 1985.
[40] S. Banerjee and A. Sharma, "Propagation characteristics of optical waveguiding structures by direct solution of the Helmholtz equation for total fields," *J. Opt. Soc. Am.*, Vol. 6, No. 12, pp. 1884–1894, 1989.
[41] T. Young, "Finite element modelling of a polarization independent optical amplifier," *IEEE J. Lightwave Technology*, Vol. LT-10, No. 5, pp. 626–633, 1992.
[42] C. N. Capsalis, J. G. Fikjoris and N. K. Uzunoglu, "Scattering from an abruptly terminated dielectric slab waveguide," *IEEE J. Lightwave Technology,* Vol. LT-3, No. 2, pp. 408–415, 1985.
[43] W. K. Uzungalu and C. N. Capsalis, "Diffraction from an abruptly terminated dielectric slab waveguide in the presence of a cylindrical scatterer," *IEEE J. Lightwave Technology*, Vol. LT-4, No. 4, pp. 405–414, 1986.
[44] E. W. Kolk, N. H. G. Baken and H. Blok, "Domain integral equation analysis of integrated optical channel and ridge waveguide in stratified media," *IEEE Trans. Microwave Theory Tech.*, Vol. 38, No. 1, pp. 78–85, 1990.
[45] H. J. W. M. Hoekstra, "An economic method for the solution of the scalar wave equation for arbitrarily shaped optical waveguides," *IEEE J. Lightwave Tech.*, Vol. 8, No. 5, pp. 789–793, 1990.
[46] G. W. Wen, "Steepest-descent approximation theory for guided modes of weakly guiding waveguides and fibres," *J. Opt. Soc. Am. A*, Vol. 8, No. 2, pp. 295–302, 1991.
[47] A. Yariv, *Quantum Electronics*, 3rd Edition (John Wiley and Sons, 1989).
[48] M. Cross and M. J. Adam, "Waveguiding properties of stripe-geometry double heterostructure injection lasers," *Solid State Electron.*, Vol. 15, pp. 919–921, 1972.
[49] C. Y. J. Chu and H. Ghafouri-Shiraz, "Structural effects on polarization sensitivity of travelling-wave semiconductor laser amplifiers," *3rd Bangor Communication Symposium*, United Kingdom, pp. 19–22, 1991.
[50] J. C. Simon, "GaInAsP semiconductor laser amplifiers for single-mode fiber communications," *J. Lightwave Technology*, Vol. LT-5, No. 9, pp. 1286–1295, 1987.
[51] Y. Mukai, Y. Yamamoto and T. Kimura, "Optical amplification by semiconductor lasers," *Semiconductor and semi-metals*, Vol. 22, Part E, pp. 265–319, Academic Press, 1985.
[52] C. Vassallo, "Theory and practical calculation of antireflection coatings on semiconductor laser diode optical amplifiers," *IEE Proc.*, Part J, Vol. 137, No. 4, pp. 193–202, 1990.
[53] C. Vassallo, "Some numerical results on polarization insensitive 2-layer antireflection coatings for semiconductor optical amplifiers," *IEE Proc.*, Part J, Vol. 137, No. 4, pp. 203–204, 1990.
[54] D. Marcuse, "Computer model of an injection laser amplifier," *IEEE J. Quantum Electronics*, Vol. QE-19, No. 1, pp. 63–73, 1983.
[55] M. J. Adam, J. V. Collins and I. D. Henning, "Analysis of semiconductor laser optical amplifiers," *IEE Proc.*, Part J, Vol. 132, pp. 58–63, 1985.

Chapter 5

Analysis and Modelling of Semiconductor Laser Diode Amplifiers: Gain and Saturation Characteristics

5.1 Introduction

In Chapter 2, a simple two-level system has been used to explain the fundamental principles and properties of an optical amplifier. The operational principles of a semiconductor laser amplifier (SLA) are more complex than that anticipated by the two-level model. For a deeper insight into SLAs, a more thorough understanding of the detailed mechanisms in SLAs is essential. Some of the basic physics of SLAs were explored in Chapter 3, where different structures and applications of SLAs have also been discussed. It was shown that to further evaluate the gain and saturation characteristics of SLAs, the following quantities have to be known:

(i) the transverse modal field distribution across the active region of the amplifier,
(ii) the material gain coefficient profile along the amplifier, and
(iii) other external factors such as pumping rate (i.e. magnitude of injection or bias current), input optical power, etc.

The first two factors above are determined by both the material characteristics and the structural parameters of the SLA. We have examined how (i) can be accomplished in the previous chapter. In this chapter, we will proceed to analyse (ii) and by incorporating (iii) into the analysis, construct a model which can be used to analyse various performance characteristics of SLAs.

In practice, the material gain coefficient in the active region cannot be determined very easily. This is because of the inter-coupling effect between the optical processes (determined by the photon rate equation) in the active p.n. junction [1, 2]. When optical signals propagate across the amplifier, it interacts with the active region via stimulated emissions and induced absorptions. The number of photons generated by stimulated emissions and lost via induced absorptions is affected by the carrier density in the conduction band (i.e. electron density) and the valence band (i.e. hole density), see Section 2.2.5, Eq. (2.14). However, such optical processes will be accompanied by recombinations of carriers, thus reducing the population inversion level in the amplifier. As we shall see later, this can be described by two simultaneous rate equations, which have to be solved in order to determine the exact material gain coefficient profile along the amplifier.

This chapter is organised as follows. First we examine how to analyse the gain and saturation characteristics of a SLA with a uniform material gain coefficient profile (hereafter for simplicity we call it a *uniform gain profile*), since this is the simplest analysis and can be used in some cases for approximation purposes [3]. Two different approaches to analysing SLAs will then be reviewed, where their merits and shortcomings will be discussed. A third approach proposed by the authors [4] is reported next, which is based on the transfer matrix method (TMM) [5]. Using this approach, a more robust and efficient technique can be used to analyse SLAs with both relatively simple (e.g. buried heterostructures) and also more complicated structures (e.g. two-section amplifiers [6]). This technique of analysis also allows us to construct an equivalent circuit model to model SLAs. Finally, we will apply this model to examine the gain and saturation characteristics of SLAs, with particular reference to structural effects on device characteristics and system consideration.

5.2 Analysis of Semiconductor Laser Diode Amplifiers with a Uniform Gain Profile

The simplest way to analyse the gain and saturation characteristics is to assume that the material gain coefficient is the same along the entire amplifier. In other words, we assume that the effect of increasing field amplitude (due to optical amplification) along z will not affect the carrier density n, and hence the material gain coefficient g_m, significantly. The validity of this commonly used assumption will be discussed later in Section 5.2.3. But first, we will see how to analyse the gain and saturation characteristics of an SLA under such an assumption. This will provide us with qualitatively accurate information about the basic characteristics of SLAs.

5.2.1 *Amplifier gain formulation in semiconductor laser amplifiers*

5.2.1.1 *Active Fabry–Perot formulation*

We have seen how the material gain coefficient g_m can be used to evaluate the single pass gain G_s of a SLA given by Eq. (4.37) of Chapter 4. To calculate the overall amplifier gain G, we must take into account the feedback, if any, provided by the finite end-facet reflectivity of the structure (see Section 3.4.1 of Chapter 3). The structure can be represented by the simplified model of a Fabry–Perot (FP) cavity illustrated in Fig. 5.1, which shows a SLA with length L and power reflection coefficients R_1 and R_2, respectively, at the end facets. The amplifier is assumed to have a uniform gain profile. The cavity resonance frequency f_0 is given by [7]

$$f_0 = \frac{mc}{2N_g L} \tag{5.1}$$

where m is an integer (i.e. the longitudinal mode number), c is the velocity of light in free space and N_g is the group effective refractive index of the active region (i.e. $N_g = v_g/c$, v_g is the group velocity) [8–10], which can be found via the modal field analysis described in the previous chapter.

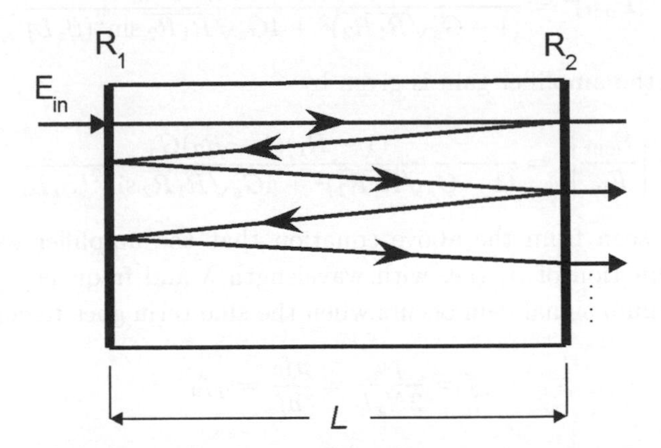

Fig. 5.1. Model used to analyse signal gain of a Fabry–Perot amplifier.

Consider an optical signal with electric field E_{in} of amplitude E_0 and angular frequency $\omega = 2\pi f_0$ incident onto the SLA. The longitudinal propagation constant β_z of the field in the active region is given by (see Section 4.2.3, Eq. (4.26))

$$\beta_z = \frac{2\pi N_{\mathrm{g}}}{\lambda} \tag{5.2}$$

where λ is the signal wavelength. The output electric field E_{out} for a uniform gain profile along the amplifier can be found as a sum of multiple reflected signals transmitted from the amplifier cavity due to reflections of R_1 and R_2, which is given by [7]

$$\begin{aligned} E_{\mathrm{out}} &= \sqrt{(1-R_1)(1-R_2)G_{\mathrm{s}}}E_0 \exp(-j\beta_z L) \\ &\quad \times [1 + \sqrt{R_1 R_2} G_{\mathrm{s}} \exp(-2j\beta_z L) + R_1 R_2 G_{\mathrm{s}}^2 \exp(-4j\beta_z L) + \cdots\cdots]. \end{aligned} \tag{5.3}$$

Using the sum for an infinite geometric progression, Eq. (5.3) can be simplified to

$$E_{\mathrm{out}} = \frac{\sqrt{(1-R_1)(1-R_2)G_{\mathrm{s}}}E_0 \exp(-j\beta_z L)}{1 - G_{\mathrm{s}}\sqrt{R_1 R_2}\exp(-2j\beta_z L)}. \tag{5.4}$$

The measured power of the amplifier output is related to the square of the magnitude of E_{out}, and from Eq. (4.18) it can be shown that

$$|E_{\mathrm{out}}|^2 = \frac{(1-R_1)(1-R_2)G_{\mathrm{s}}E_0^2}{(1-G_{\mathrm{s}}\sqrt{R_1R_2})^2 + 4G_{\mathrm{s}}\sqrt{R_1R_2}\sin^2(\beta_z L)}. \tag{5.5}$$

Therefore the amplifier gain is given by

$$G = \left|\frac{E_{\mathrm{out}}}{E_{\mathrm{in}}}\right|^2 = \frac{(1-R_1)(1-R_2)G_{\mathrm{s}}}{(1-G_{\mathrm{s}}\sqrt{R_1R_2})^2 + 4G_{\mathrm{s}}\sqrt{R_1R_2}\sin^2(\beta_z L)}. \tag{5.6}$$

It can be seen from the above equation that the amplifier gain G is a periodic function of β_z (i.e. with wavelength λ and frequency f as well). The maximum signal gain occurs when the sine term goes to zero, or

$$f = \frac{pc}{2N_{\mathrm{g}}L} = \frac{pf_0}{m} = qf_0\,. \tag{5.7}$$

Where p and q are integers. Equation (5.7) implies that G is maximum when the input signal frequency is an integral multiple of the cavity resonance frequency. The corresponding value of gain G_{max} is given by

$$G_{\mathrm{max}} = \frac{(1-R_1)(1-R_2)G_{\mathrm{s}}}{(1-G_{\mathrm{s}}\sqrt{R_1R_2})^2} . \tag{5.8}$$

The amplifier gain is minimum when the sine term is unity in Eq. (5.16). In this case, the frequency f is given by

$$f = \frac{(2p+1)c}{4N_{\mathrm{g}}L} . \tag{5.9}$$

This corresponds to a phase mismatch of π radians between signal frequency and cavity resonances. The minimum value of G is given by

$$G_{\mathrm{min}} = \frac{(1-R_1)(1-R_2)G_{\mathrm{s}}}{(1+G_{\mathrm{s}}\sqrt{R_1R_2})^2} . \tag{5.10}$$

The levels of the ripples in the amplifier gain spectrum is described by the ratio between G_{max} and G_{min}. The following parameter ξ can be used to indicate this ratio [11, 12]. Using Eqs. (5.8) and (5.10) we have

$$\xi = \left(\frac{1+G_{\mathrm{s}}\sqrt{R_1R_2}}{1-G_{\mathrm{s}}\sqrt{R_1R_2}}\right)^2 . \tag{5.11}$$

Equation (5.11) provides a method of measuring the single pass gain if the facet reflectivities are known [11]. The ripple level at the amplifier output is measured by scanning the amplified spontaneous emission spectrum, and then G_{s} can be calculated using the above equation. A knowledge of G_{s} allows one to calculate G using Eq. (5.6). To facilitate this in the analysis, Mukai and Yamamoto [12] suggested an alternative form of Eq. (5.6) by noting the periodicity of the gain function, and the fact that the term $\beta_z L$ actually corresponds to the phase mismatch between signal and cavity resonances [14]

$$G = \frac{(1-R_1)(1-R_2)G_{\mathrm{s}}}{(1-G_{\mathrm{s}}\sqrt{R_1R_2})^2 + 4G_{\mathrm{s}}\sqrt{R_1R_2}\sin^2\left(\frac{2\pi(f-f_0)N_{\mathrm{g}}L}{c}\right)} \tag{5.12}$$

where f_0 is the resonance frequency of the cavity. Notice the similarity between Eqs. (5.12) and (5.6) except for the replacement of the β_z term by a frequency term. It should be noted that from both Eqs. (5.12) and (5.6), $G = G_{\mathrm{s}}$ when $R_1 = R_2 = 0$ (i.e. when the SLA is a travelling-wave amplifier (TWA)). Measuring ξ also offers a method for measuring the facet reflectivities [15, 16]. Because G_{s} is related to the injection current i, one

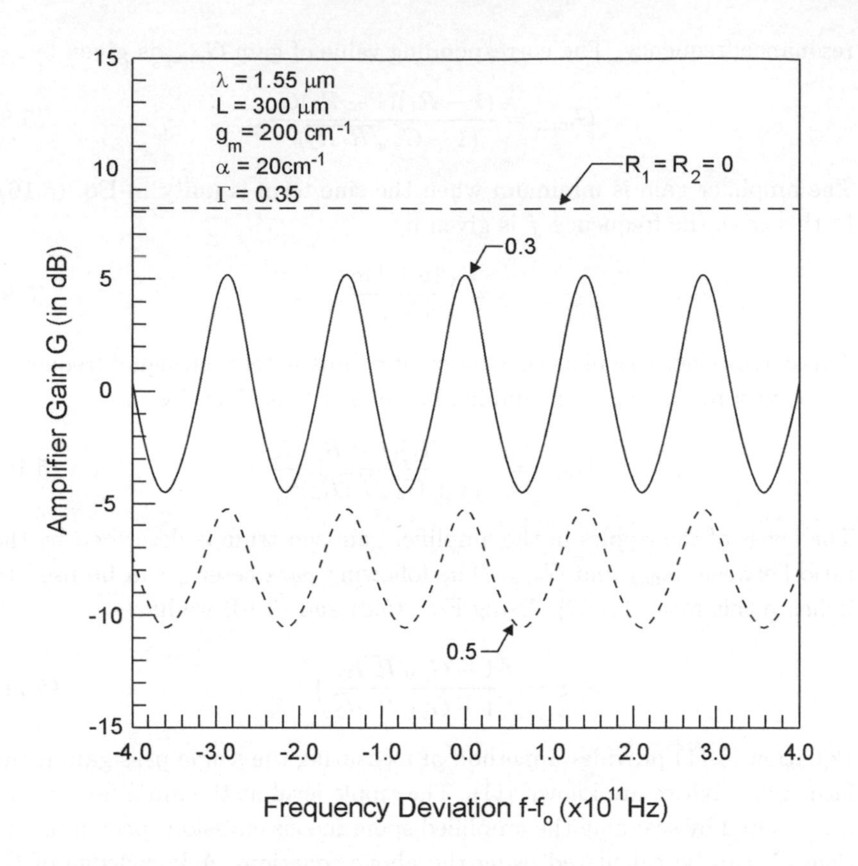

Fig. 5.2. The amplifier gain spectrum of a Fabry–Perot amplifier.

can change i to obtain a variation of ξ, and from that R_1 and R_2 can be calculated easily from Eq. (5.11) if $R_1 = R_2$.

Equation (5.11) shows that the ripple level ξ can be reduced by reducing the facet reflectivities R_1 and R_2. This is illustrated in Fig. 5.2, where we have analysed a FPA with different facet reflectivities and single pass gain G_s. A Fabry–Perot amplifier (FPA) with $L = 300$ μm, $N_g = 3.524$, $\lambda = 1.55$ μm is analysed for (i) $R_1 = R_2 = 0$, (ii) $R_1 = R_2 = 0.3$ and $R_1 = R_2 = 0.5$, respectively. The resulting signal gain is plotted against $(f - f_0)$, calculated using Eq. (5.12). In these calculations, we have neglected the dependence of material gain on the signal wavelength. The effective signal bandwidth is also increased, and a flat and uniform spectrum is obtained when $R_1 = R_2 = 0$, which corresponds to a TWA. The gain can

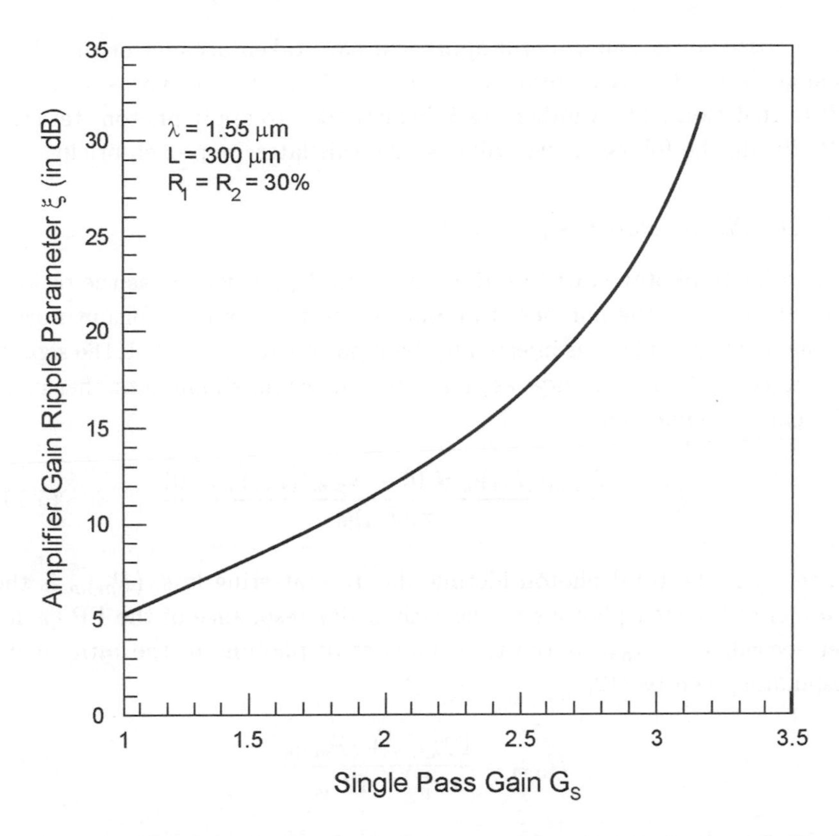

Fig. 5.3. Variation of amplifier gain ripple parameter ξ against single-pass gain G_s of a Fabry–Perot amplifier.

be maximised when both R_1 and R_2 are non-zero and the FPA is tuned to resonance. In practice, R_1 and R_2 are often finite (see the discussion in Section 3.4.1 in Chapter 3) and hence the amplifier gain spectrum is seldom flat. The gain of the amplifier can then be maximised by tuning the cavity resonance of the amplifier to match the input signal frequency. This is usually achieved by changing the temperature of the amplifier [17]. In Fig. 5.3 we have plotted the variation of ξ with G_s for $R_1 = R_2 = 0.3$. The magnitude of ξ increases rapidly as G_s increases, and hence a FPA is more sensitive to frequency mismatch if G_s is high, making it unsuitable as a stable high gain power amplifier [13]. Indeed, if G_s is increased such that the denominator of Eq. (5.11) becomes zero, the FPA reaches the lasing threshold and oscillations occur. This is the familiar threshold condition in semiconductor lasers [1, 7, 18]. The above derivations of the signal gain of

SLAs, based on a Fabry–Perot approach, have taken account of the phase mismatch due to cavity resonances. Hence, it is also known as an active FP formulation [19]. Another possible method is to apply photon statistics [12, 20]. In the following, we will describe this latter approach briefly.

5.2.1.2 *Photon statistics formulation*

According to photon statistics, the signal gain G_{co} is defined as the ratio of the increment of the number of photons extracted from the output mirror to the number of photons injected to the input mirror [19, 21]. If the signals are tuned to the mth cavity resonance to achieve maximum amplifier gain, G_{co} can be defined as

$$G_{co} = \frac{N_{p,m}((P_m)_{in} \neq 0) - N_{p,m}((P_m)_{in} = 0)}{\tau_p (P_m)_{in}} \tag{5.13}$$

where τ_p is the total photon lifetime due to scattering loss, $(P_m)_{in}$ is the number of injected photons to the mth cavity resonance of the FP cavity per second, and $N_{p,m}$ is the total number of photons in the mth cavity resonance given by [12]

$$N_{p,m} = \frac{\Gamma E_{cf,m} + (P_m)_{in}}{\tau_p^{-1} - \Gamma G_m} \tag{5.14}$$

where Γ is the optical confinement factor, G_m is the *stimulated emission coefficient*, which is equal to the stimulated emission rate, and $E_{cf,m}$ is the *spontaneous emission probability* [11, 12]. These two quantities are related to the band structure of the material. The subscript m indicates that these parameters are calculated with respect to the mth cavity resonance frequency. The numerator in Eq. (5.13) involves two terms. The first term is the total number of photons in the mth mode, including signal, stimulated emissions and amplified spontaneous emissions. The second term with $(P_m)_{in} = 0$ isolates out the signal photons, and includes those photons due to amplified spontaneous and stimulated emissions induced by amplified spontaneous emissions only. Hence, the numerator of Eq. (5.13) represents the total number of photons due to injected signal and those due to stimulated emissions induced by the signal. The form of Eq. (5.13) excludes the effect of finite coupling losses at the input and output mirror facets with power reflection coefficients of R_1 and R_2, respectively (see Fig. 5.1). To take account of these, we introduce two efficiency parameters

[12]. The injection efficiency η_{in} which is given by

$$\eta_{\text{in}} = \frac{(1 - R_1)G_s}{1 + R_1 G_s} \tag{5.15}$$

and the extraction efficiency η_{out} given by

$$\eta_{\text{out}} = \frac{\tau_p}{\left(\frac{c}{LN_g}\right) \cdot \ln\left(\sqrt{\frac{1}{R_2}}\right)} . \tag{5.16}$$

The derived signal gain G using this approach is given by [12]

$$\begin{aligned} G &= \eta_{\text{in}} G_{\text{co}} \eta_{\text{out}} \\ &= \frac{N_{p,m}((P_m)_{\text{in}} \neq 0) - N_{p,m}((P_m)_{\text{in}} = 0)}{\left(\frac{c}{LN_g}\right) \cdot \ln\left(\sqrt{\frac{1}{R_2}}\right) \cdot (P_m)_{\text{in}}} \cdot \frac{(1 - R_1)G_s}{1 + R_1 G_s} . \end{aligned} \tag{5.17}$$

5.2.1.3 *Comparisons between the two formulations*

We have described two different mathematical descriptions of G: Eqs. (5.6) and (5.12) using the active FP approach, and Eq. (5.17) using the photon statistics approach. An inspection of the different forms of Eqs. (5.17) with (5.12) reveals that the photon statistics formulation does not take into account the frequency dependence of the amplifier gain spectrum (see Fig. 5.1), and requires additional information about the amplified spontaneous emission in order to calculate $N_{p,m}$. Mukai and Yamamoto [12, 19] have found in their investigations that these two approaches have given different results in their analysis. Marcuse [3] argued that the photon statistics approach assumes that one can analyse the characteristics of a SLA accurately by ignoring the spatial variation of the number of photons along the cavity and hence the use of the total number of photons to calculate G can be justified. He pointed out that this requires a uniform electric field distribution along the cavity. Whereas this assumption is approximately true in a FP cavity, the field distribution becomes increasingly non-uniform as the facet reflectivities of the SLA reduce. Hence, the validity of Eq. (5.17) becomes questionable in analysing SLAs with low facet reflectivities (e.g. in TWA). Cassidy [22] compared these two approaches critically and found that the active FP approach is more appropriate, because the phase information of the fields inside the cavity can be retained. In order to improve the accuracy in the analysis of the characteristics of SLAs, the active FP approach should be adopted.

5.2.2 *Gain saturation formulation in semiconductor laser diode amplifiers*

In the previous section, we have examined how to analyse the amplifier gain for SLAs. However, a careful examination of the previous mathematical expressions reveals that the material gain (hence the amplifier gain) is not affected by the input signal power for a fixed injection current, which is contrary to experimental observations [19], as well as the qualitative prediction using a simple two-level system, as explained in Section 2.3.3 in Chapter 2. In SLAs, as the signal photons propagate across the amplifier, stimulated emissions can also change the population inversion level. The change in population level is dynamic, and the competition between stimulated emissions and recombinations can be described by the rate equations similar to Eqs. (2.12) and (2.15) [12, 19]. Usually these rate equations have to be solved by numerical iterations [18]. The saturation output power for a SLA with a particular bias or injection current can then be found by plotting the amplifier gain calculated from these rate equations against optical power, and noting the −3 dB point from the unsaturated gain value on the plot [23]. However, an analytical form of gain saturation in terms of the output optical intensity can be obtained under the uniform gain profile assumption [13, 24, 25]. The optical intensity $I(z)$ is allowed to vary, as in the active FP formulation. This gives the familiar expression of material gain $g_m(z)$ at position z. That is

$$g_m(z) = \frac{g_0}{1 + \frac{I(z)}{I_{\text{Sat}}}} \tag{5.18}$$

where g_0 is the unsaturated material gain coefficient and I_{Sat} is the saturation intensity [13, 26]. The details of the derivation of Eq. (5.18) can be found in Section 5.7. If g_m is measured at the output of the amplifier, i.e. $z = L$, then I_{Sat} refers to the output saturation intensity. The exact value of I_{Sat} can be found by solving the rate equations [2]. Yamamoto *et al.* [26] have found that I_{Sat} is given by

$$I_{\text{Sat}} = \frac{hf}{A\tau_{\text{sp}}} = \frac{chf}{\beta N_{\text{g}} V_0} \tag{5.19}$$

where A is the differential gain coefficient described in Eq. (3.5) of Section 3.2.2, τ_{sp} is the carrier recombination lifetime due to spontaneous emission only, N_{g} is the group effective refractive index of the amplifying medium, $V_0 = V/\Gamma$ is the optical mode volume, V is the active region actual volume, Γ is the confinement factor and β is the spontaneous emission coefficient, which is defined as the fraction of spontaneous emission power coupled to the signal [27–30]. Notice that I_{Sat} depends mainly on the material characteristics. Thus the saturation output power, given by $P_{\text{Sat}} = I_{\text{Sat}} \cdot W \cdot d$ (where W and d are, respectively, the width and thickness of the active region) can be increased by increasing W and/or d [2, 13]. Typical value of β is around 10^{-4} for 1.3 and 1.55 μm wavelength SLAs [23]. As we have been dealing with photons in solving the rate equations (see Section 5.7), we can approximate $I(L)$ by

$$I(L) = \frac{chf}{N_{\text{g}} V_0} \langle N_{\text{p}} \rangle \tag{5.20}$$

where $\langle N_{\text{p}} \rangle$ is the average number of photons at the output of the amplifier. Using Eqs. (5.19) and (5.20), one can re-express Eq. (5.18) by [13]

$$g_m(L) = \frac{g_0}{1 + \beta \langle N_{\text{p}} \rangle} \tag{5.21}$$

at the amplifier output $z = L$. Therefore, gain saturation is related to the amount of spontaneous emissions coupled to the signal as well as the signal power (i.e. $\langle N_{\text{p}} \rangle$) itself. In fact, the material gain reduces to $g_0/2$ when the number of photons $\langle N_{\text{p}} \rangle = 1/\beta$, as revealed by Eq. (5.21). It should be pointed out that the above expressions are not only valid under the assumption of a uniform gain profile along the amplifier, but also, as shown in Section 5.7 and also in Chapter 6, they can be applied at any point z in the amplifier. The SLA starts to saturate when the material gain coefficient at a particular z begins to reduce significantly from its unsaturated value.

Another interesting aspect of gain saturation in SLAs is its homogeneous nature [31]. We illustrated in Chapter 3 (Fig. 3.4) that the effect of increasing g_{peak} on the material gain spectrum is to shift the whole spectrum homogeneously upwards. The same homogeneous behaviour can be observed when the amplifier saturates. When saturation occurs for a particular wavelength in the material gain spectrum due to high incident power, the material gain coefficients for other wavelengths across the entire spectrum also saturate simultaneously. This occurs because saturation is due to the total photon density (i.e. total power of signals and spontaneous emissions inside the cavity according to Eq. (5.21)), which explains the above phenomenon: saturation occurs homogeneously because it depends on the total amount of photons for all wavelengths which exist in the cavity.

5.2.3 *Appraisal on using a uniform gain profile in analysing SLAs*

We have compared the underlying assumptions in the different approaches in analysing gain and saturation characteristics, which are described above, in Table 5.1. It can be seen that the photon statistics approach is physically unrealistic, in which both the gain profile and the signal fields (hence photon density) are assumed to be uniform. In the active FP formulation, we have assumed that the material gain g_m is independent of z, although the variation of the field has been accounted for by considering the phase of the fields at the facets [22]. It has been shown that in semiconductor lasers with Fabry–Perot cavities, the electric field distribution is fairly uniform in the cavity along z because of the relatively high Q-factor (due to the relatively high value of facet reflectivities) of the optical cavity (which creates a standing-wave pattern) [10], and hence g_m can be assumed to be z-independent without significant loss in accuracy [1, 32], but note the critical comment concerning the loss in phase information made by Cassidy [22]. On the other hand, for other laser structures (i.e. distributed feedback lasers [33]) and most laser amplifiers [3, 23], where the facet reflectivities are fairly low, this assumption will not hold. A more accurate analysis of SLAs will require one to consider the interaction between injected carrier density, material gain and electric field amplitude point by point within the cavity. Such modelling of SLAs will be discussed in the following sections.

Table 5.1. Comparisons between different approaches of analysing SLAs.

Analysis Method	Is gain coefficient independent of z?	Is field amplitude (photon density) independent of z?
Photon statistics/ Rate Equation	Yes	Yes
Active FP Formulation	Yes	No
Travelling-Wave Equations	No	No

5.3 General Analysis of Semiconductor Laser Diode Amplifiers (A Brief Review)

In the previous section, amplifier gain and the saturation characteristics (i.e. the dependence of amplifier gain on optical intensity for a SLA have been analysed by assuming a uniform gain profile. But, to calculate these quantities, the value of the material gain coefficient has first to be found. This can be obtained conveniently by solving the modified multi-mode rate equations based on photon statistics [1, 12], for which we will describe the procedures in Section 5.3.1. However, as highlighted in Section 5.2.3, using photon statistics in analysing the characteristics of SLAs assumes implicitly that both the gain and the electric field profile are independent of z [3]. This is often not true in SLAs, and hence alternative methods for analysing the gain profile and subsequently the amplifier gain and saturation characteristics have to be derived. A commonly used method, known as the travelling-wave equation method, will be described in Section 5.3.2 [3, 23].

5.3.1 *Analysis using rate equations*

This approach is a modification of the well-known laser multi-mode rate equations [1, 34] by including an input signal term into the photon density rate equation [20]. In this case, the time derivatives of the photon density S_m and carrier density n can be expressed by the following two equations;

$$\frac{dn}{dt} = P_c - R - \sum_m \Gamma G_m S_m \tag{5.22}$$

$$\frac{dS_m}{dt} = -\frac{S_m}{\tau_p} + \Gamma G_m S_m + \Gamma E_{\mathrm{cv},m} + S_{\mathrm{in},m} \,. \tag{5.23}$$

In the above equations, P_c is the pumping rate due to the injection current, R is the recombination rate of the carriers, $S_{\mathrm{in},m}$ is the photon density injected by the incident optical signal per second. Other symbols have already been defined in Section 5.2.1.2. Notice that the rate equation for the carriers, Eq. (5.22), describes fluctuations in the carrier density "n" due to all of the optical modes in the SLA, whereas that for the photon density, Eq. (5.23), describes the fluctuations in the photon density for each optical mode. Mathematical models for R can be found in Section 3.2.2 of Chapter 3. For SLAs operating with long wavelengths of 1.3 and 1.55 μm the superlinear model (Eq. (3.10)) has to be used because of the significance of Auger recombinations [15]. Then the carrier density can be solved for the above equations, either for transient analysis [35] or for steady state operation where the left hand side of Eq. (5.22) will be set to zero for all values of m [12]. In both analyses, the solutions are found by a self-consistency analysis, similar to that for lasers [18, 36]. Once the carrier density has been found, the gain coefficient g_m can be derived from Eqs. (3.5) or (3.6) in Chapter 3, and hence the signal gain can be calculated by either the photon statistics formulation (Section 5.2.1.2) by assuming both a uniform carrier and photon distribution along the amplifier, or using the active FP formulation described in Section 5.2.1.1. The gain saturation characteristics can subsequently be found following the procedures described in Section 5.2.2.

5.3.2 *Analysis using travelling-wave equations*

The above analysis using rate equations assume that both the photon density and the material gain (hence carrier density) are independent of z (see Table 5.1). In order to take into account the spatial dependence of carrier and photon densities, Marcuse [3] modified the rate equations for the mth mode as

$$\frac{dn(z,t)}{dt} = \frac{j}{ed} - R - \sum_m \Gamma G_m(z) S_m(z) \tag{5.24}$$

$$\frac{dS_m(z,t)}{dt} = -\frac{\beta n(z)}{\tau_{\mathrm{sp}}} - \frac{S_m(z)}{\tau_{\mathrm{p}}} + \Gamma G_m(z) S_m(z) \tag{5.25}$$

where S_m is the photon density of the mth guided mode, including both photons injected by the input signal, and photons generated by stimulated emissions and spontaneous emission, τ_{sp} is the radiative carrier recombination life time due to spontaneous emissions, j is the injection current density, d is the thickness of the active layer, and e is the electronic charge ($= 1.6 \times 10^{-19}$ C) and β is the fraction of spontaneous emissions coupled to the mth mode. In deriving the above equations, the current injection efficiency is assumed to be 100%. For an energy bandwidth of $\Delta E = h\Delta f$ (i.e. an optical frequency bandwidth of Δf), we have

$$\sum_m G_m S_m = \frac{c}{N_{\mathrm{g}} \Delta E} \int_0^{\Delta E} g_m(z,E) S(z,E) dE \tag{5.26}$$

where $S(z,E)$ is the total photon density at z with photon energy E and $g_m(z,E)$ is the gain coefficient of the material at z with energy E, taking into account the spectral dependence. Notice that $S(z,E)$ includes both spontaneous emissions and signal photons. By assuming a steady state operation, Eq. (5.24) can be re-written as [23]

$$\frac{j}{ed} = R(z) + \frac{c\Gamma}{N_{\mathrm{g}} \Delta E} \int_0^{\Delta E} g_m(z,E) S(z,E) dE \,. \tag{5.27}$$

The next step is to deduce the spatial dependence of photon density $S(z, E)$. We will derive this in two steps. We first consider the photons of the mth guided mode. It is described by Eq. (5.25), which can be rewritten as

$$\frac{dS_m}{dt} = \beta R_{\rm sp}(z) + \frac{c\Gamma}{N_{\rm g}}[g_m(z, E_m) - \alpha_{\rm a}(z, E_m)]S_m(z, E_m) \tag{5.28}$$

where $R_{\rm sp}$ is the radiative recombination rate of carriers due to spontaneous emissions, E_m corresponds to photon energy of the mth mode and we have replaced $S_m/\tau_{\rm p}$ by $\Gamma\alpha_{\rm a}/N_{\rm g}$ whereas $\alpha_{\rm a}$ is the loss in the active region [3]. The time derivative dS_m/dt can be regarded as a total derivative and can be expanded as;

$$\frac{dS_m^{\pm}}{dt} = \frac{\partial S_m^{\pm}}{\partial t} \pm \frac{c}{N_{\rm g}}\frac{\partial S_m^{\pm}}{\partial z} \tag{5.29}$$

provided that $S_m^{\pm}$ varies with z only (i.e. a one-dimensional structure, see Section 5.4.2). Here, the superscript $\pm$ indicates whether the photons are travelling in a forward or backward direction. A travelling-wave equation can be found by substituting Eq. (5.29) into Eq. (5.28). For the spontaneous emission photons, the phase information is not important, as their major effects are on the intensity of the signals only [22]. Hence, it is legitimate and more convenient to work with a travelling-wave equation in terms of intensity for spontaneous emissions, which is given by $I_{{\rm sp},m} = cS_{{\rm sp},m}E_m/N_{\rm g}$. Here $S_{{\rm sp},m}E_m$ is the total optical energy density of the spontaneous emissions and $c/N_{\rm g}$ takes care of the number of photons crossing the transverse $x - y$ plane per unit time. Substituting these into Eq. (5.28) and under a steady state condition we have

$$\frac{\partial I_{{\rm sp},m}^{\pm}(z, E_m)}{\partial z} = \pm\frac{\beta R_{\rm sp}(z)E_m}{2} \pm gI_{{\rm sp},m}^{\pm}(z, E_m) \tag{5.30}$$

where g is the modal gain coefficient of the amplifier, and we have replaced β by $\beta/2$ to take into account the two possible travelling directions [23]. The first term on the right hand side of the above equation represents spontaneous emissions generated and coupled to the mth guided mode at z. The second term represents amplified spontaneous emissions travelled to z. It can be shown that if $R_{\rm sp}$ and g_m are independent of z, then Eq. (5.30) can be solved analytically if the boundary conditions are also known [23]. Furthermore, if β is small and g is relatively small throughout the entire amplifier, the contribution of spontaneous emission $I_{{\rm sp},m}^{\pm}$ to the total photon density S_m will be negligible, that is, only signal photons will dominate in the interaction between photons and carriers. This is an important point which will arise later in our discussion of using transfer matrices to analyse SLAs.

With $I_{\mathrm{sp},m}$ known, $S(z, E)$ can be found from Eq. (5.26) in the absence of the signal by $S(z, E) = \sum_m S_{\mathrm{sp},m}(z, E)$. To include the signal photons, note that the signal fields satisfy a travelling-wave equation of the form [3, 23]

$$\frac{\partial F^{\pm}(z, E_m)}{\partial z} = \pm \frac{g(z, E) F^{\pm}(z, E)}{2} \mp j\beta_z F^{\pm}(z, E) \tag{5.31}$$

where β_z is the propagation constant of the field and g is the modal gain coefficient of the amplifier. It should be noted that the unit of $F(z, E)$ is $[\text{length}]^{-3/2}$, such that square of the magnitude of $F(z, E)$ represents the photon density of the signal [23]. The solutions of the above equation can be found by imposing the appropriate boundary conditions, which depend on the structure of the SLA. Since $S(z, E)$ now represents the total photon density, the total density of photons in energy bandwidth ΔE (measured from the peak gain photon energy) is the sum of the photon density, the signals and that of spontaneous emissions, which is given in a functional form by [23]

$$\begin{aligned} S(z, E) &= [|F^{+}(z, E)|^2 + |F^{-}(z, E)|^2]\delta(E - E_{\mathrm{s}})\Delta E \\ &\quad + 2[I_{\mathrm{sp}}^{+}(z, E) + I_{\mathrm{sp}}^{-}(z, E)]\delta(E - E_m)\Delta E N_{\mathrm{g}}/cE\,. \end{aligned} \tag{5.32}$$

The factor 2 in front of the spontaneous emitted photons takes into account the two possible polarisations which the spontaneous emissions can take. $I_{\mathrm{sp}}^{\pm}$ represents the envelope of the amplified spontaneous emission intensity spectrum across the m possible modes. The delta function $\delta(E - E_m)$ selects that of the signal within ΔE.

Equations (5.27), (5.30) and (5.31) are the key equations in analysing SLAs using travelling-wave equations. The procedure for solving for the gain profile is as follows. We first solve for the signal field distribution at each point along z by using Eq. (5.31) by assuming a particular form of gain profile initially, imposing suitable boundary conditions like input signal power and facet reflectivities. We then solve Eq. (5.30) by imposing these boundary conditions and the initial gain profile. Because R_{sp} is z-dependent, the corresponding field solution $F^{\pm}(z, E)$ and the spontaneous emission intensity $I_{\mathrm{sp}}^{\pm}(z, E)$ have to be substituted into Eq. (5.27) in order to obtain a solution for the carrier density and hence the corresponding gain profile g_m by iteration and self-consistency. This is a tedious process, as discovered by Marcuse [3] in his pioneering work. Adam *et al.* [23] simplify this substantially by postulating that instead of working with S_m, which is z-dependent, we can use the photon density averaged along z-direction of the cavity in solving these equations. Moreover, $R_{\rm sp}$ is assumed to be uniform along the cavity. This avoids the need for excessive iterative computation to solve for the field at each point along z [28]. With these assumptions, Eq. (5.32) is replaced by [23]

$$S(z,E) \approx S_{\rm ave} = [S_{\rm 1ave}\delta(E-E_{\rm r})\beta(E_{\rm r}) + S_{\rm 2ave}\delta(E-E_{\rm s})]\Delta E \qquad (5.33a)$$

where

$$S_{\rm 1ave} = \left\{ \frac{(\exp(gL)-1)[(1-R_2)(1+R_1\exp(gL)) + (1-R_1)(1+R_2\exp(gL))]}{gL[1-R_1R_2\exp(2gL)]} - 2 \right\} \frac{R_{\rm sp}N_{\rm g}}{gc} \qquad (5.33b)$$

$$S_{\rm 2ave} = \left\{ \frac{(1-R_1)(\exp(gL)-1)[1+R_2\exp(gL)]}{(1-\sqrt{R_1R_2}\exp(gL))^2 + 4\sqrt{R_1R_2}\exp(gL) \times \sin^2((2\pi NL/hc)(E-E_{\rm r}))} \right\} \frac{P_{\rm in}N_{\rm g}}{EWdcgL} \qquad (5.33c)$$

where N is the effective refractive index, $S_{\rm 1ave}$ and $S_{\rm 2ave}$ represent the average spontaneous emission photon density and the signal photon density along the cavity, respectively. The average spontaneous emission photons can be found by solving Eq. (5.30) alone (i.e. considering that there is no input signal to the amplifier) with $R_{\rm sp}$ independent of z and imposing appropriate boundary conditions. $\beta(E_{\rm r})$ takes into account the proportion of spontaneous emission which is actually coupled to the cavity modes. This gives a simpler way to solve the equations without involving tedious computations (since the number of equations to be solved reduces to three instead of *three for every* point along z) but it has surprisingly good accuracy [28]. In this manner, an average gain coefficient $g_{\rm ave}$ can be obtained (instead of a number of gain coefficients $g_m(z)$ along z which form the gain profile), and this can be subsequently substituted into Eq. (5.12) in Section 5.2.1.1 to find the amplifier gain for a particular input power. By proceeding to calculate the amplifier gain for different input optical power levels, the gain saturation characteristics can also be derived. We will show some results calculated by the modified method proposed by Adam *et al.* [37] (as described above in this chapter) later. Before that, we will look at an alternative method to solve for the travelling wave equations, Eqs. (5.27), (5.30) and (5.31). This involves a special technique known as the transfer matrix method (TMM) [4, 5, 71].

5.4 Analysis of Semiconductor Laser Diode Amplifiers using Transfer Matrices

5.4.1 *A brief review of matrix methods*

We have just seen in the previous section that if one has to analyse SLAs using travelling-wave equations, which have taken into account the spatial dependence of the fields and carrier density (hence the material gain profile), three unknown quantities have to be found from Eqs. (5.26), (5.30) and (5.31) *for each value of* z. These quantities are signal fields $F^{\pm}$, spontaneous emissions intensity $I_{\mathrm{sp}}^{\pm}$, and the carrier density n (we have dropped for simplicity the functional dependence of these quantities on z and E). Hence, as discussed in Section 5.3.2, the solution will require a tedious process of numerical iteration. A good example to illustrate the complexity involved in analysing SLAs with travelling-wave equations is by considering SLAs using distributed feedback (DFB) structures [38, 39]. It is well known that to find the longitudinal field $F^{\pm}(z)$ and the longitudinal propagation constant β_z, in DFB laser structures, having only a knowledge of transverse modal fields is not sufficient. This is in contrast to the solution of the longitudinal propagation constant in simple buried heterostructures, as discussed in Chapter 4, where β_z can be found by Eq. (4.26) once the effective index has been determined by the transverse modal field analysis. A coupled wave equation has to be solved numerically [33]. In addition, the longitudinal fields have to satisfy Eq. (5.31) in an amplifier. Hence, the overall analysis on the amplifier characteristics becomes much more complicated and time consuming.

One possible method of simplifying the analysis procedure, whilst increasing its flexibility and robustness, is to employ matrix methods. Matrices have been used in many engineering problems of intensive numerical nature, such as solving transverse modal fields in arbitrarily shaped dielectric waveguides [40, 41]. In microwave engineering, matrix methods have been used to solve for the electric and magnetic fields in waveguides and microwave devices [42–44]. The advantage of using matrices is that the algorithm involved in solving the fields is identical for all types of structures with different boundary conditions. Thus the method

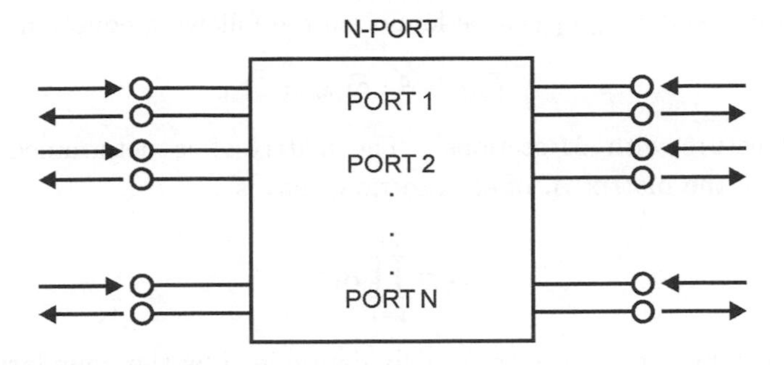

Fig. 5.4. Representation of a section in a structure analysed by matrix methods with an N-port system.

of analysis becomes more flexible. However, because matrix methods are purely numerical in nature, they do not allow one to detect whether a closed form analytical result exists for the structure. Such analytical solutions can be useful as they offer a quick way to estimate how the solutions will vary with various parameters.

In all matrix methods, the structure under analysis will be divided into a number of sections. which can be either fairly large (e.g. in the transmission-line matrix (TLM) method [42]) or as few as two or three (e.g. in scattering matrices [10]). Each section is modelled by an N-port network, with fields travelling in both directions at each port (see Fig. 5.4). The value of N will be determined by the number of interfaces joining neighboring sections. For example, for one dimensional structures, $N = 2$. There is a relation between the fields at each port for each section. At the ith section in a one-dimensional structure, this relation can be represented by the following matrix equation

$$F_{1,i} = A_i \cdot F_{2,i} \tag{5.34}$$

where $F_{1,i}$ and $F_{2,i}$ are the column vectors representing the fields at the port 1 and port 2, respectively, and A_i is the matrix which coefficients will determine the relationship between the fields $F_{1,i}$ with those in $F_{2,i}$. If the structure under analysis is one-dimensional, then at each port there can only be two fields, travelling into and out of the port, respectively [5]. Using Eq. (5.34), $F_{1,i}$ and $F_{2,i}$ can be found for each of the ith section, noting that $F_{2,i} = F_{1,i+1}$ when the entire structure is represented by cascading such two-port sections, each with a matrix A, then the relation between $F_{1,i}$ and $F_{1,M+1}$ can be found via the following equation

$$F_{1,i} = A \cdot F_{1,M+1} \tag{5.35}$$

for a structure with M-sections. The matrix A is determined by the products of the matrix A_i of each section, that is

$$A = \prod_{i=1}^{M} A_i \,. \tag{5.36}$$

Since either $F_{1,i}$ or $F_{1,M+1}$ are usually determined by the boundary conditions imposed on the structure (e.g. input power to the structure), the fields at each interface joining the ports can be found from Eq. (5.34), provided that the coefficients in A_i are known. Usually, the coefficients of A_i are determined by the structural and material characteristics of that section and can be determined for each section. If the number of sections used is sufficiently large, the continuous field profile within the structure can also be determined accurately.

The exact forms of $F_{1,i}$ and $F_{1,i+1}$ are different for different matrix methods. Some are shown in Fig. 5.5. In scattering matrix analysis, $F_{1,i}$ will

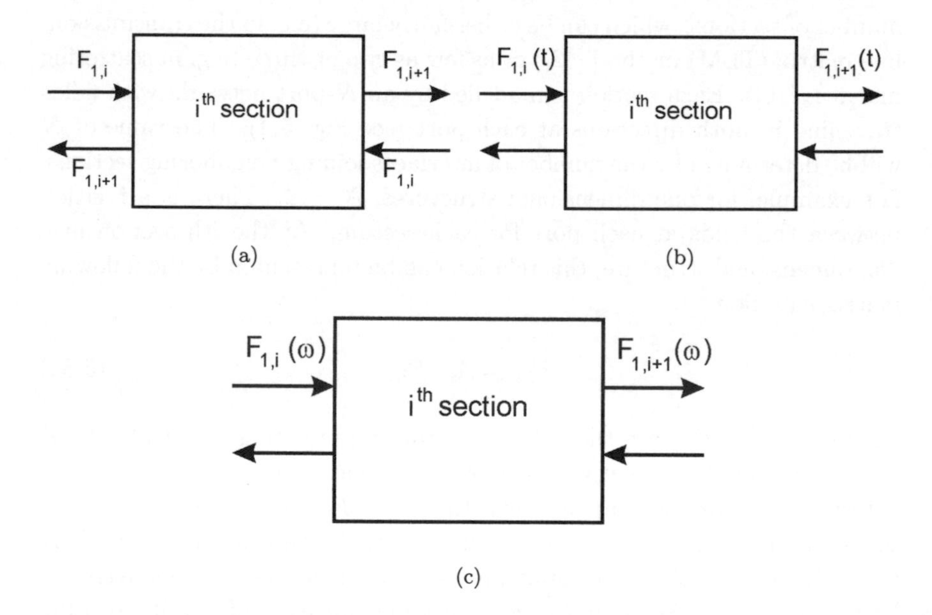

Fig. 5.5. Various matrix methods used to analyse electromagnetic fields in different structures (a) Scattering Matrix, (b) TLM and (c) TMM.

represent all the fields travelling into the port, whilst $F_{i,i+1}$ represent all the fields travelling out of the port [10]. Notice that strictly speaking the fields travelling out of the ith section should be denoted as $F_{i,i\pm1}$ because one will enter the $(i+1)$th section, whilst the other field will enter the $(i-1)$th section. In transmission line matrix (TLM) analysis, $F_{1,i}$ represents all the fields at one port, whereas $F_{i,i+1}$ represents those at the other port [42]. The difference between TLM and TMM is that the former analysis will be performed in the time domain, whereas the latter is in the frequency domain.

Equations (5.34)–(5.36) have been derived under the assumption that the structure under analysis is one-dimensional. In practice, all of these matrix methods can be extended to analyse three-dimensional structures. However, the algorithm involved is not only more complex, but is also unstable numerically [45]. In addition, as we shall see later, a three-dimensional structure can be modelled by an equivalent one-dimensional structure by dividing the structure appropriately and if some pre-analysis has been performed beforehand. This works exceptionally well in laser amplifiers, as they are essentially two-port devices with one input and one output.

Both TLM [45] and TMM [5] have been applied to analyse semiconductor laser devices. The difference in the domains of analysis of these two methods also accounts for their differences in applications. TLM works exceptionally well for analysis of the transient responses of both semiconductor lasers and laser amplifiers [35, 45, 46], although it can be used to analyse steady state performance relatively easily too. However, it is difficult to use TLM to analyse the noise performance of laser devices, e.g. spectral linewidth in semiconductor lasers, the noise figure and phase noise in laser amplifiers. This is because most of the noise-related phenomena are time-averaged stochastic processes [47]. Using TLM in the time domain to analyse noise processes will involve sampling over a very long period of time. Although this is feasible for very simple analysis (e.g. see Lowery [48]), TLM is not suitable in general to analyse noise characteristics in semiconductor laser devices.

On the other hand, because the analyses by TMM are performed in the frequency domain, both the steady state and noise characteristics (e.g. spectral linewidth) for semiconductor laser devices can be obtained accurately and efficiently [5, 38, 49]. However, the transient responses of these laser devices cannot be determined efficiently as a very wide frequency range will be required for accurate solutions in the analysis. In this book, we are primarily interested in the steady state operations of SLAs in optical communications. Hence, the TMM will be more suitable for our foregoing studies of their gain, saturation and noise characteristics. In the following section, we will examine how to implement the TMM using Eqs. (5.34) to (5.36) to analyse the longitudinal travelling fields in SLAs.

5.4.2 *Analysis of longitudinal travelling fields in SLAs using transfer matrix method*

We have seen in Section 5.3.2 that it is vital to find the longitudinal field distribution within an SLA in order to determine the material gain profile in the amplifier (Eq. (5.27)) and subsequently the overall amplifier characteristics. An efficient and robust technique of analysis which can yield this distribution will be ideal. The matrix formalism outlined in the previous discussion will be used here to develop such a technique. Most SLAs are buried heterostructures [46]. For these structures, as we examined in the previous chapter, the active region which provides gain can be represented by a rectangular dielectric waveguide of refractive index N_1 embedded in material with a different refractive index N_2. Using the transverse modal field analysis outlined in the previous chapter, one can deduce the effective group refractive index N_{eff} for the propagating signal. The $x - y$ dependence of the fields has then been eliminated. Recall that the total solution of the electric or magnetic fields in such a dielectric waveguide is given by [10]

$$E_i(x, y, z) = \phi_m(x, y)F(z) \qquad i = x, y, z\,. \tag{5.37}$$

The knowledge of $\phi_m(z, y)$ allows us to use a simple one-dimensional waveguide to represent the true three-dimensional structure. This is illustrated in Fig. 5.6. We will use this one-dimensional structure in the foregoing discussion on the implementation of TMM in analysing SLAs. Before we examine how to use TMM to analyse the longitudinal fields in the SLA, one may notice that our preceding discussion was with a BH SLA. This type of amplifier can indeed be modelled very easily by a rectangular dielectric waveguide [50]. However, modern SLAs involve more complex structures like ridge or rib waveguides [4]. In these cases, the analyses of the transverse modal fields are indeed more complicated than that outlined in Chapter 4 [51]. However, once the field solutions $\phi_m(x, y)$ and the corresponding effective refractive index N_{eff} are calculated from the transverse modal field analysis, we can use Eq. (5.37) to analyse $F(z)$ by using a one-dimensional waveguide model with a refractive index of N_{eff} as illustrated

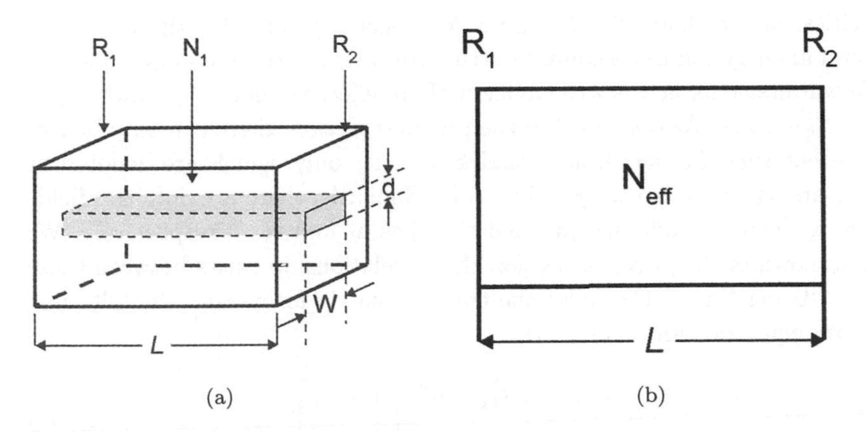

Fig. 5.6. Modelling of a buried heterostructure by a one-dimensional waveguide.

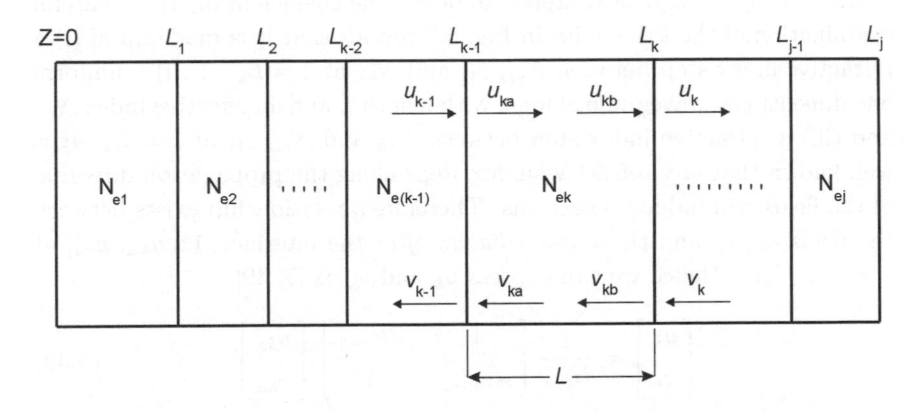

Fig. 5.7. A multi-cavity equivalent of one-dimensional waveguide used in TMM analysis.

in Fig. 5.6. Therefore the general implications on using Eq. (5.37) with the effective index remains. But one should note that the form of Eq. (5.37) assumes that $\phi_m(x, y)$ and $F(z)$ are separable. This is true for index guided structures as the propagation constant β_z is independent of x, y or z [18]. For gain guided structure (e.g. stripe geometry laser [52]), this assumption may not hold [53].

We consider an amplifier which can be modelled as a one-dimensional waveguide along z (Fig. 5.7). The structure under analysis is a general one in which we can divide the waveguide along z into a number of smaller cavities, each with an effective index N_{ei}. Such a general description will be useful in analysing more complex structures like two-section amplifiers [54]. We will first look at the kth cavity with an effective index N_{ek} and length $L = L_k - L_{k-1}$. As discussed in the previous section, there can only be two fields entering the one-dimensional kth cavity only, which are denoted as u_{k-1} and v_k, respectively (see Fig. 5.7). Similarly, there are only two fields leaving the waveguide, which are denoted as u_k and v_{k-1}, respectively. We have shown in the previous section that a relationship exists between these fields. Using Eq. (5.34) under the transfer matrix formalism, the following matrix equation can be derived.

$$\begin{bmatrix} u_k \\ v_k \end{bmatrix} = \begin{bmatrix} a_{11}^k & a_{12}^k \\ a_{21}^k & a_{22}^k \end{bmatrix} \begin{bmatrix} u_{k+1} \\ v_{k+1} \end{bmatrix} . \tag{5.38}$$

The matrix on the right hand side of Eq. (5.38) is the transfer matrix A_k of the kth cavity. Our next step is to derive the coefficient of A_k. A careful examination of the kth cavity in Fig. 5.7 reveals that it is made up of (i) a refractive index step between $N_{e(k-1)}$ and N_{ek} at $z = L_{k-1}$; (ii) a uniform one dimensional waveguide along z with length L and an effective index N_{ek} and (iii) a refractive index step between N_{ek} and $N_{e(k-1)}$ at $z = L_k$. It is well known that any refractive index steps along the propagation direction of the fields will induce reflections. Therefore a relationship exists between the fields u_k, v_k and those *immediately after* the interface, i.e u_{ka}, v_{va}, at $z = L_{k-1}$ [10]. Hence we can express u_k and v_k as [5, 49]

$$\begin{bmatrix} u_k \\ v_k \end{bmatrix} = \frac{1}{t_{k-1}} \begin{bmatrix} 1 & -r_{k-1} \\ -r_{k-1} & 1 \end{bmatrix} \begin{bmatrix} u_{ka} \\ v_{ka} \end{bmatrix} \tag{5.39}$$

where

$$r_{k-1} = \frac{N_{e(k-1)} - N_{ek}}{N_{e(k-1)} + N_{ek}} . \tag{5.40}$$

In Eq. (5.40) r_{k-1} is the *amplitude* reflectivity of the interface at $z = L_{k-1}$. The quantity t_{k-1} is the field *amplitude* transmittivity of the same interface, and is related to r_{k-1} by [10]

$$t_{k-1}^2 = 1 - r_{k-1}^2 . \tag{5.41}$$

Similarly, at the interface $z = L_k$, one can relate the fields u_{k+1} and v_{k+1} to those *immediately before* the interface by

$$\begin{bmatrix} u_{k\mathrm{b}} \\ v_{k\mathrm{b}} \end{bmatrix} = \frac{1}{t_k} \begin{bmatrix} 1 & r_k \\ r_k & 1 \end{bmatrix} \begin{bmatrix} u_{k+1} \\ v_{k+1} \end{bmatrix} \tag{5.42}$$

where

$$r_k = \frac{N_{e(k+1)} - N_{ek}}{N_{e(k+1)} + N_{ek}} . \tag{5.43}$$

In Eq. (5.43) r_k is the amplitude reflectivity of the interface at $z = L_k$. The quantity t_k is therefore the amplitude transmittivity of the same interface which is given by

$$t_k^2 = 1 - r_k^2 . \tag{5.44}$$

We have now established a relationship between (u_k, v_k) and $(u_{k\mathrm{a}}, v_{k\mathrm{a}})$ (see Eq. (5.39)), and a similar relationship between (u_{k+1}, v_{k+1}) and $(u_{k\mathrm{b}}, v_{k\mathrm{b}})$. If a relation between the internal fields $(u_{k\mathrm{a}}, v_{k\mathrm{a}})$ and $(u_{k\mathrm{b}}, v_{k\mathrm{b}})$ can be established, then the coefficients in the matrix A_k can be found. This can be achieved by noting that the rest of the kth cavity is a uniform one-dimensional waveguide. Since we have been analysing structures with index guiding, one can derive the relation between $(u_{k\mathrm{a}}, v_{k\mathrm{a}})$ and $(u_{k\mathrm{b}}, v_{k\mathrm{b}})$ by solving the one-dimensional wave equation (i.e. travelling-wave equation Eq. (5.31)) [55–58].

$$\frac{d^2F(z)}{dz^2} = -\frac{\omega^2}{c^2} N_{ek}^2 F(z) \tag{5.45}$$

where $F(z)$ can be found to have a general solution of the form

$$F(z) = u \cdot \exp(-j\beta_z z) + v \cdot \exp(j\beta_z z) . \tag{5.46}$$

Then it can be seen immediately that

$$u_{k\mathrm{a}} = u_{k_\mathrm{b}} \cdot \exp(j\beta_z L) \tag{5.47a}$$

$$v_{k\mathrm{a}} = v_{k_\mathrm{b}} \cdot \exp(-j\beta_z L) . \tag{5.47b}$$

Hence, Eq. (5.38) can be re-written as

$$\begin{bmatrix} u_k \\ v_k \end{bmatrix} = \frac{1}{t_k t_{k-1}} \begin{bmatrix} 1 & -r_{k-1} \\ -r_{k-1} & 1 \end{bmatrix} \begin{bmatrix} e^{j\beta_z L} & 0 \\ 0 & e^{-j\beta_z L} \end{bmatrix}$$

$$\times \begin{bmatrix} 1 & r_k \\ r_k & 1 \end{bmatrix} \begin{bmatrix} u_{k+1} \\ v_{k+1} \end{bmatrix} \tag{5.48}$$

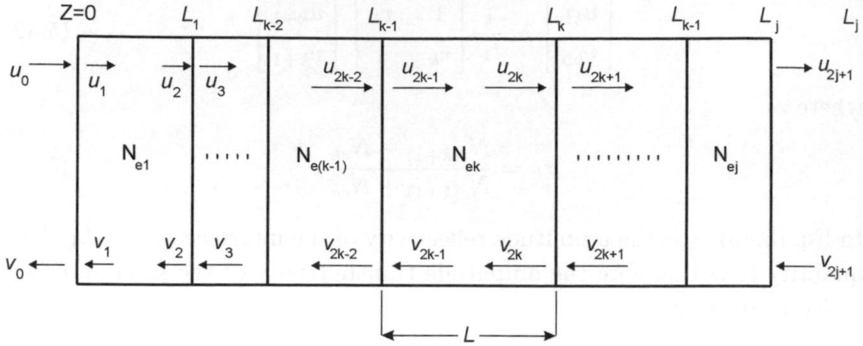

Fig. 5.8. Formalising the transfer matrix method to analyse the multi-cavity structure.

β_z can be found easily since the effective index is known (see Eq. (4.26) in Chapter 4), that is

$$\beta_z = \omega N_{ek}/c\,. \tag{5.49}$$

Hence, all the coefficients in the matrices of Eq. (5.48) are known once the structural parameters $N_{ek}, N_{e(k+1)}$ and L are known. Then if u_{k+1} and v_{k+1} are known, we can "transfer" them to form the quantities u_k and v_k easily by using Eq. (5.48). The above derivation holds for one cavity only. To formalise the analysis we should consider all K-sections shown in Fig. 5.7. In doing so, we have to redefine the internal fields as shown in Fig. 5.8. The relationship between the fields at the interface are given by Eqs. (5.39) and (5.42), respectively. The fields within the waveguide is related by Eqs. (5.47a) and (5.47b). In this manner u_{2k+1} and v_{2k+1} are related to u_0 and v_0 via the following matrix equation

$$\begin{bmatrix} u_0 \\ v_0 \end{bmatrix} = \begin{bmatrix} a_{11} & a_{12} \\ a_{21} & a_{22} \end{bmatrix} \begin{bmatrix} u_{2k+1} \\ v_{2k+1} \end{bmatrix} \tag{5.50}$$

where

$$A = \begin{bmatrix} a_{11} & a_{12} \\ a_{21} & a_{22} \end{bmatrix} = A_{01} \cdot A_{12} \cdot A_{23} \cdots A_{2k,2k+1}\,. \tag{5.51}$$

The coefficients of the above transfer matrix can be derived by using Eqs. (5.39), (5.42), (5.47a), (5.47b) and (5.48) from the structural parameters, once the fields at $z = 0$ or $z = L_k$ are known, then all the fields u_i, v_i along the structure can be calculated including the fields at $z = 0$

(if the fields at $z = L_k$ are known) or the fields at $z = L_k$ (if the fields at $z = 0$ are known), and hence the signal gain of the amplifier. One may notice from Eqs. (5.47a) and (5.47b) that in solving the fields in the one-dimensional waveguide, we have so far neglected the fact that there is optical gain in the structure (since it is a laser amplifier). To introduce this fact into the analysis, one can follow the procedures outlined in Section 4.3 of Chapter 4 by introducing an imaginary part to the longitudinal propagation constant. In this case

$$\beta_z \rightarrow \beta_z + \frac{jg}{2} \tag{5.52}$$

where g is the modal gain coefficient of the active region. Then for the kth cavity, Eq. (5.48) can be re-written as

$$\begin{bmatrix} u_k \\ v_k \end{bmatrix} = \frac{1}{t_k t_{k-1}} \begin{bmatrix} 1 & -r_{k-1} \\ -r_{k-1} & 1 \end{bmatrix} \begin{bmatrix} e^{(gL/2+j\beta_z L)} & 0 \\ 0 & e^{(gL/2-j\beta_z L)} \end{bmatrix} \times \begin{bmatrix} 1 & r_k \\ r_k & 1 \end{bmatrix} \begin{bmatrix} u_{k+1} \\ v_{k+1} \end{bmatrix} . \tag{5.53}$$

Notice that Eq. (5.53) has been derived under the assumption that g *is uniform along* L. In addition, if β_z has taken the form of Eq. (5.52), the effective indices of each cavity become complex according to Eq. (5.49). Then the corresponding reflectivities and transmittivities should also become complex. However, as discussed in Section 4.3, the imaginary part in Eq. (5.52) is much smaller compared to the real part. Hence, the imaginary parts of the effective indices will also be very small. Thus we can continue to calculate reflectivities and transmittivities at interfaces joining the cavities using the real parts of the effective indices only without significant loss in accuracy [5]. In the following section we will examine how to modify the analysis described here to consider the situation when g is not uniform along L.

5.4.3 *Analysis of SLAs with a non-uniform gain profile using transfer matrix method*

To simplify the previous discussion, we will consider SLAs with one cavity only ($K = 1$), as shown in Fig. 5.9. The analysis described here can be easily extended to SLAs having more than one cavity, following the procedures illustrated in Fig. 5.8 and Eqs. (5.50) to (5.51). As shown in Fig. 5.9,

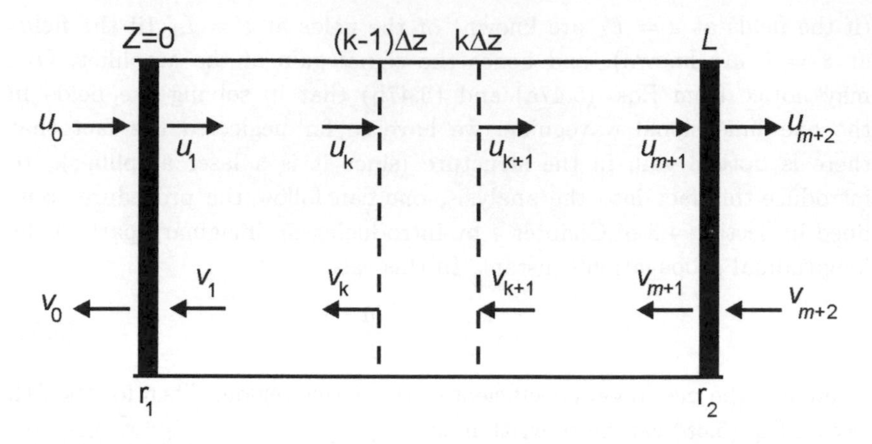

Fig. 5.9. The model for analysis of an SLA using TMM by taking into account the non-uniform gain profile.

the amplifier is modelled by a one-dimensional cavity with effective index N_{eff} and length L. We further divide the cavity into M-sections, each of length Δz. The modal gain coefficient of the kth section is given by $g(z = k\Delta z) = g_k$, and can be found from Eqs. (4.49) and (4.53) if the material gain coefficient g_m is known. For simple BH SLAs, g_k is related to $g_m(z = k\Delta z)$ by [59]

$$g_k = \Gamma(g_{mk} - \alpha_{\text{ak}}) - (1 - \Gamma)\alpha_{ck} \tag{5.54}$$

where we have used $g_{mk} = g_m(z = k\Delta z)$, and similarly α_{ak} and α_{ck} are the losses in the active region and in the claddings, respectively. As discussed in Section 5.3.2, the material gain coefficient g_{mk} and the amplitude of the electric fields are not unrelated. This is because g_{mk} is determined by the carrier density $n(z = k\Delta z) = n_k$. Under a linear model [13]

$$g_{mk} = A(n_k - n_0) \tag{5.55}$$

where n_0 is the transparency carrier density of the material (see Section 3.2.1 in Chapter 3). Notice that n_0 depends on the band structure of the material only, and is assumed to be independent of z. In addition, we have assumed that the signal wavelength is matched to that of the peak of the material gain spectrum, so that g_{mk} is the peak gain coefficient of the material. On the other hand, the carrier density n_k is related to the optical fields at $z = k\Delta z$ via the rate equation (Eq. (5.27)).

$$\frac{j}{ed} = R_k + \frac{c\Gamma}{N_g} g_{mk} S_k \,. \tag{5.56}$$

Where R_k is the recombination rate of the carrier at $z = k\Delta z$ and $S_k = S(z = k\Delta z)$ is the photon density at $z = k\Delta z$ In Eq. (5.56), we have assumed that the current density j is uniform across the active region and the contribution of spontaneous emissions are neglected by assuming that g_{mk} is small. Thus S_k consists of photons due to the input signal. If we normalise the fields u_{k+1} and v_{k+1} such that the square of their magnitudes represents the intensity of the optical signals travelling into and out of the kth section, then we have [37]

$$S_k = [|u_{k+1}|^2 + |v_{k+1}|^2] \cdot \frac{N_g}{cE} \tag{5.57}$$

where E is the energy of the signal photons. Notice that in deriving Eq. (5.57), the coupling of any spontaneous emission $I_{sp}^{\pm}$ to the guided mode of signals u_{k+1} and v_{k+1} has been neglected (i.e. $\beta \approx 0$ in Eq. (5.33)). As we shall see later in Section 5.4.4, the contribution of spontaneous emissions to the amplifier characteristics is not important in analysing SLAs, provided that the bias of the amplifier is not large and g_{mk} is small for all ks. Usually this implies that the amplifier is biased below oscillation threshold [37]. Since the performance of SLAs biased well below oscillation threshold is of most interest, the above assumption will be valid for most of our foregone analysis. Substituting Eq. (5.57) into Eq. (5.56) yields the following equation

$$\frac{j}{ed} = R_k + \frac{\Gamma g_{mk}}{E} [|u_{k+1}|^2 + |v_{k+1}|^2] \,. \tag{5.58}$$

By solving Eq. (5.58) for g_{mk} and hence n_k by using Eq. (5.55), we can deduce the coefficients of the marix A_k, which relates (u_k, v_k) and (u_{k+1}, v_{k+1})

$$\begin{bmatrix} u_k \\ v_k \end{bmatrix} = \begin{bmatrix} e^{(g_k \Delta z/2 + j\beta_z \Delta z)} & 0 \\ 0 & e^{(g_k \Delta z/2 - j\beta_z \Delta z)} \end{bmatrix} \begin{bmatrix} u_{k+1} \\ v_{k+1} \end{bmatrix} . \tag{5.59}$$

Equations (5.58) and (5.59) are the core equations in our analysis of SLAs using TMM. By solving them, we can find (i) the field, (ii) the material gain coefficient and (iii) the carrier density at $z = k\Delta z$ along the amplifier. Furthermore, the format of Eq. (5.59) suggests that the computational algorithm (see next section) will converge unconditionally, unlike solving the travelling wave equations by iteration and self-consistency [3]. Any facet reflectivities can also be accounted for with Eqs. (5.39) and (5.42) (Fig. 5.9), such that

$$\begin{bmatrix} u_{M+1} \\ v_{M+1} \end{bmatrix} = \frac{1}{t_2}\begin{bmatrix} 1 & r_2 \\ r_2 & 1 \end{bmatrix}\begin{bmatrix} u_{M+2} \\ v_{M+2} \end{bmatrix} \tag{5.60}$$

and

$$\begin{bmatrix} u_0 \\ v_0 \end{bmatrix} = \frac{1}{t_1}\begin{bmatrix} 1 & -r_1 \\ -r_1 & 1 \end{bmatrix}\begin{bmatrix} u_1 \\ v_1 \end{bmatrix}. \tag{5.61}$$

Thus by solving for (u_0, v_0) for given values of (u_{M+2}, v_{M+2}), we can calculate the amplifier gain by

$$G = \left|\frac{u_{M+2}}{u_0}\right|^2. \tag{5.62}$$

By calculating G for different levels of output optical power P_{out}, one can derive the variation of G against P_{out} and hence deduce the saturation output power.

5.4.4 *Computational considerations*

In this section, the technique described above will be used to analyse a simple Fabry–Perot amplifier (FPA), and be compared with the analysis proposed by Adams *et al.* [37]. The computation procedure with the TMM is as follows:

(i) We start with an initial value of P_{out}. This quantity is related to u_{M+2} and v_{M+2} by:

$$P_{\text{out}} = |u_{M+2}|^2 W d \tag{5.63a}$$

and

$$v_{M+2} = 0\,. \tag{5.63b}$$

(ii) Compute u_{M+1} and v_{M+1} by using Eq. (5.60).

(iii) Compute the fields u_M and v_M using Eq. (5.59) with $k = M$.
The simplest way is to express Eq. (5.58) in terms of n_k by using Eq. (5.55). The recombination rate R_k for undoped materials for

which Auger recombinations are not important is given simply by [37]

$$R_k = Bn_k^2 . \tag{5.64}$$

In long wavelength semiconductor laser devices, we have to take into account the Auger recombinations, that is the following R_k expression should be considered [72]:

$$R_k = A_{nr}n_k + Bn_k^2 + Cn_k^3 . \tag{5.65}$$

By substituting Eqs. (5.64) or (5.65) and Eq. (5.55) into Eq. (5.58), we can calculate n_k and hence g_{mk}. Then g_k can be calculated from Eq. (5.54) and the coefficients of the matrix A_k in Eq. (5.59) can be formed. Finally, u_k and v_k can be calculated. The above steps provide us with three vital pieces of information: the carrier density n_k, the material gain coefficient g_m and finally the fields u_k and v_k. If M is large enough, we can derive the material gain profile and the longitudinal field distribution along z. This is not possible with the method proposed by Adams *et al.* [37].

(iv) Repeat step (iii) until u_1 and v_1 are calculated at $M = 1$.

(v) Calculate the input fields u_0 and v_0 by using Eq. (5.61) and hence the input optical power P_{in} using the following equation:

$$P_{\text{in}} = |u_0|^2 . \tag{5.66}$$

In the absence of spontaneous emission coupled to the signal and the boundary condition of $v_{M+2} = 0$, we can show easily that $v_0 = 0$.

(vi) The amplifier gain is calculated by using Eq. (5.62).

(vii) Repeat steps (i)–(vi) for a different value of P_{out} until all the values of P_{out} in the range of interest have been covered. To compare the above technique with that proposed by Adams *et al.* [37], a simple FPA with structural parameters listed in Table 5.2 has been analysed. Auger recombinations have been neglected for simplicity.

We first compare the speed of computation between the two techniques. For $M = 300$ sections, the TMM uses approximately 0.8 s on an IBM3090 mainframe computer to calculate 50 values of G in the output power range of -60 dBm to 0 dBm for a fixed bias current. For a similar set of calculations using the technique suggested by Adams *et al.* [37], it takes approximately 1.1 s. Though the difference is not much, it should be pointed out that there are some points which do not converge using the latter method, especially when G starts to saturate. In fact, it took several days for us to

Table 5.2. SLA parameters used in calculations [16, 37].

Length of the Amplifier	$L = 500\ \mu$m
Width of the Active Region	$W = 1.5\ \mu$m
Thickness of the Active Region	$d = 0.2\ \mu$m
Optical Confinement Factor	$\Gamma = 0.5$
Total Loss Coefficient	$\alpha = 25$ cm^{-1}
Group Effective Index	$N_g = 3.5$
Reflection Coefficients	$R_1 = R_2 = 0.36\%$
Differential Gain Coefficient	$A = 4.63 \times 10^{-16}$ cm^2
Radiative Recombination Coefficient	$B = 1 \times 10^{-10}$ cm^{3}s^{-1}
Spontaneous Emission Coefficient	$\beta = 10^{-4}$

adjust the algorithm suggested by Adams *et al.* [37] properly, such that the initial trial value can lead to speedy computations with sufficient number of converged results. This critical importance of the initial value has also been discussed recently by Zaglanakis and Seeds [58]. On the other hand, the algorithm based on TMM does not need a lot of adjustment and fine-tuning (we spent only half an hour writing a successful working program for this analysis). Although the above comparisons are difficult to quantify and may depend on the experience and skills of the programmer, it clearly illustrates the ease of implementation and robustness of the TMM algorithm.

Figure 5.10 shows the results of calculated value of amplifier gain G against output optical power P_{out} using both the averaged photon density (AVPD) approximation [37] and the transfer matrix method described above. Both $\beta = 0$ and $\beta = 10^{-4}$ have been considered in the AVPD calculations. As discussed previously, the TMM approach will consider $\beta = 0$. Results shown are for three particular pumping levels of $i/i\text{th} = 0.75$, 0.81 and 0.99, respectively, where ith is the threshold current of lasing for the structure and i is the bias or injection current. There are two important points which can be observed from the diagram. Firstly, the effect of β on the calculated results using AVPD approximation is not significant until the structure approaches lasing. This has also been discovered by Adams *et al.* [37] in their studies. Hence, for normal operation of SLA, performing the analysis by taking $\beta = 0$ will be accurate enough (unless the SLA is biased close to lasing threshold, as in the optical switching application discussed in Section 3.5.3 in Chapter 3). Secondly, the

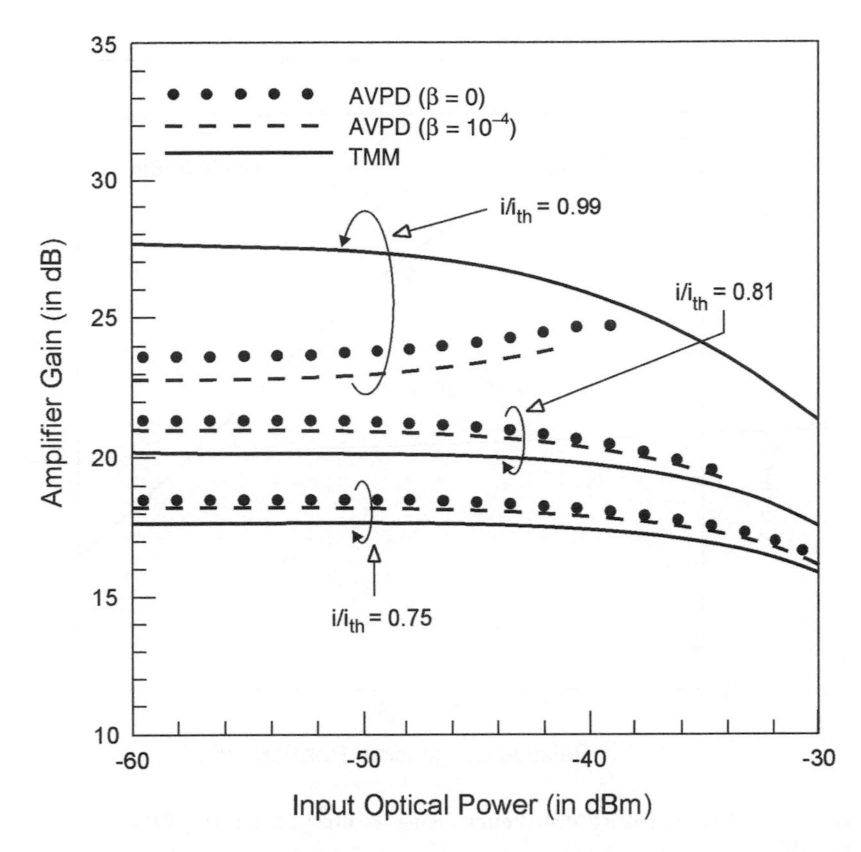

Fig. 5.10. Amplifier gain against output optical power of a FPA with parameters listed in Table 5.2 calculated by TMM and averaged photon density approximation.

differences between results calculated by TMM are not significant (around 1 dB) when the SLA is biased below threshold. A huge difference is observed when the SLA is driven close to oscillation. Thus compared with the AVPD approach, the results generated by TMM are in close agreement with those predicted by AVPD for low to moderate pumping. However, the differences between the two methods at high injection level are significant (even the shapes of the curves look different). The accuracy of both of the methods can only be assessed by comparison with experimental results. A more detailed examination of this will be given in Chapter 7.

To illustrate the benefits of using TMM, the profile of carrier density n along z and the normalised longitudinal field distribution have been calculated as shown in Figs. 5.11 and 5.12, respectively. The distribution

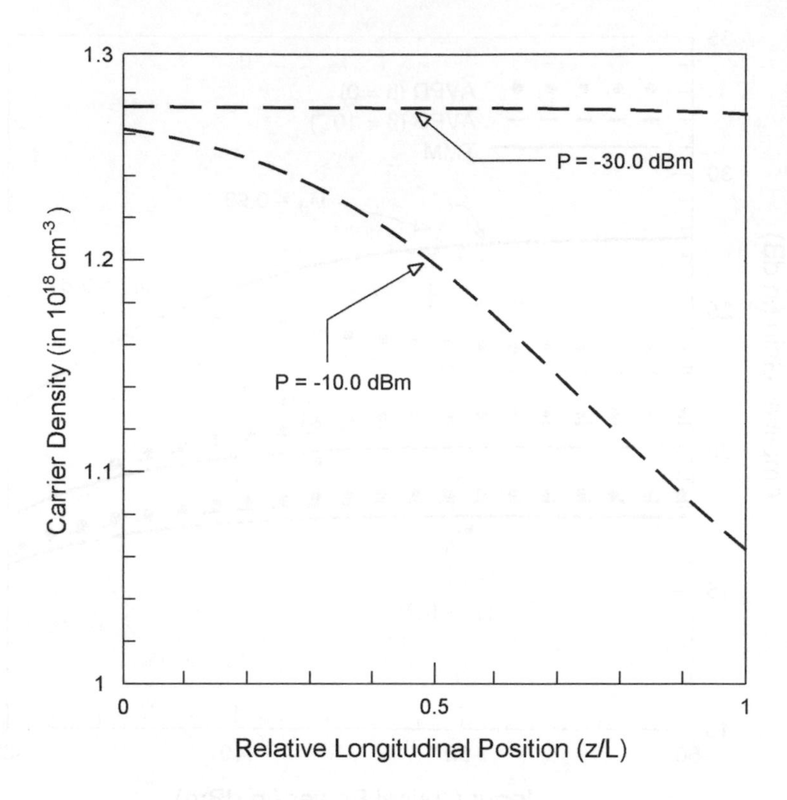

Fig. 5.11. Carrier density distribution along z direction for the FPA analysed in Fig. 5.10.

for two output powers at $i/i\text{th} = 0.81$ are shown. In Fig. 5.11, it can be seen that when the output power increases, the carrier density distribution becomes more uniform compared with the lower power level. Thus the AVPD approximation may not be accurate at high output power levels (since it assumes a uniform distribution for carrier density). This will be explored again in Chapter 7. The total fields (the two curves at the top) and the corresponding components of u and v are plotted in Fig. 5.12 with respect to z, at $i/i\text{th} = 0.81$ for two different output power levels. Interestingly, we did not get a uniform curve, as Adams *et al.* [37] obtained in their analysis. This is because the latter analysis was performed when the signal wavelength was matched to the cavity resonance of the FPA. From this, they [31] justified their AVPD approach. In the present analysis, where the signal wavelength is not matched to the cavity resonance, a strong non-uniform field distribution is observed. Thus, in using the AVPD

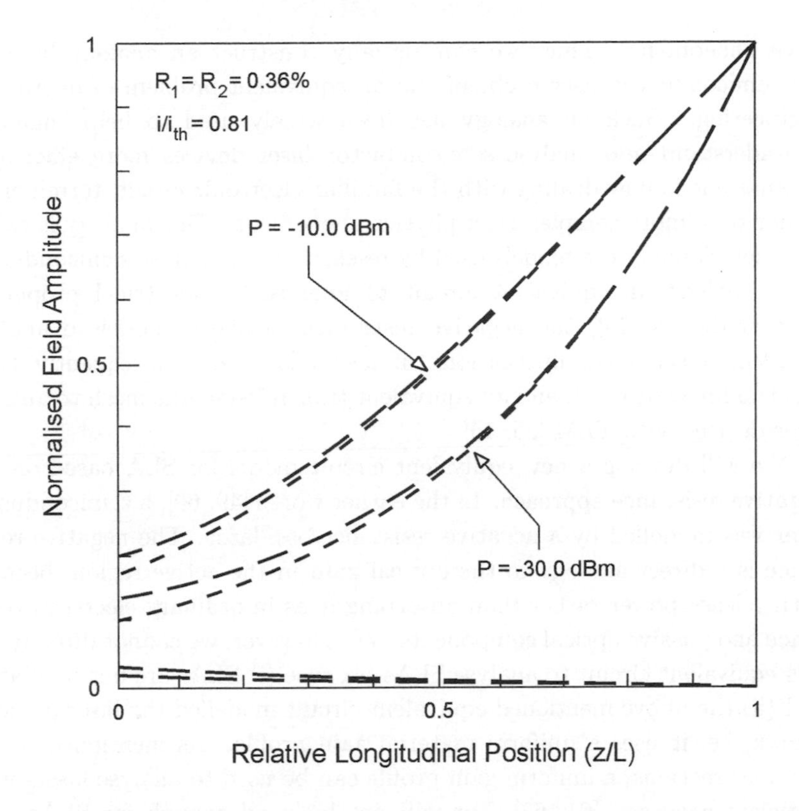

Fig. 5.12. Longitudinal field profile along z direction for the FPA analysed in Fig. 5.10.

approach, one must be careful to note that the assumptions are closest to reality only when the signal wavelength is matched with the FP cavity resonance. This illustrates the clear necessity of taking into account the non-uniform field and hence photon distribution along z in analysing SLAs.

5.5 An Equivalent Circuit Model for SLAs

The use of TMM in the analysis of SLAs was illustrated in the previous section. In this section, based on TMM, we derive an equivalent circuit to model SLAs and this equivalent circuit model will be used in the next chapter to deal with the noise performance of SLAs [4, 70–72].

As discussed in Section 5.4.1, matrix methods were originally applied to microwave engineering problems. They have also been applied to semiconductor laser devices because both methods are concerned with guided wave phenomena. Thus, we can actually construct an analogy between the semiconductor laser problem and an equivalent problem in microwave engineering. Such an analogy has been widely used to help engineers to understand and analyse semiconductor laser devices more efficiently, because one is now dealing with the familiar electronic circuit terminology rather than more complex laser physics phenomena. This analogy is called a model. Some of the models used by researchers to analyse semiconductor lasers include an equivalent circuit to analyse the electrical properties of laser diodes [59], the negative resistance oscillator model to analyse quantum mechanical fluctuations in lasers [60] and subsequently their spectral linewidth [49], and an equivalent transmission line mesh to analyse laser devices with TLM [35, 45].

We will develop a new equivalent circuit model for SLA based on the negative resistance approach. In the earlier works [49, 60], a semiconductor laser was modelled by a negative resistance oscillator. The negative resistance is a direct analogy to the optical gain in the active region, because both release power rather than absorbing it as in ordinary electrical resistance and passive optical components [60]. However, we cannot directly use this equivalent circuit to analyse SLAs because (i) SLAs are not oscillators and (ii) the above mentioned equivalent circuit modelled the laser diode as a bulk, i.e. it uses a uniform material gain profile. As mentioned in the previous sections, a uniform gain profile can be used to analyse lasers with sufficient accuracy [61, 62], but will not be good enough for SLAs. An entirely new equivalent circuit must therefore be derived for the analysis of SLAs.

As a negative resistance Z_k has been used to model the optical gain in the active region in the negative resistance oscillator analysis of semiconductor lasers, it will also be used to represent the modal gain coefficient g_k at the kth section in the one-dimensional waveguide used by TMM (Fig. 5.9). Note that instead of modelling the optical gain of the bulk active region by a single negative resistance, a single negative resistance will be used to represent a section of the amplifier under analysis only. Each section of the amplifier does not only provide the gain in field amplitude, it also introduces a phase delay due to propagation of the $\exp(j\beta_z\Delta z)$ term in A_k (see Eq. (5.59)). We can represent this by a piece of lossless transmission line of length Δz. The characteristic impedance of the line Z_0 is made such that the propagation constant in the line is identical to that in the amplifier. The equivalent circuit for a section of the amplifier under

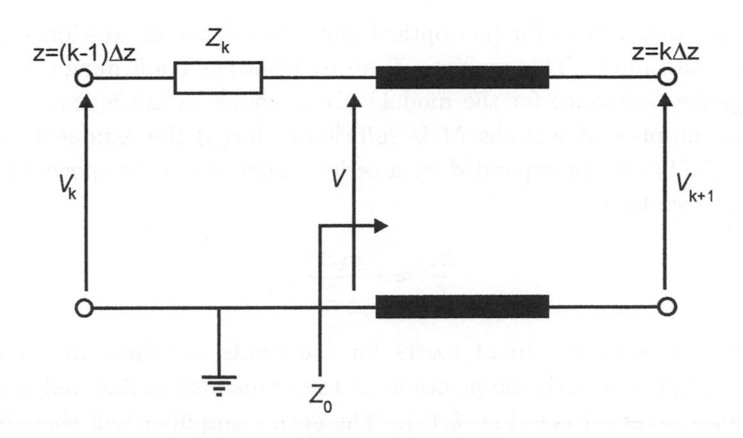

Fig. 5.13. An equivalent circuit for fields travelling in z direction in a section of an SLA.

analysis is therefore derived. This is illustrated in Fig. 5.13 which consists of two elements: a lumped negative resistance Z_k and a piece of lossless transmission line of characteristic impedance Z_0 and length Δz. We consider the voltages travelling along $+z$ direction in this equivalent circuit. By simple circuit analysis, at steady state we have:

$$V = \left(\frac{Z_0}{Z_k + Z_0}\right) \cdot V_k \tag{5.67}$$

$$V_{k+1} = V \cdot \exp(-j\beta_z \Delta z) \tag{5.68}$$

where the frequency dependence on the voltages has been assumed to be implicit. Hence:

$$\frac{V_{k+1}}{V_k} = \left(\frac{Z_0}{Z_k + Z_0}\right) \cdot \exp(-j\beta_z \Delta z)\,. \tag{5.69}$$

Comparing this with Eq. (5.59), it can be seen that

$$\frac{V_{k+1}}{V_k} \equiv \frac{u_{k+1}}{u_k} = \exp\left(\frac{g_k \Delta z}{2}\right) \exp(-j\beta_z \Delta z)\,. \tag{5.70}$$

Therefore, the following analogy can be made

$$\left(\frac{Z_0}{Z_k + Z_0}\right) \cdot = \exp\left(\frac{g_k \Delta z}{2}\right). \tag{5.71}$$

For a positive value of g (i.e. optical gain) the above expression must be greater than unity. This requires Z_k to be negative, confirming our choice of negative resistance for the model. If our choice of Δz is small enough (i.e. the number of sections M is sufficiently large) the exponential term in Eq. (5.71) can be expanded by a series. Then it can be shown that for small Δz we have

$$\frac{Z_k}{Z_0} \approx -\frac{g_k \Delta z}{2} . \tag{5.72}$$

A similar equivalent circuit exists for the fields travelling in $-z$ direction (i.e. v_k), but with the positions of the transmission line and negative resistance reversed (see Fig. 5.13). The entire amplifier will therefore be modelled by a cascade of M such equivalent circuits. The exact value of Z_k for each section has to be determined from Eq. (5.72) using TMM analysis in which g_{mk} and g_k are determined. The only remaining part of the amplifier to be modelled is the facet. This is in fact a refractive index step and can be modelled easily by two terminating loads R_{L1} and R_{L2} to account of the reflectivities r_1 and r_2, that is [10, 45]

$$r_1 = \frac{R_{L1} - Z_0}{R_{L1} + Z_0} \tag{5.73}$$

$$r_2 = \frac{R_{L2} - Z_0}{R_{L2} + Z_0} . \tag{5.74}$$

The complete equivalent circuit for a SLA is shown in Fig. 5.14. This is for a single section amplifier. To model multi-section SLAs, we can first derive an equivalent circuit for each section and then cascade them in a manner similar to cascading the one-dimensional waveguides in TMM analysis (Fig. 5.8). Notice that in theory, when M such sub-networks are cascaded, there will be reflections of voltages from later stages due to a mismatch between the characteristic impedance of the transmission line of the kth section and the negative resistance of the $(k+1)$th section. This is equivalent to the situation which we have discussed during the derivation of TMM; namely, that there should be reflections between the interfaces of each section because the optical gain is different in each section, and in theory the refractive indices (the optical gain contributes to the imaginary part; see Chapter 4) are therefore different between each section and reflection coefficients which are complex will occur. However, as discussed by Bjork and Nilsson [5], the imaginary parts of the refractive indices are

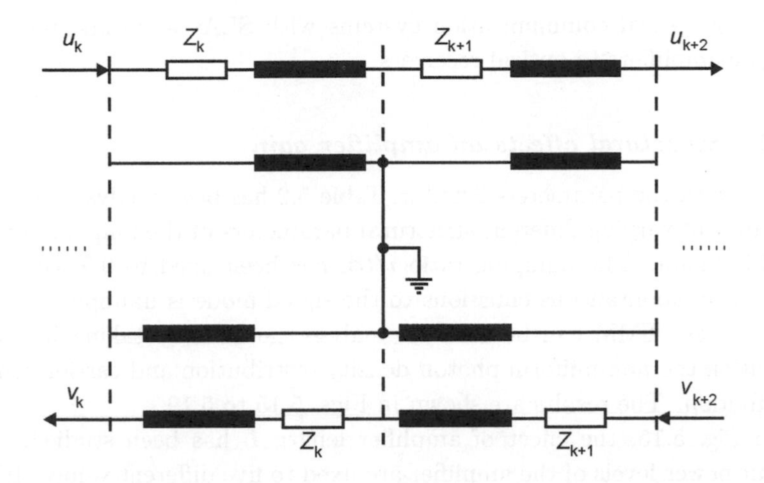

Fig. 5.14. Complete equivalent circuit model for semiconductor laser amplifier.

very small (see the discussion on the derivation of the modal gain coefficients with perturbation method in Chapter 4) and such reflections can be ignored in the TMM calculation. Therefore, in the present cascaded equivalent circuit of SLA, which is derived from the TMM analysis, the reflections because of similar mismatches between Z_0 and Z_k will also be ignored. In practice, this can be achieved by connecting stubs at the interface between kth and $(k+1)$th sub-network, such that matching between each stage occurs and no reflections are observed except at the terminating loads which model the end-facets [46].

There are a number of advantages to this equivalent circuit model. Firstly, the model can be treated as an electrical circuit, as in many microwave engineering problems. This allows us to analyse more complicated structures for SLAs. Secondly, this equivalent circuit actually allows one to include spontaneous emissions into the analysis, which has been neglected so far in the TMM analysis. This second advantage will be explored in depth in the next chapter.

5.6 Applications

In this section, the techniques developed in this chapter for analysis of SLAs will be used to study two particular aspects: (i) the effects of structural parameters on the amplifier gain and (ii) some basic considerations in designing optical communication systems with SLAs as in-line repeaters and pre-amplifiers to optical receivers.

5.6.1 *Structural effects on amplifier gain*

A SLA with the parameters listed in Table 5.2 has been analysed to study the effect of varying different structural parameters of the amplifier on the amplifier gain. The pumping ratio $i/t\text{th}$ has been fixed to 0.9 where the coupling of spontaneous emissions to the signal mode is unimportant [23]. In this case, TMM can be used to analyse the SLA by taking into consideration the non-uniform photon density distribution and carrier density distribution. The results are shown in Figs. 5.15 to 5.19.

In Fig. 5.15, the effect of amplifier length L has been studied. The output power levels of the amplifier are fixed to five different values. It can be seen that for low output power, the amplifier gain G can be maximised

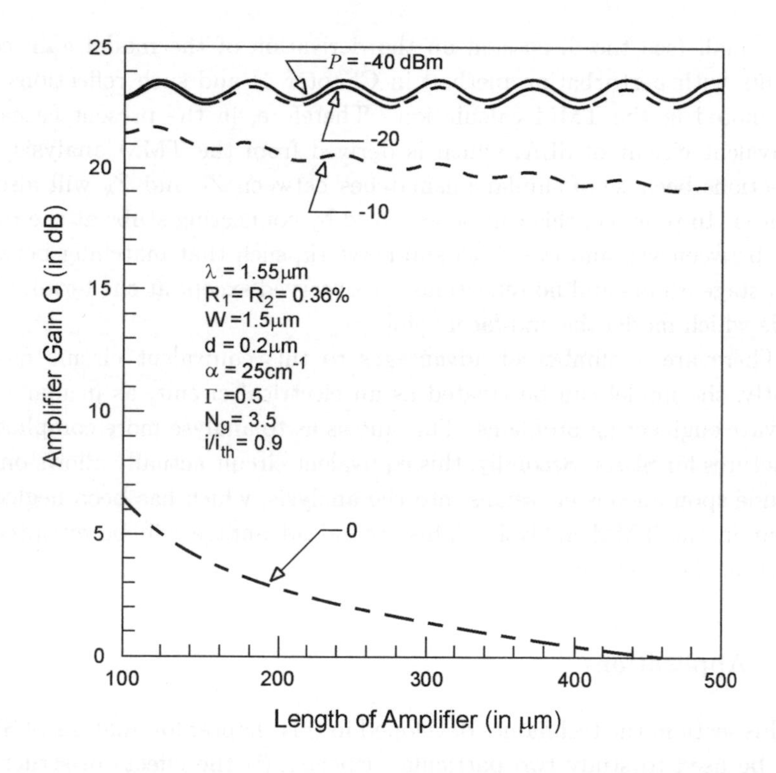

Fig. 5.15. Effects of amplifier length on the gain for a SLA analysed by TMM.

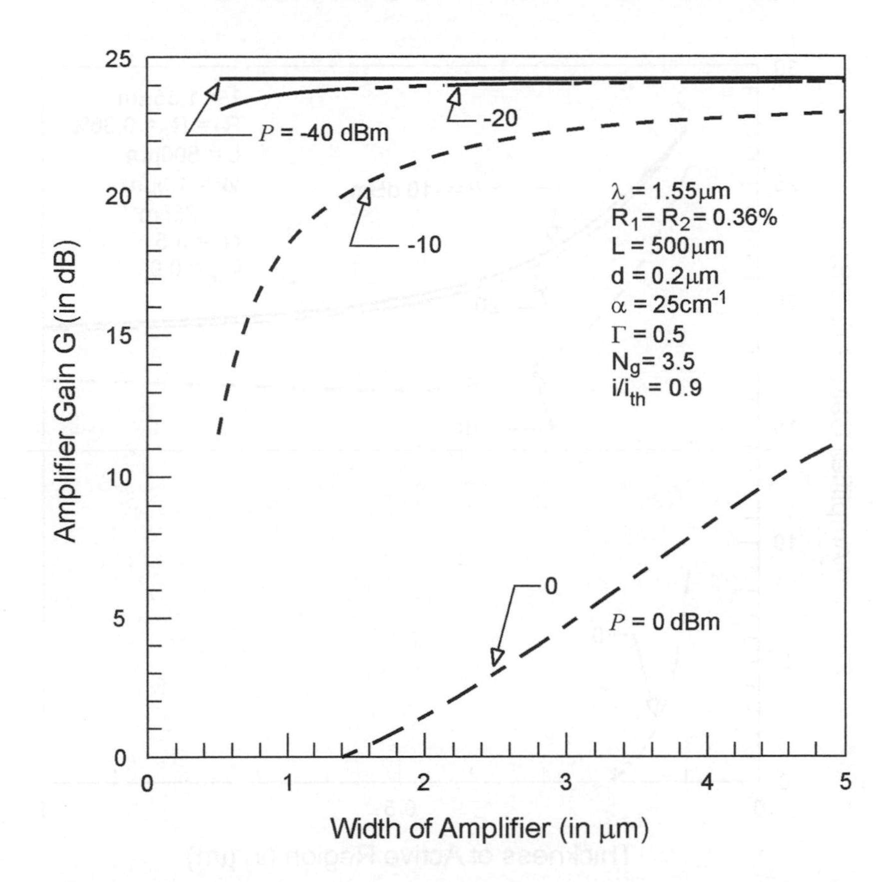

Fig. 5.16. Effect of the active layer width on the amplifier gain for a SLA analysed using TMM.

by adjusting the value of L such that the cavity resonances can match with that of signal wavelength. As the output power increases, the ripples observed in the curves for lower output power reduce and the value of G falls gradually with increasing L. This is because the amplifier starts to saturate for these values of output power. For $P = 0$ dBm, no ripples occur along the curve and the gain falls rapidly with increasing L as the amplifier is heavily saturated.

The effect of the width of the active layer W on the amplifier gain is shown on Fig. 5.16. For low output power levels, changes in W do not affect the amplifier gain significantly, although as W increases one expects that the value of threshold current simultaneously increases. As the output power increases, it can be seen that narrowing the active region will actually

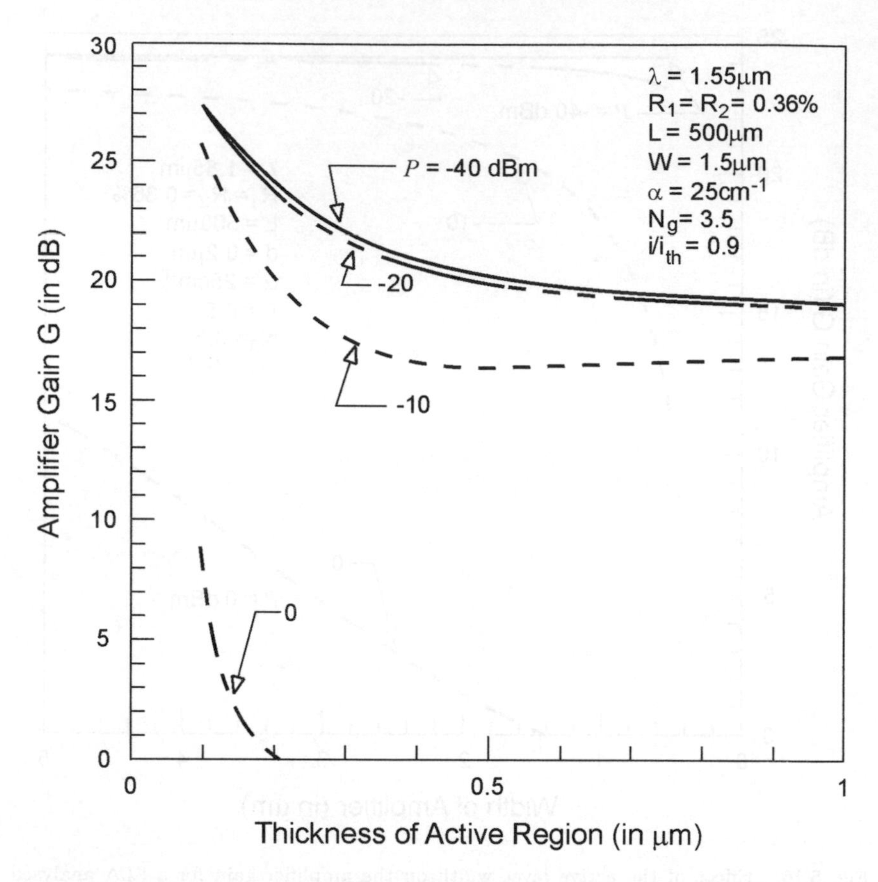

Fig. 5.17. Effect of the active layer thickness on the amplifier gain for a SLA analysed using TMM.

reduce the gain G. The reduction becomes more prominent as W is reduced beyond about 2 μm. Again for $P = 0$ dBm, where the amplifier should be saturated, the reduction in G with reducing W is very clear. In general, it can be concluded that W does not affect the unsaturated gain of a SLA significantly. This allows one to optimise the width of the active region with respect to the following three constraints: (i) the threshold current value for the SLA structure (which favours a large W), (ii) the saturation output power (which also favours a large W) [63] and (iii) the coupling efficiency [64].

In Fig. 5.17, the effect of active layer thickness d on the amplifier gain G is shown. In the calculations, the change in optical confinement factor Γ

with thickness d must be taken into account. We have used the values of Γ calculated for TE modes in Fig. 4.11 of Chapter 4 in the computation, assuming that the structure is under STM operation for all values of d. It can be seen that for all output power levels, the value of G rapidly reduces with d until around 0.35 μm, beyond which G stays fairly constant. Since a polarisation insensitive SLA will require a thicker active layer region, we can design the SLA in the region of $d > 0.3$ μm, for which the amplifier gain is fairly constant [24]. Ultimately, the upper limit on d is imposed by three constraints: (i) the value of threshold (which favours a thinner active region), (ii) the thermal resistance of the active region (which also favours a thinner active region) [63] and (iii) the coupling efficiency (which favours a thicker active region to match W in creating a more circular spot size to match that of the single-mode fibre).

Finally, the effects of facet reflectivities on the amplifier gain G are analysed and are shown in Figs. 5.18 and 5.19. In Fig. 5.18, the SLA under analysis is symmetrical (i.e. $R_1 = R_2$), whereas the results shown in Fig. 5.19 are for an asymmetric SLA. In general, it can be seen from these curves that G reduces with increasing R_1 and R_2. This is because the signal wavelength under analysis is not matched with that of the cavity resonance. Otherwise enhancement in G should be observed. Nevertheless, the amplifier becomes more and more sensitive to mismatch in frequencies, and will saturate much earlier than SLAs with lower reflectivities. Hence, it can be seen that reducing R_1 and R_2 by anti-reflection (AR) coatings is actually more beneficial than trying to match the cavity resonance with signal wavelength, as the saturation output power can also be improved by low-reflectivities.

5.6.2 *System considerations*

In practical optical fibre communication systems, the SLAs are usually employed as in-line non-regenerative repeaters and as pre-amplifiers to optical receivers [39]. In both applications, it is often necessary to design a system such that the optical power incident into the optical receiver will always be maintained above a certain minimum power level P_{min}, such that a minimum signal-to noise ratio can be maintained below a maximum bit error rate (BER) of B_e [65]. Consider such an optical communication system with M identical SLAs as in-line repeaters, each separated by a distance L_s (see Fig. 2.3 in Chapter 2). There is also an identical SLA used as a pre-amplifier to the optical receiver. Each SLA is biased at an

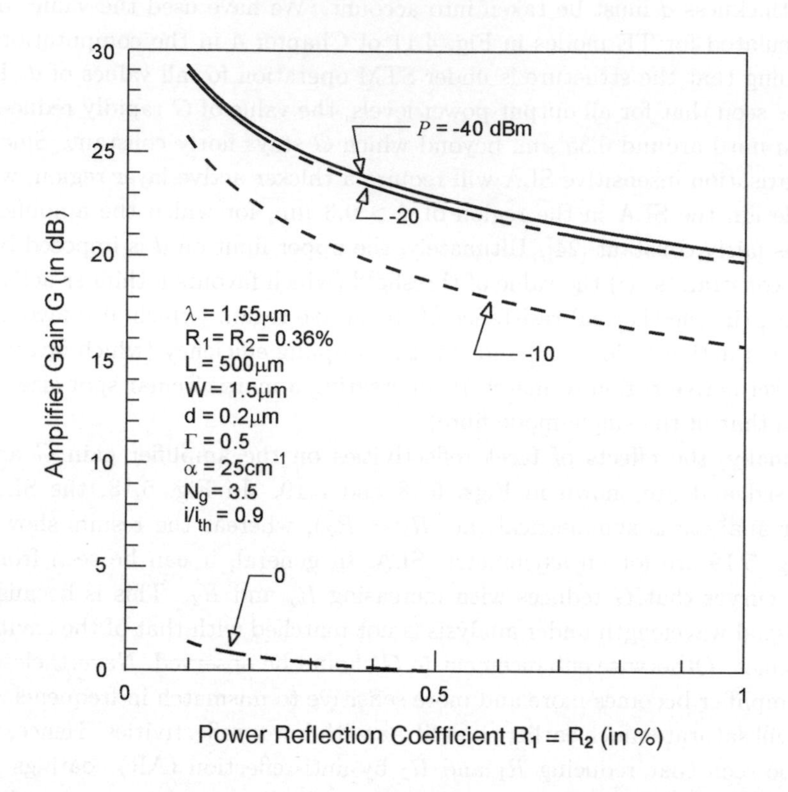

Fig. 5.18. Effect of facet reflection coefficient on the amplifier gain for a SLA analysed using TMM.

identical current i and is maintained at a constant net gain G (i.e. fibre-to-fibre gain). This can be achieved by controlling the temperature of the SLAs with feedback control by monitoring the output power from each SLA [66]. Assume the system employs intensity modulation (IM) to transmit the data. The SLAs will provide a net gain G at each stage to counteract the losses accumulated in the fibre of length L_s. If the loss of the fiber is α nep/km, then the input power at the output of the first amplifier is $[\eta_1 P_{\text{in}} \exp(-\alpha L_s)]G$ where P_{in} is the input optical power and η_1 is the coupling efficiency between the transmitter and the first fibre. Then, after M stages, the output optical power from the last section of fibre will be

$$P = \eta_1 P_{\text{in}} (G \exp(-\alpha L))^M . \tag{5.75}$$

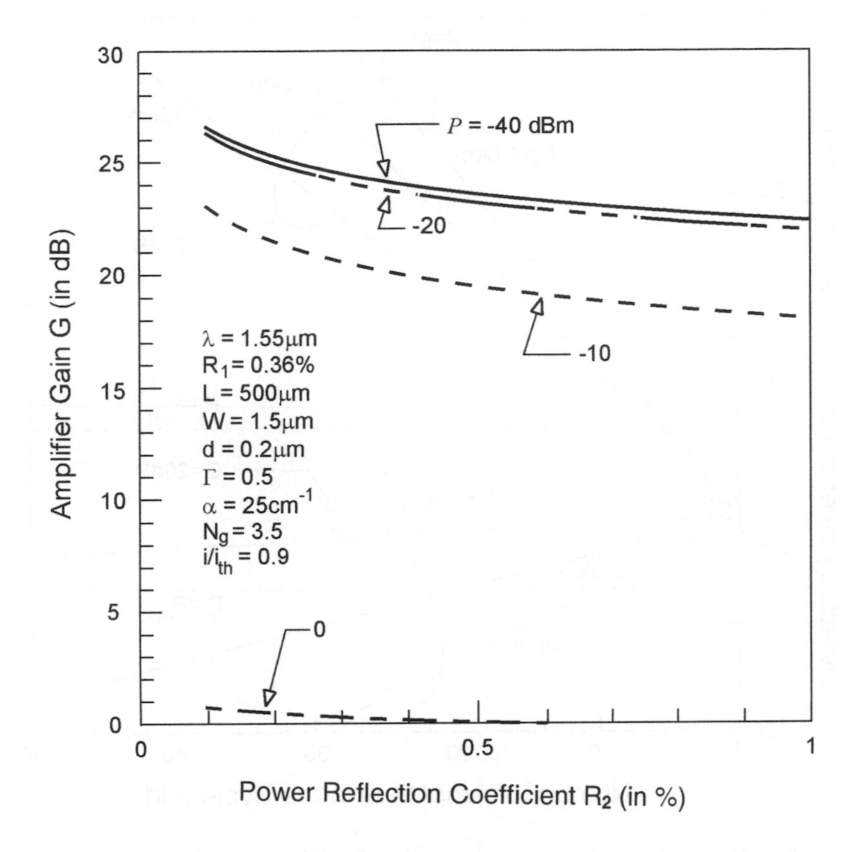

Fig. 5.19. As Fig. 5.18 but the SLA has asymmetric mirror facet configuration.

Thus the optical power incident onto the photodetector P_D is

$$P_D = \eta_2 P \tag{5.76}$$

where η_2 is the coupling efficiency between the pre-amplifier and the receiver. There are two major constraints on P_D in practical design. As we have discussed above, $P_D \geq P_{min}$ in order to maintain a minimum bit error rate in the receiver. On the other hand, P_D must be smaller than the saturation output power P_{sat} of the pre-amplifier to the receiver to avoid any saturation. Otherwise, G will becomes saturated and the overall gain is smaller than that depicted in Eq. (5.75). Such saturation will reduce P_D from the value of P_{min}, and the overall performance of the optical receiver will deteriorate.

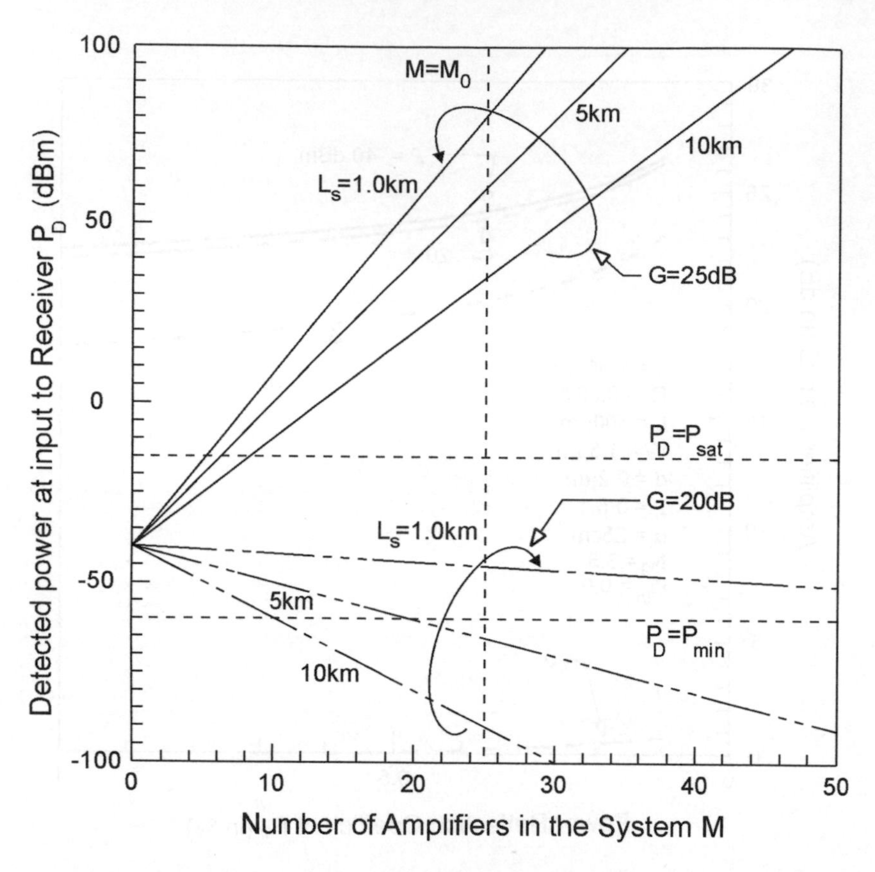

Fig. 5.20. Detected optical power at the receiver for different number of repeaters used in the system with different repeater spacing for $G = 20$ dB and 25 dB, respectively.

In Fig. 5.20, we have calculated the value of $P_{\rm D}$ against M for $L_{\rm s} = 1$ km, 5 km and 10 km, respectively, with $\eta_1 = \eta_2 = \eta_{\rm c} = -10$ dB. We have assumed that the unsaturated gain G is maintained at 20 dB and 25 dB by controlling the bias current. In some circumstances, the number of SLAs which can be used is also limited (because of, for example, cost). A third constraint that can then be added is $M \leq M_0$, as illustrated in Fig. 5.20. The bounded region indicates the possible number of amplifiers M used for different $L_{\rm s}$ at a particular value of $P_{\rm in}$, which satisfies the constraints of $P_{\rm sat} \geq P_{\rm D} \geq P_{\rm min}$ and $M \leq M_0$ (see Fig. 5.19). When $G = 20$ dB, it can be seen that the overall transmission span is lossy and no power gain is achieved at the input to the receiver. However, it thus increases the transmission span from that when $G = 0$. It should be noted also that when the system is under such circumstances, the total transmission span ML_s remains constant for all combinations of M and L_s. This is not true when $G = 25$ dB, in which there is net power gain at the input to the optical receiver. In this case, the system designer must be careful not to saturate the amplifier chain [67]. Furthermore, the total transmission span for this case is less than that in the former case when $G = 20$ dB. Thus, it appears to be desirable to operate the present system with a lower gain value. It should be mentioned, however, that there are circumstances in which it may be desirable to operate the amplifiers under saturation. This will be the case when the output power at each stage has to be maintained at a constant level. An examination of the gain, G, against the input power, P_{in}, curves calculated in the previous section reveals that when a SLA is operated under saturation condition, any sudden fall in P_{in} will actually increase the value of gain value G, hence increasing the output power P_{out}. If the design is appropriate, then a constant value of output power P_{out} can be determined in such systems.

It is apparent that the considerations which have to be taken into account in designing systems with SLAs are more complicated than may be expected by common sense, e.g. one may expect a high value of G to be beneficial to the system. In fact, examination of Fig. 5.19 indicates that different aspects of the performance of both the system and the amplifiers must be examined thoroughly and carefully.

5.7 Analysis of Gain Saturation in a SLA with a Uniform Material Gain Profile

In Section 5.2.2, an analytical formula for the gain saturation in a SLA was introduced. In this section, the detailed analysis of the formulae will be outlined [26, 30]. We start our formulation with the rate equations, which are described in Subsections 5.3.1 and 6.2.2 of Chapter 5 and Chapter 6, respectively [13]

$$\frac{dS_m}{dt} = -\frac{S_m}{\tau_p} + \Gamma G_m S_m + \Gamma E_{cf,m} + S_{in,m} \tag{5.77}$$

$$\frac{dn}{dt} = P - R - \sum_m G_m \Gamma S_m \,. \tag{5.78}$$

The definitions of the symbols have already been described in Section 5.3.1. The gain coefficient g for the number of photons can be considered as composed of two parts: an unsaturated part g_0 which is independent of photon density and a saturating part which depends on the photon density S_m. Mathematically, this can be described by;

$$g = g_0 - \frac{d\Delta g(S_m)}{dS_m} \cdot S_m \tag{5.79}$$

An inspection of Eq. (5.78) reveals that the solution of the carrier density will also consist of both an unsaturated part which is independent of the photon density and another part, which will depend on the photon density. We can attribute the unsaturated gain to the unsaturated photon density part of the carrier density. At steady state, this unsaturated part of n will be given by $P\tau_n$ if the carrier recombination rate is given by $R = n/\tau_n$ [68]. Then it can be seen that the gain in the photon number due to the amplifier is given by

$$g_0 = \frac{A\Gamma}{2}(P\tau_n - n_0) \tag{5.80}$$

which is measured in s^{-1}. If the amplifier is pumped towards threshold, the corresponding value of the threshold gain in photon number for lasing is then given by:

$$g_{0\,\mathrm{th}} = \frac{A\Gamma}{2}(P_{\mathrm{th}}\tau_n - n_0) = \frac{1}{2\tau_{\mathrm{p}}} \tag{5.81}$$

where the subscript "th" represents quantities at the threshold and τ_{p} is the lifetime of photons travelling in half of the round trip in the cavity, which are lost via the mirror facets [30]. Combining Eqs. (5.80) and (5.81) results in

$$g_0 = \frac{1}{2\tau_{\mathrm{p}}}[(1 + A\Gamma\tau_{\mathrm{p}}n_0)(P/P_{\mathrm{th}} - 1) + 1] = \frac{n_{\mathrm{sp}}P_{\mathrm{R}} + 1}{2\tau_{\mathrm{p}}} \tag{5.82}$$

where $P_{\mathrm{R}} = (P/P_{\mathrm{th}} + 1)$ is the relative pumping rate supplied to the laser, and $n_{\mathrm{sp}} = (1 + A\Gamma\tau_{\mathrm{p}}n_0)$ is the population inversion parameter. If the radiative recombinations due to spontaneous emissions are the dominating recombination mechanism, then the photon density can be expressed in terms of pumping rates below and at lasing threshold by [30]

$$S_m = (P - P_{\mathrm{th}})\tau_{\mathrm{p}} = \frac{P_{\mathrm{R}}N_{\mathrm{g}}n_{\mathrm{sp}}}{Ac\tau_{\mathrm{sp}}}\,. \tag{5.83}$$

If K is the total rate of loss of photon numbers in the laser cavity, then by using Eq. (5.79) the following inequality can be obtained if the optical amplification is to be maintained

$$\frac{n_{\text{sp}} P_{\text{R}} + 1}{2\tau_{\text{p}}} - \frac{d\Delta g}{dS_m} \cdot \frac{P_{\text{R}} N_{\text{g}} n_{\text{sp}}}{Ac\tau_{\text{sp}}} \leq K \,. \tag{5.84}$$

The amplifier gain saturates when it starts to lase, which occurs when the equality of the above equation holds. In this case, K is equal to the threshold gain value given in Eq. (5.81). Substituting this into the above equation and re-arranging gives the saturation parameter [30]

$$\frac{d\Delta g}{dS_m} = \frac{Ac\tau_{\text{sp}}}{2N_{\text{g}} V_0 \tau_{\text{p}}} \,. \tag{5.85}$$

If we substitute Eqs. (5.82), (5.83) and (5.85) into Eq. (5.79), it can be seen that the gain in photon number is halved when the intensity of the optical beam in the amplifier becomes I_{sat}, which is given by [26]

$$I_{\text{sat}} = \frac{hf}{A\tau_{\text{sp}}} \,. \tag{5.86}$$

When the gain in photon number is halved, it follows that the gain in the power of the optical field is also halved (i.e. dropped by 3 dB) as optical power is proportional to hf. The above expression has an interesting implication on the equivalent circuit representation of spontaneous emissions derived in Chapter 6. It can be shown that the spontaneous emission factor β can be derived from Eqs. (5.77) and (5.78) as [30]

$$\beta = \frac{A\Gamma c\tau_{\text{sp}}}{V_{\text{e}} N_{\text{g}}} = \frac{Ac\tau_{\text{sp}}}{V_0 N_{\text{g}}} \,. \tag{5.87}$$

If we substitute the expressions for the material gain coefficient $g_m = A(n - n_0)$ and the population inversion parameter $n_{\text{sp}} = n/(n - n_0)$ into the single-sided power spectral density of the quantum mechanical fluctuations which are used to model spontaneous emissions in Chapter 6, we can obtain the following expression using Eq. (5.87)

$$S_{\text{e}} = hf\Gamma g_m n_{\text{sp}} \Delta z = hf\Gamma An\Delta z = hf\beta R_{\text{sp}}(N_{\text{g}}/c)V_0\Delta z \tag{5.88}$$

where $R_{\text{sp}} = n/\tau_{\text{sp}}$. If we measure the power of spontaneous emission coupled to the lasing mode, it will have a bandwidth $\Delta\omega$ equal to the spectral linewidth below threshold of the laser [59]. It can be seen from the above equation that by proper adjustment of the length of each cavity Δz (i.e. the Q-factor of each individual section in the amplifier cavity), the relationship between the photon statistical interpretation of the rate equation in terms of β and R_{sp} with the spectral behaviour at the output of the amplifier (which can be measured readily) can be established correctly without the necessity of refining the definitions of terms in the rate equations (as in the critical analysis performed by Sommers [28].

5.8 Summary

In this chapter, the analysis of the signal gain of a semiconductor laser amplifier and the corresponding saturation output power have been described. Simple and basic analysis by assuming a uniform material gain profile along the amplifier has been outlined. Two different approaches to this type of analysis have been compared. The active FP formulation was found to be more suitable than the photon statistics formulation in analysing SLAs, because the phase information of the signals will be lost in the latter approach. To analyse the characteristics of SLAs in a more systematic manner, two different approaches have been reviewed. The rate equations formulation is the basic method as it describes mathematically the interaction between photons and carriers in the amplifier. However, to take into account the actual non-uniform material gain profile in the amplifier, travelling-wave equations have to be added to the basic rate equations, solutions of which will require iterations and a check for self-consistency at every point along z in the amplifier. The averaged photon density (AVPD) approximation introduced by Adam *et al.* [23] has been described, which, by using the average photon density in the rate equations after solving the travelling wave equations alone, reduces the number of equations to be solved to three instead of three at every point along z. The limitation of this approximation depends on the accuracy of using the averaged photon density instead of the actual photon density. In travelling-wave amplifiers, which have very low facet reflectivities, this approximation may not be very accurate. Moreover, the radiative carrier recombination rate due to spontaneous emissions R_{sp} has to be assumed to be independent of z. In practice, since the material gain g_m profile is not z-independent, the actual carrier density n will also be z-dependent and hence R_{sp} should also be spatial dependent. To overcome these shortcomings, an alternative solution technique using transfer matrices has been described. This allows R_{sp}, n and g_m to be z-dependent, and the field profile along the amplifier can also be solved. However, it should be noted that this technique only works for index guided SLAs for which effective indices and transverse modal fields can be calculated. Also, the carrier density n is assumed to be independent of x and y, as with all the other methods described above. Despite such assumptions, the technique gives an accurate result with a relatively simple algorithm which will always converge. It has been compared with the results calculated by AVPD approximation, and reasonable agreement has been obtained for moderate values of injection current. In this chapter, the effect of spontaneous emission on the signal has been ignored. This will be discussed in the next chapter.

Using the transfer matrix method to analyse SLAs also allows one to derive an equivalent circuit model for them. In this model, the active region of the amplifier is modelled as a cascaded network of negative resistance and lossless transmission line. This model will be extremely useful for analysis and discussion in the next chapter. Finally, the analysis of SLAs has been applied to analyse structural effects on gain characteristics of SLAs. These analyses can be used with system design constraints to derive an optimum structure for system applications. The saturation output power was shown to be extremely important in deciding the maximum number of repeaters and the spacing between them in optical communication systems. The proper use of SLAs in optical communication systems can increase the optical power incident on to an optical receiver and hence improve its sensitivity. However, just considering the signal power is not sufficient as the performance of the receiver is determined ultimately by the signal-to-noise ratio at the input of the receiver [66, 68]. In the next chapter, the noise characteristics of SLAs will be considered, so that a more complete picture can be drawn.

References

[1] G. H. B. Thompson, *Physics of semiconductor laser devices* (New York, John Wiley and Sons, 1980).

[2] J. Buus and R. Plastow, "A theoretical and experimental investigation of Fabry–Perot semiconductor laser amplifiers," *IEEE J. Quantum Electronics*, Vol. QE-21, No. 6, pp. 614–618, 1985.

[3] D. Marcuse, "Computer model of an injection laser amplifier," *IEEE J. Quantum Electronics*, Vol. QE-19, No. 1, pp. 63–73, 1983.

[4] C. Y. J Chu and H. Ghafouri-Shiraz, "Equivalent Circuit Theory of Spontaneous Emission Power in Semiconductor Laser Optical Amplifiers," *IEEE, J. Lightwave Technology*, Vol. LT-12, 1994, in press.

[5] G. Bjork and O. Nilsson, "A new exact and efficient numerical matrix theory

of complicated laser structures: properties of asymmetric phase-shifted DFB lasers," *J. Lightwave Technolog.*, Vol. LT-5, No. 1, pp. 140–146, 1987.

[6] C. Jϕrgensen *et al.*, "Two-section semiconductor optical amplifier used as an efficient channel dropping node," *IEEE Photo. Tech. Lett.*, Vol. 4, No. 4, pp. 348–350, 1992.

[7] A. Yariv, *Optical Electronics*, 3rd Edition (Holt-Saunders, 1985).

[8] M. J. Adams, *An Introduction to Optical Waveguides* (John Wiley and Sons, 1981).

[9] H. A. Haus, *Wave and fields in optoelectronics* (Prentice-Hall, 1984).

[10] C. A. Balanis, *Advanced engineering electromagnetics* (John Wiley and Sons, 1989).

[11] Y. Yamamoto, "Characteristics of AlGaAs Fabry–Perot cavity type laser amplifiers," *IEEE J. Quantum Electronics*, Vol. QE-16, No. 10, pp. 1047–1052, 1980.

[12] T. Mukai and Y. Yamamoto, "Gain, frequency bandwidth, and saturation output power of AlGaAs DH laser amplifiers," *IEEE J. Quantum Electronics*, Vol. QE-17, No. 6, pp. 1028–1034, 1981.

[13] Y. Mukai, Y. Yamamoto and T. Kimura, "Optical amplification by semiconductor lasers," *Semiconductor and semimetals*, Vol. 22, Part E, pp. 265–319, Academic Press, 1985.

[14] J. Wang, H. Olesen and K. E. Stubkjaer, "Recombination, gain and bandwidth characteristics of 1.3 μm semiconductor laser amplifiers," *IEEE J. Lightwave Technology*, Vol. LT-5, No. 1, pp. 184–189, 1987.

[15] T. Saitoh, T. Mukai and O. Mikami, "Theoretical analysis and fabrication of antireflection coatings on laser-diode facets," *J. Lightwave Technology*, Vol. LT-3, No. 2, pp. 288–293, 1985.

[16] T. Saitoh and T. Mukai, "Recent progress in semiconductor laser amplifiers," *I.EEE J. Lightwave Technology*, Vol. LT-6, No. 11, pp. 1156–1164, 1988.

[17] T. Saitoh and M. Mukai, "1.5 μm GaInAs travelling-wave semiconductor laser amplifier," *IEEE J. Quantum Electronics*, Vol. QE-23, No. 6, pp. 1010–1020, 1987.

[18] G. P. Agrawal and N. K. Dutta, *Long-wavelength semiconductor lasers* (New York, Van-Nostrad Reinhold, 1986).

[19] T. Mukai and Y. Yamamoto, "Noise in an AlGaAs semiconductor laser amplifier," *IEEE J. Quantum Electronics*, Vol. QE-18, No. 4, pp. 564–575, 1982.

[20] T. Mukai, Y. Yamamoto and T. Kimura, "S/N and error rate performance in AlGaAs semiconductor laser preamplifier and linear repeater systems," *IEEE Trans. Microwave Theory Tech.*, Vol. MTT-30, No. 10, pp. 1548–1556, 1982.

[21] R. Loudon, *The quantum theory of light*, 2nd Edition (Oxford University Press, 1983).

[22] D. T. Cassidy, "Comparison of rate-equation and Fabry–Perot approaches to modelling a diode laser," *Appl. Opt.*, Vol. 22, No. 21, pp. 3321–3326, 1983.

[23] M. J. Adams, "Time dependent analysis of active and passive optical bistability in semiconductors," *IEE Proc.*, Part J, Vol. 132, No. 6, pp. 343–348, 1985.

[24] T. Saitoh and T. Mukai, "Structural design for polarization-insensitive travelling-wave semiconductor laser amplifiers," *Opt. Quantum Electron.*, Vol. 21, pp. S47–S48, 1989.
[25] P. J. Stevens and T. Mukai, "Predicted performance of quantum-well GaAs-(GaAl)As optical amplifiers," *IEEE J. Quantum Electronics*, Vol. 26, No. 11, pp. 1910–1917, 1990.
[26] Y. Yamamoto, S. Saito and T. Mukai, "AM and FM quantum noise in semiconductor lasers-Part II: Comparison of theoretical and experimental results for AlGaAs lasers," *IEEE J. Quantum Electronics*, Vol. QE-19, No. 1, pp. 47–58, 1983.
[27] D. Marcuse, "Classical derivation of the laser rate equation," *IEEE J. Quantum Electronics*, Vol. QE-19, No. 8, pp. 1228–1231, 1983.
[28] I. D. Henning, M. J. Adams and J. V. Collins, "Performance prediction from a new optical amplifier model," *IEEE J. Quantum Electronics*, Vol. QE-21, No. 6, pp. 609-613; "amplifier model," *IEEE J. Quantum Electronics*, Vol. QE-21, No. 6, pp. 609-613.
[29] H. S. Sommers Jr., "Critical analysis and correction of the rate equation for the injection laser," *IEE Proc.*, Part J, Vol. 132, No. 1, pp. 38–41, 1985.
[30] Y. Yamamoto, "AM and FM quantum noise in semiconductor lasers-Part I: Theoretical analysis," *IEEE J. Quantum Electronics*, Vol. QE-19, No. 1, pp. 34–46, 1983.
[31] T. Mukai, K. Inoue and T. Saitoh, "Homogeneous gain saturation in 1.5 μm InGaAsP travelling-wave semiconductor laser amplifiers," *Appl. Phys. Lett.*, Vol. 51, No. 6, pp. 381–383, 1987.
[32] J. Buss, "Principles of semiconductor laser modelling," *IEE Proc.*, Part J, Vol. 132, No. 1, pp. 42–51, 1985.
[33] H. Ghafouri-Shiraz and C. Y. J. Chu, "Effect of phase shift position on spectral linewidth of the ?/2 distributed feedback laser diode," *J. Lightwave Technology*, Vol. LT-8, No. 7, pp. 1033–1038, 1990.
[34] M. I. Sargent, M. O Scully and W. E. Lamb Jr, *Laser physics* (Addison-Wesley, 1974).
[35] A. J. Lowery, "Modelling spectral effects of dynamic saturation in semiconductor laser amplifiers using the transmission-line laser model," *IEE Proc.*, Part J, Vol. 136, No. 6, pp. 320–324, 1989.
[36] J. Buss, "The effective index method and its application to semiconductor lasers," *IEEE J. Quantum Electronics*, Vol. QE–18, No. 7, pp. 1083–1089, 1982.
[37] M. J. Adam, J. V. Collins and I. D. Henning, "Analysis of semiconductor laser optical amplifiers," *IEE Proc.*, Part J, Vol. 132, pp. 58–63, 1985.
[38] T. Makino and T. Glinski, "Transfer matrix analysis of the amplified spontaneous emission of DFB semiconductor laser amplifiers," *IEEE J. Quantum Electronics*, Vol. QE-24, No. 8, pp. 1507–1518, 1988.
[39] G. Eisenstein, "Semiconductor optical amplifiers," *IEEE Circuits and Devices Magazine*, pp. 25–30, July 1989.
[40] S. Banerjee and A. Sharma, "Propagation characteristics of optical waveguiding structures by direct solution of the Helmholtz equation for total fields," *J. Opt. Soc. Am. A*, Vol. 6, No. 12, pp. 1884–1894, 1989.

[41] C. H. Henry and B. H. Verbeck, "Solution of the scalar wave equation for arbitrarily shaped dielectric waveguides by two-dimensional Fourier analysis," *IEEE J. Lightwave Technology*, Vol. LT-7, No. 32, pp. 308–313, 1989.
[42] P. B. Johns and R. L. Beurle, "Numerical solution of 2-dimensional scattering problems using a transmission-line matrix," *IEE Proc.*, Vol. 118, No. 9, pp. 1203–1208, 1971.
[43] W. J. R. Hoefer, "The transmission-line matrix method-theory and applications," *IEEE Trans. Microwave Theory Tech.*, Vol. MTT-33, No. 10, pp. 882–893, 1985.
[44] W. J. R. Hoefer, "The discrete time domain Green's function or Johns matrix-anew powerful concept in transmission line modelling (TLM)," *Int. J. Numerical Model*, Vol. 2, pp. 215–225, 1989.
[45] A. J. Lowery, "Transmission-line modelling of semiconductor lasers: the transmission-line laser model," *Int. J. Numerical Model*, Vol. 2, pp. 249–265, 1989.
[46] A. J. Lowery, "New in-line wideband dynamic semiconductor laser amplifier model," *IEE Proc.*, Part J. Vol. 135, No. 3, pp. 242–250, 1988.
[47] N. Wax, *Selected papers on noise and stochastic process* (Dover, 1954).
[48] A. J. Lowery, "Amplified spontaneous emission in semiconductor laser amplifiers: validity of the transmission-line laser model," *IEE Proc.*, Part J, Vol. 137, No. 4, pp. 241–247, 1990.
[49] G. Bjork and O. Nilsson, "A tool to calculate the linewidth of complicated semiconductor lasers," *IEEE J. Quantum Electronics*, Vol. QE-23, No. 8, pp. 1303–1313, 1987.
[50] I. Maio, "Gain saturation in travelling-wave ridge waveguide semiconductor laser amplifiers," *IEEE Photo. Tech. Lett.*, Vol. 3, No. 7, pp. 629–631, 1991.
[51] P. C. Kendall, M. J. Adam, S. Ritichie and M. J. Robertson, "Theory for calculating approximate values for the propagation constants of an optical rib waveguide by weighting the refractive indices," *IEE Proc.*, Part A, Vol. 134, No. 8, pp. 699–702, 1987.
[52] M. Cross and M. J. Adam, "Waveguiding properties of stripe-geometry double heterostructure injection lasers," *Solid State Electron.*, Vol. 15, pp. 919–921, 1972.
[53] H. A. Haus and S. Kawakami, "On the excess spontaneous emission factor in gain-guided laser amplifiers," *IEEE J. Quantum Electronics*, Vol. QE-21, No. 1, pp. 63–69, 1985.
[54] R. M. Fortenberry, A. J. Lowery and R. S. Tucker, "Up to 16 dB improvement in detected voltage using two section semiconductor optical amplifier detector," *Electronics Letters*, Vol. 28, No. 5, pp. 474–476, 1992.
[55] P. M. Morse and H. Feshbach, *Methods of theoretical physics: Part I* (MacGraw-Hill, 1953).
[56] S. G. Mikhlin and K. L. Smolitskiy, *Approximate methods for solution of differential and integral equations* (American Elsevier Publishing Co., 1967).
[57] C. Y. J. Chu and H. Ghafouri-Shiraz, "Structural effects on polarization

sensitivity of travelling-wave semiconductor laser amplifiers," *3rd IEE Bangor Symposium on Communication*, UK, pp. 19–22, 29–30 May 1991.

[58] C. D. Zaglanakis and A. J. Seeds, "Computer model for semiconductor laser amplifiers with RF intensity-modulated inputs," *IEE Proc.*, Vol. 139, Part J, No. 4, pp. 254–262, 1992.

[59] R. S. Tucker, "Circuit model of double-heterostructure laser below threshold," *IEE Proc.*, Part I, Vol. 128, No. 3, pp. 101–106, 1981.

[60] O. Nilsson, Y. Yamamoto and S. Machida, "Internal and external field fluctuations of a laser oscillator: Part II-Electrical circuit theory," *IEEE J. Quantum Electronics*, Vol. QE- 22, No. 10, pp. 2043–2051, 1986.

[61] C. H. Henry, "Theory of the phase noise and power spectrum of a single mode injection laser," *IEEE J. Quantum Electronics*, Vol. QE-19, No. 9, pp. 1391–1397, 1983.

[62] C. H. Henry, "Theory of spontaneous emission noise in open resonators and its application to lasers and optical amplifiers," *IEEE J. Lightwave Technology,* Vol. LT-4, No. 3, pp. 288–297, 1986.

[63] J. C. Simon, "GaInAsP semiconductor laser amplifiers for single-mode fiber communications," *J. Lightwave Technology*, Vol. LT-5, No. 9, pp. 1286–1295, 1987.

[64] J. John, T. S. M. Maclean, H. Ghafouri-Shiraz and J. Niblett, "Matching of single-mode fibre to laser diode by microlenses at 1.5 μm wavelength," *IEE Proc. Part J, Optoelectronics* Vol. 141, No. 3, pp. 178–184, June 1994.

[65] T. Li, *Topics in lightwave transmission system* (Academic Press, 1991).

[66] H. Kressel, *Semiconductor devices for optical communications*, 2nd Edition (Springer-Verlag, 1982).

[67] A. Elrefaire and C. Lin, "Performance degradations of multi gigabit-per-second NRZ/RZ lightwave system due to gain saturation in travelling-wave semiconductor optical amplifiers," *IEEE Photonics Technology Letters*, Vol. 1, No. 10, pp. 300–303, 1989.

[68] H. Taub and D. L. Schilling, *Principle of communication systems*, 2nd Edition (MacGraw-Hill, 1986).

[69] R. Olshansky, C. B. Su, J. Manning and W. Powazink, ' 'Measurement of radiative and nonradiative recombination rates in InGaAsP and AlGaAs light sources," *IEEE J. Quantum Electron.*, Vol. QE-20, No. 8, pp. 838–854, 1984.

[70] C. Y. J. Chu and H. Ghafouri-Shiraz, "Equivalent circuit theory of spontaneous emission power in semiconductor laser optical amplifiers," *IEEE J. Lightwave Technology*, Vol. LT-12, No. 5, pp. 760–767, May 1994.

[71] C. Y. J. Chu and H. Ghafouri-Shiraz, "Analysis of gain and saturation characteristics of a semiconductor laser optical amplifier using transfer matrices," *IEEE J. Lightwave Technology*, Vol. LT-12, No. 8, pp. 1378–1386, August 1994.

[72] C. Y. J. Chu and H. Ghafouri-Shiraz, "A simple method to determine carrier recombinations in a semiconductor laser optical amplifier," *IEEE Pholonic Technology Letter*, Vol. 5, No. 10, pp. 1182–1185, October 1993.

Chapter 6

Analysis and Modelling of Semiconductor Laser Diode Amplifiers: Noise Characteristics

6.1 Introduction

In Chapter 5, the analysis of gain and saturation characteristics of semiconductor laser amplifiers (SLAs) were discussed. The fairly complex interaction between injected carriers and photons (which include both signal and amplified spontaneous emissions) can be described by a set of rate equations and a travelling-wave equation. As discussed in Chapter 2, spontaneous emission is a random process which is statistically stationary [1, 2] and its ensemble average (i.e. power) can be measured [3]. In fact, spontaneous emission in optical devices is like shot noise in electronic circuits [4] and it is a source of noise in SLAs. In electronic amplifiers, the noise generated by the various components is measured as the output power in the absence of a signal input. As the noise processes are uncorrelated with the signal input, the noise power is directly additive to the power output due to the signal.

The above analogy between shot noise in electronic amplifiers and spontaneous emissions in SLAs reveals an additional point. That is, spontaneous emission will cause random fluctuations in both amplitude and phase of the signal and, as in electronic communication systems, this may gives rise to error in signal detection. Spontaneous emissions can influence the performance of SLAs in the following three ways [5]:

(i) Because spontaneous emission involves transition of carriers, its random nature induces fluctuations in the carrier density n.

(ii) The fluctuation in n induces fluctuations in the material gain coefficient g_m and hence affect both the amplitude and the phase of the signal, as well as the total number of photons.

(iii) The fluctuations in n also induces changes in refractive index N_g as predicted by the Kramers-Kronig relation, causing further phase changes. This is similar to the physical process behind linewidth broadening in semiconductor lasers [6, 7].

The above three aspects are fluctuations induced by spontaneous emissions via the carrier density n. In addition, the spontaneous emission photons can interact directly with the signal. As we shall see later, this interaction can be enhanced during the detection process, introducing a new type of noise in the detector known as beat noise [8].

It can be seen that the analysis of noise processes in SLAs is fairly complicated, due to the interaction between different material parameters with photons. This chapter will be devoted to such an analysis. Some conventional techniques for dealing with such problems are based on statistical methods, which will be described first. We will then utilise the equivalent circuit model developed in the previous chapter to analyse spontaneous emissions from a SLA. This approach, first reported here, allows one to investigate more complex structural effects on the spontaneous emissions from SLAs, e.g. the effects of stray reflections from ends of optical fibres. Finally, as in the previous chapter, the analysis developed here will be used to investigate its implications for the design of low-noise SLAs and the performance of optical communication systems.

6.2 Formulation of Noise in Semiconductor Laser Amplifiers

Although noise is one of the most important properties of SLAs, it is also the most complicated and difficult one to analyse. This is partly because of the mathematical complexity in describing the statistical processes which take place inside the amplifier (see Fig. 6.1). There are therefore several different approaches with different degrees of complexity in their formulation. Here we will discuss three of them.

6.2.1 *Photon statistics formulation*

Mukai and Yamamoto [8, 9] described the application of photon statistic master equations to calculate the noise properties of a SLA. This approach

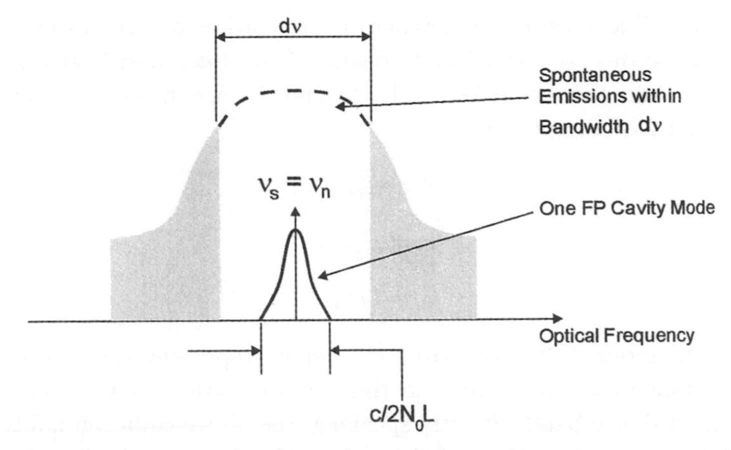

Fig. 6.1. Gain spectrum of a FPA and the corresponding spontaneous emission noise spectrum.

considers the statistical properties of the number of photons in a SLA. The statistical *momentum*, defined as the time derivative of the ensemble average of the rth power of the photon number, is evaluated by analogy with the statistics of population changes of different organisms in an ecological system due to birth and death. A simplified formulation, based on first and second order moments of the photon number N_{p} in a unit optical bandwidth has been given by [10]:

$$\frac{d\langle N_{\mathrm{p}}\rangle}{dt} = (A - B - C)\langle N_{\mathrm{p}}\rangle + A \tag{6.1}$$

$$\frac{d\langle N_{\mathrm{p}}^2\rangle}{dt} = 2(A - B - C)\langle N_{\mathrm{p}}^2\rangle + (3A + B + C)\langle N_{\mathrm{p}}\rangle + A \tag{6.2}$$

where A is the stimulated emission probability per second, B is the absorption probability per second which both describe the change in photon numbers due to these processes [11]. C will account for other loss mechanisms (e.g. scattering into claddings, intervalence band absorption [12]), which gives rise to a finite photon lifetime τ_{p}. In Eqs. (6.1) and (6.2), the total number of photons N_{p} due to coupling to output and input have been included in these equations, and hence the only possibility of change in the photon density is due to "birth" of photons due to radiative transitions and "death" of photons due to absorption and scattering (cf. compare this picture with an ecological system in which boundaries are drawn in a way that there is no net migration into and out of the boundaries, and hence the only possible reasons for change in population of organisms in the system will be due to "birth" and "death" [13]. Mukai and Yamamoto [8] postulated that the coefficients A, B and C are given by (see also Section 5.2.1.2 of Chapter 5)

$$A = E_{cv,m} \tag{6.3a}$$

$$B = E_{vc,m} \tag{6.3b}$$

$$C = 1/\tau_{\mathrm{p}} \tag{6.3c}$$

for the mth mode. The subscripts cv and vc represent the direction of radiative transitions, with c representing the conduction band, and v representing the valence band. Strictly speaking, the above equation holds only for the photon number $N_{\mathrm{p},m}$ of the mth mode. In general, the rth order momentum of $N_{\mathrm{p},m}$ is given by [8]:

$$\begin{aligned}\frac{d\langle N^r_{\mathrm{p},m}\rangle}{dt} = \sum_{j=0}^{r-1}\binom{r}{j}&[\langle(N_{\mathrm{p},m}+1)N^j_{\mathrm{p},m}\rangle E_{cv,m}\\ &+(-1)^{r+j}\langle N^{j+1}_{\mathrm{p},m}\rangle E_{vc,m} + (-1)^{r+j}\langle N^{j+1}_{\mathrm{p},m}\rangle/\tau_{\mathrm{p}}]\,.\end{aligned} \tag{6.4}$$

Physically, the photon statistics master equations (6.1) and (6.3) explain how the mean and variance of the photon population of the amplifier are affected by radiative transitions. The interaction between stimulated emissions, spontaneous emissions and absorption will result in a net optical gain, as well as fluctuations in photon number, as indicated by the finite value of the variance $\langle N^2_{\mathrm{p},m}\rangle$. Notice from Eqs. (6.3) and (6.4) that the population inversion parameter n_{sp} can be defined in terms of the coefficients A and B by [14]

$$n_{\mathrm{sp}} = \frac{A}{A-B}\,. \tag{6.5}$$

We can also derive the amplifier gain G in terms of the coefficients A, B and C [10]. Consider the number of photons $N_{\mathrm{p}}(z)$ inside the amplifier. The change in photon population for the *signal* along a distance dz is given by:

$$dN_{\mathrm{p}} = N_{\mathrm{g}}N_{\mathrm{p}}(A-B-C)(dz/c) \tag{6.6}$$

where N_{g} is the group effective refractive index of the medium and c is the velocity of light in free space. In Eq. (6.6) the quantity $(A-B-C)N_{\mathrm{p}}$ represents the net change in photon number due to stimulated emissions, absorption, and other loss mechanisms. The term dz/c represents the time elapsed as the photons travel from z to $z + dz$, i.e. the frame of reference has been transformed from a stationary one with respect to the SLA to one which is moving along the amplifier with the photons at a group velocity of $c/N_{\rm g}$. Integrating over a length L and imposing the boundary conditions of the initial number of signal photons, the gain can be derived from Eqs. (6.6) using the definition discussed in Section 5.2.1.2 as:

$$G = \frac{N_{\rm p}(t = LN_{\rm g}/c)}{N_{\rm p}(t = 0)} = \exp\left[N_{\rm g}(A - B - C)\frac{L}{c}\right]. \tag{6.7}$$

Using Eq. (6.7) we can solve the first order differential equation (6.1) to obtain the mean photon number as:

$$\langle N_{\rm p}\rangle = G\langle N_{\rm p0}\rangle + \frac{(G-1)A}{A - B - C} \tag{6.8}$$

where we have used $N_{\rm p0}$ to represent $N_{\rm p}(t = 0)$. With the assumption of $C \ll A$ and B, the term $A/(A - B - C)$ is approximately equal to $n_{\rm sp}$ (see Eq. (6.5)) and hence Eq. (6.8) can be written as:

$$\langle N_{\rm p}\rangle = G\langle N_{\rm p0}\rangle + (G-1)n_{\rm sp}\,. \tag{6.9}$$

Using a similar technique, the variance of the photon number at the output of the amplifier can be found by using Eqs. (6.2) and (6.9):

$$\begin{aligned}\langle N_{\rm p}^2\rangle &= G\langle N_{\rm p0}\rangle + (G-1)n_{\rm sp} + 2G(G-1)n_{\rm sp}\langle N_{\rm p0}\rangle \\ &\quad + (G-1)^2 n_{\rm sp}^2 + G^2[\langle N_{\rm p0}^2\rangle - \langle N_{\rm p0}\rangle^2 - \langle N_{\rm p0}\rangle]\,.\end{aligned} \tag{6.10}$$

In SLAs there can be several possible transverse modes for each longitudinal cavity resonance frequency (see Eq. (5.1) in Section 5.2.1.1 of Chapter 5). If there are $m_{\rm t}$ transverse modes for each longitudinal cavity resonance of bandwidth Δf, then the equivalent noise bandwidth of the amplifier output is given by $m_{\rm t}\Delta f$ [3, 8, 10]. Then Eqs. (6.8) and (6.10) can be written as

$$\langle N_{\rm p}\rangle = G\langle N_{\rm p0}\rangle + (G-1)n_{\rm sp}m_{\rm t}\Delta f \tag{6.11}$$

$$\begin{aligned}\langle N_{\rm p}^2\rangle &= G\langle N_{\rm p0}\rangle + (G-1)n_{\rm sp}m_{\rm t}\Delta f + 2G(G-1)n_{\rm sp}\langle N_{\rm p0}\rangle \\ &\quad + (G-1)^2 n_{\rm sp}^2 m_{\rm t}\Delta f + G^2[\langle N_{\rm p0}^2\rangle - \langle N_{\rm p0}\rangle^2 - \langle N_{\rm p0}\rangle]\,.\end{aligned} \tag{6.12}$$

In Eq. (6.11) the first term represents the average photon number due to the amplified input, whereas the second term represents that due to spontaneous emissions [8]. The five terms in Eq. (6.12) represent shot noise due to the amplified signal, shot noise due to spontaneous emissions, beat noise between signal and spontaneous emissions, beat noise between spontaneous emission components and excess noise due to incoherence of input signal when the output photons are detected by a photodetector, respectively. For a purely coherent input signal (i.e. without phase noise), the last term vanishes [10, 13, 15]. The relative noise power detected by a photodetector (PD) with a quantum efficiency of unity, unit load resistance and unit detection bandwidth can be found from the variance of the output photon number σ_{out}^2 [1, 8–16]:

$$\langle i_n^2 \rangle = e^2 \sigma_{\mathrm{out}}^2 = e^2[\langle N_{\mathrm{p}}^2 \rangle - \langle N_{\mathrm{p}} \rangle^2] \tag{6.13}$$

where e is the electron charge. The above formulation using photon statistics is general. We will illustrate briefly how to apply it to the analysis of a simple semiconductor Fabry–Perot amplifier (FPA). Analysis of more complex structures and fibre amplifiers using Eqs. (6.11) and (6.12) follow similar procedures to those outlined below.

Consider the gain spectrum of a FPA with a uniform material gain profile as illustrated in Fig. 6.1 [10]. For an optical frequency range of df, the number of spontaneous emission noise photons emitted by this FPA of length L per second is given by (see Section 6.5) [7, 17]:

$$\bar{N}_{\mathrm{p}}(f)df = \frac{(1 + R_1 G_{\mathrm{s}})(1 - R_2)(G_{\mathrm{s}} - 1)n_{\mathrm{sp}}}{|1 - G_{\mathrm{s}}\sqrt{R_1 R_2}\exp(-2\mathrm{j}\beta_z L)|^2} m_{\mathrm{t}} df \tag{6.14}$$

where $\bar{N}_{\mathrm{p}}(f)$ is the photon single-sided spectral density of photons emitted per second, G_{s} is the single pass gain of the amplifier and R_1 and R_2 are the power reflection coefficients of the facets. For one FP cavity resonance with m_{t} transverse modes, the number of spontaneous emission photons emitted per second is given by:

$$\int_{f_n - c/4L}^{f_n + c/4L} \bar{N}_{\mathrm{p}}(f)df = \frac{(1 + R_1 G_{\mathrm{s}}^n)(1 - R_2)(G_{\mathrm{s}}^n - 1)n_{\mathrm{sp}} m_{\mathrm{t}}}{1 - R_1 R_2 (G_{\mathrm{s}}^n)^2} \left(\frac{c}{2L}\right) \tag{6.15}$$

where c is the velocity of light in the amplifying medium, f_n is the nth resonance frequency, and G_{s}^n is the corresponding single pass gain of the amplifier at $f = f_n$ [18]. For a signal of frequency f_{s} matched to one of the cavity resonances, the gain of the signal is given by Eq. (5.8) (see Section 5.2.1.1 in Chapter 5):

$$G = \frac{(1 - R_1)(1 - R_2)G_{\mathrm{s}}^0}{(1 - G_{\mathrm{s}}^0\sqrt{R_1 R_2})^2} \tag{6.16}$$

where G_s^0 is the single pass gain at $f = f_s$. Then from Eq. (6.11), it can be shown that by using Eqs. (6.15) and (6.16) we can obtain [11]

$$\langle N_p \rangle = \frac{(1-R_1)(1-R_2)G_s^0}{(1-G_s^0\sqrt{R_1R_2})^2}\langle N_{p0}\rangle + \sum_n \frac{(1+R_1G_s^n)(1-R_2)(G_s^n-1)n_{sp}m_t}{1-R_1R_2(G_s^n)^2}\left(\frac{c}{2L}\right) \tag{6.17}$$

where the summation is taken over all longitudinal cavity resonances within Δv, assuming that the output of the SLA is filtered by an optical filter with a pass band bandwidth of Δv (Fig. 6.1) and $\langle N_{p0}\rangle$ is the average number of photons injected into the amplifier per second. By substituting Eqs. (6.15)–(6.17) into (6.12) and (6.13) and noticing the physical meanings of the terms in Eq. (6.12) it can be shown that

$$\langle N_p^2 \rangle - \langle N_p \rangle^2 = \langle N_p \rangle + I_1 + I_2 + \left[\frac{(1-R_1)(1-R_2)G_s^0}{(1-G_s^0\sqrt{R_1R_2})^2}\right]^2 \times (\langle N_{p0}^2\rangle - \langle N_{p0}\rangle^2 - \langle N_{p0}\rangle) \tag{6.18}$$

where $\langle N_p \rangle$ is given by Eq. (6.17). In the above derivations, we have assumed that the photodetector has low pass filter characteristics [19] such that the high frequency beating noises between the signals and that between spontaneous emission components are filtered off for the $\langle N_p \rangle^2$ term [8]. The term I_1 will involve beating between signal and spontaneous emissions [10]. Therefore, from Eq. (6.14), I_1 can be found by the following, integral:

$$\begin{aligned} I_1 &= 2G\langle N_{p0}\rangle \int_{f_n-c/4L}^{f_n+c/4L} \bar{N}_p(v)\delta(v-v_s)dv \\ &= \frac{(1-R_1)(1-R_2)^2(1+R_1G_s^0)(G_s^0-1)G_s^0 n_{sp}}{(1-G_s^0\sqrt{R_1R_2})^4}\langle N_{p0}\rangle \end{aligned} \tag{6.19}$$

where $\delta(v - v_s)$ is the Dirac delta function for $v = v_s$. In this case, $m_t = 1$ because the input signal is assumed to be in a single state of polarisation and hence the beating between the signal and spontaneous emissions can only take place along that particular transverse mode. The term I_2 can be found by considering the number of photons due to beating between spontaneous emission components within one longitudinal mode, and then summed over Δf:

$$I_2 = \sum_n \int_{f_n - c/4L}^{f_n + c/4L} \bar{N}_{\mathrm{p}}^2(v)dv\,. \tag{6.20}$$

Substituting Eq. (6.14) into the above expression, and after some algebraic manipulation, the integral can be evaluated analytically, which results in

$$I_2 = \sum_n \frac{(1+R_1G_{\mathrm{s}})^2(1-R_2)^2(G_{\mathrm{s}}-1)^2(1+R_1R_2G_{\mathrm{s}}^2)n_{\mathrm{sp}}^2 m_{\mathrm{t}}}{(1-R_1R_2G_{\mathrm{s}}^2)^3}\left(\frac{c}{2L}\right) \tag{6.21}$$

where we have dropped the superscript n on the single pass gain G_{s}^2, and hereafter, for simplicity, assuming that the dependence of G_{s} on the cavity resonance number n is implicit. The relative noise power detected at the output of the FPA is given by Eq. (6.13). By combining the above results in Eqs. (6.18) to (6.21) we can obtain the total noise power as [10]:

$$\begin{aligned}
\langle i_n^2\rangle/e^2 = \langle N_{\mathrm{p}}^2\rangle - \langle N_{\mathrm{p}}\rangle^2 &= \frac{(1-R_1)(1-R_2)G_{\mathrm{s}}^0}{(1-G_{\mathrm{s}}^0\sqrt{R_1R_2})^2}\langle N_{\mathrm{p0}}\rangle \\
&+ \sum_n \frac{(1+R_1G_{\mathrm{s}})(1-R_2)(G_{\mathrm{s}}-1)n_{\mathrm{sp}}m_{\mathrm{t}}}{1-R_1R_2G_{\mathrm{s}}^2}\left(\frac{c}{2L}\right) \\
&+ 2\frac{(1-R_1)(1-R_2)^2(1+R_1G_{\mathrm{s}}^0)(G_{\mathrm{s}}^0-1)G_{\mathrm{s}}^0 n_{\mathrm{sp}}}{(1-G_{\mathrm{s}}^0\sqrt{R_1R_2})^4}\langle N_{\mathrm{p0}}\rangle \\
&+ \sum \frac{(1+R_1G_{\mathrm{s}})^2(1-R_2)^2(G_{\mathrm{s}}-1)^2(1+R_1R_2G_{\mathrm{s}}^2)n_{\mathrm{sp}}^2 m_{\mathrm{t}}}{(1-R_1R_2G_{\mathrm{s}}^2)^3}\left(\frac{c}{2L}\right) \\
&+ \left[\frac{(1-R_1)(1-R_2)G_{\mathrm{s}}^0}{(1-G_{\mathrm{s}}^0\sqrt{R_1R_2})^2}\right]^2(\langle N_{\mathrm{p0}}^2\rangle - \langle N_{\mathrm{p0}}\rangle^2 - \langle N_{\mathrm{p0}}\rangle)\,.
\end{aligned} \tag{6.22}$$

It can be shown that by substituting Eqs. (6.11) and (6.12) into (6.13), the above equation can be written intuitively as:

$$\begin{aligned}
\langle i_n^2\rangle/e^2 &= G\langle N_{\mathrm{p0}}\rangle + (G-1)n_{\mathrm{sp}}m_{\mathrm{t}}\Delta f_1 + 2G(G-1)n_{\mathrm{sp}}\chi\langle N_{\mathrm{p0}}\rangle \\
&+ (G-1)^2 n_{\mathrm{sp}}^2 m_{\mathrm{t}}\Delta f_2 + G^2(\langle N_{\mathrm{p0}}^2\rangle - \langle N_{\mathrm{p0}}\rangle^2 - \langle N_{\mathrm{p0}}\rangle)
\end{aligned} \tag{6.23}$$

where Δf_1 is the equivalent noise bandwidth of spontaneous emission shot noise and Δf_2 is the equivalent noise bandwidth of the beat noise between spontaneous emission components. Both Δf_1 and Δf_2 can be found by comparing Eq. (6.22) with (6.23), that is [10]:

$$\Delta f_1 = \frac{\sum_n \frac{(1+R_1G_s)(1-R_2)(G_s-1)}{1-R_1R_2G_s^2}\left(\frac{c}{2L}\right)}{(G-1)} \tag{6.24}$$

$$\Delta f_2 = \frac{\sum_n \frac{(1+R_1G_s)^2(1-R_2)^2(G_s-1)^2}{(1-R_1R_2G_s^2)^2}\left(\frac{c}{2L}\right)}{(G-1)^2}\,. \tag{6.25}$$

It is interesting to see that the physical origins of Δf_1 and Δf_2 are due to the enhancement of noise by the multi-longitudinal cavity modes within the frequency band Δf. The resulting increase in spontaneous emission noise can be accounted for by a proportionate change in bandwidth, with the new net bandwidths given by Eqs. (6.24) and (6.25). The parameter χ in Eq. (6.23) is known as the excess noise coefficient [11, 20] which arises because of the enhancement of signal-spontaneous emissions beating noise by the FP cavity. It can be shown that:

$$\chi = \frac{(1-R_1)(1-R_2)(G_s^0-1)}{(1-G_s^0\sqrt{R_1R_2})^2(G-1)}\,. \tag{6.26}$$

Hence, for a travelling wave amplifier (TWA) with $R_1 = R_2 = 0$ and $G = G_s^0$ the value of $\chi = 1$. For a FPA with R_1 and $R_2 \neq 0$ the value of χ can be greater than unity, depending on the combinations of the facet reflectivities, material gain and cavity length [11]. It can also be seen that for a TWA, the corresponding noise power can be found simply by substituting $G = G_s^0$ and $\chi = 1$ into Eq. (6.23), with $\Delta f_1 = \Delta f_2 = \Delta f$. Obviously, the noise emitted from a TWA can be lower than that from a FPA [21]. The noise figure F is defined as the degradation of signal-to-noise ratio of the SLA at the output (see Section 2.3.4 in Chapter 2). As we shall see later, when G is large, the signal-spontaneous beating will dominate the overall detected noise power [11]. In this case, it can be shown that for $G \gg 1$

$$F = 2n_{sp}\chi\,. \tag{6.27}$$

For a perfect population inversion with $n_{sp} = 1$ and $\chi = 1$, we arrive at the minimum noise figure of 3 dB, which was derived in an alternative way in Section 2.5.4 of Chapter 2. The above approach of photon statistic master equations allows us to visualise the physical origins of the noise power detected by the receiver. It also allows us to separate different contributions to the total noise power by Eq. (6.12) in the analysis [11]. However, the analysis neglects the spatial dependence of spontaneous emissions and signal, as well as losing phase information by using the concept of photon counting [22–24]. Nevertheless, its relative simplicity is found to be very useful in analysing noise in SLAs, and subsequent improvements on this technique have been used. A recent application of this approach to noise analysis can be found in the work of Goldstein and Teich [13] who extended the preceding analysis to a general resonator configuration, which included coupling effects between amplifier output and PDs [25].

6.2.2 *Rate equation approach*

A more direct way to analyse noise in SLAs is to include fluctuations in the multimode rate equations used to analyse the semiconductor laser amplifier (see Section 5.3.1 in Chapter 5). Mukai and Yamamoto [8] used this approach by postulating that the carrier density n and photon number $N_{\mathrm{p},m}$ of the mth mode are related by the following equations:

$$\frac{dn}{dt} = P - R - \sum_{n} \Gamma G_m N_{\mathrm{p},m} + F_{\mathrm{e}}(t) \tag{6.28}$$

$$\frac{dN_{\mathrm{p},m}}{dt} = -\frac{N_{\mathrm{p},m}}{\tau_{\mathrm{p}}} + \Gamma G_m N_{\mathrm{p},m} + \Gamma E_{cv,m} + P_{\mathrm{in},m} + F_{\mathrm{p},m}(t) \tag{6.29}$$

where P is the pumping rate due to the injection electrical current, R is the carrier recombination rate, G_m is the stimulated emission coefficient given by $E_{cv,m} - E_{vc,m}$, with $E_{cv,m}$ being the stimulated emission probability coupled to the mth mode and $E_{vc,m}$ the corresponding induced absorption probability, τ_{p} is the photon lifetime, Γ is the optical confinement factor and $P_{\mathrm{in},m}$ is the rate of injection of signal photons of the mth mode. $F_{\mathrm{e}}(t)$ and $F_{\mathrm{p},m}(t)$ are the fluctuation operators [26] in the form of Langevin shot noise sources to account for fluctuations in n and $N_{\mathrm{p},m}$ caused by the random spontaneous emissions [7] and [17]. Note that the form of Eqs. (6.27) and (6.28) are similar to the rate equations described in Section 5.3.1 (see Chapter 5) except for the Langevin noise terms which account for fluctuations caused by spontaneous emissions [27]. The above equations can be solved by first linearising them as [8–28]:

$$n(t) = \hat{n} + \Delta n(t) \tag{6.30}$$

$$N_{\mathrm{p},m}(t) = \hat{N}_{\mathrm{p},m} + \Delta N_{\mathrm{p},m}(t)\,. \tag{6.31}$$

In the above equations, the ^ notation denotes the unperturbed steady state values for the parameters. Expressing R, G_m and $E_{cv,m}$ using these perturbed quantities with Taylor's theorem in Eqs. (6.28) and (6.29) [8]:

$$\frac{d}{dt}\Delta n(t) = \left(-\frac{\partial \hat{R}}{\partial \hat{n}}\sum_m \frac{\partial \hat{G}_m}{\partial \hat{n}} - \Gamma \hat{N}_{\mathrm{p},m}\right)\cdot \Delta n(t) - \sum_m \Gamma \hat{G}_m \Delta N_{\mathrm{p},m}(t) + F_{\mathrm{e}}(t) \tag{6.32}$$

$$\frac{d}{dt}\Delta N_{\mathrm{p},m}(t) = \Gamma\left(\hat{N}_{\mathrm{p},m}\frac{\partial \hat{G}_m}{\partial \hat{n}} + \frac{\partial \hat{E}_{cv,m}}{\partial \hat{n}}\right)\Delta n(t) + \left(\Gamma \hat{G}_m - \frac{1}{\tau_{\mathrm{p}}}\right)\Delta N_{\mathrm{p},m}(t) + F_{\mathrm{p},m}(t)\,. \tag{6.33}$$

The above equations can be solved by Fourier transform. To do this, the statistical properties of the Langevin noise sources have first to be known. This can be found by ensemble averaging Eqs. (6.28) and (6.29) in the time domain [1, 3, 17] such that [8]:

$$\langle F_{\mathrm{e}}(t)\rangle = \langle F_{\mathrm{p},m}(t)\rangle = 0 \tag{6.34}$$

$$\langle F_{\mathrm{e}}^2(\omega)\rangle = \Im\langle F_{\mathrm{e}}(t)\cdot F_{\mathrm{e}}(s)\rangle = P + \hat{R} + \sum_m [\Gamma \hat{E}_{cv,m}\hat{N}_{\mathrm{p},m} + \Gamma(\hat{E}_{cv,m} - \hat{G}_m)\hat{N}_{\mathrm{p},m}] \tag{6.35}$$

$$\langle F_{\mathrm{p},m}^2(\omega)\rangle = \Im\langle F_{\mathrm{p},m}(t)\cdot F_{\mathrm{p},m}(s)\rangle = \frac{\hat{N}_{\mathrm{p},m}}{\tau_{\mathrm{p}}} + \Gamma \hat{E}_{cv,m}(1+\hat{N}_{\mathrm{p},m}) + \Gamma(\hat{E}_{cv,m} - \hat{G}_m)\hat{N}_{\mathrm{p},m} + P_{\mathrm{in},m} \tag{6.36}$$

$$\langle F_{\mathrm{e}}(\omega)\cdot F_{\mathrm{p},m}(\omega)\rangle = \Im\langle F_{\mathrm{e}}(t)\cdot F_{\mathrm{p},m}(s)\rangle = -\Gamma[\hat{E}_{cv,m}(1+\hat{N}_{\mathrm{p},m}) + (\hat{E}_{cv,m} - \hat{G}_m)\hat{N}_{\mathrm{p},m}] \tag{6.37}$$

where $\Im$ represents the Fourier transform and $\langle AB\rangle$ represents the correlation between the two random quantities A and B [3]. Equations (6.35) and (6.36) describe the power spectrum of the Langevin noise sources $F_{\mathrm{e}}(t)$ and $F_{\mathrm{p},m}(t)$, respectively, whereas that in Eq. (6.37) represents the cross power spectrum between them. It can be seen from these equations that the postulated Langevin noise terms stem from the fluctuations in the carrier density and photon number due to spontaneous emissions (Fig. 6.1). Using Eqs. (6.34)–(6.37), Eqs. (6.32) and (6.33) can be solved in the frequency domain for $\langle \Delta n^2(\omega) \rangle$ and $\langle \Delta N_{\mathrm{p},m}^2(\omega) \rangle$ (only the mean-square values can be found because Δn and $\Delta N_{\mathrm{p},m}$ will fluctuate with time but are statistically stationary [1]). The net noise power detected by a photodetector is related to the total photon number fluctuation spectrum $\langle \Delta N_{\mathrm{p}}^2(\omega) \rangle$, which can be found to consist of two components, one due to self correlation of the photon number fluctuation in the mth mode $\langle \Delta N_{\mathrm{p},m}(\omega) \cdot \Delta N_{\mathrm{p},m}^*(\omega) \rangle$ and the other due to the cross correlation between kth and lth modes $\langle \Delta N_{\mathrm{p},k}(\omega) \cdot \Delta N_{\mathrm{p},l}^*(\omega) \rangle$ [8]. This can be found by solving Eqs. (6.32) and (6.33) for each mode m and summed over all the possible modes:

$$\langle \Delta N_{\mathrm{p}}^2 \rangle = \mathrm{Re}\left[\sum_k \sum_l \langle \Delta N_{\mathrm{p},k}(\omega) \cdot \Delta N_{\mathrm{p},l}^*(\omega) \rangle \right] \tag{6.38}$$

taking both components into consideration. The resulting noise power at the detector with load resistance R_{L} and bandwidth B_{o} with perfect coupling is given by [16]:

$$P_{\mathrm{o}}(\omega) = \frac{\langle \Delta N_{\mathrm{p}}^2(\omega) \rangle}{\tau_{\mathrm{p2}}^2} R_{\mathrm{L}} B_{\mathrm{o}} \eta_{\mathrm{D}}^2 e^2 \tag{6.39}$$

where η_{D} is the quantum efficiency of the photodetector and τ_{p2} is the photon lifetime due to output mirror loss (see Section 5.2.1.2 of Chapter 5). Compared with the photon statistics approach, the rate equation analysis is more complicated. In addition, only the total noise power can be derived using Eq. (6.38) and the corresponding noise components have to be calculated by using more indirect methods [8]. However, it is better than the photon statistics formulation in the sense that we can take into account more factors (e.g. saturation of amplifier gain due to finite population inversion: see Sections 2.3.3, 5.2.2 and 5.3) by this complex modelling. It is also interesting to note that Mukai and Yamamoto [8] discovered that the rate equation and photon statistics analysis give slightly different theoretical results. Experimental results tend to lie between the theoretical limits imposed by these two methods. The difference arises because of the difference in formulating the signal gain G. In the photon statistics analysis we calculate G by using an active FP formulation, whereas the rate equation analysis requires a gain in the format of Eq. (5.17).

6.2.3 *Travelling-wave equations formulation*

Both the rate equation and photon statistic master equation approaches ignore the spatial dependence of the signal and noise photons within the amplifier. In order to get a more complete picture of the effects of different sources of noise, as well as their effects on the spectral contents of the amplifier output, Hilton [22–24] modified the travelling-wave equations which describe SLAs (Section 5.3.2). He introduced Langevin noise terms in the travelling-wave equations for the amplitude and phase of the output signal, in addition to $F_e(t)$. The results are very complex and apparently neither analytical nor numerical solutions have been available so far. Nevertheless, he was able to predict linewidth broadening due to spontaneous emissions, which was followed by the studies carried out by Kikuchi *et al.* [5] using a similar but simpler analysis. Because of the complexity of this approach, we will not pursue this formation in any further detail.

6.3 Analysis of Noise in SLAs using the Equivalent Circuit Model

The discussion in the preceding section has illustrated the basics of analysing noise in SLAs. It can be seen that they all involve the analysis of spontaneous emissions from SLAs, and one has to understand the physical processes in semiconductor lasers clearly before such analysis can be made with the aid of probability and statistical theory. In Section 5.5 (see Chapter 5) an equivalent circuit model for an SLA has been developed. We have seen how this model can enable us to analyse the gain and saturation characteristics efficiently. In this section, we will see how this equivalent circuit model can help us to understand and analyse spontaneous emissions and the resulting noise characteristics of SLAs. This is because we can use circuit analogies, familiar to most engineers, to formulate the noise processes in the equivalent circuit instead of involving specialised theories like those used in Section 6.2. The representation of spontaneous emissions in a SLA by a circuit model will first be examined.

6.3.1 *Representation of Spontaneous Emissions in a SLA by an Equivalent Circuit*

Recall that in the equivalent circuit model, the SLA is represented by a cascade of M sub-networks, each consisting of a negative resistance (to account for net optical gain) and a small section of lossless transmission line of length $\Delta z = L/M$ with L being the length of the amplifier. For kth sub-network (corresponding to the kth section used in transfer matrix analysis (see Section 5.4.2 in Chapter 5) the negative resistance is denoted by Z_k. For a single cavity SLA, the characteristics impedance Z_0 of all the lossless transmission lines in each sub-network are the same and can be used to model real optical gain in the kth section of the SLA by the following analogy (Section 5.5, Chapter 5).

$$\frac{Z_0}{Z_0 + Z_k} = \exp\left(\frac{g_k \Delta z}{2}\right) \tag{6.40}$$

if Δz is small enough then

$$\frac{Z_0}{Z_k} \approx -\frac{g_k \Delta z}{2} \,. \tag{6.41}$$

Since the major properties of a SLA such as optical gain can be modelled by their electrical circuit analogue in this equivalent circuit model, it is also natural to seek an electrical analogue for spontaneous emissions. The similarity between spontaneous emissions in semiconductor lasers and shot noise in electrical circuits was discussed in Section 6.1. Thus, an analogy between them can be established. Consider the case when there is no signal incident onto the SLA. In the equivalent circuit model, it means that there is no signal voltage or current flowing. Then the only possible shot noise voltage source in the circuit will be due to thermal fluctuations of the lumped impedances. These thermal shot noises were analysed by Nyquist using the equipartition theorem of statistical mechanics [16]. The single-sided power spectral density (PSD) [3] of this noise voltage in a load resistance R_L is given by [16] and [29]:

$$S_{th} = 4kTR_L \tag{6.42}$$

where k is the Boltzmann constant, T is the temperature of the resistor. Louisell [30] showed that in the region where the quantum energy $\hbar\omega$ is comparable with the thermal energy kT, the zero-point fluctuations due to these quanta are also important. This can be analysed using quantum mechanics, and the corresponding noise has a single-sided PSD S_q given by [29]

$$S_q = 2\hbar\omega R_L \,. \tag{6.43}$$

If we are using the equivalent circuit to model optical frequencies, then the quantity $\hbar\omega \gg kT$ and hence the contribution of thermal fluctuation to the total noise voltage is negligible [29]. Thus, in such circumstances the PDS (single-sided) of the noise voltage in a resistance R_L which is due to quantum mechanical fluctuations is given by

$$S_e \approx S_q = 2\hbar\omega R_L . \tag{6.44}$$

The above formula is applicable for a passive resistance R_L. For a negative resistance, Bjork & Nilsson [44] suggested that Eq. (6.44) can be applied to that for a negative resistance Z if R_L is replaced by the magnitude of Z, that is:

$$S_e = 2\hbar\omega |Z| . \tag{6.45}$$

The above postulation can be justified by the following argument. Although a passive resistance R_L absorbs electrical power through dissipation, whereas a negative resistance Z releases power, the quantum mechanical fluctuations in both types of resistance are shot noise voltage sources (or current sources depending on the way of modelling the noise source [3]) which will act as a power source as well. The amount of fluctuation in both types of resistance depend on the magnitude of the power absorbed (in passive resistance) or released (in negative resistance). Hence, one can use the magnitude of the resistance to calculate the power spectrum of the quantum mechanical fluctuations in them, as in Eqs. (6.44) and (6.45).

We have now completed the mathematical representation of quantum mechanical fluctuations in the equivalent circuit of the SLA. In the optical frequency regime, which is the frequency range for our modelling parameters and of primary interest, thermal shot noise can be neglected. Instead, the major noise sources are due to quantum mechanical fluctuations. These fluctuations arise because the magnitude of the energy quanta $\hbar\omega$ are significant and the statistical uncertainty will also be significant in the measuring or observation process. In the absence of any signal voltage, these zero-point fluctuations will result in a finite reading on a root-mean-square (r.m.s.) voltmeter connected across the output of the circuit [30]. If these observations are compared with a real SLA, biased to give a finite gain, but without any incident optical signal, a strong similarity between them can be observed. In the case of a SLA, because of the population inversion created by the injection current, spontaneous emission will occur (see Eq. (2.5) in Chapter 2). The number of spontaneous emissions will depend on the level of population inversion, similar to the fact that quantum mechanical fluctuation in a resistor is proportional to the magnitude of its resistance. In the absence of any signal, the output of a biased SLA will be delivering a finite optical power (see Chapter 7), similar to the recording of zero-point fluctuations at the output of the equivalent circuit described above.

Thus, it appears that in order to model spontaneous emissions in SLAs, we can use the equivalent circuit with an added noise voltage source $e_k(t)$ with a single-sided PSD of S_{ek} for each of the sub-networks. The quantity Z in Eq. (6.45) will be replaced by Z_k, which is defined by the analogy in Eqs. (6.40) and (6.41). Notice that only the active negative resistance is noisy in each sub-network. The transmission-lines and the terminating loads are noiseless. Although it is suggested that spontaneous emissions can be modelled in this manner, because of the similarity between shot noise in electrical circuits and spontaneous emissions in SLAs [29], this argument does not justify the mathematical use of the PSD of spontaneous emission in the case of SLAs. This can be tested by using the proposed model to analyse the spontaneous emission power from a SLA with uniform material gain profile, and comparing the results with analytical ones which are well-documented in literature (e.g. see Section 6.5 [17, 27]).

6.3.2 *Validity of modeling spontaneous emissions by an equivalent circuit*

In this section, the representation of spontaneous emissions in a SLA by the quantum mechanical fluctuations in the equivalent circuit will be used to calculate the spontaneous emission power in the output of a SLA in the absence of an input signal. This information is extremely useful in the analysis of SLAs and has already been used in the photon statistics master equation in Section 6.2.1 (i.e. Eq. (6.14)). For SLAs with complex structures, their noise properties can still be analysed following the procedures discussed in Section 6.2.1, as long as the power spectral density of the spontaneous emissions (hence photon number spectral density $N_p(f)$ in Eq. (6.14)) for these structures can be found. As we shall see in this section, this can be done easily with the equivalent circuit model. Conventionally, spontaneous emission power from a SLA can be analysed using laser physics [31–32]. However, the analysis can only be applied to closed optical cavities and will not be valid for open resonator structures, such as distributed feedback (DFB) lasers and optical amplifiers. Henry [17] proposed the use of a Green function approach to solve the spontaneous emission fields via a semi-classical approach. The methodology outlined in his paper can be used to analyse DFB lasers and optical amplifiers. In fact, his method was powerful enough to analyse even more complex structures, such as coupled-cavities. However, finding the Green functions of complex optical structures is not an easy task [33]. This is illustrated by the recent work of Kahen [34]. The situation, interestingly, is quite similar to that of solving the travelling-wave equations (i.e. Eqs. (5.30) and (5.31) discussed in Section 5.5.2 in Chapter 5). The use of the equivalent circuit model together with transfer method (TMM) was proposed by Chu and Ghafouri-Shiraz [56, 57] in Chapter 5 to overcome the problem encountered with solving travelling-wave equations for optical cavities. It is therefore logical to examine whether it is also possible to analyse spontaneous emissions in SLAs by this technique. In this section the feasibility of using this technique will be demonstrated by analysing simple FPAs for which results are well known. Alternative methods of arriving at the same results using Green functions were outlined in Section 6.5.

Consider a FPA with a uniform material gain profile g_m (i.e. $g_{mk} = g_m$ for all k's) and length L. The group effective refractive index of the cavity is N_g, and the power reflection coefficients at the facets are R_1 and R_2, respectively. The equivalent circuit model for this structure is shown in Fig. 6.2, together with the proposed noise sources $e_k(t)$ added to each kth sub-network. Consider the kth section in the cavity, which is modelled by the kth sub-network in the equivalent circuit of Fig. 6.2. The values of Z_k and Z_0 of the sub-network are given by Eqs. (6.40) and (6.41). In addition, Z_0 is determined by the group velocity of the fields travelling across the

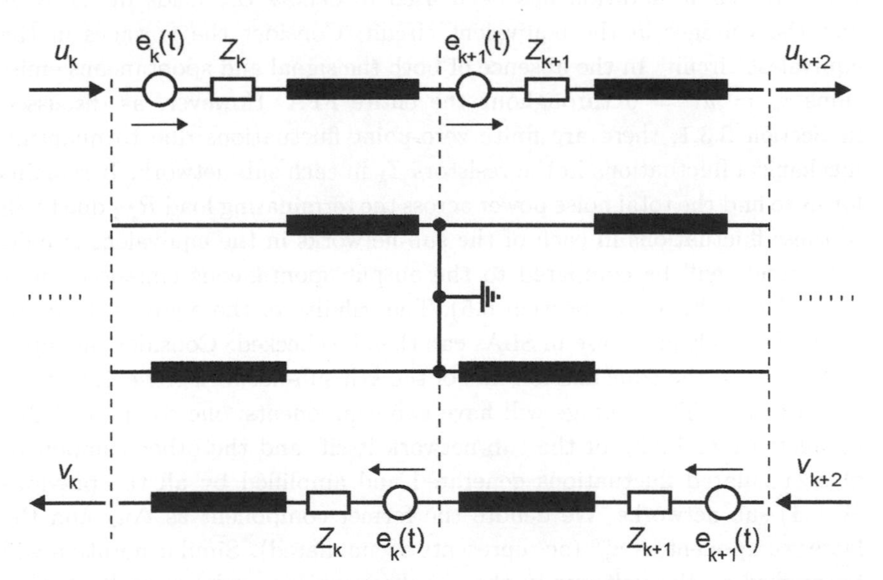

Fig. 6.2. Representation of spontaneous emissions in SLAs by the equivalent circuit model.

circuit, which is related to N_g. The modal gain coefficient g_k at the kth section for a simple buried heterostructure (BH) is given by:

$$g_k = g = \Gamma(g_m - \alpha_a) - (1 - \Gamma)\alpha_c \tag{6.46}$$

where Γ is the optical confinement factor of the active region, α_a and α_c are the losses in the active region and claddings, respectively. Note that the subscript k in Eq. (6.46) has been dropped as we have assumed that the material gain profile in the FPA is uniform. Moreover, Eq. (6.46) has assumed that a unique value of Γ can be defined. However, as discussed in Section 4.3 in Chapter 4 a unique Γ cannot be defined unambiguously in some SLA structures [35]. Hence g has to be calculated by the technique introduced in Section 4.3.1 without using Eq. (6.46). Following the notation used in Section 5.4.3 the voltages at both sides of the kth sub-network are related by the following transfer matrix expression (see Section 5.5) [56, 57]:

$$\begin{bmatrix} u_k \\ \nu_k \end{bmatrix} = \begin{bmatrix} \exp(-g_k\Delta z/2)\exp(+j\beta_z\Delta z) & 0 \\ 0 & \exp(g_k\Delta z/2)\exp(-j\beta_z\Delta z) \end{bmatrix} \times \begin{bmatrix} u_{k+1} \\ \nu_{k+1} \end{bmatrix} \tag{6.47}$$

where the same notation has been used to denote the fields in the SLA and the voltages in the equivalent circuit. Consider the voltages in the equivalent circuit. In the absence of both the signal and spontaneous emissions $v_k = u_k = 0$ throughout the entire FPA. However, as discussed in Section 6.3.1, there are finite zero-point fluctuations due to quantum mechanical fluctuations in the resistors Z_k in each sub-network. It remains for us to find the total noise power across the terminating load R_{L2} due to all of these fluctuations in each of the sub-networks in the equivalent circuit. This result will be compared to the output spontaneous emission power at R_2 for an FPA (see Section 6.5). The validity of the equivalent circuit model for analysing noise in SLAs can then be checked. Consider the input voltage u_k to the transmission line of the kth sub-network travelling in the $+z$ direction. This voltage will have two components: one component due to fluctuations in Z_k of the sub-network itself, and the other component of accumulated fluctuations generated and amplified by all the previous $(k-1)$ sub-networks. We denote the former component as Δu_k and the latter component as u_k^{ac} (ac represents accumulated). Similar notation will be applied to the voltages in the $-z$ direction (i.e. v_k's) as well. Let us consider the voltages at the termination load where $z = L = M\Delta z$ (which models the left facet of the FPA). Using Eq. (5.61) in Chapter 5, we arrive at the following expression:

$$\begin{bmatrix} u_{M+1} \\ \nu_{M+1} \end{bmatrix} = \frac{1}{t_2} \begin{bmatrix} 1 & r_2 \\ r_2 & 1 \end{bmatrix} \cdot \begin{bmatrix} u_{M+2} \\ \nu_{M+2} \end{bmatrix} \tag{6.48}$$

where $r_2 = \sqrt{R_1}$ and is given by

$$r_2 = \frac{R_{L2} - Z_0}{R_{L2} + Z_0} \tag{6.49}$$

u_{M+1} and v_{M+1} are given by

$$u_{M+1} = u_{M+1}^{\text{ac}} + \Delta u_{M+1} \tag{6.50a}$$

$$v_{M+1} = v_{M+1}^{\text{ac}} + \Delta v_{M+1} \,. \tag{6.50b}$$

Examining the meaning of each term in Eqs. (6.50a) and (6.50b), it can be seen that

$$\begin{bmatrix} u_{M+1}^{\text{ac}} \\ \nu_{M+1}^{\text{ac}} \end{bmatrix} = \begin{bmatrix} \exp(g_M \Delta z/2) \exp(-j\beta_z \Delta z) & 0 \\ 0 & \exp(-g_M \Delta z/2) \exp(j\beta_z \Delta z) \end{bmatrix} \times \begin{bmatrix} u_M \\ \nu_M \end{bmatrix} . \tag{6.51}$$

Note that Eq. (6.51) has been re-arranged based on Eq. (6.47). Notice the difference in the indices in the column vectors in the left hand side and the right hand side of Eq. (6.51). It can be seen that in general

$$\begin{bmatrix} u_{k+1}^{\text{ac}} \\ \nu_{k+1}^{\text{ac}} \end{bmatrix} = \begin{bmatrix} \exp(-g_k \Delta z/2) \exp(-j\beta_z \Delta z) & 0 \\ 0 & \exp(-g_k \Delta z/2) \exp(j\beta_z \Delta z) \end{bmatrix} \cdot \begin{bmatrix} u_k \\ \nu_k \end{bmatrix} \tag{6.52}$$

where

$$u_k = u_k^{\text{ac}} + \Delta u_k \tag{6.53a}$$

$$v_k = v_k^{\text{ac}} + \Delta v_k \,. \tag{6.53b}$$

The Δ terms in Eqs. (6.53a) and (6.53b) are fluctuations generated by the kth sub-network. Equations (6.52) and (6.53) can be used iteratively to evaluate u_{M+1} and v_{M+1} in Eqs. (6.55) and (6.51). Using Eq. (6.51) in Eq. (6.50a) we arrive at the following expression:

$$u_{M+1} = u_{M+1}^{\mathrm{ac}} = \sqrt{A_M}(u_M^{\mathrm{ac}} + \Delta u_M)\exp(-j\beta_z\Delta z) \tag{6.54}$$

where the first term on the right hand side of Eq. (6.54) is u_M^{ac} given by Eq. (6.51). The term $\sqrt{A_M}$ is used to represent the term $\exp(g_m\Delta z/2)$ for simplicity. Note that $\Delta u_{M+1} = \Delta v_{M+1} = 0$ as the terminating load R_{L2} is noiseless. However, as shown by Eq. (6.52):

$$u_M^{\mathrm{ac}} = \sqrt{A_{M-1}}(u_{M-1}^{\mathrm{ac}} + \Delta u_{M-1})\exp(-j\beta_z\Delta z)\,. \tag{6.55}$$

Then Eq. (6.54) becomes

$$\begin{aligned} u_{M+1} &= \sqrt{A_M}\{\sqrt{A_{M-1}}\exp(-i\beta_z\Delta z) \\ &\qquad \times [u_{M-1}^{\mathrm{ac}} + \Delta u_{M-1}] + \Delta u_M\}\exp(-j\beta_z\Delta z) \\ &= P_M[P_{M-1}u_{M-1}^{\mathrm{ac}} + P_{M-1}\Delta u_{M-1} + \Delta u_M] \end{aligned} \tag{6.56}$$

where we have used

$$P_M = \sqrt{A_M}\exp(-j\beta_z\Delta z)\,. \tag{6.57}$$

But from Eq. (6.52) we have:

$$u_{M-1}^{\mathrm{ac}} = \sqrt{A_{M-2}}(u_{M-2}^{\mathrm{ac}} + \Delta u_{M-2})\exp(-j\beta_z\Delta z)\,. \tag{6.58}$$

Substituting this into Eq. (6.56) gives:

$$u_{M+1} = P_M[P_{M-1}P_{M-2}(u_{M-2}^{\mathrm{ac}} + \Delta u_{M-2}) + P_{M-1}\Delta u_{M-1} + \Delta u_M]\,. \tag{6.59}$$

If we carry on the iteration with Eq. (6.52) we finally arrive at:

$$u_{M+1} = P_M\left[P_{M-1}P_{M-2}\cdots P_1u_1^{\mathrm{ac}} + \sum_{r=1}^{M-1}\left(\prod_{k=r}^{M-1}P_k\right)\Delta u_r + \Delta u_M\right] \tag{6.60}$$

u_1^{ac} can be found from Eq. (5.51) in Chapter 5 at $z = 0$:

$$\begin{bmatrix} u_0 \\ \nu_0 \end{bmatrix} = \frac{1}{t_1}\begin{bmatrix} 1 & -r_1 \\ -r_1 & 1 \end{bmatrix}\cdot\begin{bmatrix} u_1^{\mathrm{ac}} \\ u_1^{\mathrm{ac}} \end{bmatrix} \tag{6.61}$$

where $r_1 = \sqrt{R_1}$ and is given by:

$$r_1 = \frac{R_{L1} - Z_0}{R_{L1} + Z_0}\,. \tag{6.62}$$

Notice that $\Delta u_0 = \Delta v_0 = 0$ as R_{L1} is considered to be noiseless. In addition, as there is no signal coming into the amplifier, $u_0 = 0$. From Eq. (6.61), it can be shown that:

$$u_1^{\mathrm{ac}} = r_1 v_1^{\mathrm{ac}} = r_1 v_1 \,. \tag{6.63}$$

Equation (6.60) can then be re-written as:

$$u_{M+1} = P_M \left[r_1 P_{M-1} P_{M-2} \cdots P_1 v_1 + \sum_{r=1}^{M-1} \left(\prod_{k=r}^{M-1} P_k \right) \Delta v_r + \Delta v_M \right]. \tag{6.64}$$

By following similar procedures in the reverse direction for v_k, it can be shown that v_1 takes a similar form to Eq. (6.60), that is

$$v_1 = v_1^{\mathrm{ac}} = P_1 \left[P_2 P_3 \cdots P_M u_{M+1}^{\mathrm{ac}} + \sum_{r=2}^{M} \left(\prod_{k=r}^{M} P_k \right) \Delta u_{M+3-r} + \Delta u_2 \right]. \tag{6.65}$$

If the amplifier under analysis has a uniform material gain profile, then $P_1 = P_2 = \ldots P_M = P$. Equations (6.64) and (6.65) the become:

$$u_{M+1} = P[P^{M-1} u_1^{\mathrm{ac}} + P^{M-1} \Delta u_1 + P^{M-2} \Delta u_2 + \cdots + P \Delta u_{M-1} + \Delta u_M] \tag{6.66}$$

$$\nu_1 = P[P^{M-1} \nu_{M+1}^{\mathrm{ac}} + P^{M-1} \Delta \nu_{M+1} + P^{M-2} \Delta \nu_M + \cdots + P \Delta \nu_3 + \Delta \nu_2] \,. \tag{6.67}$$

Using Eq. (6.64) in Eqs. (6.66) and (6.67) we obtain the following expression:

$$\begin{aligned} u_{M+1} = P\{ & r_1 P^{M-1} P[P^{M-1} \nu_{M+1}^{\mathrm{ac}} + P^{M-1} \Delta \nu_{M+1} \\ & + P^{M-2} \Delta \nu_M + \cdots + P \Delta \nu_3 + \Delta \nu_2] \\ & + [P^{M-1} \Delta u_1 + P^{M-2} \Delta u_2 + \cdots + P \Delta u_{M-1} + \Delta u_M]\} \,. \end{aligned} \tag{6.68}$$

But from Eq. (6.48) we have

$$\nu_{M+1}^{\mathrm{ac}} = r_2 u_{M+1}^{\mathrm{ac}} = r_2 u_{M+1} \tag{6.69}$$

as $v_{M+2} = 0$ with no input signal to the FPA. Substituting Eq. (6.69) into Eq. (6.68), we obtain:

$$\begin{aligned} u_{M+1} = P\{ & r_1 r_2 P^{2M-1} \cdot u_{M+1} + [r_1 P^{2M-1} \Delta\nu_{M+1} \\ & + r_1 P^{2M-2} \Delta\nu_M + \cdots + r_1 P^{M+1} \Delta\nu_3 + r_1 P^M \Delta\nu_2] \\ & + [P^{M-1} \Delta u_1 + P^{M-2} \Delta u_2 + \cdots + P \Delta u_{M-1} + \Delta u_M]\} . \end{aligned} \tag{6.70}$$

Re-arranging Eq. (6.70) by grouping the terms for u_{M+1} by using Eq. (6.48) again for u_{M+2}, we finally arrive at the following expression:

$$\begin{aligned} u_{M+1} = t_2 u_{M+1} = & \left(\frac{t_2 P}{1 - r_1 r_2 P^{2M}} \right) \\ & \times [r_1 (P^{2M-1} \Delta\nu_{M+1} + P^{2M-2} \Delta\nu_M + \cdots + P^M \Delta\nu_2) \\ & + (P^{M-1} \Delta u_1 + P^{M-2} \Delta u_2 + \cdots + \Delta u_M)] . \end{aligned} \tag{6.71}$$

The mean-square value of voltage u_{M+1} can be found as:

$$\begin{aligned} \langle |u_{M+2}|^2 \rangle = & \frac{(1 - R_2)|P|^2}{|1 - r_1 r_2 P^{2M}|^2} \cdot \{R_1 |P|^{2M} [\langle |\Delta\nu_2|^2 \rangle + |P|^2 \langle |\Delta\nu_3|^2 \rangle \\ & + \cdots + |P|^{2M-2} \langle |\Delta\nu_{M+1}|^2 \rangle] + [|P|^{2M-2} \langle |\Delta u_1|^2 \rangle \\ & + |P|^{2M-4} \langle |\Delta u_2|^2 \rangle + \cdots + \langle |\Delta u_M|^2 \rangle]\} \end{aligned} \tag{6.72}$$

where we have assumed that the fluctuations have no spatial correlation, i.e. $\langle \Delta v_i \Delta v_j^* \rangle = 0$ and $\langle \Delta u_i \Delta u_j^* \rangle = 0$ for $i \neq j$. In addition, there is no such correlation for voltages travelling in opposite directions, i.e. $\langle \Delta u_i \Delta v_j^* \rangle = 0$. But $\langle |u_k|^2 \rangle = \langle |v_k|^2 \rangle = \langle |e(t)|^2 \rangle / (W d Z_0)$; recall that we have normalised u_k and v_k with respect to optical intensity. If the power of total fluctuation is measured at $z = L$ in a small optical bandwidth $\Delta\omega = 2\pi f$, then by Fourier analysis, it can be shown that [3]:

$$\langle |e^2(t)| \rangle = \frac{S_{\mathrm{e}} \Delta\omega}{4\pi^2} . \tag{6.73}$$

Then the power P_0 measured at the output of the equivalent circuit is given by:

$$P_0 = \frac{\Delta\omega}{4\pi^2 Z_0} \cdot S_e \cdot \frac{(1-R_2)|P|^2}{|1-r_1 r_2 P^{2M}|^2} \cdot (R_1|P|^{2M}+1) \sum_{r=0}^{M-1} P^{2r} . \tag{6.74}$$

For a FPA with a uniform material gain profile:

$$P^{2M} = \exp(gL)\exp(-2j\beta_z L) \tag{6.75}$$

where $L = M\Delta z$, then Eq. (6.74) can be re-written as:

$$P_0 = \frac{\Delta\omega}{4\pi^2 Z_0} \cdot S_e \cdot \frac{G_{FP}|P|^2}{(1-R_1)|P|^{2M}} (R_1|P|^{2M}+1) \left(\frac{|P|^{2M}-1}{|P|^2-1} \right) \tag{6.76}$$

where we have used the sum of a geometric series to evaluate $\sum P^{2r}$. The term G_{FP} is the gain of a FPA with uniform gain profile and is given by Eq. (5.4) of Section 5.2.1.1 in Chapter 5, that is:

$$G_{FP} = \frac{(1-R_1)(1-R_2)|P|^{2M}}{|1-\sqrt{R_1 R_2} P^{2M}|^2} = \frac{(1-R_1)(1-R_2)\exp(gL)}{|1-\sqrt{R_1 R_2}\exp(gL)\exp(-2j\beta_z L)|^2} . \tag{6.77}$$

Since $|P|^2 = \exp(g\Delta z) = A$ and $|P|^{2M} = \exp(gM\Delta z) = \exp(gL)$, hence Eq. (6.76) can be re-written as

$$P_0 = \frac{\Delta\omega}{4\pi^2 Z_0} \cdot S_e \cdot \left(\frac{A}{A-1} \right) \cdot \frac{G_{FP}}{(1-R_1)} (R_1 e^{gL}+1)(1-e^{-gL}) . \tag{6.78}$$

If Δz is small enough, A can be expanded by Taylor's series. Then it can be shown that:

$$\frac{A-1}{A} = 1 - \frac{1}{A} = 1 - e^{-g\Delta z} \approx g\Delta z . \tag{6.79}$$

To proceed further, we have to examine the fluctuation in Z_k for each sub-network. It was shown in Section 5.5 that Z_k represent the net optical gain. This can be split into three components [14]:

$$Z_k = Z_{1k} + Z_{2k} + Z_{ak} \tag{6.80}$$

where Z_{1k} is a negative resistance modelling the stimulated emissions, Z_{2k} is a passive resistance modelling absorption and Z_{ak} is a passive resistance modelling other optical losses like scattering. Any spontaneous emissions will add photons randomly to the stimulated emissions, and hence the major fluctuations in Z_k are due to Z_{1k} only (note that in modelling laser oscillators, quantum fluctuations in the other processes can be quite significant [14]. Hence we can write:

$$S_{\mathrm{ek}} \approx 2\hbar\omega|Z_{1k}|\,. \tag{6.81}$$

It has also been shown that g can be approximated by Eq. (6.41) for small Δz. Combining with Eq. (6.46) in an index guided structure, it can be seen that:

$$\frac{Z_{1k} + Z_{2k} + Z_{\mathrm{ak}}}{Z_0} = -\{\Gamma g_{mk} - [\Gamma\alpha_{\mathrm{ak}} + (1-\Gamma)\alpha_{ck}]\}\frac{\Delta z}{2}\,. \tag{6.82}$$

Then by comparing the analogy between each component of Z_k on the left hand side of the above equation with that on the right hand side, we arrive at the following expression:

$$Z_{1k} + Z_{2k} = -\frac{\Gamma g_{mk} Z_0 \Delta z}{2}\,. \tag{6.83}$$

This can be simplified further by noting that the population inversion parameter n_{sp} can be modelled by [14]

$$n_{\mathrm{sp}} = \frac{|Z_{1k}|}{|Z_{1k}| - Z_{2k}} = \frac{Z_{1k}}{Z_{1k} + Z_{2k}}\,. \tag{6.84}$$

Hence

$$|Z_{1k}| = \frac{\Gamma g_{mk} n_{\mathrm{sp}} Z_0 \Delta z}{2}\,. \tag{6.85}$$

Substituting into Eq. (6.81) we arrive at:

$$S_{\mathrm{ek}} = \hbar\omega\Gamma g_{mk} n_{\mathrm{sp}} Z_0 \Delta z\,. \tag{6.86}$$

The above equation is actually very interesting. It can be seen from Section 5.7 of Chapter 5 that by using the expressions for β and R_{sp} derived there, the term S_{ek} in the above equation actually will be equal to the number of spontaneous emission photons coupled to the signal mode in the optical cavity (i.e. $\beta R_{\mathrm{sp}} N_g \Delta z/c$) provided that $\Delta\omega = c/(N_g\Delta z)$. Such a correction factor for the spontaneous emission factor β was found by Sommers Jr [36], which has been attributed to the difference in using the frequency independent photon statistical interpretation of β whilst analysing the photons in the cavity which are pro-rated over a frequency range. When the equivalent circuit model is used, the amount of spontaneous emission generated at each section is controlled by Δz and g, which is what the above equation implies. Thus one can actually "fine-tune" the values of Δz such that the correction factors in rate equations applied over larger optical cavities will not be necessary, and a coherent interpretation of the physical processes described by the equivalent circuit model can be obtained [36]. The form of Eq. (6.86) suggests that the power spectrum of the fluctuations induced by spontaneous emissions in a SLA depends on g_{mk}, the material gain coefficient. In practice, because the material gain coefficient depends on emission wavelength (see Section 3.2.2 in Chapter 3), it follows that the power spectrum of the spontaneous emissions will also depend on the shape of the material gain spectrum [37–39]. Substituting this into Eq. (6.78) for an FPA with a uniform gain profile, we finally arrive at the following expression:

$$P_0 = \frac{\Delta\omega}{4\pi^2}\hbar\omega\left(\frac{n_{\mathrm{sp}}}{\eta}\right)\left(\frac{R_1 e^{gL}+1}{1-R_1}\right)(1-e^{-gL})G_{\mathrm{FP}} \tag{6.87}$$

where η is the quantum efficiency of the laser amplifier given by [14, 17]:

$$\eta = \frac{g}{\Gamma g_m} \tag{6.88}$$

for an amplifier with uniform gain profile. If we compare Eq. (6.87) with the expression for P_0 derived using Green functions in Section 6.5, it can be seen that they are identical. Hence, by adding the noise voltages $e_k(t)$ at each kth sub-network in the equivalent circuit, with a magnitude given by Eqs. (6.73) and (6.81), the equivalent circuit can also be used to model spontaneous emissions accurately. In Fig. 6.3, the variation of $P_0/(\hbar\omega\Delta\omega)$ against the pumping rate i/i_{th} for a SLA has been plotted. The structural parameters for this SLA are shown in the inset of the figure. It can be seen that the amount of spontaneous emission is not very significant for moderate values of i/i_{th}, and then increases rapidly after i/i_{th} rises beyond 0.7. This supports our proposal to ignore the coupling of spontaneous emissions to the signal mode in the TMM analysis discussed in the previous chapter.

6.3.3 *Effects of stray reflections on the spontaneous emission power from a SLA*

In this section, the equivalent circuit model will be used to analyse spontaneous emissions of SLAs subject to reflections from fibre ends. The

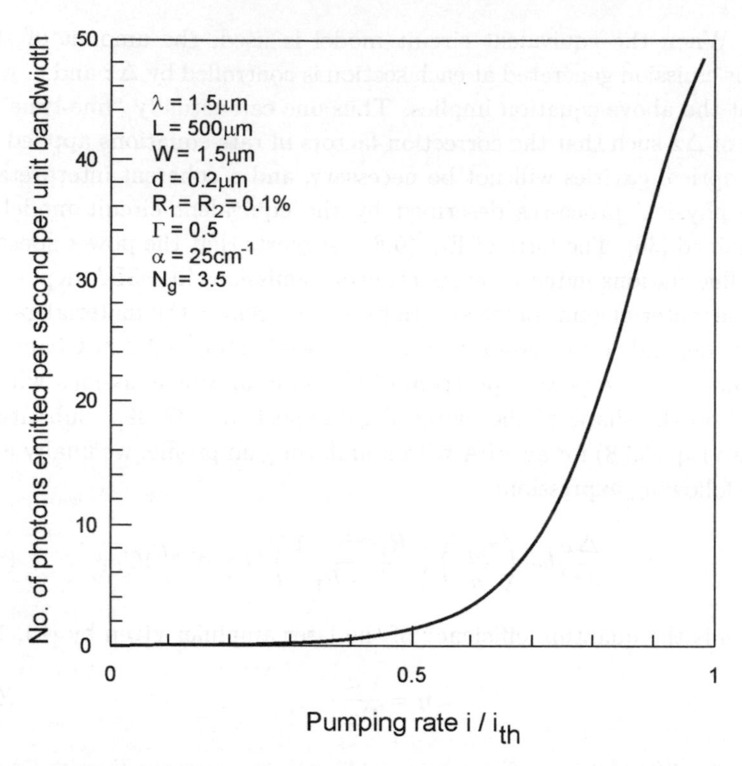

Fig. 6.3. Variation of the number of photons emitted per second per unit bandwidth due to spontaneous emissions against different pumping rate for a FPA with parameters shown in the inset.

subsequent effect on the gain characteristics will also be examined in this section. When a SLA has to be used in a system, it has to be coupled to two pieces of optical fibre, one acting as an input and the other as the output. There are discontinuities in the refractive indices between the SLA facet-air and air-fibre interfaces, which will cause reflection [33]. The effect of the reflections on the performance of the SLAs in the system are not well understood, although some work has been done with fibre amplifiers [40–41]. A formal study of the problem can be performed by the proposed method of transfer matrix analysis and equivalent circuit modelling. Here, the effect of the reflections on the spontaneous emissions from a SLA will be studied. This property is important for multi-amplifier systems, because if the spontaneous emissions are enhanced by these stray reflections (as in FPAs where the excess noise coefficient χ comes in, see Section 6.2.1), then the system may oscillate, or in some SLAs unexpected saturation will be observed [40, 42]. We simplify the analysis by assuming that the reflections from the ends of the optical fibres can be modelled by a plane mirror extending to infinity in both x and y directions (Fig. 6.5(a)) [43]. The power reflection coefficients of these mirrors are R_3 and R_4, respectively, located at $z = -L_3$ and $z = L + L_4$. In the analysis we assume the amplifier has a uniform gain profile and is divided into M sections. The relationship between U_0, V_0 and u_0, v_0 can be found by the following transfer matrix [44]:

$$\begin{bmatrix} U_0 \\ V_0 \end{bmatrix} = \frac{1}{t_3} \begin{bmatrix} 1 & -r_3 \\ -r_3 & 1 \end{bmatrix} \begin{bmatrix} \exp(jk_0 L_3) & 0 \\ 0 & \exp(-jk_0 L_3) \end{bmatrix} \begin{bmatrix} u_0 \\ \nu_0 \end{bmatrix} . \quad (6.89)$$

Similarly

$$\begin{bmatrix} u_{M+2} \\ v_{M+2} \end{bmatrix} = \frac{1}{t_4} \begin{bmatrix} \exp(jk_0 L_4) & 0 \\ 0 & \exp(-jk_0 L_4) \end{bmatrix} \begin{bmatrix} 1 & r_4 \\ r_4 & 1 \end{bmatrix} \begin{bmatrix} U_{M+2} \\ V_{M+2} \end{bmatrix} . \quad (6.90)$$

In the above equations, $r_3 = \sqrt{R_3}$ and $r_4 = \sqrt{R_4}$ are the amplitude reflectivities of the mirrors, and t_3 and t_4 are the corresponding transmittivities given by $t_4^2 = 1 - r_4^2$. The parameter k_0 is the wave number in free space which is assumed to be the medium between the mirrors and the facets of the amplifiers. From Eqs. (6.89) and (6.90) it can be shown that:

$$u_0 = r_3 v_0 \exp(-2jk_0 L_3) \quad (6.91a)$$

$$v_{M+2} = r_4 v_{M+2} \exp(-2jk_0 L_4) \, . \quad (6.91b)$$

Imposing these conditions onto Eqs. (10.48) and (10.61) we arrive at the following relations:

$$\left(\frac{u_1}{v_1} \right) = r_{\text{eff}\,1} = \frac{r_1 + r_3 \exp(-2jk_0 L_3)}{1 + r_1 r_2 \exp(-2jk_0 L_3)} \quad (6.92a)$$

$$\left(\frac{u_{M+1}}{v_{M+1}} \right) = r_{\text{eff}\,2} = \frac{r_2 + r_4 \exp(-2jk_0 L_4)}{1 + r_2 r_4 \exp(-2jk_0 L_4)} \, . \quad (6.92b)$$

Equation (6.92a) will be used in place of Eq. (6.63) whereas Eq. (6.92b) will be used in place of Eq. (6.69). With these replacements and following the procedures outlined in the previous section, we arrive at the following expression (see Eq. (6.71)):

$$u_{M+2} = \left(\frac{t_{\text{eff}\,2}P}{1 - r_{\text{eff}\,1}r_{\text{eff}\,2}P^{2M}}\right) \times [r_{\text{eff}\,1}(P^{2M-1}\Delta\nu_{M+1} + P^{2M-2}\Delta\nu_M + \cdots + P^M\Delta\nu_2) + (P^{M-1}\Delta u_1 + P^{M-2}\Delta u_2 + \cdots + \Delta u_M)] \tag{6.93}$$

where $t_{\text{eff}\,2}$ is the effective transmittance of the configuration given by:

$$|t_{\text{eff}\,2}|^2 = 1 - |r_{\text{eff}\,2}|^2\,. \tag{6.94}$$

Other terms have been defined in the previous section. An inspection of Eq. (6.93) reveals that we can replace the composite structures of Fig. 6.4(a) by an equivalent FPA in Fig. 6.4(b) where the facet reflectivities have been replaced by $r_{\text{eff}\,1}$ and $r_{\text{eff}\,2}$, respectively. Consequently, the spontaneous emission power P_0 of the structure can be found simply by an equation similar to Eq. (6.74), that is:

$$P_0 = \frac{\Delta\omega}{4\pi^2 Z_o} \cdot S_e \cdot \frac{(1 - |r_{\text{eff}\,2}|^2)}{|1 - r_{\text{eff}\,1}r_{\text{eff}\,2}P^{2M}|^2} \cdot (|r_{\text{eff}\,1}|^2|P|^{2M} + 1)\sum_{r=0}^{M-1} P^{2r} \tag{6.95}$$

which can be simplified to give:

$$P_0 = \frac{\Delta\omega}{4\pi^2}\hbar\omega\left(\frac{n_{\text{sp}}}{\eta}\right)\left(\frac{|r_{\text{eff}\,1}|^2 e^{gL} + 1}{1 - |r_{\text{eff}\,1}|^2}\right)(1 - e^{-gL})G'_{\text{FP}} \tag{6.96}$$

where G'_{FP} is given by:

$$G'_{\text{FP}} = \frac{(1 - |r_{\text{eff}\,1}|^2)(1 - |r_{\text{eff}\,2}|^2)\exp(gL)}{|1 - |r_{\text{eff}\,1}||r_{\text{eff}\,2}|\exp(gL)\exp(-2j\beta_z L)|^2}\,. \tag{6.97}$$

We calculate $P_0/(\hbar\omega\Delta\omega)$ for a FPA using Eqs. (6.96) and (6.97) with $R_1 = R_2 = 0.1\%$, $L = 500\ \mu\text{m}$, $\alpha = 25\ \text{cm}^{-1}$, $\lambda = 1.5\ \mu\text{m}$, $N_g = 3.5$, $W = 1.5\ \mu\text{m}$, $d = 0.2\ \mu\text{m}$ and $\Gamma = 0.5$ against the pumping rate $\gamma = i/i_{\text{th}}$ by considering both the case with fibre end reflections of magnitude 4% and the case where the reflections from the fibre can be neglected. The separation between the fibres and the end-facets is assumed to be the same at both the input and the output, and is fixed to 10 μm. The results are plotted in Fig. 6.5. It can be seen that the reflections actually enhance the growth of spontaneous emissions with pumping and the entire structure reaches a lasing threshold

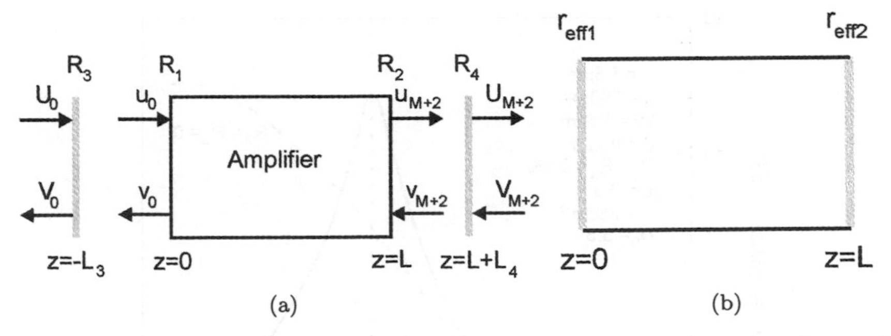

Fig. 6.4. (a) configuration used in the analysis of stray reflections from fibre ends. (b) The equivalent FPA of the configuration in (a).

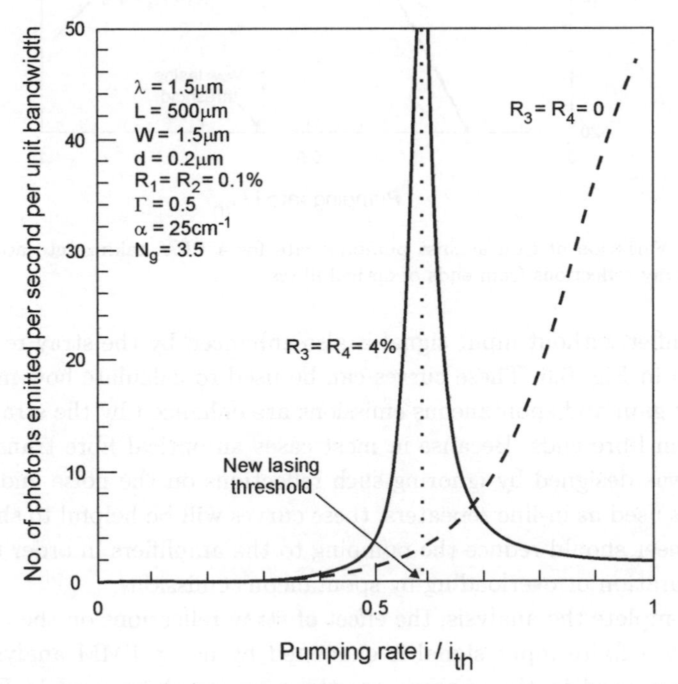

Fig. 6.5. Variation of the number of photons emitted per second per unit bandwidth against pumping rate for a FPA, taking into account the effect of stray reflections from ends of optical fibres.

earlier than that for a solitary SLA. The actual value of the new lasing threshold current and the amount of enhancement due to fibre reflections is actually determined by both the magnitude of the stray reflections as well as the distance between the facets and the fibre ends. Similarly, the gain of

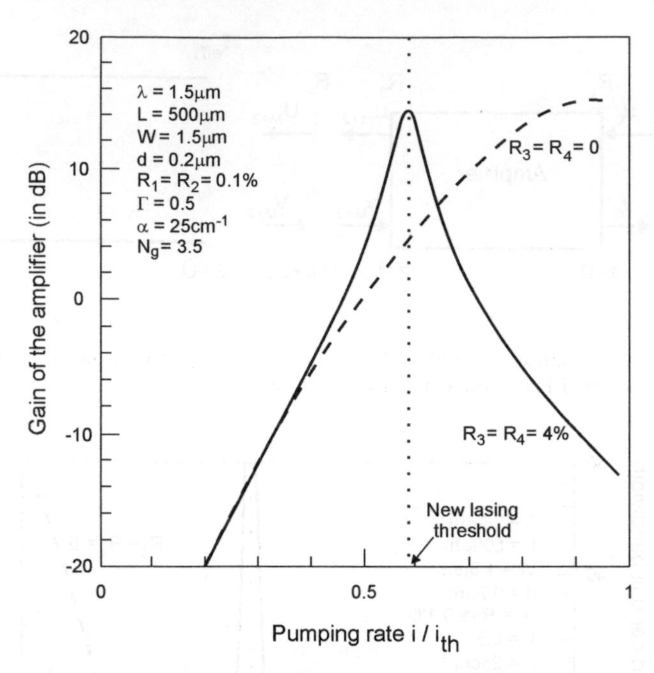

Fig. 6.6. Variation of gain against pumping rate for a FPA, taking into account the effect of stray reflections from ends of optical fibres.

the amplifier without input signal is also enhanced by the stray reflections as shown in Fig. 6.6. These curves can be used to calculate how much the amplifier gain and spontaneous emissions are enhanced by the stray reflections from fibre ends. Because in most cases an optical fibre transmission system was designed by ignoring such reflections on the noise and gain of the SLAs used as in-line repeaters, these curves will be helpful to show how the engineer should reduce the pumping to the amplifiers in order to avoid gain saturation or overloading by spontaneous emissions.

To complete the analysis, the effect of stray reflections on the amplifier gain with a finite input signal is examined by using TMM analysis. The parameters used in the analysis are the same as those used in Figs. 6.5 and 6.6. The results are shown in Fig. 6.7(a). Even though the spontaneous emissions coupling to the signal mode are ignored, the calculated results clearly illustrate the enhancement of gain due to the stray reflections which tend to drive the amplifier closer to oscillation. This can be seen more clearly when they are compared to the curves calculated for no fibre end reflections, as in Fig. 6.7(b), where it can be seen that for higher output optical power, the effect of fibre reflections becomes more important than

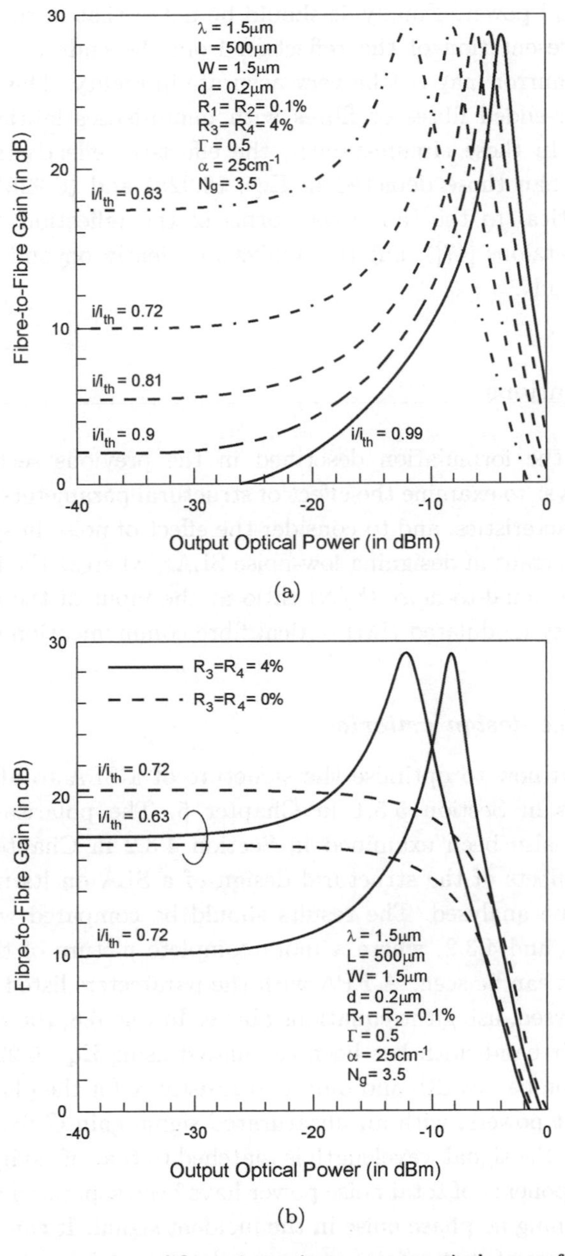

Fig. 6.7. (a) Variation of amplifier gain against output optical power for a FPA taking into account the effect of reflections from ends of optical fibres. (b) Comparison between the variation of amplifier gain against output optical power for a FPA with and without taking into account the effect of stray reflections from ends of optical fibres.

at lower optical power. Finally, it should be noted that in the above analysis, the representation of the reflection from the ends of optical fibres with a plane mirror may not be very accurate in reality. This is especially true for taper-ended fibres or fibres with micro-lenses fabricated on the tip [45, 46]. In these circumstances, the effective reflectivities are more complicated than those depicted in Eqs. (6.92a) and (6.92b) (which are actually identical to the functional forms of the reflection coefficient of Fabry–Perot etalons [47]), but the analysis is clearly beyond the scope of the present study.

6.4 Applications

We will use the formulation described in the previous sections in two particular ways: to examine the effect of structural parameters of a SLA on its noise characteristics, and to consider the effect of noise in systems. The former is important in designing low-noise SLAs, whereas the latter aspect will affect the signal-to-noise (S/N) ratio at the input of the receiver in a simple intensity modulated (IM) optical fibre communication system.

6.4.1 *Device design criteria*

We considered how to optimise the structure of a SLA to affect its gain characteristics in Section 5.5.1 in Chapter 5. The polarisation characteristics have also been examined in Section 4.3.2 in Chapter 4. In this section, the effects of the structural design of a SLA on its noise characteristics will be analysed. The results should be compared with those in Sections 5.6.1 and 4.3.2, where a more complete picture of the optimum SLA structure can be seen. A FPA with the parameters listed in Table 6.1 has been analysed, using the equations above. In Fig. 6.8, the relative noise power per unit bandwidth has been calculated using Eq. (6.22), assuming a unit detection bandwidth and unit load resistance for the photodiode, for various output powers, with an unsaturated signal gain G fixed to 20 dB, assuming that the signal wavelength is matched to that of cavity resonance. The four components of total noise power have been separated as (i)–(iv) in Fig. 6.8, assuming no phase noise in the incident signal. It can be seen that the contributions of shot noise power are relatively unimportant compared with that of beat noise, although at higher power levels they can be fairly significant [9]. The beat noise between spontaneous emissions dominate at

Table 6.1. Amplifier parameters used in noise analysis.

Unsaturated signal gain	$G = 20$ dB
Facet reflectivities	$R_1 = R_2 = 30\%$
Group index of active region	$N_g = 3.524$
Differential gain coefficient	$A = 7.2543 \times 10^{-16}$ cm^{-2}
Transparency carrier density	$N_0 = 1.5034 \times 10^{18}$ cm^{-3}
Optical confinement factor	$\Gamma = 0.52$
Amplifier length	$L = 300$ μm
Loss in active region	$\alpha = 0$ cm^{-1}

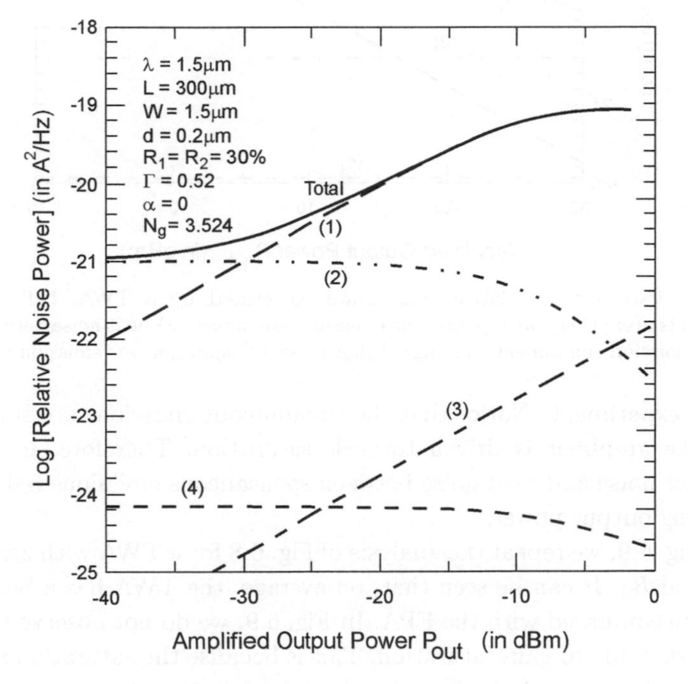

Fig. 6.8. Variation of relative noise power generated by a SLA. Different noise components are: (1) signal-spontaneous emission beat noise, (2) beat noise between spontaneous emission components, (3) signal shot noise, (4) spontaneous emission beat noise.

low power levels. As the signal power increases, the beat noises between signal and spontaneous emission become more dominant. At high output-power, where gain saturation occurs, the total noise power reduces because the beat-noises between spontaneous emission components are suppressed. This is in accord with the postulate made by Schicketanz and Zeidler [48, 50]

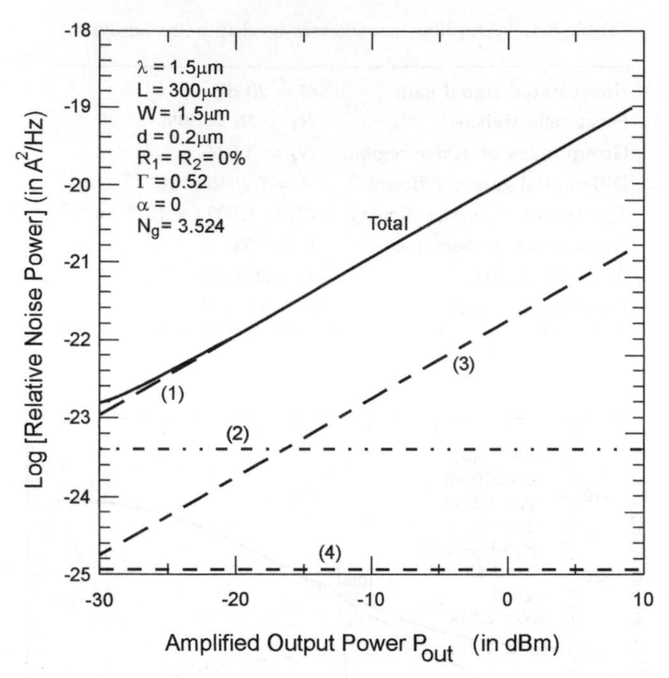

Fig. 6.9. Variation of relative noise power generated by a TWA. Different noise components are: (1) signal-sopntaneous emission beat noise, (2) beat noise between spontaneous emission components, (3) signal shot noise, (4) spontaneous emission beat noise.

in their experiment. Notice that the spontaneous emissions are suppressed when the amplifier is driven towards saturation. Therefore, in Fig. 6.8 both shot noise and beat noise between spontaneous emissions reduce with increasing output power.

In Fig. 6.9, we repeat the analysis of Fig. 6.8 for a TWA with zero values of R_1 and R_2. It can be seen that, on average, the TWA has a better performance compared with the FPA. In Fig. 6.9, we do not observe the noise suppression due to gain saturation. This is because the saturation intensity for TWAs is extremely high in the absence of cavity resonance to enhance the saturation mechanism. Hence, the noise level in TWAs can be increased quite substantially with increasing output power, before it is suppressed by saturation mechanisms. In Figs. 6.10 and 6.11, we investigate the effect of reflectivity R_1 and R_2 on the total noise emitted by the amplifier. In Fig. 6.10, we calculated the variation of relative noise power of the amplifier with varying output mirror facet reflectivity R_2 from 0 to 40% for three particular values of input mirror reflectivity R_1. The output power of the

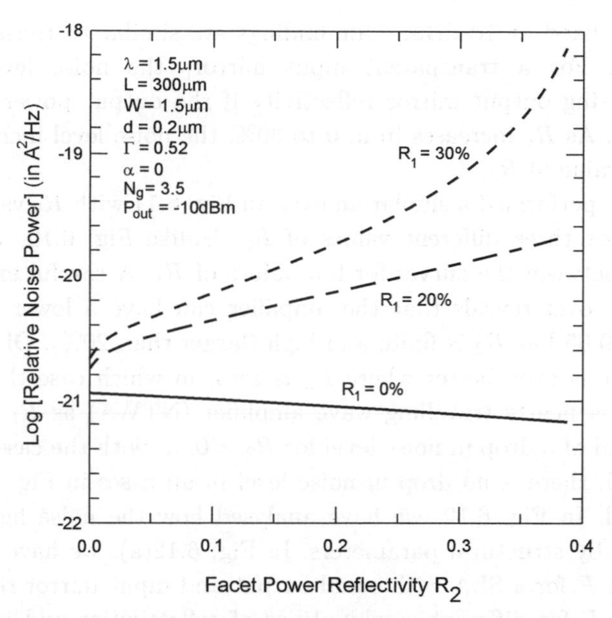

Fig. 6.10. Variation of relative noise power generated by a FPA versus R_2 with R_1 as a parameter.

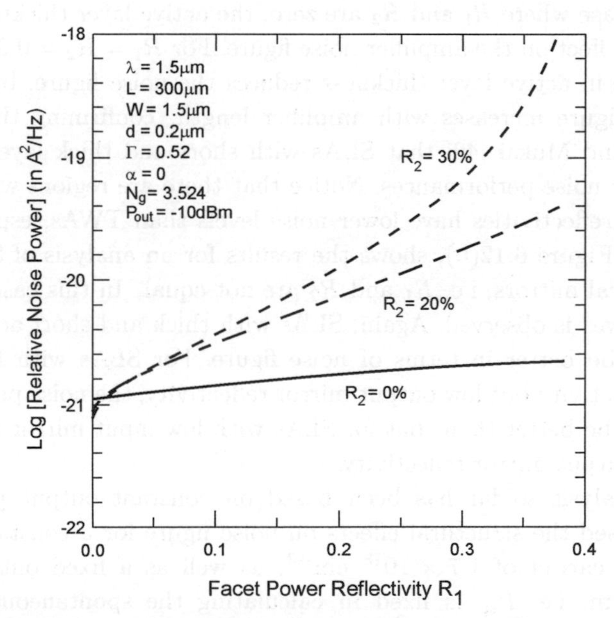

Fig. 6.11. Variation of relative noise power generated by a FPA versus R_1 with R_2 as a parameter.

amplifier is fixed at 10 dBm. Our findings are similar to those of Mukai *et al.* [18]. For a transparent input mirror, the noise level reduces with increasing output mirror reflectivity if the output power has to be maintained. As R_1 increases from 0 to 30%, the noise level increases for a particular value of R_2.

We have performed a similar analysis in Fig. 6.11 with R_1 varying from 0 to 40% for three different values of R_2. Unlike Fig. 6.10, we observe cross-over between the curves for low values of R_1. A careful examination of the cross-over reveals that the amplifier can have a lower noise level with $R_1 < 0.05$ but R_2 is finite and high (larger than 20%). Of course the performance is even better where R_2 is zero, in which case the amplifier approaches a nearly travelling wave amplifier (NTWA) as R_1 is close to zero. Instead of a drop in noise level for $R_2 = 0$, as with the case of $R_1 = 0$ in Fig. 6.10, there is no drop in noise level in all cases in Fig. 6.11 as R_1 is increased. In Fig. 6.12, we have analysed how the noise figure F can be affected by structural parameters. In Fig. 6.12(a), we have shown the variation in F for a SLA with equal output and input mirror reflectivities and length L for different combinations of reflectivities and active layer thickness d. The output power is maintained at -10 dBm as L is varied. For the TWA case where R_1 and R_2 are zero, the active layer thickness has no significant effect on the amplifier noise figure. For $R_1 = R_2 = 0.3$, however, an increase in active layer thickness reduces the noise figure. In all cases, the noise figure increases with amplifier length, confirming the findings of Saitoh and Mukai [49] that SLAs with short and thick layers tend to have better noise performances. Notice that there are regions where SLAs with finite reflectivities have lower noise levels than TWAs, especially for small L's. Figure 6.12(b), shows the results for an analysis of SLAs with asymmetrical mirrors, i.e. R_1 and R_2 are not equal. In this case, no cross over of curves is observed. Again, SLAs with thick and short active layers appear to be better in terms of noise figure. For SLAs with high input mirror reflectivity but low output mirror reflectivity, the noise performance appears to be better than that for SLAs with low input mirror reflectivity but high output mirror reflectivity.

The analysis so far has been based on constant output power. We have analysed the structural effects on noise figure for a constant density of injected carrier of 1.7×10^{18} cm^{-3}, as well as a fixed output power of -10 dBm, i.e. R_{sp} is fixed in calculating the spontaneous emission factor β. This corresponds to the situation where we have tried to maintain the output power by varying the input signal power only. Figure 6.12(c) is

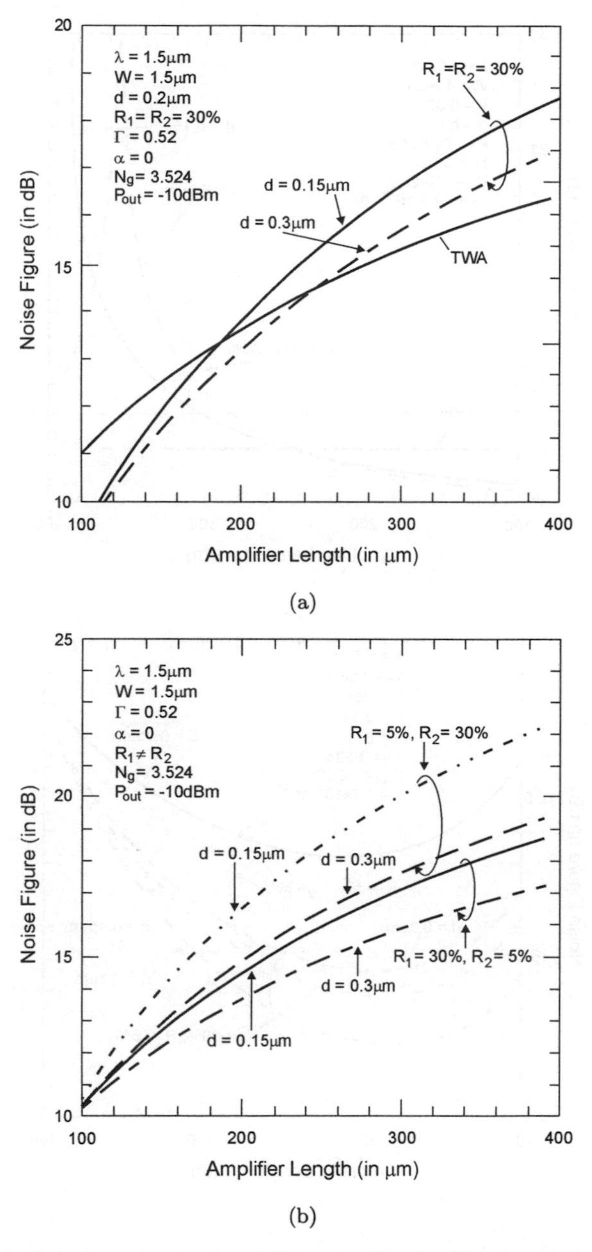

Fig. 6.12. (a) Variation of noise figure F versus the amplifier length with active layer thickness as a parameter, for both TWA and a symmetric FPA. (b) Variation of noise figure versus the amplifier length with thickness as a parameter, for both symmetric and asymmetric FPAs.

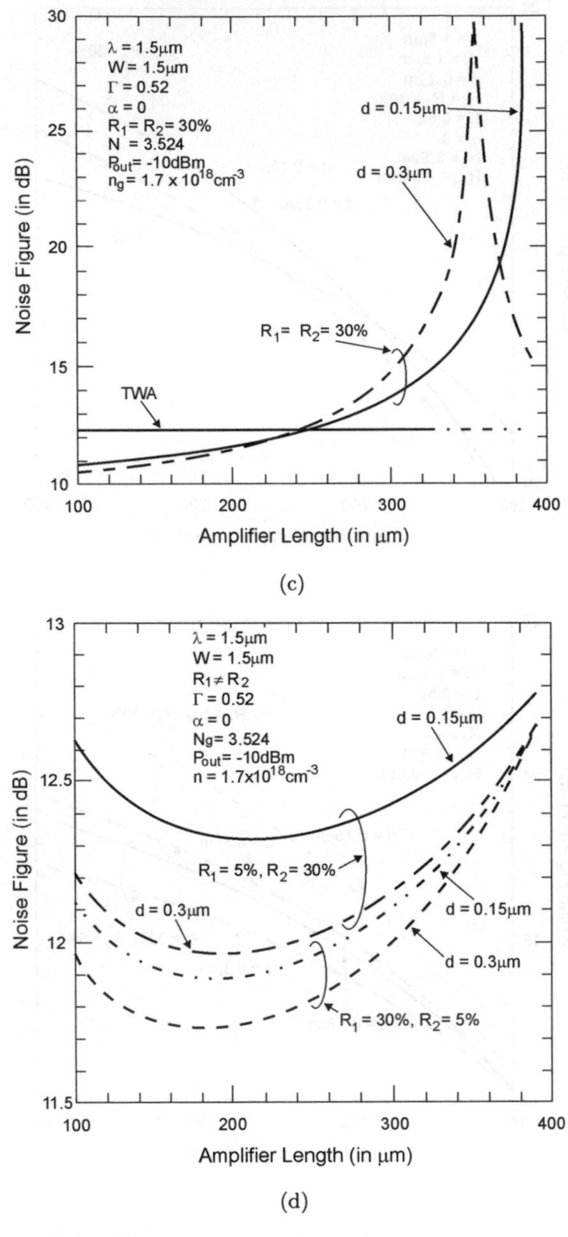

Fig. 6.12. (c) Variation of noise figure F versus the amplifier length L with active layer thickness as a parameter, for both TWA and a symmetric FPA. (d) Variation of noise figure F versus the amplifier length L with active layer thickness as a parameter, for both TWA and an asymmetric FPA.

the resulting curve with the same structural parameters as in Fig. 6.12(a), but under this new assumption. It can be seen that for a constant injection level, F is not affected by the active layer length or thickness for TWAs. On the other hand, there is dramatic variation in F with length, for SLAs with finite facet-reflectivities. For the case where $d = 0.3$ μm, a discontinuity in the curve is observed. This discontinuity represents the structural combinations of L, d and mirror reflectivities with the injection level at which oscillation occurs. A similar observation can be made for the case where $d = 0.15$ μm with R_1 and R_2 being 30%, where the structure approaches oscillation as L approaches 400 μm. Again, for shorter amplifier lengths, the noise performance is better. However, contrary to the condition where output power is fixed, the present calculations show that a thick active layer can actually have poorer noise performance when we have to hold the injection level constant.

In Fig. 6.12(d), we perform the same analysis as that for Fig. 6.12(b), but with the additional assumption of constant carrier injection. The curves show that in this situation, the noise figure exhibits a minimum as L is increased. This minimum corresponds to the best possible achievable noise figure for these configurations. Compared with SLAs with identical input and output mirrors, SLAs with asymmetrical mirror reflectivities can be more useful, as there is a definite achievable minimum value of noise figure which is of great help in designing SLA structures. In this case, a thick active layer appears to be better in terms of noise performance than the symmetrical case. As with Fig. 6.12(b), we found that high input mirror reflectivity but low output mirror reflectivity can help to reduce the noise figure of the SLA. It can be seen from the preceding discussion that the structural effects on noise figure are extremely complicated. Noise power is affected by operational conditions and hence in design it is necessary to know the operational conditions with which the SLA has to work. In addition, assessment of the best structural combinations is difficult, as they depend highly on operational conditions.

6.4.2 *System considerations*

A simple intensity modulation (IM) optical fibre communication system was analysed in Section 5.6.2 in Chapter 5. There were M identical SLAs used as in-line repeaters, each specified by an amplifier gain G, output saturation power P_{sat} and a noise figure F. The input signal power from the transmitter to the first optical fibre is P_{in}. Usually, F should depend on G.

If $G \gg 1$, then the approximation of Eq. (6.27) in Section 6.2.1 applies in the foregoing analysis. For the present study, we assume $G = 20$ dB, as in Section 5.6.2. The transmitter will also emit noise with a signal-to-noise ratio of SNR_o. The bit-error-rate of the receiver with a SLA as its pre-amplifier depends ultimately on the input signal-to-noise ratio of the receiver [3]. Let the minimum input signal-to-noise ratio of the receiver to be SNR_{min} which means that the receiver can achieve a minimum sensitivity of S [50]. The definition of noise figure F is given by [3]:

$$F = \frac{SNR_{in}}{SNR_{out}} . \tag{6.98}$$

The signal propagating across L_s km of a fibre with loss α_f/km (see Fig. 6.17) will be attenuated by $L_s\alpha_f$. Similar attenuation will be observed for the noise generated by the SLAs. Hence, the SNR for a particular stage within the transmission system is kept constant. Notice that this assumption neglects any phase noise introduced to the signal by Kerr effects along the fibre [51]. Under the above assumption, we can derive the input SNR to the optical receiver as [7, 11]:

$$SNR_d = \frac{SNR_o}{F + \frac{F}{G} + \frac{F}{G^2} + \cdots + \frac{F}{G^{M-1}}} . \tag{6.99}$$

It can be seen from the above equation that the noise figure of the first amplifier is the most significant contribution to SNR_d. Hence, for cases when different SLAs can be used, the noise figure of the first amplifier in the chain should be a minimum [11]. In Fig. 6.13, we have calculated the variation of SNR_d with M for $SNR_o = 40$ dB, and G is allowed to be either 20 dB or 25 dB. Similarly, F can be either 5 dB or 10 dB. Interestingly, for the parameters used, the value of SNR input to the receiver does not vary much with M until a critical value, where it deteriorates rapidly. This can be explained by the functional form of Eq. (6.99). Furthermore, it can be seen that increasing G for a particular value of F actually pushes this critical value forward slightly, hence reducing the transmission span. Again, as discussed in the previous chapter, optimum operation of SLAs in a system does not necessarily means a large value of G. Figure 6.13 should be used together with Fig. 5.17 in Chapter 5, because an increase in SNR_o usually requires an increase in P_{in} (in most cases the noise emitted from the transmitter is difficult to control [52]). This will increase the risk that an amplifier in the chain will become saturated.

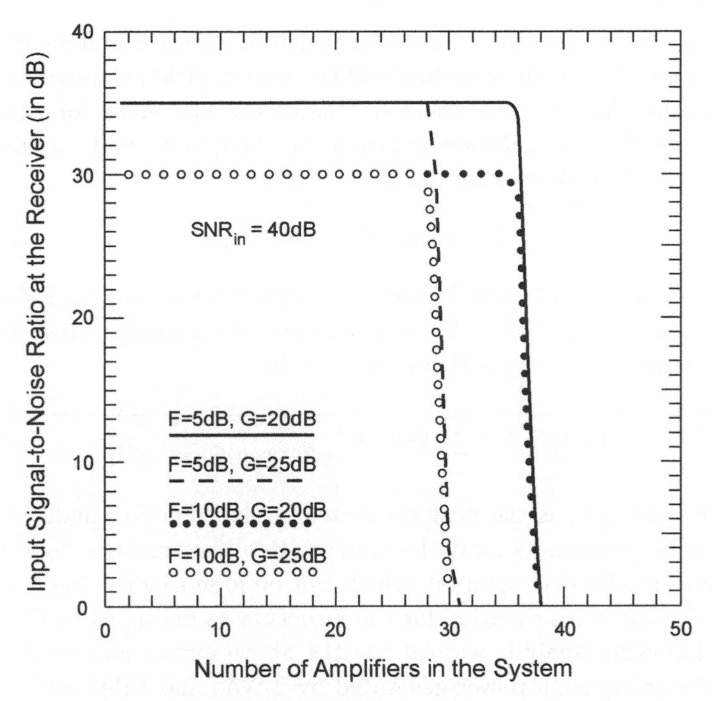

Fig. 6.12. Variation of the input signal-to-noise ratio at the optical receiver versus the number of amplifiers in an optical fibre communication link for different values of amplifier noise figure and signal gain.

6.5 Analysis of SLA Spontaneous Emission Power using Green Function Approach

In Section 6.3, the equivalent circuit model was used to derive the output spontaneous emission power generated by an SLA with a uniform material gain profile. Here we will illustrate an alternative derivation for both TWA and FPA using Green functions, first suggested by Henry [17]. Henry considered that the spontaneous emissions in an optical amplifier can be modeled as discrete Langevin noise sources distributed within the cavity. The resulting wave equation becomes an inhomogeneous one with a driving source term. This requires the use of a Green function in finding the solution of the wave equation [33]. In general, the one dimensional Green function of a one dimensional wave equation in the frequency domain is given by [53]:

$$g_n(z, z') = \frac{Z_{n^+}(z_>)Z_{n^-}(z_<)}{W_n} \tag{6.100}$$

where $z_>$ and $z_<$ are the values of z which are greater and smaller than z', respectively; z' is the coordinate of the source within the structure, and $Z_{n^+}(z_>)$ and $Z_{n^-}(z_<)$ are the solutions of the nth mode for the homogeneous one dimensional wave equation travelling in $+z$ and $-z$ directions, respectively. The Wronskian W_n is defined by:

$$W_n = Z'_{n^+}Z_{n^-} - Z_{n^+}Z'_{n^-} \tag{6.101}$$

where the dashed notation denotes derivative with respect to z. W_n is an independent function of z. The general three dimensional Green function in the frequency domain is therefore given by:

$$G_\omega(r,r') = \sum_n g_n(z,z')\frac{\Phi_n(x,y)\Phi_n(x',y')}{\langle\Phi_n(x,y)\Phi_n(x',y')\rangle} \tag{6.102}$$

where $r = (x,y,z)$ is the position vector in the structure under analysis, and r' is the position vector of the source within the structure. Φ_n is the nth transverse modal field solution, which can be found by solving the homogeneous scalar wave equation for the structure as discussed in Chapter 4. In the following analysis we will use the above expressions to derive the spontaneous emission power generated by TWAs and FPAs with uniform material gain profiles.

6.5.1 *Travelling-wave amplifier (TWA)*

For an infinite waveguide or a TWA with perfect anti-reflection coating, the function $Z_{n\pm}$ and W_n can be found as [7]:

$$Z_{n\pm} = \exp(\pm j\beta_n z) \tag{6.103}$$

$$W_n = 2j\beta_n \tag{6.104}$$

where β_n is the longitudinal propagation constant of the nth mode, and therefore Green's function for a TWA is given by:

$$G_\omega(r,r') = \sum_n \frac{\Phi_n(x,y)\Phi_n(x',y')\exp[j\beta_b|z-z'|]}{2j\beta_n\langle\Phi_n(x,y)\Phi_n(x',y')\rangle}\,. \tag{6.105}$$

The electric field in the structure can be determined from Green's function in the frequency domain by:

$$E_\omega(r) = \int dr' G_\omega(r,r')F_\omega(r')\,. \tag{6.106}$$

The subscripts indicate that the quantities are frequency dependent and the integration is taken over the entire structure. The corresponding magnetic field can be found via Maxwell's equations [33]. The term $F_\omega(r')$ is the magnitude of the Langevin noise source at r'. It satisfies the following statistical properties [7]:

$$\langle F_\omega(r)F_\omega(r')\rangle = \langle F_\omega^*(r)F_\omega^*(r')\rangle = 0 \tag{6.107}$$

$$\langle F_\omega(r)F_\omega^*(r')\rangle = 2D(r)\delta(r-r')\delta(\omega-\omega') \tag{6.108}$$

where D is the diffusion coefficient. It can be shown that by satisfying quantum statistical properties of radiation, the diffusion coefficient is given by [17, 30]:

$$2D(r) = -\frac{2\omega^4 h\varepsilon''(r)n_{\mathrm{sp}}}{\pi c^4} \tag{6.109}$$

where "ε" is the imaginary part of the dielectric constant of the amplifying medium. The noise power within the bandwidth $\Delta\omega$ can be found by first evaluating the Poynting vector and then integrating over the entire transverse section and the frequency spectrum of ω', that is:

$$P_N = \frac{c\Delta\omega}{4\pi}\iint dx\cdot dy\int_0^\infty \langle E_\omega H_\omega^* \exp[-j(\omega-\omega')t] + c.c\rangle d\omega'\,. \tag{6.110}$$

By substituting the diffusion coefficient for the Langevin force term, we obtain the following relation, using the above expressions of Eqs. (6.105) to (6.110):

$$P_N = \frac{\hbar\omega\Delta\omega g_{\mathrm{m}} n_{\mathrm{sp}}\{\exp[(g_{\mathrm{m}}-\alpha_0)L-1]\}}{4\pi^2(g_{\mathrm{m}}-\alpha_0)} \tag{6.111}$$

where g_{m} is the material gain coefficient and α_0 is the loss coefficient. The amplifier is assumed to have a length L. Notice that the power amplification factor A is identical to the exponential term in Eq. (6.111) and since the quantum efficiency of the laser amplifier is given by [14, 54]:

$$\eta = \frac{g_{\mathrm{m}}-\alpha_0}{g_{\mathrm{m}}} \tag{6.112}$$

the spontaneous emission noise power is given by:

$$P_N = \frac{\Delta\omega\hbar\omega n_{\mathrm{sp}}(A-1)}{4\pi^2\eta}\,. \tag{6.113}$$

This is a well known equation, except for the introduction of the quantum efficiency in the denominator of Eq. (6.113). Lowery [55] verified that such inclusion is necessary to obtain a correct ASE power by TLLM simulation.

6.5.2 *Fabry–Perot amplifiers*

Consider a FP cavity of length L, extending from $z = 0$ to L. The field reflectivities at the two end facets are given by r_1 and r_2 at $z = 0$ and $z = L$, respectively. Then, by considering a single transverse mode operation, the quantities $Z_{n\pm}$ and W_n for $n = 0$ are given as:

$$Z_{0-} = r_1 \exp(j\beta_0 z) + r_2 \exp(-j\beta_0 z) \tag{6.114}$$

$$Z_{0+} = r_1[\exp(j\beta_0 z) + r_2 \exp(2j\beta_0 L - j\beta_0 z)] \tag{6.115}$$

which satisfies the boundary condition. The additional factor r_1 in Eq. (6.115) is there so that at threshold $Z_{0+} = Z_{0-}$. The Wronskian is given by:

$$W_0 = 2j\beta_0 r_1[1 - r_1 r_2 \exp(2j\beta_0 L)]\,. \tag{6.116}$$

Substituting these expressions into the noise power P_N discussed above, we obtain the following expression [17]:

$$P_N = \frac{\hbar\omega\Delta\omega n_{\mathrm{sp}}}{4\pi^2\eta(1 - r_1^2)}[1 - \exp(-(g_{\mathrm{m}} - \alpha_0)L)][1 + r_1^2 \exp((g_{\mathrm{m}} - \alpha_0)L)]A \tag{6.117}$$

where the amplifier gain A is given by:

$$A = \frac{(1 - r_1^2)(1 - r_2^2)\exp[(g_{\mathrm{m}} - \alpha_0)L]}{|1 - r_1 r_2 \exp(2j\beta_0 L)|^2}\,. \tag{6.118}$$

When the FPA is driven close to threshold or for frequencies close to the cavity resonance we can approximate:

$$r_1 r_2 \exp[(g_m - \alpha_0)L] \approx 1 \tag{6.119}$$

and hence the noise power is given by

$$P_N = \frac{\hbar\omega\Delta\omega n_{\mathrm{sp}}}{4\pi^2\eta(1 - r_1^2)r_2}[1 - r_1 r_2][r_1 + r_2]A\,. \tag{6.120}$$

The above expressions are identical to those derived in the previous section using the equivalent circuit model.

6.6 Summary

In this chapter, the noise induced by random spontaneous emission events in a SLA has been analysed. Three different approaches reported in the literature have been reviewed. It was found that the approach using the photon statistics master equation was simplest to understand and apply. However, its range of applicability is fairly limited because (i) the analysis becomes more complex as the structure of the SLA becomes more complicated, and (ii) it ignores the effect of the non-uniform material gain profile on the noise characteristics. Nevertheless, this approach allows one to recognise the different components in the total output noise power detected by a photodetector, namely, shot noises due to amplified signal and amplified spontaneous emissions, and beat noise between signal and spontaneous emissions. The contribution of shot noise was found to be negligible in most cases. The other two approaches, rate equations and travelling-wave equations, attempted to overcome the short-comings of the photon statistics master equation, but the complexity grew to such a level that no simple analytical non-numerical solutions could be obtained.

Based on the equivalent circuit developed in Chapter 5 and by considering the quantum mechanical fluctuations due to the negative resistance in the equivalent circuit, the spontaneous emissions in a SLA can be modelled accurately. This has been verified by analysing the spontaneous emission power emitted from a simple FPA with a uniform gain profile. Instead of using Green's function, identical results can be achieved using the equivalent circuit model with simple circuit and signal analysis. The power of the model has been further illustrated by analysing the effect of stray reflections from fibre ends on the spontaneous emission power from the FPA. Finally, the structural design of low-noise SLAs and the effect of noise on the performance of a simple optical fibre communication system have been studied. These results may be used in conjunction with the results of Chapters 4 and 5 to devise an optimum structure for a SLA or to optimise the system configuration to achieve an acceptable performance. The equivalent circuit model proved to be extremely useful in this respect because a single model can yield all the necessary results for optimisation (cf. the separation of the analysis of gain and saturation characteristics with that of noise in most literature). A computer-aided design package, for instance, could be built based on this equivalent circuit model, together with TMM analysis.

References

[1] D. L. Snyder, "Random point processes," John Wiley and Sons, New York, 1975.
[2] R. Loudon, "The quantum theory of light," 2nd edition, Oxford University Press, Oxford, UK, 1983.
[3] H. Taub and D. L. Schilling, *Principles of communication systems*, 2nd Edition (MacGraw-Hill, 1986).
[4] N. Wax, *Selected papers on noise and stochastic process* (Dover, 1954).
[5] K. Kikuchi, C. E. Zah and T. P. Lee, "Measurement and analysis of phase noise generated from semiconductor optical amplifiers," *IEEE J. Quantum Electronics*, Vol. QE-27, No. 3, pp. 416–422, 1991.
[6] C. H. Henry, "Theory of the phase noise and power spectrum of a single mode injection laser," *IEEE J. Quantum Electronics*, Vol. QE-19, No. 9, pp. 1391–1397, 1983.
[7] C. H. Henry, "Phase noise in semiconductor lasers," *IEEE J. Lightwave Technology*, Vol. LT-4, No. 3, pp. 298–311, 1986.
[8] T. Mukai and Y. Yamamoto, "Noise in an AlGaAs semiconductor laser amplifier," *IEEE J. Quantum Electronics*, Vol. QE-18, No. 4, pp. 564–575, 1982.
[9] T. Mukai and T. Yamamoto, "Noise characteristics of semiconductor laser amplifiers," *Electronics Letters*, Vol. 17, No. 1, pp. 31–33, 1981.
[10] Y. Yamamoto, "Noise and error rate performance of semiconductor laser amplifiers in PCM-IM optical transmission systems," *IEEE J. Quantum Electronics*, Vol. QE-16, No. 10, pp. 1073–1081, 1980.
[11] Y. Mukai, Y. Yamamoto and T. Kimura, "Optical amplification by semiconductor lasers," *Semiconductor and semimetals*, Vol. 22, Part E, pp. 265–319, Academic Press, 1985.
[12] G. P. Agrawal and N. K. Dutta, *Long-wavelength semiconductor lasers* (New York, Van-Nostrad Reinhold, 1986).
[13] E. L. Goldstein and M. C. Teich, "Noise in resonant optical amplifiers of general resonator configuration," *IEEE J. Quantum Electron.*, Vol. 25, No. 11, pp. 2289–2296, 1989.
[14] G. Bjork and O. Nilsson, "A tool to calculate the linewidth of complicated semiconductor lasers," *IEEE J. Quantum Electronics*, Vol. QE-23, No. 8, pp. 1303–1313, 1987.
[15] J. A. Armstrong and A. W. Smith, "Intensity fluctuation in GaAs laser emission," *Phys. Rev.*, Vol. 140, No. 1A, pp. A155–A164, 1965.
[16] A. Yariv, *Optical Electronics*, 3rd Edition (Holt-Saunders, 1985).
[17] C. H. Henry, "Theory of spontaneous emission noise in open resonators and its application to lasers and optical amplifiers," *IEEE J. Lightwave Technology,* Vol. LT-4, No. 3, pp. 288–297, 1986.
[18] T. Mukai, Y. Yamamoto and T. Kimura, "S/N and error rate performance in AlGaAs semiconductor laser preamplifier and linear repeater systems," *IEEE Trans. Microwave Theory Tech.*, Vol. MTT-30, No. 10, pp. 1548–1556, 1982.

[19] H. Kressel, *Semiconductor devices for optical communications*, 2nd Edition (Springer-Verlag, 1982).

[20] T. Saitoh and T. Mukai, "Recent progress in semiconductor laser amplifiers," *J. Lightwave Technology*, Vol. LT-6, No. 11, pp. 1156–1164, 1988.

[21] J. C. Simon, "GaInAsP semiconductor laser amplifiers for single-mode fiber communications," *J. Lightwave Technology*, Vol. LT-5, No. 9, pp. 1286–1295, 1987.

[22] K. Hinton, "A model for noise processes in semiconductor laser amplifiers, Part 1: The travelling-wave semiconductor laser amplifier," *Opt. Quantum Electron.*, Vol. 21, pp. 533–546, 1989.

[23] K. Hinton, "Optical carrier linewidth broadening in a travelling wave semiconductor laser amplifier," *IEEE J. Quantum Electronics*, Vol. QE-26, No. 7 pp. 1176–1182, 1990.

[24] K. Hinton, "Model for noise processes in semiconductor laser amplifiers, Part 2: The Fabry–Perot semiconductor laser amplifier," *Opt. Quantum Electron.,* Vol. 23, pp. 755–773, 1991.

[25] T. J. Shepherd and E. Jakeman, "Statistical analysis of an incoherently coupled, steady-state optical amplifier," *J. Opt. Soc. Am. B*, Vol. 4, No. 11, pp. 1860–1869, 1987.

[26] K. Vahala and A. Yariv, "Semiclassical theory of noise in semiconductor lasers-Parts I and II," *IEEE J. Quantum Electronics*, Vol. QE-19, No. 6, pp. 1096–1109, 1983.

[27] K. Kikuchi, "Proposal and performance analysis of novel optical homodyne receiver having an optical preamplifier for achieving the receiver sensitivity beyond the shot-noise limit," *IEEE J. Quantum Electronics*, QE-19, No. 7, pp. 1195–1199, 1983.

[28] P. Spano, S. Piazzola and M. Tamburrini, "Phase noise in semiconductor lasers; a theoretical approach," *IEEE J. Quantum Electronics*, Vol. QE-19, No. 7, pp. 1195–1199. 1983.

[29] O. Nilsson, Y. Yamamoto and S. Machida, "Internal and external field fluctuations of a laser oscillator: Part II-Electrical circuit theory," *IEEE J. Quantum Electronics*, Vol. QE-22, No. 10, pp. 2043–2051, 1986.

[30] W. H. Louisell, *Quantum statistical properties of radiation* (New York, John Wiley and Sons, 1973).

[31] M. I. Sargent, M. O. Scully and W. E. Lamb Jr., *Laser physics* (Addison-Wesley, 1974).

[32] Y. Yamamoto, "AM and FM quantum noise in semiconductor lasers-Part I: Theoretical analysis," *IEEE J. Quantum Electronics*, Vol. QE-19, No. 1, pp. 34–46, 1983.

[33] C. A. Balanis, *Advanced engineering electromagnetics* (John Wiley and Sons, 1989).

[34] K. B. Kahen, "Green's functional approach to resonant cavity analysis," *IEEE J. Quantum Electron.*, Vol. QE-28, No. 5, pp. 1232–1235, 1992.

[35] H. Ghafouri-Shiraz and C. Y. J. Chu, "Analysis of waveguiding properties of travelling-wave semiconductor laser amplifiers using perturbation techniques," *Fiber and Integrated Optics*, Vol. 11, pp. 51–70, 1992.

[36] H. S. Sommers Jr., "Critical analysis and correction of the rate equation for the injection laser," *IEEE Proc.*, Part J, Vol. 132, No. 1, pp. 38–41, 1985.
[37] H. C. Casey and F. Stern, "Concentration-dependent absorption and spontaneous emission of heavily doped GaAs," *J. Appl. Phys.*, Vol. 47, No. 2, pp. 631–643, 1976.
[38] F. Stern, "Calculated spectral dependence of gain in excited GaAs," *J. Appl. Phys.*, Vol. 47, No. 2, pp. 5382–5386, 1976.
[39] D. Marcuse, "Classical derivation of the laser rate equation," *IEEE J. Quantum Electronics*, Vol. QE-19, No. 8, pp. 1228–1231, 1983.
[40] J. L. Gimlett and N. K. Cheung, "Effects of phase-to-intensity noise conversion by multiple reflections on giga-bit-per-second DFB laser transmission system," *J. Lightwave Technology*, Vol. LT-7, No. 6, pp. 888–895, 1989.
[41] J. L. Gimlett *et al.*, "Impact of multiple reflection noise in Gbit/s lightwave system with optical fibre amplifiers," *Electronics Letters*, Vol. 25, No. 20, pp. 1393–1394, 1989.
[42] L. Gillner, "Comparative study of some travelling-wave semicinductor laser amplifier models," *IEEE Proc.*, Part J, Vol. 139, No. 5, pp. 339–347, 1992.
[43] C. Y. J. Chu and H. Ghafouriz-Shiraz, "Effects of stray Reflections on performance of semiconductor Laer Diode Amplifiers," *Optics and Laser Technology*, In press.
[44] G. Bjork and O. Nilsson, "A new exact and efficient numerical matrix theory of complicated laser structures: Properties of asymmetric phase-shifted DFB lasers," *J. Lightwave Technolog.*, Vol. LT-5, No. 1, pp. 140–146, 1987.
[45] H. Ghafouri-Shiraz, K. Cameron and R. M. A. Fatah, "Achievement of Low Coupling Loss Between a High 'NA' MCVD-Single-Mode Fibre by Conical Microlens," *Microwave and Optical Technology Letters*, Vol. 3, No. 6, pp. 214–217, June 1990.
[46] J. John, T. S. M. Maclean, H. Ghafouri-Shiraz and J. Niblett, "Matching of single-mode fibre to laser diode by microlenses at 1.5 μm wavelength," *IEEE Proc.*, Part J, Optoelectrics, Vol. 141, No. 3, pp. 178–184, 1994.
[47] M. Born and E. Wolf, *Principle of optics*, 6th Edition (Pergamon Press, 1983).
[48] D. Schicketanz and G. Zeidler, "GaAs-double-heterostructure lasers as optical amplifiers," *IEEE J. Quantum Electron.*, Vol. QE-11, No. 2, pp. 65–69, 1975.
[49] T. Saito and T. Mukai, "1.5 μm GaInAsP travelling-wave semiconductor laser amplifier," *IEEE J. Quantum Electron.*, Vol. QE-23, No. 6, pp. 1010–1020, 1987.
[50] T. Li, *Topics in lightwave transmission system* (Academic Press, 1991).
[51] J. P. Gordon and L. F. Mollenauer, "Phase noise in photonic communications systems using linear amplifiers," *Opt. Lett.*, Vol. 15, No. 23, pp. 1351–1353, 1990.
[52] Y. Yamamoto, *Coherence amplification and quantum effects in semiconductor lasers* (John Wiley ans Sons, 1991).

[53] P. M. Morse and H. Feshbach, *Methods of theoretical physics: Part I* (MacGraw-Hill, 1953).
[54] G. H. B. Thompson, *Physics of semiconductor laser devices* (New York, John Wiley and Sons, 1980).
[55] A. J. Lowery, "Amplified spontaneous emission in semiconductor laser amplifiers: validity of the transmission-line laser model," *IEEE Proc.*, Part J, Vol. 137, No. 4, pp. 241–247, 1990.
[56] C. Y. J. Chu and H. Ghafouri-Shiraz, "Equivalent Circuit Theory of Spontaneous Emission Power in Semiconductor Laser Optical Amplifiers," *IEEE, J. Lightwave Technology*, Vol. LT-12, No. 5, pp. 760–767, May 1994.
[57] C. Y. J. Chu and H. Ghafouri-Shiraz, "Analysis of Gain and Saturation Characteristics of a Semiconductor Laser Optical Amplifier using Transfer Matrices," *IEEE J. Lightwave Technology*, Vol. LT-12, No. 8, pp. 1378–1386, August 1994.

Chapter 7

Experimental Studies on Semiconductor Laser Diode Amplifiers

7.1 Introduction

In Chapters 4–6, several characteristics of semiconductor laser amplifiers (SLAs) were studied theoretically. These include: the effective refractive index of the active region and the polarisation dependence on modal gain coefficients (Chapter 4); amplifier gain and output saturation power (Chapter 5); spontaneous emissions from a SLA and their effect on the noise characteristics of the amplifier (Chapter 6). An equivalent circuit consisting of negative resistances and lossless transmission lines has been developed, which can be used as a basis for computer-aided design and analysis of SLAs, even with fairly complex structures. This has been illustrated by using it to analyse the effects of stray reflections from the ends of coupling fibres on the spontaneous emissions generated by SLAs, and on their gain characteristics (see Subsection 6.3.3 in Chapter 6). The investigation so far has been purely theoretical. Before this theory can be used for the design of devices or systems, it is necessary to check the validity of its predictions by experimental studies. These experimental studies are arranged in three parts: (i) characterisation of the recombination mechanism in the SLA by measuring its output spontaneous emission power against injection or bias current; (ii) measurement of the gain characteristics; and (iii) measurement of the noise generated by a SLA in the presence of an input signal. These results will be compared with those predicted by theory. The experimental set-up will also demonstrate some of the practical aspects of operating SLAs.

7.2 Basic Set-up for Measurements

Experimental work has been based on a SLA supplied by British Telecom Research Laboratories (BTRL). The centre wavelength of the gain spectrum is and the principal structural parameters are given in Table 7.1. The amplifier is a **MOVPE** grown buried heterostructure laser coated with multi-layer anti-reflection (AR) coatings, and is designed to be polarisation insensitive. In order to test such a laser amplifier, it is necessary to set up a test rig, which includes a laser source to provide input signals to the amplifier, the SLA for test, and a photo-detector to measure the light output from the amplifier [1]. The details of mounting these components and the electronics for biasing and for temperature control are described in the following subsections.

Table 7.1. Structural parameters of the SLA used in the experiment.

Length of amplifier	$L = 500\ \mu\text{m}$
Width of active region	$W = 2\ \mu\text{m}$
Thickness of active region	$d = 0.1\ \mu\text{m}$
Optical confinement factors	$\Gamma_{\text{TE}} = 0.16$
Optical confinement factors	$\Gamma_{\text{TM}} = 0.13$
Facet reflection coefficients	$R_1 = R_2 = 10^{-4}$

7.2.1 *The semiconductor laser diode source*

The laser source is a semiconductor distributed feedback (DFB) laser diode, again supplied by BTRL. The emission wavelength is 1.514 μm, and the emission spectrum of this type of laser diode is very stable against small fluctuations in temperature and bias current [2].

As with all types of semiconductor laser diodes, a stable current source is required to bias the device [3]. This current source must be variable to control the output power emitted by the laser. For the initial trial, the circuits were designed to bias the laser diode both in a pulsed regime (this will reduce heating of the *p.n.* junction [4]) and in a continuous wave (CW) regime, which requires a constant current source. A simple circuit can be used to achieve this, which consists of two parts. We first examine the constant current source shown in Fig. 7.1. The two *n.p.n* transistors form a two-stage amplifier which provides the constant current required to bias

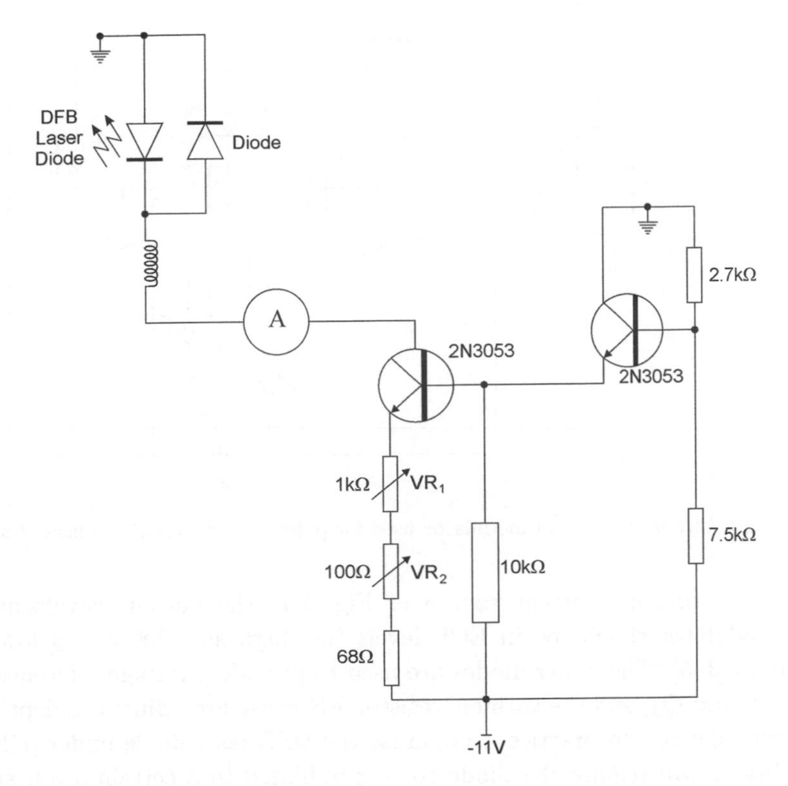

Fig. 7.1. Circuit diagram of a current source bias in both the laser diode and laser amplifier.

the laser diode. The two variable resistors VR_1 and VR_2 are used to adjust the emitter current of the transistor, which will in turn vary the collector current flowing in the transistor. This collector current flows across the laser diode, which provides the necessary bias current. Resistors VR_1 and VR_2 provide coarse and fine control of the laser diode bias current, respectively. The inductor in series with the DFB laser diode provides protection against spurious voltage spikes which may occur during, for instance, turning on the circuit [2]. Also an anti-parallel junction diode is connected across the laser diode to protect it against any accidental reversal of voltage across the laser.

The second part of the circuit is shown in Fig. 7.2, which is the modulator circuit used for pulsed operation of the DFB laser diode. The circuit is based on a simple differential amplifier using a long-tail pair with high frequency transistors, such as the BFY90 series. For compatibility

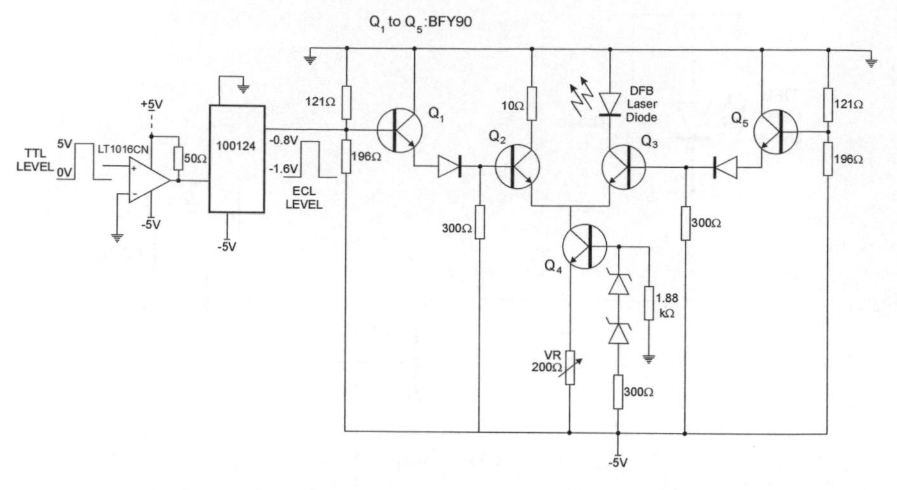

Fig. 7.2. Circuit diagram of modulator used for pulsed operations of the laser diode.

with the constant current source of Fig. 7.1, the output waveform of the modulator should be in ECL levels (i.e. high at −0.8 V and low at −1.6 V) [3–5]. The Zener diodes are used to provide a voltage reference to bias Q_2 and Q_3, and the variable resistor VR is used to adjust the depth of the modulation. In practice, to operate the DFB laser diode under pulsed conditions will require the diode to be pre-biased to a certain level, such that the effect of relaxation oscillation can be minimised [3, 6]. Thus the modulator in Fig. 7.2 is connected to the DFB laser diode with the constant current shown in Fig. 7.1. Furthermore, most signal generators produce signals with levels between +5 V and −5 V. These signals are converted to ECL levels in the modulator by passing first through a LT1016CN comparator to convert to TTL levels and then through a 100124 TTL-ECL converter. For pulsed operation the laser is first biased to its nominal operating current with the controlled current source; the modulator is then adjusted to produce the required depth of modulation. The current source and the modulator are shown in Fig. 7.3, where the metal case was used to shield the circuit from external interference. Although the emission spectra of DFB laser diodes are more stable than FP laser diodes, their operating temperature has to be stabilised. There are a number of different ways to achieve this goal [3, 7]. One way is to detect the output power of the laser diode by a photo-detector, which is used in a feedback controller to control the bias current of the laser diode. This will eliminate any increase in optical power due to temperature drift, which would subsequently cause

Fig. 7.3. Current source to bias DFB laser diode.

an increase in bias current, resulting eventually in *catastrophic breakdown* [2, 6]. However, the feedback has to be calibrated very carefully to maintain the stability of the entire circuit.

A simpler way is to use an open-loop control. A comparison between the open-loop and closed-loop feedback controls is illustrated in Figs. 7.4(a) and 7.4(b). A sensor that converts heat to electrical current is necessary. A common choice is a Peltier heat pump [8], which is placed between the heat sink and the device that is to be temperature controlled. The heat extracted by the pump is controlled by a bias current (see Fig. 7.5). The heat pump is biased at a fixed current. The device under control will be emitting an amount of heat energy Q. According to the magnitude of the bias current, the heat pump will "pump" away excessive heat (by converting it into an electrical current) such that the temperature of the device falls to a designated value T. At equilibrium, therefore, the device can maintain a fixed temperature by controlling the current flowing in the heat pump. The current required to control the temperature for a particular device can be found from a typical set of calibration curves, such as the ones shown in Fig. 7.5, which are usually applied by the manufacturer. The temperature

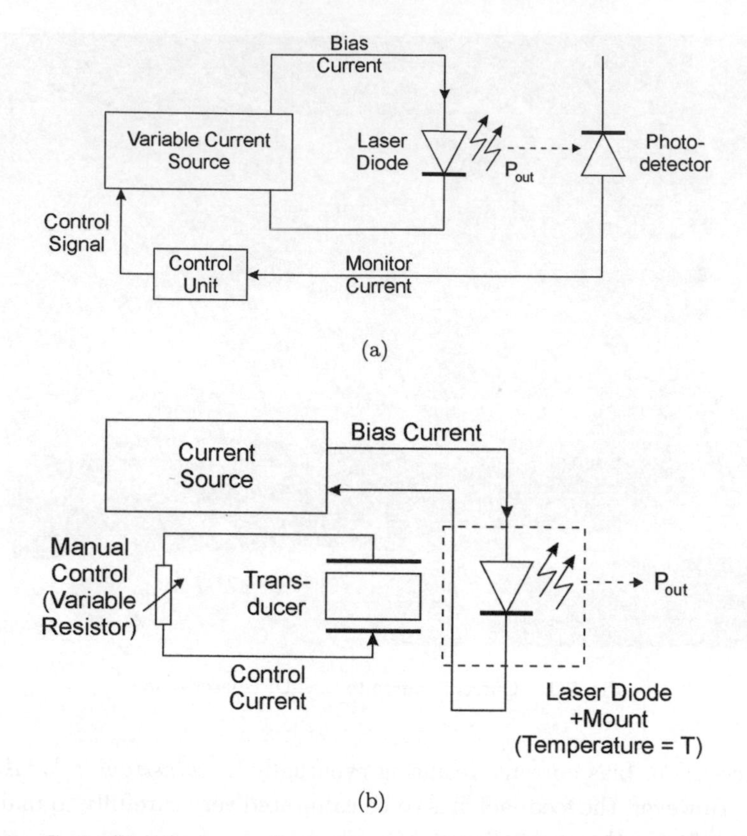

Fig. 7.4. Controlling the temperature of the laser diode using: (a) A closed loop system. (b) An open loop system.

control system shown in Fig. 7.4(b), is an open-loop types in the sense that the current flowing into the heat pump has to be controlled manually. However, in practice, whilst closed-loop control is necessary in unmanned optical communication systems, for experimental work an open-loop control is adequate.

To supply current to the heat pump, we can use a similar circuit to that used to bias the laser diode in Figure 7.1. However, although the heat pump requires a very small bias voltage ($V_{\max} \leq 0.5$ V), it needs a fairly large amount of current ($I_{\max} \approx 1.2$ A). Hence, the circuit in Fig. 7.1 is not suitable for this purpose. Instead, the circuit shown in Fig. 7.6 (see also Fig. 7.7) was used where the VR controls the pump current and is adjusted manually to maintain the desired temperature at the laser. Figure 7.8

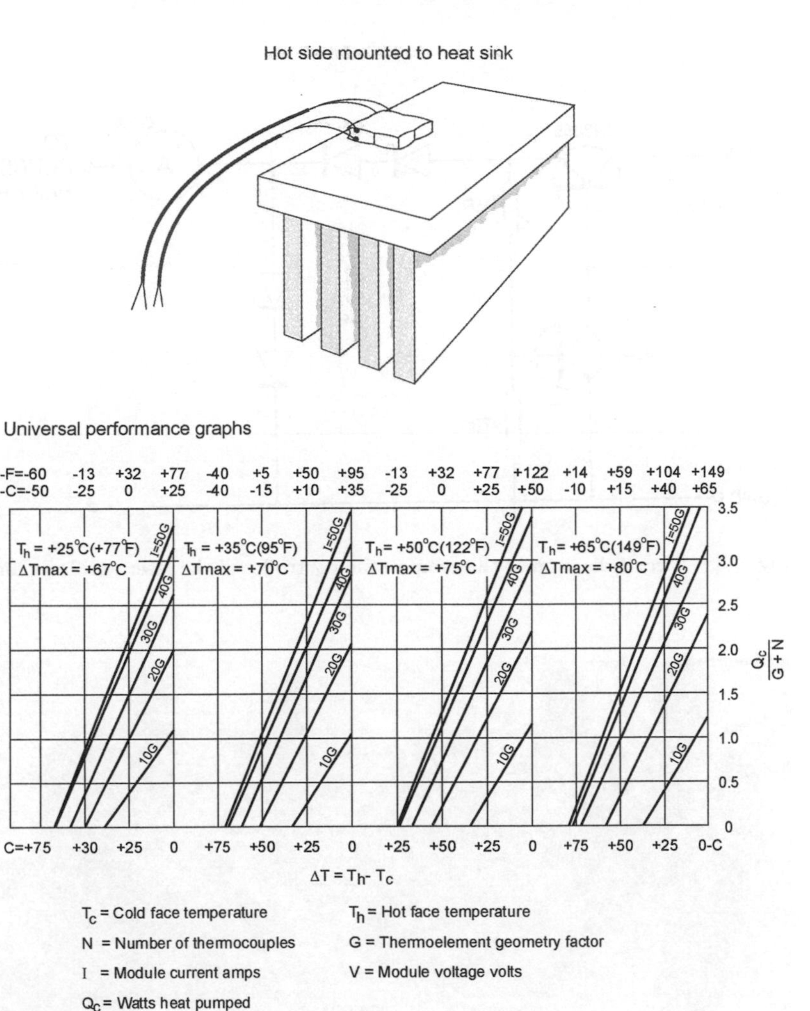

Fig. 7.5. A sketch of the Peltier heat pump used as temperature controller, together with calibration curves.

shows how the laser diode is mounted on a micro-positioner. The laser is fixed to a small copper plate attached to the Peltier heat pump, which is isolated thermally by a Perspex block, so that it is only the excess heat from the laser that is pumped away. The mount is sprayed black to minimise stray reflections, as such reflections can affect the stability of the emission spectrum [9].

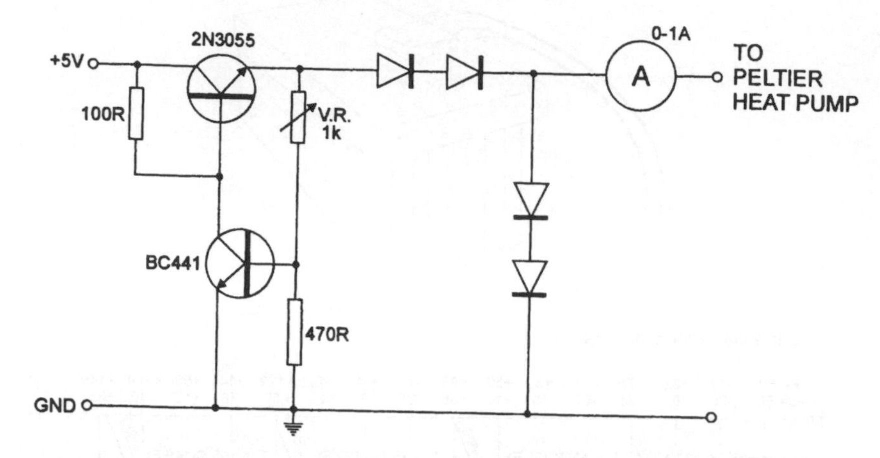

Fig. 7.6. Circuit diagram of the temperatue controller of the laser and laser amplifier.

Fig. 7.7. Temperature control for laser diode and amplifier.

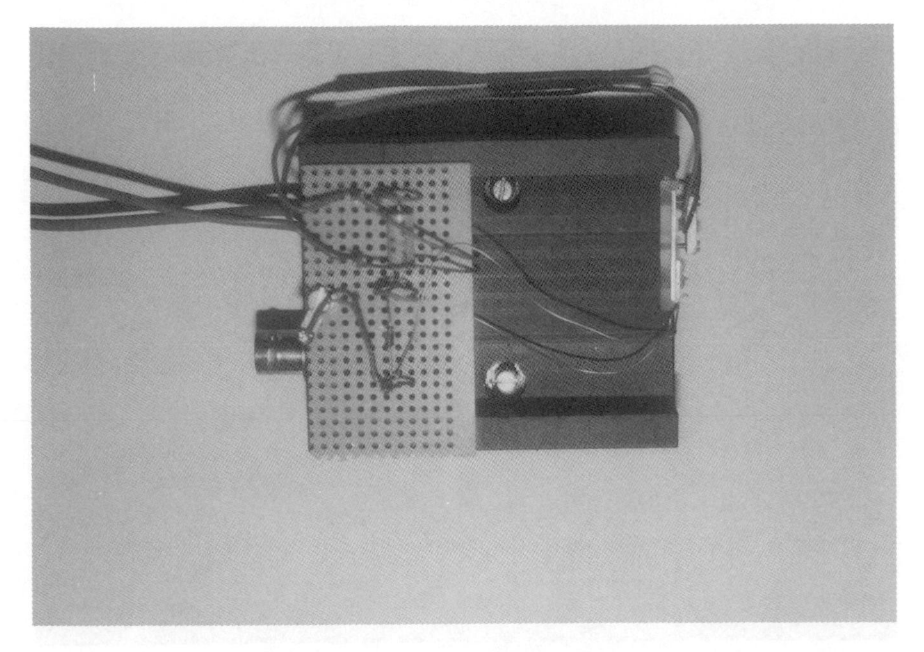

Fig. 7.8. Mount for the DFB laser diode.

7.2.2 *Semiconductor laser diode amplifier*

The SLA used in the tests has to be biased with a constant current and also has to be stabilised in temperature. The controlled current source is the same as in Fig. 7.1 and the Peltier heat pump circuit is as in Fig. 7.4. The actual circuit constructed for the current source is as shown in Fig. 7.9. The mounting is similar to the laser diode (see Fig. 7.10) differing only in that the SLA is raised to facilitate the coupling of the input and output signals.

7.2.3 *Detection circuit*

Two types of detector were required for the experimental study. A precision optical power meter, by Anritsu, with its own optical detector head, was used for most of the measurements. This detector head has a relatively large surface area, which improves its sensitivity and coupling efficiency. For noise measurements it is necessary to use a simple detection circuit consisting of a photo-detector connected to a load resistor [3]. The photo-detector is an InGaAsP pin photo-diode supplied by British Telecom Research Laboratories. This device has a quantum efficiency approaching 95%.

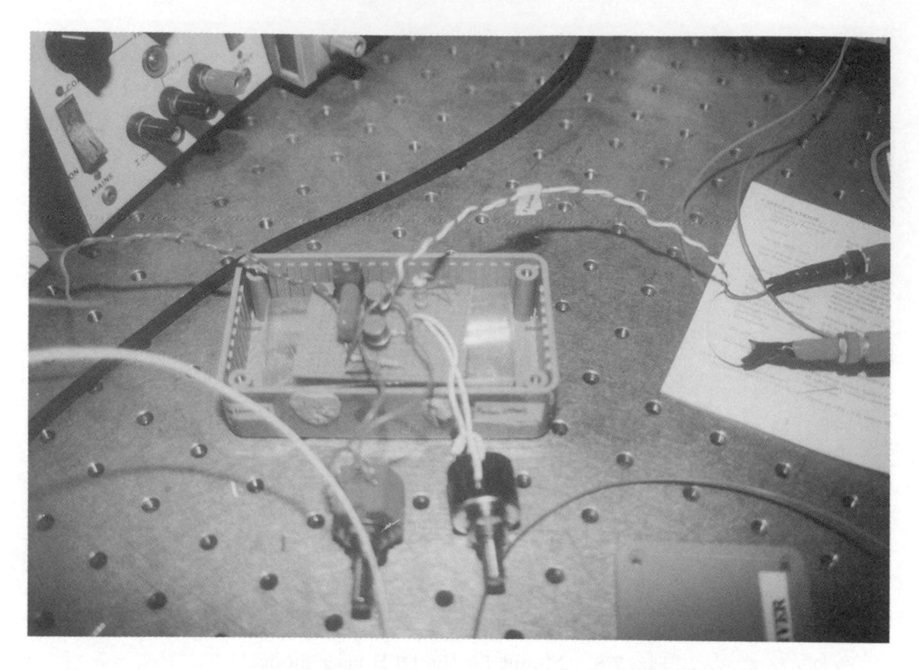

Fig. 7.9. Constant current source used to bias the laser amplifier.

Fig. 7.10. Mount used for the semiconductor laser amplifier.

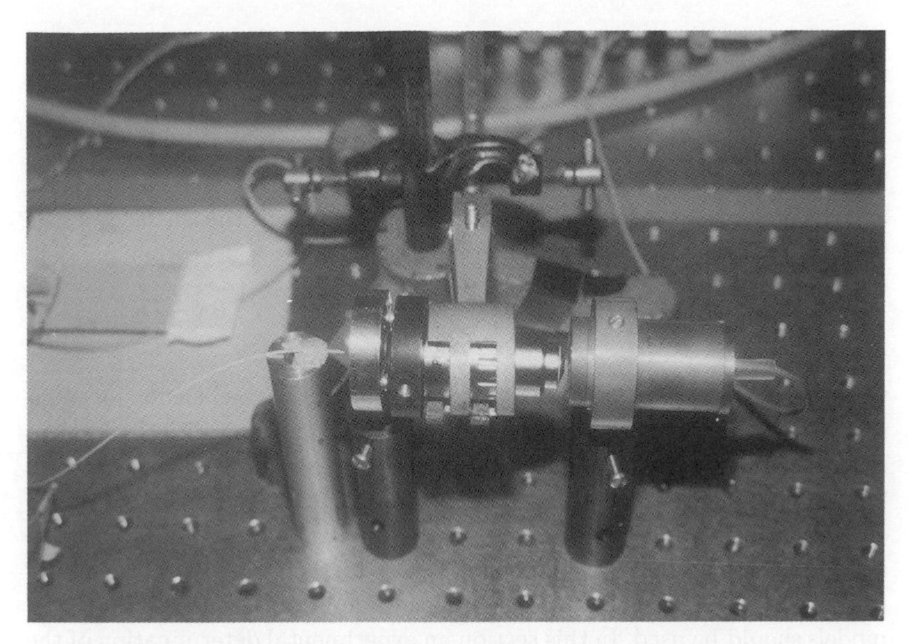

Fig. 7.11. Detector set-up using the pin photo-diode supplied by BTRL, showing also the simple detection circuit at the back.

One advantage in using this type of photo-diode is that the shot noise due to thermal fluctuations, and the dark current in the load resistor, are low [6, 7]. However, the surface area of this photo-diode is quite small (the diameter of the active area is approximately 50 μm), therefore care has to be taken to obtain good coupling efficiency between the signal and the photo-diode. This photo-diode is mounted on a copper cylinder, and connected as shown in Fig. 7.11. Two additional optical components are included in this assembly; a narrow band optical filter with a bandwidth of around 10 nm and a lens. The filter is used to minimise unwanted spontaneous emission noise generated by the amplifier, whilst the lens is used to improve the coupling efficiency between the output from the filter to the photo-detector.

7.3 Experimental Studies on Recombination Mechanisms

The objective of the first set of experiments is to investigate the recombination mechanisms in the SLA. This is necessary because we saw in Chapter 3 that there are two possible models for the carrier recombination rate R. The bimolecular model which takes into account radiative recombination due to spontaneous emissions only; if the active region is undoped,

$$R = B_{\text{rad}}\, n^2 \tag{7.1}$$

The other possible model will include the effects of Auger recombination and recombination with defects and impurities near the surface of the material as well [2, 10] (see also the discussion in Subsection 3.2.2 in Chapter 3) for which

$$R = A_{\text{nr}}\, n + B_{\text{rad}}\, n^2 + C_{\text{Aug}}\, n^3 \tag{7.2}$$

where A_{nr}, B_{rad} and C_{Aug} are the recombination coefficients for non-radiative recombination of carriers with defects and impurities, radiative recombination owing to spontaneous emissions, and non-radiative recombination of carriers owing to Auger processes, respectively. In the above equations, n is the injected carrier density, and we have assumed that the active region of the SLA is undoped. Before further analysis of the device under study can be made, it is necessary to determine which of these two models should be used. In this section an experimental determination of the recombination mechanisms will be described such that we can determine whether we should use Eq. (7.1) or (7.2) in subsequent analyses.

7.3.1 *Principles of the experimental measurement*

As discussed in the previous chapter (Subsection 6.3.2), if a SLA is biased by a finite injection current in the absence of an input signal, a finite amount of optical power will be detected at the output owing to amplified spontaneous emissions (ASEs). This is because the random spontaneous emissions generated by the amplifier are amplified by the optical gain provided by the injection current. The amount of spontaneous emission power detected will be proportional to the optical gain in the active region. On the other hand, the magnitude of the optical gain will also depend on the injection current. We have seen in Chapter 5, Subsection 5.3.2, that at a distance z within the active region, the modal gain coefficient $g(z)$ is related to the injection current density j via the rate equation [11]

$$\frac{j}{ed} = R(z) + \frac{c\Gamma}{N_{\text{g}}\Delta E}\int_0^{\Delta E} g(z,E)S(z,E)dE \tag{7.3}$$

where d is the active layer thickness, E is the energy of the photon, Γ is the optical confinement factor, N_{g} is the group effective index of the active region, S is the photon density at z, and $\Delta E = h\Delta f$ where Δf is the optical bandwidth of the material. The carrier recombination rate R can take the form of either Eq. (7.1) or (7.2), depending whether Auger recombination is important. Therefore Eq. (7.3) can be solved in terms of the carrier density n for a particular injection current density j, in which the modal gain coefficient can be found from the material gain coefficient g_{m}:

$$g(z,E) = \Gamma[g_{\mathrm{m}}(z,E) - \alpha_{\mathrm{a}}(z,E)] - (1-\Gamma)\alpha_{\mathrm{c}}(z,E) \tag{7.4}$$

where we have assumed that the optical confinement factor Γ can be well defined ([12]; see also Chapter 4). This is true in the present case, as the SLA supplied by BTRL is a buried heterostructure. The solution procedures of these equations have already been outlined in Subsection 5.3.2 and Section 5.4. The peak-gain of material gain coefficient for a particular value of n can be found either by the well-known linear model

$$g_{\mathrm{m}} = A(n - n_{\mathrm{o}}) \tag{7.5}$$

or by the parabolic model suggested by Ghafouri-Shiraz [13] (Subsection 3.2.2 in Chapter 3):

$$g_{\mathrm{m}} = an^2 + bn + c \tag{7.6}$$

Knowing n for a particular injection current and hence the peak material gain g (from either Eq. (7.5) or (7.6)) and the peak modal gain coefficient g allows us to calculate the output spontaneous emission power P_{out} generated by the amplifier at the peak wavelength of the material gain spectrum (known hereafter as the *peak-gain wavelength*). Obviously the amount of P_{out} generated by a SLA biased at particular injection current density j will then be determined by the model used for the recombination rate R, as discussed by Wang *et al.* [14]. Hence, by measuring the actual value of P_{out} at the peak-gain wavelength for the device with parameters listed in Table 7.1 [15–16], and comparing the measurements with the theoretical values predicted by the two recombination models, we can then see which model, Eq. (7.1) or (7.2), is the appropriate one to be used in subsequent analysis [31].

7.3.2 *Experimental procedures*

The laser amplifier with parameters tabulated in Table 7.1 was mounted and connected to the current source and the temperature controller. The temperature was maintained at room temperature (20°C) by biasing the

Peltier heat pump at a fixed current (around 400 mA), which can be determined from the calibration curves in Fig. 7.5. A fixed injection current was used to bias the amplifier. The peak-gain wavelength of the amplifier supplied by the manufacturer is 1.51 μm, and the output power of the amplifier was measured at this wavelength by the Anritsu optical power meter. To minimise coupling loss between the amplifier output and the detection head of the power meter, the detector had to be as close as possible to the amplifier *without* hitting the facet and damaging the device. This was done by aligning it with a micropositioner and the aid of a microscope. The output spontaneous emission power is measured by varying the injection current from 10 mA to 80 mA in 5 mA steps.

7.3.3 *Results and discussions*

The measured results are shown as the solid circles in Fig. 7.12 [31]. They are compared with the theoretical spontaneous emission power outputs for the two recombination models given by Eqs. (7.1) and (7.2). The theoretical results calculated by considering radiative recombination owing to

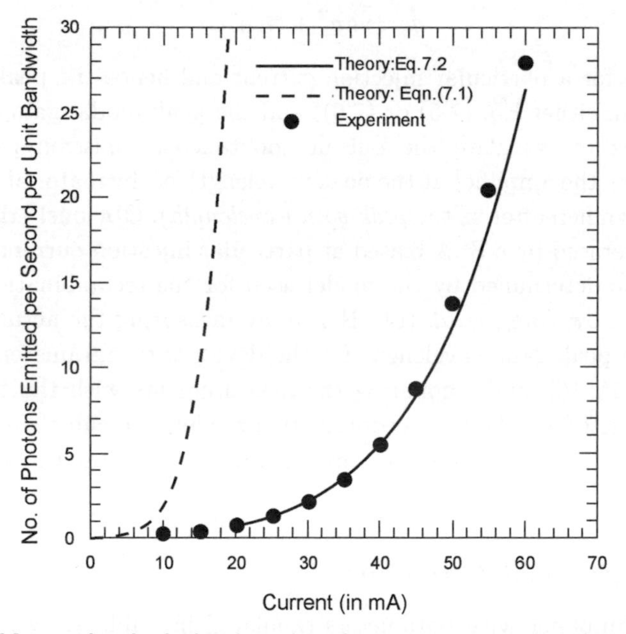

Fig. 7.12. Measured and calculated number of spontaneous emission photons generated by the SLA at different bias current levels. Two recombination models have been used for comparison.

Table 7.2. Material parameters used in the analysis of the SLA [1, 8, 15–16].

Differential gain coefficient	$A = 4.63 \times 10^{-16}$ cm^2
Transparency carrier density	$n_0 = 8.31 \times 10^{17}$ cm^{-3}
Non-radiative recombination constant	$A_{\mathrm{nr}} = 10^{-8}$ s
Spontaneous recombination constant	$B_{\mathrm{rad}} = 10^{-10}$ cm^3/s
Auger recombination constant	$C_{\mathrm{Aug}} = 7.5 \times 10^{-29}$ cm^6/s
Loss coefficient in active region	$\alpha_{\mathrm{a}} = 25$ cm^{-1}
Loss coefficient in claddings	$\alpha_{\mathrm{c}} = 50$ cm^{-1}

spontaneous emissions only are shown as the solid line in Fig. 7.12, whereas the results calculated by including Auger recombination and recombination with defects are shown as the dashed curve. They are calculated by using the averaged photon density (AVPF) approximation proposed by Adams [11] with $\beta = 10^{-4}$. The additional parameters required for the calculations are tabulated in Table 7.2. Sources for some of these parameters are shown in the table. The experimental results agree well with those predicted by the recombination rate described by Eq. (7.2), i.e. with non-radiative recombination included. It can be seen that if the carrier recombination is due to radiative spontaneous emissions only, then the theoretical results will overestimate the amount of spontaneous emission generated by the amplifier for each particular value of injection current. In fact, this model will predict an onset of oscillation when the amplifier is biased at around 10 mA, which does not happen at all in this experiment. On the other hand, the full recombination model, taking into account Auger recombination and recombination with defects, gives a much more accurate prediction over the range of measurement (20–60 mA), therefore we will use Eq. (7.2) for the carrier recombination rate R.

A note about the peak-gain coefficient model. Two models are possible, which have been discussed in Eqs. (7.5) (the linear model) and (7.6) (the parabolic model) already. Both have been used in analysing the above measurements. However, there are no significant differences between using either model. For simplicity, the theoretical results plotted on Fig. 7.12 are calculated using the simple linear model of Eq. (7.5), and this model will be used hereafter for other analyses. This does not imply that there is no difference between the two models. As seen from Fig. 3.2, Subsection 3.2.2 in Chapter 3, there is a significant difference between the values of the peak-gain coefficient calculated by these two models when the injection carrier density n is close to the transparency carrier density n_o. However, because of the losses in the active region and claddings in actual devices, the value of n required to create a significant value of optical gain in the active region will be well in the region of n where both models do not differ significantly.

7.4 Measurement of Gain Characteristics

7.4.1 *Experimental set-up*

The next set of measurements is to determine the gain of the amplifier for different input signal levels and different magnitudes of injection current. The DFB laser source used has to be set up as discussed in Subsection 7.2.1. A measurement of the output power of the DFB laser diode against its bias current has first been taken, which is plotted on Fig. 7.13. It can be seen that the stability of the DFB laser diode is extremely good, with no "kinks" observed on the light power-current curve (i.e. no mode jumping) [3, 7]. The threshold current of the laser diode is around 15 mA at room temperature. By varying the bias current to the DFB laser diode, its output power can be changed as described by the curve in Fig. 7.13. This is used

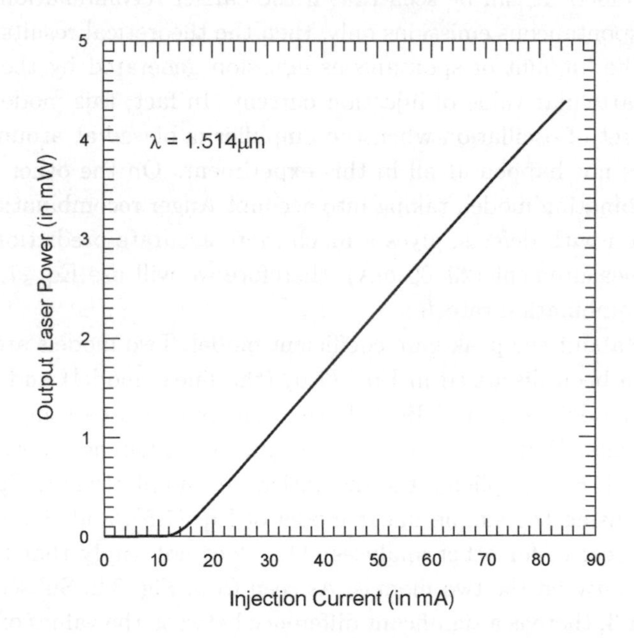

Fig. 7.13. Measured output DFB laser diode output power against bias current.

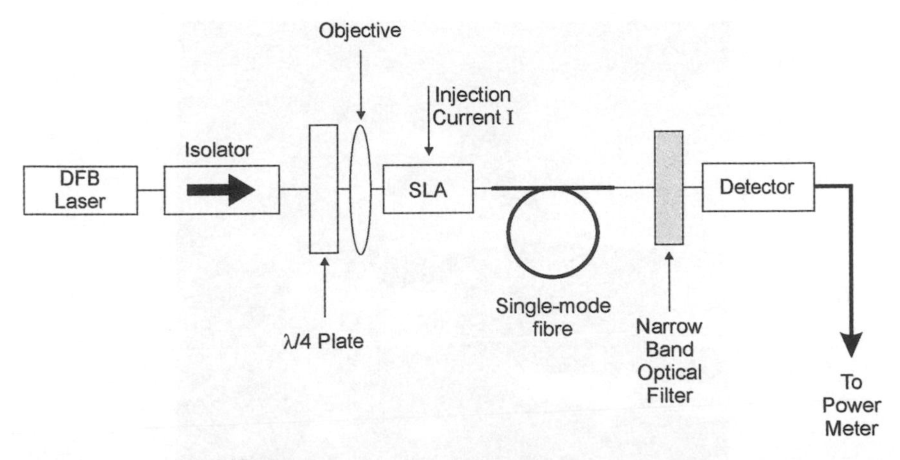

Fig. 7.14. Experimental set-up for measuring gain and output saturation power in an SLA.

as the variable input optical signal source to the amplifier for the following measurements.

The experimental set-up for measuring the amplifier gain is shown in Fig. 7.14 [1, 8]. The output of the DFB laser diode, which is used as an input signal source for the amplifier, is coupled to an isolator. This isolator, shown in Fig. 7.15, is a polarisation beam-splitting cube followed by a quarter-wave plate. The principle of this isolator is illustrated in Fig. 7.16. Any reflected signals towards the DFB laser will be converted, after the quarter-wave plate, to one which has orthogonal polarisation to that of the input signal to the isolator. The polarisation beam-splitting cube will ensure that this signal will not be able to return to the laser diode, thereby isolating the emitted beam from reflected beam. Because the signal emerging after the isolator is circularly polarised [17], it has to be converted back to a linearly polarised signal before coupling into the SLA. This can be achieved by an additional quarter-wave plate, as shown in Figs. 7.14 and 7.15. By rotating this quarter-wave plate we can control the plane of polarisation of the linearly polarised light emerging from the plate [17]. This will be used to control the state of polarisation of the input signal to the amplifier when the polarisation dependence of the amplifier gain is to be measured [18]. An objective is used to couple the linearly polarised light to the laser amplifier, as shown in Fig. 7.15.

The output of the amplifier is butt coupled to a piece of single-mode fibre having a length of 2 m (Fig. 7.17). This fibre was supplied by BTRL and

Fig. 7.15. Isolator and retardation plates used to couple signal to the amplifier.

was designed specially with a high numerical aperture [7, 19] to improve the butt coupling efficiency between the amplifier and the optical fibre [20–21]. The loss of the optical fibre is 0.25 dB/km, according to the manufacturer. The coupling efficiency between the output signal and the detector head can be improved by the insertion of this length of low loss optical fibre because the matching of the spot sizes can be improved. Finally, before detection, the signals are passed through a narrow band optical filter of bandwidth 10 nm centred at 1.50 μm to eliminate excessive spontaneous emission power, which would otherwise affect the measurements. Alternatively, a lock-in amplifier can be used for this purpose, but its sensitivity is not as good as the Anritsu power meter. Therefore the latter equipment has been used. This is shown in Fig. 7.18. The entire setup is displayed in Figs. 7.19 and 7.20.

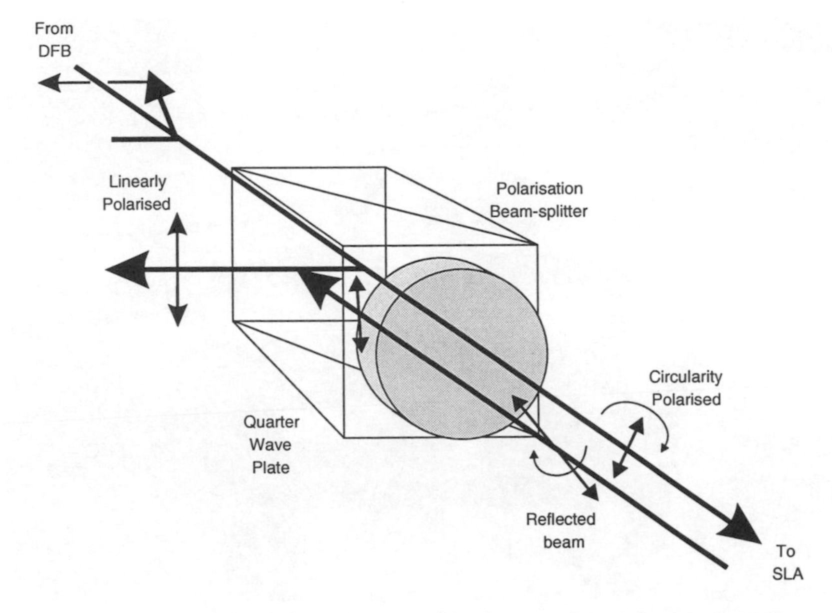

Fig. 7.16. Principle of the isolator used in the experimental measurements.

Fig. 7.17. Fibre coupling to the output of the amplifier.

Fig. 7.18. The signals are passed through a narrow band optical filter (10 nm bandwidth centred at 1.5 μm wavelength).

Fig. 7.19. The complete experimental set-up used to measure amplifier characteristics.

Fig. 7.20. A close-up of the laser source–laser amplifier used in the experimental set-up.

7.4.2 *Experimental procedures*

7.4.2.1 *Determination of coupling losses in the set-up*

The first part of this set of measurements will involve determination of the coupling losses within the set-up shown in Fig. 7.14, which will be required for the measurement of gain in the following measurements. We will assume that the coupling losses are independent of the signal power such that one set of measurements based on a particular optical power level will be sufficient. This measurement has to be performed simultaneously when the alignment and coupling of the devices and optical components is undertaken. The amplifier is connected first. The SLA is biased at 40 mA at a temperature of 20°C (i.e. room temperature), which generates spontaneous emission power at the output. The power emitted immediately after the facet of the SLA has been measured in Subsection 7.3.3, and is plotted in Fig. 7.12. The optical components can then be added one by one: first the retardation plate/objective combination, secondly the isolator, thirdly the fibre, and then the optical filter/detector head combination. When each component is added, alignment with the three-axes micro-positioners is made to ensure maximum coupling (in the case of coupling between the fibre and the amplifier output, additional help from a microscope may be necessary). This is done by measuring the optical power after adding each optical component and adjusting the alignment until a maximum optical power is recorded. This measured maximum power will also be used to calculate the coupling loss between the optical components. If the previously measured maximum optical power before addition of the extra optical component is P_2, then the coupling loss η_c between these optical components is simply given by:

$$\eta_c = \frac{P_2}{P_1} \tag{7.7}$$

In this manner, the coupling loss between different optical components in the experimental set-up shown in Fig. 7.14 can be determined. The measured coupling losses are tabulated in Table 7.3. These values will be used in the following experimental measurements.

Table 7.3. Results of coupling efficiencies measureents.

Coupling efficiency between DFB laser and isolator	$\eta_1 = -6.17$ dB
Coupling efficiency between isolator and quarter-wave plate/objective combination	$\eta_2 = -3.75$ dB
Coupling efficiency between objective (lens) and SLA	$\eta_{c1} = -7.64$ dB
Coupling efficiency between fibre and SLA	$\eta_{c2} = -11.37$ dB
Coupling efficiency between fibre and filter/detector	$\eta_3 = -14.63$ dB

(N.B. The losses in the low-loss single-mode fibre has been neglected)

7.4.2.2 *Measurement of amplifier gain*

The output of the SLA detected by the optical power meter will consist of both the power of the spontaneous emissions that pass through the optical filter as well as the signal. Therefore before any measurement is taken, the amplifier is set up without any input signal, and the corresponding values of the spontaneous emissions reaching the optical power meter are noted. To measure the amplifier gain, the following procedures are followed. First, the DFB laser diode is biased by a particular current, with its temperature maintained at room temperature. Similarly, the SLA will also be biased at a fixed current. The coupling of the DFB laser diode with the SLA is maximised by proper alignment with the three-axes micro-positioner (Fig. 7.20). The power P emitted by the DFB laser diode for a given current can be found from the curve in Fig. 7.13. The true input power P_{in} to the optical amplifier will be $\eta_1\eta_2\eta_{c1}P$, where η_1, is the coupling efficiency between the isolator and the DFB laser, η_{c1} is the coupling efficiency between the quarter-wave plate/objective with the facet of the laser amplifier. The temperature of the amplifier is then adjusted until a maximum power is detected by the detector head of the optical power meter. This is the case when the amplifier's peak-gain wavelength is matched with that from the DFB laser and the centre wavelength of the optical filter. The quarter-wave plate/objective combination (Fig. 7.15) is then rotated. Rotating the quarter-wave plate/objective combination will not affect P_{in}, only the state of polarisation of the input signal. This will affect the amplifier gain G as discussed in Chapter 4. The value of G is a maximum when the state of polarisation of the input signal is TE, whereas when the input signal is TM polarised, G will become a minimum. Thus the state of polarisation of the input signal to the SLA can be adjusted by rotating the quarter-wave plate/objective combination until the detected power is a maximum (i.e. TE mode) or minimum (i.e. TM mode). If the detected power of the signal is P_{D} (i.e. after the spontaneous emission power has been deducted), the true output power of the SLA P_{out} is given by $P_{\text{D}}/(\eta_{c2}\eta_3)$ assuming that the loss in the optical fibre is negligible, where η_{c2} is the coupling efficiency between the SLA and single-mode fibre, and η_3 is the coupling efficiency between the optical fibre and the narrow band filter. The values of all of these coupling efficiencies have already been tabulated in Table 7.3. Then the amplifier gain G is simply given by the ratio $P_{\text{out}}/P_{\text{in}}$. The resulting values of are measured by the optical power meter. The bias current to the DFB laser diode is varied such that the output power of the amplifier is incremented from -40 dBm in steps of -5 dBm until a maximum value, which corresponds to the maximum power input to the amplifier. By noting the currents required to bias the DFB laser diode to achieve these output optical powers, P can be found from Fig. 7.13. Using the arguments above, the values of P_{out} and P_{in} can be determined, and hence G can be found for various output power levels. The measurements are repeated for different values of the bias current to the SLA ranging from 30 mA to 80 mA in 5 mA steps. Hence the variation of G with the injection current i to the SLA for a particular input power level can also be measured.

7.4.3 *Results and discussions*

The measured values of the TE mode gain G versus P_{out} are plotted in Fig. 7.21 for i = 20, 30, 40, 50 and 60 mA, and similar results are plotted in Fig. 7.22 for the TM mode. The theoretical results (solid curves)

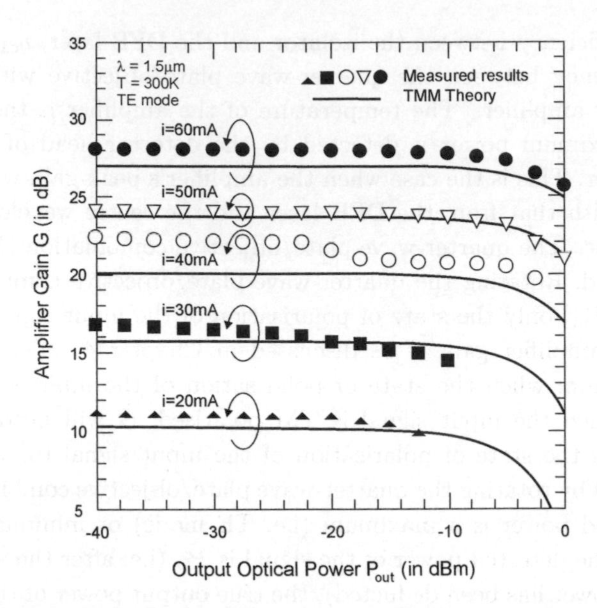

Fig. 7.21. Measured and calculated amplifier gain for TE mode versus output optical power from the SLA with bias current as a parameter. The solid curves represent results calculated by the transfer matrix method (TMM).

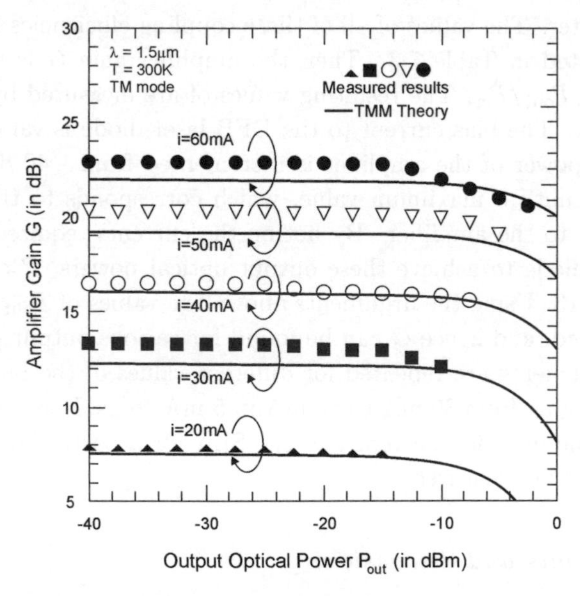

Fig. 7.22. Measured and calculated amplifier gain for TM mode versus output optical power from the SLA with bias current as a parameter. The solid curves represent results calculated by the transfer matrix method (TMM).

are calculated from the equivalent circuit model using transfer matrix analysis. This theoretical model, as discussed in Chapters 5 and 6, will be able to take account of the non-uniform material gain profile in the SLA because of the increasing photon density with z along the amplifier [22]. To assess the accuracy of this model, the amplifier gain G is also calculated using the AVPD approximation proposed by Adams [11] as discussed in Subsection 5.3.2 in Chapter 5. The results calculated by this method are plotted as the dashed lines in Fig. 7.23 for i = 40 mA. The structural parameters used in the analysis have already been tabulated in Table 7.1. In the theoretical analyses, the optical confinement factors shown in Table 7.1 are obtained from the curves shown in Fig. 4.10(b) in Chapter 4. The confinement factors for the slab TE and TM modes can be used here because the dimensions of the active region, also shown in Table 7.1, satisfy the single-transverse mode condition (i.e. the field is cut off along the x-direction) [23]. As discussed in Section 7.3 above, the carrier recombination rate used in these analyses will include Auger recombination and recombination with defects. Moreover, in the calculations we found that the theoretical results agree better with the measured ones if the depen-

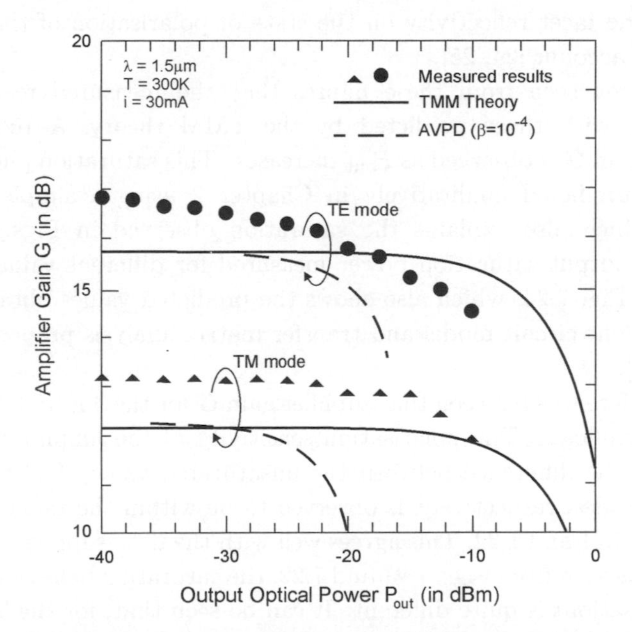

Fig. 7.23. Comparison between amplifier gain of versus its output optical power calculated by TMM and the average photon density (AVPD) approximation.

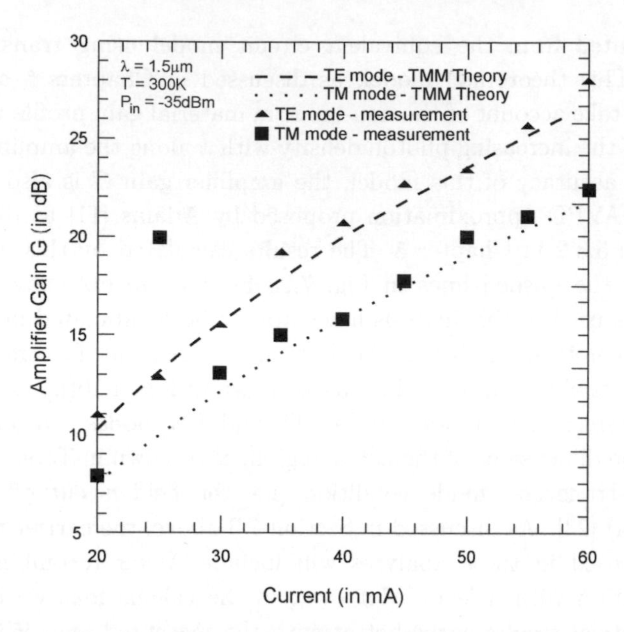

Fig. 7.24. Variation of the amplifier gain versus bias current.

dence of the facet reflectivity on the state of polarisation of the signal is taken into account [24, 25].

It can be seen from these figures that the measured results agree reasonably with those predicted by the TMM theory. A reduction in amplifier gain G is observed as P_{out} increases. This saturation phenomenon has been predicted qualitatively in Chapter 2 using a simple two-level system, which also explains the saturation observed in Figs. 7.21 and 7.22. The output saturation power measured for different values of i are plotted in Fig. 7.24, which also shows the predicted values obtained from the equivalent circuit model and transfer matrix analysis proposed by the author [32].

The differences between the amplifier gain G for the TE and TM modes are not significant. The polarisation sensitivity of the amplifier, which is defined as the difference between the unsaturated value of G for the TE and TM modes, respectively, is observed to be within the range of 3.4 dB from Figs. 7.21 and 7.22. This agrees well with the data supplied by BTRL. However, as seen from Figs. 7.21 and 7.22, the saturation behaviour of these two polarisations is quite different. It can be seen that, for the TM mode, saturation occurs earlier because the optical confinement factor for this mode is smaller [26].

It is interesting to compare the experimental measurements with the theoretical results predicted by the TMM and the AVPD approximation in Fig. 7.23. It can be seen that the AVPD gives an unsaturated gain which is close to that predicted by the TMM. However, the AVPD predicts saturation output powers which are much smaller than those obtained from the measurements and from the theoretical results predicted by the TMM. This can be due to the fact that the facet reflectivity of the amplifier under test is relatively low, and the photon density distribution inside the amplifier becomes increasingly non-uniform. The AVPD approximation will not work well under such circumstances, whereas the TMM, by taking care of this variation, can be more accurate. This supports the use of the more accurate analysis using the TMM proposed by the present author.

7.5 Measurement of Noise Characteristics

7.5.1 *Experimental set-up*

This is the most delicate set of measurements to be made. There are two reasons for this; first, the alignment between the optical components and devices has to be optimised carefully stage by stage to obtain maximum coupling efficiency, as with the previous set of measurements. In the present measurement of noise properties we have to measure the r.m.s. values of the fluctuations in the detector current flowing across the load resistor in the photo-detector circuit. Thus, additional electronic and optical components will be involved, as discussed in Subsection 7.2.3 (Fig. 7.11).

The experimental set-up for noise measurements is shown in Fig. 7.25. This set-up was used by Mukai and Yamamoto [27, 28] in their noise measurements, which was modified from the set-up used to measure intensity fluctuations of a GaAs diode laser by Armstrong and Smith [29]. An optical signal has to be coupled into the SLA, and both the bias current to the SLA and the input optical power (i.e. the output power from the DFB laser source) has to be varied. Thus most of the set-up is identical to that used in gain measurements. The major difference is in the detection scheme. In the present case, the output of the SLA will pass through the single-mode optical fibre and the optical filter, which is then incident on the InGaAsP pin photo-diode. An objective is placed between the photo-diode and the output of the filter to improve coupling. The resulting current generated by the detected optical power, which includes both the amplified signals and the spontaneous emissions generated from the SLA, will pass

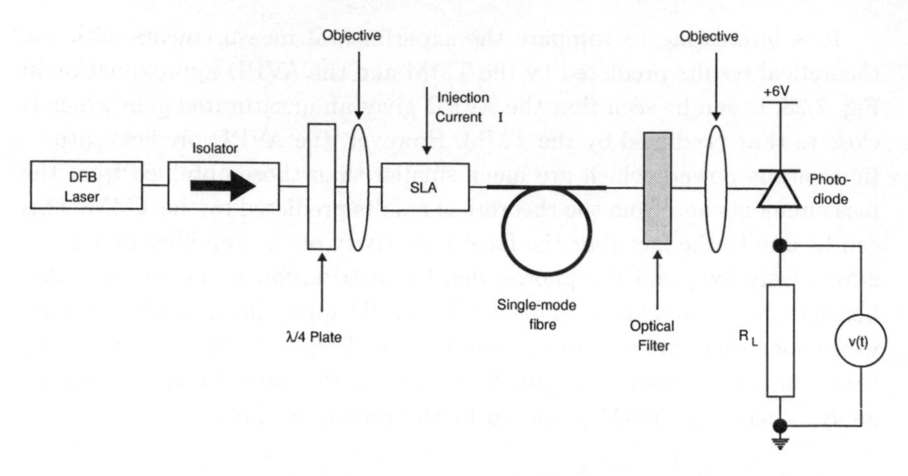

Fig. 7.25. Experimental set-up for measurement of noise characteristics of SLA.

through the load resistor RL. The fluctuations in the detected power will then result in fluctuations in the voltage across the load resistor, and this can be measured by a high sensitivity r.m.s. voltmeter.

7.5.2 *Experimental procedures*

Because we have changed from the detector head to a pin photo-diode, the coupling efficiency between the output of the fibre and the photo-diode has to be re-measured. This follows the same procedures as in the previous measurement. The SLA is biased with no input signal, and the power of the spontaneous emissions after the fibre and before the photo-diode are measured. The coupling efficiency depends strongly on the alignment between the photo-diode and the fibre. They have to be aligned both horizontally and vertically (to focus the beam from the objective onto the active surface of the photo-diode). The maximum coupling efficiency between the fibre and the photo-diode is achieved when the measured spontaneous emission power after the objective is maximised. This was measured to be −27.68 dB. Compared with the previous set-up, there is more coupling loss in the present set-up because there are more optical components. This introduces more misalignment into the set-up. Although such misalignments can be minimised by proper alignment with the micro-positioners and optical mounts, it cannot be totally eliminated. The more optical components the set-up possesses, the more difficult it is to eliminate misalignment. Furthermore, there are finite losses (not owing to coupling but because of, for instance, scattering and absorption) in the optical components. An increase in the number of optical components will definitely increase the total loss in the system.

As before, the DFB laser diode is biased at different currents, with its temperature stabilised at room temperature to give a variable optical output power. The SLA is biased at a constant level. As with the previous set of measurements, the quarter-wave plate/objective combination before the input to the SLA is used to adjust the state of polarisation of the input light signal to the amplifier. The temperature of the SLA will be adjusted such that its peakgain wavelength will coincide with that of the DFB laser source and that of the centre wavelength of the optical filter. The bias current to the DFB laser source will then be varied as in the previous gain measurements to give an output optical power from −40 dBm to about 0 dBm in steps of 5 dBm. The corresponding input power to the SLA can be found from the curve in Fig. 7.13 and the coupling efficiency data tabulated in Table 7.3. The r.m.s. voltage across the load resistor R_{L} is then measured for each of these input power levels. The measurement is then repeated for different SLA bias current levels, ranging again from 20 mA to 80 mA in steps of 10 mA. The mean square voltage recorded across the load resistor R_{L} in an *electrical* bandwidth of $\Delta\omega$ will be given by [26, 28]

$$\begin{aligned}\langle v^2(t)\rangle = {} & [\eta_{\mathrm{c2}}^2\eta_3^2\eta_{\mathrm{D}}^2\langle i_{\mathrm{beat}}^2\rangle + 2e\eta_{\mathrm{c2}}\eta_3\eta_{\mathrm{D}}\langle i_{\mathrm{pho}}\rangle + 2e\eta_{\mathrm{D}}\langle i_{\mathrm{dark}}\rangle] \\ & \times R_{\mathrm{L}}\Delta\omega + 4kT\Delta\omega/R_{\mathrm{L}} \end{aligned} \tag{7.8}$$

In the above equation, $\langle i_{\mathrm{beat}}^2\rangle$ is the beat noise power per unit bandwidth generated in the photo-diode with $\eta_{\mathrm{D}} = 1$ by the beating between the signal and the spontaneous emission components; and between the spontaneous emission components *after filtering* by the optical filter with bandwidth $\Delta\omega \cdot \langle i_{\mathrm{pho}}\rangle$ is the steady-state photo-current generated by the photo-diode after detecting the output from the SLA. $\langle i_{\mathrm{dark}}\rangle$ is the dark current of the photo-diode if its quantum efficiency is unity [6]. η_{D} is the actual quantum efficiency of the photo-diode, and the product $\eta_{\mathrm{c2}}\eta_3$ is the total coupling efficiency between the SLA output and the photo-diode. R_{L} is the load resistance of the photo-detector [30]. Thus, the four terms in the above equation represent the voltage fluctuation induced by (i) the beat noise current generated by the SLA output; (ii) the shot noise current generated by the SLA output; (iii) the shot noise current generated by the dark current of the photo-diode; and (iv) the shot noise current generated by the thermal noise of the resistive part of the detection circuit, respectively. It can be seen that the choice of the value of the load resistor R_L is vital in the measurement. If the selected value is too small, then the r.m.s voltage becomes too small to measure accurately, but the noise power owing to the dark current and thermal noise from the resistive part of the circuit will also be very small. The sensitivity of the measurement can be improved by increasing the value of R_L, but the noise power due to the dark current and thermal noise becomes very large. However, we can measure the noise power owing to the shot noise generated by the dark current and thermal fluctuations by measuring the r.m.s. voltage across the load resistor with the photo-diode shielded from any stray fight. Therefore, for sensitivity reasons (remember that there is quite a significant loss between the output of the SLA and the detector), a relatively large resistance of $R_L = 100$ kΩ has been used in the measurements, with the resulting noise power induced by the dark current and thermal fluctuations measured before the output from the SLA is coupled to the photo-diode. The noise power which is due to the SLA can be obtained by subtracting the mean-square voltage across the load resistor R_L, without a signal incident on the photo-diode from the total mean-square voltage when the signal is present.

7.5.3 *Results and discussions*

The measured noise power per unit electrical bandwidth has been plotted in Figs. 7.26(a) and 7.26(b) versus the input optical signal power for different bias currents, for both the TE and TM modes. This quantity has been estimated from the measured voltage fluctuations across the load resistor by considering its frequency components lie within the bandwidth of the r.m.s. voltmeter. This has been done under two assumptions: (i) that the bandwidth of the r.m.s. voltmeter is smaller than that of the detection circuit (according to the manufacturer, the bandwidth of the r.m.s. voltmeter is approximately 1 MHz, whereas for the load resistance we have been using the bandwidth of the detection circuit estimated to be 1 GHz if the capacitance of the photo-diode is similar to other pin photo-diodes of around 1 pF [3]); and (ii) the frequency responses for both the r.m.s. voltmeter and the detection circuit can be assumed to be that of simple first-order systems with ideal low-pass characteristics. Under these assumptions, only the power of those frequency components of the total fluctuations of the voltage across the load resistor that fall within the pass band of the frequency response of the r.m.s. voltmeter will be measured.

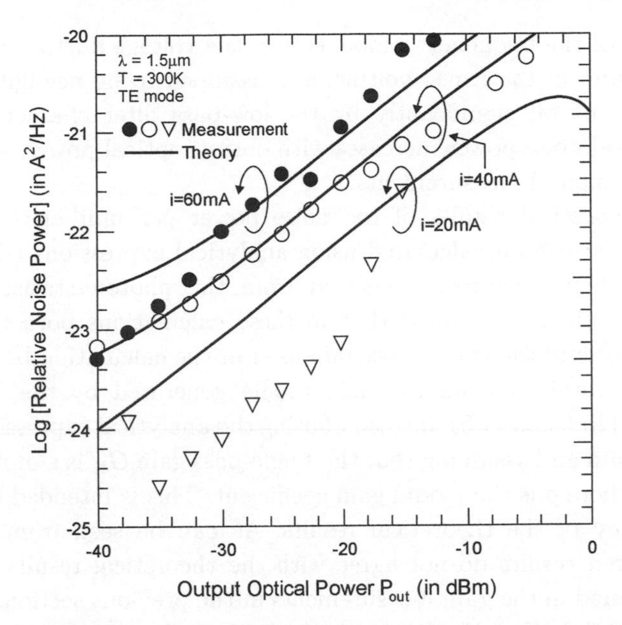

Fig. 7.26. (a) Calculated and measured relative noise powers detected by the photodetector for the TE mode.

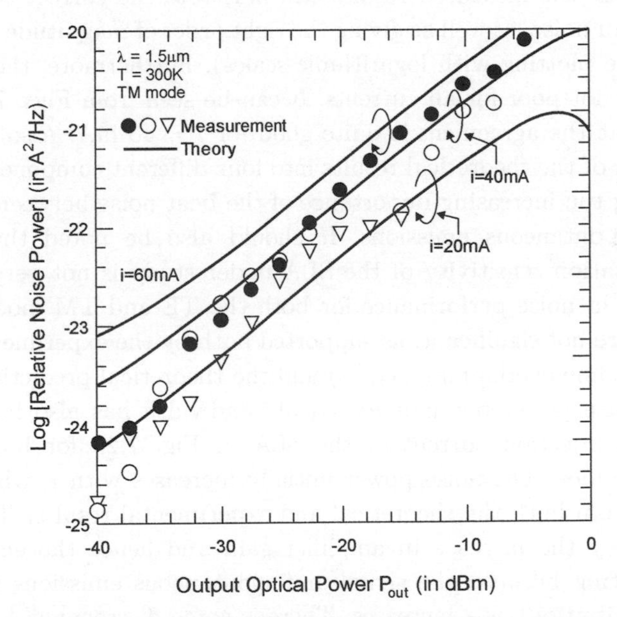

Fig. 7.26. (b) Calculated (solid curves) and measured relative noise power for the TM signal mode.

The power of the frequency components of the voltage fluctuations outside the pass band of the r.m.s. voltmeter is assumed to be negligible as they have been cut off significantly by the low-pass filter characteristics. In general, total noise power increases with output optical power, as observed from experimental measurements.

The theoretical results of the noise power per unit electrical bandwidth have also been calculated using analytical expressions (Eq. (6.22)), Section 6.2.1 in Chapter 6, derived from the photon statistics master equation. It should be noted that in these calculations both the gain of the amplifier and the single-pass gain used in the calculation by Eq. (6.22) have been obtained from the field profile generated by the TMM (see Fig. 5.16(b) in Chapter 5), instead of using the analytical expressions for the amplifier gain and assuming that the single-pass gain G_s, is simply given by $\exp(gL)$, where g is the modal gain coefficient. This is intended to improve the accuracy of the theoretical results. As can be seen from Fig. 7.26, the measured results do not agree with the theoretical results as well as those obtained in the gain measurements in the previous section. This may be because of the fact that the estimation of the frequency response of the r.m.s. voltmeter may not be very accurate. However, with these rather crude assumptions, the measured results still illustrate the correct trend in the behaviour of noise, as well as giving the right order of magnitude (remember that we are plotting with logarithmic scales). Furthermore, the measured results are not poor for all currents. It can be seen from Figs. 7.27(a) and 7.27(b) that the agreement is quite good for $i = 40$ mA. Also shown is a breakdown of the theoretical results into four different components, clearly illustrating the increasing importance of the beat noise between the signal and the spontaneous emissions. It should also be noted that because the polarisation sensitivity of the SLA under study is not very high, the differences in noise performance for both the TE and TM modes for this amplifier are not significant, as supported both by the experimental results (in which some overlapping occurs) and the theoretical predictions.

The noise power per unit electrical bandwidth has also been plotted versus the injection current of the SLA in Fig. 7.28 for both the TE and TM modes. The noise power initially increases with i, which can be observed from both the theoretical and experimental results. This can be explained by the increase in amplifier gain and hence the enhancement of the beating between the signal and spontaneous emissions (the major noise contribution) as i increases. There is some discrepancy between the measured results and the theoretical results for the TE measurements, but

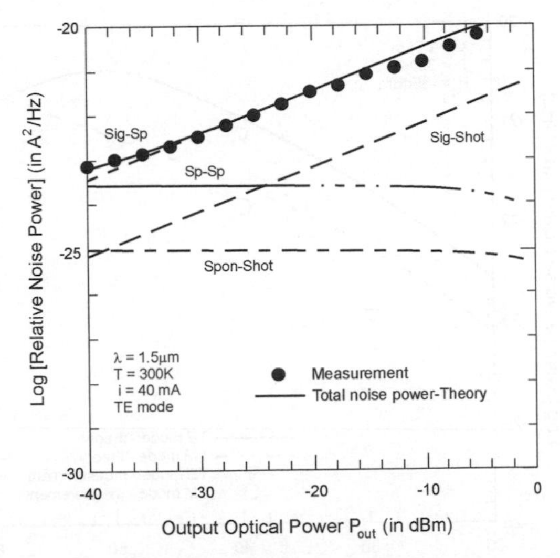

Fig. 7.27. (a) Calculated (solid curves) and measured relative noise power for the TE signal mode. Also shown are different components contributing to the total noise power, that is; (i) signal to spontaneous emission beat noise, (ii) spontaneous to spontaneous noise due to beating between different spontaneous emission components, (iii) signal shot noise and (iv) shot noise due to spontaneous emissions.

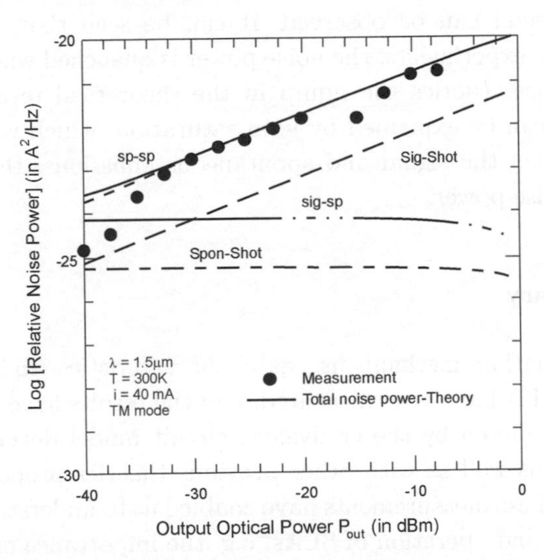

Fig. 7.27. (b) Calculated (solid curves) and measured relative noise power for the TM signal mode. Also shown are different components contributing to the total noise power, that is; (i) signal to spontaneous emission beat noise, (ii) spontaneous to spontaneous noise due to beating between different spontaneous emission components, (iii) signal shot noise and (iv) shot noise due to spontaneous emission.

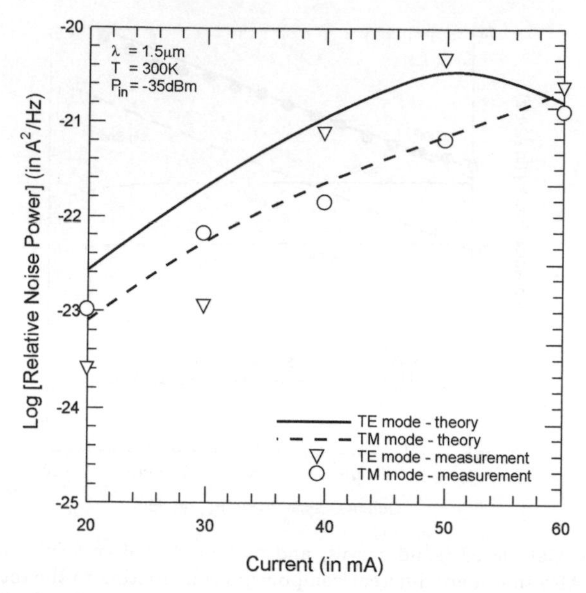

Fig. 7.28. Calculated and measured relative noise power of SLA against bias current for both TE and TM modes.

the general trend can be observed. It can be seen that, as observed in many previous experiments, the noise power is quenched when the injection current increases (notice the hump in the theoretical results for the TE mode). This can be explained by gain saturation, which will suppress the beating between the signal and spontaneous emissions, thereby reducing the overall noise power.

7.6 Summary

The recombination mechanisms, gain characteristics and noise characteristics of a SLA have been measured, and the results have been compared with those predicted by the equivalent circuit model developed in Chapters 5 and 6, as well as with other previous theories proposed by various researchers. These measurements have enabled us to understand more about the principles and operation of SLAs, e.g. the importance of Auger recombination, the effect of beating noise generated by the output of a SLA in a photo-diode, the saturation characteristic dependent on the bias current, etc. In addition, it gives some support for the accuracy, applicability and power of the equivalent circuit model proposed by the author in Chapters 5 and 6, which encourages further use of this theoretical model in designing and analysing SLAs with more complicated structures for use in future optical fibre communications systems.

References

[1] T. Saitoh and T. Mukai, "Recent progress in semiconductor laser amplifiers," *IEEE J. Lightwave Technol.*, LT-6, No. 11, pp. 1156–1164, 1988.

[2] G. P. Agrawal and N. K. Dutta, *Long-wavelength Semiconductor Lasers* (New York, VanNostrand Reinhold, 1986).

[3] H. Kressel, *Semiconductor Devices for Optical Communications*, 2nd edition (Springer-Verlag, Berlin, 1982).

[4] T. Li, *Topics in Lightwave Transmission System* (London, Academic Press, 1991).

[5] Y. Imai, E. Sano and K. Asai, "Design and performance of wideband GaAs MMICs for high-speed optical communication systems," *IEEE Trans. Microwave Theory Technol.*, 40, No. 2, pp. 185–189, 1992.

[6] A. Yariv, *Optical Electronics*, 3rd edition (Holt-Saunders, 1985).

[7] J. M. Senior, *Optical Fibre Communications: Principles and Practice*, 2nd edition (New York, Prentice-Hall, 1992).

[8] T. Saitoh and M. Mukai, "1.5 μm GaInAs travelling-wave semiconductor laser amplifier," *IEEE J. Quantum Electron.*, QE-23, No. 6, pp. 1010–1020, 1987.

[9] Y. Yamamoto, *Coherence Amplification and Quantum Effects in Semiconductor Lasers* (New York, John Wiley and Sons, 1991).

[10] R. Olshansky, C. B. Su, J. Manning and W. Powazink, "Measurement of radiative and nonradiative recombination rates in InGaAsP and AlGaAs light sources," *IEEE J. Quantum Electron.*, QE-20, No. 8, pp. 838–854, 1984.

[11] M. J. Adams, "Time dependent analysis of active and passive optical bistability in semiconductors," *IEEE Proc.*, 132, Part J, No. 6, pp. 343–348, 1985.

[12] C. Y. J. Chu and H. Ghafouri-Shiraz, "Structural effects on polarization sensitivity of travelling-wave semiconductor laser amplifiers," *3rd Bangor Communication Symposium*, United Kingdom, pp. 19–22, 1991.

[13] H. Ghafouri-Shiraz, "A model for peak-gain coefficient in lnGaAsP/lnP semiconductor laser diodes," *Opt. Quantum Electron.*, 20, pp. 153–163, 1988.

[14] J. Wang, H. Olesen and K. E. Stubkjaer, "Recombination, gain and bandwidth characteristics of 1.3 μm semiconductor laser amplifiers," *IEEE J. Lightwave Technol.*, LT-5, No. 1, pp. 184–189, 1987.

[15] I. D. Henning, M. J. Adams and J. V. Collins, "Performance prediction from a new optical amplifier model," *IEEE J. Quantum Electron.*, QE-21, No. 6, pp. 609–613, 1985.

[16] H. Ghafouri-Shiraz, Unpublished work, 1992.

[17] A. Yariv and P. Yeh, *Optical Waves in Crystals* (New York, John Wiley and Sons, 1984).

[18] R. M. Jopson *et al.*, "Polarisation-dependent gain spectrum of a 1.5 pm travelling-wave optical amplifier," *Electron. Lett.*, 22, No. 21, pp. 1105–1107, 1986.
[19] A. W. Snyder and J. D. Love, *Optical Waveguide Theory* (London, Chapman & Hall, 1983).
[20] J. John, T. S. M. Maclean, H. Ghafouri-Shiraz and J. Niblett, "Matching of single-mode fibre to laser diode by microlenses at 1.5 μm wavelength," *JEE Proc.*, 141, Part J, No. 3, pp. 178–184, 1994.
[21] H. Ghafouri-Shiraz and C. Y. J. Chu, "Effect of phase shift position on spectral linewidth of the $\pi/2$ distributed feedback laser diode," *J. Lightwave Technol.*, LT-8, No. 7, pp. 1033–1038, 1990.
[22] D. Marcuse, "Computer model of an injection laser amplifier," *IEEE J. Quantum Electron.*, QE-19, No. 1, pp. 63–73, 1983.
[23] H. Ghafouri-Shiraz, "Single transverse mode condition in long wavelength SCH semiconductor laser diode," *Trans. IEICE Jpn.*, E70, No. 2, pp. 130–134, 1987.
[24] C. Vasallo, "Rigorous and approximate calculations of antireflection layer parameters for travelling-wave diode laser amplifiers," *Electron. Lett.*, 21, No. 8, pp. 333–334, 1985.
[25] C. Vassallo, "Theory and practical calculation of antireflection coatings on semiconductor laser diode optical amplifiers," *JEE Proc.*, 137, Part J, No. 4, pp. 193–202, 1990.
[26] Y. Mukai, Y. Yamamoto and T. Kimura, "Optical amplification by semiconductor lasers," *Semiconductors and Semimetals*, 22, Part E, pp. 265–319, Academic Press, London, 1985.
[27] T. Mukai and T. Yamamoto, "Noise characteristics of semiconductor laser amplifiers," *Electron. Lett.*, 17, No. 1, pp. 31–33, 1981.
[28] T. Mukai and Y. Yamamoto, "Noise in an AlGaAs semiconductor laser amplifier," *IEEE J. Quantum Electron.*, **QE-18**, No. 4, pp. 564–575, 1982.
[29] J. A. Armstrong and A. W. Smith, "Intensity fluctuation in GaAs laser emission," *Phys. Rev.*, 140, No. 1A, pp. A155–A164, 1965.
[30] H. Taub and D. L. Schilling, *Principles of Communication Systems*, 2nd edition (New York, McGraw-Hill, 1986).
[31] C. Y. J. Chu and H. Ghafouri-Shiraz, "A simple method to determine carrier recombinations in a semiconductor laser optical amplifier," *IEEE Photon. Technol. Lett.*, 5, No. 10, pp. 1182–1185, 1993.
[32] C. Y. J. Chu and H. Ghafouri-Shiraz, "Analysis of gain and saturation characteristics of a semiconductor laser optical amplifier using transfer matrices," *IEEE J. Lightwave Technol.*, LT-12, No. 8, pp. 1378–1386, 1994.

Chapter 8

Novel Semiconductor Laser Diode Amplifier Structure

8.1 Introduction

Semiconductor laser diode amplifiers (SLDAs) have been developed mainly to replace complex optoelectronic regenerating systems. Some of their main functions are to act as: (i) post-amplifiers for the optical transmitter, (ii) in-line amplifiers and (iii) preamplifier for optical receivers [1]. Recently, the development of the erbium doped fibre-amplifiers (EDFAs) has challenged many of the applications of SLDAs. The limitation of the SLDA is the difficulty in achieving high gain, low polarisation sensitivity, low noise figure and low coupling loss at the same time [2]. However, SLDAs have the advantage of small size, low power consumption and low cost (when mass produced), and can be integrated monolithically with other photonic devices like lasers or detectors. Moreover, they are also available at a wavelength of 1.3 μm and perform other functions such as switching, pulse-shape regeneration and wavelength conversion.

Combinations of antireflection coating, window facet and angle facet structures [3] have made it possible to reduce the facet reflectivity of the SLDA to a negligible value, hence resulting in an almost ideal *travelling wave semiconductor laser amplifier* (TW-SLDA). Compared with a Fabry–Perot laser diode amplifier, which has a facet reflectivity of about 30% [4], a TW-SLDA has a wider bandwidth, higher gain and is less sensitive to (i) state of polarisation of the input signal and (ii) fluctuations in current and temperature.

Gain saturation in TW-SLDAs severely limits many of its applications. In a wavelength division multiplexing system, the gain saturation is determined by the total power of all the channels involved. As a result, the number of channels in a WDM system for a given gain is limited by the gain saturation. This also results in undesirable effects, such as gain saturation induced channel cross-talk [5], and nearly degenerates four-wave mixing [6]. When SLDAs are used as in-line amplifiers, the gain provided by each amplifier (which also determines the number of required amplifiers) is limited by the gain saturation of the amplifier. Many other applications, such as booster amplifiers, also require high gain saturation.

One method of obtaining high gain saturation in a SLDA is to increase the cross-sectional area of its active region. There are two aspects that result in an improvement in gain saturation. Firstly, by increasing the cross-sectional area of the active region, the intensity is reduced for a given optical power. Secondly, there are more carriers available for providing gain since the active volume is bigger. The cross-sectional area is enhanced by increasing the width of the active region. The cross-sectional area of conventional TW-SLDA is limited by the condition for single-mode propagation, which is desirable for high coupling efficiency to single-mode fibres [7]. To achieve both high gain saturation and single-mode propagation simultaneously, the width of the TW-SLDA can be increased gradually from the input to the output [8]. The performance of tapered TWSLAs has been investigated in [9–11]. It was found that a linearly tapered TW-SLDA gave a better gain saturation performance and a higher normalized fundamental mode output power, whereas an exponentially tapered one gave a lower relative amplified spontaneous emission noise.

In this chapter, we have investigated the performance characteristics of a novel SLDA device where its active region has a semi-exponential-linear tapered (SELT) waveguide structure (see Fig. 8.1). This device offers a good compromise between the exponentially and linearly waveguide structures in terms of saturation power and amplified spontaneous emission noise. The chapter is organized as follows. In Section 8.2 the wave propagation, gain and amplified spontaneous emissions of the SELT waveguide structure are studied. The numerical results are discussed in Section 8.3 and, finally, in Section 8.4 conclusions are summarised.

8.2 Theoretical Model

8.2.1 *The normalised power of the fundamental mode*

The SELT waveguide structure under consideration is shown in Fig. 8.1. We have applied the step-transition method (STM) [12, 13] to analyse this

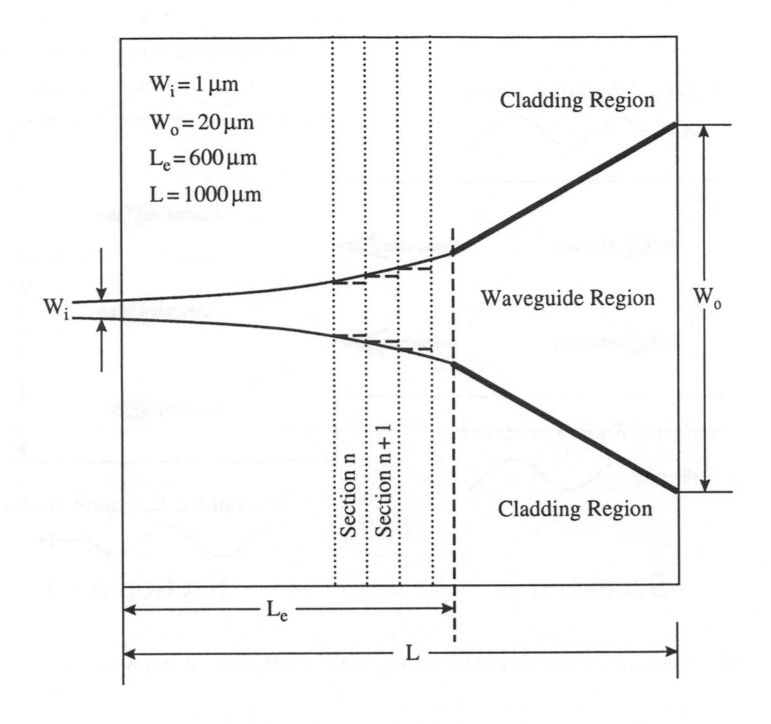

Fig. 8.1. Configuration of a semi-exponential-linear tapered (SELT) waveguide.

structure. In this method, the waveguide is divided into a large number of small rectangular sections, and then by using the effective index method [14] the local modes can be obtained. The step transition method is used as it requires less simulation time than the beam propagation method [15]. In the following analysis we have assumed that (i) the fundamental mode is the only guided mode at the input of this waveguide and (ii) the power of this mode is normalised to unity. Let A_n and A_{n+1} be the complex electric field amplitude coefficients of the nth and $(n+1)$th sections normalised such that $|A| = 1$. In Fig. 8.2 we show two incident $(E_{i,n}, E_{j,n})$, two reflected $(E_{i,n}^R, E_{j,n}^R)$ and three transmitted $(E_{i,n+1}, E_{j,n+1}, E_{k,n+1})$ local guided modes in the nth and $(n+1)$th sections. In addition to these modes, reflected radiation modes (RRM) and transmitted radiation modes (TRM) also exist. By applying the boundary conditions at the interface between the adjacent sections n and $n+1$ (see Fig. 8.2) we have

$$A_{j,n+1} = \sum_{i=1}^{M_n} c_{ij} A_{i,n} \tag{8.1}$$

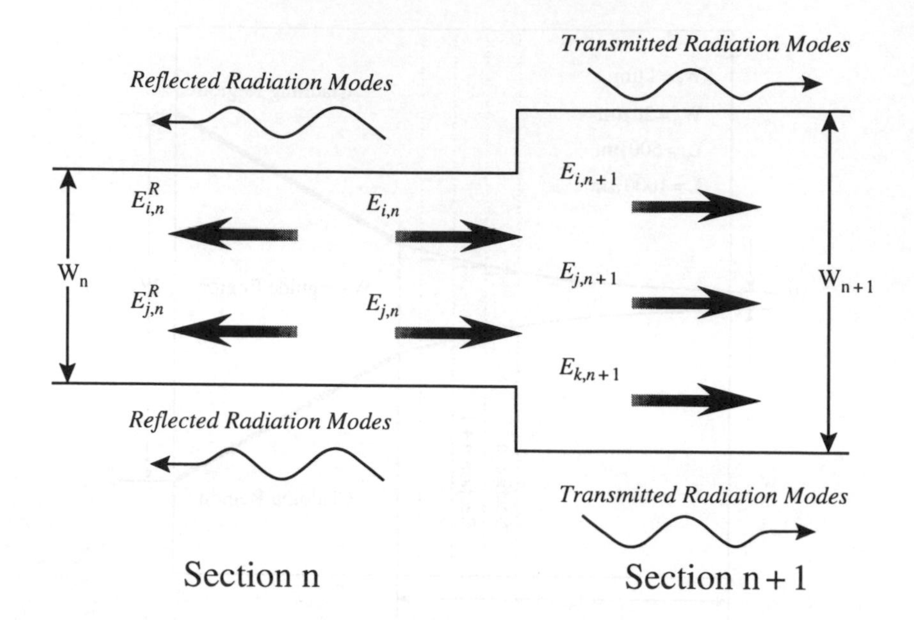

Fig. 8.2. An approximation of the actual SELT waveguide, using a series of steps.

where M_n is the maximum higher-order local guided mode number in the nth section. c_{ij} is the coupling coefficient which relates the complex amplitude coefficient of mode j of section $(n+1)$ to mode i of section n. The c_{ij} coefficient in Eq. (8.1) can be evaluated for a TE polarised mode by matching of the transverse field components at the interface between the adjacent sections. That is [10, 16]:

$$c_{ij}(\mathrm{TE}) = \begin{cases} \dfrac{2\sqrt{\beta_{i,n}\beta_{j,n+1}}(\beta_{j,n}+\beta_{i,n+1})I_{i,n;j,n+1}}{(\beta_{j,n}+\beta_{j,n+1})(\beta_{i,n}+\beta_{i,n+1})\sqrt{I_{i,n;i,n}I_{j,n+1;j,n+1}}} & \text{for } j \le M_n \\ \dfrac{2\sqrt{\beta_{i,n}\beta_{j,n+1}}I_{i,n;j,n+1}}{(\beta_{i,n}+\beta_{i,n+1})\sqrt{I_{i,n;i,n}I_{j,n+1;j,n+1}}} & \text{for } j > M_n \end{cases} \tag{8.2a}$$

where

$$I_{m,a;n,b} = \int_{-\infty}^{\infty} E_{m,a}E_{n,b}dx\,. \tag{8.2b}$$

In Eq. (8.2) $\beta_{p,q}$ ($p = m, n$ and $q = a, b$) denotes the propagation constant of the pth mode of the qth section and $E_{p,q}$ is the associated lateral electric field solution.

8.2.2 *The gain saturation performance*

The rate equation for the carrier density in the TW-SLDA can be expressed as [17]:

$$\frac{\partial N}{\partial t} = D\frac{\partial^2 N}{\partial x^2} + \frac{J}{ed} - \frac{N}{\tau_s(N)} - g_m \frac{i(x, y, z, t)}{hf} \tag{8.3}$$

where N is the carrier density, J is the injection current density, e is the electron charge, d is the thickness of the active region, D is the diffusion constant, h is Planck's constant and f is the optical frequency, $\tau_s(N)$ is the carrier lifetime, $g_m(N)$ is the material gain and i is the instantaneous intensity. The carrier recombination lifetime $\tau_s(N)$ [18] and the material gain $g_m(N)$ [19] are given by:

$$\tau_s(N) = \frac{1}{A_{nr} + B_{rad}N + C_{Aug}N^2} \tag{8.4}$$

$$g_m(N) = a(N - N_0) \tag{8.5}$$

where A_{nr} is the non-radiative recombination coefficient, B_{rad} is the spontaneous recombination coefficient, C_{Aug} the Auger recombination coefficient, a is the gain efficient constant and N_0 is the carrier density at transparency. The average optical intensity $I(z, t)$ and the carrier density $N(x, z)$ may be expressed as [16]:

$$I(z, t) = \frac{\beta}{2\omega\mu_0}|E(x, y, z, t)|^2 \tag{8.6}$$

$$N(x, z) = \begin{cases} N_{av}(z) - N_1(z)\cos\left(\dfrac{2\pi x}{W(z)}\right) & \text{for } x < |W(z)|/2 \\ 0 & \text{otherwise} \end{cases} \tag{8.7}$$

where $E(x, y, z, t) = E_x(x, z)E_y(y, z)E_z(z)\exp[j(\omega t - \beta z)]$ is the optical electric field distribution, $E_z(z)\exp[j(\omega t - \beta z)]$ is the field solution along the propagating direction, $E_z(z)$ is the complex amplitude, and $E_x(x, z)$ and $E_y(y, z)$ are the lateral field solutions along the x and y directions respectively. Also $N_{av}(z)$ is the average carrier density along the x direction,

$N_1(z)$ is the spatial carrier density distribution along the lateral direction and W is the width of the active region. The exact solution of the carrier density distribution along the lateral direction has been proposed by Yamada and Suematsu [20]. Alternatively, Eq. (8.7) has been suggested [8, 9], which takes into account the effects of both hole burning and carrier diffusion along the lateral direction. The exact solution requires extensive computation; however, Eq. (8.7) is substantially simplified whilst maintaining sufficient accuracy. In a given section (see Fig. 8.2), we assume that there is no z dependence, hence $E_z(z)$ is constant and both $E_x(x,z)$ and $E_y(y,z)$ are only functions of x and y, respectively. Let us substitute Eq. (8.7) into Eq. (8.3) and integrate the result over the active layer cross-sectional area. This gives the following equation for the average carrier density N_{av}:

$$\frac{\partial N_{\text{av}}}{\partial t} = \frac{J}{ed} - A_{\text{nr}}N_{\text{av}} - B_{\text{rad}}\left(N_{\text{av}}^2 + \frac{N_1^2}{2}\right) - C_{\text{Aug}}\left(N_{\text{av}}^3 + \frac{3}{2}N_{\text{av}}N_1^2\right)$$
$$- \Gamma[g_{\text{m}}(N_{\text{av}}) - \eta a N_1]\frac{I}{hf}\,. \tag{8.8}$$

If we again substitute Eq. (8.7) into Eq. (8.3) but now multiply the result by $\cos(2\pi x/W)$ before integrating over the active layer cross-sectional area, we obtain the following equation for N_1:

$$\frac{\partial N_1}{\partial t} = -D\left(\frac{2\pi}{W}\right)^2 N_1 - A_{\text{nr}}N_1 - 2B_{\text{rad}}N_{\text{av}}N_1 - C_{\text{Aug}}\left(3N_1N_{\text{av}}^2 + \frac{3}{4}{N_1}^3\right)$$
$$+ 2\Gamma[\eta g_{\text{m}}(N_{\text{av}}) - \sigma a N_1]\frac{I}{hf}\,. \tag{8.9}$$

In the above equations η and σ take into account the lateral hole burning and diffusion and Γ is the confinement factor. That is:

$$\eta = \frac{\int_{-W/2}^{W/2}\cos(2\pi x/W)E_x^2 dx}{\int_{-W/2}^{W/2}E_x^2 dx} \tag{8.10}$$

$$\sigma = \frac{\int_{-W/2}^{W/2}\cos^2(2\pi x/W)E_x^2 dx}{\int_{-W/2}^{W/2}E_x^2 dx} \tag{8.11}$$

$$\Gamma = \frac{\int_{-W/2}^{W/2}E_x^2 dx\int_{-d/2}^{d/2}E_y^2 dy}{\iint E_x^2E_y^2 dx dy}\,. \tag{8.12}$$

In the steady state, we have $\partial N_{\text{av}}/\partial t = \partial N_1/\partial t = 0$ and hence for a given intensity the set of Eqs. (8.8) and (8.9) can be solved to obtain the average carrier density N_{av} and its deviation N_1. The amplification rate of the optical intensity along the SELT waveguide structure can be expressed as [8]:

$$\frac{\partial I(z,t)}{\partial z} = [\Gamma(g_{\text{m}}(N_{\text{av}}) - a\eta N_1) - \alpha_{\text{loss}} - \alpha_{\text{tap}}]I(z,t) \tag{8.13}$$

where

$$\alpha_{\text{loss}} = \Gamma\alpha_{\text{act}} + (1-\Gamma)\alpha_{\text{clad}} \tag{8.14}$$

$$\alpha_{\text{tap}} = -W(z)\frac{d}{dz}\left(\frac{1}{W(z)}\right). \tag{8.15}$$

In the above equations α_{loss} is the material loss coefficient, α_{tap} accounts for the variation of intensity due to the tapering shape of the waveguide, α_{act} is the loss coefficient of the active layer and α_{clad} is that of the cladding layer. Equation (8.14) can be obtained as follows. For a given power flow $P(z,t)$, we have $P(z,t) = I(z,t)A(z)$ where $A(z) = W(z)d$ is the cross-sectional area of the active layer. Since the active layer thickness d is constant, then for a fixed power we have:

$$\frac{\partial}{\partial z}(I(z,t)W(z)) = 0 \tag{8.16a}$$

or

$$I(z,t)\frac{\partial}{\partial z}W(z) + W(z)\frac{\partial}{\partial z}I(z,t) = 0\,. \tag{8.16b}$$

Equation (8.16b) shows how the light intensity varies with the width of the active layer. That is:

$$\frac{\partial I(z,t)}{\partial z} = I(z,t)\left(-\frac{1}{W(z)}\frac{\partial}{\partial z}W(z)\right) = -\alpha_{\text{tap}}I(z,t)\,. \tag{8.17}$$

If the cavity is divided into a large number of small sections, then we may assume that in each section the carrier density, gain and loss factor are constants. Hence, the general solution for Eq. (8.13) is given by

$$I(z+\Delta z) = I(z)\exp\{[\Gamma(g_{\text{m}}(N_{\text{av}},\lambda) - a\eta N_1) - (\alpha_{\text{loss}} + \alpha_{\text{tap}})]\Delta z\} \tag{8.18}$$

where Δz is the length of a section (i.e. $\Delta z = L/M$ where L is the cavity length and M is the number of sections). Equation (8.16) is used to calculate the intensity of the $(n+1)$th section given the average carrier density

N_{av}, its deviation N_1 and the intensity of the nth section. Equations (8.8), (8.9) and (8.16) are the main equations which can be used to analyse performance characteristics of the semi-exponential-linear tapered laser diode amplifier.

8.2.3 *The relative amplified spontaneous emission*

The average ASE power, P_n, at the output of a semiconductor laser diode amplifier can be expressed as [21]

$$P_n = \hbar f n_{\mathrm{sp}} m_t (G-1)\Delta f_1 \tag{8.19}$$

where $\hbar f$ is the photon energy, n_{sp} is the population inversion parameter of the amplifier medium, m_t is the number of effective transverse modes, G is the power gain and Δf_1 is the equivalent noise bandwidth. From Eq. (8.17), by assuming that n_{sp} and Δf_1, are constants throughout the cavity, the normalised amplified spontaneous emission power P_{ASE} of the SELT waveguide structure can be expressed as [10]:

$$P_{\mathrm{ASE}} = \frac{\sum_{m=0}^{M} \exp(\Gamma g_{m0} - \alpha_{\mathrm{tot}}) L_m}{\exp(\Gamma g_{m0} - \alpha_{\mathrm{tot}}) L} \tag{8.20}$$

where g_{m0} is the unsaturated gain coefficient, α_{tot}, is the total effective loss coefficient, L_m, is the length at which the mth mode propagates and L is the total length or the propagating distance of the fundamental mode. It should be noted that the normalised P_{ASE} is a measure of the ASE power in the SELT waveguide TW-SLDA as compared with that of the conventional single-mode TW-SLDA.

8.3 Analysis, Results and Discussions

8.3.1 *The shape of the taper structure*

The structure of the device is shown in Fig. 8.1. The expression for the variation of the width $W(z)$ of this semi-exponential-linear tapered waveguide structure as a function of z is given by:

$$W(z) = \begin{cases} W_{\mathrm{i}} \exp(\alpha_{\mathrm{e}} z) & \text{for } 0 < z \le L_{\mathrm{e}} \\ W_{\mathrm{e}} + (W_{\mathrm{o}} - W_{\mathrm{e}}) \dfrac{z - L_{\mathrm{e}}}{L - L_{\mathrm{e}}} & \text{for } L_{\mathrm{e}} < z \le L \end{cases} \tag{8.21}$$

where

$$\alpha_e = \frac{\ln(W_o/W_i)}{L} \tag{8.22}$$

$$W_e = W_i \exp(\alpha_e L_e)\,. \tag{8.23}$$

In the above equations, W_i and W_o are the widths of the active region at the input and output respectively, and L_e is the length of the exponential taper profile within the cavity. Note that when L_e is equal to zero the exponential part of the taper profile is zero and the taper profile is completely linear, and when L_e is equal to the total length, the tapering profile of the whole structure is exponential. Table 8.1 lists the parameters that have been used in calculating the results shown in Figs. 8.3 to 8.8.

Table 8.1. List of parameters and their values.

Description	Values/Units
Active layer thickness	$d = 0.1\ \mu$m
Active region width at the input	$W_i = 1\ \mu$m
Active region width at the output	$W_o = 1$ to $30\ \mu$m
Length of amplifier	$L = 1000\ \mu$m
Non-radiative recombination constant	$A_{nr} = 0$ s^{-1}
Spontaneous recombination constant	$B_{rad} = 8 \times 10^{-11}$ cm^{3}s^{-1}
Auger recombination constant	$C_{Aug} = 9 \times 10^{-29}$ cm^{6}s^{-1}
Active region loss coefficient	$\alpha_{act} = 1.2 \times 10^2$ cm^{-1}
Loss coefficient of the cladding region	$\alpha_{clad} = 5.6$ cm^{-1}
Gain coefficient	$a = 1.64 \times 10^{-16}$ cm^2
Carrier density at transparency	$N_0 = 10^{18}$ cm^{-3}
Current density	$J = 10$ kAcm^{-2}
Diffusion coefficient	$D = 10$ cm^{-2}s^{-1}
Confinement factor	$\Gamma = 0.17$

8.3.2 *Adiabatic single-mode condition*

One of the main reasons for using a tapered waveguide TW-SLDA rather than a broad area TW-SLDA is that the tapered profile allows most of the power to be confined in the fundamental mode. Figure 8.3 shows the fraction of the power carried by the local fundamental mode (TE_{00}) as a

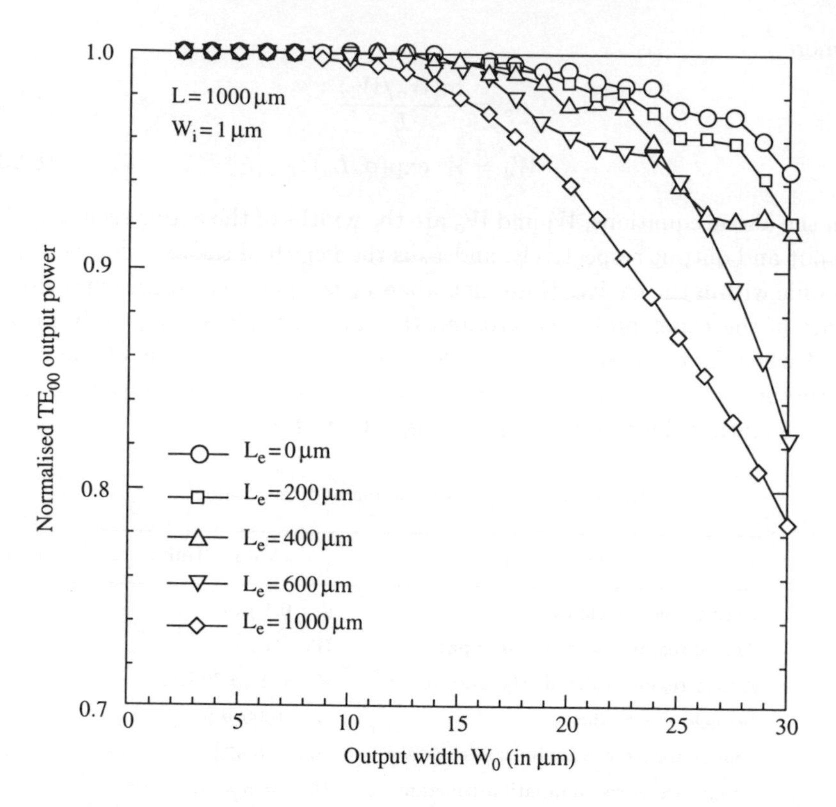

Fig. 8.3. Normalized TE_{00} output power as a function of the output width for various taper profiles.

function of output width for various values of L_e. The input width W_{in} is fixed at 1 μm to satisfy the single-mode condition and the total length of the device is 1000 μm. The equivalent indices of the active and cladding regions are 3.27 and 3.17, respectively, which are typical values for buried heterostructure index guided devices. In the calculations the length of each section was set to 2 μm. The number of sections used in analysing various taper structures was made large enough to ensure the numerical convergence of the solutions. The computation is carried out as follows. The total power at the input, which is carried by the fundamental mode, is set to 1. The coupling coefficient between each mode of the nth and $(n+1)$th sections is then computed by using equation (8.2). The complex amplitude coefficients of the various modes of the $(n+1)$th section can then be calculated by using Eq. (8.1) (see Fig. 8.2). This process is repeated for the subsequent sections until we reach the output end of the cavity.

The normalised output power of the fundamental mode is the square of the complex amplitude coefficient divided by the sum of the square of the complex amplitude coefficient of all modes.

As shown in Fig. 8.3, the effect of increasing the length of the exponential taper L_e is reduction in the normalised output power of the fundamental mode. When L_e increases, the linear taper part of the SELT waveguide becomes steeper, which results in more conversion of power from the fundamental mode to the higher-order modes. However, for $L_e < 400$ μm and output width of 30 μm over 90% of the output power remains within the fundamental mode.

8.3.3 *The intensity and carrier distributions*

In order to present a clearer picture of the amplification process within the amplifier, the intensity distribution and the carrier density are plotted against the propagation distance z. These are shown in Figs. 8.4 and 8.5

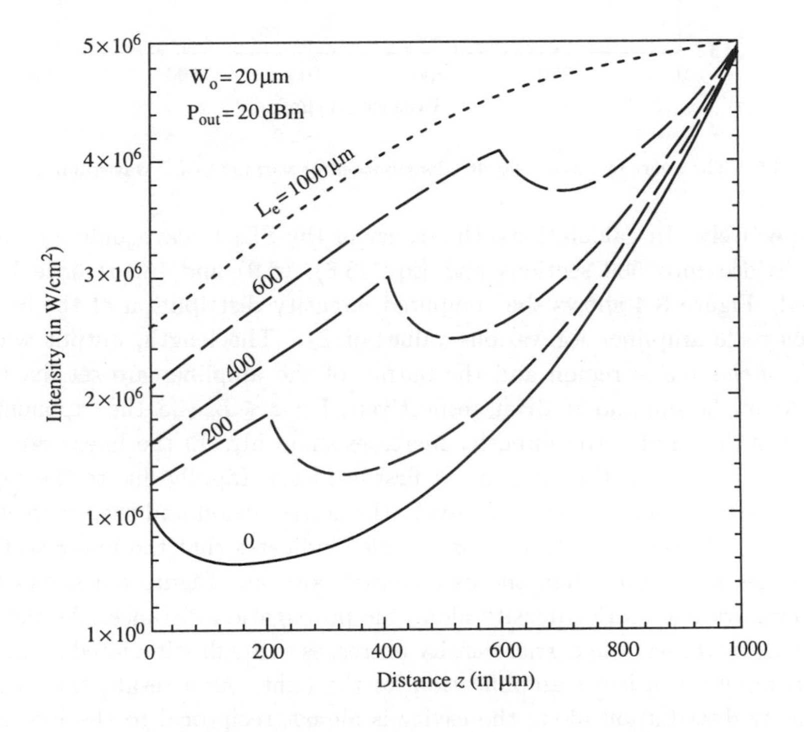

Fig. 8.4. The intensity distribution for various SELT waveguides.

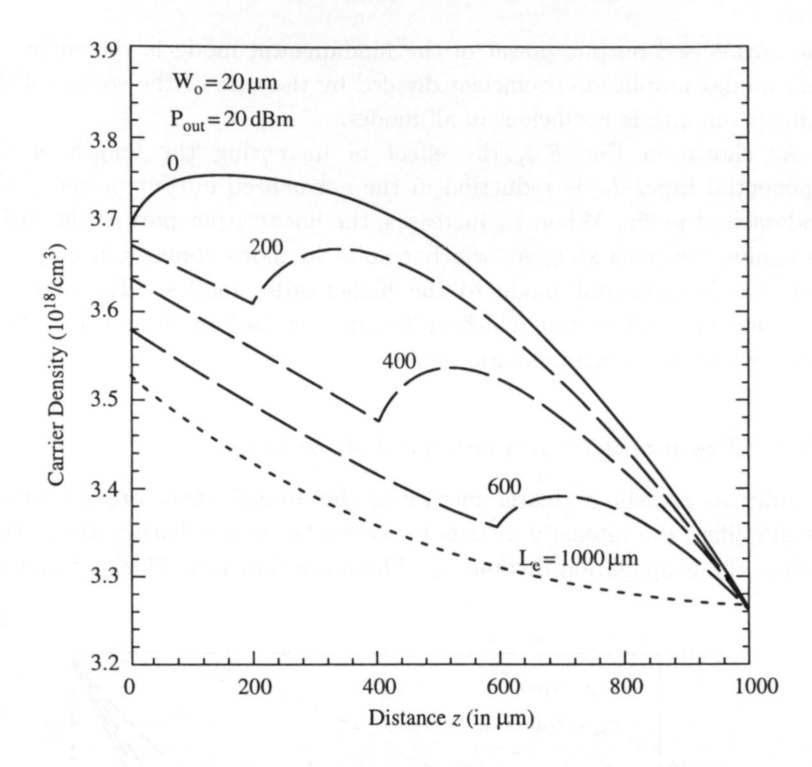

Fig. 8.5. The average carrier density distribution for various SELT waveguide profiles.

respectively. In calculations, the cavity of the SELT waveguide amplifier is divided into 500 sections and Eqs. (8.8), (8.9) and (8.16) have been used. Figure 8.4 shows the computed intensity distribution of the SELT waveguide amplifier for various values of L_e. The length, output width W_o of the active region and the output of the amplifier are set fixed at 1000 μm, 20 μm and 20 dBm, respectively. For $z \leq L_e$ (i.e. the exponential part of the cavity) the intensity increases smoothly. In the linear portion (i.e. $L_e < z \leq L_e$) the intensity at first decreases rapidly due to the rapid increase in the cross-sectional area of the active region and then increases. The gain decreases as L_e increases, which indicates that the linear section provides more gain than the exponential portion. Figure 8.5 shows the lateral average carrier density along the propagation distance. As the intensity increases, the carrier density decreases through stimulated emission and results in a lower amplification for the light. As a result, the carrier density distribution along the cavity is almost reciprocal to the intensity distribution.

8.3.4 *The gain saturation and relative amplified spontaneous emission*

The total power gain of SELT waveguide amplifier for $L_e = 400$ μm is plotted against output power for various output width W_o as shown in Fig. 8.6. As the output width increases, a higher saturation output power is obtained. Shown in Figs. 8.7 is the 3 dB saturation output power versus W_o for different L_e. For a given W_o the saturation output power decreases as L_e increases. The difference in saturation output power is small for smaller W_o, but becomes bigger as the output width increases. The improvement of the saturation output power slowly reduces as W_o increases. The relative amplified spontaneous emission (ASE) versus W_o with L_e as a parameter is calculated as shown in Fig. 8.8. The figure clearly shows that the relative

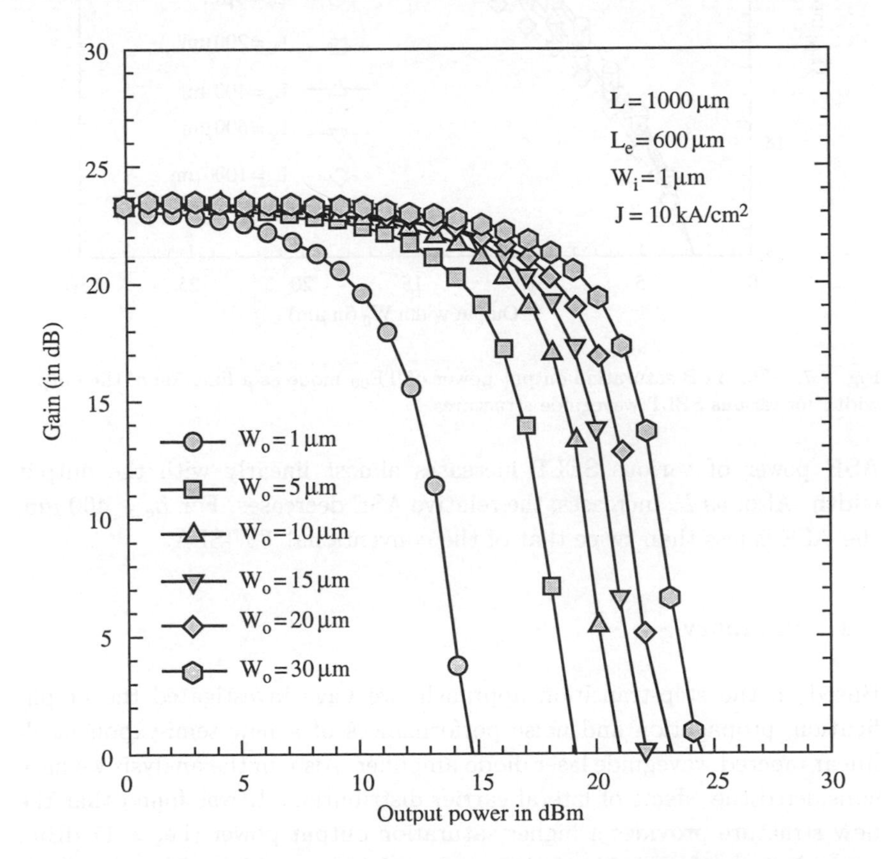

Fig. 8.6. The signal gain as a function of output power for various SELT waveguide structures.

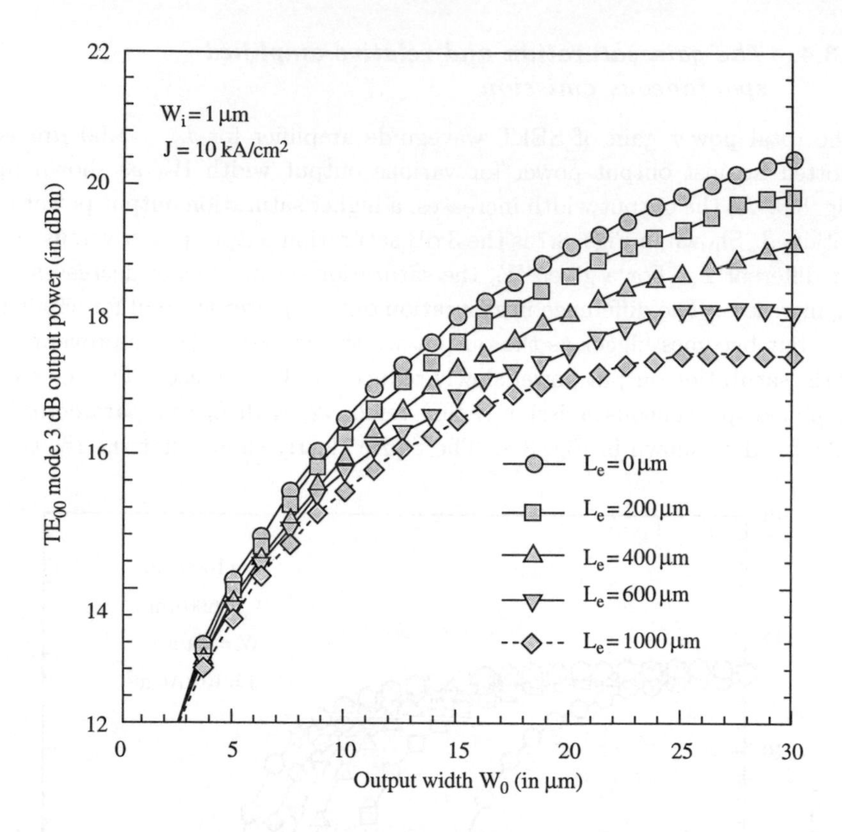

Fig. 8.7. The 3 dB saturation output power of TE_{00} mode as a function of the output width for various SELT waveguide structures.

ASE power of various SELT increases almost linearly with the output width. Also, as L_e increases, the relative ASE decreases. For $L_e \geq 400$ μm, the ASE is less than twice that of the conventional TW-SLA.

8.4 Summary

Based on the step-transition approach, we have investigated the amplification, propagation and noise performances of a new semi-exponential-linear tapered waveguide laser diode amplifier. Also, in the analysis we have considered the effects of lateral carrier distribution. It was found that the new structure provides a higher saturation output power (i.e. > 17 dBm, depending on the length of the exponential waveguide) with a small increase in ASE noise compared with the conventional travelling wave laser

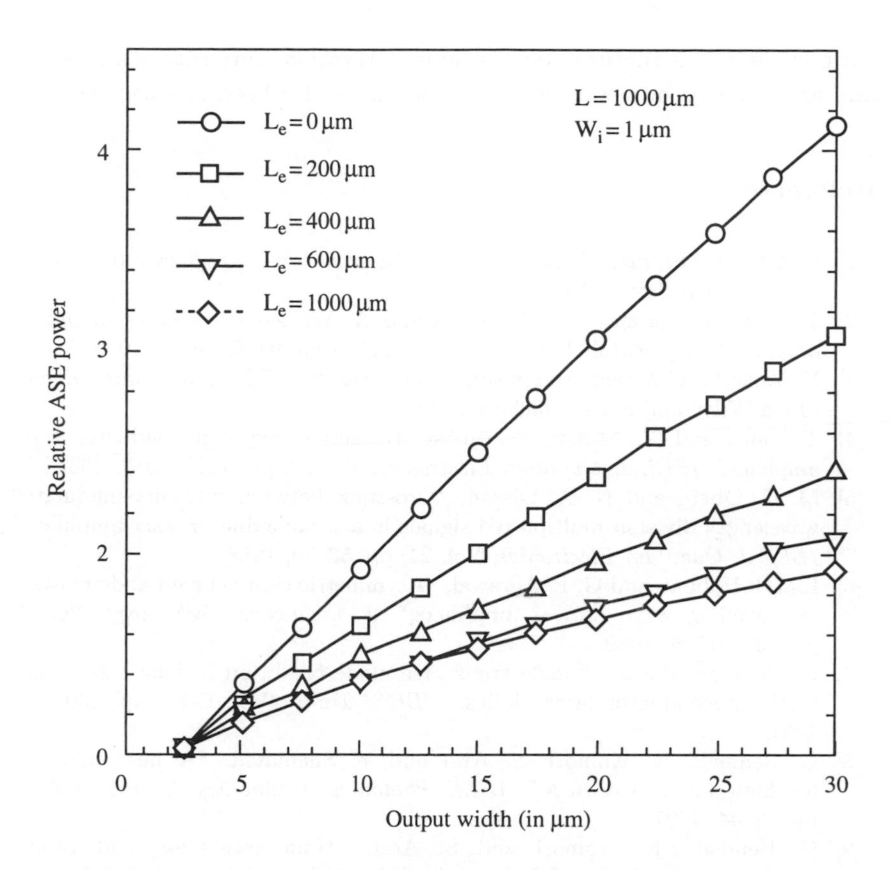

Fig. 8.8. The relative amplified spontaneous emission of exponential-linear tapered waveguide TW-SLA.

diode amplifiers. The effect of various structures on the variation of the intensity and carrier distribution along the cavity has been demonstrated. The coupling efficiency between this new device and single-mode fibres can be improved considerably by using either a vertical graded index lens to increase the beam width along the y direction [22] or a monolithically integrated waveguide lens to reduce the beam width along the x direction [23].

In conclusion, in this chapter a new semiconductor laser diode amplifier structure has been introduceed, which consists of a semi-exponential-linear tapered waveguide active layer to improve the saturation output power. Various characteristic performances of this device have been analysed based on the step-transition approach, taking into account the effects of the spatial distributions of the carrier density and the optical field. The effects of various tapered structures on the gain saturation, intensity distribution and amplified spontaneous emission noise have also been investigated.

References

[1] H. Ghafouri-Shiraz, *Fundamentals of Laser Diode Amplifiers* (John Wiley and Sons, Chichester 1996).
[2] J. Auge, J. Chesnoy, P. M. Gabla and A. Weygang, *Progress in optical amplification* (Alcatel) Electrical Communication 4th Quarter, 1992.
[3] Y. Yamada, *Coherent Amplification and Quantum Effects in Semiconductor* (John Wiley and Sons, Chichester 1991).
[4] T. Saito and T. Mukai, "GaInAsP travelling-wave semiconductor laser amplifier," *IEEE J. Quantum Electronics*, Vol. 23, pp. 1010–1019, 1987.
[5] M. G. Oberg and N. A. Olsson, "Crosstalk between intensity-modulated wavelength division multiplexed signals in a semiconductor laser amplifier," *IEEE J. Quantum Electronics*, Vol. 24, pp. 52–59, 1988.
[6] I. M. I. Habbab and G. P. Agrawal, "Asymmetric channel gain and crosstalk in travelling wave optical amplifiers," *J. Lightwave Technology*, Vol. 7, pp. 1351–1359, 1989.
[7] H. Ghafouri-Shiraz, "Single transverse mode condition in long wavelength SCH semiconductor laser diodes," *IEICE Trans.*, Vol. E-70, pp. 130–134, 1994.
[8] G. Bendelli, K. Komori, S. Arai and Y. Suematsu, "A new structure for high-power TW-SLA," *IEEE Photonics Technology Letters*, Vol. 3, pp. 42–44, 1991.
[9] G. Bendelli, K. Komori and S. Arai, "Gain saturation and propagation characteristics of index-guided tapered waveguide travelling-wave semiconductor laser amplifiers (TTW-SLA's)," *IEEE J. Quantum Electron.*, Vol. 28, pp. 447–454, 1992.
[10] E. El. Yumin, K. Komori, S. Arai and G. Bendelli, "Taper-shape dependence of tapered-waveguide wave semiconductor laser amplifier (T'IV-SLA)," *IEICE Trans. Electron.*, Vol. E77-C, pp. 7–15, 1994.
[11] H. Ghafouri-Shiraz and P. W. Tan, "Analysis of a semi-linear tapered-waveguide laser-diode amplifier," *Microwave and Optical Technology Letters*, Vol. 12, pp. 53–56, 1996.
[12] A. R. Nelson, "Coupling optical waveguides by tapers," *Applied Optics*, Vol. 14, pp. 3012–3015, 1975.
[13] A. F. Milton and W. K. Bums, "Mode coupling in optical waveguide horns," *IEEE J. Quantum Electronics*, Vol. 13, pp. 828–835, 1977.
[14] G. P. Agrawal and N. K. Dutta, *Long Wavelength Semiconductor Lasers* (New York, Van Nostrand, 1986).
[15] F. Gonthier, A. Henault, S. Lacroix, R. J. Black and J. Bures, "Mode coupling in nonuniform fibers: comparison between coupled mode theory and finite difference beam propagation method simulations," *J. Optical Society of America B*, Vol. 8, pp. 416–421, 1991.

[16] W. K. Bums and A. F. Milton, "Mode conversion in planar-dielectric separating waveguides," *IEEE J. Quantum Electronics*, Vol. 11, pp. 32–39, 1975.

[17] G. P. Agrawal, "Lateral analysis of quasi-index guided injection lasers: transition from gain to index guiding," *J. Lightwave Technology*, Vol. 2, pp. 537–543, 1984.

[18] R. Olshansky, C. B. Su, J. Manning and W. Powazinik, "Measurement of radiative and non-radiative recombination rates in InGaAsP and AlGaAs light sources," *IEEE J. Quantum Electronics*, Vol. 20, pp. 837–854, 1984.

[19] J. Wang , H. Olesen and K. E. Stubkjear, "Recombination, gain and bandwidth characteristics of semiconductor laser amplifiers," *J. Lightwave Technology*, Vol. 5, pp. 184–189, 1987.

[20] M. Yamada and Y. Suematsu, "Analysis of gain suppression in undoped injection lasers," *J. Appl. Phys.*, Vol. 52, pp. 2653–2664, 1981.

[21] Y. Yamamoto, "Noise and error rate performance of semiconductor laser amplifiers in PCM-IM optical transmission systems," *IEEE J. Quantum Electronics*, Vol. 16, pp. 1073–1081, 1980.

[22] S. El. Yumin, K. Komori and S. Arai, "GalnAsP/lnP semiconductor vertical GRIN lens for semiconductor optical devices," *IEEE Photonics Technology Letters*, Vol. 5, pp. 601–604, 1994.

[23] F. Koyama, K. Y. Liou, A. G. Dentai, T. Tanbun-ek and C. A. Barrus, "Multiple-quantum-well GaInAs/GaInAsP tapered broad-area amplifiers with monolithically integrated waveguide lens for high-power applications," *IEEE Photonics Technology Letters*, Vol. 5, pp. 916–919, 1993.

Chapter 9

Picosecond Pulse Amplification in Tapered-Waveguide Semiconductor Laser Diode Amplifiers

9.1 Introduction

Currently due to potential applications in high-speed optical fibre communication and logic systems, amplification of ultrashort optical pulses in laser diode amplifiers have been the subject of considerable research interest. The large amplification bandwidth of the laser diode amplifier (LDA) which is about 6000 GHz (50 nm) at 1.55 μm offers the possibility of amplifying ultrashort pulses without distortion [1].

In a conventional straight cavity LDA, the carrier density reduces as the light intensity increases. This results in gain saturation of the amplifier, which severely limits its applications. The saturation output power can be increased by providing a larger cross-sectional area for the active region which provides a larger number of carriers and a lower optical intensity for a given optical power. However, a larger active-region cross-sectional area results in more than one transverse mode propagation. Bendelli *et al.* [2] proposed a tapered-waveguide structure for the LDA active region to overcome this problem. In their structure the taper width increases gradually from the input to the output of the amplifier. This allows most of the power to be carried by the fundamental mode, and provides high saturation output power. It has been shown theoretically [2–4] and experimentally [5–10] that a tapered-waveguide type LDA provides higher gain saturation compared to a conventional non-tapered travelling-wave LDA. Various tapered waveguides such as linear, exponential, quadratic and Gaussian have been proposed to improve the saturation output power. It has been shown in [11] that both linear and Gaussian tapered waveguides have higher saturation output power, whereas the exponential one has lower amplified spontaneous emission noise.

Ultrashort-pulse amplification in LDAs has received significant attentions for its various applications in fibre-optic systems [12–15]. LDAs can exhibit either temporal pulse compression (TPC) or broadening, where it has been found that in both cases the amplifier gain saturation is the dominant mechanism. The amount of pulse compression or broadening depends on the input pulse shape [16]. If the optical input power to a LDA increases, then the stimulated recombination reduces the carrier concentration, hence decreasing the amplifier gain [17]. In such a case, when a picosecond pulse propagates through the amplifier its trailing edge receives less gain than its leading edge. This results in pulse distortion which may also lead to pulse compression [12]. Recently this effect has been used to explain the TPC in LDAs [18–19] which is an alternative to complex mode-locked laser transmitters for the generation of ultrashort pulses for optically time division multiplexed systems [20].

In this chapter we have studied the performance characteristics of an ultrashort Gaussian pulse propagation in both linearly- and exponentially-tapered LDA structures, taking into account the variation of carrier density along the lateral direction.

9.2 Theory

The exponential and linear tapered waveguide LDA structures considered in this study are shown in Fig. 1. The z dependence of the width $W(z)$ of these structures can be expressed as:

$$W(z) = W_{\text{in}} \exp\left[\frac{z}{L} \ln\left(\frac{W_{\text{out}}}{W_{\text{in}}}\right)\right] \quad \text{for Fig. 9.1(a)} \tag{9.1a}$$

and

$$W(z) = W_{\text{in}} + \left(\frac{W_{\text{out}} - W_{\text{in}}}{L}\right) z \quad \text{for Fig. 9.1(b)} \tag{9.1b}$$

where W_{in} and W_{out} are the input and output widths, respectively. The rate equation for the carrier density in the TW-LDA can be expressed as [14] (see also Chapter 8)

$$\frac{\partial N}{\partial t} = D\frac{\partial^2 N}{\partial x^2} + \frac{J}{ed} - \frac{N}{\tau_{\text{s}}(N)} - g_{\text{m}}(N)\frac{i}{hf} \tag{9.2a}$$

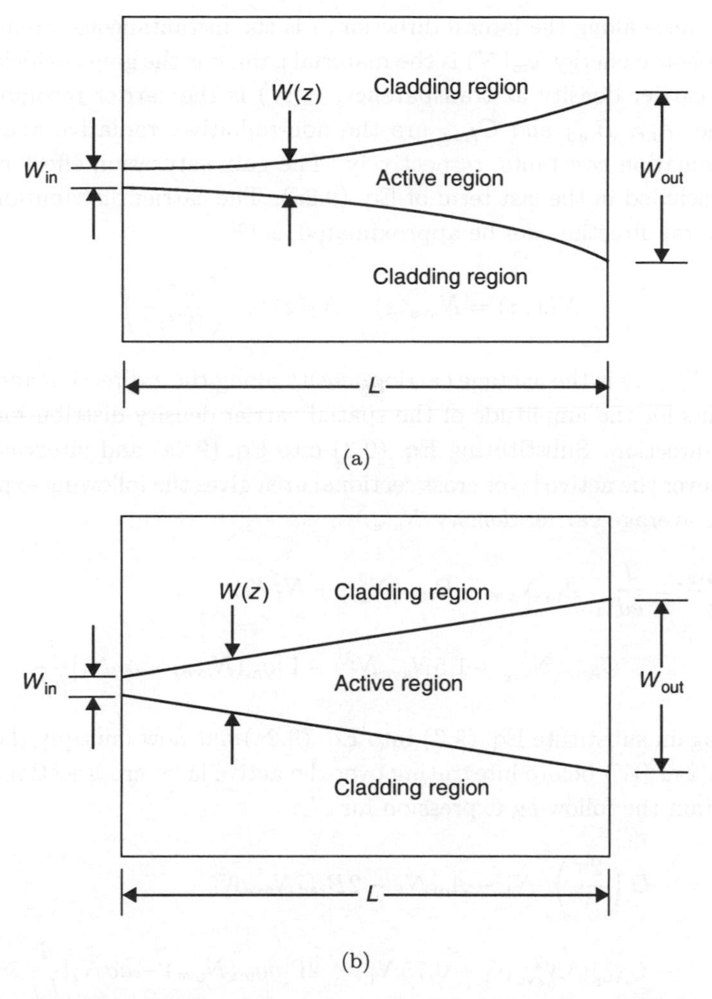

Fig. 9.1. Schematic diagram of the active region of a tapered-waveguide laser diode amplifier (a) exponential and (b) linear.

where

$$\tau_s(N) = \frac{1}{A_{nr} + B_{rad}N + C_{Aug}N^2} \tag{9.2b}$$

$$g_m(N) = a(N - N_0)\,. \tag{9.2c}$$

In Eq. (9.2), N is the carrier density, D is the diffusion coefficient, J is the current density, e is the electron charge, d is the active layer thickness, x is the distance along the lateral direction, i is the instantaneous intensity, hf is the photon energy, $g_m(N)$ is the material gain, a is the gain coefficient, N_0 is the carrier density at transparency, $\tau_s(N)$ is the carrier recombination lifetime, A_{nr}, B_{rad} and C_{Aug} are the non-radiative, radiative and Auger recombination constants, respectively. The gain saturation effect has also been included in the last term of Eq. (9.2a). The carrier distribution along the lateral direction can be approximated as [2]:

$$N(x,z) = N_{ave}(z) - N_1(z)\cos\left(\frac{2\pi x}{W(z)}\right) \tag{9.3}$$

where $N_{ave}(z)$ is the average carrier density along the z direction and $N_1(z)$ accounts for the amplitude of the spatial carrier density distribution along the x direction. Substituting Eq. (9.3) into Eq. (9.2a) and integrating the result over the active layer cross sectional area gives the following expression for the average carrier density N_{ave}.

$$\begin{aligned}\frac{\partial N_{ave}}{\partial t} &= \frac{J}{ed} - A_{nr}N_{ave} - B_{rad}(N_{ave}^2 + N_1^2/2) \\ &\quad - C_{Aug}(N_{ave}^3 + 1.5N_{ave}N_1^2) - \Gamma[g_m(N_{ave}) - a\eta N_1]\frac{I}{hf}\,.\end{aligned} \tag{9.4}$$

If we again substitute Eq. (9.3) into Eq. (9.2a) but now multiply the result by $\cos(2\pi x/W)$ before integrating over the active layer cross sectional area, we obtain the following expression for N_1:

$$\begin{aligned}\frac{\partial N_1}{\partial t} &= -D\left(\frac{2\pi}{W}\right)^2 N_1 - A_{nr}N_1 - 2B_{rad}N_{ave}N_1 \\ &\quad - C_{Aug}(3N_{ave}^2N_1 + 0.75N_1^3) + 2\Gamma[\eta g_m(N_{ave}) - a\sigma N_1]\frac{I}{hf}\,.\end{aligned} \tag{9.5}$$

In the above equations, η and σ take into account for the lateral hole burning and diffusion and Γ is the confinement factor. They are given by the following expressions:

$$\eta = \frac{\int_{-W/2}^{W/2}\cos(2\pi x/W)E_x^2dx}{\int_{-W/2}^{W/2}E_x^2dx} \tag{9.6a}$$

$$\sigma = \frac{\int_{-W/2}^{W/2}\cos^2(2\pi x/W)E_x^2dx}{\int_{-W/2}^{W/2}E_x^2dx} \tag{9.6b}$$

$$\Gamma = \frac{\int_{-W/2}^{W/2} E_x^2 dx \int_{-W/2}^{W/2} E_y^2 dy}{\iint E_x^2 E_y^2 dx dy}. \tag{9.6c}$$

In Eq. (9.6), E_x and E_y are the electric field solutions along the lateral and transverse directions, respectively, and I is the total average intensity. Equations (9.4) and (9.5) can be solved to obtain the steady state values of N_{ave} and N_1 [21]. The amplification rate of the optical intensity along the waveguide structure can be expressed as

$$\frac{\partial I(z,t)}{\partial z} = \{\Gamma[g_{\text{m}}(N_{\text{ave}}) - a\eta N_1] - \alpha_{\text{loss}} - \alpha_{\text{tap}}\} I(z,t) \tag{9.7a}$$

$$\alpha_{\text{loss}} = \Gamma\alpha_{\text{a}} + (1-\Gamma)\alpha_{\text{c}} \tag{9.7b}$$

$$\alpha_{\text{tap}} = -W(z)\frac{d}{dz}\left(\frac{1}{W(z)}\right) \tag{9.7c}$$

where α_{loss} is the material loss coefficient, α_{tap} accounts for the variation of intensity due to the tapering shape of the waveguide, α_{a} and α_{c} are the active and cladding layers loss coefficients, respectively. As we have not considered the effect of dispersion on the output pulse hence the variation of the refractive index has no effect on the shape of the output pulse [12]. For an unchirped Gaussian input pulse, the power $P_{\text{in}}(\tau)$ can be expressed as:

$$P_{\text{in}}(\tau) = \frac{E_{\text{in}}}{\tau_{\text{o}}\sqrt{\pi}} \exp\left(-\frac{\tau^2}{\tau_0^2}\right) \tag{9.8}$$

where E_{in} is the input pulse energy and τ_o is related to pulse width τ_{m} which is the full width at half maximum (FWHM) as $\tau_{\text{m}} = 1.665\tau_{\text{o}}$. The time parameter $\tau = t - z/v_{\text{g}}$ is measured in a reference frame moving with the pulse. Parameters t, z and v_{g} are, respectively, actual time, propagation distance and group velocity. The saturation energy of the amplifier at the input is given by

$$E_{\text{sat(in)}} = \frac{hfW_{\text{in}}d}{a\Gamma} \tag{9.9}$$

where $E_{\text{sat(in)}}$ is the saturation energy for the conventional laser amplifier where $W_{\text{out}} = W_{\text{in}}$. For the parameters used in calculations (see Table 1) $E_{\text{sat(in)}} = 4.75$ pj. Our analysis is based on the step-transition method (as explained in Chapter 8) [21] where the laser cavity is divided into 200 sections and the input pulse is divided into 1600 steps over the interval of 40 ps. The intensity at the input of the cavity is first calculated from the given input energy and hence the input power. The steady state value of the carrier density is calculated by setting $\partial N_{\text{ave}}/\partial t = \partial N_1/\partial t = 0$. Then this value is used as the initial condition for Runge–Kutta numerical simulation of the pulse dynamics. The corresponding carrier density is then obtained by solving Eqs. (9.4) and (9.5) using the Runge–Kutta (RK) method. The intensity of subsequent sections is then obtained by solving Eq. (9.7), again using the RK method. This process is repeated until the output of the amplifier is reached.

9.3 Results and Discussions

The pulse amplification in the conventional (i.e. non-taper) LDA, exponential and linear TW-LDAs has been studied using the carrier rate equations given by Eqs. (9.4) and (9.5). In the following study, the input pulse is an unchirped Gaussian with pulse width $\tau_{\text{m}} = 10$ ps. For this picosecond pulse the ultrafast non-linearities such as carrier heating and spectral hole burning have negligible effects on the performance of the amplifier [14–22]. The amplifier is a highly index guided InGaAsP buried heterostructure device operating at a wavelength of 1.55 μm. The parameters used in the analysis are listed in Table 1. The conventional LDA has a straight cavity, with equal input and output widths of 1 μm (i.e. $W_{\text{in}} = W_{\text{out}} = 1$ μm). Its active layer width and thickness are designed so as to satisfy the single

Table 9.1. Parameters used in the simulations.

Description	Values
Active layer thickness	$d = 0.1$ μm
Input active layer width	$W_{\text{i}} = 1$ μm
Length of Amplifier	$L = 900$ μm
Non-radiative recombination coefficient	$A_{\text{nr}} = 0$ s^{-1}
Spontaneous recombination constant	$B_{\text{rad}} = 8 \times 10^{-11}$ cm^3/s
Auger recombination constant	$C_{\text{Aug}} = 9 \times 10^{-29}$ cm^6/s
Active region loss coefficient	$\alpha_{\text{a}} = 1.2 \times 10^2$/ cm
Cladding layer loss coefficient	$\alpha_{\text{c}} = 5.6$/cm
Carrier density at transparency	$N_0 = 10^{18}$cm^{-3}
Gain coefficient	$a = 1.64 \times 10^{-16}$ cm^2
Current density	$J = 10$ kA/cm^2
Diffusion coefficient	$D = 10$ cm^2/s
Confinement factor	$\Gamma = 0.17$

transverse mode condition [23] and to optimise its polarisation sensitivity [24]. In a tapered waveguide amplifier, the width increases gradually along the cavity from the input to the output. Equations (9.1a) and (9.1b) describe the tapering profile of the exponential and linear TW-LDA structures, respectively.

Figure 9.2 shows the evolution of the input pulse shape within the 900 μm long conventional non-taper TW-LDA for an unchirped Gaussian input pulse. We have considered the case where the input pulse energy is comparable to the saturation energy of the amplifier (i.e. $E_{\text{in}} = E_{\text{sat(in)}} =$ 4.75 pj). As the figure clearly shows, the amplified pulse shape becomes asymmetric and its leading edge is sharper compared with its trailing edge. This sharpening occurs because the leading edge of the pulse saturates the amplifier and reduces the gain available for the trailing edge. Moreover, the pulse also experiences broadening as it propagates towards the output facet of the amplifier.

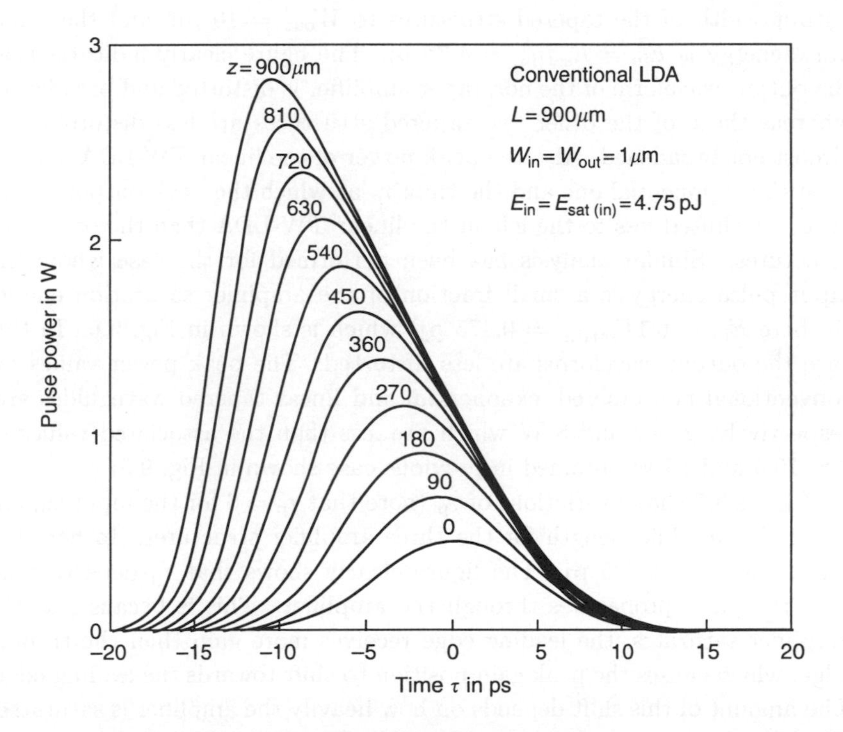

Fig. 9.2. Evolution of an unchirped Gaussian input pulse inside a conventional non-tapered laser diode amplifier in *saturated* regime (i.e. $E_{\text{in}} = E_{\text{sat(in)}} = 4.75$ pj).

The input pulse evolution for the exponentially-tapered TW-LDA is also shown in Fig. 9.3 where the output width of the active layer is $W_{\text{out}} = 10\ \mu\text{m}$ (see Fig. 9.1(a)). In this case the input pulse receives more amplification as it propagates towards the output of the amplifier cavity and is less distorted as compared with the results of the non-tapered TW-LDA structure shown in Fig. 9.2. This improvement is due to the broader cross-sectional area of the structure which improves the saturation performance and hence makes the difference between amplifications of the leading and trailing edges of the pulse small. The results for the case of the linearly-tapered TW-LDA with $W_{\text{out}} = 10\ \mu\text{m}$ (see Fig. 9.1(b)) are also shown in Fig. 9.4 which again shows the pulse is less distorted as compared with Fig. 9.2. However, compared with Fig. 9.3 the pulse receives more amplification (i.e. peak powers are larger). This is because the cross sectional area of the linear TW-LDA is larger than that of the exponential one.

To compare the three LDA's performances, we have plotted the input and output pulse waveforms as shown in Fig. 9.5. In this analysis we set the output width of the tapered structures to $W_{\text{out}} = 10\ \mu\text{m}$ and the input pulse energy is $E_{\text{in}} = E_{\text{sat(in)}} = 4.75$ pj. The figure clearly indicates that the output waveform of the non-taper amplifier is distorted and broadened, whereas those of the other two tapered structures are less distorted and almost not broadened. Also the peak power of the linear TW-LDA is more than the exponential one and the time τ_{p} at which the peak output power occurs is shifted less to the left in the linear TW-LDA than the other two structures. Similar analysis has been performed for the case where the input pulse energy is a small fraction of the amplifier saturation energy (in here $E_{\text{in}} = 0.1E_{\text{sat(in)}} = 0.475$ pj) which is shown in Fig. 9.6. In this case the output waveforms are less distorted. The peak power values for conventional non-tapered, exponential and linear tapered waveguides are, respectively, 2, 6.7 and 8 W which are less than the associated values of 2.8, 10.4 and 13 W obtained in previous case shown in Fig. 9.5.

Figure 9.7 shows variations of τ_{p} (note that $\tau_{\text{p}} = 0$ for the input signal) along the amplifier length for the three amplifier structures. In here the input energy is 0.475 pj. The figure clearly shows that τ_{p} departs from 0 as the pulse propagates through the amplifier. This is because, as the amplifier saturates, the leading edge receives more gain than the trailing edge, which causes the peak gain position to shift towards the leading edge. The amount of this shift depends on how heavily the amplifier is saturated. The shift from 0 is larger in conventional non-tapered LDA compared with the tapered ones and is least in the linear TW-LDA. Large value of τ_{p} may

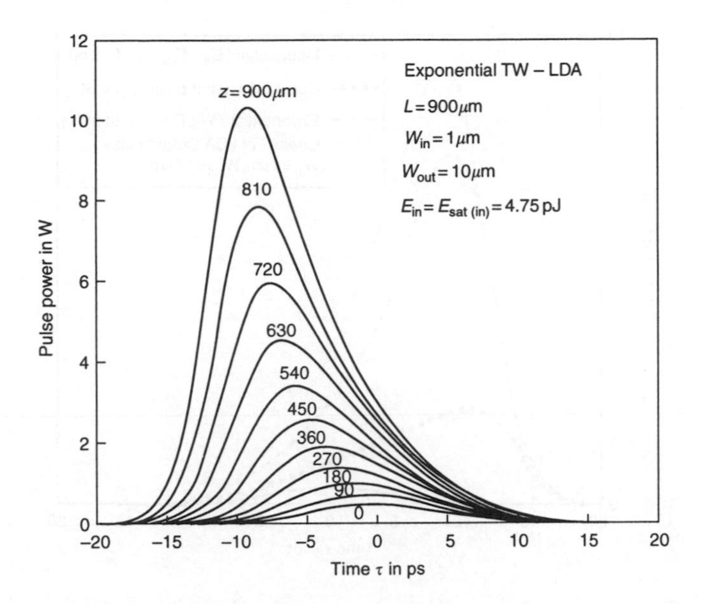

Fig. 9.3. Evolution of an unchirped Gaussian input pulse inside an exponential tapered-waveguide laser diode amplifier in *saturated* regime (i.e. $E_{\text{in}} = E_{\text{sat(in)}} = 4.75$ pj).

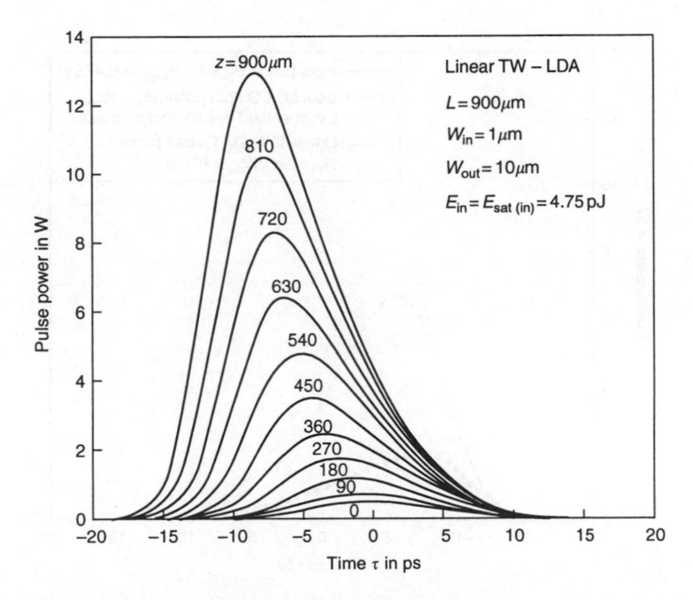

Fig. 9.4. Evolution of an unchirped Gaussian input pulse inside a linear tapered-waveguide laser diode amplifier in *saturated* regime (i.e. $E_{\text{in}} = E_{\text{sat(in)}} = 4.75$ pj).

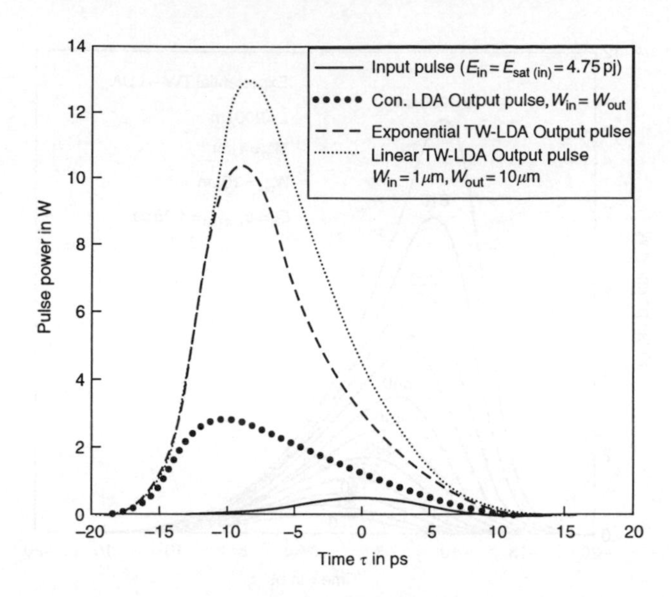

Fig. 9.5. The output pulse waveforms in conventional non-tapered, exponential- and linear-tapered-waveguide amplifier structures in *saturated* regime ($E_{\text{in}} = E_{\text{sat(in)}} = 4.75$ pj).

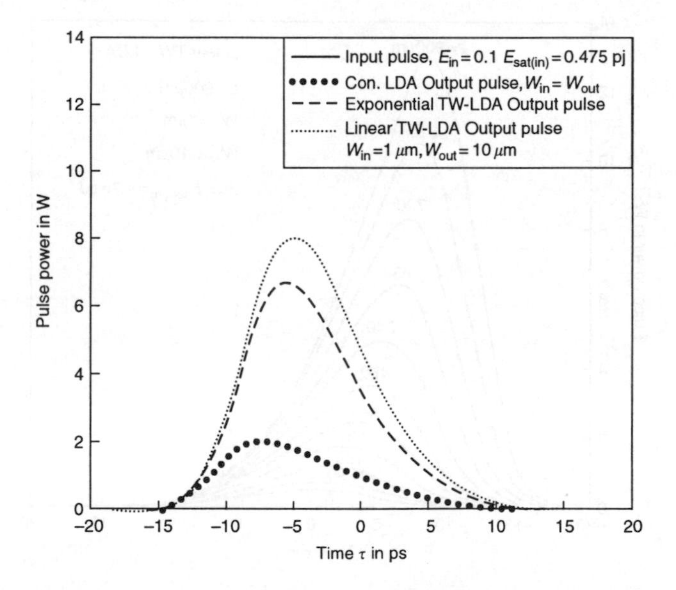

Fig. 9.6. The output pulse waveforms in conventional non-tapered, exponential- and linear-tapered-waveguide amplifier structures in *unsaturated* regime ($E_{\text{in}} = 0.1E_{\text{sat(in)}} = 0.475$ pj).

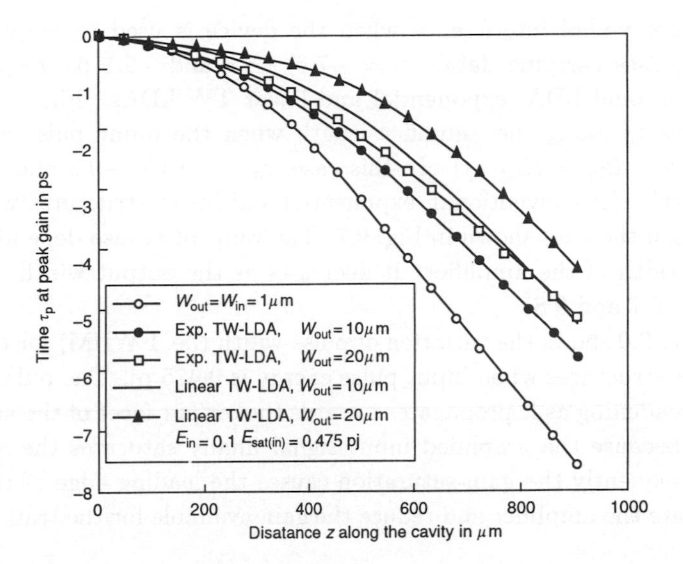

Fig. 9.7. Temporal shift in peak gain along the amplifier cavity length for the conventional non-tapered, exponential- and linear-tapered-waveguide amplifier structures in *unsaturated* regime ($E_{\rm in} = 0.1\, E_{\rm sat(in)} = 0.475$ pj).

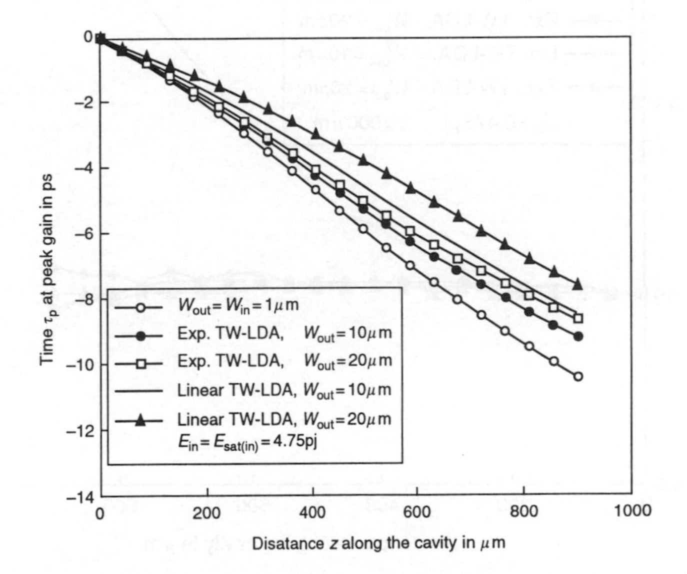

Fig. 9.8. Temporal shift in peak gain along the amplifier cavity length for the conventional non-tapered, exponential- and linear-tapered-waveguide amplifier structures in *saturated* regime ($E_{\rm in} = E_{\rm sat(in)} = 4.75$ pj).

cause intersymbol interference when the device is used to amplify picosecond pulses carrying data. $\tau_p = -7.6$, -5.8 and -5.1 ps, respectively, in conventional LDA, exponential and linear TW-LDAs. Figure 8 shows variations τ_p along the amplifier length when the input pulse energy is 4.75 pj (i.e. $E_{in} = E_{sat(in)}$). In this case, $\tau_p = -10.5$, -9.3 and -8.8 ps, respectively, for conventional, exponential and linear structures which are larger than the ones shown in Fig. 9.7. The value of τ_p also depends on the output width of the amplifier. It decreases as the output width increases (see Figs 9.7 and 9.8).

Figure 9.9 shows the variation of pulse-width (i.e. FWHM) for the three amplifier structures when input pulse energy is 0.475 pj. The pulse experiences broadening as it propagates towards the output facet of the amplifier. This is because the amplified input signal finally saturates the amplifier and consequently the gain-saturation causes the leading edge of the pulse to saturate the amplifier and reduce the gain available for the trailing edge,

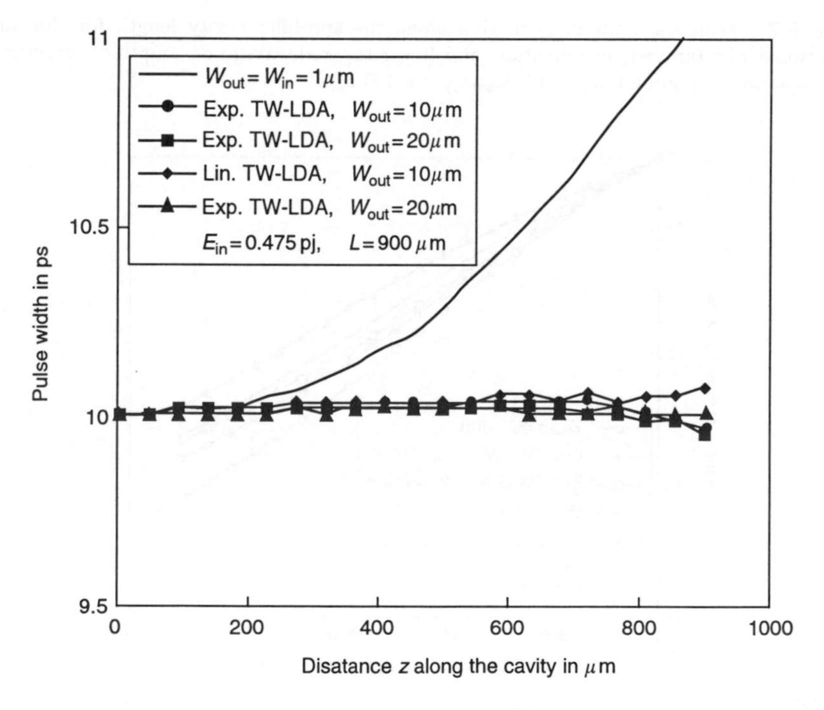

Fig. 9.9. Variations of the FWHM pulse width along the amplifier cavity length for the conventional non-tapered, exponential- and linear-tapered-waveguide amplifier structures in *unsaturated* regime ($E_{in} = 0.1\,E_{sat(in)} = 0.475$ pj).

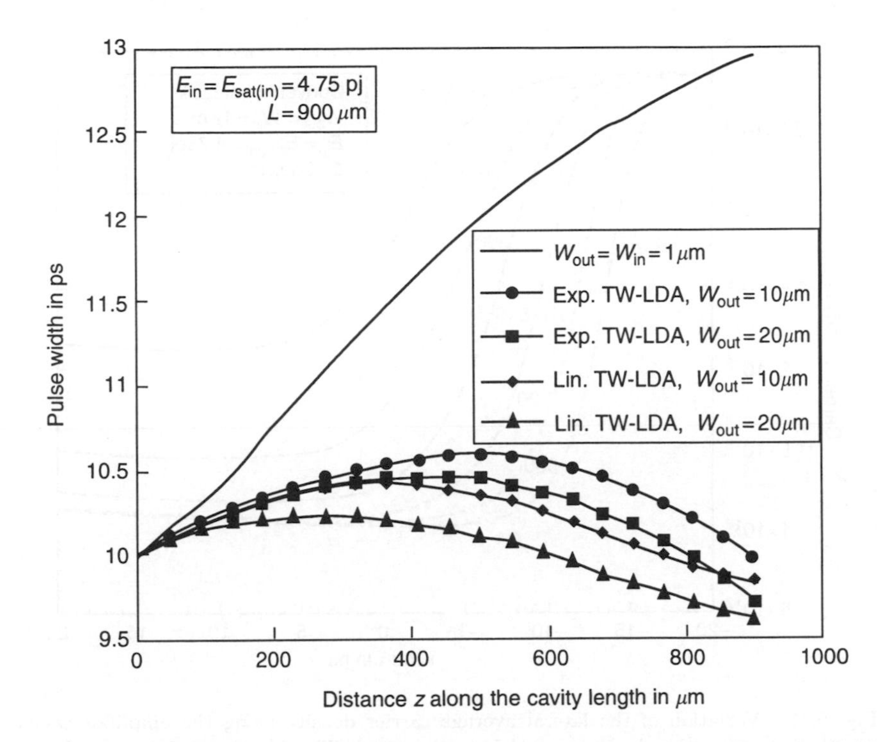

Fig. 9.10. Variations of the FWHM pulse width along the amplifier cavity length for the conventional non-tapered, exponential- and linear-tapered-waveguide amplifier structures in *saturated* regime.

which results in pulse broadening. In case of the tapered structure, the amplified input pulse does not saturate the amplifier as it propagates within the amplifier. This is because the tapered structure has a broader active layer area and hence the change in the output pulse width is insignificant. Both the exponential and linear tapered waveguide amplifiers have almost negligible pulse broadening as compared with the conventional type.

Figure 9.10 shows similar calculations for the case where the input pulse energy is 4.75 pj (i.e. $E_{\text{in}} = E_{\text{sat(in)}}$). Again the non-tapered LDA output pulse width is broadened significantly, whereas when the input pulse propagates within the exponential or linear tapered amplifier initially it is slightly broadened and then at the later stage it is compressed, which results in the output pulse width being slightly smaller. The compression is because the leading edge of the pulse receives more amplification than the trailing edge, which results in the trailing edge being slightly trimmed. The amount of pulse broadening in both exponential and linear TW amplifier structures

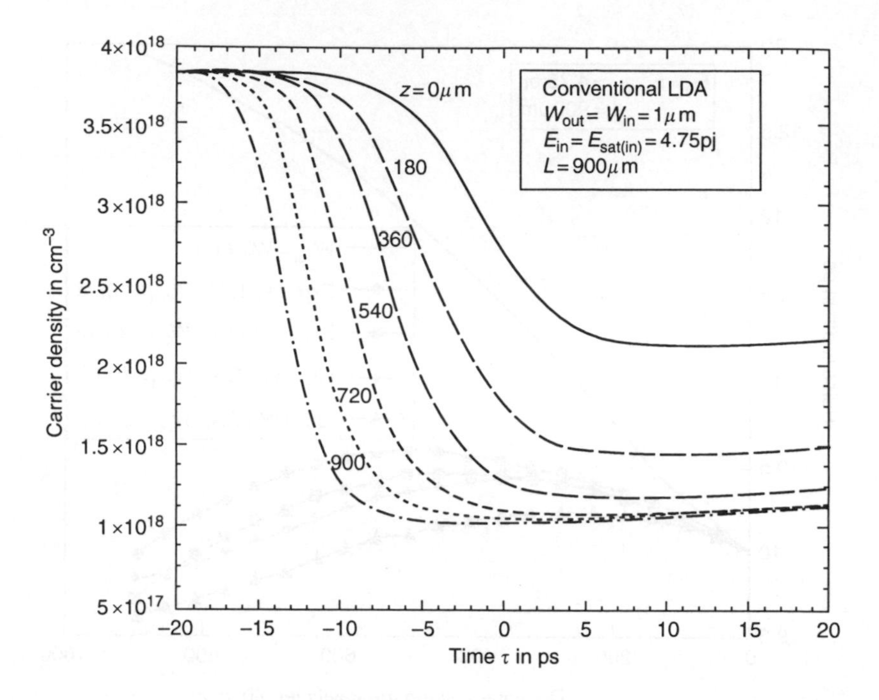

Fig. 9.11. Variation of the lateral average carrier density along the amplifier cavity length in a conventional non-tapered laser diode amplifier in *saturated* regime ($E_{\text{in}} = E_{\text{sat(in)}} = 4.75$ pj).

is almost the same for $z < 250$ μm, however, thereafter is lesser in linear TW-LDA. Also the amplified pulse is more compressed in linear TW-LDA than in exponential one. The results shown in both Figs. 9.9 and 9.10 are for both $W_{\text{out}} = 10$ and 20 μm.

Figure 9.11 shows variations of the average carrier density N_{ave} along the amplifier length for the case shown in Fig. 9.2. When the input pulse is amplified at the amplifier output the associated N_{ave} decreases through stimulated emission rapidly and approaches its carrier transparency value of $N_{\text{o}} = 10^{18}\text{cm}^{-3}$ where there is either very little or no amplification. The rate of gain recovery is characterised by the slope at the trailing edge of the amplified pulse, which in the case of 10 ps pulse is not significant.

Figure 9.12 shows variations of N_{ave} for the case of exponential TW-LDA. In this case, also as the pulse is amplified, N_{ave} decreases; however, since the active layer width increases along the amplifier cavity length the rate of reduction of the carrier density is slower than the non-taper case although the pulse receives more amplification. Moreover, the N_{ave} at

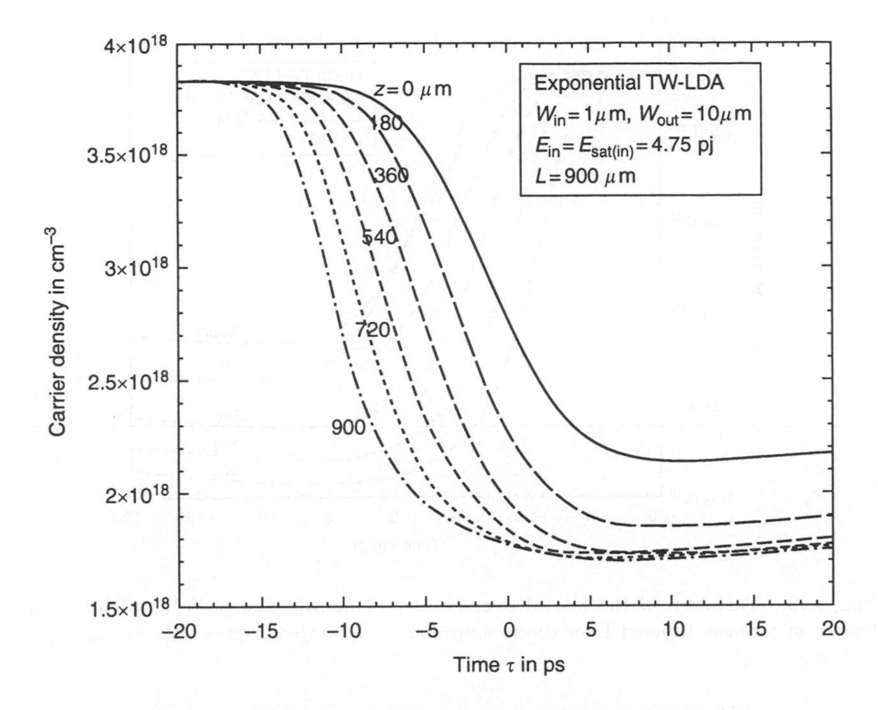

Fig. 9.12. Variation of the lateral average carrier density along the amplifier cavity length in an exponential tapered laser diode amplifier in *saturated* regime ($E_{\rm in} = E_{\rm sat(in)} = 4.75$ pj).

the output does not approach $N_{\rm o}$, which indicates that the amplifier has not yet saturated. Similar calculations have been performed for the case of linear TW-LDA which is shown in Fig. 9.13.

Figure 9.14 shows the gain saturation performance of both 600 and 900 μm long exponential-tapered-waveguide as compared with the non-taper conventional LDA. For $L = 600$ μm, the improvement in the 3 dB input pulse saturation energy over that of the conventional non-tapered one are 6 and 7.7 dB for the output widths of 10 and 20 μm, respectively, whereas for $L = 900$ μm these values are 6.8 and 9 dB, respectively. A similar gain saturation performance for the linear TW-LDA is shown in Fig. 9.15. For $L = 600$ μm the improvement in the 3 dB input pulse saturation energy is 7 and 9.8 dB, respectively, for the output width of 10 and 20 μm which are higher than the exponential TW-LDA. For $L = 900$ μm, these values are 7.5 and 10.5 dB, respectively. In all cases, the linear TW-LDA provides higher gain saturation performance than the

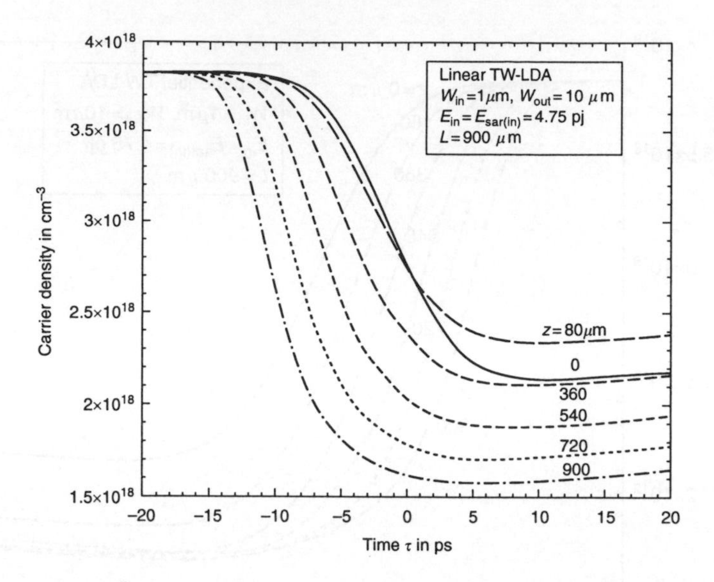

Fig. 9.13. Variation of the lateral average carrier density along the amplifier cavity length in a linear tapered laser diode amplifier in *saturated* regime ($E_{\rm in} = E_{\rm sat(in)} =$ 4.75 pj).

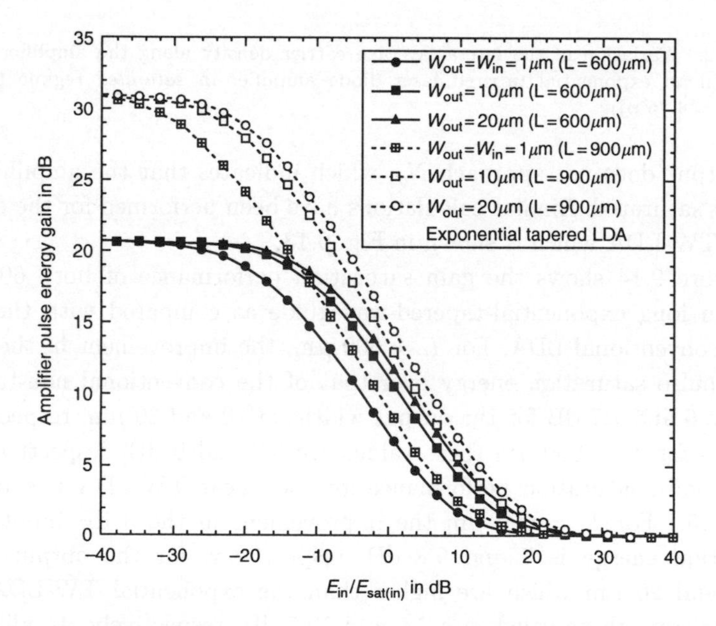

Fig. 9.14. Variations of pulse energy gain with $E_{\rm in}/E_{\rm sat(in)}$ in the conventional non-tapered, and *exponential* tapered-waveguide laser diode amplifier with cavity length as a parameters.

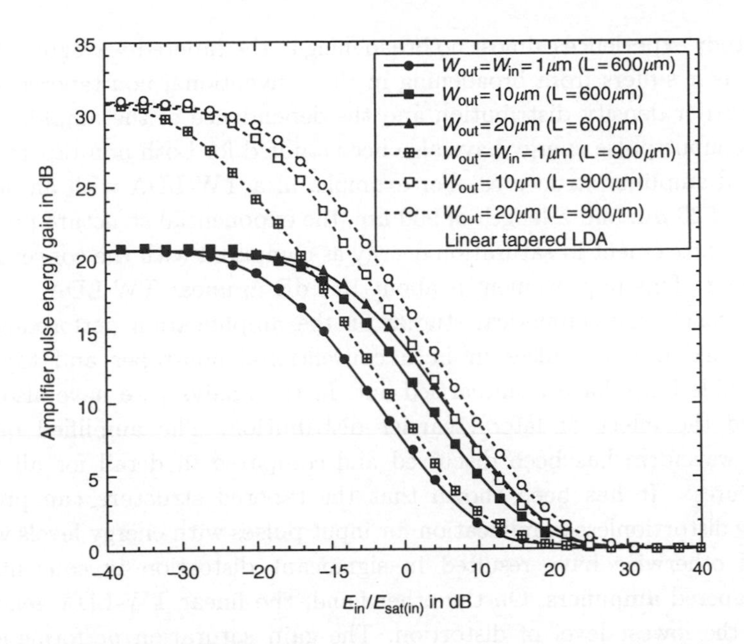

Fig. 9.15. Variations of pulse energy gain with $E_{\text{in}}/E_{\text{sat(in)}}$ in the conventional non-taperad, and linear tapered-waveguide laser diode amplifier with cavity length as a parameter.

exponential one, which is due to a larger cross sectional area of the linear taper.

9.4 Summary

The amplification characteristics of picosecond Gaussian pulses in conventional non-tapered and both linear and exponential tapered-waveguides (TWs) laser diode amplifier (LDA) structures have been studied. The analysis is based on numerical simulation of the rate equation which also takes into account the effect of lateral carrier density distribution. The amount of pulse distortion experienced within the amplifier for input pulses having energies $E_{\text{in}} = 0.1E_{\text{sat(in)}} = 0.475$ pj (where $E_{\text{sat(in)}}$ is the input saturation energy of the amplifier) and $E_{\text{in}} = E_{\text{sat(in)}} = 4.75$ pj have been analysed for each structure which has a length of 900 μm and an input width of 1 μm. It has been found that the TW-LDA provides higher gain saturation and hence imposes less distortion on the amplified pulse as compared with a conventional non-tapered LDA. The amplified 10 ps pulse used in this study experiences almost no broadening in the tapered-waveguide LDA, whereas it suffers from broadening in the conventional non-tapered LDA. The carrier density distribution and the dependence of the amplifier gain on the input pulse energy have also been studied for both non-tapered and tapered amplifier structures. For example, in a TW-LDA with an output width of 20 μm and a length of 900 μm, the exponential structure provides 9 dB improvement in saturation energy as compared with the conventional amplifier. This improvement is about 10.5 dB in linear TW-LDA.

In conclusion, numerical studies on the amplification performance of picosecond optical pulses in both conventional non-taper and tapered-waveguide LDA have been carried out. In the analysis we have also considered the effect of lateral carrier distribution. The amplified output pulse waveform has been discussed and compared in detail for all three structures. It has been shown that the tapered structure can provide nearly distortionless amplification for input pulses with energy levels which would otherwise have resulted in significant distortion in conventional non-tapered amplifiers. On the other hand, the linear TW-LDA generally gives the lowest level of distortion. The gain saturation performance of picosecond pulses improve significantly in both tapered-waveguide LDA.

References

[1] H. Ghafouri-Shiraz, *Fundamentals of Laser Diode Amplifiers* (John Wiley and Sons, ISBN 0471958727, Feb. 1996).

[2] G. Bendelli, K. Komori and S. Arai, "Gain Saturation and Propagation Characteristics of Index-Guided Tapered-Waveguide Travelling-Wave Semiconductor Laser Amplifiers (TTW-SLAs)," *IEEE J. Quantum Electronics*, Vol. QE-28, No. 2, pp. 447–458, 1992.

[3] E. El. Yumin, K. Komori, S. Arai and G. Bendelli, "Taper-Shape Dependence of Tapered-Waveguide Travelling Wave Semiconductor Laser Amplifier(TTW-SLA)," *IEICE Trans. Electron.*, Vol. E77-C, No. 4, pp. 7–15, April 1994.

[4] H. Ghafouri-Shiraz, P. W. Tan and T. Aruga, "Analysis of a Semi-Linear Tapered-Waveguide Laser Diode Amplifier," *Microwave and Optical Technology Letters*, Vol. 12, No. 2, pp. 53–56, June 1996.

[5] F. Koyama, K. Y. Liou, A. G. Dentai, T. Tanbun-ek and C. A. Burrus, "Multiple-Quantum-Well GaInAs/GaInAsP Tapered Broad-Area Amplifiers with Monolithically Integrated Waveguide Lens for High-Power Applications," *IEEE Photonics Technology Letters*, Vol. 5, No. 8, pp. 916–919, Aug. 1993.

[6] L. Goldberg, D. Mehuys and D. Welch, "High Power Mode-Lock Compound

Laser Using a Tapered Semiconductor Amplifier," *IEEE Photonics Technology Letters*, Vol. 6, No. 9, pp. 1070–1072, 1994.

[7] B. Zhu, I. H. White, K. A. Williams, F. R. Laughton and R. V. Penty, "High-Peak Power Optical Pulse Generation from Q-switched Bow-Tie Laser with a Tapered Travelling-wave Amplifier," *IEEE Photonics Technology Letters*, Vol. 8, No. 4, pp. 503–505, 1996.

[8] T. Toyonaka, S. Tsuji, K. Hiramoto, S. Aoki and H. G. Shiraz, "Gain Saturation Characteristics of 1.55 μm MQW Amplifiers and the Bit Error Performance at 2.5 Gb/s-120 km Transmission," *IEEE/OSA*, *IEE/IEICE 1993 Technical Digest Series*, Vol. 14, pp. 162–165, Paper No. MD4, Optical Amplifiers and Their Applications, July 1993, Yokohama, Japan.

[9] A. Mar, R. Helkey, J. Bowers, D. Mehuys and D. Welch, "Mode-Locked Operation of a Master Oscillator Power Amplifier," *IEEE Photonic Technology Letters*, Vol. 6, No. 9, pp. 1067–1069, Sept. 1994.

[10] D. J. Bossert, J. R. Marciante and M. W. Wright, "Feedback Effects in Tapered Broad Area Semiconductor Lasers and Amplifiers," *IEEE Photonic Technology Letters*, Vol. 7, No. 5, pp. 470–472, May 1995.

[11] S. E. L. Yumin, K. Komori, S. Arai and G. Bendelli, "Taper-Shape Dependence of Tapered-Waveguide Traveling Wave Semiconductor Laser Amplifier (TTW-SLA)," *IEICE Trans. Electronics*, Vol. E77-C, No. 4, pp. 7–15, 1994.

[12] G. P. Agrawal and N. A. Olsson, "Amplification and compression of weak picosecond optical pulses by using semiconductor laser amplifiers," *Opt. Lett.*, Vol. 14, pp. 500–502, May 1989.

[13] A. J. Lowery, "Pulse Compression mechanisms in semiconductor laser amplifiers," *IEE Proceedings Pt. J*, Vol. 136, No. 3, pp. 141–146, June 1989.

[14] G. P. Agrawal and N. A. Olsson, "Self-Phase Modulation and Spectral Broadening of Optical Pulses in Semiconductor Laser Amplifiers," *IEEE J. Quantum Electronics*, Vol. 25, pp. 2297–2306, Nov. 1989.

[15] T. Saitoh and T. Mukai, "Gain Saturation Characteristics of Travelling-Wave Semiconductor Laser Amplifiers in Short Optical Pulse Amplification," *IEEE J. Quantum Electronics*, Vol. 26, pp. 2086–2094, Dec. 1990.

[16] A. J. Lowery, "Explanation and Modelling of Pulse Compression and Broadening in Near-Traveling-Wave Laser Amplifiers," *Electronics Letters*, Vol. 24, No. 18, pp. 1125–1126, 1988.

[17] J. C. Simon, "Semiconductor Laser Amplifier for Single-Mode Optical Fiber Communications," *J. Opt. Commun.*, Vol. 4, pp. 51–62, 1983.

[18] I. W. Marshall and D. M. Spirit, "Observation of Large Pulse Compression by a Saturated Travelling-Wave Semiconductor Laser Amplifier," *CLEO'88*, Anaheim, California, USA, Paper TuM64, April 1988.

[19] I. W. Marshall, D. M. Spirit and M. J. O'Mahony, "Picosecond Pulse Response of a Travelling-Wave Semiconductor Laser Amplifier," *Electronics Letters*, Vol. 23, No. 16, pp. 818–819, 1987.

[20] J. M. Weisenfeld, G. Eisenstein, R. S. Tucker, G. Raybon and P. B. Hansen, "Distortionless Picosecond Pulse Amplification and Gain Compression in a Travelling-Wave InGaAsP Optical Amplifier," *Applied Physics Letters*, Vol. 53, pp. 1239–1241, 1988.

[21] H. Ghafouri-Shiraz and P. W. Tan, "Study of a Novel Laser Diode Amplifier Structure," *Semiconductor Science Technology*, Vol. 11, August 1996, in press.

[22] C. B. Kin, En. T. Peng, C. B. Su, W. Rideout and G. H. Cha, "Measurement of 50 fs Nonlinear Gain Time Constant in Semiconductor Lasers," *IEEE Photonics Technology Letters*, Vol. 4, No. 9, Sept. 1992.

[23] H. Ghafouri-Shiraz, "Single transverse mode condition in long wavelength SCH semiconductor laser diodes," *Trans. IEICE*, Vol. E70, No. 2, pp. 130–134, 1987.

[24] T. Saitoh and T. Mukai, "Structural design for polarization-insensitive travelling-wave semiconductor laser amplifiers," *Optical and Quantum Electronics,* Vol. 21, pp. S47–S48, 1989.

Chapter 10

Sub-picosecond Gain Dynamic in Highly-Index Guided Tapered-Waveguide Semiconductor Laser Diode Optical Amplifiers

10.1 Introduction

Ultrafast gain dynamics of laser diode amplifiers have been widely studied due to some emerging ultrafast non-linear applications [1–6]. Ultra-short pulses can be generated either by using a Q-switching laser or external cavity mode locking technique [7–11]. Amplification is often desirable for such pulses. Sub-picosecond pulse amplification in a laser diode amplifier is affected by many factors. With regard to the gain dynamics, it is determined by carrier density depletion, spectral hole burning, carrier heating effects and two-photon absorption. There are various sources which result in carrier heating. Firstly, injected carriers need excess energy to overcome the barrier at the junction, to reach the active layer, and the release of this energy results in injection heating. On the other hand, recombination heating which results from the stimulated recombination process depletes carriers with below average energy, and hence leaves the system with those carriers having energies higher than average. In the process of Auger recombination, the carrier is excited to a higher energy state via the recombination of another carrier, resulting in an increase in the carrier energy. In addition, free-carrier absorption refers to the absorption of the photons by phonons or by impurities which results in a higher temperature.

The cooling is mainly due to inelastic longitudinal optical (LO) phonon scattering. In the case of spectral hole burning, it has been shown in a detailed calculation that the spectral hole burning effect is very closely represented by the non-linear model $1/(1 + \varepsilon_{\text{shb}}S)$ where S is the intensity and ε_{shb} is the spectral hole burning parameter [1]. The study of carrier dynamics is mainly carried out using the pump-probe technique. Many sub-picosecond pump-probe experiments have been carried out to study the ultrafast gain dynamics in a semiconductor optical amplifier [12–15]. The typical relaxation time of the carrier temperature is 0.5–1 ps. The carrier-carrier scattering time responsible for spectral hole burning is about 50–100 fs.

Apart from gain dynamics, there are a number of other factors which contribute to the distortion of the pulse in time and spectral domains [16–19]. The refractive index of the semiconductor material is affected by the carrier density, the carrier temperature and the instantaneous intensity, and is characterised by associated linewidth enhancement factors in [16–18]. These changes in the refractive index cause self-phase modulation. If dispersion were significant, the change in the refractive index would result in the pulse at various wavelengths travelling at a different velocity, with further distortion of the pulse both in the time and frequency domains. If the pulse shape is severely altered, the amplification characteristics of the pulse can be significantly affected.

As mentioned in Chapter 8 the highly index guided tapered-waveguide semiconductor laser amplifier is known to provide significant improvement in terms of gain saturation for the semiconductor laser amplifier [20–23]. It has been utilised in a number of experiments involving the generation of ultrashort pulses using Q-switching or mode locking techniques. The model introduced in Chapter 9 to analyse the picosecond pulse amplification in tapered waveguide amplifiers has only considered the carrier depletion. However, this model is only applicable for pulses having a duration of more than a few picoseconds. In this chapter, the effects of both carrier heating and spectral hole burning have been included in the analysis of sub-picosecond pulse gain dynamics within a highly index guided tapered-waveguide semiconductor laser amplifier.

However, in the model described in this chapter, the lateral distribution of carrier and carrier temperature have also been taken into account. A few assumptions have been made in using this model. Firstly, it is assumed that an adiabatic single lateral mode condition remains throughout the cavity. A number of theoretical and experimental works have shown that with a gently tapering profile, most of the power would remain in the fundamental mode. This is due to the single mode condition at the input end and the low level of conversion into higher order modes [20–25]. The pulse distortion as a result of self-phase modulation, dispersion and memory effects are beyond the scope of this chapter. For simplicity, the variation in the energy density with the carrier density and carrier temperature have been modeled linearly.

This chapter studies the effects of lateral carrier and temperature distributions on the amplification of sub-picosecond pulses in both conventional and tapered-waveguide laser diode amplifiers with the effect of lateral distribution both included and excluded. The saturation characteristics of picosecond and sub-picosecond pulses in such amplifiers are also explained.

10.2 Theoretical Model

The device considered here is a double heterostructure highly index guided travelling-wave semiconductor laser amplifier operating at 1.55 μm, as shown in Fig. 10.1. We have assumed that the device has negligible facet reflectivities. The active region at the input has a smaller width, which satisfies the single mode condition. The width is gradually increased from the input end to the output end to preserve most of the power in the fundamental mode. The analysis of the passive tapered waveguide has been covered by a number of authors [20–25]. Hence, in this chapter we have only investigated the amplification performance of sub-picosecond pulses. The carrier rate equation, which determines the response of the carrier density in a semiconductor to an incoming optical signal, is given by:

$$\frac{\partial N}{\partial t} = D\frac{\partial^2 N}{\partial x^2} + \frac{J}{ed} - \frac{N}{\tau_s(N)} - g\frac{i(x,y,z,t)}{hf} \tag{10.1}$$

where N is the carrier density, D is the diffusion constant, x is the axis along the lateral direction, J is the current density, e is the electron charge, d is

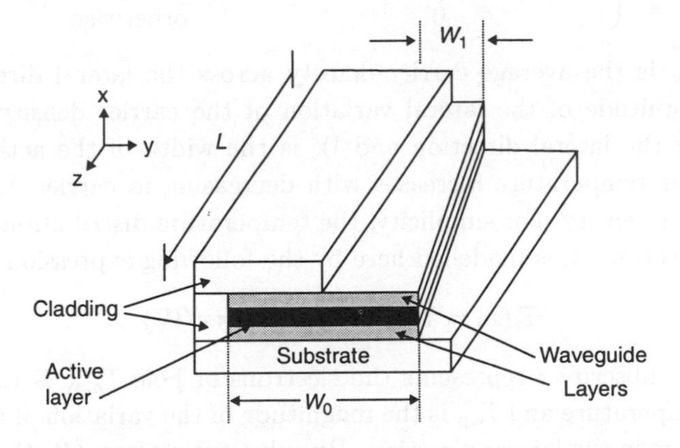

Fig. 10.1. Structure of tapered-waveguide laser diode amplifier.

the thickness of the active region, τ_s is the carrier recombination lifetime, i is the instantaneous intensity, h is the plank's constant, f is the optical frequency and g is the material gain which can be expressed as [5]:

$$g \approx (g_0 + g_N \Delta N + g_{Tc} \Delta T_c + g_{Tv} \Delta T_v) \frac{1}{1 + \varepsilon_{shb} S} . \tag{10.2}$$

In Eq. (10.2), g_0 is the gain coefficient constant, g_N g_{Tc} and g_{Tv} are parameters for the variation of the material gain with respect to the variation of the carrier density ΔN and carrier temperatures ΔT_c and ΔT_v. The rate equations for the carrier temperature can be derived from the density matrix as [5]:

$$\frac{\partial T_i}{\partial t} = h_i \left(\mu_i - E_i + \frac{\beta_i}{g} hf \right) v_g \, gS - \frac{T_i - T_L}{\tau_{h,i}} \quad i = c, v \tag{10.3}$$

where T_i is the temperature of the carriers, E_i is the carrier energy, h_i is defined as $h_i^{-1} = (\partial U_i / \partial T_i)_N$, μ_i, is defined as $\mu_i = (\partial U_i / \partial N_i)_{T_i}$, U_i is the carrier energy density, β is the free carrier absorption coefficient, $\tau_{h,i}$ is the intraband scattering time, v_g is the group velocity, $S(z, t)$ is the photon density and T_L is the lattice temperature. The subscript $i = c, v$ represents the carrier temperature in the conduction and valence bands, respectively. Assuming an adiabatic single lateral mode condition in the laser diode amplifier, the carrier density distribution across the lateral direction can be represented by the following equation [20–23]:

$$N(x) = \begin{cases} N_{av} - N_1 \cos\left(\dfrac{2\pi x}{W}\right) & \text{for} \quad N(x) < |W(z)/2| \\ 0 & \text{otherwise} \end{cases} \tag{10.4}$$

where N_{av} is the average carrier density across the lateral direction, N_1 is the magnitude of the lateral variation of the carrier density, x is the axis along the lateral direction and W is the width of the active region. The carrier temperature increases with decreasing in carrier density and increasing intensity. For simplicity, the temperature distribution along the lateral direction, x, is modelled here by the following expression [25]:

$$T_i(x) = T_{i,av} + T_{i,p} \cos(2\pi x / W) \tag{10.5}$$

where the subscript i represents the electrons or hole $T_{x,av}$ is the average carrier temperature and $T_{i,p}$ is the magnitude of the variation of the carrier temperature in the lateral direction. By substituting Eqs. (10.4) and (10.5) into the rate Eqs. (10.1) and (10.3), and separating the averaged and the cosine parts, the following rate equations for the carrier density can be obtained.

$$\frac{\partial N_{\rm av}}{\partial t} = \frac{J}{ed} - A_{\rm nr}N_{\rm av} - B_{\rm rad}\left(N_{\rm av}^2 + \frac{N_1^2}{2}\right) - C_{\rm Aug}\left(N_{\rm av}^3 + \frac{3N_{\rm av}N_1^2}{2}\right) - \Gamma g_1\left(\frac{I}{hf}\right) \quad (10.6a)$$

$$\frac{\partial N_1}{\partial t} = -DN_1\left(\frac{2\pi}{W}\right)^2 - A_{\rm nr}N_1 - 2B_{\rm rad}N_{\rm av}N_1 - C_{\rm Aug}\left(3N_1N_{\rm av}^2 + \frac{3N_1^3}{4}\right) - 2\Gamma g_2\left(\frac{I}{hf}\right) \quad (10.6b)$$

where $A_{\rm nr}$ is the non-radiative recombination coefficient, $B_{\rm rad}$ is the spontaneous recombination coefficient, $C_{\rm Aug}$ is the Auger recombination coefficient, I is the total intensity averaged over the active region. The rate equations for the carrier temperature are obtained as:

$$\frac{\partial T_{i,\rm av}}{\partial t} = \Gamma(h_i\mu_i - h_iE_i)g_1\frac{I}{hf} + \Gamma h_i\beta I - \frac{T_{i,\rm av} - T_{\rm L}}{\tau_{{\rm h},i}} \quad (10.7a)$$

$$\frac{\partial T_{i,\rm p}}{\partial t} = 2\Gamma(h_i\mu_i - h_iE_i)g_2\frac{I}{hf} + 2\Gamma h_i\beta\eta I - \frac{T_{i,\rm p}}{\tau_{{\rm h},i}}. \quad (10.7b)$$

The gain g_1 and g_2 is defined as:

$$g_1 = \{g_0 + g_N(N_{\rm av} - N_0) + g_{T{\rm c}}(T_{\rm c,av} - T_{\rm L}) + g_{T{\rm v}}(T_{\rm v,av} - T_{\rm L}) - \eta g_N N_1 + \eta(g_{T{\rm c}}T_{\rm c,p} + g_{T{\rm v}}T_{\rm v,p})\}\frac{1}{1+\varepsilon_{\rm shb}S} \quad (10.8a)$$

$$g_2 = \{\eta[g_0 + g_N(N_{\rm av} - N_0) + g_{T{\rm c}}(T_{\rm c,av} - T_{\rm L}) + g_{T{\rm v}}(T_{\rm v,av} - T_{\rm L})] - \sigma g_N N_1 + \sigma(g_{T_{\rm c}}T_{\rm c,p} + g_{T{\rm v}}T_{\rm v,p})\}\frac{1}{1+\varepsilon_{\rm shb}S} \quad (10.8b)$$

where N_0 is the carrier density at transparency. In these equations Γ is the optical confinement factor, and both η and σ are parameters take into account the effects of lateral distribution. The latter two are given by the following expressions:

$$\eta = \frac{\int_{-W/2}^{W/2}\cos(2\pi x/W)X^2dx}{\int_{-W/2}^{W/2}X^2dx} \quad (10.9a)$$

$$\sigma = \frac{\int_{-W/2}^{W/2} \cos^2(2\pi x/W) X^2 dx}{\int_{-W/2}^{W/2} X^2 dx} \tag{10.9b}$$

with X being the field solution along the lateral direction. Pump-probe experiments are commonly applied to investigate the ultrafast dynamic response of semiconductor laser amplifiers. If $P(z,t)$ is the intensity of the pump pulse and $p(z,t)$ is the intensity of the probe pulse, then the two photon absorption (TPA) is proportional to the square of the photon density. The equation for the photon density in a conventional laser diode amplifier is given as:

$$\frac{\partial S}{\partial z} + \frac{1}{v_g}\frac{\partial S}{\partial t} = (\Gamma g_1 - \alpha_{\text{loss}} - \alpha_{\text{tap}})S - \beta_2 S \tag{10.10}$$

where α_{loss} is the material loss, β_2 is the parameter accounting for the two-photon absorption. The parameter α_{tap} in the above equation takes into account the reduction of the intensity as a result of the tapering of the waveguide and is given by:

$$\alpha_{\text{tap}} = -W(z)\frac{d}{dz}\left(\frac{1}{W(z)}\right). \tag{10.11}$$

It should be noted that $W_i = W(0)$ and $W_o = W(L)$. Using Eqs. (10.10) and (10.11) and the transform for localised time $\tau = t - z/v_g$, the equations for the pump and probe photon density at the localised time τ are given by:

$$\begin{aligned}\frac{\partial P(z,\tau)}{\partial z} &= \Gamma g_1 P(z,\tau) - \alpha_{\text{loss}} P(z,\tau) - \alpha_{\text{tap}} P(z,\tau) \\ &\quad - \beta_2 P^2(z,\tau)\end{aligned} \tag{10.12a}$$

$$\begin{aligned}\frac{\partial p(z,\tau)}{\partial z} &= \Gamma g_1 p(z,\tau) - \alpha_{\text{loss}} p(z,\tau) - \alpha_{\text{tap}} p(z,\tau) \\ &\quad - 2\beta_2 I(z,\tau) p(z,\tau)\,.\end{aligned} \tag{10.12b}$$

10.3 Results and Discussions

In [26], Hansen *et al.* have investigated the effects of periodic input pulse trains on the amplifier gain saturation for conventional laser diode amplifiers. However, in this section we only consider Gaussian input pulses for both the pump and probe with similar duration. The analysis is carried

Table 10.1. Parameter used in simulations.

Description	Values
Active layer thickness	$d = 0.1\ \mu\text{m}$
Input active region width	$W_i = 2\ \mu\text{m}$
Length of Amplifier	$L = 600\ \mu\text{m}$
Non-radiative recombination constant	$A_{\text{nr}} = 0\ \text{s}^{-1}$
Spontaneous recombination constant	$B_{\text{rad}} = 8 \times 10^{-11}\ \text{cm}^3/\text{s}$
Auger recombination constant	$C_{\text{Aug}} = 9 \times 10^{-29}\ \text{cm}^6/\text{s}$
Material loss coefficient	$\alpha_{\text{loss}} = 20\ \text{cm}^{-1}$
Carrier density at transparency	$N_0 = 10^{18}\ \text{cm}^{-3}$
Current density	$J = 1.67\ \text{kA/cm}^2$
Diffusion coefficient	$D = 2\ \text{cm}^2/\text{s}$
Confinement factor	$\Gamma = 0.3$
Gain coefficient	$g_N = 4.31 \times 10^{-16}\ \text{cm}^2$
$(\partial U_i/\partial T_i)_N$	$h_{\text{c}} = 6.26 \times 10^{-21}\ \text{m}^3\text{K/eV}$
	$h_{\text{v}} = 3.41 \times 10^{-21}\ \text{m}^3\text{K/eV}$
$(\partial U_i/\partial N_i)_{T_i}$	$\mu_{\text{c}} = 0.145\ \text{eV}$
	$\mu_{\text{v}} = 0.0431\ \text{eV}$
Carrier energies affected by stimulated emission	$E_{\text{c}} = 0.0453\ \text{eV}$
	$E_{\text{v}} = 0.00469\ \text{eV}$
Free carrier absorption coefficient	$\beta_{\text{c}} = 800\ \text{m}^{-1}$
	$\beta_{\text{v}} = 0\ \text{m}^{-1}$
Temperature relaxation time	$\tau_{\text{h,c}} = 700\ \text{fs}$
	$\tau_{\text{h,v}} = 200\ \text{fs}$
Loss coefficient for carrier temperature	$g_{T\text{c}} = -108\ \text{m}^{-1}\text{K}^{-1}$
	$g_{T\text{v}} = -224\ \text{m}^{-1}\text{K}^{-1}$
Spectral Hole Burning non-linear coefficient	$\varepsilon_{\text{shb}} = 4.1 \times 10^{-24}\ \text{m}^3$
Lattice Temperature	$T_{\text{L}} = 300^{\circ}\ \text{K}$

out by dividing the amplifier cavity into 200 sections and the pulse in the time domain into 800 steps. The responses of the carrier density and carrier temperature within each section are calculated using Eqs. (10.6) and (10.7). The pulse amplification in a particular section is calculated using Eq. (10.12a) for the pump, and Eq. (10.12b) for the probe when the probe pulse is involved. The parameters used in these simulations are given in Table 10.1.

Figures 10.2 and 10.3 show variations of the amplified pulse energy with distance z along the amplifier cavity for both conventional and tapered laser diode amplifiers with output widths, W_{o} of 2 μm, 10 μm, and 20 μm. The

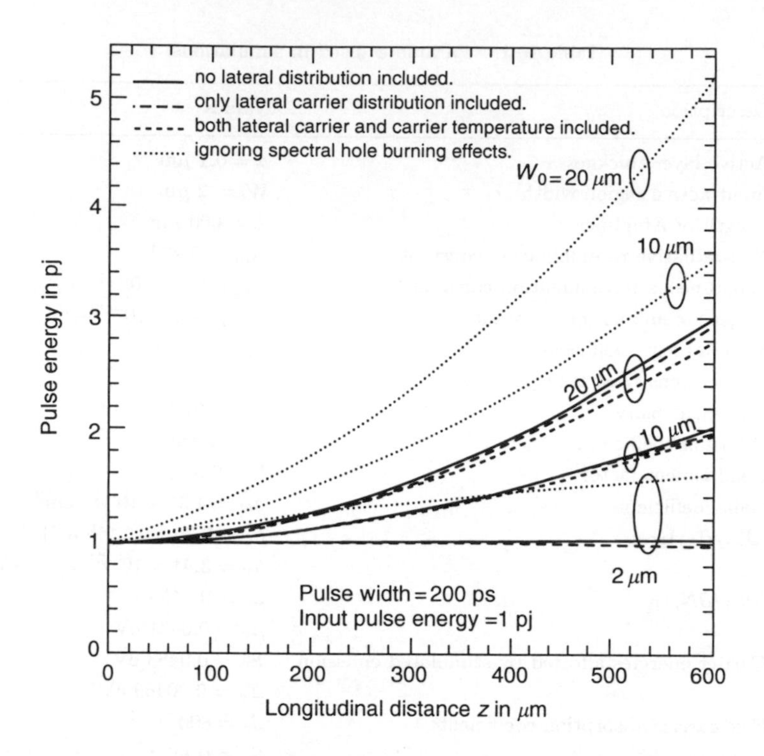

Fig. 10.2. Amplification of the pulse energy along the cavity of various amplifiers for a 200 fs pulse.

input pulse energy is 1 pj, and the FWHM pulsewidths for Figs. 10.2 and 10.3 are 200 fs and 2 ps, respectively. For comparison we have also included the results for the case where $\varepsilon_{\text{shb}} = 0$. As indicated in the figures, the lateral distribution has negligible effect on the amplification of pulses in a conventional non-tapered amplifier. However, with the increase of the taper width, the effect of the lateral distribution becomes stronger in reducing the gain of the amplifier. The analysis indicates that the lateral distribution of carrier density and carrier temperature has a significant effect. Moreover, comparison of Figs. 10.2 and 10.3 shows that the effect of lateral carrier temperature is much more significant for 200 fs pulse as compared to the 2 ps pulse, and that the effect of lateral carrier distribution also increases. This is because a pulse with shorter duration has higher intensity, which results in further uneven lateral distribution. Also, as the pulse duration becomes smaller, the effect of carrier diffusion and cooling on flattening the lateral distribution reduces. For the 200 fs pulse, the lateral effect is

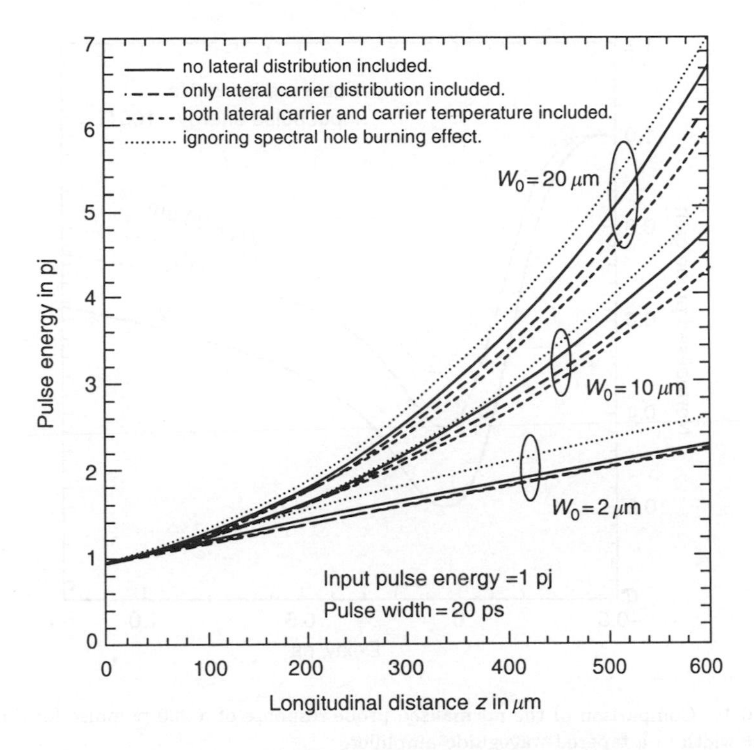

Fig. 10.3. Amplification of the pulse energy along the cavity of various amplifiers for a 2 ps pulse.

relatively small as compared with the effect of the $\varepsilon_{\rm shb}$ parameter. However, for the 2 ps pulse, the effect of lateral distribution is comparable with the $\varepsilon_{\rm shb}$ parameter.

The pump-probe responses for both 200 fs and 2 ps pulses having an input pulse energy of 100 fj are calculated and shown in Figs. 10.4 and 10.5, respectively, for both conventional laser diode amplifiers and tapered waveguide amplifiers. The vertical axis is the amplification gain of the probe normalised to the gain when the pump pulse does not exist. The horizontal axis is the delay between the pump and the probe pulses in picoseconds. The normalised probe response for a pulse with a FWHM of 200 fs is shown in Fig. 10.4. It can be seen clearly that the gain depletion due to both carrier depletion and carrier heating reduces significantly as the output width of the amplifier increases. The effect of spectral hole burning is reduced in the tapered waveguide laser diode amplifier as well. This is the result of the lower intensity for a pulse of given energy in a tapered waveguide amplifier.

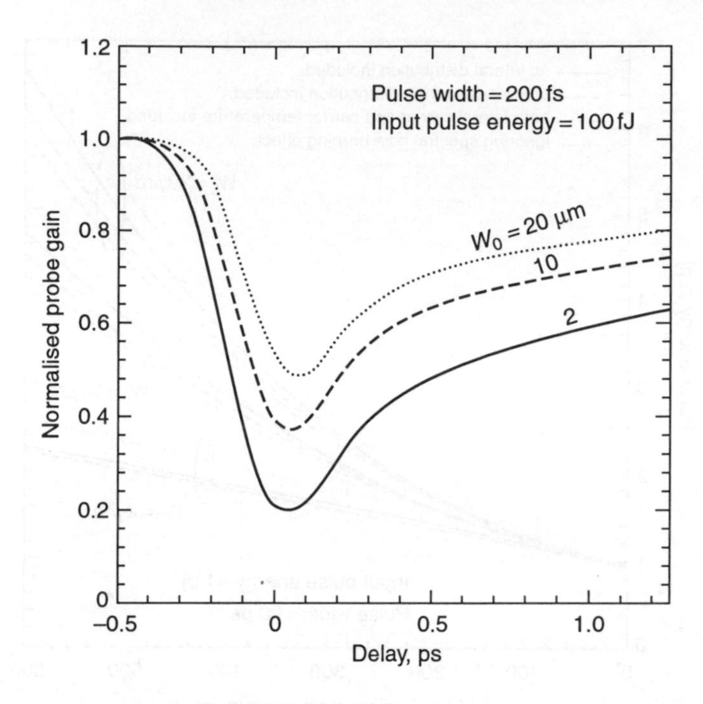

Fig. 10.4. Comparison of the normalised probe response of a 200 ps pulse for different output width in a tapered-waveguide amplifier.

Similarly, the probe response for a pulse with a FWHM of 2 ps is shown in Fig. 10.5. In both cases, the tapering of the cavity does not alter the shape of the gain depletion, even though the amount of gain depletion is greatly reduced. Hence, it can be concluded that the lateral distribution contributes very little to the gain recovery characteristic time.

The carrier temperature variation within a conventional laser diode amplifier for a 200 fs pulse is shown in Fig. 10.6. As the pulse propagates towards the amplifier output, the carrier temperature increases with the pulse intensity as it approaches the output of the amplifier. Figure 10.7 shows similar plots for a tapered-waveguide semiconductor laser amplifier having an output width of 10 μm. It can be seen that, as the width of the amplifier increases from the input to the output, the change in the electron temperature is not significant. Similarly, the variation in electron temperature in a 20 μm output width amplifier is shown in Fig. 10.8, and is lower compared to that shown in Fig. 10.7. This reveals that the tapering of the active region has greatly reduced the effect of carrier heating in laser diode amplifiers. Moreover, as shown in Fig. 10.8, the electron temperature (ET)

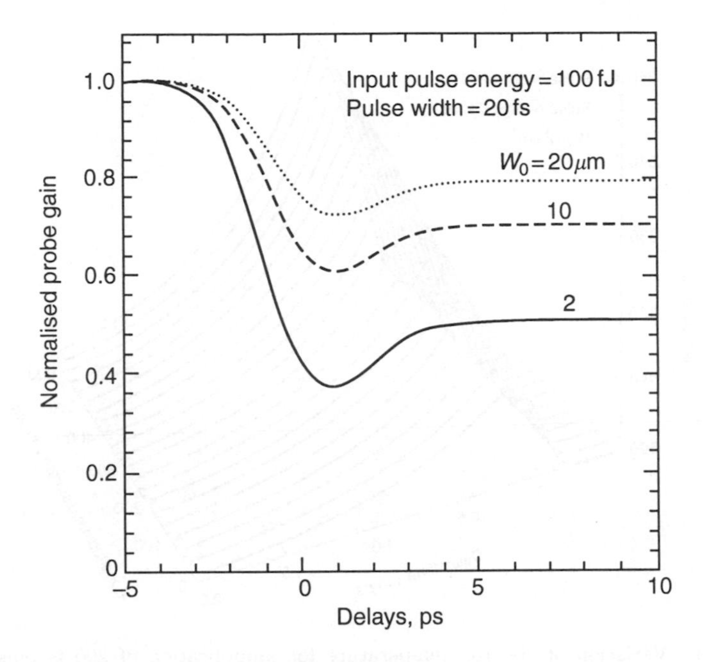

Fig. 10.5. Comparison of the normalised probe response of a 20 ps pulse for different output width in a tapered-waveguide amplifier.

first decreases when $z < 90$ μm. This is because the amplifier gain in this region is not high enough to amplify the light intensity significantly, and hence ET decreases. For $z > 90$ μm, however, the light intensity increases as the amplifier gain increases with z and hence ET increases.

The amplifier gain is defined as $G = E_{\text{out}}/E_{\text{in}}$, where E_{out}, is the energy of the output pulse and E_i, is the energy of the input pulse. The initial peak power $P_{\text{in-pk}}$ is related to the initial pulse energy as $P_{\text{in-pk}} = 1.665\ E_{\text{in}}/(\sqrt{\pi}\tau_{\text{in}})$, where τ_{in} is the pulse width. The gain dependence of the amplifier with E_{out} for both 200 fs and 20 ps pulses and amplifier output widths of 2 μm, 10 μm, and 20 μm is shown in Fig. 10.9. The amplifier operates in the saturation region for the values of $E_{\text{out}}(\text{sat})$ which reduce the amplifier gain to half its value at $E_{\text{out}} = 0.01$ pj. As Fig. 10.9 shows for both 200 fs and 20 ps pulses, $E_{\text{out}}(\text{sat})$ increases with the amplifier output width. However, this increase depends on the pulsewidth and is smaller for shorter pulsewidth. The model was compared with the experimental data in [13] for the case of a conventional non-tapered amplifier (i.e. $W_i = W_o$, and assuming a coupling loss of 12.5 dB), where we have included the carrier diffusion to study its effect on a sub-picosecond

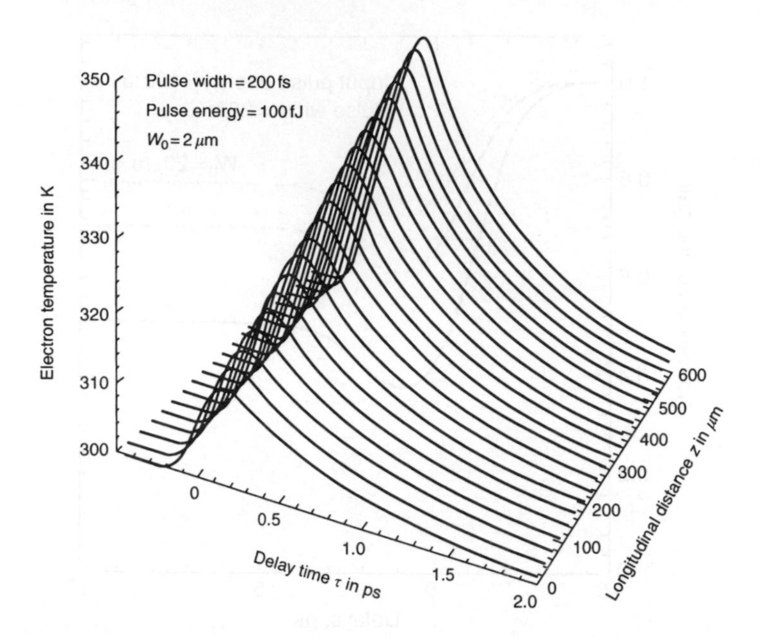

Fig. 10.6. Variation of electron temperature for amplification of 200 fs pulse. The tapered-waveguide laser diode amplifier output width is 2 μm.

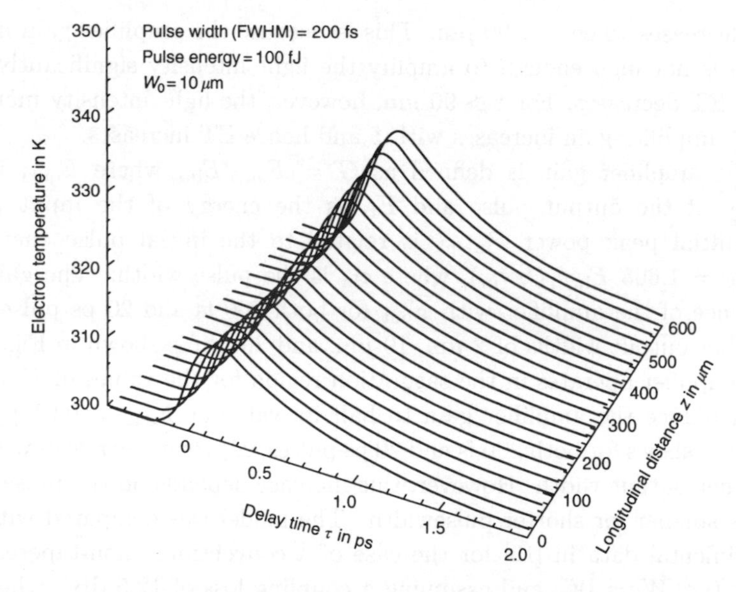

Fig. 10.7. Variation of electron temperature for amplification of 200 fs pulse. The tapered-waveguide amplifier output width is 10 μm.

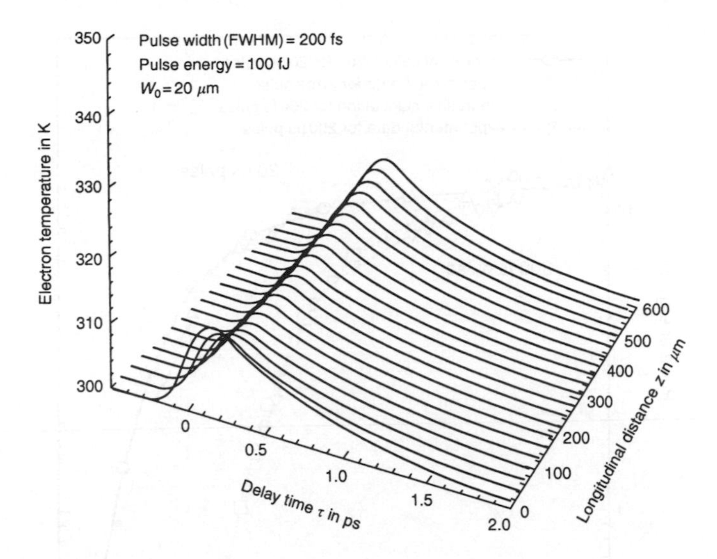

Fig. 10.8. Variation of electron temperature for amplification of 200 fs pulse. The tapered-waveguide amplifier output width is 20 μm.

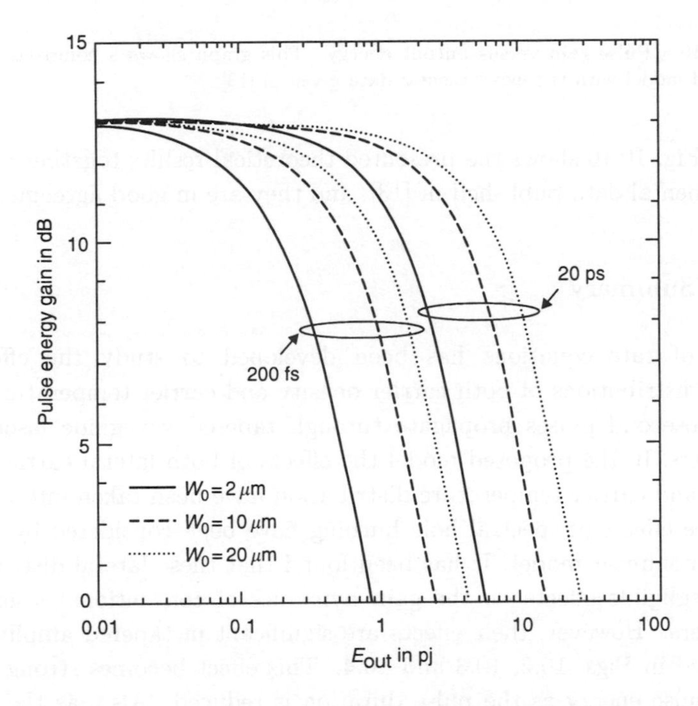

Fig. 10.9. Saturation characteristics with various widths for diffferent output widths.

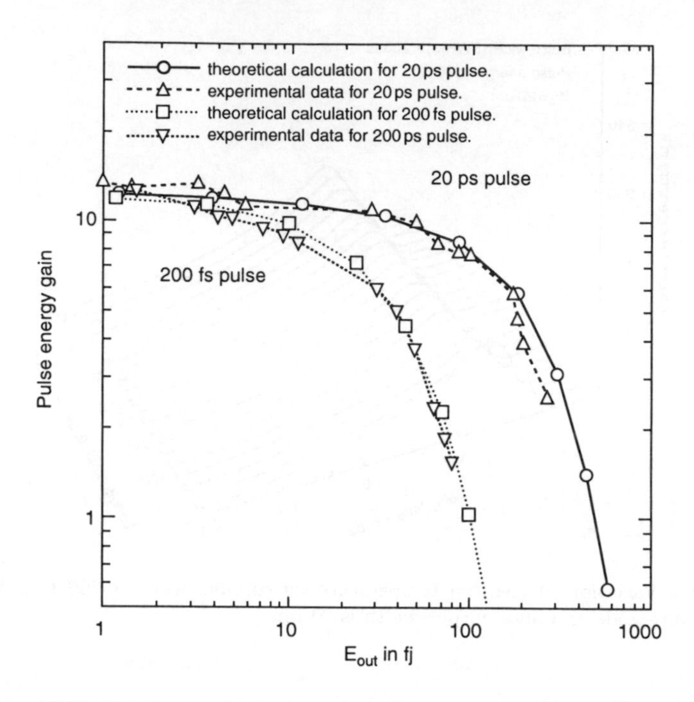

Fig. 10.10. Pulse gain versus output energy. This graph shows a comparison of the proposed model with the experimental data given in [13].

pulse. Fig. 10.10 shows the presented theoretical results together with the experimental data published in [13], and they are in good agreement.

10.4 Summary

A set of rate equations has been developed to study the effects of lateral distributions of both carrier density and carrier temperature when sub-picosecond pulses propagate through tapered-waveguide laser diode amplifiers. In the proposed model the effects of both lateral carrier distribution and carrier temperature distribution have been taken into account. Also the effects of spectral hole burning have been considered by using a simple nonlinear model. It has been found that these lateral distributions have negligible effects on the gain dynamics of conventional laser diode amplifiers. However, their effects are significant in tapered amplifiers, as indicated in Figs. 10.2, 10.3 and 10.4. This effect becomes stronger for a given pulse energy as the pulse duration is reduced. Also, as the output width of the tapered amplifier increases, the carrier heating effect reduces significantly, as shown in Figs. 10.6, 10.7 and 10.8. It was found that by changing the output width of the amplifier from 2 μm to 20 μm, the saturation energy of a 20 ps pulse is nearly 8 times higher than that of a 200 fs pulse. It should be mentioned that the proposed model is restricted to (i) the region where the amplifier gain constant varies linearly with the carrier density and the carrier temperature, and (ii) highly index-guided tapered-waveguide amplifiers, where the power is mostly confined to the fundamental mode.

In summary, the sub-picosecond gain dynamic in a highly index-guided tapered-waveguide laser diode amplifier is studied in this chapter. The analysis is based on a set of rate equations for the carrier density N and carrier temperature T, where the effects of lateral distributions of both N and T are considered. It is found that both N and T have significant effects on the gain dynamic of the tapered-waveguide laser diode amplifier. The gain saturation of the tapered-waveguide amplifier for both 20 ps and 200 fs pulses is also calculated.

References

[1] C. Y. Tsai, R. M. Spencer, Y. H. Lo and L. F. Eastman, "Nonlinear Gain Coefficients in Semiconductor-Lasers - Effects of Carrier Heating," *IEEE Journal of Quantum Electronics*, Vol. 32, No. 2, pp. 201–212, 1996.

[2] V. I. Tolstikhin and M. Willander, "Carrier Heating Effects in Dynamic Single- Frequency GaInAsP-InP Laser-Diodes," *IEEE Journal of Quantum Electronics*, Vol. 31, No. 5, pp. 814–833, 1995.

[3] J. Mork, J. Mark and C. P. Seltzer, "Carrier Heating in InGaAsP Laser-Amplifiers due to 2-Photon Absorption," *Applied Physics Letters*, Vol. 64, No. 17, pp. 2206–2208, 1994.

[4] M. Willatzen, J. Mark, J. Mork and C. P. Seltzer, "Carrier Temperature and Spectral Hole Burning Dynamics in InGaAsP Quantum-Well Laser-Amplifiers," *Applied Physics Letters*, Vol. 64, No. 2, pp. 143–145, 1994.

[5] A. Uskov, J. Mork and J. Mark, "Theory of Short-Pulse Gain Saturation in Semiconductor-Laser Amplifiers," *IEEE Photonics Technology Letters*, Vol. 4, No. 5, pp. 443–446, 1992.

[6] R. Schnabel, R. Ludwig, W. Pieper, S. Diez and H. G. Weber, "Ultrafast Optical Signal-Processing Using Semiconductor-Laser Amplifiers," *Philosophical Transactions of the Royal Society of London Series A-Mathematical Physical and Engineering Sciences*, Vol. 354, No. 1708, pp. 733–744, 1996.

[7] B. Zhu, I. H. White, K. A. Williams, F. R. Laughton and R. V. Penty, "High-Peak-Power Picosecond Optical Pulse Generation from Q-Switched

Bow-Tie Laser with a Tapered Travelling Wave Amplifier," *IEEE Photonics Technology Letters*, Vol. 8, No. 4, pp. 503–505, 1996.

[8] Peter P. Vasil'ev, "High Power High-Frequency Picosecond Pulse Generation by Passively Q-Switched Diode Lasers," *IEEE Journal of Quantum Electronics,* Vol. 29, No. 6, pp. 1687–1692, 1993.

[9] P. J. Delfyett *et al.*, "High-Power Ultrafast Laser-Diodes," *IEEE Journal of Quantum Electronics*, Vol. 28, No. 10, pp. 2203–2219, 1992.

[10] A. Mar, R. Helkey, J. Bowers, D. Mehuys and D. Welch, "Mode-locked Operation of a Master Oscillator Power Amplifier," *IEEE Photonic Technology Letters*, Vol. 6, No. 9, pp. 1067–1069, 1994.

[11] L. Goldberg, D. Mehuys and D. Welch, "High Power Mode-Locked Compound Laser Using a Tapered Semiconductor Amplifier," *IEEE Photonic Technology Letters*, Vol. 6, No. 9, pp. 1070–1072, 1994.

[12] K. L. Hall, J. Mark, K. P. Ippen and G. Eisenstein, "Femtosecond Gain Dynamics in InGaAsP Optical Amplifiers," *Applied Physics Letters*, Vol. 56, No. 18, pp. 1740–1742, 1990.

[13] Y. Lai, K. L. Hall, E. P. Ippen and G. Eisenstein, "Short Pulse Gain Saturation in InGaAsP Diode Laser Amplifiers," *IEEE Photonics Technology Letters*, Vol. 2, No. 10, pp. 711–713, OCT 1990.

[14] J. Mark and J. Mork, "Subpicosecond Gain Dynamics in InGaAsP Optical Amplifiers — Experiment and Theory," *Applied Physics Letters*, Vol. 61, No. 19, pp. 2281–2283, 1992.

[15] K. L. Hall, G. Lenz, A. M. Darwish and E. P. Ippen, "Subpicosecond Gain and Index Nonlinearities in InGaAsP Diode Lasers," *Optics Communications*, 111, 589–612, 1994.

[16] A Dienes, J. P. Heritage, M. Y. Hong and Y. H. Chang, "Time- and spectral- domain evolution of subpicosecond pulses in semiconductor optical amplifiers," *Optics Letters*, Vol. 17, No. 22, pp. 1602–1604, 1992

[17] M. Y. Hong, Y. H. Chang, A. Dienes, J. P. Heritage and P. J. Delfyett, "Subpicosecond Pulse Amplification in Semiconductor Laser Amplifiers: Theory and Experiment," *IEEE Journal of Quantum Electronics*, Vol. 30, No. 4, pp. 1122–1131, 1994.

[18] A Dienes, J. P. Heritage, C. Jasti and M. Y. Hong, "Femtosecond optical pulse amplification in saturated media," *J. Optical Society of America B.*, Vol. 13, No. 4, pp. 725–734, 1996.

[19] R. A. Indik *et al.*, "Role of Plasma Cooling, Heating, and Memory Effects in Subpicosecond Pulse-Propagation in Semiconductor Amplifiers," *Physical Review A*, Vol. 53, No. 5, pp. 3614–3620, 1996.

[20] G. Bendelli, K. Komori and S. Arai, "Gain Saturation and Propagation Characteristics of Index-Guided Tapered-Waveguide Travelling-Wave Semiconductor-Laser Amplifiers (TTW-SLAS)," *IEEE Journal of Quantum Electronics*, Vol. 28, No. 2, pp. 447–458, 1992.

[21] S. El Yumin, K. Komori, S. Arai and G. Bendelli, "Taper-Shape Dependence of Tapered-Wave-Guide Travelling-Wave Semiconductor-Laser Amplifier (TTW-SLA)," *IEICE Transactions on Electronics*, Vol. E77C, No. 4, pp. 624–632, 1994.

[22] H. Ghafouri Shiraz, P. W. Tan and T. Aruga, "Analysis of a Semi-linear Tapered-Wave-Guide Laser-Diode Amplifier," *Microwave and Optical Technology Letters*, Vol. 12, No. 2, pp. 53–56, 1996.

[23] H. Ghafouri Shiraz and P. W. Tan, "Study of a Novel Laser-Diode Amplifier Structure," *Semiconductor Science and Technology*, Vol. 11, No. 10, pp. 1443–1449, 1996.

[24] S. H. Yang, S. Smith, J. Fitz and C. H. Lee, "Generation of High-Power Picosecond Pulses from a Gain-Switched Two-Section Quantum-Well Laser with a Laterally Tapered Energy-Storing Section," *IEEE Photonic Technology Letters*, Vol. 8, No. 3, pp. 337–339, Mar 1996.

[25] A. G. Plyavenek, "Influence of Electron Heating on the Dynamic Behaviour of Semiconductor Injection Lasers," *Optics Communications* 87, pp. 115–121, 1994.

[26] P. B. Hansen, J. M. Wiesenfeld, G. Eisenstein, R. S. Tucker and G. Raybon, "Repetition-rate dependence of gain compression in InGaAsP optical amplifiers using picosecond optical pulses," *IEEE Journal Quantum Electronics*, Vol. 25, No. 12, pp. 2611–2619, 1989.

Chapter 11

Saturation Intensity of InGaAsP Tapered Travelling-Wave Semiconductor Laser Amplifier Structures

11.1 Introduction

In recent years, semiconductor laser amplifiers (SLA) with linear [1] or exponential tapers [1–2] and other combinations [3–4] have been studied intensively as they provide higher saturation output power and better quantum efficiency. Semiconductor laser amplifiers are known to exhibit different values of gain in different states of polarisation. This is due to the difference in values of residual facet reflectivity, optical confinement factor and effective index of the two orthogonal polarisation states (i.e. TE and TM modes). The amplifier gain dependence upon the state of polarisation needs to be reduced as it degrades the performance of optical communication systems using coherent detection. Although several approaches have been proposed to compensate for the polarisation dependence of signal gain, polarisation-insensitive optical amplifiers are more favourable. Therefore, it is useful to investigate the polarisation dependence of signal gain in tapered travelling-wave SLA structures as a preliminary step to achieve a structural design for polarisation-insensitive tapered amplifier. In this chapter, polarisation sensitivity of two tapered amplifier structures will be presented based on a simple approximate expression for the amplifier saturation intensity.

11.2 An Analytical Expression of Saturation Intensity of a Tapered TW-SLA Structure

In previous chapters the step-transition approach has been used to analyse the field propagation in the tapered TW-SLA. In the following analysis we have considered an active waveguide. The tapered structure is approximated by a series of stair-like rectangular sections. The coupling between the incident and the transmitted local guided modes of two adjacent sections can be expressed as [1]:

$$A_{j,n+1}e^{-j\beta_{j,n+1}z} = \sum_{i=1}^{M_n} C_{ij} G_s^{1/2} A_{i,n} e^{-j\beta_{i,n}z} \tag{11.1}$$

where

$$G_s = \exp[(\Gamma g_m - \alpha_{loss})\Delta z] \tag{11.2}$$

$$\alpha_{loss} = \Gamma\alpha_{act} + (1 - \Gamma)\alpha_{clad}\,. \tag{11.3}$$

Here, G_s is the single-pass gain, A is the complex field amplitude normalised such that a unit power is carried by the associated mode (i.e. $|A| = 1$), M_n is the maximum local guided mode, β is the propagation constant, i and j are integers labelling the guided modes in the nth and $(n+1)$th section, respectively, z is the distance away from the input, C_{ij} is the coupling coefficient [1], α_{loss} is the conventional average loss coefficient, α_{act} is the active region loss coefficient, α_{clad} is the cladding region loss coefficient, Γ is the optical confinement factor, g_m is the material gain coefficient, Δz is the length of each section. Both guided and radiation modes have been considered in the analysis [5]. The amplifier structure is assumed to be a symmetric separate confinement heterostructure (SCH). The material gain coefficient, g_m is approximated by the familiar closed form expression given as [6]:

$$g_m = \frac{g_{m0}}{1 + \frac{I(z)}{I_s}} \tag{11.4}$$

where g_{m0} is the unsaturated material gain coefficient, $I(z)$ is the optical intensity and $I_s = hv/(\Gamma a\tau)$, where a is the gain coefficient, τ is the carrier lifetime and hv is the photon energy, I_s is the saturation intensity defined as the optical intensity that reduces g_{m0} to $g_{m0}/2$. In a conventional laser amplifier where the input and output widths are equal, the saturation intensity I_s is constant. However, in a tapered amplifier structure this is not the case and the saturation intensity I_{sat} also depends on the active layer width, that is:

$$I_{sat}(w) = I_s f(w) \tag{11.5}$$

where the function $f(w)$ represents the effect of active layer width on I_{sat}. This function can be found by analysing the variation of g_{m} along the active region. By following the approach given in [1] we can calculate the values of material gain coefficient, g_{m} along the tapered TW-SLA. This is because the optical intensity distribution is known for a given input power level, that is:

$$\frac{dI(z)}{dz} = (\Gamma(g_{\text{m}}(N_{\text{av}}) - a\eta N_1) - \alpha_{\text{loss}} - \alpha_{\text{tap}})I(z) \tag{11.6}$$

where α_{tap} is the loss coefficient introduced by the taper, N_1 is the first Fourier term of the spatial carrier density distribution and η takes into account lateral hole burning which is expressed by,

$$\eta(z) = \frac{\int_{-w/2}^{w/2} \cos\left(\frac{2\pi x}{W}\right) X^2(x,z)dx}{\int_{-w/2}^{w/2} X^2(x,z)dx} \tag{11.7}$$

where $X(x,z)$ is the lateral optical field distribution and W is the width of the active layer which is z dependent. We considered three taper geometries, linear [1], exponential [1–2], and semi-exponential-linear tapers [4] to investigate the variation of saturation intensity along the tapered amplifier structures. In the following analysis we have assumed that the optical power is confined mainly in the fundamental mode. For the device parameters listed in Table 11.1, we have calculated variation of the intensity at $g_{\text{m}} = g_{\text{m0}}/2$ (i.e. saturation intensity) for various width. This is shown by solid circles in Fig. 11.1. An empirical expression for $I_{\text{sat}}(w)$ based on this data has also been obtained by using fitting technique (see the solid line in Fig. 11.1). This approximate empirical expression is given by,

$$I_{\text{sat}}(w) = 2.75 I_{\text{s}} e^{\left(\frac{-1.19}{w^{1.5}}\right)} \tag{11.8}$$

where in the above expression the unit of w is in μm.

Equation (11.8) can be used for all three amplifier structures discussed in this letter. Figure 11.1 clearly shows that for $w \leq 10$ μm, I_{sat} increases rapidly and then increases at a slower rate. For a large value of w, I_{sat} approaches its maximum limit of $2.75 I_{\text{s}}$. The above expression provides a simple approximation to complex numerical solution of the rate equations of tapered amplifier structures.

Table 11.1. Device parameters used in calculations.

Input active region width	$W_i = 1\ \mu m$
Output active region width	$W_o = 5$ to $50\ \mu m$
Amplifier length	$L = 1000\ \mu m$
Active region thickness	$d = 0.1\ \mu m$
Waveguide thickness	$t_w = 0.2\ \mu m$
Cladding thickness	$t_{clad} = 4\ \mu m$
Active region refractive index	$n_{act} = 3.553$
Cladding refractive index	$n_{clad} = 3.169$
Peak wavelength at transparency	$\lambda_p = 1.55\ \mu m$
Active region loss coefficient	$\alpha_{act} = 1.2 \times 10^4\ m^{-1}$
Cladding loss coefficient	$\alpha_{cladd} = 5.6 \times 10^2\ m^{-1}$
Unsaturated power gain	$G_0 = 23.5$ dB
TE optical confinement factor	$\Gamma_{TE} = 0.17$
Unsaturated gain coefficient	$4.64 \times 10^4\ m^{-1}$
Photon energy	$hv = 1.324 \times 10^{-19}$ Joule
Gain coefficient	$a = 1.64 \times 10^{-20}\ m^2$
Spontaneous carrier lifetime	$\tau = 610$ ps
Length of each section	$\Delta z = 2\ \mu m$

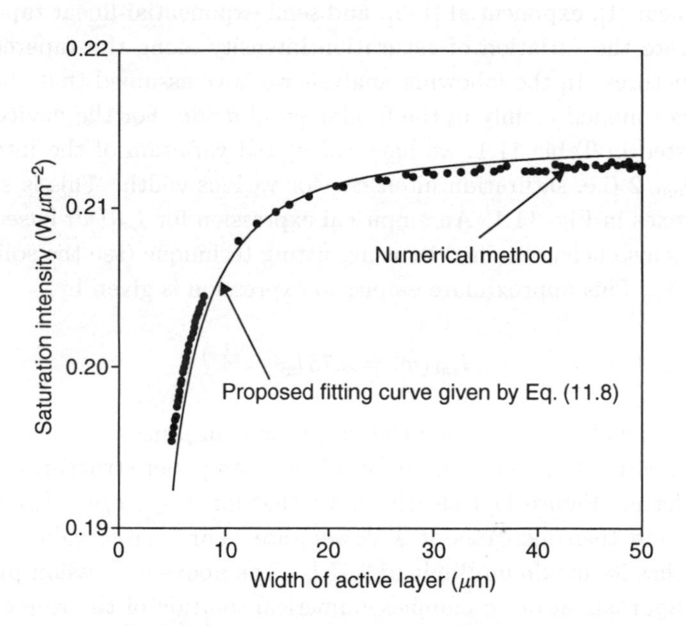

Fig. 11.1. The analytical expression of saturation intensity of tapered TW-SLA structures. The output width range from 5 μm to 50 μm.

11.3 Effects of Gain Saturation on Polarisation Sensitivity

The polarisation sensitivity (PS) of a tapered TW-SLA is defined as:

$$PS(dB) = 10\log_{10}\left(\frac{TE_{00}\,gain}{TM_{00}\,gain}\right). \tag{11.9}$$

11.3.1 *Polarisation sensitivity of tapered TW-SLA structures*

Two structures have been investigated, the linear and exponential tapers. We have investigated the polarisation sensitivity of the amplifier by taking into account mode conversion in both saturated and unsaturated conditions.

Figure 11.2 shows the variation of polarisation sensitivity of the structures when they are operated in the unsaturated region. Polarisation sensitivity is almost constant at smaller output widths. When output width, W_o exceeds 20 μm, polarisation sensitivity of the linear taper begins to fluctuate dramatically. In comparison, fluctuation of polarisation sensitivity

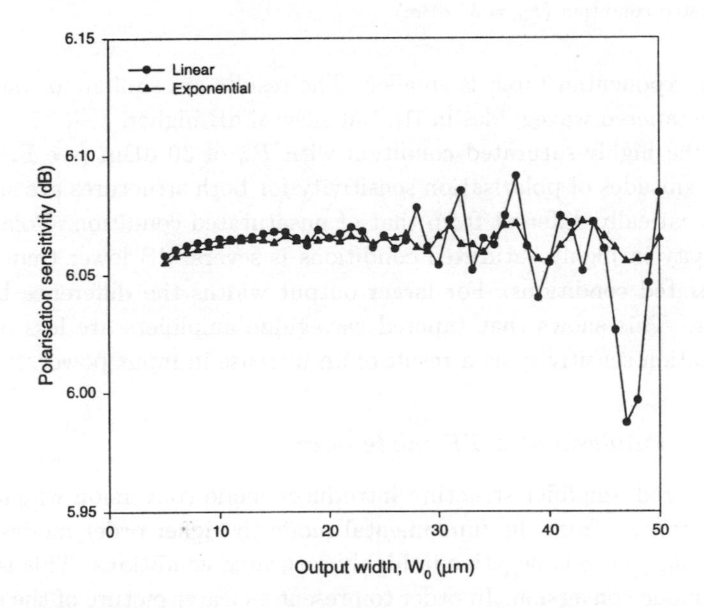

Fig. 11.2. Polarisation sensitivity of the linear and exponential tapers with $L = 1000$ μm in unsaturated condition ($P_{\text{in}} = -30$ dBm).

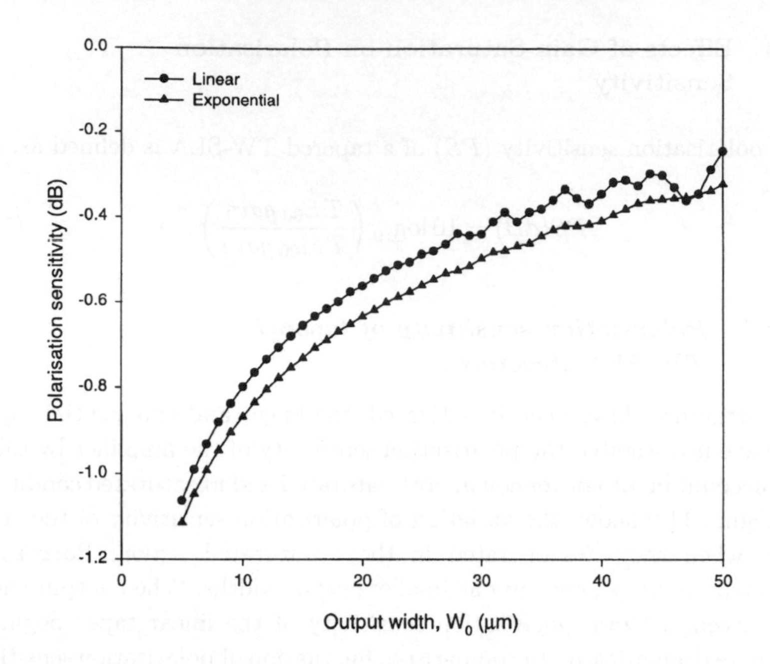

Fig. 11.3. Polarisation sensitivity of the linear and exponential taper with $L = 1000$ μm in saturated condition ($P_{\text{in}} = 30$ dBm).

for the exponential taper is smaller. The results are similar to that of the passive tapered waveguides in [1], but several dB higher.

In the highly-saturated condition with P_{in} of 30 dBm (see Fig. 11.3), the magnitudes of polarisation sensitivity for both structures are shown to be dramatically different from that of unsaturated condition. Polarisation sensitivity in highly saturated conditions is several dB lower than that in unsaturated conditions. For larger output widths the difference becomes smaller. This shows that tapered waveguide amplifiers are less prone to polarisation sensitivity as a result of an increase in input power.

11.3.2 *Fundamental TE mode gain*

The tapered amplifier structure introduces mode conversion which causes power transfer from the fundamental mode to higher order modes. TE_{00} (and TM_{00}) *gain* is negative in highly saturated conditions. This is due to severe mode conversion. In order to present a clearer picture of the effect of mode conversion in gain saturation conditions, TE_{00} *gain* is plotted against output width.

In the highly saturated condition (see Fig. 11.5), the TE_{00} *gain* is negative and hence losses occur. A wider linear taper will have higher

TE_{00} *gain* when it is highly saturated. The TE_{00} *gain* of the exponential taper initially increases with output width and later drops. Beyond output width of 25 μm, severe mode conversion occurs and TE_{00} *gain* reduces further.

Figures 11.4 and 11.5 show the TE_{00} *gain* of the tapered amplifier structures. In the unsaturated condition shown in Fig. 11.4, TE_{00} *gain* of the linear taper remains almost constant when output width, W_o is less than 20 μm. In the same graph, TE_{00} *gain* of the exponential taper is almost constant below output width of 13 μm. Beyond these respective values of W_o, a fall in TE_{00} *gain* is observed. This is due to mode conversion in wider tapers. The fall in TE_{00} *gain* is more drastic for the exponential taper than for the linear taper.

11.4 Summary

In this chapter an analytical expression for saturation intensity of tapered travelling-wave semiconductor laser amplifier (TW-SLA) structures

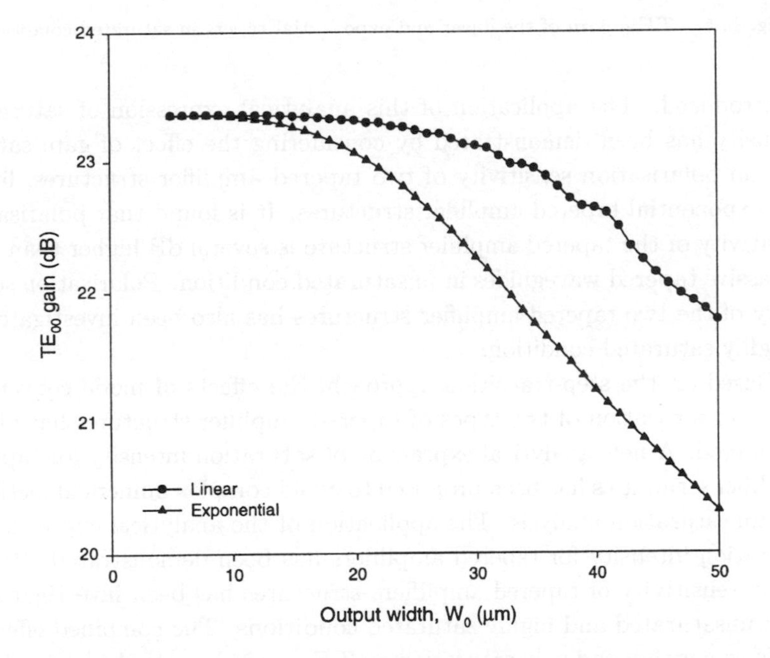

Fig. 11.4. TE_{00} *gain* of the linear and exponential tapers in unsaturated condition.

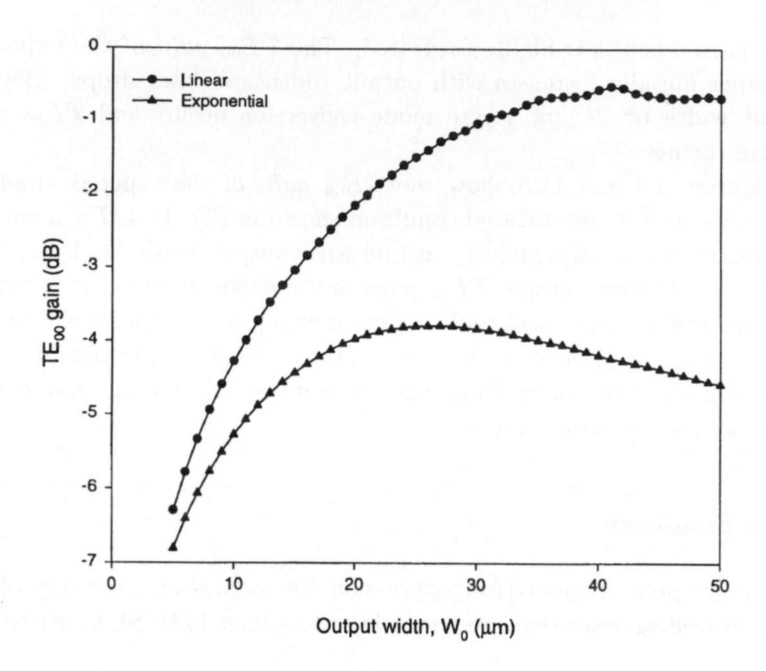

Fig. 11.5. TE_{00} *gain* of the linear and exponential tapers in saturated condition.

is introduced. The application of this analytical expression of saturation intensity has been demonstrated by considering the effect of gain saturation on polarisation sensitivity of two tapered amplifier structures, linear and exponential tapered amplifier structures. It is found that polarisation sensitivity of the tapered amplifier structure is several dB higher than that of passive tapered waveguides in unsaturated condition. Polarisation sensitivity of the two tapered amplifier structures has also been investigated in a highly saturated condition.

Based on the step-transition approach, the effects of mode conversion and gain saturation of two types of tapered amplifier structures have been considered. A new analytical expression of saturation intensity for tapered amplifier structures has been proposed to avoid complex numerical methods in gain saturation analysis. The application of the analytical expression of saturation intensity for tapered amplifiers has been demonstrated. Polarisation sensitivity of tapered amplifiers structures has been investigated in both unsaturated and highly saturated conditions. The combined effect of mode conversion and gain saturation on TE_{00} *gain* has also been simulated using the analytical expression of saturation intensity.

References

[1] G. Bendelli, K. Komori and S. Arai, "Gain Saturation and Propagation Characteristics of Index-Guided Tapered Waveguided Travelling Wave Semiconductor Laser Amplifiers (TTW-SLA's)," *IEEE J. Quantum Electron.*, Vol. 28, No. 2, Feb. 1992, pp. 447–458.

[2] G. Bendelli, K. Komori, S. Arai and Y. Suematsu, "A New Structure for High- Power TW-SLA," *IEEE Photonics Tech. Lett.*, Vol. 3, No. 1, Jan. 1991, pp. 42–44.

[3] H. Ghafouri-Shiraz and P. W. Tan, "Study of a novel laser diode amplifier structure," *Semicond. Sci. Technol.*, 11 (1996), pp. 1443–1449.

[4] H. Ghafouri-Shiraz, P. W. Tan and T. Aruga, "Analysis of a Semi-Linear Tapered Waveguide Laser Diode Amplifier," *Microwave and Optical Technology Letters*, Vol. 12, No. 2, June 5 1996, pp. 53–?.

[5] D. Marcuse, "Radiation Losses of Tapered Dielectric Slab Waveguides," *Bell System Tech. Journal*, Feb. 1970, pp. 273–290.

[6] T. Saitoh and T. Mukai, "1.5 μm GaInAsP Travelling-Wave Semiconductor Laser Amplifier," *IEEE J. Quantum Electron.*, QE-23, No. 6, June 1987, pp. 1010–1019.

Chapter 12

Wavelength Conversion in Tapered-Waveguide Laser Diode Amplifiers Using Cross-Gain Modulation

12.1 Introduction

Wavelength conversion has received much attention recently due to its potential to increase the capacity of a wavelength division multiplexing (WDM) communication system by assigning dynamic links between channels. A semiconductor laser amplifier (SLA) with high non-linearity and short carrier lifetime has the potential to achieve all-optical wavelength conversion. The wide gain bandwidth travelling-wave SLA provides the necessary broad wavelength conversion span and the lifetime of the carriers provides the speed necessary for high bit-rate wavelength conversion. The three main techniques of achieving wavelength conversion in a SLA are (i) Cross-Gain Modulation (XGM) [1–4], (ii) Cross-Phase Modulation (XPM) [5–8] and (iii) Four Wave Mixing (FWM) [9–11]. The prime objective of a wavelength converter is to achieve conversion rate for a signal of 10 Gb/s and above with a high extinction ratio, low noise factor and high conversion efficiency. Wavelength conversion using cross gain saturation in SLAs is the most straightforward method among the various proposed methods. It is simple to implement. Polarisation insensitive amplifiers can be employed in order to minimize the polarisation effects of the device. Recently, experimental and theoretical results have shown that wavelength conversion using XGM in SLAs has the potential to handle data transmission rates ranging from 10 to 40 Gb/s [12–16]. However, the disadvantage of the XGM is the extinction ratio degradation for conversion

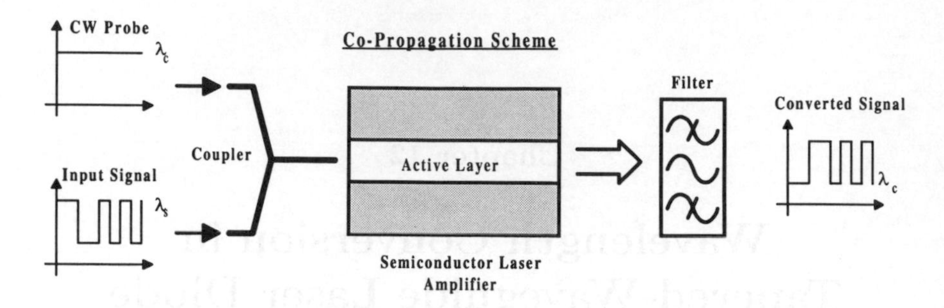

Fig. 12.1. Block Diagram of Wavelength Conversion by Cross-Gain Modulation Technique with Co-Propagation Scheme.

to longer wavelengths. The output extinction ratio for shorter wavelengths is higher, and extinction ratio degradation is unavoidable for conversion to longer wavelengths. The extinction ratio is vital to the performance of the XGM wavelength converter, as it has been shown to be the main cause of signal deterioration when used in a multistage system [17]. To date, most of the experimental and theoretical works are carried out on the XGM conversion in conventional SLA as shown in Fig. 12.1. This figure shows the cross-gain modulation technique using co-propagation scheme. A continuous-wave (CW) probe signal together with an input signal is injected into the SLA. The input signal beam carrying the information at λ_s modulates the gain of the SLA by depleting the carriers. The probe beam at wavelength λ_c encounters the modulated gain and the probe amplitude is changed by the input signal. Thus, the gain experienced by the probe is reduced when the input signal is high and conversely increased when the input signal goes to its minimum. So after the SLA, the probe beam carries the inverted information of the input signal with the wavelength of λ_c. In this chapter, we explain the XGM in a linear tapered-waveguide SLA. Several important parameters, such as extinction ratio, output signal power and conversion speed have been investigated.

12.2 Theoretical Method

The carrier rate equations in a linear tapered-waveguide SLA for the average carrier density N_{ave} and the amplitude of lateral carrier density variation N_1 are given as follows [18–21]:

$$\frac{\partial N_{\mathrm{av}}}{\partial t} = \frac{J}{ed} - A_{\mathrm{nr}}N_{\mathrm{av}} - B_{\mathrm{rad}}\left(N_{\mathrm{av}}{}^2 + \frac{N_1{}^2}{2}\right) - C_{\mathrm{aug}}\left(N_{\mathrm{av}}{}^3 + \frac{3}{2}N_{\mathrm{av}}N_1{}^2\right)$$
$$- \Gamma\{g_{\mathrm{m}}(N_{\mathrm{av}}, \lambda_{\mathrm{s}}) - \eta a N_1\}\frac{I_{\mathrm{s}}}{hv_{\mathrm{s}}} - \Gamma\{g_{\mathrm{m}}(N_{\mathrm{av}}, \lambda_{\mathrm{p}}) - \eta a N_1\}\frac{I_{\mathrm{p}}}{h\nu_{\mathrm{p}}} \tag{12.1a}$$

$$\frac{\partial N_1}{\partial t} = -DN_1\left(\frac{2\pi}{W}\right)^2 - A_{\mathrm{nr}}N_1 - 2B_{\mathrm{rad}}N_{\mathrm{av}}N_1 - C_{\mathrm{aug}}\left(3N_{\mathrm{av}}^2N_1 + \frac{3}{4}N_1{}^3\right)$$
$$+ 2\Gamma\{\eta g_{\mathrm{m}}(N_{\mathrm{av}}, \lambda_{\mathrm{s}}) - \sigma a N_1\}\frac{I_{\mathrm{s}}}{hv_{\mathrm{s}}} + 2\Gamma\{\eta g_{\mathrm{m}}(N_{\mathrm{av}}, \lambda_{\mathrm{p}}) - \sigma a N_1\}\frac{I_{\mathrm{p}}}{h\nu_{\mathrm{p}}}. \tag{12.1b}$$

In Eq. (12.1), J is the current density, e is the electron charge, d is the active layer thickness, A_{nr}, B_{rad} and C_{aug} are the non-radiative, radiative and Auger recombination constants, respectively. N_{p} is the carrier density at transparency, W_i is the active layer input width, D is the diffusion coefficient, hv_{s} is the signal photon energy, hv_{p} is the probe photon energy, a is the gain coefficient, I_{s} and I_{p} are the intensities of the input and continuous probe signals, respectively. Γ, σ and η are parameters given in Eqs. (12.5) to (12.7). The material gain $g_{\mathrm{m}}(N)$ is given by:

$$g_{\mathrm{m}}(N, \lambda) = a(N - N_{\mathrm{g}}) - \zeta(\lambda - \lambda_{\mathrm{pk}})^2. \tag{12.2}$$

The peak wavelength is assumed to vary linearly with the carrier density:

$$\lambda_{\mathrm{pk}} = \lambda_0 - g_{\mathrm{p}}(N - N_{\mathrm{p}}). \tag{12.3}$$

The equation for the amplification of the signal and probe is given by:

$$\frac{\partial I_x(z,t)}{\partial z} = [\Gamma(g_{\mathrm{m}}(N_{\mathrm{av}}, \lambda_x) - a\eta N_1) - \alpha_{\mathrm{loss}} - \alpha_{\mathrm{tap}}]\, I_x(z,t) \tag{12.4}$$

where the subscript "x" denotes pump and probe signal, α_{loss} is the material loss and α_{tap} is the reduction of the intensity as a result of the tapering. The confinement factor is given as:

$$\Gamma = \frac{\int_{-w/2}^{w/2} X^2 dx \int_{-d/2}^{d/2} Y^2 dy}{\iint X^2 Y^2 dx dy} \tag{12.5}$$

η and σ are defined as

$$\eta = \frac{\int_{-w/2}^{w/2} \cos(2\pi x/W) X^2 dx}{\int_{-w/2}^{w/2} X^2 dx} \tag{12.6}$$

$$\sigma = \frac{\int_{-w/2}^{w/2} \cos^2(2\pi x/W) X^2 dx}{\int_{-w/2}^{w/2} X^2 dx} \tag{12.7}$$

where "X" is the lateral electric field distribution. The definitions of other parameters in the above equations are defined in Table 1.

Table 12.1. Parameters used in simulations [16, 26].

Description	Values/Units
Active layer thickness	$d = 0.15$ μm
Input active region width	$W_i = 1.35$ mum
Nonradiative recombination constant	$A_{\mathrm{nr}} = 1 \times 10^8$ s^{-1}
Spontaneous recombination constant	$B_{\mathrm{rad}} = 2.5 \times 10^{-11}$ cm^{3}s^{-1}
Auger recombination constant	$C_{\mathrm{aug}} = 9.4 \times 10^{-29}$ cm^{6}s^{-1}
Gain coefficient	$a = 4.63 \times 10^{-16}$ cm^2
Carrier density at transparency	$N_{\mathrm{p}} = 1.0 \times 10^{18}$ cm^{-3}
Diffusion coefficient	$D = 10$ cm^{2}s^{-1}
Confinement factor	$\Gamma = 0.2$
Length of amplifier	$L = 450/1200$ μm
Effective loss coefficient	$\alpha = 40$ cm^{-1}
Bandwidth coefficient	$\zeta = 1.5 \times 10^{-5} \mu$m^{-1}nm^{-2}
Wavelength at transparency	$\lambda_0 = 1.55$ μm
Peak wavelength shift coefficient	$g_{\mathrm{p}} = 2.7 \times 10^{-17}$ cm^{3}nm

12.3 Simulation Results

All the following simulations are for wavelength conversion using cross-gain modulation technique with co-propagation scheme. The parameters used for the simulation are given in Table 1. By maintaining a similar current density, as in both 450 μm and 1200 μm, the effect of both conventional and linear tapered-waveguide SLAs have been studied. In the following simulation, unless otherwise stated, we have used the optimal input signal and CW probe powers of -3 dBm and -15 dBm, respectively, and on input extinction ratio of 10 dB.

12.3.1 *Extinction ratio for up and down conversion*

An important feature for a wavelength converter is equal performance for both up and down conversion. However, the major limitation of the XGM conversion is the different performance for up and down conversion. Figure 12.2 shows the dependence of extinction ratio for both up and down conversion in a conventional SLA with the cavity length of 450 μm. The extinction ratio is higher for the shorter continuous probe wavelengths, and the conversion to the same or longer wavelengths gives a reduced extinction ratio. The larger the difference between the input signal wavelength and that of the probe signal wavelength for down-conversion, the higher the extinction ratio that can be achieved. The change in the extinction ratio with wavelength is caused by the variation of the differential gain with the signal and CW probe wavelength. Better extinction ratios can be obtained by shorter wavelengths for the CW probe signal. This is due to the gain peak of the amplifier shifting towards the longer wavelengths when the gain is saturated, resulting in a higher slope on the shorter wavelength side of

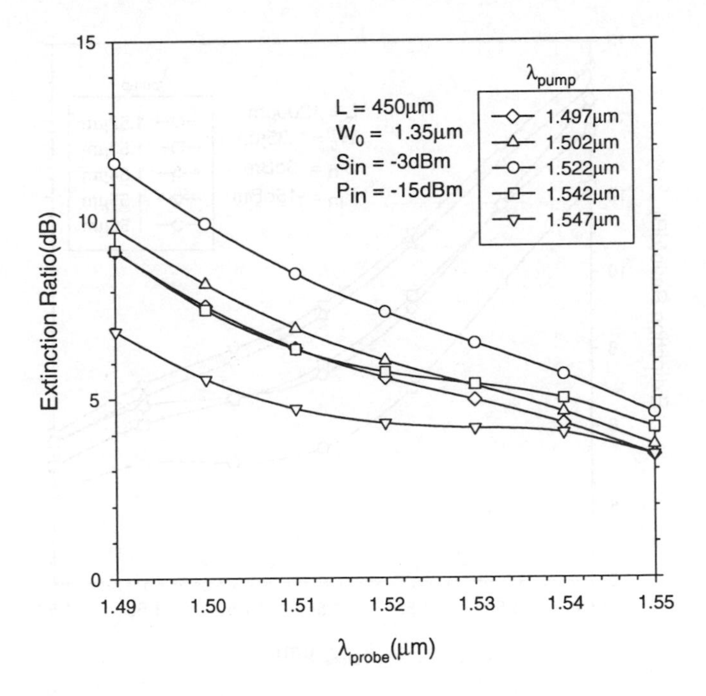

Fig. 12.2. Extinction ratio of converted signal vs probe wavelength for various input signal wavelengths.

the gain peak. Therefore, the probe channels at the shorter wavelengths will experience larger gain variations.

The dependence of the extinction ratio for both up and down conversions in a conventional SLA having the cavity length of 1200 μm is shown in Fig. 12.3. A large extinction ratio of the converted signal is expected for the long SLA since a smaller bandwidth gives larger differential gain at the short wavelength side of the gain peak. However, the difference between up and down conversion for the longer SLA becomes greater. The peak gain wavelength varies along the length of the SLA as longitudinal amplification of the beams causes a carrier density gradient. As the alignment of gain peak and signal wavelength is carried out for longer SLA, the converted signal power is reduced due to the gain peak being a long way away from the signal wavelength, which causes the signal to experience lower gain. For the shorter SLA, saturation occurs only a short distance down the SLA, due to the high gain and the gain peak which remains at the signal wavelength over the remaining length of the SLA. The depen-

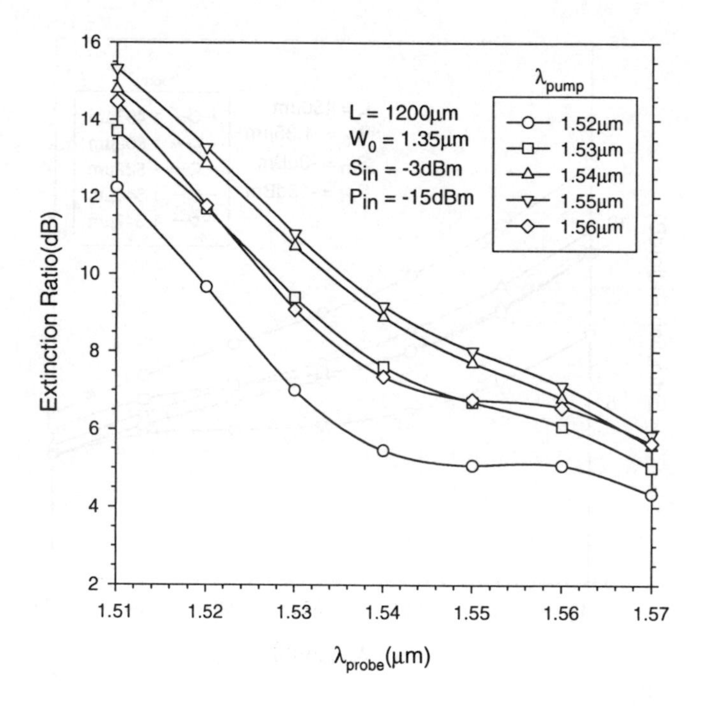

Fig. 12.3. Extinction ratio of converted signal vs probe wavelength for various input signal wavelengths.

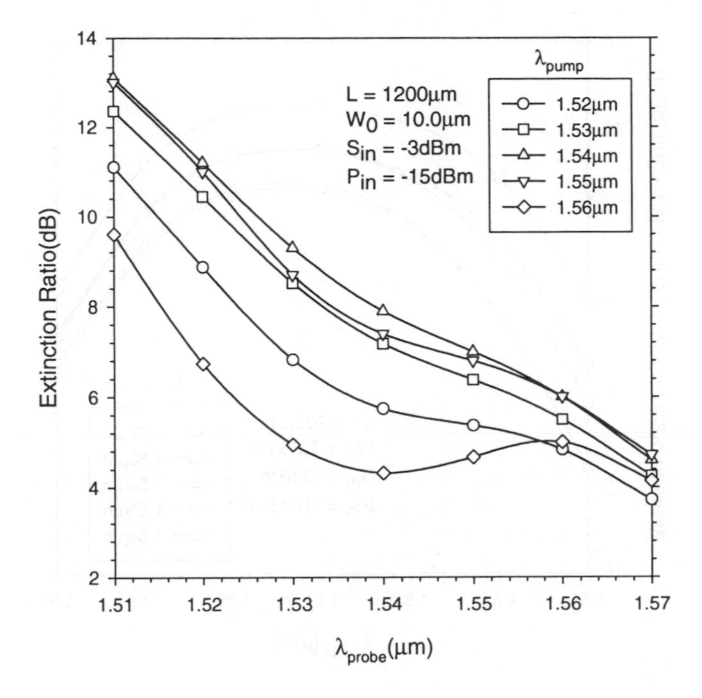

Fig. 12.4. Extinction ratio of converted signal vs probe wavelength for various input signal wavelengths.

dence of the extinction ratio on the wavelength of the signal and CW probe for a linear tapered-waveguide SLA with an output width of 10 μm is shown in Fig. 12.4. The difference between up and down conversion is less severe than for the conventional SLA. The up and down conversions for input signals with wavelengths ranging from 1.53 to 1.55 μm are similar.

12.3.2 *Dependence of signal converted power on signal and probe wavelength*

The average output power of the converted signal is shown in Figs. 12.5 and 12.6. The peak wavelength of the amplifier is 1.54 μm. Figure 12.5 shows that the highest average output power occurs when the input signal (pump) wavelength is 1.52 μm. Signal wavelengths around the peak wavelength have a lower average output power compared to the signal wavelengths away from the peak wavelength. The reason behind the performance for different input signal wavelengths is the result of the saturation imposed by the input

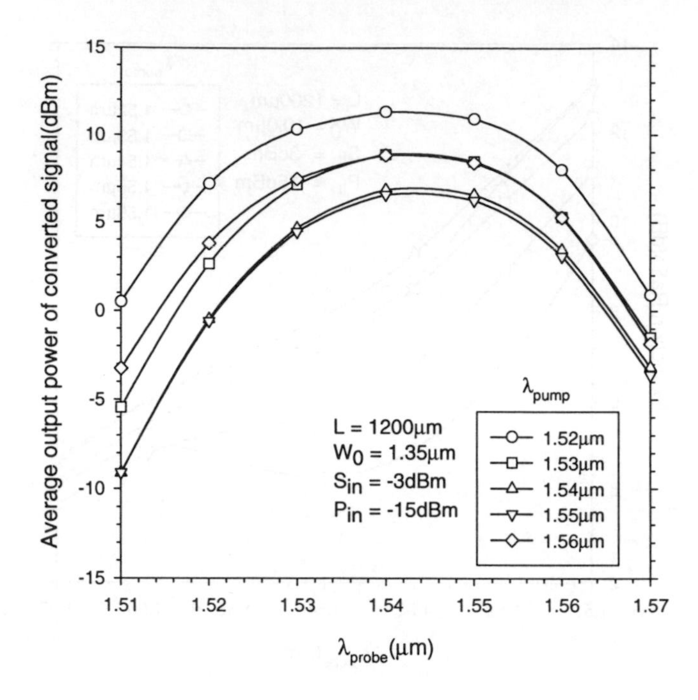

Fig. 12.5. Extinction ratio of converted signal vs probe wavelength for various input signal wavelengths.

signal. For the input signal with a shorter wavelength, the carrier density is higher than that of the longer wavelength. This would result in lower saturation and hence higher gain for the converted signal. When the input signal has a wavelength close to the peak wavelength, it receives greater amplification and causes a higher level of saturation, hence the average power of the converted signal is lower. In all cases, the average output power of the converted signal falls rapidly when the probe signals deviate from the peak wavelength. An increase of probe signal wavelength would result in a subsequently smaller gain suppression, when the probe signal wavelength is at the gain peak, it will experience maximum gain which would give a higher average output power. After the peak wavelength, the amplifier gain is saturated, which causes the average output power to fall rapidly. The average output power of the converted signal in the case of linear tapered-waveguide SLA is shown in Fig. 12.6. The average output power of the converted signal in all cases is 15 dBm to 19 dBm as compared to 6 dBm to 11 dBm for the cases of non-tapered SLAs. The power of the

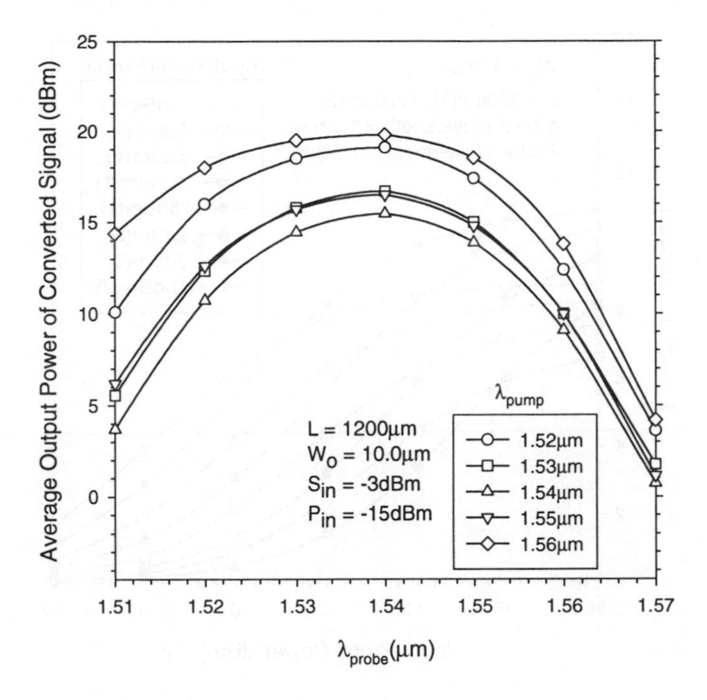

Fig. 12.6. Extinction ratio of converted signal vs probe wavelength for various input signal wavelengths.

converted signal falls more rapidly towards the longer wavelength side as compared to the conventional SLA. This is because the gain peak varies along the length of the SLA as longitudinal amplification of the beams causes a carrier density gradient. A higher slope at the longer wavelength side at the gain peak means that higher saturation output power can be achieved.

12.3.3 *Dependence of extinction ratio on input signal and probe power*

Figure 12.7 illustrates the dependence of extinction ratios on the power of the input signal, probe continuous-wave and the length of the device. The conversion is performed on wavelengths from 1.52 μm to 1.53 μm and the extinction ratio for the input signal is 10 dB. For a given input probe power, the extinction ratio improves with increasing input signal powers. With the higher input signal power, the gain decreases hence

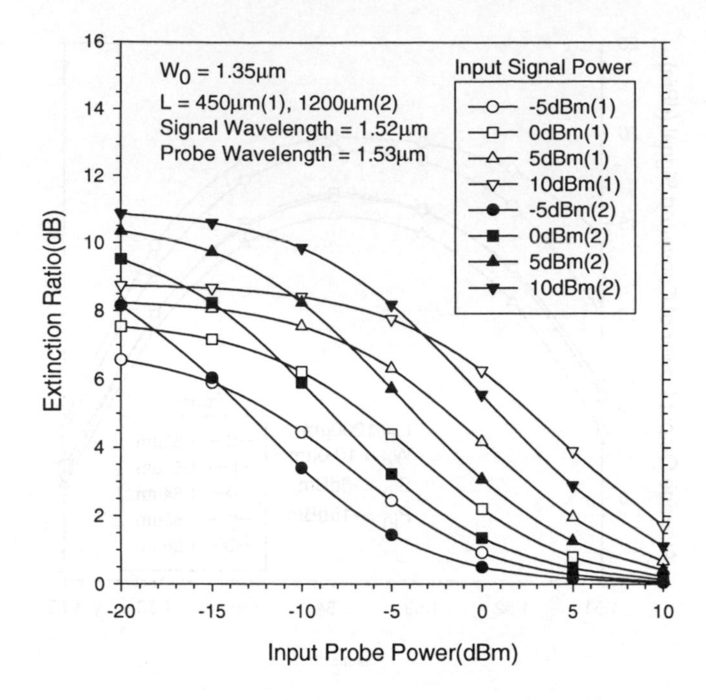

Fig. 12.7. Dependence of extinction ratio on input signal and probe power in a Conventional TW-SLA.

the gain peak shifts to a longer wavelength that gives a longer differential gain at the short wavelength side of the gain peak. A smaller probe power would also result in a higher extinction ratio. When the probe power increases, its contribution to the gain saturation also increases, hence reducing the extinction ratio for the converted signal. A higher extinction ratio can be achieved for longer devices. The improvement in the extinction ratio is expected for the longer SLA since a smaller bandwidth gives a larger differential gain, at the short wavelength side of the gain peak. Figure 12.8 shows the dependency of the extinction ratio on the power of the input and probe signal. By comparing Figs. 12.7 and 12.8, it can be seen that for a linear tapered-waveguide SLA with shorter length, a higher input signal power is required to obtain a similar extinction ratio to the one achieved by the conventional SLA. However, for the longer length, the extinction ratio is similar for both SLAs. With a shorter length, the saturation effect is concentrated near to the output end of the amplifier. When a linear tapered-waveguide SLA is

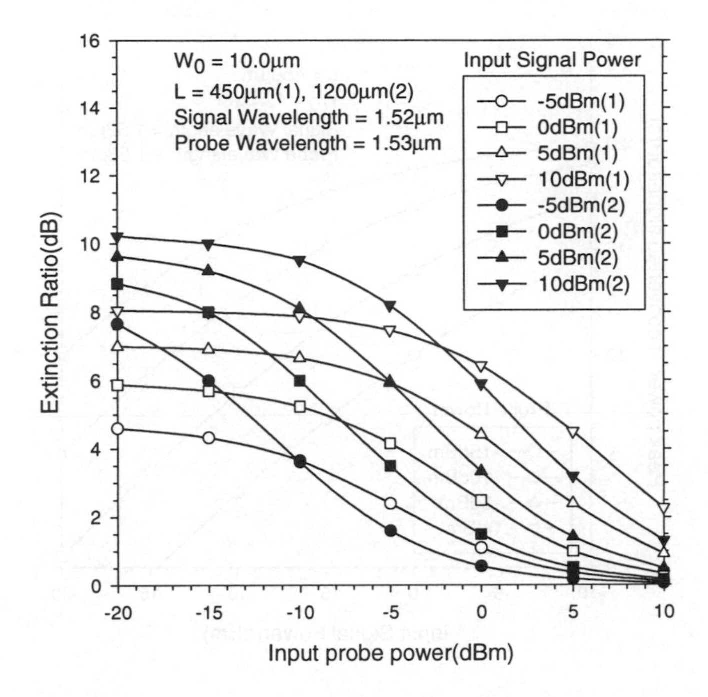

Fig. 12.8. Dependence of extinction ratio on input signal and probe power in a tapered waveguide TW-SLA.

used, a higher input signal power is required to achieve a similar degree of saturation at the output end as the conventional SLA experiences. For a longer length, the saturation effect is spread over a longer distance. When a linear tapered-waveguide SLA with a longer length is used, the conversion still occurs along most parts of the amplifier, resulting in a similar extinction ratio.

12.3.4 *Effect of signal power and probe power on the peak power of the converted signal*

Figures 12.9 and 12.10 show the effects of the input signal and probe power on the peak power of the converted signal for a conventional SLA with different cavity lengths. It shows that power of the converted signal reduces rapidly with the input signal power, especially for the input probe with lower power. When the input signal power increases, the signal is intense enough to compress the gain significantly at the amplifier, thus the gain

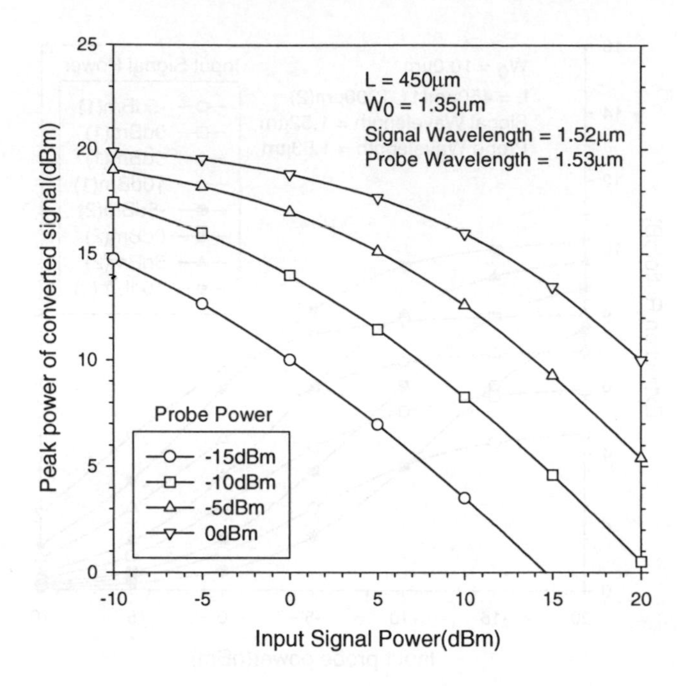

Fig. 12.9. The Peak power of converted signal vs probe power for various input signal power.

experienced by the probe signal is reduced. Therefore the peak power also reduces accordingly. Another reason behind it is that when the input signal power is high, it drives the amplifier deeper into saturation, resulting in a correspondingly severe depletion of carriers. This causes an excess penalty due to an increased turn-on delay for the converted pulse. For a probe signal with lower power, the effect of the input signal is less significant, thus a higher level of spontaneous emission is present. The dependency of the peak output power on the input signal and probe signal power for the linear tapered-waveguide SLAs with cavity lengths of 450 and 1200 μm are shown in Figs. 12.11 and 12.12, respectively. It can be seen that the peak output powers have increased approximately by 6 dBm to 10 dBm as compared tothe conventional SLA. In a conventional SLA the output power is limited by the gain saturation, whereas in a linear tapered-waveguide SLA, since the power is distributed over a larger active region, the gain saturation is higher. Comparison of 450 μm and 1200 μm linear tapered-waveguide SLAs results indicates that the shorter device has a steeper slope and hence reaches saturation earlier than the longer device.

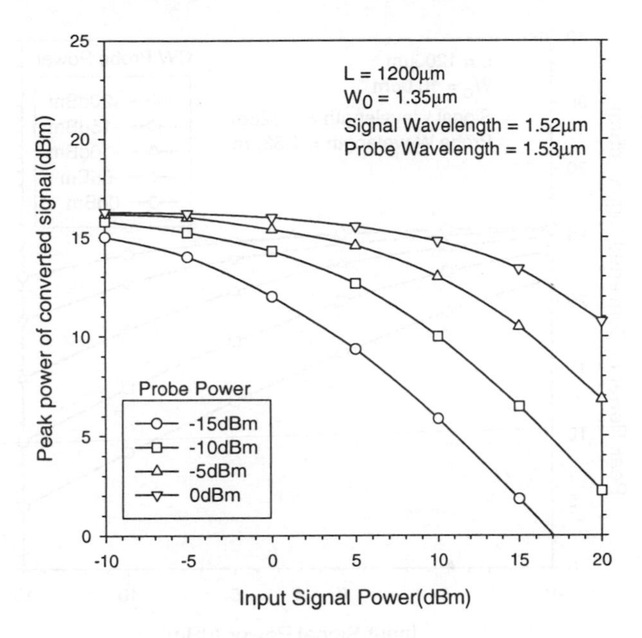

Fig. 12.10. The Peak power of converted signal vs probe power for various input signal power.

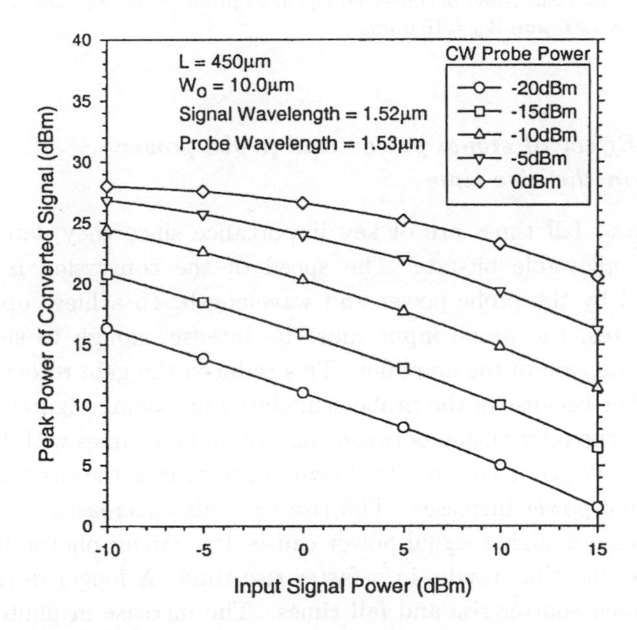

Fig. 12.11. The Peak power of the converted signal vs the probe power for various input signal power.

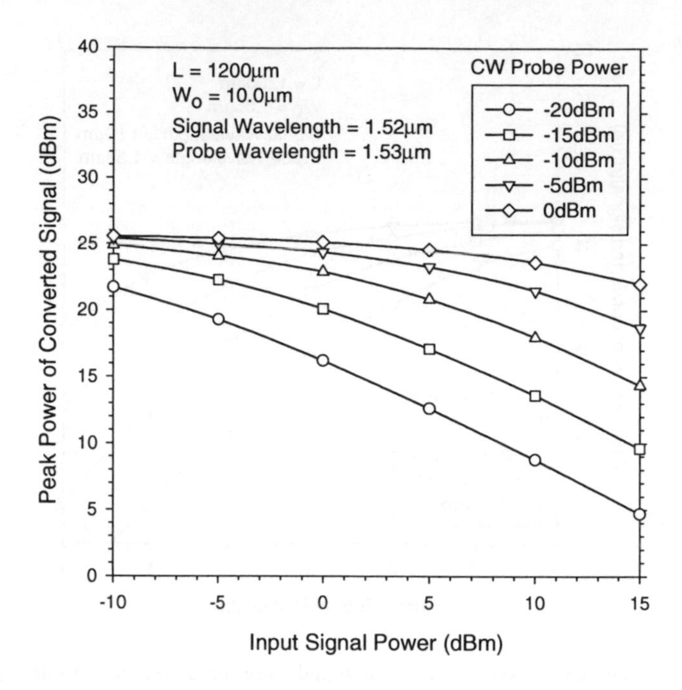

Fig. 12.12. The Peak power of converted signal vs probe power for various input signal power for $L = 1200$ μm, $W_0 = 10.0$ μm.

12.3.5 *Effect of signal power and probe power on the rise time*

The rise and fall times are of key importance since they determine the maximum allowable bit-rate. The speed of the conversion is primarily determined by the probe power and wavelength. To achieve operation at high bit-rates, the probe input must be intense enough to significantly compress the gain of the amplifier. This reduced the gain recovery time of the amplifier because of the probe stimulated emission. Figures 12.13 and 12.14 show the relationship between the rise and fall times with both input signal and CW probe powers. As shown in the figures, the rise time reduces as the probe power increases. The rise time also decreases for increasing signal power. A higher signal power causes the carrier-photon interaction to increase and thus results in a faster rise time. A longer device is seen to have much shorter rise and fall times. The increase in photon density for the longer device enhances the carrier-photon interaction resulting in

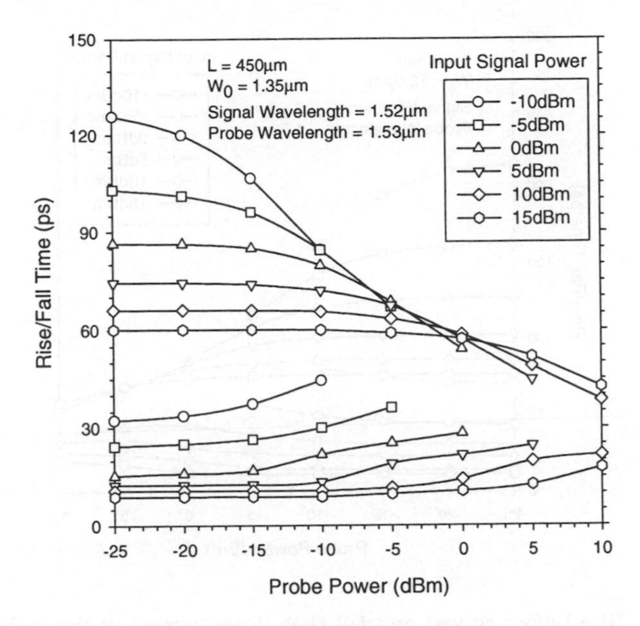

Fig. 12.13. Rise (upper curves) and fall time (lower curves) vs the probe power for various input signal power.

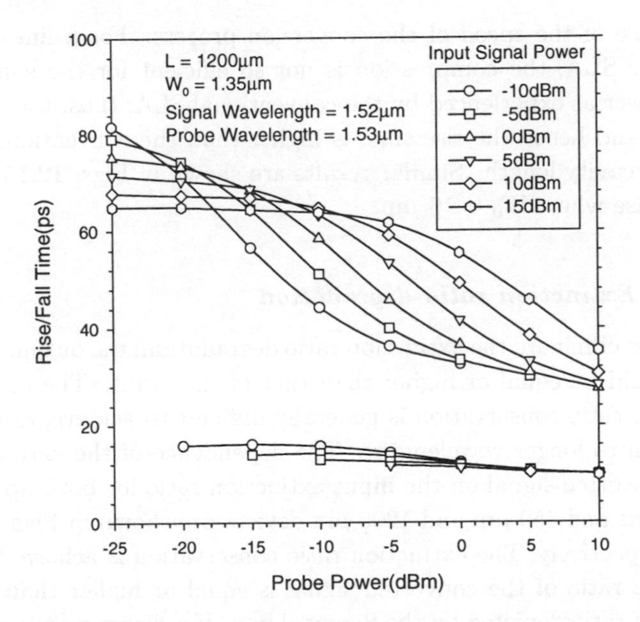

Fig. 12.14. Rise (upper curves) and fall time (lower curves) vs the probe power for various input signal power.

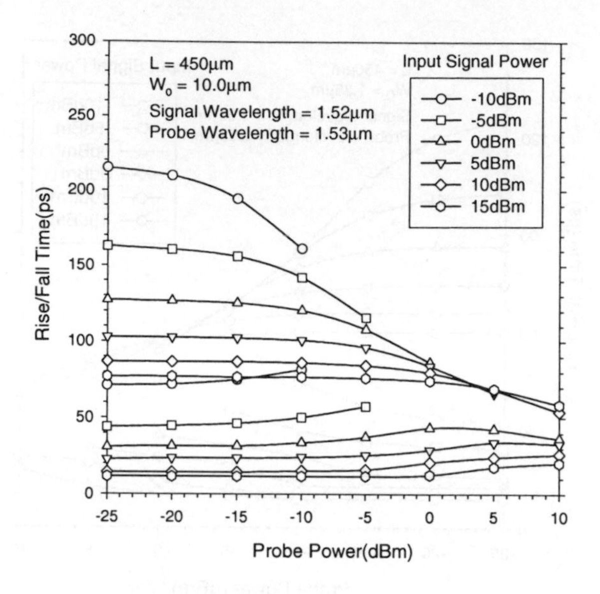

Fig. 12.15. Rise (upper curves) and fall time (lower curves) vs the probe power for various input signal power.

an increase of the speed of the conversion process. For a linear tapered-waveguide SLA, the compression is not so efficient for the same level of probe power as experienced by the conventional SLA, thus, the gain recovery time and hence the rise time is higher than the conventional SLA for the same cavity length. Similar results are shown in Figs. 12.15 and 12.16 for the case where $W_0 = 10\ \mu$m.

12.3.6 *Extinction ratio degradation*

In order to eliminate the extinction ratio degradation, the output extinction ratio should be equal or higher than that of the input. The condition for extinction ratio conservation is generally difficult to achieve, especially for conversion to longer wavelengths. The dependence of the extinction ratio of the converted signal on the input extinction ratio for both up and down conversions and 450 μm and 1200 μm devices are shown in Figs. 12.17 and 12.18, respectively. The extinction ratio conservation is achieved when the extinction ratio of the converted signal is equal or higher than the input extinction ratio depicted by the diagonal line. If a device possesses the property of extinction ratio conservation, it is possible to prevent the extinction

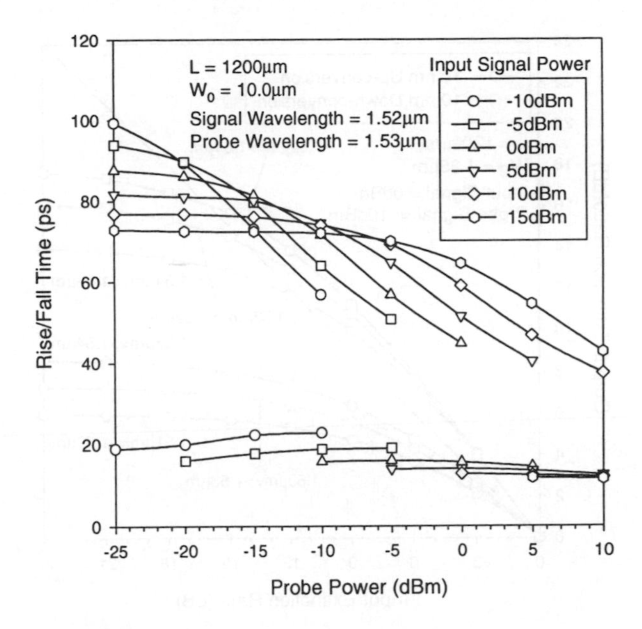

Fig. 12.16. Rise (upper curves) and fall time (lower curves) vs the probe power for various input signal power.

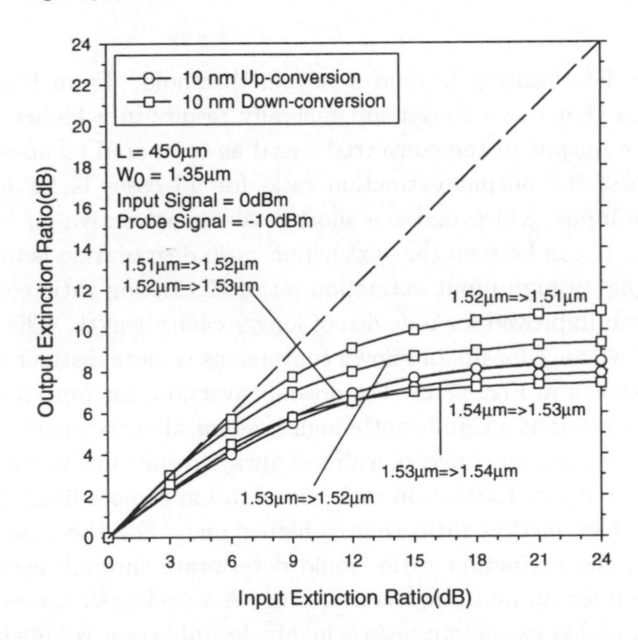

Fig. 12.17. The Extinction ratio of converted signal vs the extinction ratio of input signal.

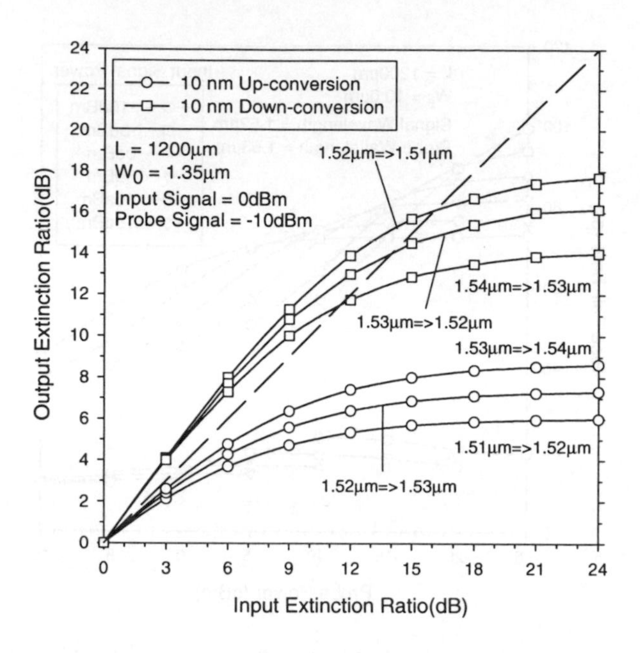

Fig. 12.18. The Extinction ratio of converted signal vs the extinction ratio of input signal.

ratio from deteriorating beyond a certain threshold. From Fig. 12.17, it can be seen that down-conversion generally results in a higher extinction ratio at the output of the converted signal as compared to up-conversion. Nevertheless, the output extinction ratio for all cases is far lower than that of the input, which makes a short device unattractive for wavelength conversion. It can be seen that extinction ratio degradation is more severe for the signal of high input extinction ratio. Extinction ratio conservation is very much improved for a device of longer cavity length. The difference in the performance for up and down conversions is more distinct in a longer device as shown in Fig. 12.18. For down-conversion, the output extinction ratio is conserved at a significantly high level for all wavelengths indicated. On the other hand, up-conversion almost always results in a lower extinction ratio at the output. Extinction ratio conservation is more likely for a lower input signal extinction ratio than a higher one. For the case shown in Fig. 12.17, the extinction ratio would deteriorate through each stage of conservation for up and down conversions. A wavelength converter which results in a higher extinction ratio is highly desirable as it results in a higher signal to noise ratio. The length of the SLA was extended to 2400 μm to

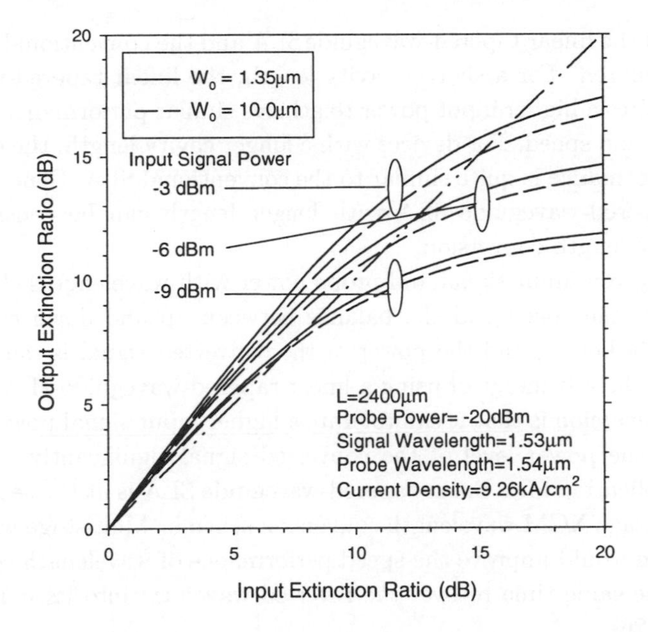

Fig. 12.19. Comparison of Extinction Ratio in 2400 μm long Semiconductor Laser Amplifier for output widths of $W_0 = 1.35$ μm and $W_0 = 10.0$ μm.

compare the extinction ratio conservation between conventional and linear tapered-waveguide devices for a 10 Gb/s data stream as shown in Fig. 12.19. The input pump and probe power were close to the optimum value. It can be seen that with such a long cavity, a low probe power and a typical input signal power, it is possible to preserve the extinction ratio over 10 dB. The output extinction ratio of the linear tapered-waveguide SLA may be slightly lower, but its output power is significantly higher.

12.4 Summary

The performance of XGM wavelength converter is closely related to the power of both input and probe signals, as well as their respective wavelengths. For a given signal power, the optimal extinction ratio is determined by the difference between the input signal power and probe power. The extinction ratio increases with the input signal power. However, the input signal power should not be too high in order to avoid an excess penalty caused by an increased turn-on delay for the converted pulses. The performances of the linear tapered-waveguide SLA and the conventional SLA have been compared. For a shorter cavity length, the linear tapered-waveguide SLA requires a higher input power to achieve similar performance in extinction ratio and speed. For devices with a longer cavity length, the extinction ratio performance is quite similar to the conventional SLA. Hence, only the linear tapered-waveguide SLA with longer length can be considered for XGM wavelength conversion.

For a given input signal and probe power with wavelengths close to the peak gain, the speed and the balance between up and down conversions are slightly better, and the power of the converted signal is much higher. However, the advantage of using a linear tapered-waveguide SLA for wavelength conversion is that it can tolerate a higher input signal power without reducing the power level of the converted signal significantly. The most likely application for a linear tapered-waveguide SLA is at the second stage of a two-stage XGM wavelength conversion scheme. Multistage wavelength conversion would improve the speed performance of wavelength converters, and at the same time recover the inverted waveform into its non-inverted form [24–26].

In this chapter, wavelength conversion using cross-gain modulation (XGM) in linear tapered-waveguide semiconductor laser amplifiers (SLAs) has been studied and compared with the non-tapered semiconductor laser amplifier. For example, we have found that for the linear tapered-waveguide SLA with a 450 μm cavity length, a higher signal power is required to achieve a similar extinction ratio as obtained by the conventional SLA with the same cavity length. However, a similar extinction ratio is observed for both conventional and linear tapered-waveguide with the cavity length of 1200 μm. The difference between the extinction ratio for up and down conversions is less severe in a linear tapered-waveguide SLA as compared to that for a conventional SLA. The rise time for linear tapered-waveguide is observed to be higher than that for conventional SLA. It is possible to attain extinction ratio conservation using XGM for both up and down conversions in a SLA with a very long cavity length of 2400 μm.

References

[1] B. Glance *et al.*, "High Performance Optical Wavelength Shifter," *Electronics Letters*, Vol. 28, No. 18, pp. 1714–1715, 1992.

[2] I. Valiente, J. C. Simon and M. Le Ligne, "Theoretical Analysis of Semicon-

ductor Optical Amplifier Wavelength Shifter," *Electronics Letters*, Vol. 29, No. 5, pp. 502–503, 1993.

[3] D. A. O. Davis, "Small Signal Analysis of Wavelength Conversion in Semiconductor Laser Amplifiers via Gain Saturation," *IEEE Photonics Technology Letters*, Vol. 7, No. 6, pp. 617–619, 1995.

[4] C. Joergensen, T. Durhuus, C. Braagaard, B. Mikkelsen and K. E. Stubkjaer, "4 Gb/s Optical Wavelength Conversion Using Semiconductor Optical Amplifiers," *IEEE Photonics Technology Letters*, Vol. 5, No. 6, pp. 657–659, 1993.

[5] T. Durhuus, C. Joergensen, B. Mikkelsen, R. J. S. Pedersen and K. E. Stunjaer, "All Optical Wavelength Conversion by SOA's in a Mach-Zehnder Configuration," *IEEE Photonics Technology Letters*, Vol. 6, No. 1, pp. 53–55, 1994.

[6] F. Ratovelomanana *et al.*, "An All-Optical Wavelength Converter with Semiconductor Optical Amplifiers Monolithically Integrated in an Asymmetric Passive Mach-Zehnder Interferomater," *IEEE Photonics Technology Letters*, Vol. 7, No. 10, pp. 992–994, 1995.

[7] X. Pan and J. M. Wiesenfeld, "Dynamic Operation of a Three-Port, Integrated Mach-Zehnder Wavelength Converter," *IEEE Photonics Technology Letters*, Vol. 7, No. 9, pp. 995–997, 1995.

[8] W. Idler *et al.*, "10 Gb/s Wavelength Conversion with Integrated Multiquantum-Well-Based 3-Port Mach-Zehnder Interferometer," *IEEE Photonics Technology Letters*, Vol. 8, No. 9, pp. 1163–1165, 1996.

[9] J. Zhou *et al.*, "Efficiency of Broadband Four-Wave Mixing Wavelength Conversion Using Semiconductor Travelling-Wave Amplifiers," *IEEE Photonics Technology Letters*, Vol. 6, No. 1, pp. 50–52, 1994.

[10] A. D'Ottavi *et al.*, "Efficiency and Noise Performance of Wavelength Converters Based on FWM in Semiconductor Optical Amplifiers," *IEEE Photonics Technology Letters*, Vol. 7, No. 4, pp. 357–359, 1995.

[11] G. P. Bava, P. Debernardi and Osella, "Frequency Conversion in Travelling Wave Semiconductor Laser Amplifiers with Bulk and Quantum-Well Structures," *IEE Proc.-Optoelectronics*, Vol. 7, No. 9, pp. 995–997, 1995.

[12] J. M. Wiesenfeld, B. Glance, J. S. Perino and A. H. Gnauck, "Wavelength Conversion at 10 Gb/s using Semiconductor Optical Amplifier," *IEEE Photonics Technology Letters*, Vol. 5, No. 11, pp. 1300–1303, 1993.

[13] J. M. Wiesenfeld, B. Glance, J. S. Perino and A. H. Gnauck, "Bit Error Rate Performance for Wavelength Conversion at 20 Gb/s," *IEEE Photonics Technology Letters*, Vol. 30, No. 9, pp. 720–721, 1994.

[14] W. Shieh and A. E. Willner, "Optimal Conditions for High-Speed All-Optical SOA-Based Wavelength Shifting," *IEEE Photonics Technology Letters*, Vol. 7, No. 11, pp. 1273–1275, 1995.

[15] T. Durhuus, B. Mikkelsen, C. Joergensen, S. L. Lykke Danielsen and K. E. Stubkjaer, "All-Optical Wavelength Conversion by Semiconductor Optical Amplifiers," *Journal of Lightwave Technology*, Vol. 14, No. 6, pp. 942–954, 1996.

[16] M. Asghari, I. H. White and R. V. Penty, "Wavelength Conversion using Semiconductor Optical Amplifiers," *Journal of Lightwave Technology*, Vol. 15, No. 7, pp. 1181–1190, 1997.

[17] M. E. Bray and M. J. O'Mahony, "Cascading Gain-Saturation Semiconductor Laser Amplifier Wavelength Translators," *IEE Proc.-Optoelectonics*, Vol. 143, No. 1, pp. 1–6, 1996.

[18] P. A. Yazaki, K. Komori, S. Arai and Y. Suematsu, "A GaInAsP/InP Tapered Wave-Guide Semiconductor Laser Amplifier Integrated with a 1.5?m Distributed Feedback laser," *IEEE Photonics Technology Letters*, Vol. 3, No. 12, pp. 1060–1063, 1991.

[19] G. Bendelli, K. K. Komori and S. Arai, "Gain Saturation and Propagation Characteristics of Index-Guided Tapered-Waveguide Travelling-Wave Semiconductor Laser Amplifiers(TTW-SLA's)," *IEEE J.Quantum Electronics,* Vol. 28, No. 2, pp. 447–454, 1992.

[20] E. El. Yumin, K. Komori, S. Arai and G. Bendelli, "Taper-Shape Dependence of Tapered-Waveguide Travelling Wave Semiconductor Laser Amplifier (TTW-SLA)," *IEICE Trans.Electron*, Vol. E77-C, No. 4, pp. 7–15, 1994.

[21] H. Ghafouri Shiraz, P. W. Tan, T. Aruga, "Picosecond Pulse Amplification in Tapered-Waveguide Laser Diode Amplifiers," *IEEE Selected Topic in Quantum Electronics*, Vol. 3, No. 2, pp. 210–217, 1997.

[22] K. Inoue and K. Oda, "Noise Suppression in Wavelength Conversion Using a Light-Injected Laser Diode," *IEEE Photonics Technology Letters*, Vol. 7, No. 5, pp. 500–501, 1995.

[23] K. Inoue, "Noise Transfer Characteristics in Wavelength Conversion Based on Cross-gain Saturation in a Semiconductor Optical Amplifier," *IEEE Photonics Technology Letters*, Vol. 87, No. 7, pp. 888–890, 1996.

[24] J. M. Wiesenfeld and B. Glance, "Cascadibility and Fanout of Semiconductor Optical Amplifier Wavelength Shifter," *IEEE Photonics Technology Letters*, Vol. 4, No. 10, pp. 1168–1171, 1996.

[25] D. Marcenac and A. Mecozzi, "Switches and Frequency Converters Based on Cross-gain Modulation in Semiconductor Optical Amplifiers," *IEEE Photonics Technology Letters*, Vol. 9, No. 6, pp. 749–751, 1997.

[26] P. W. Tan and H. Ghafouri-Shiraz, "Sub-picosecond Gain Dynamic in Highly-Index Guided Tapered-Waveguide Laser Diode Optical Amplifiers," *IEE Proc. Opto-electronics*, Vol. 146, No. 2, pp. 83–88, April 1999.

Chapter 13

The Semiconductor Laser: Basic Concepts and Applications

13.1 Introduction

The fundamentals of the semiconductor laser are presented in this chapter, which provide the foundation for understanding the time-domain laser models in Chapters 14 and 15. The basic concepts here also supplement the additional theory provided in Chapter 16 that is required to model the tapered structure semiconductor laser. In addition, sufficient background in ultrashort pulse generation schemes using laser diodes will be given, with emphasis on active mode-locking, the technique used in the novel multisegment mode-locked laser design in Chapter 17.

The first laser, based on the ruby rod, was demonstrated by Maiman in 1960 [1]. The dawn of optoelectronics began in 1962 when gallium arsenide (GaAs) and gallium arsenide phosphide (GaAsP) semiconductor lasers were independently demonstrated by four groups led by Hall, Nathan, Rediker, and Holonyak [2]. The semiconductor laser developed slowly from a pulse- operated simple *p.n.* homojunction to continuously operated (300 K) double heterostructure (DH) in 1970 [3]. Research and development of the semiconductor laser, together with the introduction of low-loss optical fibres, anticipated as early as 1966 by Kao and Hockham [4], have opened up the door for optical communications.

The semiconductor laser diode was chosen to play a key role in optical communications because of its compact-size, suitable wavelength range, and high reliability. The direct energy transition of the III-V semiconductor compounds means that we can achieve high quantum efficiency in the conversion of electron-hole pairs to photons, providing a clear advantage over indirectly excited lasers [5]. Furthermore, the compact laser diode can be integrated monolithically with other circuit elements in optoelectronic integrated circuits (OEIC) and photonic integrated circuits (PIC) [6].

Generally, semiconductor lasers that operate in the near-infrared region fall into two main categories [7]:

(i) Gallium aluminium arsenide (GaAlAs) on gallium arsenide (GaAs) for emission from 780–850 nm (short wavelength)
(ii) Indium gallium arsenide phosphide (InGaAsP) on indium phosphide (InP) for emission from 1100–1650 nm (long wavelength)

The latter category is favoured in long-haul optical communication systems because they cover the zero dispersion (1300 nm) and minimum loss wavelengths (1550 nm) of the optical fibre (see Fig. 13.1). In addition, there is also the exclusive 980 nm pump sources used in erbium-doped fibre amplifiers, which have indium gallium arsenide (InGaAs) active layers on gallium arsenide (GaAs) substrates.

Research in semiconductor laser diodes for communications has traditionally focused on achieving stable single frequency operation, high output power and increasing direct modulation bandwidth. However, as

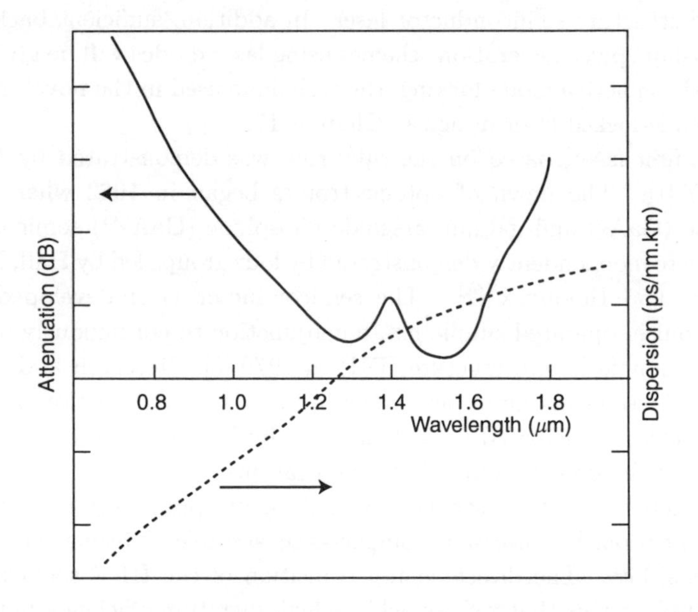

Fig. 13.1. Attenuation and dispersion characteristics of the standard silica glass fibre.

optical communications evolve into more complex systems and networks, the additional functionality of laser diodes such as wavelength conversion, routing, regeneration, multiplexing/demultiplexing and tunability become just as important [8–10].

13.2 Fundamental Optical Processes

To illustrate the principle behind the fundamental processes that occur during light emission from the laser diode, a simple two-level atomic system of Fig. 13.2 is used [6]. In Fig. 13.2, the energy level, E_1, corresponds to the ground state and E_2 corresponds to the excited state. If a photon is incident on the atomic system, its energy may be absorbed, causing an electron in the ground state, E_1, to rise to the excited state, E_2. This is referred to as *stimulated absorption.* If an electron is initially in the excited state, E_2 (known as *inversion*), it may suddenly make a transition to the ground state, E_1, and its energy is released in the form of a photon. This is referred to as *spontaneous emission.* The random nature of spontaneous emission results in incoherent radiation, i.e. the photons emitted have no phase relationship between them and propagate in random directions.

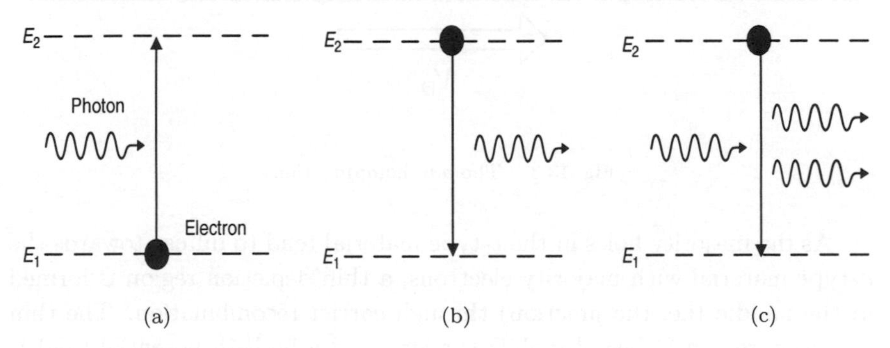

Fig. 13.2. Fundamental optical processes (a) absorption (b) spontaneous emission and (c) stimulated emission.

The light emitted from the laser is predominantly caused by *stimulated emission.* This is when an existing photon interacts with an excited electron, causing it to return to the ground state, and energy is released in the form of a second photon. The second photon has the same phase, energy, polarisation, and propagation direction as the original photon. This accounts for the coherent radiation from the laser.

13.3 Homojunction and Double Heterojunction

In a semiconductor material, the presence of valence and conduction bands separated by a forbidden energy bandgap, E_g, resembles the two distinct energy states of the simplified two-level atomic system. There are generally three types of semiconductor material $-i$ (intrinsic), n-type and p-type. In the intrinsic semiconductor containing no impurities, the Fermi level, E_f, is located at the centre of the bandgap. By doping the semiconductor material with donor and acceptor impurities, the Fermi level can be raised above and lowered below the centre of the bandgap to create n-type or p-type semiconductors, respectively. The n-type has more free electrons and the p-type has more free holes. The most basic structure of a semiconductor light source is the $p.n.$ homojunction formed by joining a p-type and n-type semiconductor together as shown in Fig. 13.3.

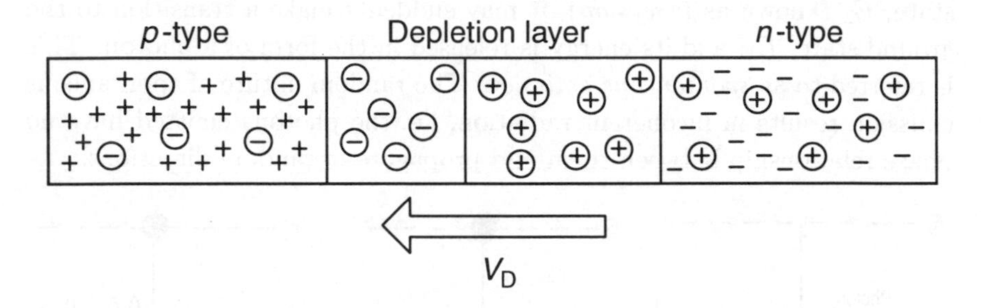

Fig. 13.3. The $p.n.$ homojunction.

As the majority holes in the p-type material tend to diffuse towards the n-type material with majority electrons, a thin depletion region is formed in the middle (i.e. the junction) through carrier recombination. The thin depletion region is free of mobile carriers, and a built-in potential (V_D) is created, which is directed from the n-type toward the p-type to prevent further diffusion of mobile carriers. When an external voltage is applied in the same direction as the built-in potential (Fig. 13.4(a)), i.e. *reverse-biased*, the width of the depletion region increases resulting in a higher potential barrier. Any carriers present in the depletion region are immediately separated by the electric field, with electrons going towards the n-type region and holes towards the p-type region.

On the other hand, if an external voltage is applied in the opposite direction of the built-in potential (Fig. 13.4(b)), i.e. *forward-biased*, the

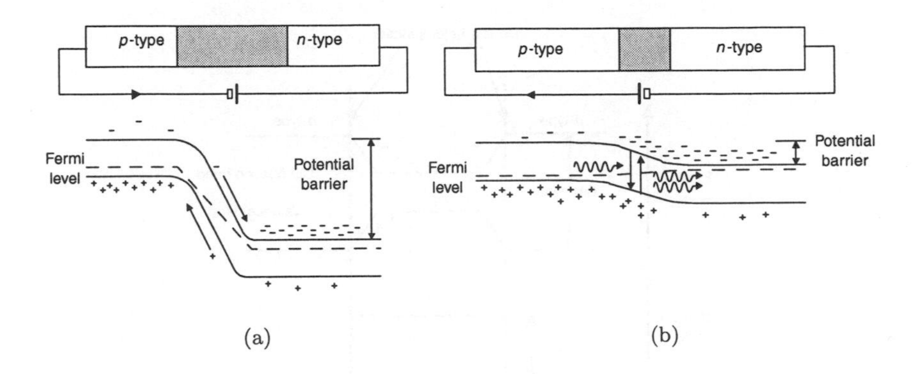

Fig. 13.4. Homojunction (a) Reverse-biased and (b) forward-biased.

width of the depletion region shrinks and the potential barrier is reduced. Electrons from the n-type region and holes from p-type region can flow more easily across the junction. As a result, both free electrons and holes are present within the depletion region, leading to a high inversion level. When these carriers recombine, energy is released in the form of photons (i.e. light is emitted).

The problem with $p.n.$ homojunctions is that it is difficult to realise high carrier densities in the active region, resulting in low efficiency for light generation. By using semiconductor materials with different bandgaps, a *heterojunction* is formed and better carrier confinement can be achieved. The original research work carried out by Z. I. Alferov made a major contribution to heterojunction semiconductor lasers and their continuous-wave (CW) operation at room temperature. By using two such heterojunctions, the double heterojunction (DH) is formed with a thin active layer sandwiched between p-type and n-type semiconductor layers with larger bandgaps (see Fig. 13.5). The active layer, where carrier recombination and light generation take place, also has a slightly higher index of refraction and acts as a dielectric waveguide for light confinement. It is the DH structure that has made the semiconductor laser practical for its role in optical communication systems.

13.4 Lasing Condition

Besides optical gain, the semiconductor light source also requires feedback to achieve self-sustained oscillation known as *lasing*. The simplest laser cav-

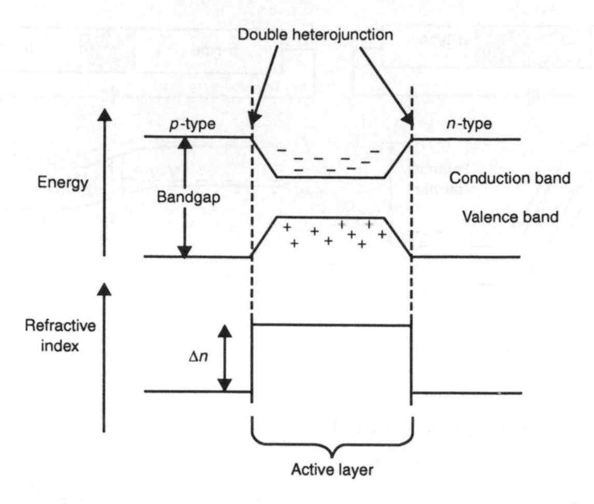

Fig. 13.5. The double heterojunction (DH) laser for carrier and light confinement.

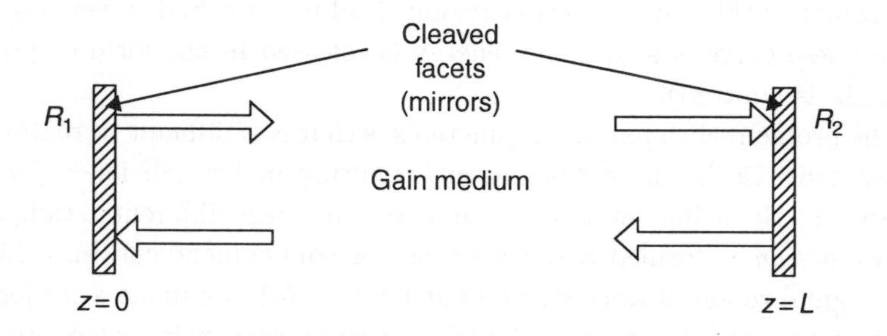

Fig. 13.6. The Fabry Perot (FP) laser cavity.

ity is the Fabry–Perot (FP) laser with cleaved facets providing the optical feedback as shown in Fig. 13.6. We can gain considerable insight into the threshold condition required for lasing by assuming plane-wave solutions of the time-independent wave equation expressed by:

$$\nabla^2 \tilde{E} + \varepsilon k_0^2 \tilde{E} = 0 \tag{13.1}$$

where $\tilde{E}$ is the complex amplitude of the electric field, ε is complex dielectric constant, and k_0 is the free space wavenumber. The plane wave approximation is then given by:

$$\tilde{E} = \hat{x} E_0 \exp(i\tilde{\beta} z) \tag{13.2}$$

where $\hat{x}$ is the polarisation unit vector, and E_0 is a constant amplitude. It is worth pointing out that the laser modes are never plane-waves but are of transverse electric (TE) and transverse magnetic (TM) types, resulting from dielectric waveguiding [11]. However, the plane-wave approximation is reasonably accurate in helping us understand the essential physics of the lasing process. The complex propagation constant ($\tilde{\beta}$) in Eq. (13.2) takes the following form:

$$\tilde{\beta} = nk_0 + i\frac{g_{\text{net}}}{2} \tag{13.3}$$

where n is the index of refraction, and g_{net} is the net power gain coefficient [6]. In order to obtain the threshold condition, we require that the optical field reproduce itself after each round trip in the steady-state continuous wave (CW) operating condition [6, 11–12]. If R_1 are R_2 the cleaved facet reflectivities, the net change in the field amplitude after one round trip must be set to unity at the laser threshold, that is:

$$\sqrt{R_1R_2}\exp(2i\tilde{\beta}L) = 1 \tag{13.4}$$

where L is the total length of the laser cavity. By substituting Eq. (13.3) into Eq. (13.4) and equating their real and imaginary parts we obtain the following equations:

$$\text{amplitude condition:} \qquad \sqrt{R_1R_2}\exp(-g_{\text{net}}L) = 1 \tag{13.5}$$

$$\text{phase condition:} \qquad \cos(2nk_0L) = 1\,. \tag{13.6}$$

In the threshold condition, the mirror loss, α_{m}, due to optical radiation escaping from the laser cavity must be equal to the net gain, g_{net}, that is:

$$\alpha_{\text{m}} = g_{\text{net}} = g - \alpha_{\text{int}} \tag{13.7}$$

where g is the gain coefficient, and α_{int} is the internal loss due to several mechanisms, such as free carrier absorption and scattering loss. From Eq. (13.5), the mirror loss, α_{m}, may also be defined in terms of cavity length (L) and facet reflectivities (R_1 and R_2) as [6]:

$$\alpha_{\text{m}} = \frac{1}{2L}\ln\left(\frac{1}{R_1R_2}\right). \tag{13.8}$$

The threshold gain coefficient in Eq. (13.7) may be rewritten as:

$$g = \alpha_{\text{m}} + \alpha_{\text{int}}\,. \tag{13.9}$$

This means that in order to achieve threshold condition, the total gain due to external electrical pumping must balance the total loss (mirror loss and internal loss).

For practical interest, the threshold current, $I_{\rm th}$, is usually used to define the threshold condition instead of the threshold gain coefficient, $g_{\rm th}$. The threshold current is defined as [6]:

$$I_{\rm th} = \frac{q v_{\rm a} N_{\rm th}}{\tau_n} + I_{\rm L} \,. \tag{13.10}$$

where $v_{\rm a}$ is the active region volume, q is the electronic charge, $N_{\rm th}$ is the threshold carrier density, and $I_{\rm L}$ is the leakage current. In the general case, the carrier recombination lifetime, τ_n, is a function of carrier density. For long-wavelength semiconductor lasers, an increase in temperature leads to a decrease in τ_n due to Auger recombination [6]. At threshold, the net rate of stimulated emission (G) must equal the net rate of photon loss due to photons escaping the cavity and other internal loss, that is:

$$G = \frac{1}{\tau_{\rm p}} = \Gamma v_{\rm g} a (N - N_{\rm tr}) \tag{13.11}$$

where Γ is the optical confinement factor, $v_{\rm g}$ is the group velocity, a is the differential gain coefficient, $N_{\rm tr}$ is transparency carrier density, and the photon lifetime ($\tau_{\rm p}$) is defined by:

$$\frac{1}{\tau_{\rm p}} = v_{\rm g}(\alpha_{\rm m} + \alpha_{\rm int}) \,. \tag{13.12}$$

By rearranging Eq. (13.11), the threshold carrier density, $N_{\rm th}$, can be found from:

$$N_{\rm th} = N_{\rm tr} + \frac{1}{\Gamma v_{\rm g} a \tau_{\rm p}} \,. \tag{13.13}$$

This value of threshold carrier density ($N_{\rm th}$) may eventually be used to find the rough estimate of the threshold current ($I_{\rm th}$) of the laser diode. Since photons escape from the laser cavity at a rate of $v_{\rm g}\alpha_{\rm m}$, the optical power emitted by each facet is related to the photon density inside the cavity by:

$$P_{\rm out} = \frac{1}{2} h f v_{\rm g} \alpha_{\rm m} v_s S \tag{13.14}$$

where h is Planck's constant, f is the optical frequency, and S is the photon density. In the above, the amplitude condition (see Eq. (13.5) is used to obtain the threshold current defined by Eq. (13.10). On the other hand, the phase condition of Eq. (13.6) is used to obtain the resonant frequencies of the laser cavity. The periodic nature of the trigonometric function in Eq. (13.6) leads to a multiple solution of the form:

$$2nk_0L = 2m\pi \tag{13.15}$$

where m is any integer. The frequency of the mth longitudinal mode of the FP cavity is thus defined as:

$$f_m = \frac{mc}{2nL}\,. \tag{13.16}$$

The multiple longitudinal modes correspond to the peaks (or troughs) of the standing waves produced by the optical waves running back and forth with the FP cavity as shown in Fig. 13.7. Which one and how many of these longitudinal modes are lasing depends on the gain spectrum of the active semiconductor material. The spectral dependence of material gain selects the mode closest to its gain peak as the lasing mode as shown in Fig. 13.8.

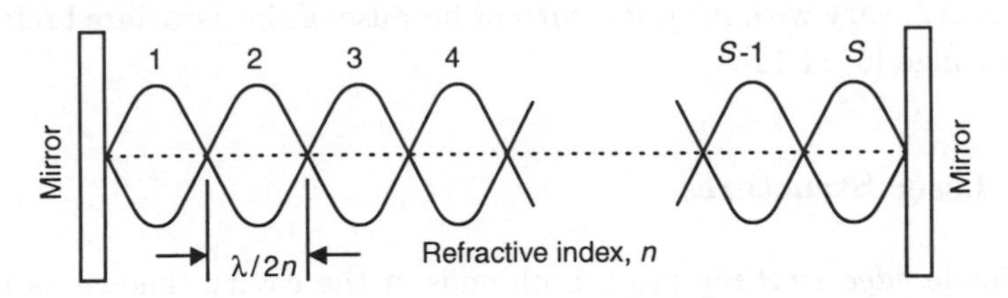

Fig. 13.7. Standing waves of the longitudinal modes.

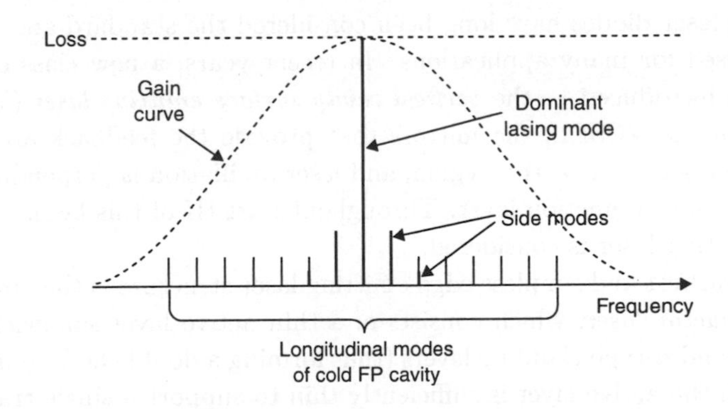

Fig. 13.8. The wavelength selective mechanism of the gain spectrum.

In practice, neighbouring modes on each side of the dominant lasing mode may also overcome the threshold condition and carry a significant portion of the total power [13]. This is typical for FP lasers, which are sometimes known as multimode lasers.

As the refractive index n of the semiconductor laser varies with frequency due to the dispersive semiconductor material, the intermode spacing (also known as the free spectral range, FSR) is given as:

$$\Delta f = \frac{c}{2n_g L} \tag{3.17a}$$

where

$$n_g = n + f\frac{\partial n}{\partial f}\,. \tag{3.17b}$$

The term n_g is known as the group refractive index, which takes into account material dispersion. An important and unique feature of semiconductor lasers is that the longitudinal mode frequencies, f_m, and their intermode spacing, Δf, vary with injection current because of the associated refractive index change [6–11,12].

13.5 Laser Structures

In a simple *edge-emitting* laser, both ends of the cavity (known as facets) are cleaved to act as partially reflective mirrors, and light is emitted parallel to the junction plane. It is the optical feedback provided by the reflective facets that sustains the laser oscillation inside the resonant cavity. Edge-emitting laser diodes have long been considered the standard and are still widely used for many applications. In recent years, a new class of lasers has been introduced — the *vertical cavity surface emitting laser* (VCSEL) [6, 7]. In the VCSEL, the mirrors that provide the feedback are placed above and below the active region, and laser oscillation is perpendicular to the plane of the junction layer. Throughout Part III of this book, only the edge-emitting laser is considered.

The earliest and simplest edge-emitting laser structure is the *broad-area semiconductor laser*, which consists of a thin active layer sandwiched between p- and n-type cladding layers (thus forming a double heterojunction). Although the active layer is sufficiently thin to support a single transverse mode, there is no light confinement mechanism in the lateral direction, resulting in highly elliptical beam and unacceptably high threshold currents

($\sim$1A) [6, 7]. Therefore, improved laser structures have been proposed to achieve lateral mode confinement, which are explained as follows.

13.5.1 *Lateral mode confinement*

Lateral mode confinement is achieved by using either *gain-guided* or *index-guided* lasers. In the gain-guided laser, the injected current is restricted to flow only within a central narrow region in the active region, and lateral optical confinement is provided by the optical gain itself. An example is the junction stripe-geometry laser [6], as shown in Fig. 13.9(a). Diffusion of Zn over the central region of the top n-type layer converts it to a p-type providing the current path into the active region. The current is blocked in the remaining regions on the top n-type layer because reverse-biased $p.n.$ junctions are formed. Gain-guided lasers typically have threshold currents of 10–100 mA [6, 14].

Index-guided lasers are further subcategorised as either *weakly index-guided* or *strongly index-guided.* An example of a weakly index-guided laser is the ridge waveguide laser [15]. A ridge is formed by etching away parts of the p-InP cladding layer, and a SiO_2 or Si_3N_4 layer is then deposited as insulating layers, which provides current confinement to the central region of the active layer. Since the ridge has a larger refractive index than that of the insulating layer, a lateral index step of $\sim$0.01–0.02 can be achieved, providing optical waveguiding. The ridge waveguide laser structure and its parasitics are shown in Fig. 15.1 of Chapter 15. Ridge waveguide lasers have typical threshold currents of 40–60 mA [6].

In the case of strongly index-guided lasers, the active region of dimensions $\approx 0.1 \times 1$ μm is buried on all sides by cladding regions of lower refractive indices. A typical example is the etched mesa buried heterostructure (EMBH) laser [15], which is shown in Fig. 13.9(b). The lateral index step of these laser structures is $\sim$0.2–0.3, which is about two orders of magnitude higher than that provided by carrier-induced effects in gain-guided lasers. EMBH lasers have internal threshold currents of 10-15 mA but leakage currents are significant [6]. Important issues in these laser structures are control of leakage current and difficulty of the fabrication process [6, 15]. Nevertheless, the emitted light from strongly index-guided lasers supports a single spatial mode and is inherently stable. Index guided structures are invariably used as long-wavelength semiconductor lasers in most lightwave systems because of their superior performance over gain-guided lasers.

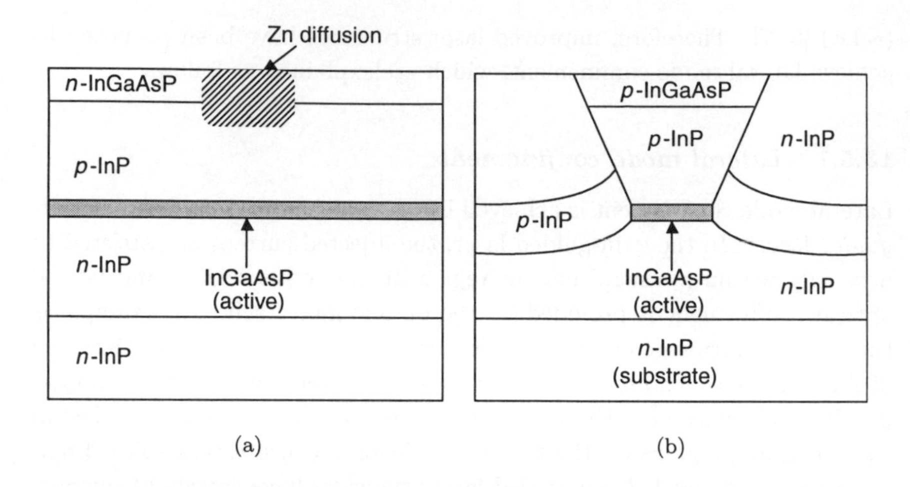

Fig. 13.9. (a) Gain-guided laser: The junction stripe geometry laser (b) Index-guided laser: The etched mesa buried heterostructure (EMBH) laser.

The simple Fabry Perot (FP) that is formed by cleaved facets at both edges of the laser chip exhibits multimode characteristics and this leads to undesirable effects, such as mode-hopping and pulse broadening, which ultimately limit the bit rate-distance (BL) product due to fibre dispersion [16]. In order to minimise power penalty due to dispersion and maximise the BL product, stable single frequency diode lasers are required. Several laser designs have been proposed for longitudinal mode control, as explained in the following.

13.5.2 *Longitudinal mode control*

This can be achieved by providing feedback only at a certain frequency instead of multiple frequencies at the resonant modes of the FP cavity. One successful method is to use corrugated gratings on the semiconductor substrate to provide feedback at a chosen wavelength, called the *Bragg wavelength* (λ_B), which may be expressed as [6, 14]:

$$\lambda_B = \frac{2\bar{n}\Lambda}{m} \tag{3.18}$$

where $\bar{n}$ is the effective mode index, Λ is the grating period, and m is the integer that represents the order of Bragg diffraction. In these corrugated grating structures, the feedback is not localised at the end facets but

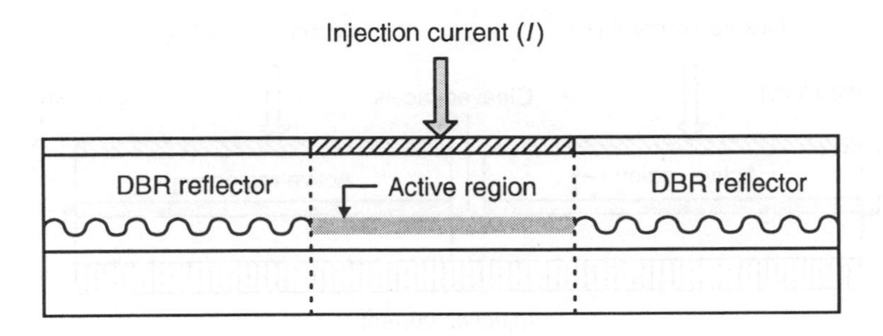

Fig. 13.10. Distributed Bragg reflector (DBR) laser.

distributed all along the cavity length. Hence, it is called the *distributed feedback* (DFB) laser [6, 17], where the corrugated grating is etched along the entire length of the laser that is pumped by electrical current. Two examples of the DFB laser (i.e. uniform DFB and quarter-wave-shifted DFB lasers) are given in Chapter 17. Another variant is the *distributed Bragg reflector* (DBR) laser [6, 17], where the grating is formed outside the active region, as shown in Fig. 13.10. The unpumped DBR regions then act as effective mirrors, which are wavelength dependent.

Single longitudinal mode (SLM) emission can also be achieved by using *coupled-cavity* lasers [6]. In a coupled-cavity laser design, such as the *external* cavity (EC) laser, light is emitted into an external cavity but a portion of reflected light from a mirror-like element (e.g. diffraction grating) at the end of the external cavity is fed back into the laser. Since in-phase feedback occurs only for those laser modes whose wavelength nearly coincides with resonant modes of the external cavity, the laser facet facing the external cavity sees an effective reflectivity which is wavelength dependent. The laser mode that is closest to the gain peak, and which coincides with a trough in the loss profile (corresponding to in-phase feedback) becomes the dominant lasing mode.

Besides single wavelength selection and linewidth reduction, EC lasers can also be used for active mode-locking, which will be discussed later in this chapter. In another variant of coupled-cavity lasers, a conventional FP laser is cleaved in the middle, so that two active sections are separated by a narrow air gap (~ 1 μm) forming the *cleaved coupled cavity* (C^3) laser (Fig. 13.11). Both sections are electrically separated and can be independently pumped to control the behaviour of the device for applications such as wavelength tuning and optical bistabillity [6].

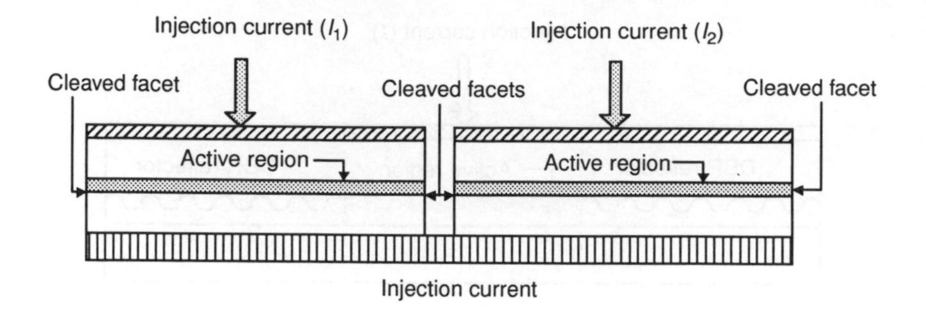

Fig. 13.11. Cleaved coupled cavity (C^3) laser.

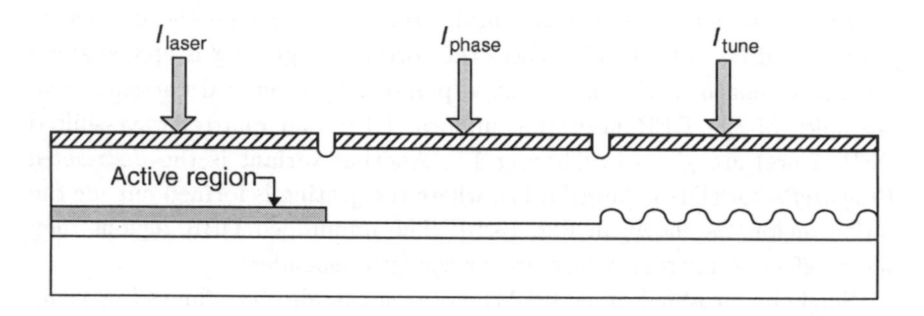

Fig. 13.12. Multielectrode DBR laser.

With the advent of wavelength division multiplexed (WDM) transmission system, wavelength tunability is becoming increasingly important [18]. In order to achieve both SLM stability and tuning, low chirp in intensity modulation (IM) systems or low amplitude modulation in frequency modulation (FM) systems, novel multielectrode DFB and DBR lasers can be used [19–21]. A multielectrode DBR laser is shown in Fig. 13.12 where the active section current (I_{laser}), phase-control current (I_{phane}), and Bragg section current (I_{tune}) can be controlled independently for continuous wavelength tuning.

13.6 Rate Equations

The laser rate equations describe the exchange of energy between electrons and photons in a resonant cavity through the three fundamental optical

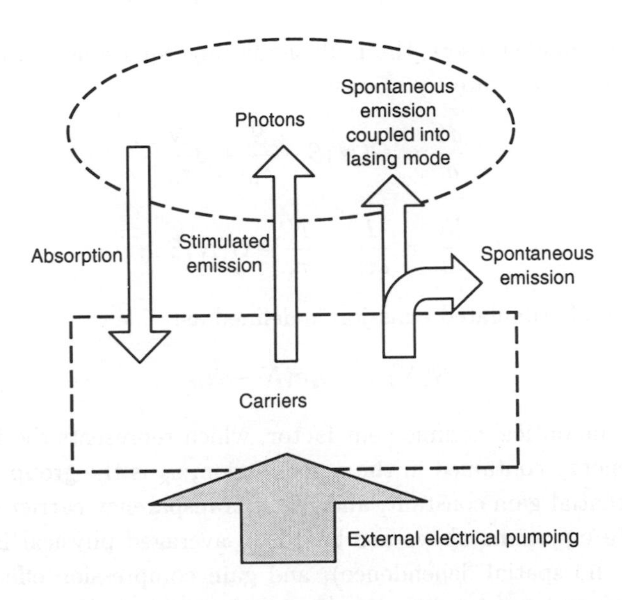

Fig. 13.13. Simple illustration of the laser rate equations.

processes: stimulated emission, spontaneous emission and absorption. The semiconductor material is electrically pumped into a forward-biased heterojunction diode, and charge carriers that are injected into the thin active layer provide the optical gain. Much of the understanding of high-speed modulation of semiconductor lasers comes from either a small signal or large-signal analysis of the laser rate equations, which is basically a bookkeeping of the rate of supply, annihilation, and creation of carriers and photons inside the laser cavity (see Fig. 13.13). During the switch-up of the laser, the dynamic interplay between the photons and carriers gives rise to transient ringing effects known as *relaxation oscillations* [22–23]. The general *multimode* rate equations are given as [24]:

$$\begin{aligned} \frac{dS_m}{dt} &= G_m(N)S_m - \frac{S_m}{\tau_p} + \beta_m \frac{N}{\tau_n} \\ \frac{dN}{dt} &= \frac{I}{qv_a} - \frac{N}{\tau_n} - \sum_m G_m(N)S_m \end{aligned} \tag{3.19}$$

where S_m is the photon density of the mth longitudinal mode, G_m is the stimulated emission rate of the corresponding mode, β_m is the fraction of spontaneous emission coupled into the mth mode, N is the carrier density, I is the injection current, and v_a is the active region volume. For predominantly single-moded lasers, the multimode rate equations can be reduced into the *single-mode* rate equations:

$$\frac{dS}{dt} = G(N)S - \frac{S}{\tau_p} + \beta\frac{N}{\tau_n} \tag{3.20}$$

$$\frac{dN}{dt} = \frac{I}{qv_a} - \frac{N}{\tau_n} - G(N)S\,. \tag{3.21}$$

The net rate of stimulated emission is defined as:

$$G(N) = \Gamma v_g a(N - N_{tr}) \tag{3.22}$$

where Γ is the optical confinement factor, which represents the fraction of the mode energy contained in the active region, v_g is the group velocity, a is the differential gain constant, and N_{tr} is transparency carrier density.

The rate equations above are based on averaged physical laser parameters (i.e. no spatial dependence), and gain compression effects due to hole-burning and carrier diffusion [5, 14], have not been included. The photon rate equation, i.e. Eq. (13.20), may be derived from Maxwell's equations, while the carrier rate equation, i.e. Eq. (13.21), may be derived from a quantum-mechanical approach for the induced polarisation [6]. The advantage of these rigorous approaches is that various approximations that were made can clearly be identified. Alternatively, the rate equations can be obtained heuristically by considering the fundamental optical processes (stimulated emission, spontaneous emission, and absorption) and the current continuity equation, which describe how the photons and carriers change with time.

In the photon rate equation, i.e. Eq. (13.20), the first term on the right hand side (RHS) accounts for the growth of photons due to stimulated emission, the second term represents annihilation of photons due to absorption and mirror loss, while the third term is the fraction (denoted by the factor β) of the total photons created by spontaneous emission that couples into the lasing mode. Although stimulated emission supplies energy only to the modes of the resonant cavity, spontaneous emission is omnidirectional, therefore only a small fraction of all spontaneously emitted photons feeds energy into the resonant modes [25]. In the carrier rate equation, i.e. Eq. (13.21), carriers are supplied by the injection current, which is represented by the first term on the RHS, but are depleted through carrier recombination during spontaneous and stimulated emissions, which are represented by the second and third terms on the RHS, respectively.

13.7 Laser Linewidth and Chirping

The linewidth of an individual FP longitudinal mode is not negligible, as in the ideal case shown in Fig. 13.8, but has values between 10 MHz and 100 MHz (see Fig. 13.14(a) for example) when operating at a power level below 10 mW, and saturates to a range between 1 and 10 MHz at powers above 10 mW [16, 26]. The first careful measurement of linewidth of an AlGaAs semiconductor laser was reported by Fleming and Mooradian [27], where the lineshape was found to be Lorentzian. The full width at half maximum (FWHM) of the linewidth, Δf, was also confirmed to be inversely proportional to output power, P_0.

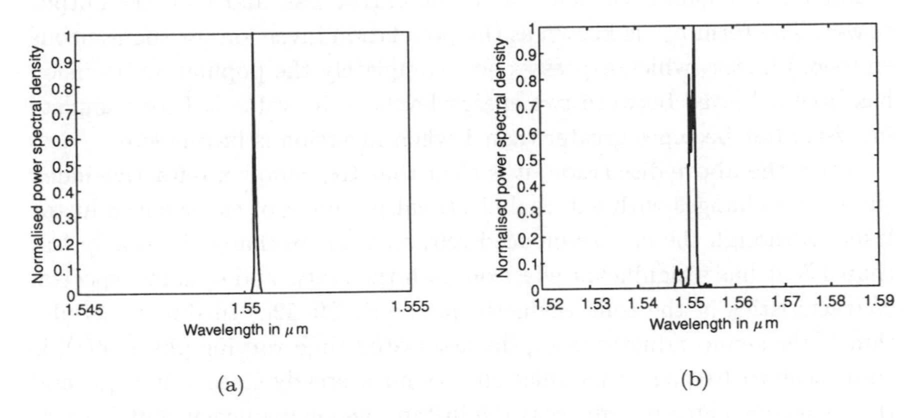

Fig. 13.14. (a) Static linewidth due to intensity fluctuations (b) dynamic linewidth due to transient chirping.

However, the linewidth of semiconductor lasers was found to be a factor of 5 to 40 times greater than that predicted by using the modified Schawlow-Townes formula [8]. This linewidth enhancement phenomenon was explained by Henry [28], which he attributed to the field amplitude-phase coupling effect. Briefly, this is explained as the following. Random spontaneous emission induces both phase and intensity changes in the laser field. To restore the steady-state intensity, the laser intensity undergoes relaxation oscillations, which changes the imaginary part of the refractive index, $\Delta n''$ (gain). This will also alter the real part of the refractive index, $\Delta n'$ (phase), through the relationship given by [6]:

$$\alpha_{\mathrm{H}} = \frac{\Delta n'}{\Delta n''} = -2k_0 \left(\frac{\partial n/\partial N}{\partial g/\partial N} \right) \tag{3.23}$$

which is known as Henry's linewidth enhancement factor, α_{H}, usually taken as a constant. From Eq. (13.23), it is clear that gain modulation $(\partial g/\partial N)$ will also lead to phase modulation $(\partial n/\partial N)$ which exhibits itself in the spectral broadening of the laser linewidth, known as frequency *chirping.* Consequently, the modified Schawlow-Townes formula may be rewritten as [28]:

$$\Delta f = \frac{R}{4\pi I}(1+\alpha_{\mathrm{H}}) = \frac{v_{\mathrm{g}}^2 hfgn_{\mathrm{sp}}\alpha_{\mathrm{m}}(1+\alpha_{\mathrm{H}})}{8\pi P_0} \tag{3.24}$$

where R is the spontaneous emission rate, I is the number of photon population, g is the gain coefficient, α_{m} is the mirror loss, and P_0 is the output power. The term n_{sp} is known as the population inversion (or spontaneous emission) factor, which expresses how completely the population inversion has been achieved between two energy levels — its value is 1 for complete inversion but becomes greater than 1 when inversion is incomplete.

From the above discussion, it is clear that the complex refractive index $(n' + jn'')$ changes with external electrical pumping of the semiconductor laser. Although the carrier-induced refractive index change is usually less than 1%, it has a significant effect on both the static and dynamic spectral characteristics of the semiconductor laser [26, 29–32]. In direct modulation of the semiconductor laser, the associated time-varying phase, $\phi(t)$, is equivalent to frequency modulation around a steady-state value, f_0, and the frequency chirp is defined as the instantaneous frequency shift from f_0 in the form of [6, 8]:

$$\delta f(t) = \frac{1}{2\pi}\frac{d\phi}{dt} = \frac{\alpha_{\mathrm{H}}}{4\pi}\left[\left(\frac{d}{dt}\ln P(t)\right) + xP(t)\right] \tag{3.25}$$

where $P(t)$ is the time variation of the optical power, and x is a constant. The first term inside the square brackets in Eq. (13.25) is referred to as the *transient chirp* due to relaxation oscillations (see for example Fig. 13.14(b)), and the second term is referred to as the *adiabatic chirp* at steady-state.

The adiabatic chirp is useful in coherent optical systems that use frequency shift keying (FSK) modulation [16]. The desired frequency shift for FSK modulation is produced by changing the bias current between two distinct levels above threshold. In intensity modulation direct detection (IM/DD) systems [16], spectral broadening owing to frequency chirp affects the pulse shape at the fibre output because of group velocity dispersion, which consequently degrades the system performance.

13.8 Laser Noise

The output of semiconductor lasers exhibits fluctuations in its intensity, phase, and frequency even when the laser is biased at a constant current with negligible current fluctuations. The two fundamental noise mechanisms are spontaneous emission and electron-hole recombination (shot noise) [16, 33] of which spontaneous emission is the dominant mechanism. The presence of noise perturbs both amplitude and phase of the coherent field (established by stimulated emission) by adding to it a small field component whose phase is random. Noise characteristics of laser diode modulation will be discussed in the following.

13.8.1 *Relative Intensity Noise (RIN)*

Intensity fluctuations in the modulated laser diode lead to a degradation in the signal-to-noise ratio (SNR). In general, intensity noise peaks in the vicinity of laser threshold, and decreases rapidly with higher bias levels. This amplitude fluctuation is characterised by the relative intensity noise (RIN), which is defined as the Fourier transform of the intensity-autocorrelation function [16]:

$$RIN(\omega) = \int_{-\infty}^{\infty} C_{pp}(\tau) \exp(-i\omega\tau) d\tau \tag{3.26}$$

where ω is the angular frequency, and τ is the time delay. The term C_{pp} is the intensity-autocorrelation function, which is defined as:

$$C_{pp}(\tau) = \frac{\langle \delta P(t+\tau)\delta P(t)\rangle}{\bar{P}^2} \tag{3.27}$$

where $\bar{P}$ is the average power, and $\delta P(t) = P(t) - \bar{P}$ represents a fluctuation. The RIN spectrum peaks near the relaxation oscillation frequency, ω_{R}, because of the laser's intrinsic resonance [6], and drops sharply when $\omega \gg \omega_{\mathrm{R}}$ because the laser is not able to respond to such high frequencies. The RIN spectrum characteristics show that the semiconductor laser acts like a bandpass filter to spontaneous emission noise, as shown by the solid curve in Fig. 13.15. RIN is directly proportional to the P^{-3} at low power levels but changes to a P^{-1} dependence at higher power levels.

13.8.2 *Mode partition noise*

The small net gain difference between neighbouring longitudinal modes allows the position of the dominant lasing mode to shift randomly in time

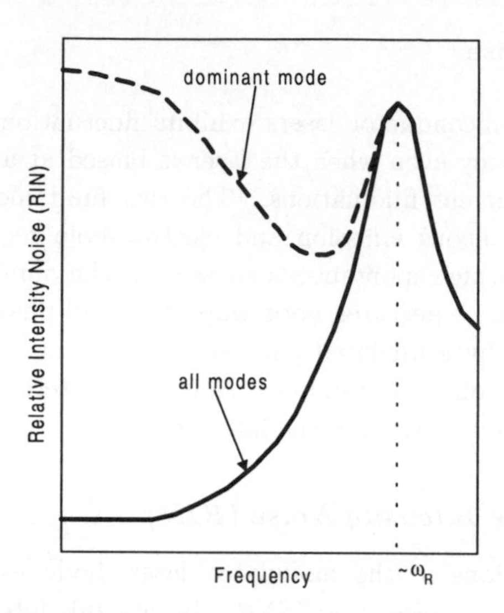

Fig. 13.15. Relative intensity noise (RIN) spectrum (log scale in both axes).

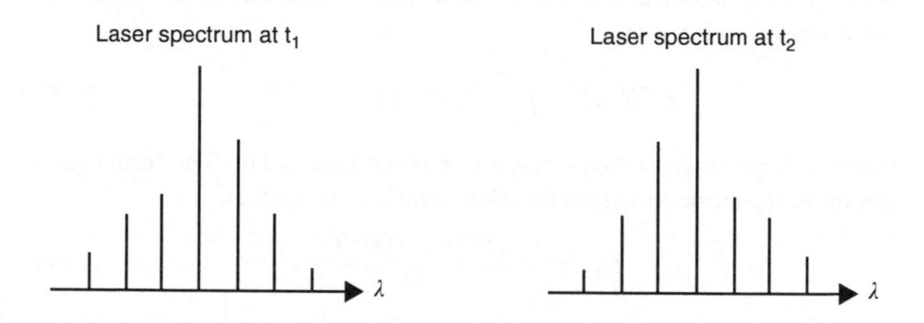

Fig. 13.16. Optical spectrum at different time t_1, t_2 illustrating mode partition noise.

due to spontaneous emission. The main and side modes fluctuate in such a way as to leave the total intensity relatively constant, as illustrated in Fig. 13.16. Therefore, the intensity noise of the superposition of all modes (solid curve in Fig. 13.15) is usually much smaller than the noise for a single mode alone (dashed curve in Fig. 13.15), a phenomenon know as mode partition noise. If only one dominant lasing mode is detected, a very high RIN would be measured, which can be several orders of magnitude higher than the RIN of the total light emission [13]. Mode partition noise causes the multimoded spectrum to vary randomly from pulse to pulse, or even within the duration of a single pulse. Due to its interaction with fibre dispersion [34], the successive pulses travel at different speeds, leading to intersymbol interference and SNR degradation at the receiver. Fabry–Perot (FP) lasers with only a few dominant modes above threshold have been shown to exhibit a higher level of mode partition noise than truly single or truly multimoded lasers [13].

13.8.3 *Phase Noise*

Phase fluctuations lead to a finite spectral linewidth, which is an important issue in coherent optical communication systems [35]. The broadening of the semiconductor laser linewidth due to the amplitude-phase coupling effect has already been explained in Section 13.7. The modified Schawlow-Townes formula, together with Henry's linewidth enhancement factor [28], describes the semiconductor laser linewidth (FWHM) at low power levels above threshold. In addition, the spectrum can be found from the Fourier transform of the field-autocorrelation function, that is [16]:

$$S(\omega) = \int_{-\infty}^{\infty} \Gamma_{\mathrm{EE}}(\tau) \exp(-i\omega\tau) d\tau \tag{3.28}$$

where the field-autocorrelation function is:

$$\Gamma_{\mathrm{EE}}(\tau) = \langle E^*(t+\tau)E(t)\rangle \tag{3.29}$$

and the optical field is represented in phasor form as:

$$E(t) = [\bar{P} + \delta P(t)]^{\frac{1}{2}} \exp[-i\{\omega_0 + \bar{\phi} + \delta\phi(t)\}] \tag{3.30}$$

where ω_0 is the central frequency of the single longitudinal mode under consideration, $\bar{P}$ and $\bar{\phi}$ are the average power and phase, respectively. Fluctuations are denoted by $\delta P(t)$ and $\delta\phi(t)$. The spectrum consists of a dominant central peak located at ω_0, and satellite peaks located at multiples of the relaxation oscillation, ω_{R} [6]. A narrow linewidth laser is a strict requirement in coherent optical communication systems [35]. The linewidth can be reduced by adding an external passive cavity with a reflector (e.g. diffraction grating) to the laser chip [26]. It has been shown that the laser linewidth narrows approximately as the square of the external reflector separation, L_{p} [26]. From the modified Schawlow-Townes formula of Eq. (13.24), a long passive section will decrease the average spontaneous emission rate ($R \propto L_{\mathrm{p}}^{-1}$), and increase the average photon population ($I \propto L_\mathrm{p}$), resulting in an overall linewidth reduction proportional to L_p^{-2}.

13.9 Modulation Behaviour

The simplest technique for impressing a signal on the output of a semiconductor laser is by direct modulation. This is achieved by superimposing a modulation signal on the bias current. The total drive current is then of the form:

$$I(t) = I_\mathrm{b} + I_m(t) \tag{3.31}$$

where I_b is the DC bias current, and $I_m(t)$ is the time-varying modulation current. In this way, the time-varying light output waveform follows that of the modulation current waveform, provided that the laser diode is operated in the linear region. There are two regimes of laser diode modulation, i.e. (i) small-signal and (ii) large-signal, which are discussed briefly in the following.

13.9.1 *Small-signal modulation*

In small-signal modulation, the bias current, I_b, is well above threshold level, I_th, and the amplitude of the modulation current, $I_m(t)$, is small enough to ensure that the laser is never below threshold, $I_{m(t)} \ll [I_\mathrm{b} - I_\mathrm{th}]$. If the small-signal analysis assumes that the modulation current waveform is sinusoidal, i.e. $I_m(t) = I_\mathrm{p}\sin(2\pi f_m t)$, then we can linearise the rate equations to find analytical expressions for the laser diode modulation response [6]. The terms I_p and f_m denote the peak current amplitude and modulation frequency, respectively. The small-signal analysis is only valid if the modulation depth, m, is small, that is:

$$m = \frac{[I_m(t)]_\mathrm{max}}{I_\mathrm{b} - I_\mathrm{th}} \ll 1\,. \tag{3.32}$$

The 3-dB modulation bandwidth can be found analytically from [16, 36]:

$$f_\mathrm{3dB} = \frac{\sqrt{3}f_\mathrm{r}}{2\pi} = \frac{\sqrt{3}}{2\pi}\sqrt{\frac{\Gamma v_\mathrm{g} a S}{\tau_\mathrm{p}}} \tag{3.33}$$

where f_r is the relaxation oscillation frequency of the laser, Γ is the optical confinement factor, v_g is the group velocity, a is the differential gain coefficient, S is the average photon density in the laser cavity, and τ_p is the photon lifetime. Equation (13.33) implies that there are three important requirements for providing a wide bandwidth, which are (i) large photon density (or equivalently output power) (ii) high differential gain, and (iii) short photon lifetime. To achieve high power levels, we can simply increase the bias current level. From Eq. (13.14), we also know that for a given output power, the average photon density in the active region can be increased by using a smaller active region width, such as in *constricted mesa buried heterostructure* (CMBH) lasers [15]. High differential gain can be achieved by cooling the laser [37] or by using quantum well lasers [38]. Alternatively, reducing the length of the laser cavity reduces the photon lifetime, see Eqs. (13.8) and (13.12).

Besides a large resonance frequency, a wide-band laser should also have low parasitics (to be discussed in Chapter 15), and strong damping to provide a flat response (see Fig. 13.17). This flat response is to avoid nonlinear distortion during modulation of the laser diode [13, 39]. When the laser is operated near threshold, the damping mechanism is dominated by spontaneous emission. However, at high bias levels, damping is dominated by the gain compression effect due to spatial or spectral hole-burning [13, 15]:

A characteristic feature of semiconductor lasers is that intensity modulation (IM) leads simultaneously to phase or frequency modulation (FM). The interdependence of IM and FM is due to the field amplitude-phase

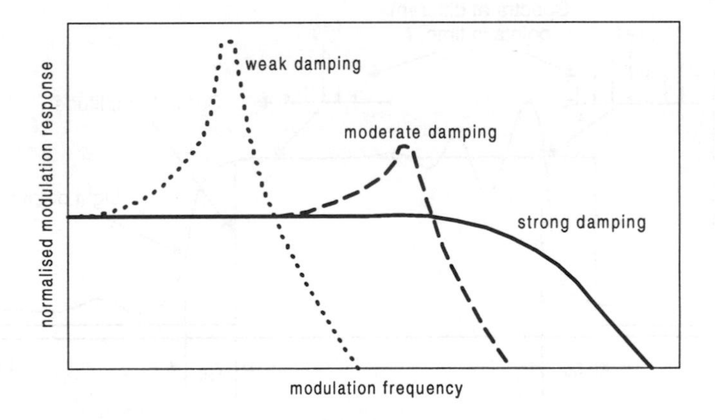

Fig. 13.17. Small signal modulation response with different levels of damping (both axes are in log scale.

coupling effect that is governed by the linewidth enhancement factor, α_H (Section 13.7). For sinusoidal modulation, the maximum frequency chirp can be approximated as the following [6]:

$$\delta f_{\max} = \frac{1}{2}\alpha_H f_m m \,. \tag{3.34}$$

Where f_m is the modulation frequency, and m is the modulation depth.

13.9.2 *Large-signal modulation*

In pulse coded modulation (PCM) for digital communication systems, the laser is typically biased below or close to threshold, and is modulated considerably above and below threshold [16]. In this case, the modulation depth does not satisfy Eq. (13.32), and nonlinear effects play an important role in determining the large-signal modulation characteristics. In the simple two-level PCM, the laser is modulated by a rectangular current waveform such that an optical pulse is produced to give the "on" state, and the absence of a pulse denotes an "off" state. The speed of transmission is therefore limited by the speed at which the laser can be switched between "on" and "off" states.

In practice, the optical pulse is not an exact replica of the current waveform because of ringing transient effects (i.e. relaxation oscillations) at the leading and trailing edges of the pulse (see Fig. 13.18). The associated

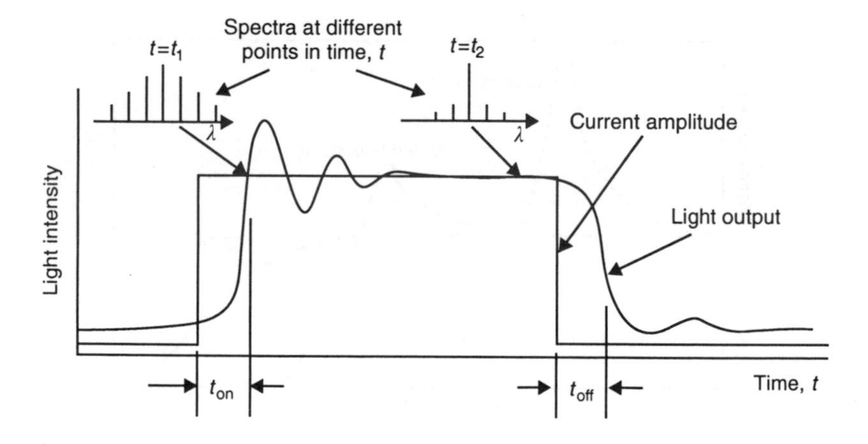

Fig. 13.18. The light output pulse during large-signal modulation (rectangular current pulse).

"turn-on" (t_{on}) and "turn-off" (t_{off}) delay times also limit the transmission speed [13, 15]. Furthermore, a long bit sequence of "on" states in non-return-to-zero (NRZ) modulation format results in different spectral width at different points in time (see Fig. 13.18). This leads to different parts of the train of "on" bits being transmitted at different speeds due to fibre dispersion [34].

During PCM, the mode frequency shifts toward the blue side near the leading edge, and toward the red side near the trailing edge of the optical pulse. This transient chirping leads to additional spectral broadening, which may contribute to the pulse broadening effect. In contrast to small-signal modulation, the maximum transient chirp, $\delta f_{\max}$, in large-signal modulation is independent of modulation depth, m. For a Gaussian pulse shape proportional to $\exp(-t^2/T^2)$, the chirp can be approximated by the following [6]:

$$\delta f_{\max} \approx \frac{1}{2\pi T}\alpha_{\text{H}} \tag{3.35}$$

where the parameter, T, is related to the FWHM of the pulse (t_{p}) by $t_{\text{p}} = 1.665T$. If standard fibres are to be used in 1.55 μm lightwave systems for transmission between 80–100 km, dynamic frequency chirping limits the bit rates to values below 2 Gb/s [32, 40]. Chirp reduction can be achieved by coupling together optical cavities such as lateral coupled waveguide (LCW) lasers, cleaved coupled cavity (CCC) lasers, and lasers coupled to external Bragg reflectors [13]. Frequency chirp can also be reduced by using quantum well devices, which exhibit lower values of linewidth enhancement factor, α_{H} [41].

13.9.3 *Nonlinear distortion*

In multichannel fibre-optic transmission of analogue microwave signals, good system linearity is required to avoid channel interference due to nonlinear distortion [13, 39 and 42]. For example, if two modulation frequencies, ω_1 and ω_2, are transmitted within a channel, the second order distortions are $2\omega_1$, $2\omega_2$, $\omega_2 \pm \omega_1$ while the third order distortions are $3\omega_1$, $3\omega_2$, $2\omega_1 \pm \omega_2$, and $2\omega_2 \pm \omega_1$ as shown in Fig. 13.19. If the transmission channel is carefully chosen, harmonic distortions $(2\omega_1, 2\omega_2, 3\omega_1, 3\omega_2, \ldots)$ are of little concern since they generally do not fall within the channel. However, the third order intermodulation (TOI) products at $2\omega_1 - \omega_2$ and

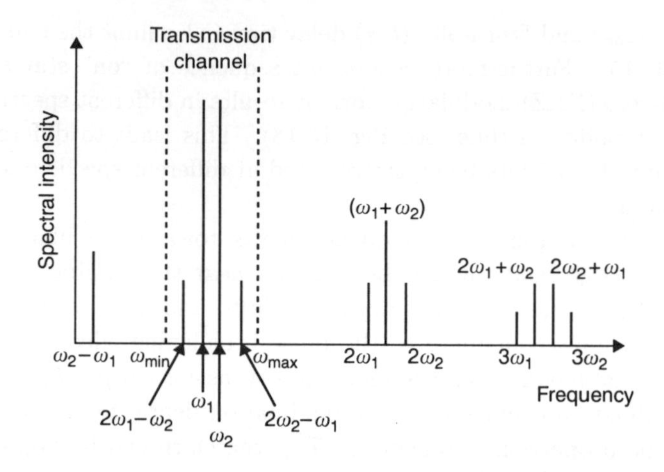

Fig. 13.19. Harmonic and intermodulation distortion for two-tone modulation.

$2\omega_2 - \omega_1$ usually lie within the channel and must be carefully considered (see Fig. 13.19).

At low modulation frequencies (few tens of MHz and below), harmonic distortions are introduced due to the imperfect linearity of the light-current (L-I) characteristics. Generally, the second harmonics are proportional to m (i.e. modulation depth), and third harmonics proportional to m^2 [13]. Gain-guided lasers typically exhibit a slight curvature in their L-I characteristics above threshold and therefore have relatively higher second order harmonic distortion than index-guided lasers.

At high modulation frequencies, harmonic and intermodulation distortions mainly depend on the relaxation oscillation phenomenon. For sufficiently low modulation frequencies below the relaxation oscillation frequency (f_r), the intermodulation products increase at a rate of 40dB/decade, then reache a plateau at $f_r/2$ before resuming the 40dB/decade increment as the frequency exceeds f_r [39]. The nonlinear distortion level also depends on the damping of relaxation oscillations, where strong damping is desirable to achieve low nonlinear distortions (see Fig. 13.17).

13.10 Short Pulse Generation Schemes

To realise multi Gb/s PCM systems, short optical pulses are required to encode the digital data stream [16]. In order to increase the transmission capacity, optical time division multiplexed (OTDM) systems can be used where picosecond optical pulses from several lower bit-rate channels are interleaved to achieve higher aggregate bit-rates [43]. Besides its role in digital modulation, short optical pulses can produce usable frequency harmonics up to the millimetre-wave (mm-wave) region for microwave applications such as phased array radars and radio-over-fibre systems [44].

Picosecond laser sources are also required in *soliton* transmission systems [34]. Soliton propagation relies on the interplay between group velocity dispersion (GVD) and self phase modulation (SPM) in the anomalous dispersion regime so that the optical pulse can propagate undistorted in long-haul transmission systems. In addition, picosecond optical pulses are important instrumentation tools for studying ultrafast physical and chemical phenomena. Compact and reliable semiconductor picosecond laser sources are also highly desirable for instrumentation applications such as noninvasive characterisation and testing of high-speed electronic circuits, from semiconductor transistors to photoconductive switches [45].

According to Fourier analysis, a very short temporal pulse requires a very wide frequency bandwidth to describe it. The ideal case is the impulse, which requires an infinite bandwidth to describe it. In general, the minimum pulsewidth, Δt, is inversely proportional to its spectral bandwidth, Δf, that is [46]:

$$\Delta t = \frac{k}{\Delta f} \tag{3.36}$$

where k is a constant determined by the pulse shape. The time-bandwidth product of an arbitrary pulse shape depends on the exact shape of the pulse (Gaussian, square, exponential, sech, etc.) and the amount of chirp in the pulse [5]. Pulses with time-bandwidth product close to 0.5 are often referred to as *transform-limited.* For example, the transform-limited Gaussian pulse has a time-bandwidth product of 0.44. The presence of chirp tends to increase the time-bandwidth product, which is an indication of the amount of pulse broadening to expect when the pulse propagates in the optical fibre in the presence of group velocity dispersion.

There are three important techniques for short pulse generation using semiconductor lasers: (i) Q-switching (ii) gain-switching, and (iii) mode-locking. The shortest pulses so far have been generated by the mode-locking techniques [46]. *Active mode-locking* gives relatively lower timing jitters compared to gain-switching and other mode-locking techniques. Ultrashort pulse generation techniques are briefly discussed in the following, with emphasis given to the mode-locking techniques.

(*A*) *Q-switching*: Briefly, the cavity loss is raised (i.e. low cavity-Q) by an intracavity element to inhibit lasing so that the laser can be pumped to a very high inversion level. As the inversion reaches its peak value, the high cavity-Q condition is abruptly resumed so that laser suddenly finds itself with a huge amount of excess gain, and responds by emitting a short intense optical pulse to dissipate the excess gain [5]. The laser diode can be divided into separate sections and independently biased so that the absorber section may be incorporated into the same semiconductor material [46].

(*B*) *Gain-switching*: The basic idea is to modulate the laser in such a way as to excite the first relaxation oscillation overshoot by raising the current level above threshold, and suppress the onset of any subsequent overshoot. This short pulse generation technique is attractive for its simplicity. Either a comb generator [47] or an RF sine-wave generator [48] can be used as the electrical current source for gain-switching the laser diode.

(*C*) *Mode-locking*: In general, mode-locking techniques can be classified into (i) active mode-locking, (ii) passive mode locking, and (iii) hybrid mode-locking. Hybrid mode-locking is simply a combination of techniques (i) and (ii) [49]. Therefore, only techniques (i) and (ii) will be explained in the following.

(*i*) *Active mode-locking*: In the time-domain description of active mode-locking, the optical pulse circulates in the laser cavity (which usually includes a longer passive cavity) in synchronism with the RF drive signal that provides the gain modulation. The pulse arrives in the laser cavity at the instant the gain modulation reaches its maximum value, thus sharpening the pulse at every round trip. The amplitudes of the RF modulation current and bias current are chosen to create a very short time-window of net gain, as depicted in Fig. 13.20. The best pulse stability occurs when the RF drive frequency is exactly equal to the inverse of round-trip time.

Usually, the short cavity length of diode lasers (100–300 μm) results in large roundtrip frequencies (125–375 GHz), making it very difficult to modulate the gain at such high frequencies. The solution is to place an external passive cavity (see Fig. 13.21) to increase the cavity round-trip time, hence decreasing the roundtrip frequencies to around 0.5–20 GHz. As the conventional bulk optics set-up is prone to excessive loss and poor stability, a monolithic design is usually preferred. In a monolithic device, the external cavity can be replaced by an extended waveguide section in the same semiconductor material [50–51]. A passive cavity is normally used instead

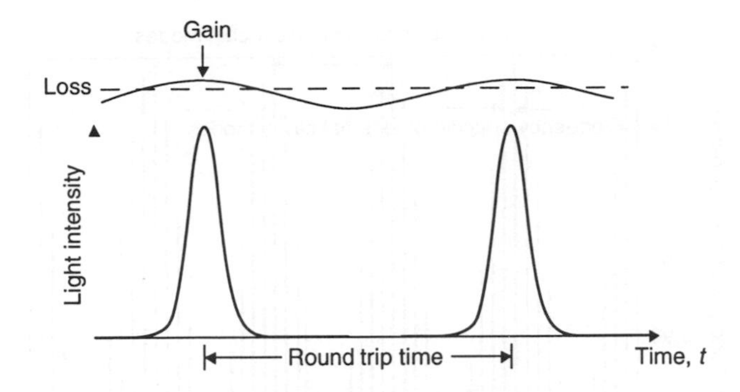

Fig. 13.20. Time domain picture of active mode-locking.

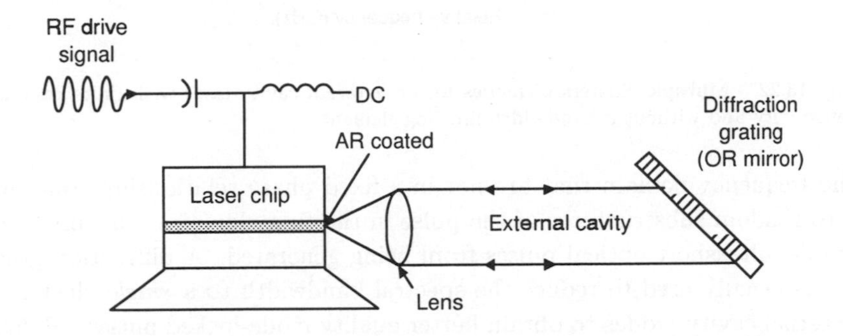

Fig. 13.21. Active mode-locking using an external cavity (EC) laser.

of an active cavity to avoid the dynamic gain saturation effects that will lead to pulse broadening and instability. High quality mode-locked pulses require (i) a wide gain bandwidth (ii) efficient and stable coupling between the passive and active sections, and (iii) a low-loss passive waveguide section.

In the presence of finite reflectivity at the laser facet facing the external cavity, clusters of external cavity modes begin to appear, due to the formation of a second resonant cavity within the laser chip itself. Consequently, the clusters are separated from one another by the frequency spacing of the resonant Fabry–Perot (FP) modes of the laser chip, as shown in Fig. 13.22. The RF modulation results in phase-locking of external cavity modes *within* each cluster but modes from one cluster may not be in phase with those of adjacent clusters. The external cavity modes from different clusters in

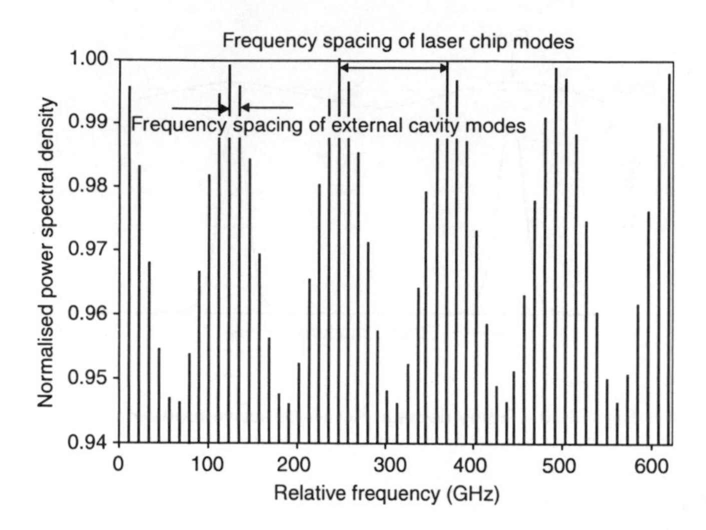

Fig. 13.22. Multiple clusters of modes for an external cavity laser with finite residual reflectivity and without a bandwidth-limiting element.

the frequency domain that are not in a fixed phase relationship translate into random substructures of the pulse in the time domain. This prevents stable, ultrashort optical pulses from being generated. A diffraction grating is usually used to reduce the spectral bandwidth to a *single* cluster of external cavity modes to obtain better quality mode-locked pulses [46, 52].

The general principle behind mode-locked operation is to induce all the longitudinal modes of the resonant cavity to oscillate in a fixed phase relation between them [5, 45–46]. Therefore, it is vital to have many longitudinal modes to achieve mode-locking. The greater the number of modes that are locked in phase, the narrower the resultant optical pulses will be. The time interval between successive pulses (inverse of repetition rate) is equal to the round trip time of the total cavity length.

In *active* mode-locking, the spectrum is forced to be in phase by active RF modulation of the laser chip using the correct round-trip frequency of the total cavity length, $f_{\text{cav}} = c/(2nL)$. Among all the pulse amplitude profiles due to the different phase relationships between the longitudinal modes, the one in which the phases add, so as to create a pulse that arrives at the laser chip when the gain is maximum (and loss is minimum) will be preferentially amplified (i.e. in-phase). One situation when this criterion is satisfied is when the phases between all of the longitudinal modes are zero. In the frequency-domain picture of active mode-locking, each spectral mode is "locked" (the origin of the term *mode-locking*) by the modulation side-bands of its adjacent spectral mode. In this case, the total field amplitude, E_{total} is given by [53]:

$$E_{\text{total}} = \sum_m E_m e^{j2\pi f_m t + \Phi_m} \tag{3.37}$$

where E_m is the field amplitude, f_m is the frequency, ϕ_m is the phase of the mth-mode, and t denotes time. The frequencies of the longitudinal modes are given as (see Eq. (13.16)):

$$f_m = f_0 + \frac{mc}{2nL} \, . \tag{3.38}$$

Where f_0 is the central frequency, c is the speed of light, n is the group refractive index, and L is the total cavity length. For simplicity, we neglect material dispersion, and the intermodal spacing is taken as $\Delta f = c/(2nL)$. By setting the phases $\phi_m = 0$, and amplitudes $E_m = E_0$, we have:

$$\begin{aligned} E &= E_0 \sum_m e^{j2\pi f_m t} = E_0 \sum_m e^{j2\pi (f_0 + \frac{mc}{2nL})t} \\ &= E_0 e^{j2\pi f_0 t} \sum_m e^{j2\pi (\frac{mc}{2nL})t} \, . \end{aligned} \tag{3.39}$$

By letting $\varphi = \frac{\pi ct}{nL}$, we can simplify the summation of Eq. (13.39) into the following form:

$$E = E_0 e^{j2\pi f_0 t}(1 + e^{i\varphi} + e^{i2\varphi} + \cdots) = E_0 e^{j2\pi f_0 t} \frac{\sin \frac{M\varphi}{2}}{\sin \frac{\varphi}{2}} \, . \tag{3.40}$$

The function of Eq. (13.40) has a period of $T = 2nL/c$, and its magnitude peaks at the value of ME_0. If we normalise the field amplitudes to unity, i.e. $E_m = E_0 = 1$, then the maximum field amplitude is 10 for 10 modes, and its intensity (i.e. $I \propto |E^2|$) will be 100 (see Fig. 13.23). There will be (M-1) minima points in between successive pulses. However, if the relative amplitudes of the modes are arranged in a Gaussian fashion, then the signal only has one minimum between pulses. From Fig. 13.23, it is clear that the higher number of modes produces shorter mode-locked pulses.

(ii) Passive mode-locking: In *passive* mode-locking, the selective amplification of the circulating pulse in the laser cavity does not require any external modulation current but depends solely on the saturable absorber. The saturable absorber that saturates easier than the gain medium serves

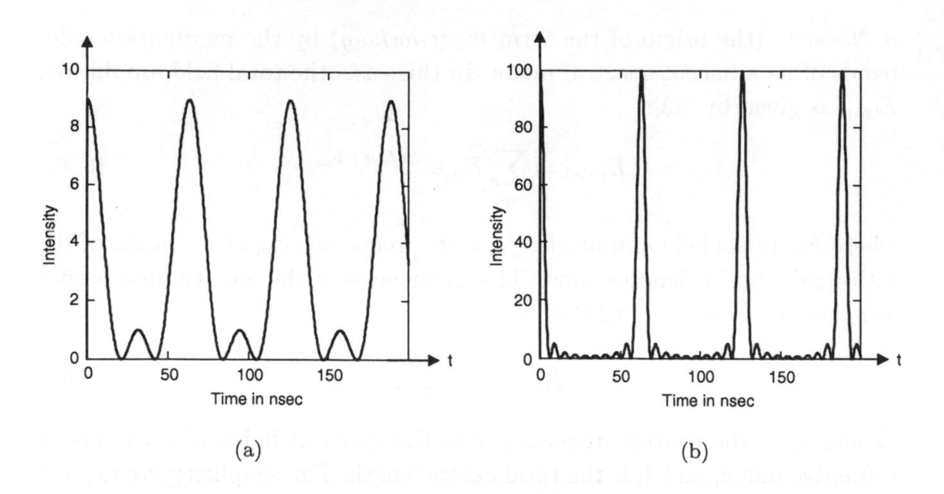

Fig. 13.23. Multiple clusters of modes for an external cavity laser with finite residual reflectivity and without a bandwidth-limiting element.

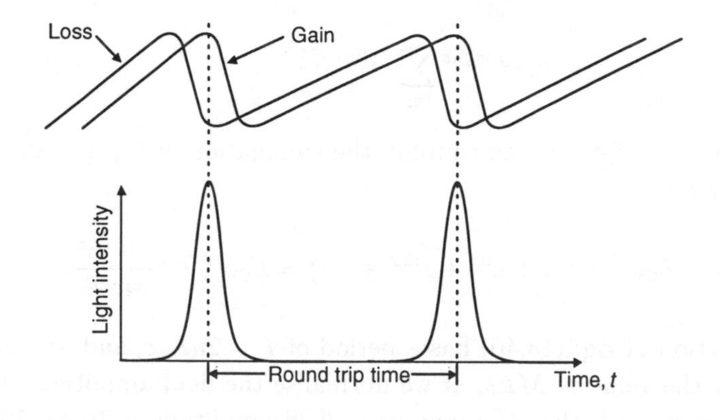

Fig. 13.24. The time-domain picture of passive mode-locking.

to sharpen up the leading edge of the pulse while the subsequent gain saturation sharpens the trailing edge. Furthermore, the loss is greater than the gain at every point in time except at the central part of the pulse which experiences a net gain as shown in Fig. 13.24.

On each round trip in the laser cavity, the combined action of gain and loss saturation shortens the optical pulse by amplifying the peak of the pulse while attenuating the leading and trailing edges. Passive mode-locking can be improved with colliding pulse effects in saturable absorbers such as self colliding pulse mode-locking (SCPM) and colliding pulse mode-locking (CPM) [46, 49, 54].

13.11 Summary

Since its beginning in the early 1960s, the semiconductor laser has played a key role in optical communication systems. Today, the laser diode provides many additional functions, such as wavelength conversion, optical switching, regeneration, demultiplexing, and optical logic besides its early function as an optical source for carrying modulated signals. In this chapter, the fundamental optical processes leading to the emission of light, and the semiconductor basics for understanding the confinement of externally injected carriers were explained first. Then, the threshold condition for lasing action, and the existence of longitudinal modes in the optical spectrum were discussed.

Typical laser structures for lateral carrier and optical confinement, as well as single longitudinal mode (SLM) control, were also presented. Next, the rate equations that describe the laser dynamics were explained intuitively. We explained how the static and dynamic frequency chirp of the laser was attributed to the carrier induced refractive index change, which is governed by Henry,s linewidth enhancement factor, $\alpha_{\rm H}$. Noise characteristics of the laser due to intensity and phase fluctuations are important factors to consider because they affect the bit error rate (BER) and signal-noise-ratio (SNR) of fibre-optic transmission systems. The modulation behaviour of the laser diode for small-signal and large-signal modulation was then explained qualitatively, and the contributing factors leading to signal degradation were also discussed. Finally, the underlying principles behind ultrashort pulse generation in semiconductor laser diodes were presented.

References

[1] T. H. Maiman, "Stimulated optical radiation in ruby masers," *Nature*, Vol. 187, p. 493, 1960.

[2] N. Holonyak Jr, "The Semiconductor Laser: A Thirty-Five-Year Perspective," *Proc. IEEE*, Vol. 85, pp. 1678–1693, 1997.

[3] I. Hayashi, M. B. Panish, P. W. Foy, and A. Sumski, "Function lasers which operate continuously at room temperature," *Appl. Phys. Lett.*, Vol. 17, p. 109, 1970.
[4] K. C. Kao and G. A. Hockham, "Dielectric-fiber surface waveguides for optical frequencies," *Proc. IEEE*, Vol. 113, p. 1151, 1966.
[5] A. E. Siegman, LASERS. Mill Valley, California: University Science Books, 1986.
[6] G. P. Agrawal and N. K. Dutta, *Semiconductor lasers*, 2nd ed. (New York: Van Nostrand Reinhold, 1993).
[7] J. Hecht, *Understanding fiber optics*, 3rd ed. (Upper Saddle River, NJ: Prentice Hall, 1999).
[8] D. J. G. Mestdagh, *Fundamentals of Multiaccess Optical Fiber Networks* (Boston: Artech House, 1995).
[9] M. J. O'Mahoney, "Optical multiplexing in fiber networks: progress in WDM and OTDM," *IEEE Comm. Mag.*, Dec., pp. 82–88, 1995.
[10] R. Ramaswami and K. N. Sivarajan, *Optical Networks: A Practical Perspective* (San Francisco, CA: Morgan Kaufmann Publishers, Inc., 1998).
[11] H. Kressel and J. K. Butler, *Semiconductor Lasers and Heterojunction LEDs* (New York: Academic Press, 1977).
[12] H. C. J. Casey and M. B. Panish, *Heterostructure Lasers: Part A, Fundamental Principles* (New York: Academic Press, 1978).
[13] K. Petermann, *Laser Diode Modulation and Noise* (Dordrecht: Kluwer Academic Publishers, 1991).
[14] G. H. B. Thompson, *Physics of semiconductor laser devices*, (Chichester: John Wiley & Sons, 1980).
[15] I. P. Kaminow and R. S. Tucker, "Mode-controlled semiconductor lasers," in *Guided-Wave Optoelectronics*, 2nd ed., T. Tamir, Ed. (Berlin: Springer-Verlag, 1990), p. Ch. 5.
[16] G. P. Agrawal, *Fiber-Optic Communication Systems* (New York: John Wiley & Sons, Inc., 1992).
[17] T. L. Koch and U. Koren, "Semiconductor Lasers for Coherent Optical Fiber Communications," *IEEE J. Lightwave Technol.*, Vol. LT-8, No. 3, pp. 274–293, 1990.
[18] D. J. G. Mestdagh, *Fundamentals of Multiaccess Optical Fiber Networks* (Boston: Artech House, 1995).
[19] H. Ishii, Y. Kondo, F. Kano, and Y. Yoshikuni, "A Tunable Distributed Amplification DFB Laser Diode (TDA-DFB-LD)," *IEEE Photon. Tech. Lett.*, Vol. 10, No. 1, pp. 30–32, 1998.
[20] H. Ishii, H. Tanobe, Y. Kondo, and Y. Yoshikuni, "A tunable interdigital electrode (TIE) DBR laser for single-current continuous tuning," *IEEE Photon. Tech. Lett.*, Vol. 7, No. 11, pp. 1246–1248, 1995.
[21] Y. Yoshikuni and G. Motosugi, "Multielectrode Distributed Feedback Laser for Pure Frequency Modulation and Chirping Suppressed Amplitude Modulation," *IEEE J. Lightwave Technol.*, Vol. LT-5, No. 4, pp. 516–522, 1987.

[22] M. J. Adams, "Rate equations and transient phenomena in semiconductor lasers," *Opto-electronics*, Vol. 5, pp. 201–215, 1973.

[23] D. Marcuse and T. P. Lee, "On Approximate Analytical Solutions of Rate Equations fo Studying Transient Spectra of Injection Lasers," *IEEE J. Quant. Electron.*, Vol. QE-19, No. 9, pp. 1397–1406, 1983.

[24] T. P. Lee, C. A. Burrus, J. A. Copeland, A. G. Dentai, and D. Marcuse, "Short-Cavity InGaAsP Injection Lasers: Dependence of Mode Spectra and Single-Longitudinal-Mode Power on Cavity Length," *IEEE J. Quant. Electron.,* Vol. QE-18, No. 7, pp. 1101–1113, 1982.

[25] D. Marcuse, "Classical derivation of the laser rate equation," *IEEE J. Quant. Electron.*, Vol. QE-19, pp. 1228–1231, 1983.

[26] C. Henry, "Phase Noise in Semiconductor Lasers," *IEEE J. Lightwave Technol.,* Vol. LT-4, No. 3, pp. 298–311, 1986.

[27] M. W. Fleming and A. Mooradian, "Fundamental line broadening of single-mode (GaAl)As diode lasers," *Appl. Phys. Lett.*, Vol. 38, p. 511, 1981.

[28] C. H. Henry, "Theory of the Linewidth of Semiconductor Lasers," *IEEE J. Quant. Electron.*, Vol. QE-18, No. 2, pp. 259–264, 1982.

[29] B. W. Hakki, "Optical and microwave instabilities in injection lasers," *J. Appl. Phys.*, Vol. 51(1), pp. 68–73, 1980.

[30] N. A. Olsson, N. K. Dutta, and K. Y. Liou, "Dynamic Linewidth of Amplitude- Modulated Single-Longitudinal-Mode Semiconductor Lasers Operating at 1.5 mum Wavelength," *Electron. Lett.*, Vol. 20, No. 3, pp. 121–122, 1984.

[31] M. Osinski and M. J. Adams, "Intrinsic Manifestation of Regular Pulsations in Time-Averaged Spectra of Semiconductor Lasers," *Electron. Lett.*, Vol. 20, No. 13, pp. 525–526, 1984.

[32] R. A. Linke, "Transient chirping in single-frequency lasers: Lightwave systems consequences," *Electron. Lett.*, Vol. 20, No. 11, pp. 472–474, 1984.

[33] A. Yariv, Optical Electronics, 4th ed. (Fort Worth: Saunders College Publishing, 1991).

[34] G. P. Agrawal, *Nonlinear Fiber Optics* (Boston: Academic Press Inc., 1989).

[35] T. Okoshi and K. Kikuchi, *Coherent optical fiber communications* (Dordrecht: Kluwer Academic Publishers, 1988).

[36] K. Y. Lau and A. Yariv, "Ultra-High Speed Semiconductor Lasers," *IEEE J. Quant. Electron.*, Vol. QE-21, No. 2, pp. 121–138, 1985.

[37] K. Y. Lau, C. Harder and A. Yariv, "Direct modulation of semiconductor lasers at $>$ 10 GHz by low-temperature operation," *Appl. Phys. Lett.*, Vol. 44(3), pp. 273–275, 1984.

[38] J. Peter S. Zory, Ed., *Quantum Well Lasers* (San Diego: Academic Press, Inc., 1993).

[39] K. Y. Lau and A. Yariv, "Intermodulation distortion in a directly modulated semiconductor injection laser," *Appl. Phys. Lett.*, Vol. 45, pp. 1034–1036, 1984.

[40] R. A. Linke, "Modulation Induced Transient Chirping in Single Frequency Lasers," *IEEE J. Quant. Electron.*, Vol. QE-2,1 No. 6, pp. 593–597, 1985.

[41] Y. Arakawa and A. Yariv, "Quantum well lasers — gain, spectra, dynamics," *IEEE J. Quant. Electron.*, Vol. QE-22, No. 9, pp. 1887–1899, 1986.

[42] W. I. Way, "Large Signal Nonlinear Distortion Prediction for a Single-Mode Laser Diode Under Microwave Intensity Modulation," *IEEE J. Lightwave Technol.*, Vol. LT-5, No. 3, pp. 305–315, 1987.

[43] R. S. Tucker, G. Eisenstein, and S. K. Korotky, "Optical Time-Division Multiplexing For Very High Bit-Rate Transmission," *IEEE J. Lightwave Technol.*, Vol. LT-6, No. 11, pp. 1737–1749, 1988.

[44] A. S. Daryoush, "Special Issue on Microwave Photonics," *Microw. Opt. Tech. Lett.*, Vol. 6, No. 1, pp. 1-?, 1993.

[45] W. M. Robertson, *Optoelectronics Techniques for Microwave and Millimeter-Wave Engineering* (Boston: Artech House, 1995).

[46] P. P. Vasil'Ev, "Ultrashort pulse generation in diode lasers," *Opt. Quant. Elec.*, Vol. 24, pp. 801–824, 1992.

[47] P. L. Liu, C. Lin, I. P. Kaminow, and J. J. Hsieh, "Picosecond pulse generation from InGaAsP lasers at 1.25 and 1.3 mum by electrical pulse pumping," *IEEE J. Quant. Electron.*, Vol. QE-17, pp. 671–674, 1981.

[48] N. Onodera, H. Ito, and H. Inaba, "Generation and Control of Bandwidth-Limited, Single-Mode Picosecond Optical Pulses by Strong RF Modulation of a Distributed Feedback InGaAsP Diode Laser," *IEEE J. Quant. Electron.*, Vol. QE-21, No. 6, pp. 568–575, 1985.

[49] D. J. Derickson *et al.*, "Short Pulse Generation using Multisegment Mode-Locked Semiconductor Lasers," *IEEE J. Quant. Electron.*, Vol. QE-28, No. 10, pp. 2186–2202, 1992.

[50] R. S. Tucker *et al.*, "40 GHz active mode-locking in a 1.5 mum monolithic extended-cavity laser," *Electron. Lett.*, Vol. 25 No. 10, pp. 621–622, 1989.

[51] P. B. Hansen *et al.*, "5.5 mum long InGaAsP Monolithic Extended-Cavity Laser with an Integrated Bragg-Reflector for Active Mode Locking," *IEEE Photon. Tech. Lett.*, Vol. 4, No. 3, pp. 215–217, 1992.

[52] L. Zhai, A. J. Lowery, and Z. Ahmed, "Diffraction grating model for transmission-line laser model of actively mode-locked semiconductor lasers," *IEE Proceedings Pt. J.*, Vol. 141, No. 1, pp. 21–26, 1994.

[53] G. H. C. New, "The generation of ultrashort laser pulses," *Rep. Prog. Phys.*, Vol. 46, pp. 877–971, 1983.

[54] D. J. Jones, L. M. Zhang, J. E. Carroll, and D. D. Marcenac, "Dynamics of Monolithic Passively Mode-Locked Semiconductor Lasers," *IEEE J. Quant. Electron.*, Vol. QE-31, No. 6, pp. 1051–1058, 1995.

Chapter 14

Microwave Circuit Techniques and Semiconductor Laser Modelling

14.1 Introduction

Although today microwave and optical engineering appear to be separate disciplines, there has been a tradition of interchange of ideas between them. In fact, many traditional microwave concepts have been adapted to yield optical counterparts. The laser, as an optical device that plays a key role in optoelectronics and fibre-optic communications, grew from the work of its microwave predecessor, the maser (microwave amplification by stimulated emission of radiation) [1]. The operating principle behind the laser is very similar to that of the microwave oscillator. In a semiconductor laser, the required feedback may either be provided by the cleaved facets of Fabry–Perot (FP) lasers or by a periodic grating in distributed feedback (DFB) lasers. Certain optical techniques, such as injection locking of lasers by external light [2], are ideas borrowed from the phenomenon of injection locking of microwave oscillators by an external electronic signal [3]. The close relationship between optical and microwave principles suggests that it may be advantageous to apply microwave circuit techniques in modelling of semiconductor lasers.

Engineers work best when using tools they are familiar with. In particular, electrical and electronic engineers are familiar with well-established electrical circuit models as tools to aid themselves in the understanding and prediction of behaviour of electrical machines or electronic devices. Since the early days of radio frequency (RF) and microwave engineering, microwave circuit theory has allowed us to explore fundamental properties of electromagnetic waves by giving us an intuitive understanding of electromagnetic waves without the need to invoke detailed and rigorous electromagnetic field theories [4, 5]. In the same spirit, microwave circuit formulation of the semiconductor laser diode enhances our understanding of the device, which is otherwise obscured by hard-to-visualise mathematical formulations. Complex mathematical models are too sophisticated to be desirable for engineers, especially those who are not specialists in the field of laser physics but would like to have an easy-to-digest method of understanding and designing semiconductor laser devices. It is far more convenient to work in terms of voltages, currents, and impedances. In fact, electromagnetic field theory and distributed-element circuits (transmission-lines) give identical solutions when we are dealing with transverse electro-magnetic (TEM) fields, where voltages and currents in the transmission-lines are uniquely related to the transverse electric and magnetic fields, respectively.

The attractiveness of using equivalent circuit models for semiconductor laser devices stems from their ability to provide an analogy of laser theory in terms of microwave circuit principles. In addition, microwave circuit models of laser diodes are compatible with existing circuit models of microwave devices such as heterojunction bipolar transistors (HBTs) and field-effect transistors (FETs) — an attractive feature for optoelectronic integrated circuit (OEIC) design [6]. Equivalent circuit models have effectively helped many to understand, design, and optimise integrated circuits (IC's) in the microelectronics industry and they have the potential to do the same for the optoelectronics industry. The main theme of this chapter is microwave circuit modelling techniques applied to semiconductor laser devices.

In part III of this book, two types of microwave circuit models for semiconductor lasers have been investigated: (i) the simple *lumped element* model based on low frequency circuit concepts and (ii) the more versatile *distributed element* model based on transmission-line modelling. The former (lumped-element circuit models) is based on the simplifying assumption that the phase of current or voltage across the dimension of the components has little variation. This is true when considering only the modulated signal instead of the optical carrier signal. In this case, Kirchoff's law can be applied, which is nothing more than a special case of Maxwell's equations [7]. Strictly speaking, laser devices have dimensions in the order of the operating wavelength, thus lumped-element models may not be suitable in ultrafast applications where propagation plays an important role, such as in active mode-locking [8]. However, the lumped-element circuit is reasonably accurate for microwave applications if all the important processes and effects are modelled accordingly by the circuit on an equivalence basis. The lumped-element circuit model will be discussed in Chapter 15.

The latter of the two circuit modelling techniques (i.e. transmission-line modelling) is a more powerful circuit model that includes distributed effects, which will be discussed in detail in this chapter. In Chapter 15, a comparison will be made between the two types of circuit models. It is worth pointing out that at microwave frequencies and above, voltmeters and ammeters for direct measurement of voltages and currents do not exist, where voltage and current waves are only introduced conceptually in the microwave circuit to make optimum use of the low frequency circuit concepts.

14.2 The Transmission-Line Matrix (TLM) Method

The transmission-line matrix (TLM) was originally developed to model passive microwave cavities by using meshes of transmission-lines [9–10]. The numerical processes involved in TLM resemble the mechanism of wave propagation but they are discretised in both time and space [10–11]. A lot of work has been carried out using the TLM method for analysis of passive microwave waveguide structures (see [12] and references therein). Most of the work done involved two-dimensional and three-dimensional TLM, with the exception of application to lumped networks [13–14], the heat diffusion problem [15], and semiconductor laser modelling [16]. Although the TLM is unconditionally stable when modelling passive devices, the semiconductor laser is an active device and therefore requires more careful consideration. The basics of the one-dimensional (1-D) TLM will be presented in the following section, which forms the basis of the transmission-line laser model (TLLM) [17].

14.3 TLM Link-Lines and Stub-Lines

The TLM is a discrete-time model of wave propagation simulated by voltage pulses travelling along transmission-lines. The medium of propagation is represented by transmission-lines — a general or lossy transmission-line consists of series resistance, shunt admittance, series inductance, and shunt capacitance per unit length, whereas an ideal or lossless transmission-line has reactive elements only. The transmission-line may be described by a set of *telegraphist equations* [7], which can be shown to be equivalent to the Maxwell's equations. There are two types of TLM elements that can be used as building blocks of a complete TLM network — they are the TLM (i) stub-lines and (ii) link-lines [13].

14.3.1 *TLM link-lines*

For a lossless transmission-line, the velocity of propagation is expressed by:

$$v_p = \frac{1}{\sqrt{L_\mathrm{d} C_\mathrm{d}}} = \frac{\Delta l}{\Delta t} \tag{14.1}$$

where L_d is inductance per unit length, C_d is capacitance per unit length, Δl is the unit section length, and Δt is the model timestep. In Fig. 14.1, it is shown that a lumped series inductor (L) is equivalent to a transmission-line

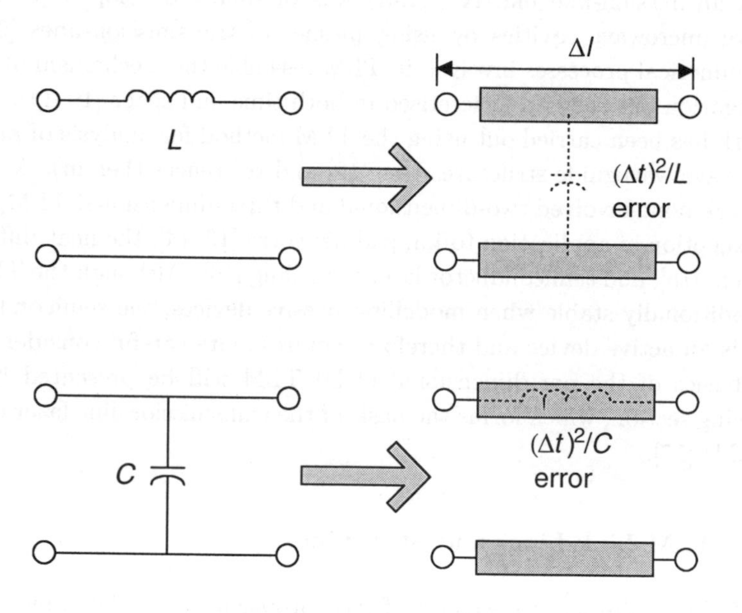

Fig. 14.1. TLM limk-lines.

with inductance per unit length of L_d, where [13]:

$$L = L_\mathrm{d} \Delta l\,. \tag{14.2}$$

The characteristic impedance (Z_0) of the transmission-line can be found from:

$$Z_0 = \sqrt{\frac{L_d}{C_d}} = \frac{L}{\Delta t} . \tag{14.3}$$

However, there is a small error associated with the shunt capacitance of the transmission-line, which can be expressed as:

$$C_e = \frac{(\Delta t)^2}{L} . \tag{14.4}$$

Similarly, the lumped shunt capacitor (C) is equivalent to a transmission-line with capacitance per unit length of C_d (Fig. 14.1) where:

$$C = C_d \Delta l . \tag{14.5}$$

The characteristic impedance (Z_0) of the line can be found to be expressed by [13]:

$$Z_0 = \frac{\Delta t}{C} \tag{14.6}$$

and the associated error in the form of a series inductor is given by:

$$L_e = \frac{(\Delta t)^2}{C} . \tag{14.7}$$

The errors C_e and L_e are of the order of $(\Delta t)^2$ and can be reduced by using a smaller model timestep. In practice, there is no component that is purely inductive nor purely capacitive. The parasitic errors can therefore be adjusted by changing the timestep (Δt) to model stray inductance or capacitance. If two adjacent reactive elements are required, then the parasitic error from one line can be "absorbed" into its adjacent line so that the parasitic error may be eliminated for at least one of the lines.

14.3.2 *TLM stub-lines*

In the preceding section, we have seen how lumped reactive elements can be simulated by TLM link-lines. The lumped reactive elements may also be modelled by TLM stub-lines, as shown in Fig. 14.2. The lumped inductor L is also equivalent to a short-circuit stub with a characteristic impedance of [13]:

$$Z_0 = \frac{2L}{\Delta t} \tag{14.8}$$

and has a parasitic capacitance expressed by:

$$C_{\mathrm{e}} = \frac{(\Delta t)^2}{4L} \,. \tag{14.9}$$

On the other hand, the lumped capacitor is equivalent to an open-circuit stub with characteristic impedance of [13]:

$$Z_0 = \frac{\Delta t}{2C} \tag{14.10}$$

and has a parasitic inductance expressed by:

$$L_{\mathrm{e}} = \frac{(\Delta t)^2}{4C} \,. \tag{14.11}$$

For TLM stub-lines, the length of the transmission-line is chosen so that it takes half a model timestep ($\Delta t/2$) for the pulse to travel from one end to another (see Fig. 14.2). The reason is to allow the voltage pulses to propagate to the termination of the stub and back again at the scattering node in one complete time iteration (Δt). This way, all incident voltage pulses will arrive at their scattering nodes in exactly the same amount of time, irrespective of stub-lines or link-lines, i.e. the voltage pulses are *synchronised.*

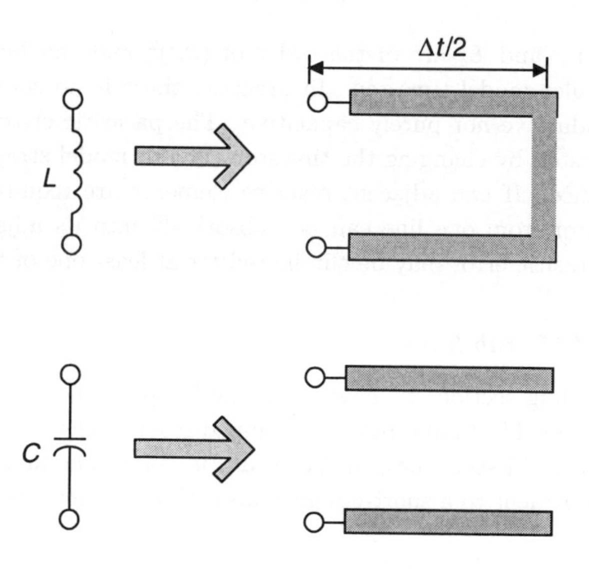

Fig. 14.2. TLM stub-lines.

14.4 Scattering and Connecting Matrices

The most basic algorithm of TLM involves two main processes: (i) scattering and (ii) connecting. When the incident voltage pulses, $\mathbf{V^i}$, arrive at the scattering node, they are operated by a *scattering matrix* and reflected voltage pulses, $\mathbf{V^r}$, are produced. These reflected pulses then continue to propagate along the transmission-lines and become incident pulses at adjacent scattering nodes — this process is described by the *connecting matrix*. Formally, the TLM algorithm may be expressed as:

$$
\begin{aligned}
{}_k\mathbf{V}^{\mathbf{rT}} &= \mathbf{S}_k \mathbf{V}^{\mathbf{iT}} \qquad \text{[SCATTERING]} \\
{}_{k+1}\mathbf{V}^{\mathbf{iT}} &= \mathbf{C}_k \mathbf{V}^{\mathbf{rt}} \qquad \text{[CONNECTING]}\,.
\end{aligned}
\tag{14.12}
$$

The terms $\mathbf{V^{iT}}$ and $\mathbf{V^{rT}}$ are the transpose matrix of the incident and reflected pulses, respectively. The terms k and $k+1$ denote the kth and $(k+1)$th time iterations, respectively. The scattering and connecting matrices are denoted by $\mathbf{S}$ and $\mathbf{C}$, respectively. As the matrices involved in Eq. (14.12) depend on the type of TLM subnetwork, a worked example is given in the following based on the TLM subnetwork of Fig. 14.3.

The TLM subnetwork consists of three "branches" of lossy transmission-lines as shown in Fig. 14.3, where scattering and connecting of the voltage pulses are clearly described pictorially. The *normalised* impedances are unity for the two lines connected to adjacent nodes and Z_s for the remaining branch (open–circuit stub line). The associated *normalised* resistances are R, and R_s, respectively. The matrices $\mathbf{V^{iT}}$ and $\mathbf{V^{rT}}$ are given as:

$$
\mathbf{V}^{\mathbf{rT}} = \begin{bmatrix} V_1 \\ V_2 \\ V_3 \end{bmatrix}^r \qquad \mathbf{V}^{\mathbf{iT}} = \begin{bmatrix} V_1 \\ V_2 \\ V_3 \end{bmatrix}^i
\tag{14.13}
$$

where V_1, V_2 and V_3 are the voltage pulses on ports 1, 2 and 3, respectively, on each scattering node (see Fig. 14.3). The superscripts r and i denote reflected and incident pulses, respectively. It is convenient to break up the scattering matrix $\mathbf{S}$ and express it as [15–18]:

$$
\mathbf{S} = \mathbf{p} \times \mathbf{q} - \mathbf{r}
\tag{14.14}
$$

where the matrices $\mathbf{p}$, $\mathbf{q}$, and $\mathbf{r}$ are defined in the following. For the type of TLM subnetwork in Fig. 14.3, we have:

$$
\mathbf{q} = [q_1 \quad q_2 \quad q_3]
$$

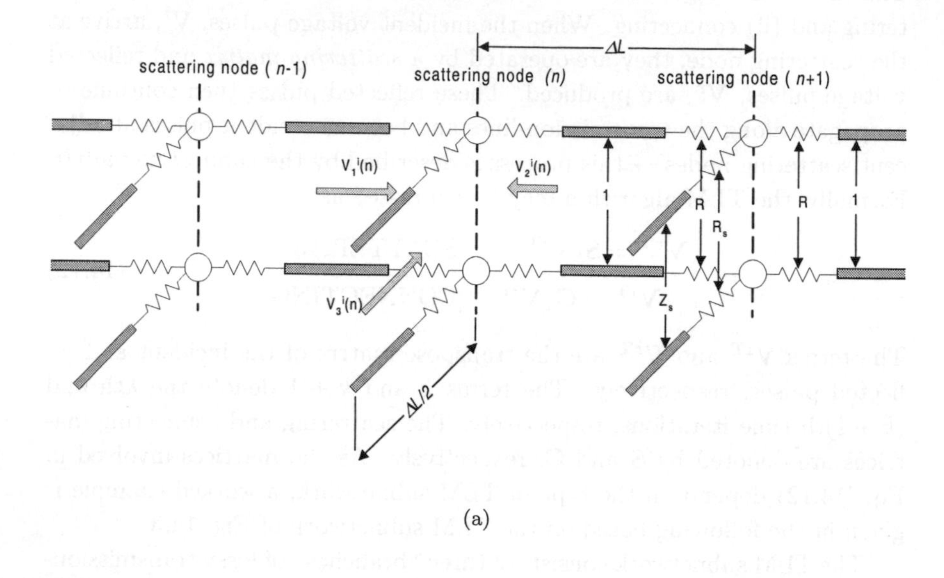
ΔL
scattering node (n-1)
scattering node (n)
scattering node (n+1)
V1i(n)
V2i(n)
V3i(n)
1
R
Rs
R
1
Zs
ΔL/2

(a)

(b)

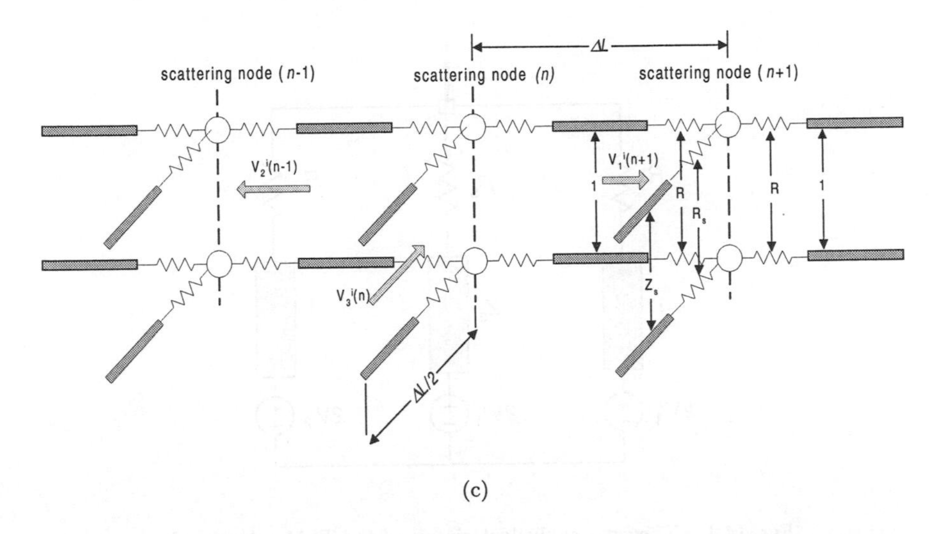

(c)

Fig. 14.3. The TLM stub-line: (a) incident pulses arriving at scattering node (b) incident pulses scattered into reflected pulses (scattering) (c) reflected pulse arriving at adjacent nodes (connecting).

where

$$q_i = \begin{cases} \dfrac{2(R_s + Z_s)}{1 + R + 2(R_s + Z_s)} & i = 1, 2 \\ \dfrac{2(1 + R)}{1 + R + 2(R_s + Z_s)} & i = 3 . \end{cases} \tag{14.15}$$

The matrix $\mathbf{q}$ may be found by replacing the subnetwork by its Thevenin equivalent circuit, which is shown in Fig. 14.4. By definition of the voltage pulse, V_x^i [5], its generator or source require a value of $2V_x^i$, where x denotes the port number (1, 2 or 3). The nodal voltage ($\mathbf{v}$) can be defined as:

$$\mathbf{v} = \mathbf{q}_k \mathbf{V}^{\mathbf{iT}} . \tag{14.16}$$

This will be further explained by using two simple TLM subnetworks as examples later. The matrix $\mathbf{p}$ is found by applying the voltage division rule and is given as:

$$\mathbf{p} = [p_1 \quad p_2 \quad p_3]^{\mathbf{T}}$$

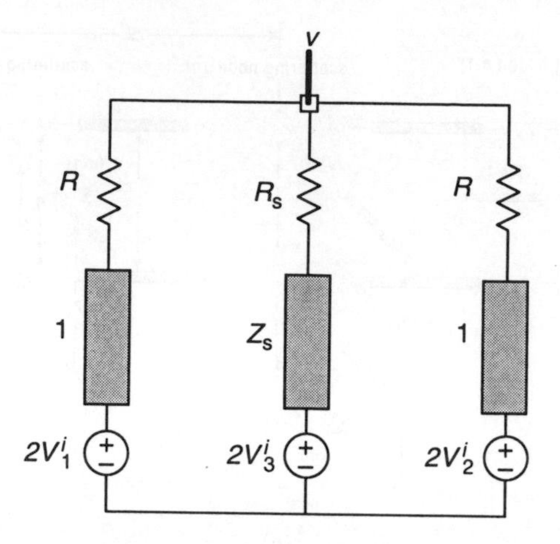

Fig. 14.4. Thevenin equivalent circuit of the TLM subnetwork.

where

$$p_i = \begin{cases} \dfrac{1}{1+R} & i = 1,2 \\[2ex] \dfrac{Z_s}{R_s + Z_s} & i = 3 \end{cases} \tag{14.17}$$

and the matrix $\mathbf{r}$ in Eq. (14.14) is given by:

$$\mathbf{r} = \begin{bmatrix} r_{11} & 0 & 0 \\ 0 & r_{22} & 0 \\ 0 & 0 & r_{33} \end{bmatrix}$$

where

$$r_{ii} = \begin{cases} \dfrac{1-R}{1+R} & i = 1,2 \\[2ex] \dfrac{Z_s - R_s}{R_s + Z_s} & i = 3\,. \end{cases} \tag{14.18}$$

If lossless transmission-lines are used, we have $R = R_s = 0$, and Eqs. (14.15), (14.17) and (14.18) become:

$$\mathbf{q} = [q_1 \quad q_2 \quad q_3]$$

$$q_i = \begin{cases} \dfrac{2Z_\mathrm{s}}{1+2Z_\mathrm{s}} & i = 1,2 \\ \dfrac{2}{1+2Z_\mathrm{s}} & i = 3 \end{cases}$$

$$\mathbf{p} = \begin{bmatrix} 1 \\ 1 \\ 1 \end{bmatrix} \tag{14.19}$$

$$\mathbf{r} = \begin{bmatrix} 1 & 0 & 0 \\ 0 & 1 & 0 \\ 0 & 0 & 1 \end{bmatrix}.$$

The connecting matrix depends on how the transmission-lines are connected, that is which port(s) of one scattering node is connected to which port(s) of other adjacent scattering node(s). The matrix element in the connecting matrix is unity only if a connection allows a pulse to travel from port (i) of node (m) to port (j) of node (n). In the example given in Fig. 14.3, the connecting matrix $\mathbf{C}$ may be expressed by 3×3 elements, where we have pulses travelling from:

(i) port 2 of node $(n-1)$ to port 1 of node (n)
(ii) port 3 of node (n) to port 3 of node (n) (open circuit stub)
(iii) port 1 of node $(n+1)$ to port 2 of node (n).

Thus, the connecting matrices for node (n) and its adjacent nodes $(n-1)$ and $(n+1)$ are given as:

$$\mathbf{C}_{n-(n-1)} = \begin{bmatrix} 0 & 0 & 0 \\ 1 & 0 & 0 \\ 0 & 0 & 0 \end{bmatrix}$$

$$\mathbf{C}_{n-n} = \begin{bmatrix} 0 & 0 & 0 \\ 0 & 0 & 0 \\ 0 & 0 & 1 \end{bmatrix}$$

$$\mathbf{C}_{n-(n+1)} = \begin{bmatrix} 0 & 1 & 0 \\ 0 & 0 & 0 \\ 0 & 0 & 0 \end{bmatrix}. \tag{14.20}$$

A simple example of another TLM sub-network is shown in Fig. 14.5(a), which consists of a resistor "sandwiched" between two lossless transmission-lines. In the Thevenin equivalent circuit shown in Fig. 14.5(b), each transmission-line is replaced by its characteristic impedance in series with a voltage generator of twice the incident voltage pulse. The *incident* voltage pulses are denoted as V_1^i and V_2^i, while the *reflected* pulses are V_1^r and V_2^r. The impedances of the lossless transmission-lines are Z_1 and Z_2. From Fig. 14.5(b), the *nodal voltages* (v_1 and v_2) may be expressed as:

$$v_1 = \frac{i_1}{Y_1}, \qquad v_2 = \frac{i_2}{Y_2} \tag{14.21}$$

where

$$i_1 = \frac{2V_1^i}{Z_1} + \frac{2V_2^i}{(R+Z_2)}, \qquad i_2 = \frac{2V_1^i}{(R+Z_1)} + \frac{2V_2^i}{Z_2} \tag{14.22}$$

and

$$Y_1 = \frac{1}{Z_1} + \frac{1}{R+Z_2}, \qquad Y_2 = \frac{1}{Z_1} + \frac{1}{R+Z_2}. \tag{14.23}$$

By substituting Eqs. (14.22) and (14.23) into (14.21), we have:

$$\begin{aligned} v_1 &= \left(\frac{1}{R+Z_1+Z_2}\right)\left[2(R+Z_2)V_1^i + 2Z_1V_2^i\right] \\ v_2 &= \left(\frac{1}{R+Z_1+Z_2}\right)\left[2Z_2V_1^i + 2(R+Z_1)V_2^i\right]. \end{aligned} \tag{14.24}$$

By the original definition of the nodal voltage [5], we have the relationships:

$$\begin{aligned} v_1 &= V_1^i + V_1^r \\ v_2 &= V_2^i + V_2^r. \end{aligned} \tag{14.25}$$

Now we can write down the reflected voltage pulses in terms of the incident pulses as:

$$\begin{aligned} V_1^r &= v_1 - V_1^i \\ V_1^r &= \left(\frac{1}{R+Z_1+Z_2}\right)\left[(R-Z_1+Z_2)V_1^i + 2Z_1V_2^i\right] \quad \text{and} \\ V_2^r &= v_2 - V_2^i \\ V_2^r &= \left(\frac{1}{R+Z_1+Z_2}\right)\left[2Z_2V_1^i + (R-Z_2+Z_1)V_2^i\right]. \end{aligned} \tag{14.26}$$

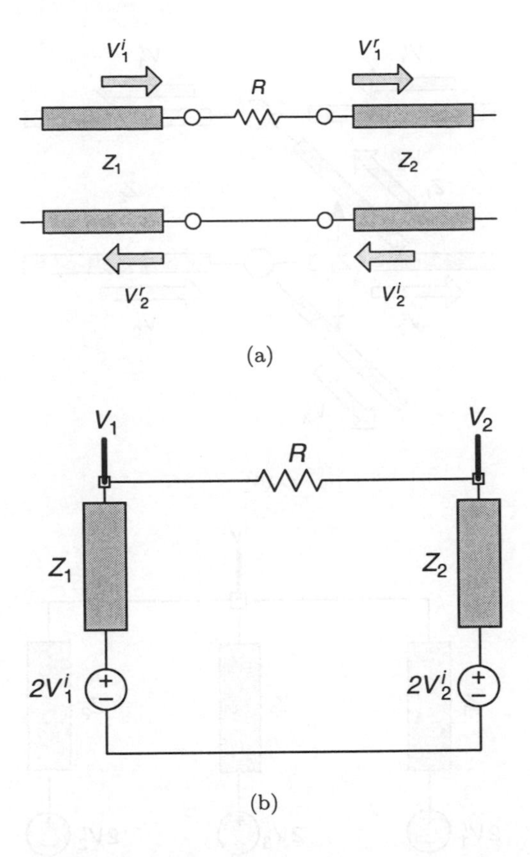

Fig. 14.5. (a) TLM subnetwork (b) its Thevenin equivalent circuit.

Finally, the complete scattering matrix of the TLM network in Fig. 14.5 at the kth iteration is expressed by [13]:

$$S_k = \left(\frac{1}{R + Z_1 + Z_2}\right)\begin{bmatrix}(R - Z_1 + Z_2) & 2Z_1 \\ 2Z_2 & (R - Z_2 + Z_1)\end{bmatrix}. \qquad (14.27)$$

Another simple but useful TLM subnetwork is shown in Fig. 14.6(a) and its Thevenin equivalent is given in Fig. 14.6(b). This is similar to one of the subnetworks in Fig. 14.3 but it is lossless in this case and each branch has a different value of line impedance. Now, the common *nodal current* is defined by:

$$i = \frac{2V_1^i}{Z_1} + \frac{2V_3^i}{Z_3} + \frac{2V_2^i}{Z_2}\,. \qquad (14.28)$$

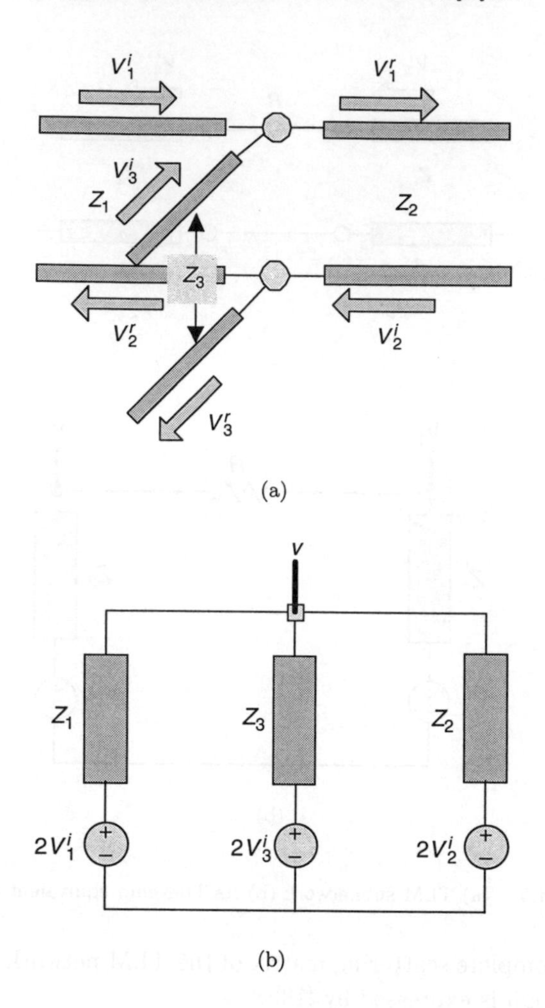

Fig. 14.6. (a) TLM subnetwork (b) its Thevenin equivalent circuit.

From Fig. 14.6(b), the total admittance at the scattering node is expressed as:

$$
\begin{aligned}
Y &= \left(\frac{1}{Z_1} + \frac{1}{Z_3} + \frac{1}{Z_2}\right) \\
Y &= \frac{Z_2 Z_3 + Z_1 Z_2 + Z_1 Z_3}{Z_1 Z_2 Z_3} \,.
\end{aligned}
\tag{14.29}
$$

From Millman's thereom [19] the common nodal voltage can be found from:

$$v = \frac{i}{Y} = \frac{2Z_2Z_3V_1^i + 2Z_1Z_2V_3^i + 2Z_1Z_3V_2^i}{Z_2Z_3 + Z_1Z_2 + Z_1Z_3}. \tag{14.30}$$

Since the sum of incident voltage pulse and reflected pulse gives the nodal voltage, see Eq. (14.25), the reflected pulse may again be found by subtracting the incident pulse from the nodal voltage, that is:

$$\begin{aligned} V_1^r &= v - V_1^i \\ V_1^r &= \frac{2Z_2Z_3V_1^i + 2Z_1Z_2V_3^i + 2Z_1Z_3V_2^i}{Z_2Z_3 + Z_1Z_2 + Z_1Z_3} - V_1^i \\ V_1^r &= \frac{(Z_2Z_3 - Z_1Z_2 - Z_1Z_3)V_1^i + 2Z_1Z_2V_3^i + 2Z_1Z_3V_2^i}{Z_2Z_3 + Z_1Z_2 + Z_1Z_3}. \end{aligned} \tag{14.31}$$

In a similar manner, the reflected pulses, V_2^r and V_3^r can also be found, and the complete scattering matrix of the sub-network in Fig. 14.6 is expressed by:

$$S_k = \frac{\begin{bmatrix} Z_2Z_3 - Z_1Z_2 - Z_1Z_3 & 2Z_1Z_2 & 2Z_1Z_3 \\ 2Z_2Z_3 & Z_1Z_2 - Z_2Z_3 - Z_1Z_3 & 2Z_1Z_3 \\ 2Z_2Z_3 & 2Z_1Z_2 & Z_1Z_3 - Z_2Z_3 - Z_1Z_2 \end{bmatrix}}{Z_1Z_2 + Z_1Z_3 + Z_2Z_3}. \tag{14.32}$$

The TLM subnetwork in Fig. 14.6(a) is the basis of the matching network that will be discussed in Chapter 15. There are many other TLM subnetworks that can be formed by using series and shunt resistors together with the TLM link-lines and stub-lines. For example, periodically unmatched boundaries of TLM link-lines can be used to mimic corrugated gratings [20] and shunt conductances can be included to model gain-coupled DFB lasers [21]. A TLM subnetwork, which consists of a series resistor and two reactive stub-lines, will be used to model the wavelength dependence of semiconductor laser gain in the transmission- line laser model (TLLM).

14.5 Transmission-Line Laser Modelling (TLLM)

The transmission-line laser model (TLLM) is a wide-bandwidth dynamic laser model that takes into account important considerations such as inhomogeneous effects, multiple longitudinal modes, spectral dependence of gain, carrier-induced refractive index change, and spontaneous emission noise. The TLLM is very flexible and has successfully been used to model a wide range of laser devices, including Fabry–Perot (FP) lasers [16], DFB (index- coupled and gain-coupled) and DBR lasers [21–23], quantum-well (QW) lasers [24], cleaved-coupled-cavity (CCC) lasers [25] external-cavity (EC) mode-locked lasers [26], fibre grating lasers [27], and laser amplifiers [28]. Electrical parasitics and matching circuits can also be included as part of the model [29].

The TLLM can be thought of as a pedagogical model to aid in the physical understanding of the laser in terms of more familiar circuit techniques. The topology of the model closely mimics the physical structure of the laser. It was first developed by Lowery [16] based on the 1-D TLM concepts discussed in the preceding section. Since the TLLM is a time-domain model, its main application is transient analysis, where laser nonlinearity is important.

14.6 Basic Construction of the Model

In the TLLM, the laser cavity is divided into many smaller sections, as shown in Fig. 14.7, hence longitudinal distribution of carrier and photon density can be modelled. The individual sections are connected to one another by dispersionless transmission-lines (i.e. group velocity equals to phase velocity) with characteristic impedance of Z_0. Voltage pulses travel along these transmission-lines in both forward and backward directions analogous to the propagation of optical waves inside the laser cavity. The model timestep, (Δt), describes the time required for the voltage pulse to propagate from one section to the next adjacent section. At the centre of each section is the scattering node, where the TLM scattering matrix is placed. When incident voltage pulses arrive at the scattering nodes, the TLM scattering matrix operates on them to produce reflected voltage pulses (see Section 14.4).

The fundamental processes of the laser, such as gain (stimulated emission), loss (absorption), and noise (spontaneous emission) are contained in the scattering matrix. Spectral dependence of the material gain, which the optical waves experience, is also included. Coupling between counter-propagating waves that occurs in DFB lasers can also be modelled. At the laser facets, the reflections that provide feedback to achieve lasing action are simulated by unmatched terminal loads.

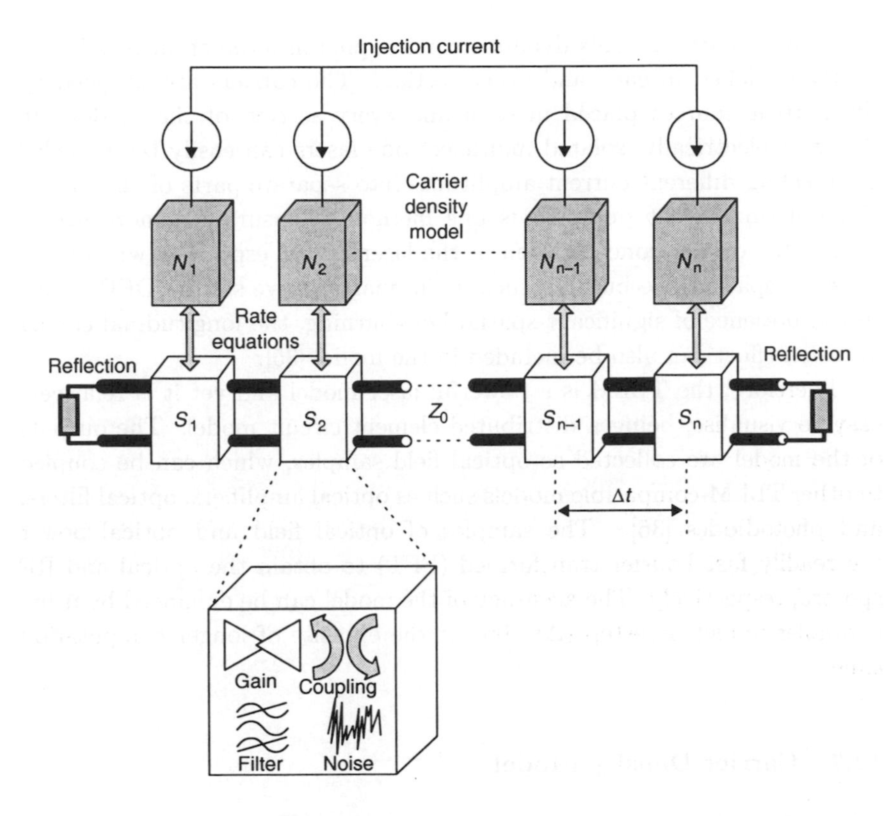

Fig. 14.7. The transmission-line laser model (TLLM) and its components.

The model assumes single transverse mode operation of the laser diode, and that variations of the carrier and photon densities in the lateral and transverse dimensions are not significant, except for broad area lasers, such as the tapered waveguide structure [30]. Today, advanced fabrication techniques allow single transverse and longitudinal mode laser devices to be achieved, typical of strongly index guided laser structures such as buried heterostructure (BH) lasers. By spatially averaging the lateral-x and transverse-y dependence of the optical field amplitude, the model is simplified into a 1-D model [16]. The 1-D TLLM is a reasonable approach, which has been supported by many other 1-D dynamic laser models such as the transfer matrix model (TMM) [31], time-domain model (TDM) [32–33], and power matrix model (PMM) [34–43].

The reservoir of carriers (N_n where n is the section number) in the carrier density model interacts with the photon density (S_n) through the laser rate equations. This dynamic carrier- photon interaction is independently modelled in each and every section. The carriers are supplied by the current sources placed in each and every section of the model. In this way, electrically-isolated multielectrode lasers can easily be modelled by injecting different current amplitudes into separate parts of the model. Nonuniform current pumping is one method of ensuring a more evenly distributed carrier concentration in the laser cavity, especially when there is severe spatial hole-burning such as in quarter-wave shifted DFB lasers. In the presence of significant spatial hole-burning, the longitudinal carrier diffusion effect can also be included in the model [35].

Therefore, the TLLM is a powerful laser model and yet it is relatively easy to visualise, being a distributed element circuit model. The outputs of the model are collected as optical field samples, which can be coupled to other TLLM-compatible models such as optical amplifiers, optical filters, and photodiodes [36]. The samples of optical field and optical power are readily fast Fourier transformed (FFT) to obtain the optical and RF spectra, respectively. The accuracy of the model can be enhanced by using a smaller model timestep (Δt) but at the expense of longer computation time.

14.7 Carrier Density Model

The carrier-photon resonance that leads to the transient phenomenon of relaxation oscillations is governed by the laser rate equations. The carrier rate equation is given as:

$$\frac{dN}{dt} = \frac{I}{qv_{\mathrm{a}}} - \frac{N}{\tau_{\mathrm{n}}} - G(N)S \tag{14.33}$$

where N is the carrier density, I is the injected current, v_{a} is the active layer volume of the unit section, q is the electronic charge, τ_{n} is the carrier lifetime, $G(N)$ is the stimulated recombination rate, and S is the photon density. The carrier rate equation can be modelled by an equivalent circuit (see Fig. 14.8) at each and every section of the TLLM, where the carrier density is represented by the voltage, V. In the equivalent circuit of Fig. 14.8, the current, I_{inj}, represents injected current into the active region, the storage capacitor, C, represents carrier build-up/depletion, the resistor, R_{sp}, represents spontaneous emission rate, and the current, I_{stim}, represents stimulated emission rate. The corresponding circuit equation

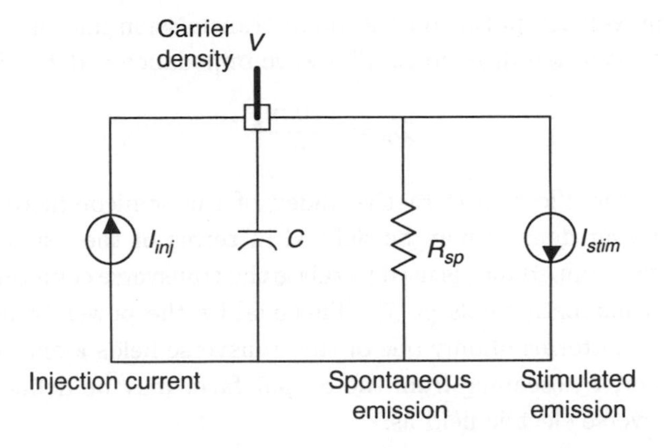

Fig. 14.8. Equivalent circuit model of the carrier density rate equation.

is given as:

$$\frac{dQ}{dt} = I_{\text{inj}} - \frac{V}{R_{\text{sp}}} - I_{\text{stim}} \tag{14.34a}$$

where

$$Q = Nqv_{\text{a}} = CV\,. \tag{14.34b}$$

Comparing the carrier rate equation from (14.33) and the circuit equation from (14.34), the equivalence between the recombination terms and the circuit variables is found by inspection as:

$$\begin{cases} I_{\text{inj}} = I \\ R_{\text{sp}} = \dfrac{\tau_{\text{n}}}{qv_{\text{a}}} \\ I_{\text{stim}} = qv_{\text{a}}G(N)S\,. \end{cases} \tag{14.35}$$

Moreover, the photon density of the local section, S_n, may be found from [17]:

$$S_n = \frac{n_{\text{g}}(|F_M^i(n)|^2 + |B_M^i(n)|^2)}{Z_p h f_0 c_0 m^2} \tag{14.36}$$

where n_g is the effective group index, h is Planck's constant, f_0 is the lasing frequency, m is a unity constant with dimension of length, F_M^i and B_M^i are the incident voltage pulses on the main transmission-line of section-n in opposite propagation directions. The wave impedance is defined by:

$$Z_{\mathrm{p}} = \frac{120\pi n_{\mathrm{g}}}{n_{\mathrm{e}}} \tag{14.37}$$

where n_{e} is the effective refractive index of the semiconductor material. Since the power density may be defined in terms of the transverse fields alone, the wave impedance is used to relate the transverse components of the electric and magnetic fields [5, 7]. This enables the power transmitted to be expressed in terms of only one of the transverse fields alone. Hence, the optical power, P, escaping from the output facet may be defined in terms of the transverse electric field as:

$$P = |F_M^r(S)|^2(1 - R_2)\frac{Wd}{Z_{\mathrm{p}}} \tag{14.38}$$

where W is the output width, d is the active layer thickness, R_2 is the output facet reflectivity, and F_M^r is the forward travelling voltage pulse (transverse electric field) on the main transmission-line at the output section (S).

14.8 Laser Amplification

The amplification rate of the optical intensity is assumed to be of the form:

$$\begin{aligned} \frac{dI}{dz} &= (\Gamma g - \alpha_{\mathrm{sc}})I \\ \int_{I=I_{(n-1)\Delta L}}^{I_{n\Delta L}} \frac{dI}{I} &= \int_{z=(n-1)\Delta L}^{n\Delta L} (\Gamma g - \alpha_{\mathrm{sc}})dz \\ I_{n\Delta L} &= I_{(n-1)\Delta L}\exp[(\Gamma g - \alpha_{\mathrm{sc}})\Delta L] \end{aligned} \tag{14.39}$$

where $I_{(n-1)}\Delta L$ is the initial optical intensity, $I_{n\Delta L}$ is the intensity after propagating a distance of ΔL, Γ is the optical confinement factor, g is the gain coefficient that is *frequency-dependent*, and α_{sc} is the loss factor. As the optical field amplitude is proportional to the square root of the intensity, we can write:

$$E_{n\Delta L} = E_{(n-1)\Delta L}\exp[(\Gamma g - \alpha_{\mathrm{sc}})\Delta L/2] \tag{14.40}$$

where E represents the optical field amplitude. The gain term, $\exp(\Gamma g\Delta L/2)$, in Eq. (14.40) may be approximated by a Taylor expansion

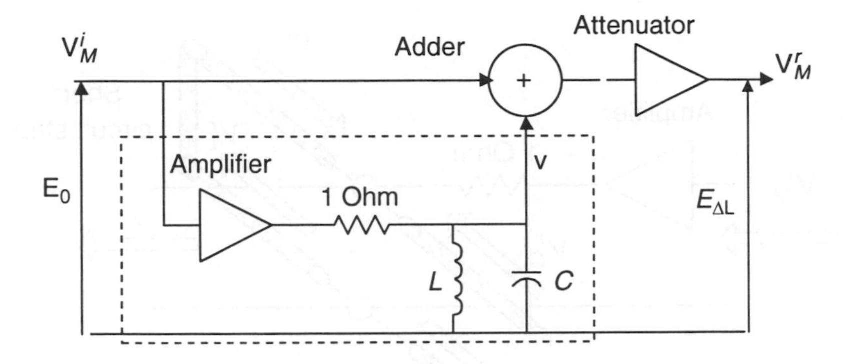

Fig. 14.9. Block diagram of the gain filter.

as the first bracketed terms in the following expression [16]:

$$E_{n\Delta L} = E_{(n-1)\Delta L}\left(1 + \frac{\Delta L\Gamma g}{2}\right)\exp\left(-\frac{\alpha_{\mathrm{sc}}\Delta L}{2}\right). \tag{14.41}$$

Meanwhile, the loss term remains as an exponential term in Eq. (14.41) and is assumed to be independent of frequency. In modelling the laser gain spectrum, a combination of TLM and digital signal processing techniques are involved. The block diagram equivalent of Eq. (14.41) is shown in Fig. 14.9. The frequency (wavelength) dependence of the small amplified signal term, $(\Gamma g\Delta L/2)E_0$, is modelled by the RLC bandpass filter, which is then added back to the incoming signal (E_0) and attenuated by the factor $\exp(-\alpha_{\mathrm{sc}}\Delta L/2)$ to produce the output amplified signal $(E_{\Delta L})$. The lumped circuit representation of the second order bandpass filter in Fig. 14.9 (rectangular dashed lines) may be converted into its TLM counterpart, as shown in Fig. 14.10. The bandpass filter has a Lorentzian response, assuming that the laser is a homogeneously broadened two-level system [6, 37]. The Thevenin equivalent circuit of Fig. 14.10 is also shown in Fig. 14.11. The lumped inductance and capacitance of the bandpass filter are modelled as short-circuit and open- circuit TLM stubs, respectively. From the Thevenin equivalent circuit the common nodal voltage (v) can be expressed as:

$$v = \frac{(gV_{\mathrm{M}}^i + 2V_{\mathrm{L}}^iY_{\mathrm{L}} + 2V_{\mathrm{C}}^iY_{\mathrm{C}})}{(1 + Y_{\mathrm{L}} + Y_{\mathrm{C}})} \quad \text{where} \quad Y_{\mathrm{L}} = \frac{1}{Z_{\mathrm{L}}} \quad \text{and} \quad Y_{\mathrm{C}} = \frac{1}{Z_{\mathrm{C}}} \tag{14.42}$$

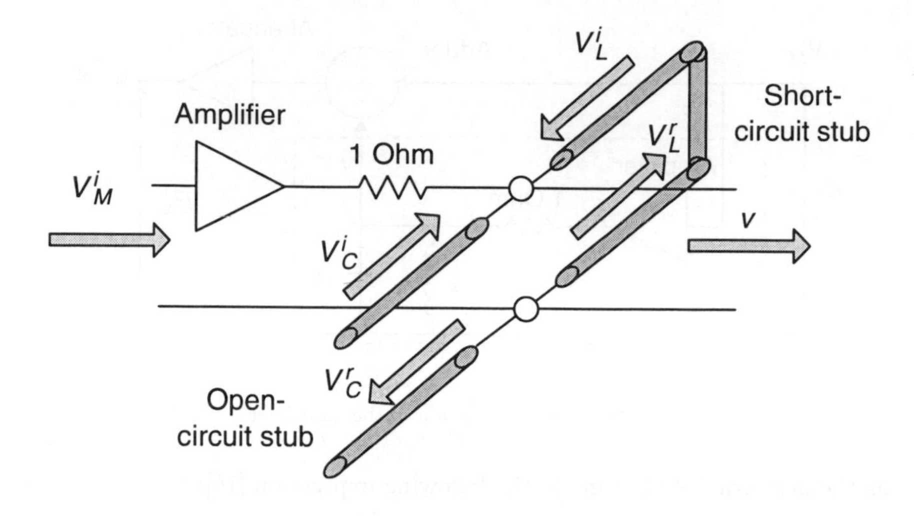

Fig. 14.10. The TLM stub filter.

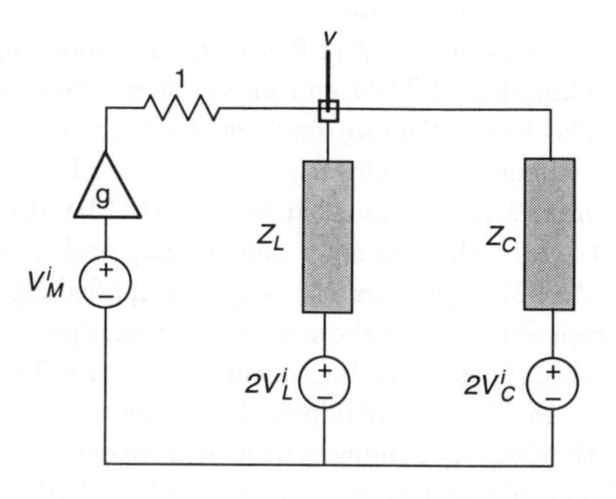

Fig. 14.11. Thevenin equivalent circuit of The TLM stub filter.

Z_{L} and Z_{C} are the impedance values of the inductive and capacitive stub-lines, respectively, V_{M}, V_{L} and V_{c} are the voltage pulses on the main transmission-line, inductive stub-line (short-circuit), and capacitance stub-line (open-circuit), respectively. The superscripts i and r are used to denote incident and reflected pulses, respectively. The reflected voltage pulses are found from the following relationships:

$$\begin{cases} V_{\rm M}^{\rm r} = V_{\rm M}^{\rm i} + v \\ V_{\rm L}^{\rm r} = v - V_{\rm L}^{\rm i} \\ V_{\rm C}^{\rm r} = v - V_{C}^{\rm i} \end{cases} \tag{14.43}$$

From Eq. (14.43), the scattering matrix may be found as:

$$\begin{gathered} {}_k\begin{bmatrix} V_{\rm M} \\ V_{\rm C} \\ V_{\rm L} \end{bmatrix}^{\rm r} = \frac{1}{y} \begin{bmatrix} t(g+y) & 2tY_{\rm C} & 2tY_{\rm L} \\ g & 2Y_{\rm C} - y & 2Y_{\rm L} \\ g & 2Y_{\rm C} & 2Y_{\rm L} - y \end{bmatrix}_k \begin{bmatrix} V_{\rm M} \\ V_{\rm C} \\ V_{\rm L} \end{bmatrix}^{\rm i} \\ y = 1 + Y_{\rm C} + Y_{\rm L} \\ t = \exp\left(\frac{-\alpha_{\rm sc}\Delta L}{2}\right). \end{gathered} \tag{14.44}$$

The scattering mechanism of Eq. (14.44) is implemented for pulses travelling in two opposite propagating directions inside the laser cavity, i.e. forward and backward. If the gain coefficient is less than 10^{-10} the gain model may be assumed linear as expressed in the following form:

$$g = av_{\rm g}(N - N_{\rm tr})\frac{S}{(1+\varepsilon)S} \tag{14.45a}$$

where

$$v_{\rm g} = \frac{c_0}{n_{\rm g}} \tag{14.45b}$$

where S is the photon density of the local model section, a is the differential gain constant, $n_{\rm g}$ is the effective group index, and ε is the gain compression factor. However, if the gain is greater than 10^{-10}, the gain model is assumed to be logarithmic [38] in the form of:

$$g = av_{\rm g}\ln\left(\frac{N}{N_{\rm tr}}\right)\frac{S}{(1+\varepsilon)S}\,. \tag{14.46}$$

The linear gain model is assumed throughout this book. As the model bandwidth is limited by the sampling rate used, the TLM stub filter response falls to zero at both edges of the model bandwidth. However, there is a good fit between the stub filter response and the actual Lorentzian lineshape near the peak, as shown in Fig. 14.12. The stub filter response was found by performing an FFT on the time-domain impulse response.

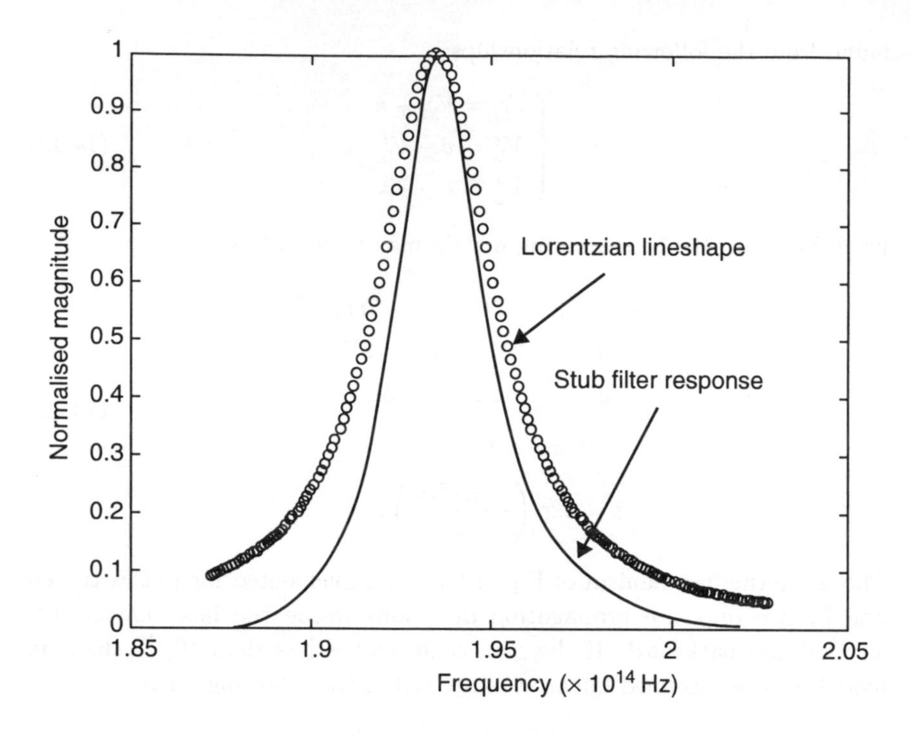

Fig. 14.12. Comparison between the stub filter response and actual Lorentzian shape.

The bandwidth of laser gain spectrum is governed by the Q-factor of the stub filter, which is related to the admittances Y_c and Y_L of the stub filter. The Q-factor is defined as the ratio of centre frequency to its the bandwidth. We shall now find the relationship between the Q-factor and the admittances of the stub filter. From transmission-line theory, the input impedance, Z_{in}, is expressed by [5]:

$$Z_{\text{in}} = Z_0 \frac{Z(l) + jZ_0 \tan(k_0 l)}{Z_0 + jZ(l) \tan(k_0 l)} \tag{14.47}$$

where k_0 is the free-space wavenumber, l is the distance from load (e.g. open circuit and short-circuit), and Z_0 is the characteristic impedance of the line. When an open circuit stub is used to simulate a capacitive stub line, $Z(l)$ approaches infinity, and we have:

$$Z_{\text{iC}} = \frac{Z_{\text{C}}}{j \tan(k_0 l)} = -jZ_{\text{C}} \cot(k_0 l) \tag{14.48}$$

where Z_C is the impedance of the capacitive stub-line. When a short circuit stub is used to simulate an inductive stub line, $Z(l)$ is zero, and we have:

$$Z_{iL} = jZ_L \tan(k_0 l) \tag{14.49}$$

where Z_L is the impedance of the inductive stub-line. In the laser cavity, the free space wavenumber (k_0) should be replaced by the propagation constant of the semiconductor material, which is given as:

$$\beta = k_0 n_g = \frac{2\pi f_0 n_g}{c_0} \tag{14.50}$$

where c_0 is the speed of light in free space. Since the model is discretised in time (Δt) and space (Δl), and to comply to the synchronisation criterion, the length of the stub must be half of a unit section length, i.e. $\Delta l/2$. Hence, the input impedance of the stubs can be expressed by:

$$\begin{aligned} Z_{iC} &= -jZ_C \cot(\pi f_0 \Delta t) \\ Z_{iL} &= jZ_L \tan(\pi f_0 \Delta t)\,. \end{aligned} \tag{14.51}$$

In a parallel RLC filter with unity resistance ($R = 1\Omega$), the Q-factor is given as:

$$Q = \frac{R}{\omega_0 L} = \sqrt{\frac{C}{L}} \quad \text{where} \quad \omega_0 = \frac{1}{\sqrt{LC}}\,. \tag{14.52}$$

By making use of Eqs. (14.8) and (14.10), (14.52) can be rewritten in terms of transmission-line impedances:

$$Q = \sqrt{\frac{1}{Z_C Z_L}}\,. \tag{14.53}$$

At resonance, the parallel combination of stub admittance is zero ($Y_{iC} + Y_{iL} = 0$), leading to the following relationship:

$$\sqrt{\frac{Z_C}{Z_L}} = \tan(\pi f_0 \Delta t)\,. \tag{14.54}$$

From Eqs. (14.53) and (14.54), we can define the admittances of the stub filter as the following:

$$\begin{aligned} Y_L &= Q \tan(\pi f_0 \Delta t) \\ Y_C &= \frac{Q}{\tan(\pi f_0 \Delta t)}\,. \end{aligned} \tag{14.55}$$

For the purpose of computational efficiency, the baseband Q-factor (Q_{dc}) is used in the model instead of the actual Q-factor. The relationship between them is given as [16]

$$Q_{\text{dc}} = Q\left(1 - \frac{Bf_{\text{samp}}}{f_0}\right) \quad \text{where} \quad f_{\text{samp}} = \frac{1}{\Delta t} \tag{14.56}$$

f_0 is the actual lasing frequency, and B is the bandnumber, which will be discussed in Section 14.11. For the same reason, the baseband lasing frequency (f_{dc}) replaces the actual lasing frequency (f_0). The baseband lasing frequency is usually placed at the centre of the model bandwidth, i.e. $f_{\text{dc}} = 1/4\Delta t$.

As mentioned earlier, the model bandwidth is determined by the sampling rate used. Therefore, the gain filter response will fall to zero at the edges of the available bandwidth as shown in Fig. 14.13 by the dotted lines. This corresponds to model timestep of $\Delta t = 50$ fs that gives a total

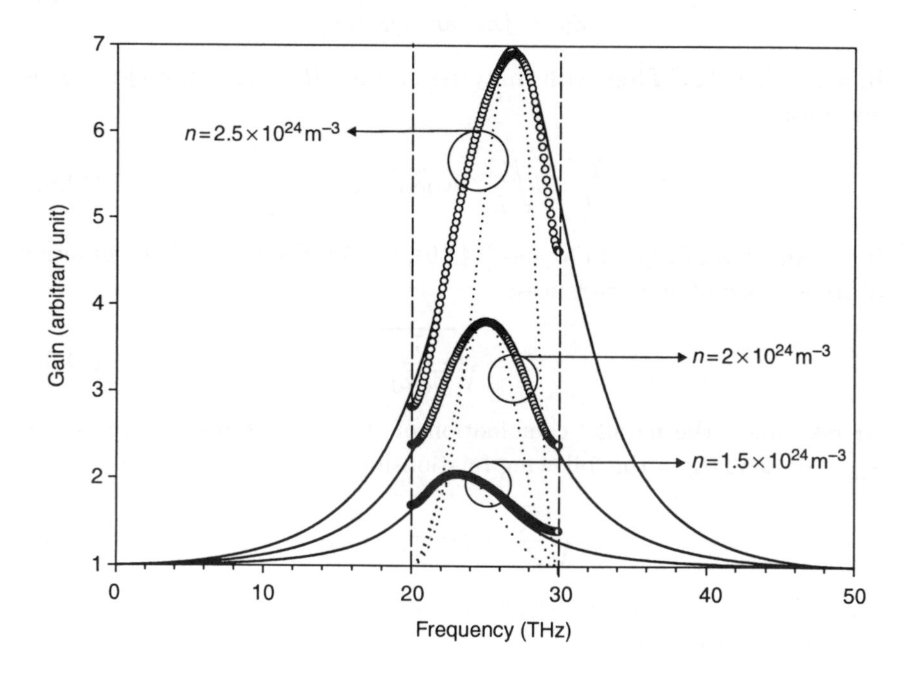

Fig. 14.13. Improvement in gain spectrum by using additional TLM stub-filters. Carrier dependence of the gain is included. Solid lines: $\Delta T = 10$ fs (simple gain stub filter); Dotted lines: $\Delta T = 50$ fs (simple gain stub filter); Circles: $\Delta T = 50$ fs (improved gain stub filter including HP and LP stub-filters).

bandwidth of only 10 THz, as shown by the dashed lines. By using a smaller timestep $\Delta t = 10$ fs, the total available bandwidth increased by five times to 50 THz but at the expense of lower computational speed. The runtime is now 25 times slower because the computational task is proportional to S^2, where S is the total number of sections in the model. One method to increase computational speed and yet achieve reasonably accurate gain spectra is to use first-order TLM stub filters to perform some shaping to the spectral response in addition to some discrete-time signal processing [39]. The improvements are shown as circular symbols in Fig. 14.13. Notice that now the spectral responses do not fall to zero at the band edges. The dynamic gain peak shift can also be modelled by simply including the carrier dependence of the lasing frequency, f_{dc} [35].

The progression of the amplified and filtered voltage pulses (i.e. connection) along the main transmission-lines are described by the following connecting relationship:

$$\begin{aligned} {}_{k+1}F_{\mathrm{M}}^{\mathrm{i}}(n) &= {}_{k}F_{\mathrm{M}}^{\mathrm{r}}(n-1) \\ {}_{k+1}B_{\mathrm{M}}^{\mathrm{i}}(n) &= {}_{k}B_{\mathrm{M}}^{\mathrm{r}}(n+1) \end{aligned} \tag{14.57}$$

where $F_{\mathrm{M}}(n)$ and $B_{\mathrm{M}}(n)$ are the forward and backward travelling voltage pulses of section-n, respectively. At the facets, the connecting stems from the fact that the pulses are reflected into the opposite direction due to the impedance mismatch between transmission-line and the terminal load, that is:

$$\begin{aligned} {}_{k+1}F_{\mathrm{M}}^{\mathrm{i}}(1) &= \sqrt{R_1}\,{}_{k}B_{\mathrm{M}}^{\mathrm{r}}(1) \\ {}_{k+1}B_{\mathrm{M}}^{\mathrm{i}}(S) &= \sqrt{R_2}\,{}_{k}F_{\mathrm{M}}^{\mathrm{r}}(S) \end{aligned} \tag{14.58}$$

where R_1 and R_2 are the power reflectivities at the left (section 1) and right end (section S) facets, respectively. This is assuming that the voltage pulses travel from left to right for the forward direction, and vice versa for the backward direction. Finally, the connecting algorithm for the stub lines is given as,

$$\begin{aligned} {}_{k+1}V_{\mathrm{C}}^{\mathrm{i}}(n) &= {}_{k}V_{\mathrm{C}}^{\mathrm{r}}(n) \quad \text{(open circuit stub)} \\ {}_{k+1}V_{\mathrm{L}}^{\mathrm{i}}(n) &= -{}_{k}V_{\mathrm{L}}^{\mathrm{r}}(n) \quad \text{(short circuit stub)} \end{aligned} \tag{14.59}$$

where the voltage pulse, V, is applicable for both the forward and backward direction (F or B).

14.9 Carrier-Induced Frequency Chirp

The consequence of carrier dependence of the active layer refractive index is a dynamic spectral shift, commonly known as chirping. To simulate this refractive index change, the phase length is altered by phase (*adjusting*)-stubs, which can either be placed in (i) each and every section of the model or (ii) only at the facets of the laser cavity. For the time being, we will only discuss case (ii), which is known as the *stub-attenuator* model [40]. The former (i.e. case (i)) takes into account inhomogeneous distribution of the carrier density and will only be discussed in Chapter 16, where inhomogeneous effects are important, owing to the travelling-wave condition.

In the stub-attenuator model of Fig. 14.14, the voltage pulses get reflected back into the opposite direction due to the impedance mismatch between main transmission-lines and terminal loads. The voltage pulses then enter the phase-stubs to be delayed before they continue propagating along the main transmission-line. The three-port circulator at the left facet (section 1) shown in Fig. 14.14 ensures that the delayed pulses from the phase-stubs that are incident at port 2 continue to propagate into port 3 and not port 1 (e.g. see Fig. 14.15). The scattering matrix of a three-port

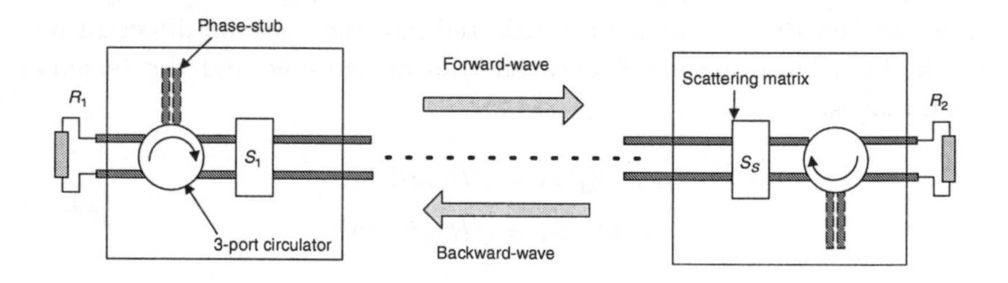

Fig. 14.14. Stub-attenuator model.

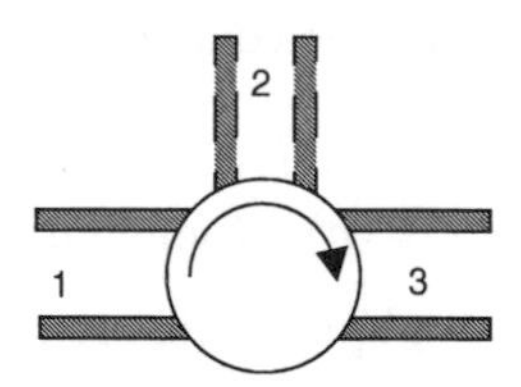

Fig. 14.15. Three-port circulator.

circulator is given as [4]:

$$\begin{bmatrix} V_1 \\ V_2 \\ V_3 \end{bmatrix}^{\mathrm{r}} = \begin{bmatrix} 0 & 0 & 1 \\ 1 & 0 & 0 \\ 0 & 1 & 0 \end{bmatrix} \begin{bmatrix} V_1 \\ V_2 \\ V_3 \end{bmatrix}^{\mathrm{i}} . \tag{14.60}$$

Taking the left facet (Section 1) as example, the backward travelling wave (i.e. a train of voltage pulses) enters port 3 of the circulator and gets transferred into port 1, where it gets reflected into the phase-stubs (port 2). The reflected wave then becomes incident wave at port 2 of the circulator at the next time step. The delayed wave now entering port 2 gets diverted into port 3 and continues to propagate along the main transmission-lines. The scattering matrix that describes the stub-attenuator model at the left facet is given by:

$$\begin{bmatrix} V_1(1) \\ V_2(1) \\ V_3(1) \end{bmatrix}^{\mathrm{r}} = \begin{bmatrix} 0 & 0 & 1 \\ \dfrac{2Z_{\mathrm{s}}}{(1+Z_{\mathrm{s}})} & \dfrac{(1-Z_{\mathrm{s}})}{(1+Z_{\mathrm{s}})} & 0 \\ \dfrac{(Z_{\mathrm{s}}-1)}{(1+Z_{\mathrm{s}})} & \dfrac{2}{(1+Z_{\mathrm{s}})} & 0 \end{bmatrix} \begin{bmatrix} V_1(1) \\ V_2(1) \\ V_3(1) \end{bmatrix}^{\mathrm{i}} \tag{14.61}$$

where all impedances are normalised to the characteristic impedance of the main transmission-lines (Z_0), Z_{s} is the phase-stub impedance, and R_1 is the power reflectivity of the left facet (1st section). A similar scattering matrix is needed at the right end facet (Sth section). On first thought, the phase-stub length should be varied to adjust the total phase-length of the laser cavity. However, the need for synchronisation of pulses in the TLLM methodology only allows us to vary the stub impedance. This is an equivalent approach, since the stub impedance is related to the stub length. To find this relationship, the input impedance (Z_{in}) of the *fixed-length* stub (variable line impedance) is equated to the input impedance (Z_{in}) of the *adjustable-length* open-circuit stub (fixed line impedance). The input impedance of the adjustable open-circuit stub is:

$$Z_{\mathrm{in}} = \frac{Z_0}{j\tan(\beta \Delta l_{\mathrm{ph}})} \tag{14.62}$$

where Δl_{ph} is the desired extension/contraction. The fixed-length stubs can either be capacitive (open-circuit) or inductive (short-circuit), as defined in Eq. (14.51), this depends on the average carrier concentration level. This will be explained as follows.

Case 1: When the change in carrier concentration (ΔN) is negative, the change in phase length is positive, and the fixed-length capacitive stub is used. Comparing Eq. (14.62), and Eq. (14.51), for the capacitive stub, we have:

$$Z_{\text{in}} = \frac{Z_0}{j\tan(\beta\Delta l_{\text{ph}})} = -Z_{\text{s}} j\cot(\pi f_0 \Delta t) \tag{14.63a}$$

or

$$Z_{\text{s}} = \frac{Z_0 \tan(\pi f_0 \Delta t)}{\tan(\beta\Delta l_{\text{ph}})} \tag{14.63b}$$

where Z_{s} is the fixed-length stub's *variable* impedance.

Case 2: On the other hand, when the change in carrier concentration is positive, the change in phase length is negative, and the fixed-length inductive stub is used. Comparing Eqs. (14.62) and (14.51) for the inductive stub, we have:

$$Z_{\text{in}} = \frac{Z_0}{j\tan(\beta\Delta l_{\text{ph}})} = jZ_{\text{s}}\tan(\pi f_0 \Delta t) \tag{14.64a}$$

or

$$Z_z = -\frac{Z_0}{\tan(\pi f_0 \Delta t)\tan(\beta\Delta l_{\text{ph}})}\,. \tag{14.64b}$$

The variable phase length (Δl_{ph}) can have values above and below zero depending on whether the carrier concentration drops below or rises above the reference value (N_{ref}), corresponding to a red and blue chirp, respectively. Since the change in resonant frequency is the same whether it is due to a change in (i) effective refractive index (Δn) or (ii) variable phase length (Δl_{ph}), the relationship between Δn and ΔI_{ph} can be expressed by [40]:

$$\begin{aligned}\Delta l_{\text{ph}} &= \frac{\Gamma L \Delta n}{n_{\text{g}}}\\ \Delta n &= \frac{dn}{dN}(N'_{\text{av}} - N_{\text{ref}})\end{aligned} \tag{14.65}$$

where N'_{av} is the carrier density averaged over the entire cavity, and dn/dN is the rate of refractive index change with carrier density. The dispersionless transmission-lines are based on the assumption that group index (n_{g}) equals to phase index (n). The term N_{ref} is the reference value of carrier density for zero phase shift, usually set to the threshold level. The rate of change of n with N is defined as [40]:

$$\frac{dn}{dN} = -\frac{\alpha_{\rm H} c_0 a}{4\pi f_0} \tag{14.66}$$

where $\alpha_{\rm H}$ is Henry's linewidth enhancement factor [41] and a is the differential gain constant.

14.10 Spontaneous Emission Model

The TLLM is a stochastic model that takes noise, which is inevitable in a real laser device, into account. Consideration of noise in a laser model is important for the following reasons. In ultrashort pulse generation using gain-switched and mode-locked lasers, the emitted optical pulses do not exactly have regularly-spaced intervals but noise-induced timing jitter exists [42–43]. This may lead to increased bit error rates (BER) when used as pulse sources in optical time-division multiplexed (OTDM) communication systems [44]. Salathe *et al.*, found that spontaneous emission noise in the laser rate equation smoothens the transition between non-lasing to lasing condition [45]. The presence of spontaneous emission has also been found to dampen relaxation oscillations during pulse code modulation (PCM) [46–47]. In addition, spontaneous emission contributes to phase noise and determines the continuous (CW) linewidth of each longitudinal mode. The CW linewidth of the laser is crucial when used as local oscillator in coherent optical systems [48].

The spontaneous emission noise is modelled by current sources in TLLM, which will be shown equivalent to the free current density term in Maxwell's equations [49]. The presence of a current source along the lossless transmission-lines (see Fig. 14.16) can be described mathematically

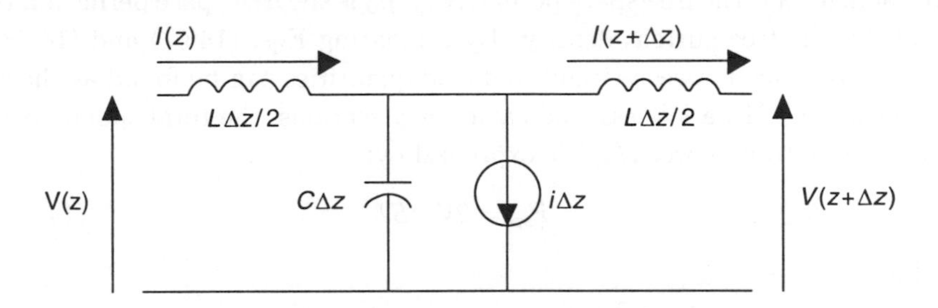

Fig. 14.16. Equivalent circuit diagram of the lossless transmission-line including a current source.

Table 14.1. Equivalence between circuit and field quantities.

Circuit quantities	Field quantities
V/m	E_x
I/m	$-H_y$
L	μ_0
C	$n^2\varepsilon_0$
i/m	j_x

by the telegraphist equations [7], given as:

$$-\frac{\partial I}{\partial z} = C\frac{\partial V}{\partial t} + i \tag{14.67a}$$

and

$$\frac{\partial V}{\partial z} = -L\frac{\partial I}{\partial t} \tag{14.67b}$$

where C is capacitance per unit length, L is inductance per unit length, V is the voltage wave, I is the current wave, and i is the external current source per unit length. For TEM (plane) waves ($E_x = H_z = 0$), Maxwell's equations can be written as:

$$\frac{\partial H_y}{\partial z} = n^2\varepsilon_0\frac{\partial E_x}{\partial t} + j_x \tag{14.68a}$$

and

$$\frac{\partial E_x}{\partial z} = \mu_0\frac{\partial H_y}{\partial t} \tag{14.68b}$$

where E is the electric field, H is the magnetic field, n is the index of refraction, ε_0 is the free-space permittivity, μ_0 is the free-space permeability, and j is the free current density. By comparing Eqs. (14.67) and (14.68), the equivalence between circuit and field quantities can be found as shown in Table 14.1. In a lossless cold cavity of S sections, the total spontaneous emission output power (P_{sp}) is expressed by:

$$P_{\text{sp}} = 2WdSI \tag{14.69a}$$

where

$$I = \frac{1}{Z_{\text{p}}}\left(\frac{V}{m}\right)^2 \tag{14.69b}$$

and

$$V^2 = i_{\rm rms}^2 \left(\frac{Z_{\rm p}\Delta l}{2} \right)^2 \tag{14.69c}$$

d is the active layer thickness, W is the active layer width, I is the output intensity of a unit section due to spontaneous emission, and m is a unity constant of unit length to achieve dimensional correctness. With a spontaneous emission current source placed at each and every node of a S-section model and split into two propagating directions (forward and backward), the output power per longitudinal mode out of both facets may be expressed by:

$$P_m = \frac{Wd\langle i^2 \rangle Z_{\rm p}(\Delta l)^2}{2m^2} \tag{14.70}$$

where $\langle i^2 \rangle$ is the mean square value of the current source. Alternatively, from the photon rate equation, the spontaneous emission output power per longitudinal mode can be written as:

$$P_m = WdLhf_0\beta R(N) \tag{14.71a}$$

where

$$L = S\Delta l \tag{14.71b}$$

is the spontaneous emission coupling factor and $R(N)$ is the recombination rate. By comparing Eqs. (14.70) and (14.71), the mean squared current is found to be given as [49]:

$$\langle i^2 \rangle = \frac{2\beta R(N) hf_0 m^2 S}{Z_p \Delta l}\,. \tag{14.72}$$

As spontaneous emission is a random process, only its mean squared value could be defined. The spontaneous emission current source is simulated by a Gaussian random number generator [50] with mean square of $\langle i^2 \rangle$ and mean of zero. Since the mean value of white Gaussian noise is zero, the mean square value equals to its variance [51]. In order to model the distributive nature of the spontaneous emission noise, the current sources are placed along the TLLM and the mean square current is calculated using local carrier concentration rather than just the average value of the entire cavity. The equivalent circuit diagram of the distributed current source model of spontaneous emission is shown in Fig. 14.17. In each section

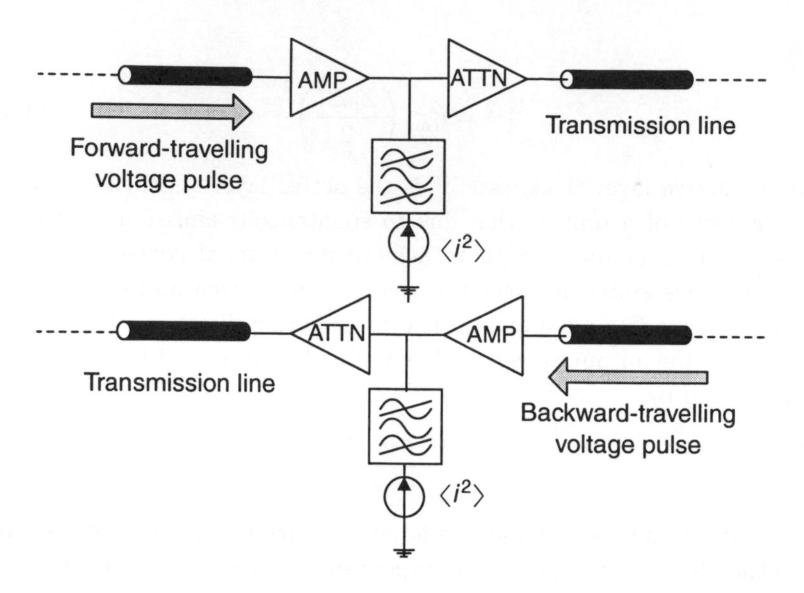

Fig. 14.17. One section of the distributed current source model of spontaneous emission.

of the TLLM, the incoming voltage pulses are amplified by wavelength-selective amplifiers (AMP) before spontaneous emission noise is added to it. Note that the filtered spontaneous emission current must first be converted to voltage before being added to the main voltage pulses. Along the transmission-lines, the attenuators (ATTN) simulate the material and scattering losses in the optical cavity.

The spontaneous emission spectrum has Gaussian noise statistics and is not white noise (frequency independent) but coloured noise (frequency dependent). Therefore, the TLM stub filter (see Section 14.8) is placed at the output of the spontaneous emission current source to model the frequency dependence. The resonant cavity modes collect all the power that is spontaneously emitted into the frequency intervals between adjacent cavity modes [52].

14.11 Computational Efficiency-Baseband Transformation

Modelling with digital computers requires discretisation of signals, meaning that real continuous waveform must be sampled at equally spaced points in time. Since there is a close relationship between time and frequency, the need for sampling in time leads to a signal of finite frequency bandwidth.

According to Nyquist's sampling theorem, an accurate reconstruction of the original signal from the sampled version can only be obtained if the sampling rate is twice the bandwidth of the original signal. Equivalently, the unit section length, Δl, must be at most one half of the group wavelength, λ_g, in the laser cavity. Therefore, the minimal number of sections S allowed in the laser model with length of L is:

$$S \geq \frac{2L}{\lambda_g} \tag{14.73a}$$

where

$$\lambda_g = \frac{c_0}{f_0 n_g} \,. \tag{14.73b}$$

Since laser devices operate at the near-infrared region, this will lead to a huge amount of computation if the Nyquist criterion is strictly followed for the optical carrier frequency (f_0). For example, a 300 μm laser chip with effective group index of 4 operating at 1.55 μm would require at least 1549 sections! This is too computationally intensive considering the amount of processing tasks (that is, the laser physical processes that must be taken into account) at each and every section at each time iteration, which may take up weeks to process on a Pentium^{TM} processor. For reasonable computation times, the total number of sections (S) used in TLLM is usually between 10 and 100. Interestingly, the value of S also corresponds to the number of longitudinal modes that are taken into account by the model [16]. Computation time is proportional to S^2, which is why S must not be too large.

Usually, the modulated light signal is at microwave or millimetre-wave frequencies, which has a bandwidth at least a few thousand times smaller than its optical carrier frequency. Therefore, the TLLM makes use of a *baseband transformation* method to shift the optical carrier down to the DC level to speed up the computational time [16], which is illustrated in Fig. 14.18. Only the positive frequencies are considered here because the model uses real signals, even though complex signals can also be used but at the expense of lower computational efficiency. The baseband transformation method has also been applied to the quantum-mechanical model in [32].

The baseband transformation method depicted in Fig. 14.18 is acceptable because the semiconductor laser has a relatively narrow *signal* bandwidth (THz region or less) compared to its optical *carrier* frequency

(10^{14} Hz and above). For example, in Fig. 14.18 the signal bandwidth is given as $(f_{\max} - f_{\min})$ while the optical frequency is at a much higher value, denoted by f_0. The concept here is to neglect the spectral regions that are not of interest lying between DC and the optical carrier frequency (f_0) known as the *deadband* (the lighter lines in Fig. 14.18). Hence, the optical spectrum can be down-converted to baseband without any signal distortion, if Nyquist's sampling criterion is satisfied for the signal bandwidth (f_{band}). Now, the lower limit of the sampling rate (f_{samp}) is only restricted by:

$$f_{\text{samp}} \geq 2 f_{\text{band}} \,. \tag{14.74}$$

The above restriction ensures that there is no overlap between the alias bands, which would otherwise lead to signal lost and distortion. The degree of scaling down of the actual lasing frequency can be described by the *bandnumber* (B), which is expressed by:

$$B = \text{int}\left(\frac{f_0}{f_{\text{samp}}}\right) = \text{int}\left(\frac{f_0}{2 f_{\text{band}}}\right) \tag{14.75}$$

where the function "int" rounds off to the nearest but lower integer. In the example of Fig. 14.18, the bandnumber is $B = 4$. The down-converted lasing frequency, f_{dc}, is usually chosen to be at the cavity resonant frequency nearest to the centre of the model bandwidth $(1/\Delta t)$ if a symmetrical gain spectrum is assumed.

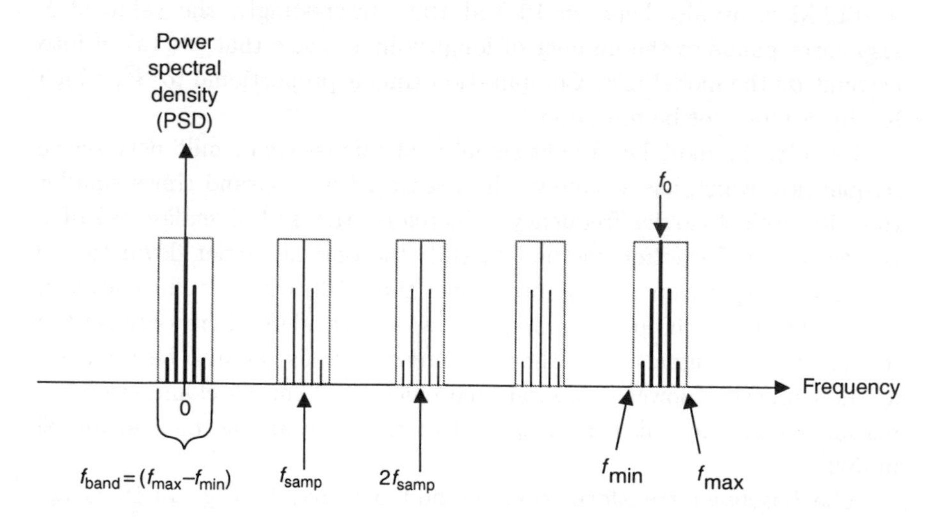

Fig. 14.18. Baseband transformation technique to reduce sampling rate.

14.12 Signal Analysis–Post-Processing Methods

In order to analyse the performances of the laser device, some form of signal processing must be applied on the laser output. This closely mimics the laboratory instrumentation for signal measurement and analysis such as optical spectrum analysers, sampling oscilloscopes, and streak cameras. The post-processing techniques included in the TLLM are explained in the following.

(1) *The Fourier Transform*: The output samples of the TLLM are voltage pulses in time domain. In order to find the frequency spectrum of the output signal, the discrete Fourier transform (DFT) or the more computationally efficient fast Fourier transform (FFT) can be used [53]. The information content in the frequency domain will always be less than that of the time domain data since it must be truncated at some point — a full frequency domain description can only be obtained if the time samples extend to infinity. The FFT produces uniformly distributed data points over a frequency bandwidth extending from $-f_{\text{samp}}/2$ to $f_{\text{samp}}/2$. However, only the *relative* frequency with respect to the lasing optical frequency (f_0) is important. The optical spectrum obtained will be a *continuous* spectrum displaying linewidth broadening effects due to noise fluctuations associated with spontaneous emission [54]. This is in contrast to the unrealistic *discrete* optical spectra produced from multimode rate equations [55–56].

(2) *Moving Average Filter*: The TLLM is a highly realistic model, where the output samples are voltage pulses that represent optical field amplitudes. The fast varying field amplitudes can be converted to instantaneous power by using Eq. (14.38). In practical measurements, the time-averaged power or optical intensity is usually observed. In the TLLM, digital filtering can be used to remove the high frequency oscillations of the optical carrier but at the expense of time resolution. For example, taking the average of a group of 10 samples causes the resolution to become 10 times coarser (ten points become one averaged point). To solve this problem, the moving average digital filter is used. This is briefly illustrated in Fig. 14.19.

The advantage is that the resolution of the filter does not depend on the size of the average moving window. However, for a window containing a large number of time samples, there is no sharp cutoff — significant attenuation can occur within the frequency range of interest. In order to achieve better cutoff characteristics, a weighted moving average window

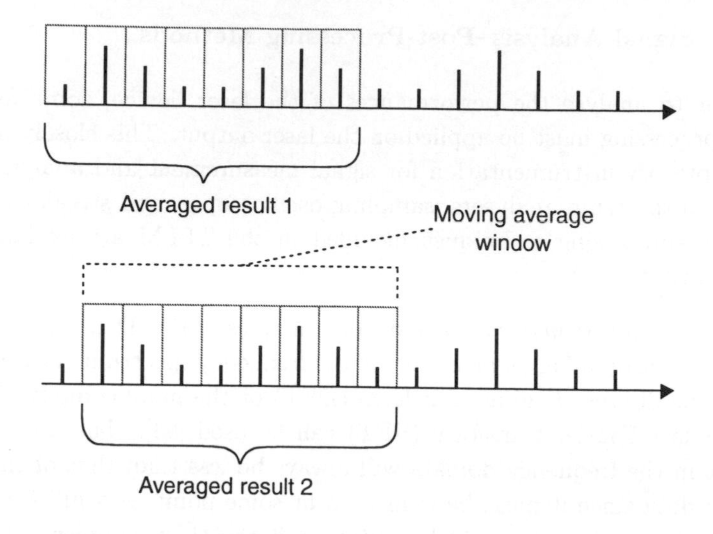

Fig. 14.19. Moving average digital filter.

such as a Gaussian response or binomial series can be used, but at the expense of longer computation time.

(3) *Stable Averaging Method*: When investigating short pulse generation in TLLM such as gain-switching and mode-locking, the optical pulses grow from spontaneous emission. Since spontaneous emission is a random process, this leads to the existence of timing jitter and random substructures within the pulse. As a result, the optical spectrum changes with time owing to random fluctuations in turn-on delay time, pulsewidth, and dynamic spectral shift (chirp). Even in CW operation, noise fluctuations are unavoidable unless the noise can be switched off at steady-state [57]. This makes it impossible to define a single value of optical pulsewidth or spectral width without using signal averaging techniques.

In practical measurements, stable pulses can be observed by using digital sampling oscilloscopes in averaging mode. The stable averaging algorithm [58] is implemented in TLLM to obtain *time-averaged* optical pulse and spectrum by filtering out any noise due to spontaneous emission and its beating effects. The stable averaging algorithm can be expressed as [58]:

$$P_m^i = P_{m-1}^i + \frac{p_m^i - P_{m-1}^i}{m} \tag{14.76}$$

where P_m^i is the averaged power of the ith time-slot at the mth sweep, and p_m^i is the instantaneous power from TLLM. A predetermined number of time-slots in memory are used to store the averaged results. To display the power envelope instead of the fast-varying instantaneous power, the term p_m^i can be replaced by the power that has been averaged by the moving-average filter. In this way, a single average value of pulsewidth and spectral width (FWHM) can be found. In addition, the random noise will be reduced by a factor of $\sqrt{m}$ for m number of sweeps used [58].

(4) *Sample and overlay method*: Sometimes it is helpful to be able to see how the repetitive optical pulses deviate from one another. By using the stable averaging technique any pulse instability is eliminated and thus cannot be seen in the final averaged pulse. One way of observing the pulse timing jitter is by overlaying samples of sequential pulses in a manner similar to a digital sampling oscilloscope in infinite persistence mode without signal averaging. Any pulse instability will show up as broadening of the traces since sequential pulses will not lie on top of one another exactly while a thin trace indicates good pulse stability.

(5) *Smoothing algorithm*: In order to find the spectral width of the time-averaged spectrum, we need to use both stable averaging and smoothing algorithms. The stable averaging method is used to obtain the time-averaged spectrum. However, in a multimode spectrum with mutiple peaks and substructures, especially during short pulse generation, it is difficult to define the full width at half maximum (FWHM). Smoothing is thus required before finding the FWHM of the optical spectrum. The smoothing algorithm used in TLLM is based on FFT-filtering [59]. The first step is to set-up a smoothing kernel, k, given as:

$$\begin{aligned} k &= \frac{1-r^2}{1-2r\cos(t)+r^2} \quad \text{where} \\ t &= 2\pi \times \left(0 \cdots \left[1-\frac{1}{2N_\mathrm{s}}\right] \text{ in steps of } \frac{1}{N_\mathrm{s}}\right) \end{aligned} \tag{14.77}$$

N_s is the total number of samples used (size), and r is the smoothing control ($0 < r < 1$). Next, we take the FFT of the original time-averaged spectrum and of the smoothing kernel to convert them into time domain, then we multiply them together. In the frequency domain, this is equivalent to a convolution between the original spectrum and the kernel. The result from this multiplication is then transformed back into the frequency domain to obtain the smoothed optical spectrum. Finally, we normalise the smoothed result to its size, N_s. Formally, the smoothing algorithm can be summarised as follows:

$t_k = fft(k)$	; FFT of kernel
$t_x = fft(x)$	; FFT of data
$t_d = \text{mulcpx}(t_k, t_x)$	; convolve kernel and data
$s_d = invfft(t_d)$	; transform back to frequency domain
$t_{sd} = \text{real}(s_d)/\text{size}[\text{real}(s_d)]$	; normalise the data
note : $\text{size}(s_d) = 2x\ \text{size}[\text{real}(s_d)]$	

where x is the data (array) we want smoothed, and t_{sd} is the final smoothed data. The function fft is the FFT, mulcpx is the multiplication in time domain (convolution in frequency domain), invfft is the inverse FFT, real is taking real values of the complex data, and **size** finds the total number of data samples. The above smoothing method can also be used for finding the FWHM of highly unstable pulses but usually the pulsewidth is not of interest outside the stable region of operation in ultrashort pulse generation.

14.13 Summary

Microwave circuit techniques when applied to semiconductor laser modelling provide us with additional insight into the operation of the device. The transmission-line laser model (TLLM) can be classified as a distributed element circuit model, which is based on the 1-D transmission-line matrix (TLM) method. The building blocks of a TLM network are the TLM link-lines and stub-lines. It has been shown how the scattering matrices of several TLM subnetworks may be derived by using Thevenin equivalent circuits. *Scattering* and *connecting* are the two main processes that form the basis of TLM. The scattering matrix at a TLM node takes incident voltage pulses and operates on them to produce reflected pulses that travel away from the node. The connecting matrix then directs the reflected pulses from one TLM node to adjacent TLM nodes, where they become incident pulses of the adjacent nodes in the next time iteration.

In TLLM, the voltage pulses represent the optical waves that circulate inside the laser cavity. All the important optical processes in the laser are taken into account, such as the spectrally-dependent gain of stimulated emission, material and scattering loss, spontaneous emission, carrier-photon interaction, and carrier-dependent phase shift. The microwave circuit elements of TLLM are used to describe these laser processes on an equivalence basis. The baseband transformation method is used to enhance the computational efficiency by down-converting from the true optical carrier frequency to its equivalent baseband value. The TLLM is a stochastic laser model because random noise effects are included, making it a highly realistic model compared to deterministic laser models [60–61]. However, intensive time averaging and smoothing techniques are required to obtain the wanted signal, which may otherwise be masked by noise.

References

[1] E. I. Gordon, "Optical maser oscillators and noise," *Bell Syst. Tech. J.*, Vol. 43, pp. 507–539, 1964.

[2] S. Kobayashi, "Injection-locked semiconductor laser amplifiers," in *Coherence, Amplification, and Quantum Effects in Semiconductor Lasers*, Y. Yamamoto, Ed. (New York: John-Wiley & Sons, Inc., 1991), p. 646.

[3] R. Adler, "A study of locking phenomena in oscillators," *Proc. I.R.E & Wav. Elec.*, June, pp. 351–357, 1946.

[4] J. L. Altman, Microwave Circuits (Princeton, New Jersey: D. Van Nostrand Company, Inc., 1964).

[5] R. E. Collins, *Foundations of Microwave Engineering* (New York: McGraw-Hill, 1966).

[6] G. P. Agrawal and N. K. Dutta, *Semiconductor lasers*, 2nd ed. (New York: Van Nostrand Reinhold, 1993).

[7] S. Ramo, J. R. Whinnery, and T. Van Duzer, *Fields and waves in communication electronics*, 3rd ed. (New York: John Wiley & Sons, 1994).

[8] P. P. Vasil'Ev, "Ultrashort pulse generation in diode lasers," *Opt. Quant. Elec.*, Vol. 24, pp. 801–824, 1992.

[9] P. B. Johns and R. L. Beurle, "Numerical solution of 2-dimensional scattering problems using a transmission-line matrix," *Proc. IEEE*, Vol. 188, No. 9, pp. 1203–1208, 1971.

[10] P. B. Johns, "A new mathematical model to describe the physics of propagation," *Radio Electron. Eng.*, Vol. 44, pp. 657–666, 1974.

[11] C. Christopoulos, *The Transmission-Line Modeling Method TLM* (New York: The IEEE/OUP Series on Electromagnetic Wave Theory, 1995).

[12] W. J. R. Hoefer, "The Transmission-Line Matrix Method-Theory and Applications," *IEEE Trans. Microwave Theory Tech.*, Vol. MTT-33, No. 10, pp. 882–893, 1985.

[13] J. W. Bandler, P. B. Johns, and M. R. M. Rizk, "Transmission-Line Modeling and Sensitivity Evaluation for Lumped Network Simulation and Design in the Time Domain," *Jour. Frank. Inst.*, Vol. 304, pp. 15–32, 1977.

[14] P. B. Johns and M. J. O'Brien, "Use of transmission-line modelling (t.l.m) method to solve non-linear lumped networks," *Radio Electron. Eng.*, Vol. 50, pp. 59–70, 1980.

[15] P. B. Johns, "A simple explicit and unconditionally stable numerical routine for the solution of the diffusion equation," *International Journal for Numerical Methods in Engineering*, Vol. 11, pp. 1307–1328, 1977.
[16] A. J. Lowery, "New dynamic semiconductor laser model based on the transmission-line modelling method," *IEE Proceedings Pt. J*, Vol. 134, pp. 281–289, 1987.
[17] A. J. Lowery, "Transmission-line modelling of semiconductor lasers: The transmission-line laser model," *Int. Jour. Num. Model.*, Vol. 2, pp. 249–265, 1989.
[18] P. B. Johns, "On the Relationship Between TLM and Finite-Difference Methods for Maxwell's Equations," *IEEE Trans. Microwave Theory Tech.*, Vol. MTT-35, No. 1, pp. 60–61, 1987.
[19] S. A. Boctor, *Electric Circuit Analysis* (Englewood Cliffs, NJ: Prentice-Hall, 1987).
[20] A. J. Lowery, "Dynamic Modelling of Distributed-Feedback Lasers Using Scattering Matrices," *Electron. Lett.*, Vol. 25, No. 19, pp. 1307–1308, 1989.
[21] A. J. Lowery and D. F. Hewitt, "Large-signal dynamic model for gain-coupled DFB lasers based on the transmission-line model," *Electron. Lett.*, Vol. 28, No. 21, pp. 1959–1960, 1992.
[22] A. J. Lowery, "Integrated mode-locked laser design with a distributed Bragg reflector," *IEE Proceedings Pt. J*, Vol. 138, No. 1, pp. 39–46, 1991.
[23] A. J. Lowery, "New dynamic model for multimode chirp in DFB semiconductor lasers," *IEE Proceedings Pt. J*, Vol. 137, No. 5, pp. 293–300, 1990.
[24] L. V. T. Nguyen, A. J. Lowery, P. C. R. Gurney, and D. Novak, "A Time Domain Model for High-Speed Quantum-Well Lasers Including Carrier Transport Effects," *IEEE Jour. Select. Top. Quant. Elec.*, Vol. 1, No. 2, pp. 494–504, 1995.
[25] A. J. Lowery, "New dynamic multimode model for external cavity semiconductor lasers," *IEE Proceedings Pt. J*, Vol. 136, No. 4, pp. 229–237, 1989.
[26] A. J. Lowery, "New time-domain model for active mode locking based on the transmission line laser model," *IEE Proceedings Pt. J*, Vol. 136, No. 5, pp. 264–272, 1989.
[27] L. Zhai, A. J. Lowery, and Z. Ahmed, "Locking bandwidth of actively mode-locked semiconductor lasers using fiber-grating external cavities," *IEEE J. Quant. Electron.*, Vol. 31, No. 11, pp. 1998–2005, 1995.
[28] A. J. Lowery, "New inline wideband dynamic semiconductor laser amplifier model," *IEE Proceedings Pt. J*, Vol. 135, No. 3, pp. 242–250, 1988.
[29] W. M. Wong and H. Ghafouri-Shiraz, "Integrated semiconductor laser-transmitter model for microwave-optoelectronic simulation based on transmission-line modelling," *IEE Proceedings Pt. J*, Vol. 146, No. 4, pp. 181–188, 1999.
[30] W. M. Wong and H. Ghafouri-Shiraz, "Dynamic model of tapered semiconductor lasers and amplifiers based on transmission line laser modeling," *IEEE Jour. Select. Top. Quant. Elec.*, Vol. 6, No. 4, pp. 585–593, 2000.
[31] M. G. Davis and R. F. Dowd, "A Transfer Matrix Method Based Large-

Signal Dynamic Model for Multielectrode DFB Lasers," *IEEE J. Quant. Electron.*, Vol. 30, No. 11, pp. 2458–2466, 1994.

[32] D. D. Marcenac and J. E. Carroll, "Quantum-mechanical model for realistics Fabry–Perot lasers," *IEE Proceedings Pt. J*, Vol. 140, No. 3, pp. 157–171, 1993.

[33] C. F. Tsang, D. D. Marcenac, J. E. Carroll, and L. M. Zhang, "Comparison between "power matrix mode" and "time domain model" in modelling large signal response of DFB lasers," *IEE Proceedings Pt. J*, Vol. 141, No. 2, pp. 89–96, 1994.

[34] L. M. Zhang and J. E. Carroll, "Large-Signal Dynamic Model of the DFB Laser," *IEEE J. Quant. Electron.*, Vol. 28, No. 3, pp. 604–611, 1992.

[35] A. J. Lowery, "A Study of the Static and Multigigabit Dynamic Effects of Gain Spectra Carrier Dependence in Semiconductor Lasers Using a Transmission-Line Laser Model," *IEEE J. Quant. Electron.*, Vol. 24, pp. 2376–2385, 1988.

[36] A. J. Lowery, P. C. R. Gurney, X. H. Wang, L. V. T. Nguyen, Y. C. Chan, and M. Premaratne, "Time-domain simulation of photonic devices, circuits, and systems," *Proceedings of SPIE*, Vol. 2693, pp. 624–635, 1996.

[37] A. E. Siegman, *LASERS* (Mill Valley, California: University Science Books, 1986).

[38] P. W. Mc Ilroy, A. Kurobe, and Y. Uematsu, "Analysis and application of theoretical gain curves to the design of multi-quantum-well lasers," *IEEE J. Quant. Electron.*, Vol. 21, pp. 1958–1963, 1985.

[39] L. V. T. Nguyen, A. J. Lowery, P. C. R. Gurney, D. Novak, and C. N. Murtonen, "Efficient Material-Gain Models for the Transmission-Line Laser Model," *Int. Jour. Num. Model.*, Vol. 8, pp. 315–330, 1995.

[40] A. J. Lowery, "Model for multimode picosecond dynamic laser chirp based on transmission line laser model," *IEE Proceedings Pt. J*, Vol. 135, No. 2, pp. 126–132, 1988.

[41] C. H. Henry, "Theory of the Linewidth of Semiconductor Lasers," *IEEE J. Quant. Electron.*, Vol. QE-18, No. 2, pp. 259–264, 1982.

[42] A. Valle, M. Rodriguez, and C. R. Mirasso, "Analytical calculation of timing jitter in single-mode semiconductor lasers under fast periodic modulation," *Opt. Lett.*, Vol. 17, No. 21, pp. 1523–1525, 1992.

[43] D. J. Derickson, P. A. Morton, J. E. Bowers, and R. L. Thorton, "Comparison of timing jitter in external and monolithic cavity mode-locked semiconductor lasers," *Appl. Phys. Lett.*, Vol. 59 (26), pp. 3372–3374, 1991.

[44] R. S. Tucker, G. Eisenstein, and S. K. Korotky, "Optical Time-Division Multiplexing For Very High Bit-Rate Transmission," *IEEE J. Lightwave Technol.*, Vol. LT-6, No. 11, pp. 1737–1749, 1988.

[45] R. Salathe, C. Voumard, and H. Weber, "Rate equation approach for diode lasers. Part 1: Steady-state solutions for a single diode," *Opto-electronics*, Vol. 6, pp. 451–456, 1974.

[46] M. Danielson, "A theoretical analysis for Gb/s pulse code modulation of semiconductor lasers," *IEEE J. Quant. Electron.*, Vol. QE-12, pp. 657–659, 1976.

[47] P. M. Boers, M. T. Vlaardingerbroek, and M. Danielsen, "Dynamic Behaviour of Semiconductor Lasers," *Electron. Lett.*, Vol. 11, No. 10, pp. 206–208, 1975.

[48] T. Okoshi and K. Kikuchi, *Coherent optical fiber communications* (Dordrecht: Kluwer Academic Publishers, 1988).

[49] A. J. Lowery, "A new time-domain model for spontaneous emission in semiconductor lasers and its use in predicting their transient response," *Int. Jour. Num. Model.*, Vol. 1, pp. 153–164, 1988.

[50] "Chapter G05 Random Number Generators," NAG Fortran Library (Mark 16).

[51] E. Kreyszig, *Advanced engineering mathematics*, 7th Ed. (New York: John Wiley & Sons, Inc., 1993).

[52] D. Marcuse, "Classical derivation of the laser rate equation," *IEEE J. Quant. Electron.*, Vol. QE-19, pp. 1228–1231, 1983.

[53] "Chapter C06 Summation of Series," NAG Fortran Library (Mark 16).

[54] Y. C. Chan, M. Premaratne, and A. J. Lowery, "Semiconductor laser linewidth from the transmission-line laser model," *IEE Proceedings Pt. J*, Vol. 144, No. 4, pp. 246–252, 1997.

[55] D. Marcuse and T. P. Lee, "On Approximate Analytical Solutions of Rate Equations fo Studying Transient Spectra of Injection Lasers," *IEEE J. Quant. Electron.*, Vol. QE-19, No. 9, pp. 1397–1406, 1983.

[56] T. P. Lee, C. A. Burrus, J. A. Copeland, A. G. Dentai, and D. Marcuse, "Short-Cavity InGaAsP Injection Lasers: Dependence of Mode Spectra and Single-Longitudinal-Mode Power on Cavity Length," *IEEE J. Quant. Electron.*, Vol. QE-18, No. 7, pp. 1101–1113, 1982.

[57] A. J. Lowery and Y. C. Chan, "Deterministic spectrum simulation using the transmission line laser model," *Electron. Lett.*, Vol. 20, No. 2, pp. 134–136, 1994.

[58] J. E. Deardorff and C. R. Trimble, "Calibrated Real-Time Signal Averaging," *HP Journal*, April 1968, pp. 8–13, 1968.

[59] SPSS Inc., "Sigmaplot 4.0," smooth.xfm.

[60] D. Marcuse, "Computer model of an injection laser amplifier," *IEEE J. Quant. Electron.*, Vol. QE-19, No. 1, pp. 63–73, 1983.

[61] G. Bendelli, K. Komori, and S. Arai, "Gain Saturation and Propagation Characteristics of Index-Guided Tapered-Waveguide Traveling-Wave Semiconductor Laser Amplifiers (TTW-SLA's)," *IEEE J. Quant. Electron.*, Vol. 28, No. 2, pp. 447–457, 1992.

Chapter 15

Microwave Circuit Models of Semiconductor Lasers

15.1 Introduction

Microwave signals are traditionally transmitted via coaxial cables and waveguides, which are very lossy and suffer from electromagnetic interference. Fibre-optic links offer a good alternative because of low-loss, high bandwidth, and immunity to electromagnetic interference. In recent years, much effort has been devoted to optical generation of millimetre-wave (mm-wave) signals for applications such as fibre-fed wireless communication systems and phased array antennas [1, 2]. In novel radio-over-fibre systems, integration between optical and microwave devices may be required [3–5]. Therefore, equivalent circuit models of optical devices are handy simulator tools that can be combined with existing circuit models of microwave devices for design and optimisation in such systems.

Microwave circuit techniques are well-established in the microwave research community and equivalent circuit models of microwave devices have been extensively used for design and modelling purposes for many years now [6, 7]. However, well-accepted circuit models of laser diodes are still rare compared to that of microwave devices. In this chapter, an introduction of a *lumped-element* circuit model, based on the laser rate equations, is first presented. Lumped-element circuit models of laser diodes have been used for many different applications in recent years [8–21]. These equivalent circuit models based on lumped-elements are useful in the fast emerging research area of microwave photonics because of the insight they provide and for their simplicity [22, 23].

The main advantage of lumped-element circuit models of laser diodes is that they are compatible with general purpose circuit analysis programs such as PSPICE or Hewlett-Packard's Microwave Design System (MDS) software. However, the usefulness of these models is limited by many simplifying assumptions as they are based on low-frequency concepts, where the operating wavelength is much larger than the dimensions of the device. These limitations can be overcome by using the *distributed-element* circuit model based on transmission-line laser modelling (TLLM), which is introduced in Chapter 15. The TLLM is a powerful and flexible equivalent circuit modelling technique to simulate various types of multimoded laser devices, where reflections and feedback are important [24]. Random spontaneous emission noise is also included in the stochastic TLLM model [25–26], making it possible to study signal-to-noise ratio (SNR) and bit-error-rates (BER) in amplified optical systems as well as timing jitter during ultrashort pulse generation.

Later in this chapter, a comprehensive laser diode transmitter model referred to as the *integrated TLLM model* [27] that was developed in this book will be presented. The integrated TLLM model is suitable for a microwave optoelectronic simulation environment because it includes practical microwave considerations such as electrical parasitics, a matching network as well as the intrinsic laser diode. The proposed model is a realistic microwave-optoelectronic laser diode model that is able to conveniently include both the electrical (matching circuit and parasitics) and optical (travelling-wave rate equations) characteristics of the laser diode transmitter within the same source code. This is possible because the integrated TLLM model is entirely based on microwave circuit modelling techniques.

15.2 Electrical Parasitics

Electrical characterisation through the measurement of current-voltage ($I - V$) derivatives was used as an early method to understand current paths in the injection laser and its lasing behaviour [28–30]. For the stripe-geometry laser, the circuit equivalent below threshold is just an ideal diode with a series resistance R. The $I - V$ characteristics of the diode can be expressed by the Shockley equation [31]:

$$I = I_{\rm s} \exp\left[\left(\frac{qV_{\rm d}}{\eta kT}\right) - 1\right] \tag{15.1}$$

where I_s is the saturation current, k is the Boltzmann constant, T is the absolute temperature, V_d is the voltage across the diode, and η is the ideality factor. *Above threshold*, the voltage across the laser diode saturates, and the measured voltage across the laser diode is simply given as:

$$V = V_d + IR. \tag{15.2}$$

The threshold point can be identified by observation of a "kink" in the $I(dV/dI)$ versus I characteristics [30].

In practice, a DC shunt path exists diverting the current away from the active region, which can be approximated as a simple shunt resistor [28]. During high-speed modulation, however, frequency-dependent shunt paths exist, which are more complicated than just a simple shunt resistor. These shunt paths also depend on the type of the laser structure used (including geometry and doping density) [32]. They are known as the *electrical parasitics* of the laser, which limit the extrinsic modulation bandwidth [33–34].

To illustrate the structural dependence of electrical parasitics, the cross-section of the *ridge waveguide laser* is shown in Fig. 15.1, with the parasitic paths identified and clearly labelled [35]. The total series resistance (R_s) consists of the ridge resistance (R_{SR}) and substrate resistance (R_{SS}), directly under the active region, which is typically several ohms. The *p.n.* junction formed between the InGaAsP layers is modelled by a heterojunction diode (D_L) and its associated space-charge capacitance (C_L). The large resistance provided by R_Q prevents excessive current leakage from the active region to the heterojunction diode (D_L).

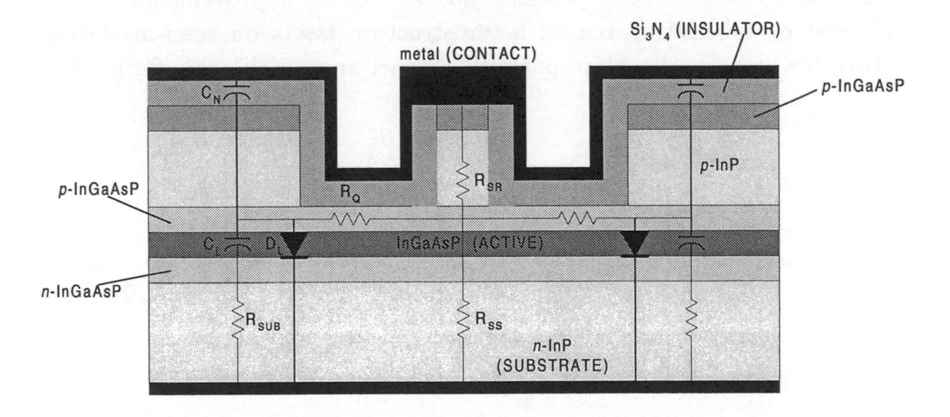

Fig. 15.1. Schematic cross-section of ridge-waveguide laser with parasitics included.

In series with C_L is the metal-insulator-semiconductor (MIS) capacitance (C_N) and substrate resistance (R_{SUB}). The capacitance C_N depends on the thickness of the insulator layer while C_L depends on the doping density of the InGaAsP layers. Typically, the total shunt capacitance (C_S) is dominated by the smaller of the two, that is $C_N \approx 10$ pF, this follows from circuit analysis that total capacitance of several capacitors in series is dominated by the smallest capacitance value. The resistance, R_{SUB}, depends on the doping density of the substrate layer, and is usually small (1-2 ohms) for large doping densities (in excess of $10^{24}/\mathrm{m}^3$). All the parasitic elements above are referred to as the *chip* parasitics.

When the laser chip is mounted on a metal stud, there are contributions from the *package* parasitics [36] such as standoff shunt capacitance (C_P several pF order), bondwire inductance (L_P several nH), and bondwire resistance (R_P several ohms). In order to reduce the contributions from the package parasitics, monolithically integrated laser devices can be used [37–38]. The equivalent circuit model used to represent the overall parasitics is shown in Fig. 15.2, where R_{IN} is the source resistance and I_L is the leakage current. The principal factor limiting the laser diode bandwidth is the shunt parasitic capacitance, C_S, and the series contact resistance, R_S, which forms a low-pass filter. The RC product corresponds to a cutoff frequency (f_c) given by:

$$f_c = \frac{1}{2\pi R_S C_S}\,. \tag{15.3}$$

The *etched mesa buried heterostructure* (EMBH) laser typically exhibits larger values of the $R_S C_S$ product [35, 39]. Some improvements can be achieved by fabricating buried heterostructure lasers on semi-insulating substrates, where the chip parasitic capacitance is in the form of a

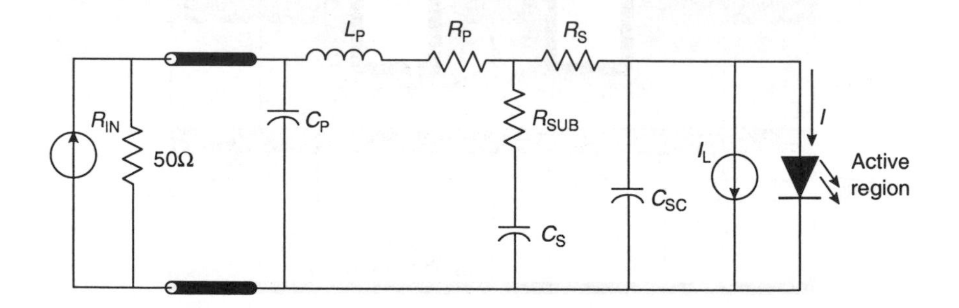

Fig. 15.2. The equivalent lumped circuit model of the electrical parasitics.

distributed RC network [40]. At high frequencies, the capacitance elements far away from the lasing junction do not contribute to the effective total capacitance. Furthermore, the coupling resistances reduce the amount of current diverted into the parasitic capacitor network.

15.3 Lumped-Element Circuit Models

15.3.1 *Large signal model*

The large-signal circuit model is derived from the rate equation by translating physical parameters of the laser, such as photons and carriers, into circuit variables, such as voltages and currents. A detailed explanation of the derivation is given in Section 15.14. The large-signal circuit model used in this book is shown in Fig. 15.3.

The circuit model of package and chip parasitics can be combined with the laser circuit model at its input port. The injection current reaching the active region, after passing the parasitics model, is denoted by I. The term $C_{\rm sc}$ denotes the space-charge capacitance, $I_{\rm sp}$ is the spontaneous emission current term, $I_{\rm g}$ is the stimulated emission current term, and $R_{\rm ph}$ is the photon-decay (absorption and scattering loss) resistance. The term $\beta I_{\rm sp}$ is the spontaneous emission current coupled in the lasing mode, the term $\tau_{\rm ns}(dI_{\rm sp}/dt)$ represents the carrier build-up/depletion effect, V_j is the junction voltage, and $S_{\rm n}$ is the normalised output photon density

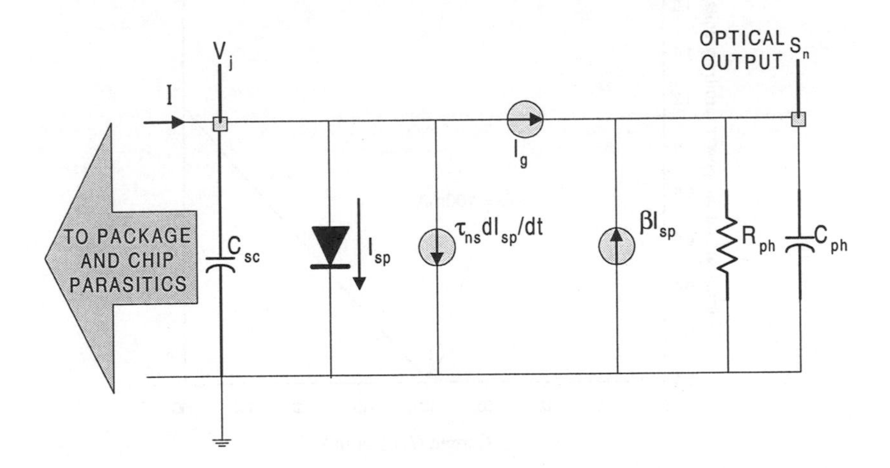

Fig. 15.3. Tucker's large-signal circuit model of the laser diode.

($\propto$ intensity and power). The capacitor C_{ph} is the energy storage element that describes the photon build-up/depletion effect. This large-signal circuit model is constructed by using symbolically-defined devices (SDD) in HP-MDS [41], which allows nonlinear dependent sources to be implemented. Several application examples of how the large-signal circuit model can be used are presented in the following.

15.3.2 *Light-current characteristics*

By injecting DC current into the large signal circuit model of Fig. 15.3, and taking the optical output given by S_{n}, the light-current ($L - I$) characteristics of three different types of laser structures may be obtained, and these are plotted in Fig. 15.4. The parameter values of these lasers are taken from [35, 42] but can also be found in Section 15.14. The threshold currents were found to be approximately 100 mA, 45 mA, and 30 mA for the stripe-geometry, ridge waveguide, and EMBH lasers, respectively. *Gain-guided* lasers (stripe-geometry) are less attractive than *index-guided* lasers because they have relatively high threshold currents, increased Auger recombination [31], and increased index antiguiding (due to carrier-induced refractive index change). Strongly index-guided lasers (e.g. buried heterostructure) have lateral refractive index step of about two orders of magnitude higher

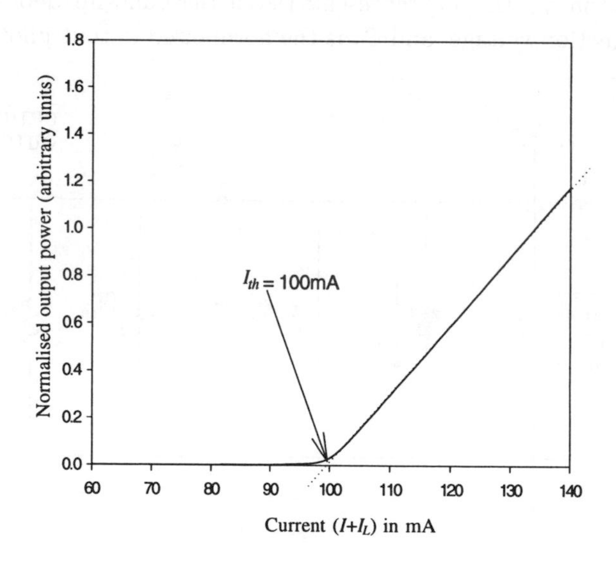

(a)

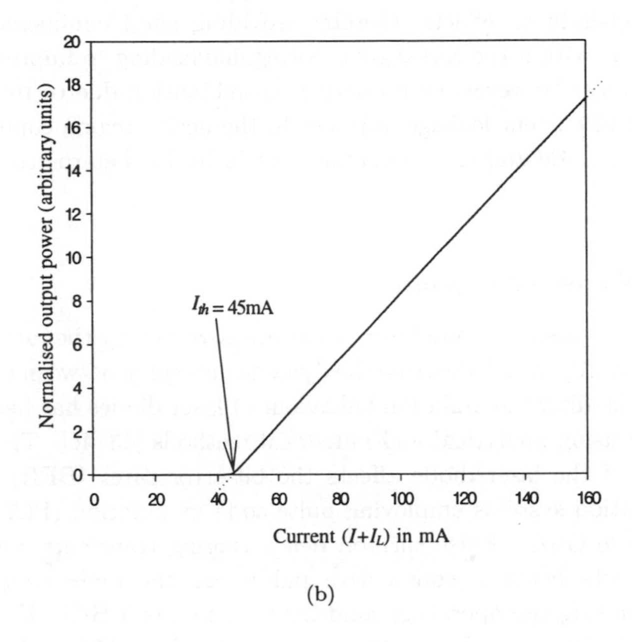

(b)

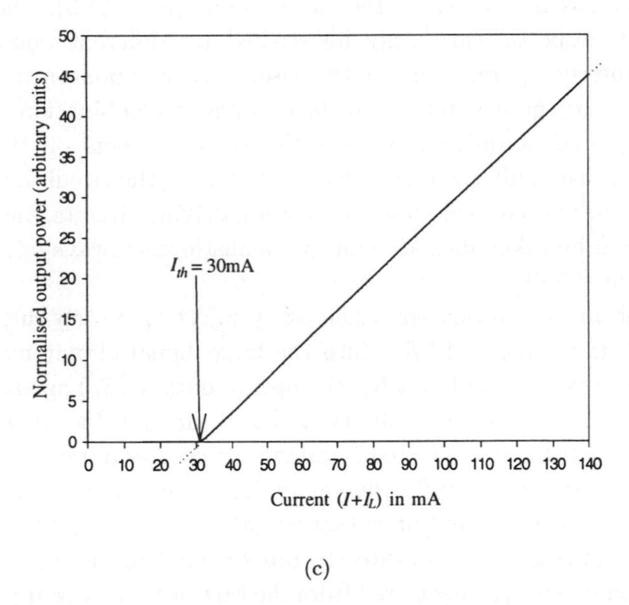

(c)

Fig. 15.4. Light-current ($L-I$) characteristics of (a) gain-guided laser (stripe-geometry) (b) weakly index-guided laser (ridge waveguide) and (c) strongly index-guided laser (EMBH).

than carrier-induced effects, thereby providing good confinement of the optical mode within the rectangular waveguide leading to improved, lasing characteristics. However, nonradiative recombination due to regrowth defects [31] and current leakage adjacent to the active region counteract the good lateral mode and carrier confinement in buried heterostructure (BH) lasers.

15.3.3 *Transient response*

The transient phenomena in laser diodes are governed by the rate equations (see Chapter 13), which describe the dynamic interplay between photon and carriers. The direct modulation behaviour of laser diodes has been studied intensively using analytical and numerical methods [43–46]. The dynamic behaviour of the laser diode affects the bit-error rates (BER) of optical communication systems employing pulse code modulation (PCM) at high speeds above Gb/s. Pulse turn-on delay, ringing transients, and charge-storage effects between consecutive pulses are the main considerations when optimising the operating conditions for minimal BER. By using the large-signal circuit model of the laser diode (Fig. 15.3), the dynamic modulation response can easily be studied for different operating conditions. Internal parameters of the laser such as spontaneous emission coupling factor, gain compression factor, and recombination coefficients can be adjusted accordingly to investigate their effects on the dynamic behaviour. Most of all, the main advantage of using the circuit model is that external circuits such as electrical parasitics, driving circuits, and matching networks can be taken into account in a realistic microwave-optoelectronic environment [47–48].

Relaxation oscillations are observed by injecting a step current (I_{step}) of 150 mA (i.e. $I_{\mathrm{step}} = 1.5 I_{\mathrm{th}}$) into the large signal circuit model of the stripe-geometry laser and plotting the optical output (S_{n}) against time (t), as shown in Fig. 15.5. It is observed that a larger value of spontaneous emission coupling factor ($\beta = 0.05$) provides a greater damping effect. The value of β is usually larger for the gain-guided lasers than for index guided lasers due to their wavefront curvature [31, 49]. The junction voltage, V_{j} (defined in Fig. 15.3), is directly proportional to the carrier density. The turn-on delay (t_{on}) measured from the first overshoot is approximately 1.8 ns for both cases, which can be improved (i.e. reduced) by biasing the device high above threshold.

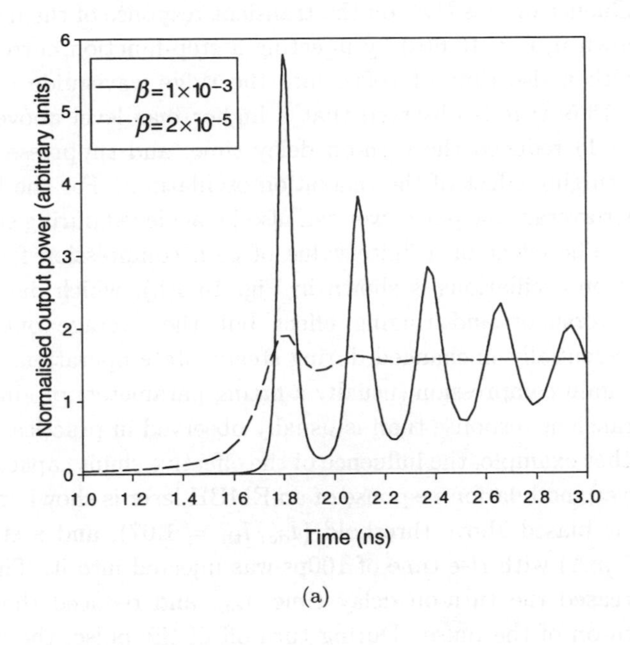

(a)

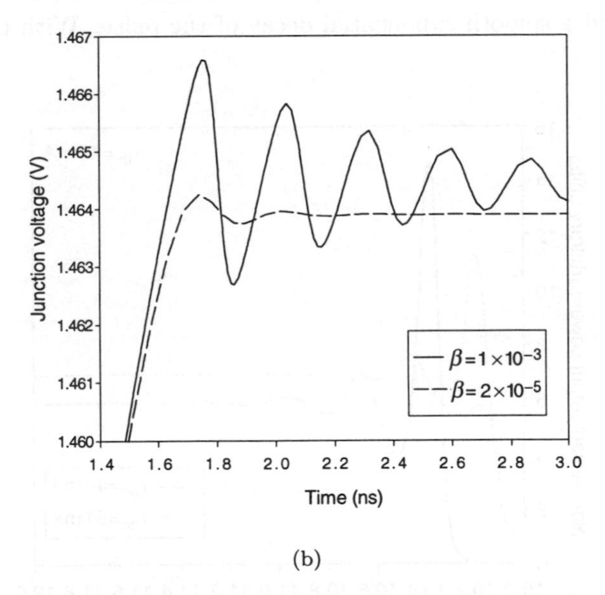

(b)

Fig. 15.5. (a) Light output response for step function drive current for two differe values of spontaneous emission coupling factor (β), and (b) the corresponding junctic voltage response.

The influence of bias level on the transient response of the index-guided laser is shown in Fig. 15.6(a) by injecting a step-function current ($I_{\text{step}} =$ 40 mA) with a rise time of 100ps into the ridge waveguide laser model. From Fig. 15.6(a) it is observed that a higher bias level above threshold ($I_b/I_{\text{th}} = 1.1$) reduced the turn-on delay time, and suppressed the overshoot and ringing effect of the relaxation oscillations. For the higher bias level, higher average output power can also be achieved during steady-state operation. The effect of a finite value of gain compression factor (ε) on the relaxation oscillations is shown in Fig. 15.6(b), which dampened the transient overshoot and ringing effect but the average output power remained essentially unchanged during steady-state operation. The inclusion of the gain compression (usually a fitting parameter) produces a more realistic transient response than is usually observed in practice.

In another example, the influence of the effective shunt capacitance (C_{S}) on the pulsed modulation response of an EMBH laser is shown in Fig. 15.7. The laser is biased above threshold ($I_{\text{dc}}/I_{\text{th}} = 1.07$), and a step current ($I_{\text{step}} = 35$ mA) with rise time of 100ps was injected into it. The presence of C_{S} increased the turn-on delay time, t_{on}, and reduced the overshoot during turn-on of the pulse. During turn-off of the pulse, the presence of C_{S} produced a smooth exponential decay of the pulse. With the presence

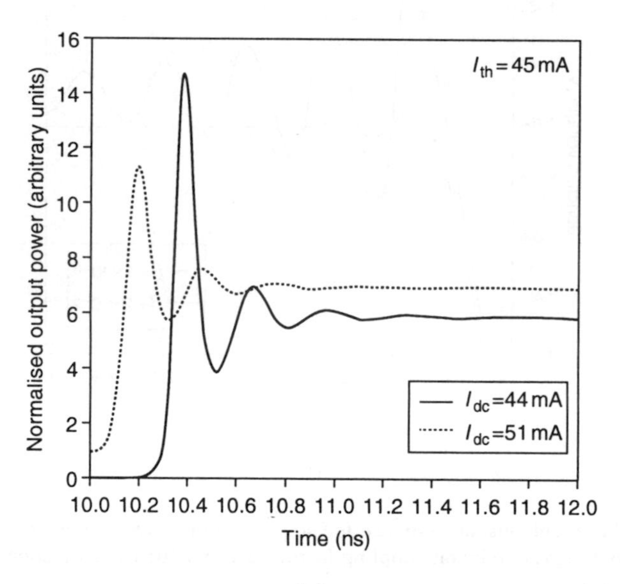

(a)

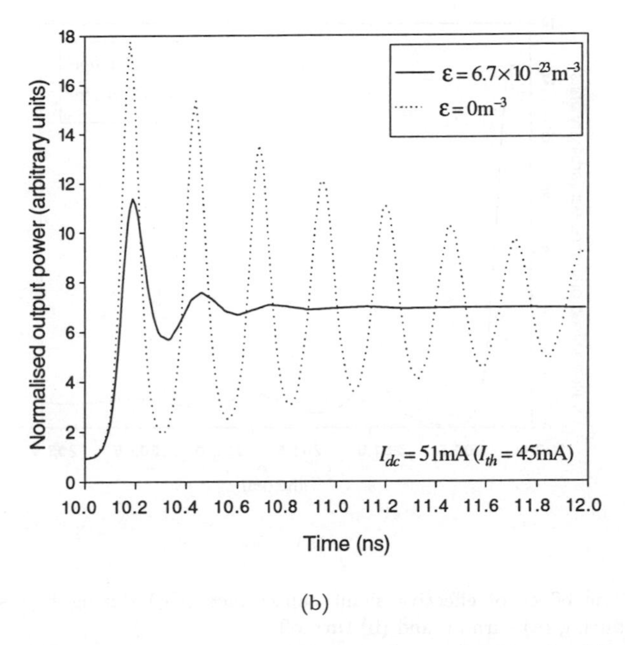

(b)

Fig. 15.6. Transient response of ridge waveguide laser showing the (a) effect of bias level and (b) gain compression.

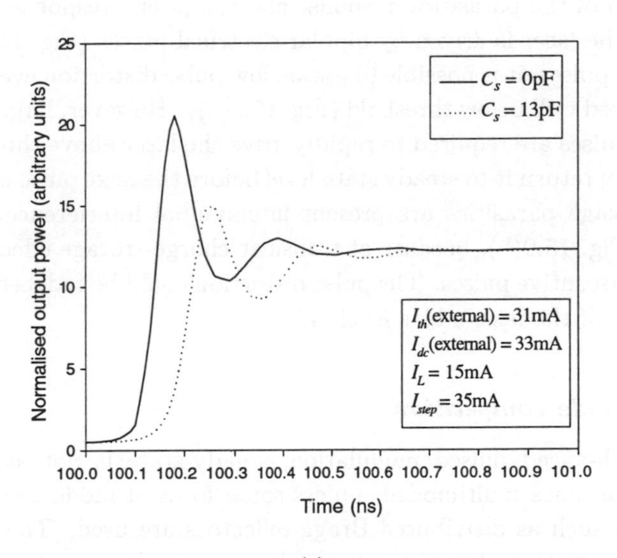

(a)

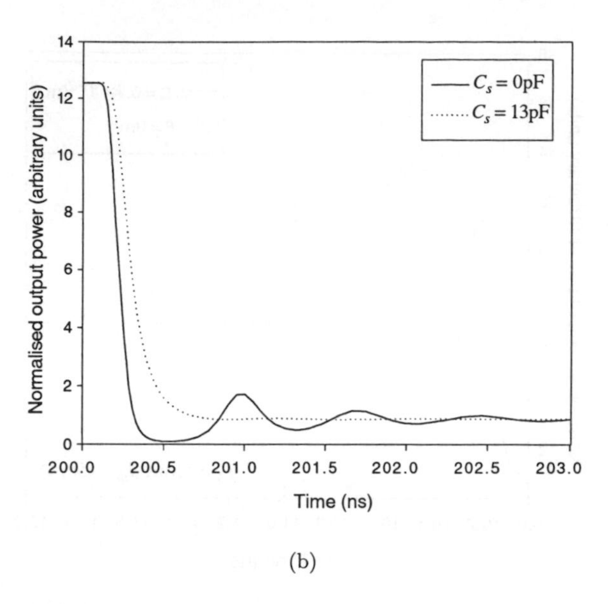

(b)

Fig. 15.7. The effect of effective shunt capacitance (C_S) during high-speed pulsed modulation during (a) turn-on and (b) turn-off.

of parasitics, the overall transient response of the laser is given by the convolution of the parasitics' response and the laser's response [50].

When the laser is driven by bipolar electrical pulses (Fig. 15.8) instead of unipolar pulses, it is possible to obtain low pulse distortion even when the laser is biased well below threshold (Fig. 15.9(a)). However, large amplitude electrical pulses are required to rapidly drive the laser above threshold, and then quickly return it to steady state level before the next pulse commences. When package parasitics are present intersymbol interference occurs, as shown in Fig. 15.9(b), because of transient charge-storage effects between the two consecutive pulses. The pulse distortion could be reduced by careful adjustment of the laser's bias level [42].

15.3.4 *Mode competition*

During high-speed pulsed modulation, steady-state is not attained and the laser becomes multimoded, unless some form of mode discrimination mechanism such as distributed Bragg reflectors are used. The broadened spectrum limits the achievable bit rate-distance product due to fibre dispersion. In order to take into account multiple longitudinal modes of the laser,

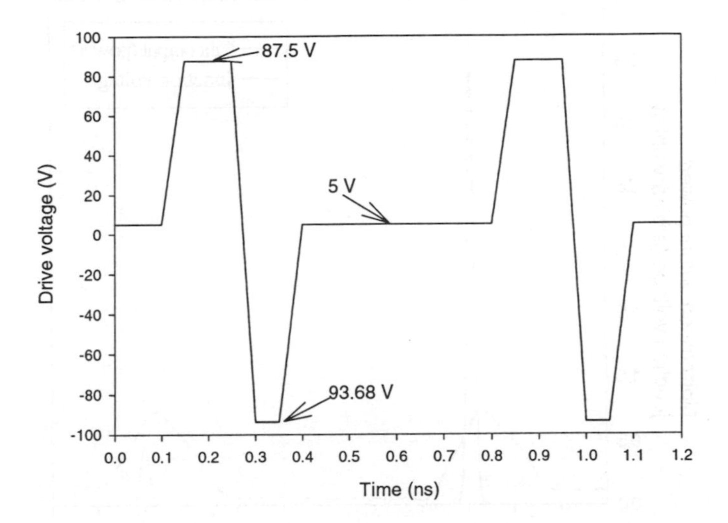

Fig. 15.8. Bipolar pulse modulation of the stripe-geometry laser: drive voltage.

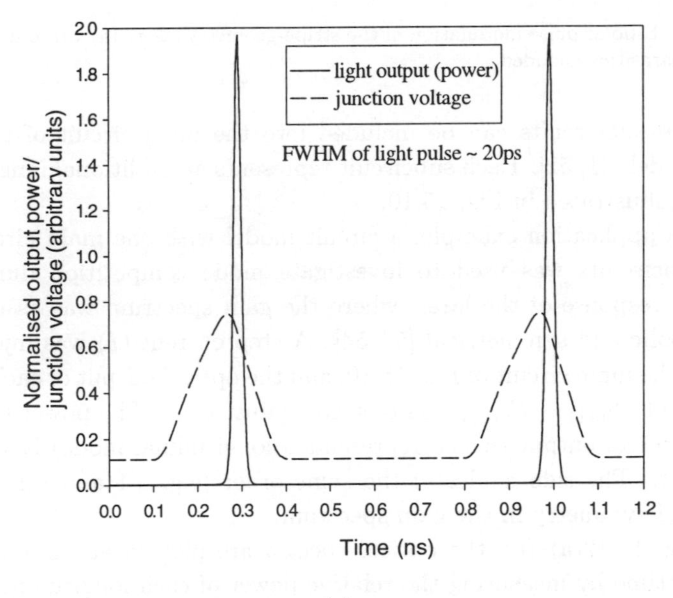

(a)

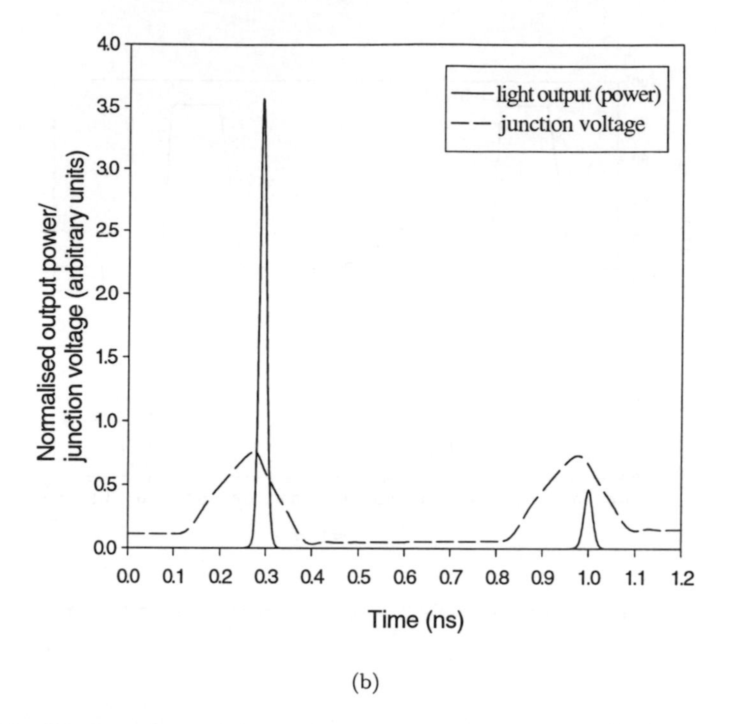

(b)

Fig. 15.9. Bipolar pulse modulation of the stripe-geometry laser: (a) without parasitics (b) with parasitics included.

additional subcircuits can be included into the main circuit of the large signal model [51, 52]. Each subcircuit represents an additional longitudinal mode, as illustrated in Fig. 15.10.

As an application example, a circuit model with one main circuit and eight subcircuits was used to investigate mode competition during the transient response of the laser, where the gain spectrum was assumed to be parabolic and symmetrical [53–54]. A step current (I) was injected as input to the main circuit of Fig. 15.10, and the optical output of each individual circuit ($S_{\mathrm{n}(1)} \cdots S_{\mathrm{n}(\mathrm{n})}$) was plotted against time. The power transient of each optical output (which represents a longitudinal mode) is shown in Fig. 15.11. The side modes of the same order (e.g. +1 and −1) overlap because of symmetry in the gain spectrum.

In Fig. 15.12(a)–(c), the optical spectra are plotted at three different points in time by measuring the relative power of each longitudinal mode. The carrier dependence of the gain peak and its displacement from the dominant lasing ($\Delta\lambda$) could also be included in the circuit model to demonstrate

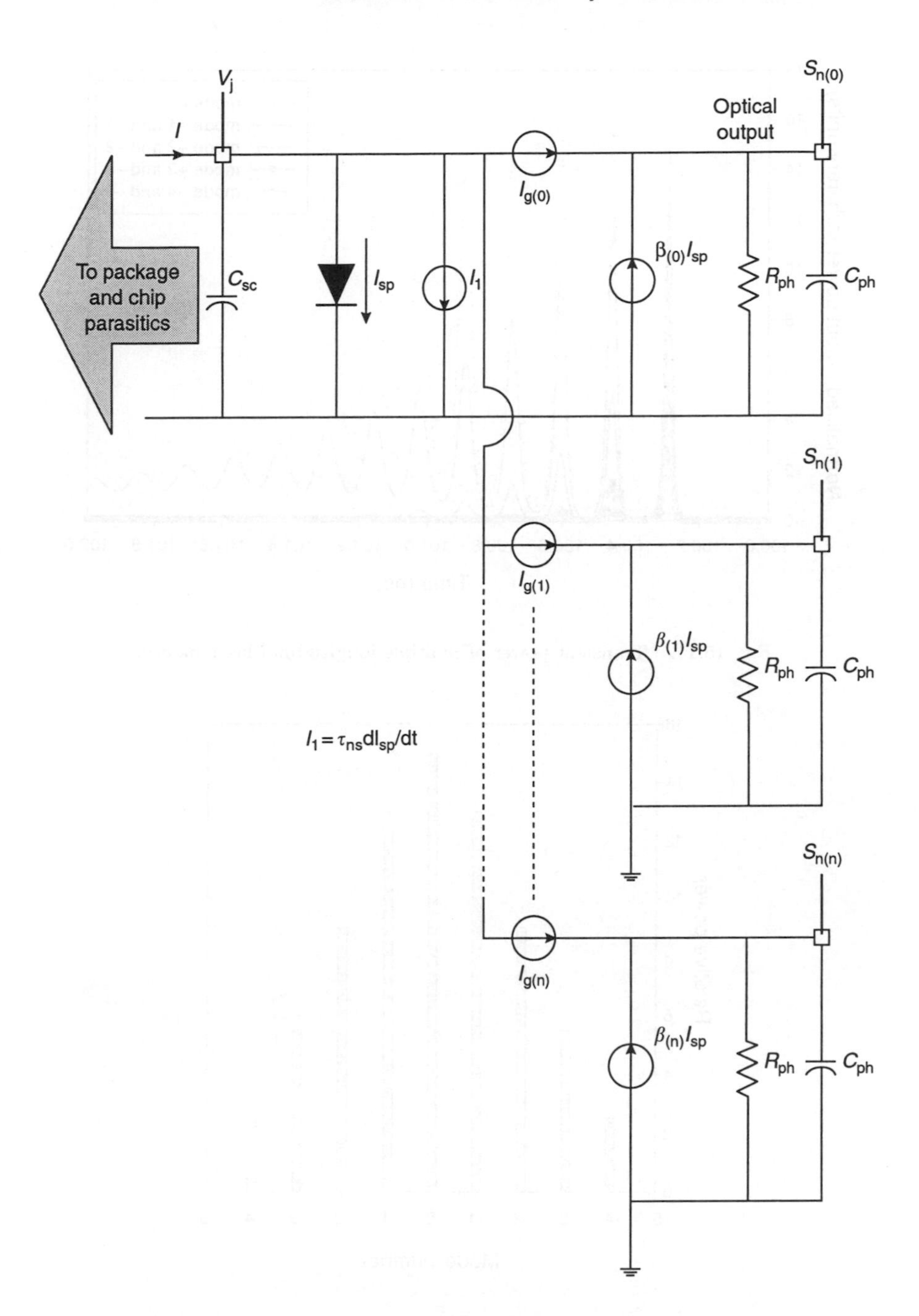

Fig. 15.10. Multimode circuit model of the Fabry Perot laser diode.

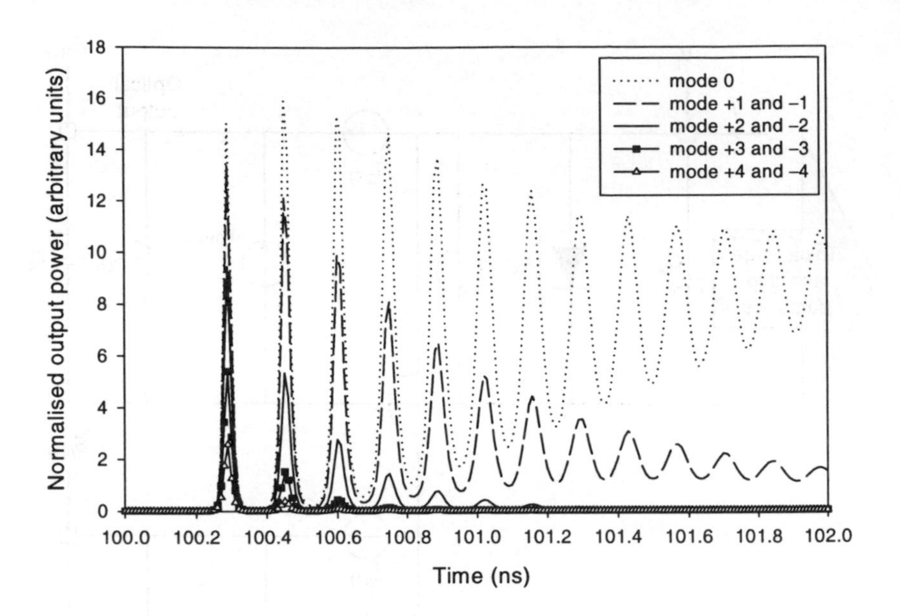

Fig. 15.11. Transient power of multiple longitudinal laser modes.

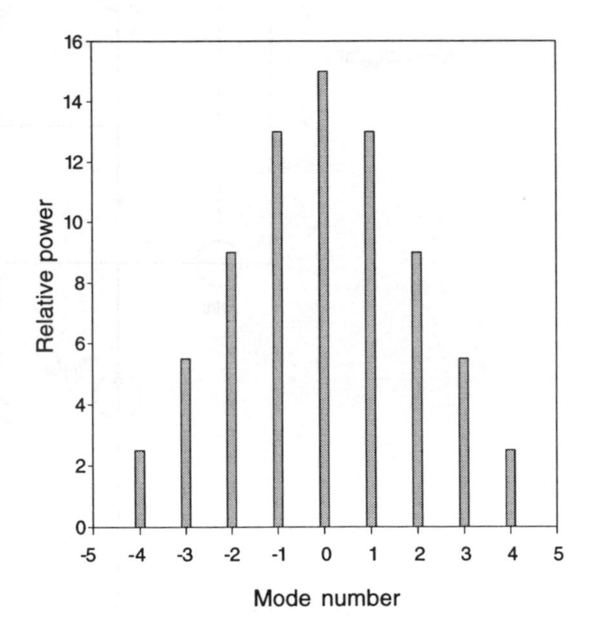

(a)

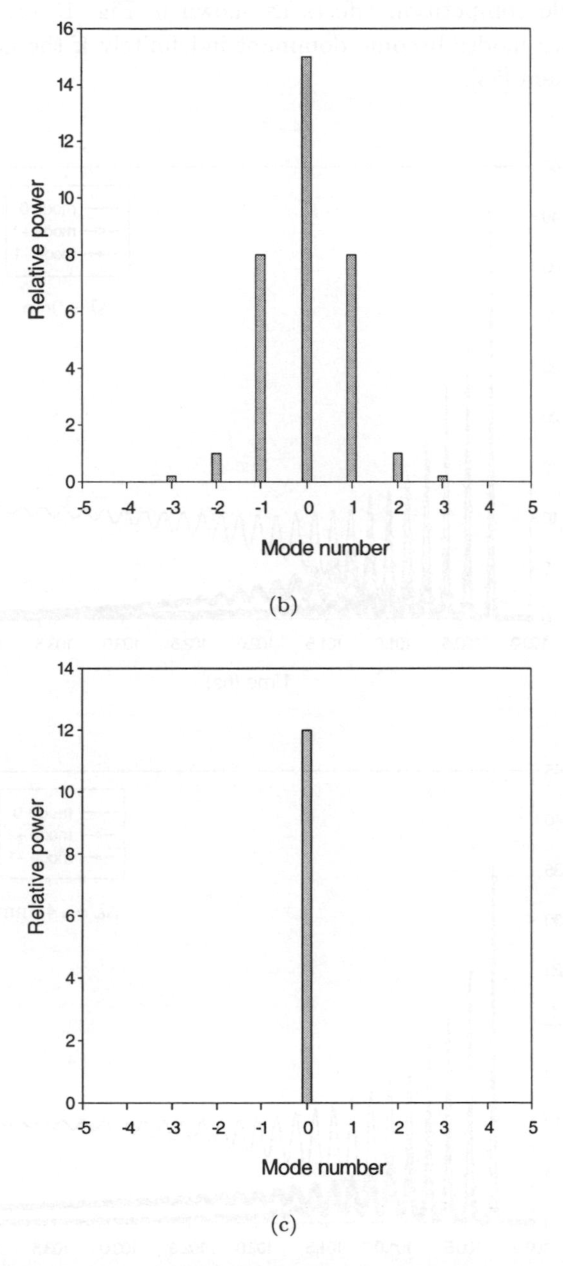

Fig. 15.12. Evolution of the transient spectra at (a) first pulse (b) fourth pulse and (c) steady state (side modes could be seen more clearly if log scale is used).

dynamic mode competition effects as shown in Fig. 15.13. It has been found that two modes become dominant indefinitely if the gain peak falls in between them [53].

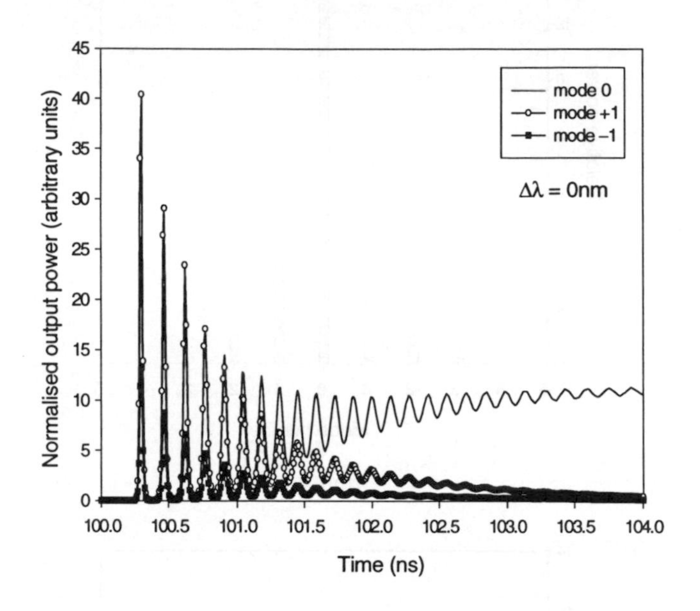

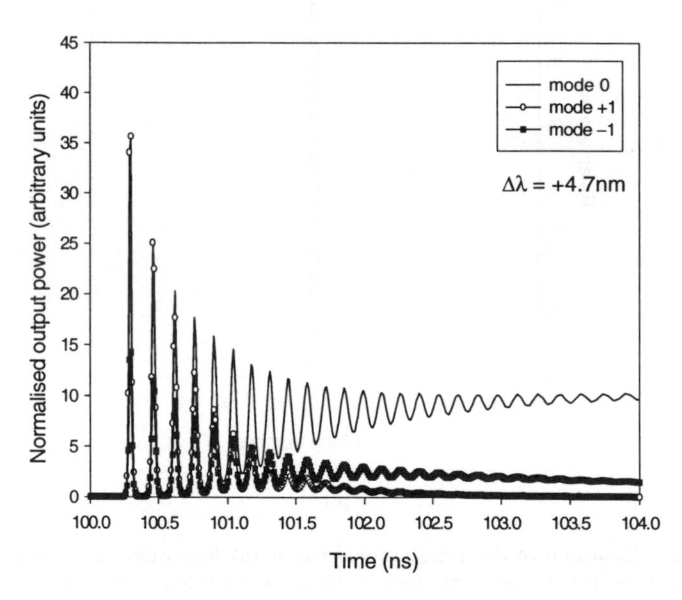

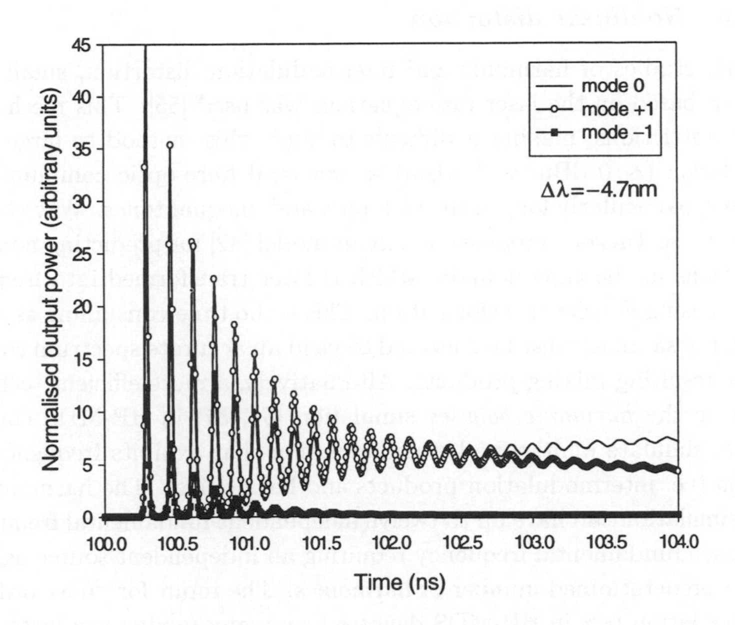

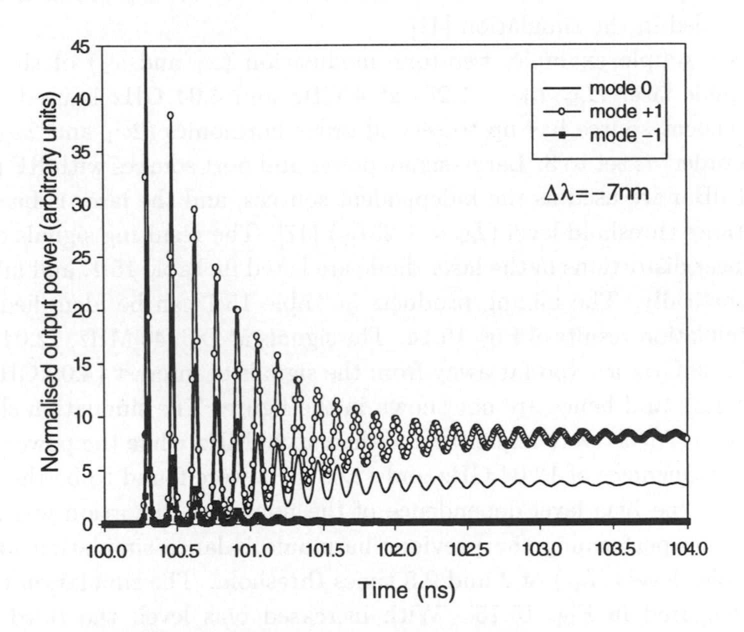

Fig. 15.13. Output power transient response for different gain peak displacement ($\Delta\lambda$) ($I_{\mathrm{dc}}/I_{\mathrm{th}} = 0.9$ and $I_{\mathrm{dc}}/I_{\mathrm{th}} = 1.5$). This shows modal competition effects clearly.

15.3.5 *Nonlinear distortion*

In early studies of harmonic and intermodulation distortion, small-signal analysis based on the laser rate equations was used [55]. This method has many restrictions, making it difficult to apply this method to large-signal modulation ($\sim$ 0 dBm to 5 dBm) in practical fibre-optic communication systems, particularly for multitone inputs and unequal tones. Way [47] proposed to use Tucker's large-signal circuit model [42] for predicting nonlinear distortions in the time domain, which is later transformed into frequency domain using Fourier transformation. This is too time-consuming as a large number of samples must be collected to yield an accurate spectrum covering all the resulting mixing products. Alternatively, a more efficient technique based on the *harmonic balance* simulation [56, 57] in HP-MDS could be used to simulate multi-tone laser modulation that exhibits frequency conversion (i.e. intermodulation products and harmonics). The harmonic balance simulation can have up to twelve independent fundamental frequencies with each fundamental frequency requiring an independent source assigned with a predetermined number of harmonics. The input for "max order" in the simulation box in HP-MDS denotes how many mixing products are to be included in the simulation [41].

As a simple example, two-tone modulation (ω_1 and ω_2) of the ridge waveguide laser ($I_{dc}/I_{th} = 1.25$) at 4 GHz and 4.04 GHz is used. Each independent source has up to second order harmonics ($2\omega_1$ and $2\omega_2$) and "max order" is set to 3. Large-signal power and port sources with RF power of -1 dBm are used as the independent sources, and the laser is biased at 1.25 times threshold level ($I_{dc} = 1.25I_{th}$) [47]. The resulting signals due to nonlinear distortions in the laser diode are listed in Table 15.1, and labelled alphabetically. The mixing products in Table 15.1 can be identified from the simulation results of Fig. 15.14. The signals at DC, 40 MHz, 12.04 GHz, and 12.08 GHz are too far away from the signals of interest (4.00 GHz and 4.04 GHz), and hence are not shown in the figure. The simulation showed that power levels were higher at DC than at 40 MHz while the power levels at the frequencies of 12.04 GHz and 12.08 GHz were found to be the same.

Next, the bias level dependence of the nonlinear distortion was investigated by performing the previous harmonic balance simulation for two other bias levels (I_{dc}) of 2 and 2.5 times threshold. The simulation results are compared in Fig. 15.15. With increased bias level, the third order intermodulation products of (c) and (f) have become less pronounced, while there is almost no change in the output power at the signal frequencies

Table 15.1. The resulting mixing products (up to order of 3) of two-tone modulation (ω_1, ω_2).

(a)	DC	Always present
(b)	$\omega_2 - \omega_1 = 40$ MHz	Intermodulation
(c)	$2\omega_1 - \omega_2 = 3.96$ GHz	Intermodulation
(d)	$\omega_1 = 4$ GHz	Signal frequency 1
(e)	$\omega_2 = 4.04$ GHz	Signal frequency 2
(f)	$2\omega_2 - \omega_1 = 4.08$ GHz	Intermodulation
(g)	$2\omega_1 = 8$ GHz	Harmonic
(h)	$\omega_1 + \omega_2 = 8.04$ GHz	Intermodulation
(i)	$2\omega_2 = 8.08$ GHz	Harmonic
(j)	$2\omega_1 + \omega_2 = 12.04$ GHz	Intermodulation
(k)	$2\omega_2 + \omega_1 = 12.08$ GHz	Intermodulation

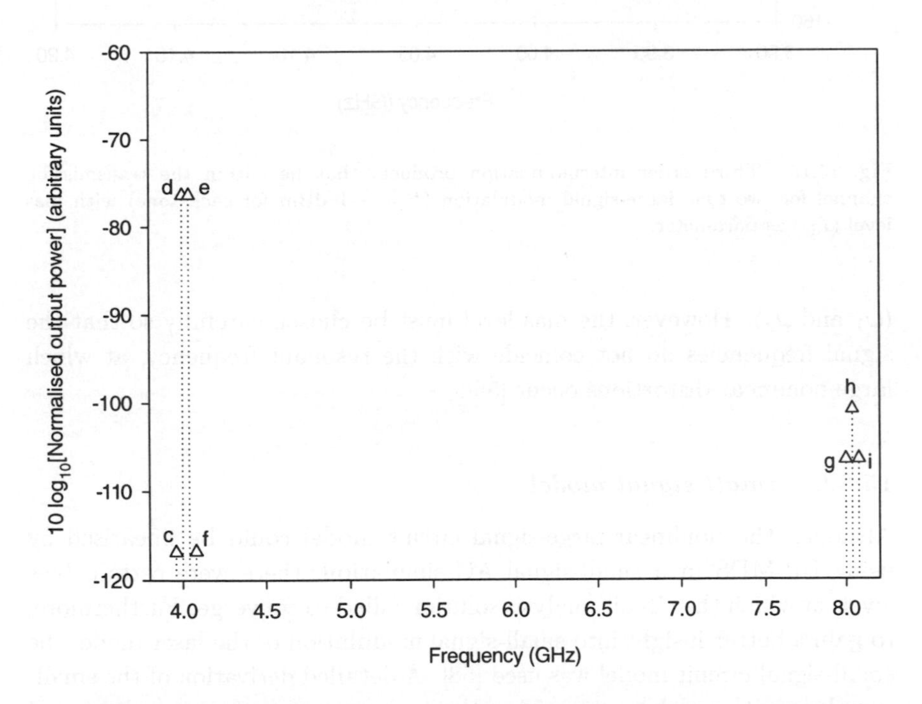

Fig. 15.14. Intermodulation and harmonic distortion due to laser nonlinearity (compared with Fig. 13.19 in Chapter 13).

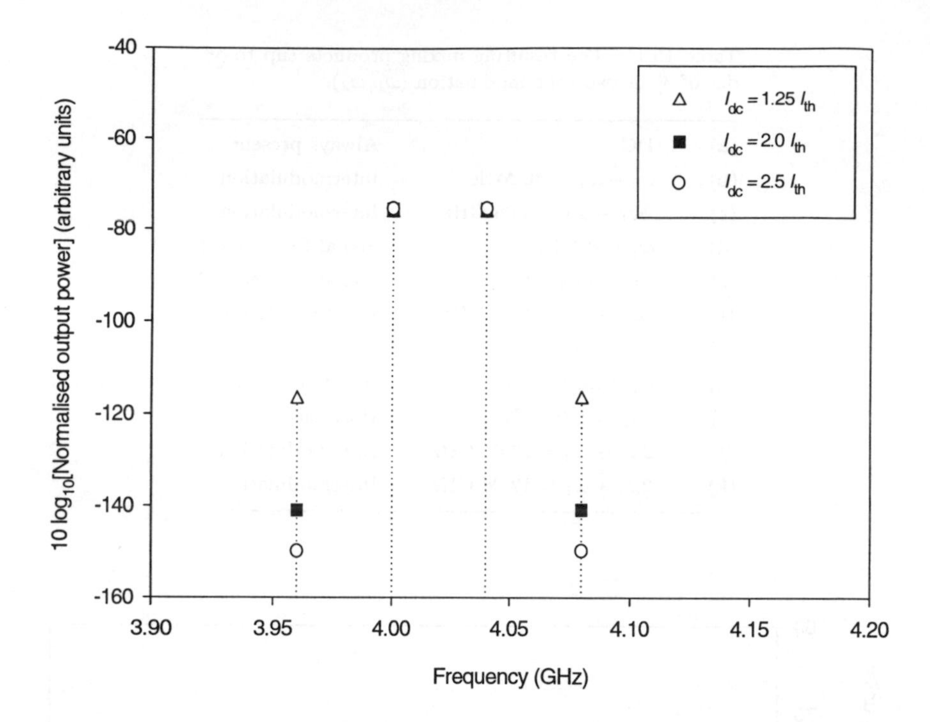

Fig. 15.15. Third order intermodulation products that lie within the transmission channel for two-tone large-signal modulation ($P = -1$ dBm for each tone) with bias level ($I_{\rm dc}$) as parameter.

(ω_1 and ω_2). However, the bias level must be chosen carefully so that the signal frequencies do not coincide with the resonant frequency, at which large nonlinear distortions occur [55].

15.3.6 *Small signal model*

Although the nonlinear large-signal circuit model could be linearised by using HP-MDS in a small-signal AC simulation, there were certain bias levels at which the circuit analysis solution failed to converge. Furthermore, to gain a better insight into small-signal modulation of the laser diode, the small-signal circuit model was used [58]. A detailed derivation of the small-signal circuit model by linearising the laser rate equations can be found in Section 15.15. The final schematic layout of the small-signal circuit model of the laser diode (including the gain compression effect) is shown

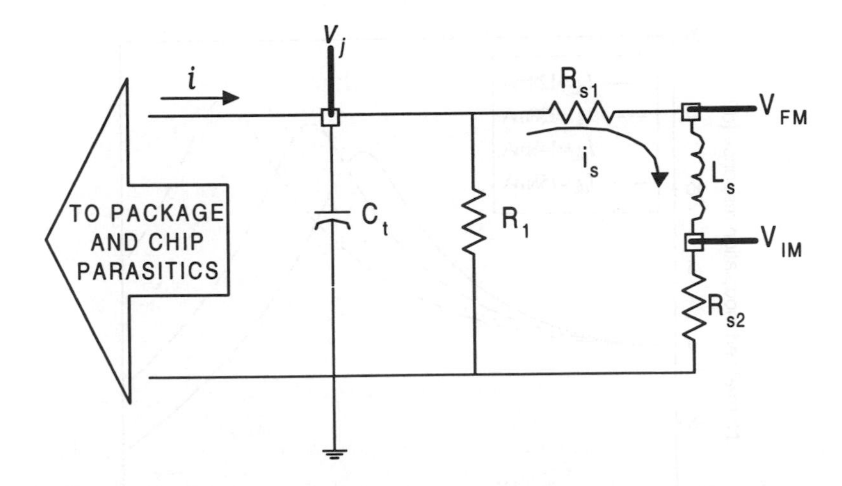

Fig. 15.16. Small-signal circuit model of the laser diode.

in Fig. 15.16. The physical significance of each circuit element is briefly explained in the following.

Laser oscillation arises from the transient charge-discharge effect (equivalent to the dynamic exchange of energy between carriers and photons) between the effective capacitance ($C_t = C_d + C_{sc}$) and the inductance (L_s). Damping of the resonance is determined by the resistances R_1, R_{s1} and R_{s2}. The resistance, R_1, models the damping due to spontaneous and stimulated recombination, while the resistance, R_{s1}, models damping due to spontaneous emission coupled into the lasing mode. The resistance, R_{s2}, provides the same effect as the spontaneous emission coupling factor (β) used in the large-signal circuit model (see Fig. 15.5). On the other hand, the resistance, R_{s1}, allows additional damping (due to lateral carrier diffusion or other gain compression effects) to be modelled without relying on using an unrealistically large value of β [59]. For intensity modulation (IM) response, the voltage, V_{IM}, is taken while the voltage, V_{FM}, represents the frequency modulation (FM) response.

15.3.7 *Intensity modulation (IM) response*

As an application example of the small-signal circuit model, the modulation responses of three different types of laser structures have been investigated. First, the modulation response of the stripe-geometry laser with negligible parasitics is plotted in Fig. 15.17(a). It is observed that the resonant peak

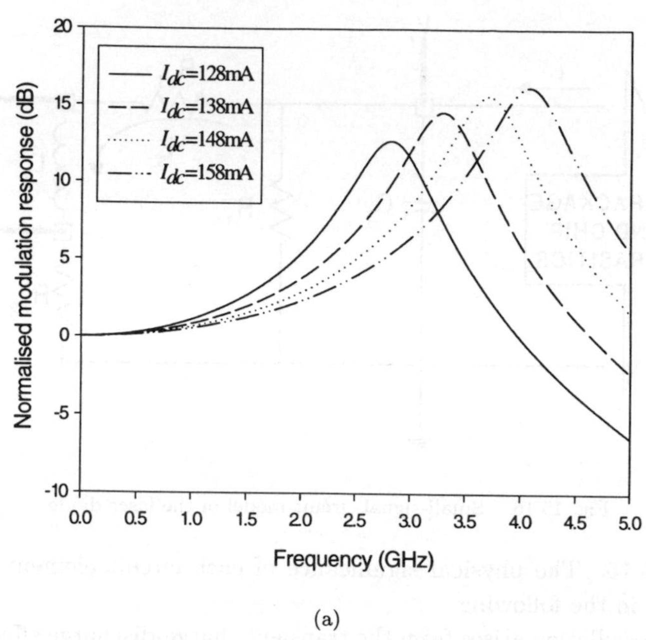

(a)

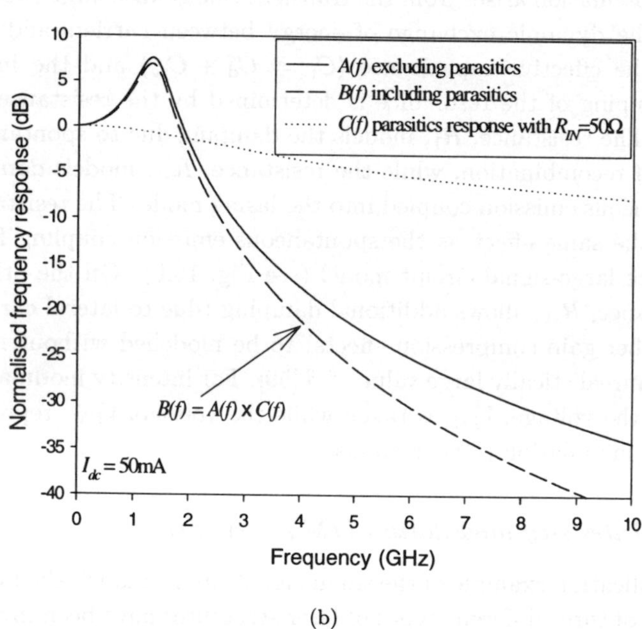

(b)

Fig. 15.17. (a) Intrinsic modulation response of stripe-geometry laser and (b) the effect of parasitics.

shifts to higher frequencies as the bias level is increased. The magnitude of the resonant peak also becomes greater with increasing bias level. In practice, the electrical parasitics discussed in Sec. 15.2 play a very important role in limiting the modulation bandwidth of the laser diode.

With the presence of parasitics, the *extrinsic* modulation response, $B(f)$, of the laser diode is given by the multiplication of the *intrinsic* modulation response, $A(f)$, and the parasitics response, $C(f)$. This is illustrated in Fig. 15.17(b). In other words, the modulation bandwidth of the laser cannot be increased indefinitely by increasing the bias level. Neglecting other degradation and saturation effects, the modulation bandwidth is ultimately limited by the parasitics roll-off at high frequencies (>1 GHz). Realistic modulation responses of the ridge-waveguide laser and EMBH laser are shown in Fig. 15.18(a) and (b), respectively. Consideration of the parasitics response is essential for the EMBH laser, which typically exhibits a dip in the modulation response before the resonant peak. This is because the EMBH laser has a larger $R_S C_S$ product than the ridge-waveguide laser.

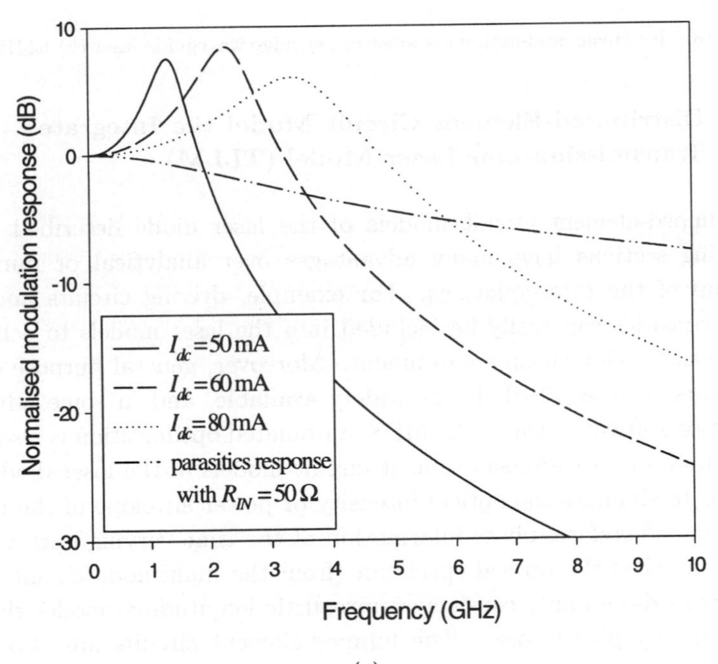

(a)

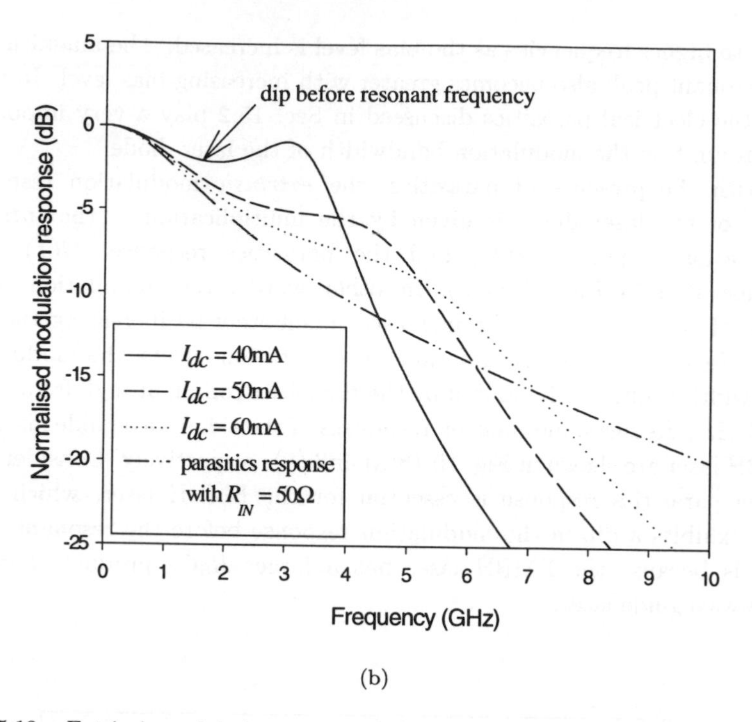

(b)

Fig. 15.18. Extrinsic modulation response of (a) ridge waveguide laser (b) EMBH laser.

15.4 Distributed-Element Circuit Model the Integrated Transmission-Line Laser Model (TLLM)

The lumped-element circuit models of the laser diode described in the preceding sections have many advantages over analytical or numerical solutions of the rate equations. For example, driving circuits and electrical parasitics can easily be included into the laser models to achieve a more realistic simulation environment. Moreover, general purpose circuit simulators such as PSPICE are widely available, and in some advanced simulation software such as HP-MDS, automated optimisation is also available. However, the lumped-element circuit models of the laser diode only allow us to simulate the optical intensity or power envelope of the optical radiation. Therefore, phase information of the time-varying optical fields is lost and thus the optical spectrum (from the multimode circuit model of the laser diode) only consists of unrealistic longitudinal modes that are free from any phase noise. The lumped-element circuits are also based on average values of photon and carrier densities, thereby neglecting any inhomogeneous distribution along the laser cavity. In other words, there is no spatial correspondence between the circuit model and the physical device.

All these limitations can be overcome by using the *distributed-element* circuit model — the transmission-line laser model (TLLM). In this case, the operating wavelength is much smaller than the device elements, and propagation effects can be taken into account. The basic construction of the TLLM was explained earlier in Chapter 14. Further comparisons between the lumped-element circuit model and the distributed-element TLLM are given in Table 15.2. In a combined microwave optoelectronic modelling approach proposed by Sum and Gomes [48, 60], overall simulation of a laser transmitter is carried out in two separate steps. First, the electrical characteristic of the laser is simulated by an equivalent circuit based on lumped-elements. This includes the electrical parasitics, and possibly, the matching circuit. Then, the output from that simulation is used as input to the time-domain model (TDM) [61], which is the intrinsic laser model. In this book we have included the matching circuit, electrical parasitics and intrinsic laser diode in an *integrated* TLLM model, which is based on distributed circuit elements (see Fig. 15.19). The optical characteristics of the intrinsic laser are modelled by the standard TLLM [62]. In addition, the electrical parasitics and matching networks are simulated using the transmission-line matrix (TLM) link-lines and stub-lines [63].

Previously, an integrated equivalent circuit model based on lumped elements, including electrical parasitics, and matched by a single-stub microstrip circuit, was also proposed by Sum and Gomes [64]. However, their lumped circuit model was based on many simplifying assumptions that limited the number of applications for the model. For example, their model assumed (i) single-mode operation, and neglected (ii) carrier-induced refractive index change, (iii) random spontaneous emission noise, (iv) reflections, (v) optical field phase information, and (vi) inhomogeneous spatial distribution of carrier density and photon flux. The proposed integrated TLLM laser transmitter model in this book takes into consideration all these effects.

First, we briefly review the intrinsic laser model based on the standard TLLM model. Next, we will explain how lumped networks can be simulated by TLM link-lines and stub-lines so that the electrical characteristics of the laser transmitter can be included for a more complete microwave-optoelectronic simulation. Finally, an application example of the integrated

Table 15.2. Comparison between two equivalent circuit models of the semiconductor laser.

Questions	Lumped-element circuit model	Distributed-element circuit-model (TLLM)
Specialised knowledge required?	No, only require general purpose circuit simulator software	Yes, custom-written program
Interconnectable to external circuit/other models?	Yes	Yes
Wide optical bandwidth?	No, usually single-moded but multiple modes possible (< 10 modes)	Yes, determined by sampling rate (10–100 modes)
Random spontaneous emission noise?	No, deterministic model	Yes, stochastic model
Optical spectrum generated?	Yes, but discrete spectrum only	Yes, continuous spectrum (includes phase noise)
Inhomogeneous carrier density distribution?	No	Yes
Flexible to simulate novel devices?	No, entirely new circuit for different laser devices	Yes, only slight modifications to existing model
Excessive averaging?	None required	Yes, signal is masked by noise
Optical waves simulated?	No, only optical intensity	Yes, phase information of optical field included
Large-signal?	Yes	Yes
Carrier-induced dynamic chirp?	No	Yes
Computationally efficient?	Yes, electrical circuit analysis	Yes, baseband transformation method
Optimisation?	Yes, included in HP-MDS	No, but possible

TLLM model is presented through a microwave signal generation technique known as harmonic generation by *gain-switching* (see Chapter 13).

15.5 Intrinsic Laser Model

The intrinsic laser diode of the integrated TLLM model is based on the standard TLLM model [62], where the laser cavity is divided into S sections each having a length of ΔL. Each section is modelled by a scattering matrix

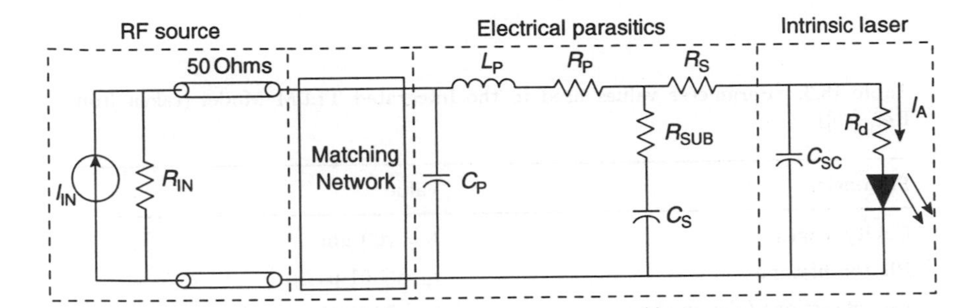

Fig. 15.19. Integrated laser transmitter model for microwave optoelectronic simulation.

that includes the stimulated gain, which is wavelength-dependent, and the facets are simply modelled by unmatched terminal loads. The interaction between photon density (proportional to the square of the optical field) and carrier density is described by the laser rate equations at each and every local section so that inhomogeneous effects can be considered.

Two processes called *scattering* and *connecting* are the main algorithms that form TLLM [24]. Scattering nodes take incoming (incident) waves and scatter them to produce outgoing (reflected) waves. After a delay, the reflected waves will impinge on adjacent scattering nodes, becoming incident waves. The optical waves are represented as *voltage* pulses on dispersionless transmission-lines. All voltage pulses must arrive at scattering nodes at the same time; i.e. they must be synchronised.

The TLLM is a stochastic model, where optical power builds up from spontaneous emission noise, represented as current sources of random magnitude along the transmission-lines. Each spontaneous emission current source has a root-mean-square (RMS) value that is dependent upon the *local* carrier density [25]. The parameter values used in the intrinsic laser of the integrated TLLM model are listed in Table 15.3.

The laser cavity is divided into 23 sections to allow a reasonable simulation time, giving a timestep (Δt) of 116 fs. As such, the resolution in frequency domain is poor at microwave frequencies unless a large number of samples are collected. The number of samples collected is $2^{18} = 262144$ iterations, equivalent to a period of 30.4 ns. This gives a frequency resolution of 32.9 MHz, which is reasonable since linewidth measurement is not required here. Another advantage of TLLM is that computational efficiency can be controlled by adjusting the timestep (Δt) used. When used for device optimisation, accuracy can be traded off for efficiency during the early stages.

Table 15.3. Parameter values used in the Integrated TLLM Model (taken from Ref. [65]).

Parameter	Value
Cavity length	$L = 200\ \mu\text{m}$
Photon lifetime	$\tau_\text{p} = 1.01$ ps
No. of sections (also modes)	$S = 23$ section/mode
Active region thickness	$d_\text{a} = 0.2\ \mu\text{m}$
Active region width	$W = 5\ \mu\text{m}$
Effective index	$n_\text{eff} = 3.3$
Group index	$n_\text{g} = 4$
Confinement factor	$\Gamma = 0.239$
Wave impedance	$Z_\text{p} = 138.472\ \Omega$
Free-space lasing wavelength	$\lambda_\text{o} = 1.325\ \mu\text{m}$
Power facet reflectivity	$R = 0.32$
Linewidth enhancement factor	$\alpha_\text{H} = 5.6$
Carrier lifetime	$\tau_\text{n} = 10$ ns
Radiative recombination coefficient	$B_0 = 6.0 \times 10^{-17}\ \text{m}^3\text{s}^{-1}$
Radiative recombination coefficient	$B_1 = 1.1 \times 10^{-41}\ \text{m}^6\text{s}^{-1}$
Auger recombination coefficient	$C_\text{aug} = 4.0 \times 10^{-41}\ \text{m}^6\text{s}^{-1}$
Internal attenuation factor	$\alpha_\text{sc} = 75\ \text{cm}^{-1}$
Spatial gain per unit inversion	$a = 4.2 \times 10^{-16}\ \text{cm}^{-2}$
Gain compression factor	$\varepsilon = 6.7 \times 10^{-23}\ \text{m}^3$
Times step	$\Delta t = 115.942$ fs
Gain spectrum Q-factor	40
Spontaneous emission spectrum Q-factor	15
Spontaneous emission coupling factor	$\beta = 1.0 \times 10^{-5}$
Transparency carrier density	$N_\text{tr} = 1.1 \times 10^{24}\ \text{m}^{-3}$
Threshold carrier density (calculated)	$N_\text{th} = 2.417 \times 10^{24}\ \text{m}^{-3}$
Initial carrier density (time-saving)	$N_\text{init} = 2.417 \times 10^{24}\ \text{m}^{-3}$
Internal threshold current (active layer)	$I_\text{th} = 32.111$ mA
Bias current	$I_\text{dc} = 1.3\ I_\text{th} = 41.745$ mA
Modulation current	$I_\text{m} = 14.2$ mA (+ 1 dB max. power)

15.6 Electrical Parasitics Model

High-frequency limitations due to electrical parasitics have been analysed using lumped-element circuit models [35]. At high frequencies, (in the GHz range), electrical parasitics cause a roll-off in the laser's modulation response. This means that the design of matching networks is affected as frequency increases. Therefore, when designing matching networks, the contribution of electrical parasitics should be taken into consideration.

Due to the incompatibility between non-circuit based numerical models of semiconductor lasers and electrical circuit networks, the laser parasitics and matching circuit designs are rarely considered. An exception to this is the TLLM, which can simulate the laser parasitics network based on TLM lumped elements [63]. A two-port model with bidirectional interfaces for both electrical and optical ports was first proposed by Lowery [66].

The electrical parasitics and matching network used in the integrated TLLM were modelled by TLM stub-lines and link-lines (see Chapter 14). The TLM representation of the electrical parasitics network of Fig. 15.19 is shown in Fig. 15.20. The equivalence between the lumped elements and TLM components (distributed elements) are also listed in Table 15.4.

The parasitic components values used in the integrated TLLM model for the ridge-waveguide laser structure are listed in Table 15.5 [35]. Alternatively, a parameter optimisation software such as HP-MDS could be used to find the effective parasitic values that match experimentally measured S_{11} plots [58]. The bondwire inductance (L_p) and space-charge capacitance

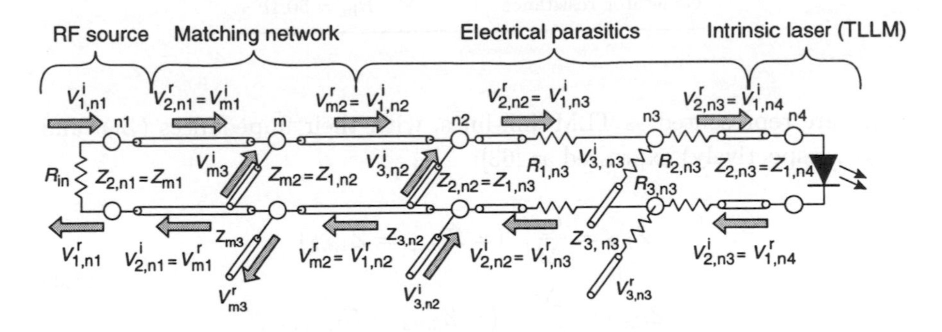

Fig. 15.20. The TLM network representation of the integrated TLLM model (compare with Fig. 15.19).

Table 15.4. Equivalence between the lumped elements and TLM components.

Lumped element	Equivalent TLM impedances
Stand-off shunt capacitance, C_p	$Z_{3,n2}$
Bondwire inductance, L_p	$Z_{2,n2} = Z_{1,n3}$
Bondwire resistance, R_p	$R_{1,n3}$
Chip substrate resistance, R_{sub}	$R_{3,n3}$
Shunt parasitic capacitance, C_s	$Z_{3,n3}$
Chip series resistance, R_s	$R_{2,n3}$
Space-charge capacitance, C_{sc}	$Z_{2,n3} = Z_{1,n4}$

Table 15.5. Parameter values used in the electrical parasitics network.

Parasitic element	Value
Bondwire resistance	$R_p = 1.0\ \Omega$
Bondwire inductance	$L_p = 0.63$ nH
Stand-off shunt capacitance	$C_p = 8.0$ pF
Chip substrate resistance	$R_{sub} = 1.5\ \Omega$
Shunt parasitic capacitance	$C_s = 0.23$ pF
Chip series resistance	$R_s = 5.5\ \Omega$
Space-charge capacitance	$C_{sc} = 2.0$ pF
Forward bias resistance	$R_d = 2.0\ \Omega$
Generator resistance	$R_{in} = 50\ \Omega$

(C_{sc}) are represented as TLM link-lines, with their impedances (Z_{LP} and Z_{CSC}, respectively) expressed as [63]:

$$
\begin{aligned}
Z_{LP} &= \frac{L_p}{\Delta t} \quad (= Z_{2,n2} = Z_{1,n3}) \\
Z_{CSC} &= \frac{\Delta t}{C_{sc}} \quad (= Z_{2,n3} = Z_{1,n4})\,.
\end{aligned}
\tag{15.4}
$$

The stand-off shunt capacitance (C_p) and chip shunt capacitance (C_s) are modelled as TLM stub-lines with their impedances (Z_{CP} and Z_{SC}, respectively) expressed as [63]:

$$\begin{aligned} Z_{\mathrm{CP}} &= \frac{\Delta t}{2C_{\mathrm{p}}} \quad (= Z_{3,n2}) \\ Z_{\mathrm{CS}} &= \frac{\Delta t}{2C_{\mathrm{s}}} \quad (= Z_{3,n3})\,. \end{aligned} \tag{15.5}$$

The TLM network "sees" the intrinsic laser as a small series resistance when it is forward biased [40, 67, 68]. At above threshold, the carrier density was clamped and a constant value of 2.0 Ω was used for the series resistance, R_{d} [69], while the associated space-charge capacitance, C_{sc}, was taken as 2.0 pF [35]. Using a constant linear load for the intrinsic laser improves computational efficiency of our model. Strictly speaking, the junction impedance of the intrinsic laser is carrier dependent [66, 70]. The only effect of this carrier dependence is a turn-on delay of the current flowing into the active region during switch up.

The frequency response of the ridge waveguide laser parasitics with the RF source resistance (R_{IN}) as parameter was simulated using the TLM parasitics model of Fig. 15.20. This was done by injecting an impulse voltage ($V^i_{1,n1}$) into the TLM network, and collecting the output voltage pulse samples in time domain ($V^i_{1,n4}$) before finally applying an FFT to convert them into frequency domain. For the sake of comparison, the parasitics response was also simulated using the circuit simulator in HP-MDS. In HP-MDS, a small-signal AC current source was injected in the lumped circuit model of the laser parasitics network, and then plotting the active region current (see current in Fig. 15.2) against frequency. Both simulation results are plotted in Fig. 15.21, showing good agreement. Note that the frequency responses are normalised to their DC level. Interestingly, for large values of R_{IN}, the resonance between the bond-wire inductance and stand-off shunt capacitance causes an increase in the frequency response [35]. The RF source resistance used throughout this book was chosen to be 50 Ω, which is typically used in practical measurements.

15.7 Matching Considerations

Recently, there have been proposals for using semiconductor lasers as sources of millimetre-wave signal for mobile broadband communication systems [2, 3, 71]. Increasing interest in millimetre-wave fibre-radio systems

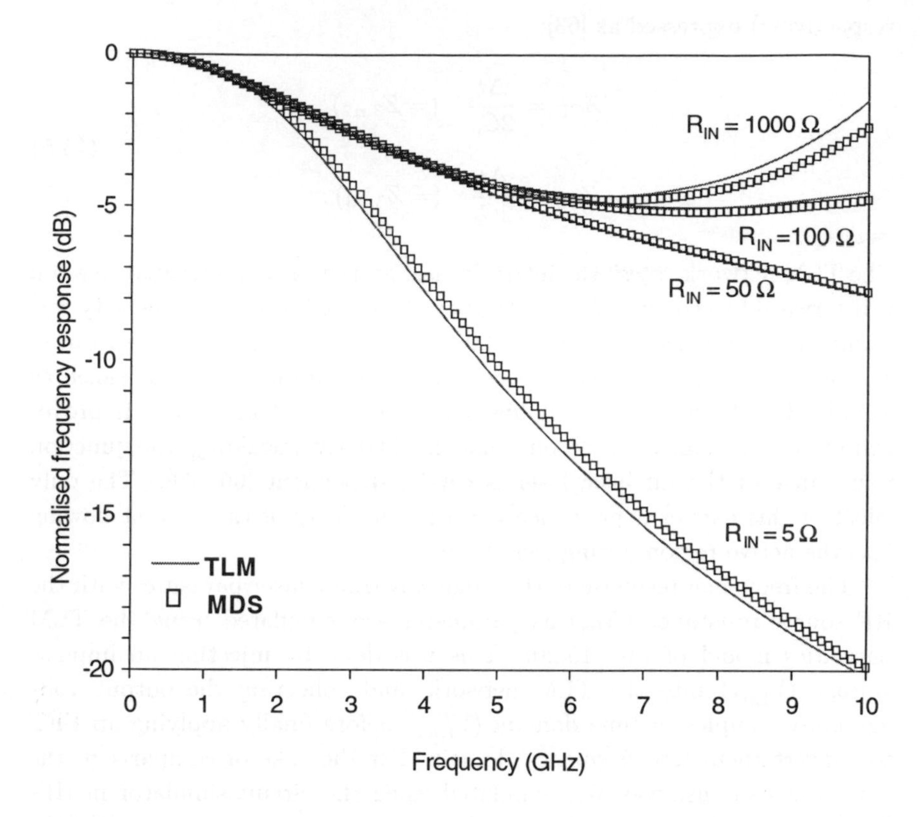

Fig. 15.21. Frequency response of the ridge waveguide laser parasitics for different source resistances (comparison between TLM and MDS).

has also lead to the investigation of optical signal-processing techniques for microwave and millimetre-wave signal generation by Lowery and Gurney [72]. In microwave and millimetre-wave applications, the laser diode should be matched to the signal generator to minimise reflections due to impedance mismatch [73]. In addition, high levels of RF power are required to overcome the losses associated with the electrical cable, bias tee, and laser mount.

Matching networks are often preferred to the conventional method of connecting a chip resistor (43–48 Ω) in series with the laser diode [67]. The reason is matching networks provide better power transfer between the signal generator and laser diode. Broadband operation, however, becomes more difficult to achieve. There have been numerous reports of matching techniques for laser diode transmitters. Narrowband techniques (typically 10–15% of signal bandwidth) include quarter-wave transformers (QWT) [74] and stub tuning [64]. Broadband techniques (100% or more of signal bandwidth) include the pseudo-bandpass LC ladder network [75] and the resonant circuit step transformer [69]. In practice, the matching is degraded due to the parasitics, and there are also constraints imposed by the matching elements themselves, e.g. realisation of transmission-line segments in microstrip are limited to impedances between 10 and 120 Ω [76] while lumped elements become unmanageably small at millimetre-wave frequencies [77]. The matching circuit has never been considered in any previous TLLM simulations, and in this book it has been incorporated into the integrated TLLM model [27]. The matching network chosen for this book is based on lumped-elements for its simplicity. At frequencies above 1 GHz, lumped elements are difficult to realise since they must be made very small relative to the operating wavelength. However, Maricot *et al.* [78] have successfully implemented a passive lumped-element matching network at 5.6 GHz based on monolithically integrated metal-insulator-metal (MIM) capacitors and spiral inductors. The monolithic implementation is preferred to hybrid matching [69] as it eliminates any stray inductance from bond wires used to attach the lumped and distributed elements together.

The matching network is an L-section type [79] as shown in Fig. 15.22(a), which consists of two reactive lumped elements, i.e. a capacitor and an inductor. For compatibility with the transmission-line laser model, TLM link-lines and stub-lines [63] are used to model the lumped-element matching network. The general transmission-lines used to simulate the reactive lumped matching elements are shown in Fig. 15.22(b) with the scattering directions of voltage pulses and their notations clearly labelled. On the matching node, V^{i}_{mi} and V^{r}_{mi} are the incident and reflected voltage pulses of the ith transmission-line, R_{mi} and Z_{mi} are the series resistances and characteristic impedances of the ith transmission-line. Since the matching network is purely reactive, the resistances can be neglected here (i.e. $R_{m1} = R_{m2} = R_{m3} = 0$). The reactive matching network in Fig. 15.22 may be incorporated into the overall integrated TLLM model as shown in Fig. 15.20. Referring to Fig. 15.20, the first scattering node from the left is the *input node* (n_1), the second is the *matching node* (m), and the remaining three scattering nodes will be referred to as the *parasitics nodes* $n_2 - n_4$).

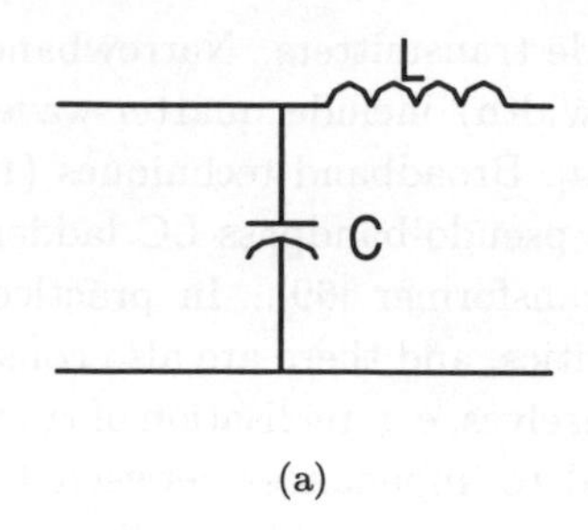

(a)

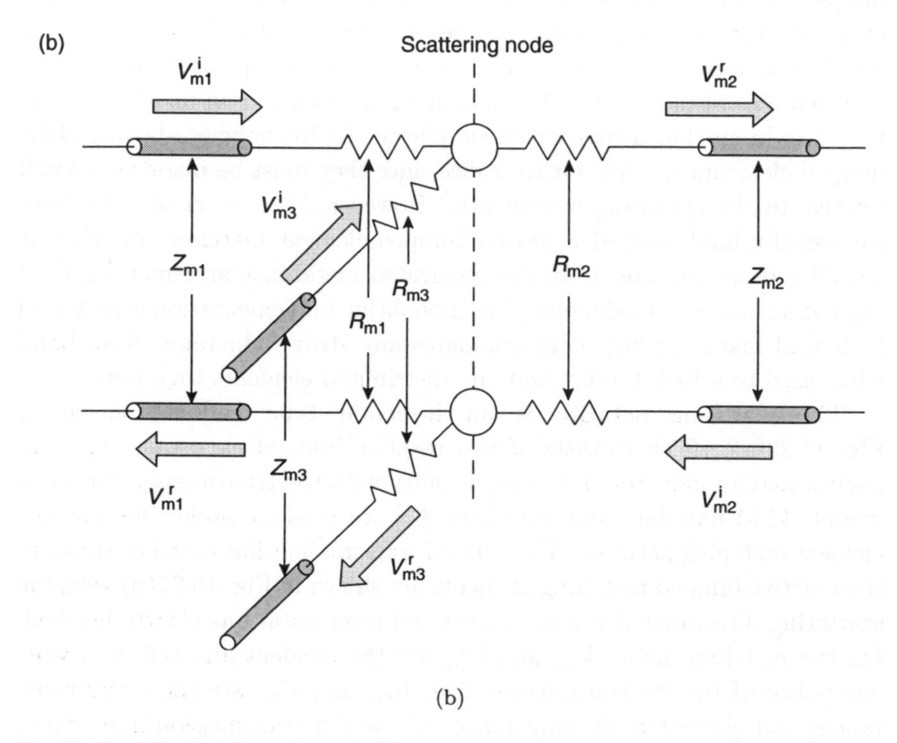

(b)

Fig. 15.22. The matching network (a) Lumped representation (b) TLM representation.

15.8 Small-Signal Modulation

In order to check for improvement in the emitted light signal as a result of matching, the small-signal modulation responses of the laser for both the unmatched and matched cases were plotted and shown in Fig. 15.23. These results were obtained by applying the FFT to the impulse response of the integrated TLLM model, similar to the method of obtaining the

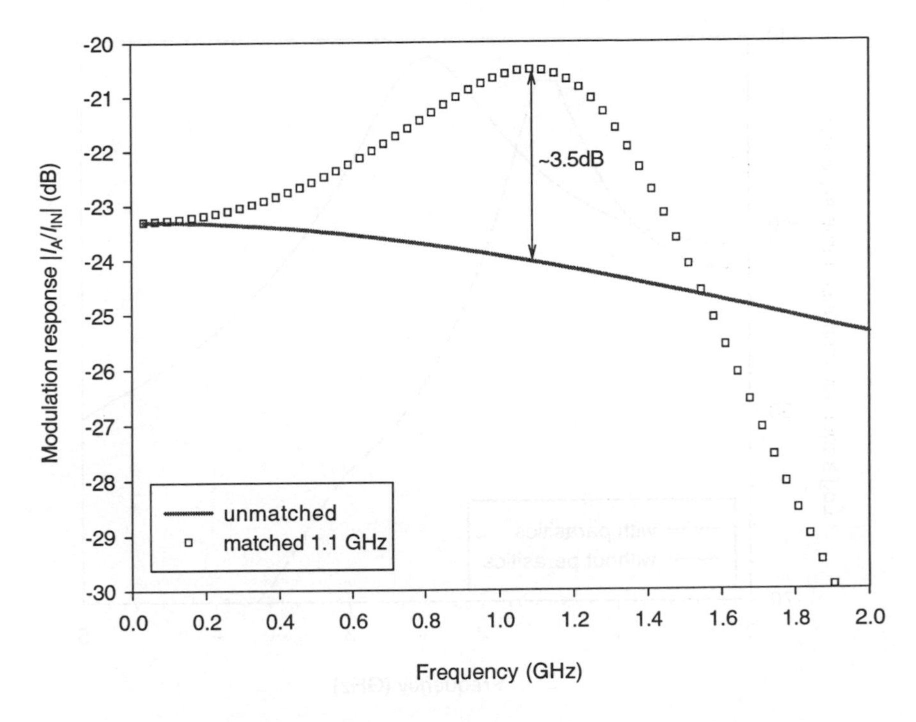

Fig. 15.23. Improvement in the integrated TLLM modulation response, matched at 1.1 GHz.

frequency response of the TLM parasitics network. The results show an improvement in emitted signal of about 3.5 dB, in agreement with the results of [80] for a matched laser transmitter. A resonant peak is absent for the unmatched case in Fig. 15.23 because the intrinsic laser impedance was simply modelled by a resistor (R_d) in parallel with the space-charge capacitance (C_{sc}) in the TLM lumped-network [80].

Furthermore, the *intrinsic* and *extrinsic* relaxation oscillation frequencies were found to be 2.4 GHz and 1.5 GHz, respectively. These values were found by taking the current flowing across the intrinsic laser diode (seen as a linear load by the TLM network), and feeding it directly (i.e. within the same source code) into the nonlinear laser model (TLLM). The -3 dB modulation bandwidth ($f_{-3\text{ dB}}$) of the laser transmitter is usually slightly higher than its relaxation frequency (f_r). However, when large electrical parasitics are present, their low-pass transfer function will significantly reduce $f_{-3\text{ dB}}$ to below f_r. This has been observed in EMBH lasers (see Fig. 15.18(b)). From Fig. 15.24, the extrinsic (i.e. parasitics included)

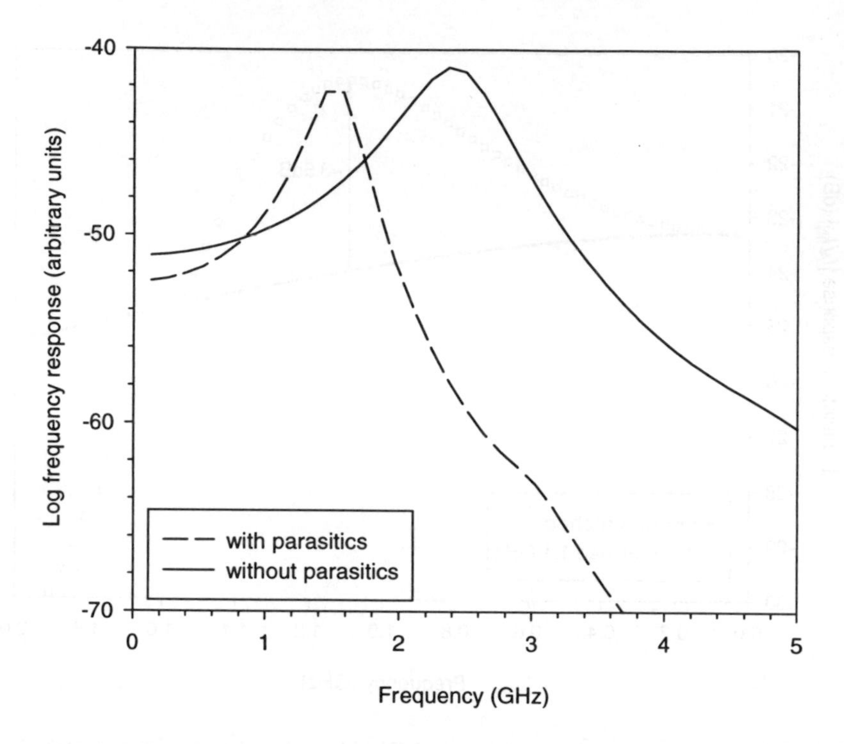

Fig. 15.24. Intrinsic and extrinsic resonant frequencies of the integrated TLLM model.

of the integrated TLLM model is about 1.5 GHz. Also notice that the inclusion of parasitics has weakened the resonant peak and shifted it to a lower frequency. By using the expression for intrinsic f_r (i.e. without parasitics) given as [31]:

$$f_r = \frac{1}{2\pi}\sqrt{\frac{1 + N_{tr}\Gamma v_g a \tau_p}{\tau_n(N)\tau_p} \cdot \left(\frac{I}{I_{th}} - 1\right)}$$

where

$$\tau_n(N_{th}) = \frac{1}{\frac{N_{th}}{\tau_n} + (B_0 - B_1 N_{th})N_{th} + C_{aug}N_{th}^2} \tag{15.6}$$

the calculated value of intrinsic f_r was found to be 2.4 GHz. The parameters in Eq. (15.6) have been defined in Table 15.3. This analytical result agrees well with the simulation result of the integrated TLLM model (parasitics neglected), which gave a value of 2.5 GHz. The error is therefore within only 5% of the calculated result.

15.9 Large-Signal Modulation

The laser diode provides a compact, reliable, and potentially low-cost method for the optical generation and distribution of mm-wave signals [2, 19]. By large-signal direct modulation of the laser diode, such as gain-switching or mode-locking, harmonic powers can be generated up to millimetre-wave frequencies. Direct modulation is attractive because of its simplicity. In the following large-signal application example, the integrated TLLM model was gain-switched at different subharmonic frequencies to generate a RF signal at 6.6 GHz, useful for locking a GaAs-based monolithic microwave integrated circuit voltage controlled oscillator (MMIC VCO) [82].

A multimoded 1.3 μm InGaAsP ridge-waveguide laser diode model was used in the simulations, with a comprehensive list of parameter values taken from [65], also listed in Table 15.3. The modulation bandwidth of the laser model is well below 6.6 GHz, hence direct modulation at this frequency is highly inefficient. Therefore, a simple solution is to gain-switch the laser at subharmonic frequencies of 6.6 GHz such as 1.1 GHz, 1.32 GHz, 1.65 GHz, 2.2 GHz, and 3.3 GHz so that the 6.6 GHz component is included as part of the comb of harmonic frequencies. Furthermore, by using subharmonic frequencies, higher drive power is available since RF amplifiers at high frequencies are difficult to realise.

Matching circuits should be used in large-signal modulation of semiconductor lasers to achieve higher RF power and better modulation depth. Matching techniques such as microstrip stub matching circuits [64] and the quarter-wave transformer can be used to match the laser transmitter, improving the peak optical power that can be achieved through gain-switching [74]. Alternatively, monolithically integrated lumped networks can be used to achieve impedance matching [78]. In the following, the design of a matching circuit based on monolithically integrated reactive elements is explained.

15.10 Design of the Matching Circuit

The matching network is designed for five different RF frequencies that correspond to the subharmonic frequencies of 6.6 GHz. Depending on the

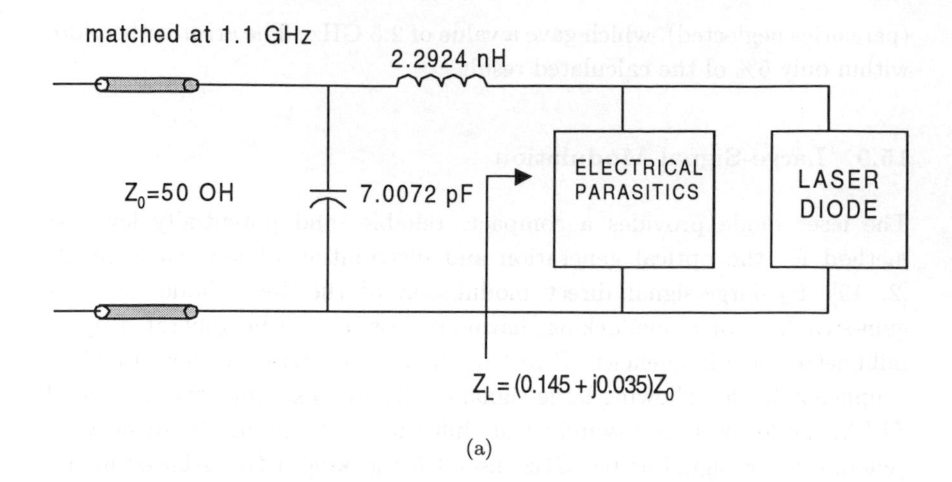

(a)

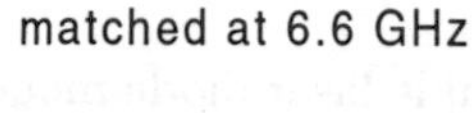

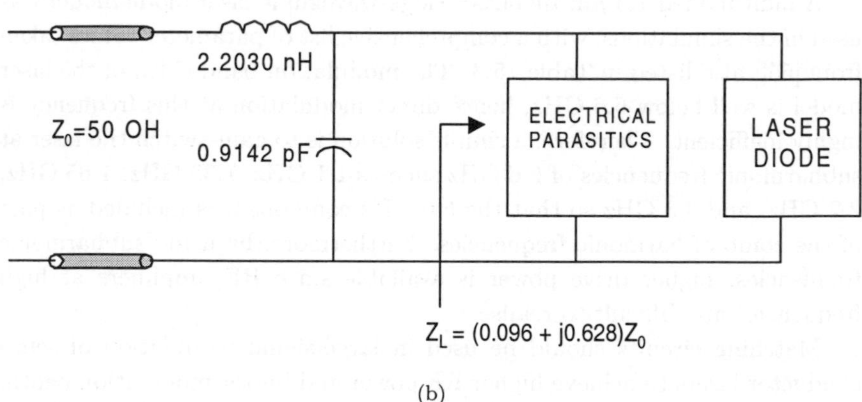

(b)

Fig. 15.25. Two circuit configurations for the matching circuit of the integrated TLLM model.

desired matching frequency, there are two circuit configurationsas shown within the frequency range of interest (1.1 GHz–6.6 GHz), in Figs. 15.25(a) and (b). It should be noted that Fig. 15.25(a) is applicable for subharmonic frequencies of 1.1 GHz, 1.32 GHz, 1.65 GHz, 2.2 GHz, and 3.3 GHz, whereas Fig. 15.25(b) is only applicable for the fundamental frequency of 6.6 GHz. Note that in Fig. 15.25(b) the positions of both are interchanged from their position in Fig. 15.25(a). From Fig. 15.20, the scattering at the input node ($n1$) can be described by [7]:

$$\begin{bmatrix} V^{\mathrm{r}}_{1,n1} \\ V^{\mathrm{r}}_{2,n1} \end{bmatrix} = \begin{bmatrix} \dfrac{Z_{\mathrm{m1}} - R_{\mathrm{in}}}{R_{\mathrm{in}} + Z_{\mathrm{m1}}} & \dfrac{2R_{\mathrm{in}}}{R_{\mathrm{in}} + Z_{\mathrm{m1}}} \\ \dfrac{2Z_{\mathrm{m1}}}{R_{\mathrm{in}} + Z_{\mathrm{m1}}} & \dfrac{R_{\mathrm{in}} - Z_{\mathrm{m1}}}{R_{\mathrm{in}} + Z_{\mathrm{m1}}} \end{bmatrix} \begin{bmatrix} V^{\mathrm{i}}_{1,n1} \\ V^{\mathrm{i}}_{2,n1} \end{bmatrix} \tag{15.7}$$

where R_{in} is the source resistance, Z_{m1} is the link-line impedance, $V^{i}_{1,n1}$ is the input signal (port 1 of node-$n1$) comprising the bias current and an RF signal superimposed on it, $V^{\mathrm{r}}_{2,n1}$ is the voltage pulse transmitted into the matching circuit (port 2 of node-$n1$), $V^{\mathrm{i}}_{2,n1}$ is the incident voltage pulse from the matching circuit, and $V^{\mathrm{r}}_{1,n1}$ is the voltage pulse transmitted back to the source.

The TLM link-line at port 1 of the matching node is labelled as Z_{m1}, which is the transmission-line feeding the RF signal into the matching circuit. When matched at subharmonic frequencies below 6.6 GHz, Z_{m1} is taken to be equal to the source resistance (R_{in}) of 50 Ω. The capacitive and inductive matching elements are respectively modelled as TLM stub-lines and link-lines expressed by:

$$Z_{\mathrm{m3}} = \frac{\Delta t}{2C_{\mathrm{match}}} \quad \text{and} \quad Z_{\mathrm{m2}} = \frac{L_{\mathrm{match}}}{\Delta t}\,. \tag{15.8}$$

On the other hand, when the laser is to be matched at 6.6 GHz, Z_{m1} is used to model the inductive matching element instead, that is:

$$Z_{\mathrm{m1}} = \frac{L_{\mathrm{match}}}{\Delta t}\,. \tag{15.9}$$

Now the inductive element (TLM link-line) comes before the capacitive element (TLM-stub-line). In this case, the matched transmission-line feeding the RF signal is made redundant. This does not affect the simulation results since there is no reflection before the matching circuit anyway. An important point to note is that the link-line of port 2 of the matching node (node-m) must now be made part of the stand-off shunt capacitance (C_{p}) of the adjacent parasitics node (node-$n2$). This is done by splitting the stand-off shunt capacitance equally as a TLM stub-line and a TLM link-line, where their corresponding impedances can be expressed by:

$$Z_{1,n2} = \frac{2\Delta t}{C_{\mathrm{p}}} \quad \text{and} \quad Z_{3,n2} = \frac{\Delta t}{C_{\mathrm{P}}} \tag{15.10}$$

where $Z_{1,n2}$ is the impedance of the TLM link-line connecting the matching node (node-m) and the first parasitics node (see Fig. 15.20), and $Z_{3,n2}$ is the impedance of the TLM stub-line at the first parasitics node (node-$n2$). Therefore, the total stand-off shunt capacitance remains unchanged.

The scattering matrix of the TLM matching network in Fig. 15.22 is given as:

$$\begin{bmatrix} V^{\mathrm{r}}_{\mathrm{m1}} \\ V^{\mathrm{r}}_{\mathrm{m2}} \\ V^{\mathrm{r}}_{\mathrm{m3}} \end{bmatrix} = C_{\mathrm{m}} \begin{bmatrix} 2P_{\mathrm{m1}}Z_{\mathrm{s2}}Z_{\mathrm{s3}}+R_{\mathrm{m11}} & 2P_{\mathrm{m1}}Z_{\mathrm{s1}}Z_{\mathrm{s3}} & 2P_{m1}Z_{s1}Z_{s2} \\ 2P_{\mathrm{m2}}Z_{\mathrm{s2}}Z_{s3} & 2P_{\mathrm{m2}}Z_{s1}Z_{\mathrm{s3}}+R_{\mathrm{m22}} & 2P_{\mathrm{m2}}Z_{\mathrm{s1}}Z_{\mathrm{s2}} \\ 2P_{\mathrm{m3}}Z_{\mathrm{s2}}Z_{\mathrm{s3}} & 2P_{\mathrm{m3}}Z_{\mathrm{s1}}Z_{s3} & 2P_{\mathrm{m3}}Z_{\mathrm{s1}}Z_{\mathrm{s2}}+R_{\mathrm{m33}} \end{bmatrix} \times \begin{bmatrix} V^{i}_{\mathrm{m1}} \\ V^{i}_{\mathrm{m2}} \\ V^{i}_{\mathrm{m3}} \end{bmatrix}$$

where

$$\begin{aligned} C_m &= \frac{1}{Z_{\mathrm{s1}}Z_{\mathrm{s2}} + Z_{\mathrm{s1}}Z_{s3} + Z_{\mathrm{s2}}Z_{\mathrm{s3}}} \qquad (15.11) \\ R_{mii} &= \frac{R_{mi} - Z_{mi}}{R_{mi} + Z_{mi}} \qquad \text{where} \quad i = 1,2,3 \\ Z_{\mathrm{s}i} &= R_{mi} + Z_{mi} \qquad \text{where} \quad i = 1,2,3 \\ P_{mi} &= \frac{Z_{mi}}{Z_{mi} + R_{mi}} \qquad \text{where} \quad i = 1,2,3\,. \end{aligned}$$

The scattering matrix above was found by referring to Eq. (14.14) in Chapter 14, and with the help of the Thevenin equivalent of the TLM network of Fig. 15.22. Since the metal-insulator-metal (MIM) capacitor and spiral inductor used for matching were purely reactive elements, the resistances of the lines (R_{mi} where $i = 1,2,3$) were set to zero. The scattering matrix of Eq. (15.10) is thus defined for the most general case. Similar scattering matrices may be used for the parasitics nodes, each with their respective values for series resistances ($R_{x,n(y+1)}$) and line impedances ($Z_{x,n(y+1)}$) where x (either 1, 2 or 3) denotes one of the three ports of the yth parasitics node.

Next, we determined the initial values for the reactive matching elements of Fig. 15.25 with the help of the Z-Smith chart. Then, the initial values were used as inputs into a parameter optimisation software such as HP-MDS. The matching elements were then finely tuned to achieve insertion

Table 15.6. Optimised values of the L-matching network for different subharmonic frequencies.

nth subharmonic frequency	L_{match} (nH)	C_{match} (pF)
6th (1.1 GHz)	2.2924	7.0072
5th (1.32 GHz)	1.7953	6.0429
4th (1.65 GHz)	1.2677	5.0663
3th (2.2 GHz)	0.7330	4.1072
2nd (3.3 GHz)	0.1822	3.0424
Fundamental (6.6 GHz)	2.2030	0.9142

losses up to -50 dB. The optimised values obtained for the matching elements are listed in Table 15.6, while the S_{11} plots (matched at 1.1 GHz and at 6.6 GHz) are shown in Fig. 15.26. The results obtained from TLM and MDS are compared in Fig. 15.26, and are found to be in good agreement. The magnitude of S_{11} for the case of 1.1 GHz only managed to reach -30 dB for the TLM model because of its lower frequency resolution. This can easily be solved by collecting a larger number of time-domain samples.

15.11 Harmonic Generation by Gain-Switching

In this section, the integrated TLLM model (parasitics and matching network included) was used to study a microwave signal generation technique known as harmonic generation by gain-switching. The simulation results obtained from the integrated TLLM model were then compared with the results from a lumped-element integrated laser model by Sum and Gomes [64]. In contrast to the noiseless lumped-circuit models, the stochastic integrated TLLM model may be useful for investigating the effects of up-conversion of low-frequency noise in gain-switching [83, 84].

The 6th subharmonic frequency (1.1 GHz) with +1 dBm power was injected into the integrated laser transmitter model to generate gain-switched optical pulses, which subsequently produced a comb of RF frequencies at harmonics of 1.1 GHz owing to laser nonlinearity. The RF power at these harmonic frequencies for both the matched and unmatched cases are compared in Fig. 15.27. When matched, the RF power increased for all harmonic frequencies and an improvement of 2 dB was achieved for the 6.6 GHz component. It can also be observed in Fig. 15.27 that the higher harmonics had greater improvement than the lower harmonics.

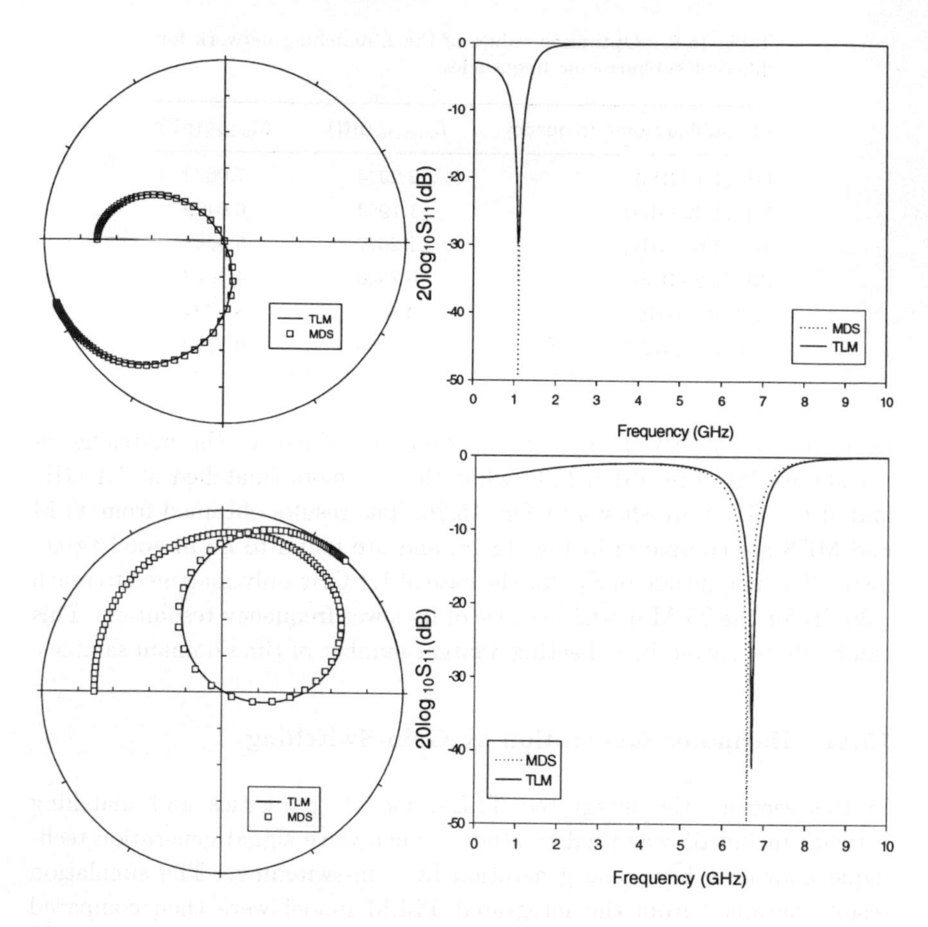

Fig. 15.26. S_{11} plots (left: Z-smith; right: magnitude) when matched at 1.1 GHz (top) and 6.6 GHz (bottom).

An analytical approach to investigate the effect of parasitics on the injected electrical pulse has been presented by Liu *et al.* [85]. The presence of electrical parasitics tends to widen the input electrical pulse and reduces its peak amplitude. In the approach in [85], the analytical form of the pulse shape after passing the parasitics has to be estimated beforehand. With the integrated TLLM model, no such estimation is necessary. In fact, there is no restriction as to the type of input modulation waveform used in the integrated TLLM model, since the electrical waveform after passing the parasitics need not be estimated beforehand.

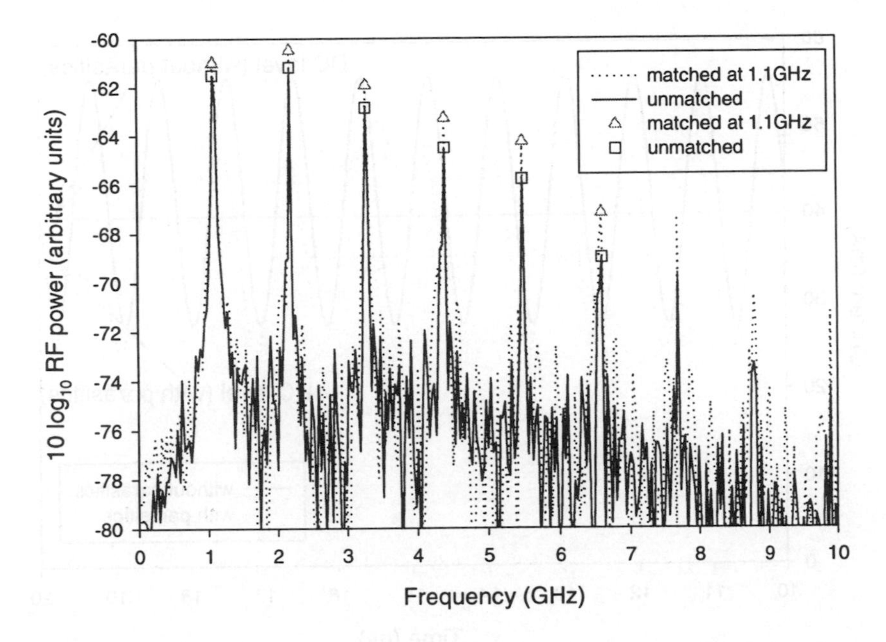

Fig. 15.27. The RF spectrum: comparison between matched and unmatched power.

A sinusoidal RF signal was used here in contrast to the electrical pulses from a comb generator that were used in [85]. The injected RF modulation current before and after the electrical parasitics are compared in Fig. 15.28. From Fig. 15.28, the presence of parasitics reduced the DC bias level, as well as peak-to-peak (p.p.) amplitude of the RF modulation current. There was also a small time delay in the current reaching the active region. As a result, the gain-switched optical pulses were broadened and had lower peak powers.

When the matching circuit was included, narrower optical pulses with higher peak powers could be achieved. Figure 15.29(a) shows the train of optical pulses that were emitted from the gain-switched laser, where the optical power was time-averaged over every 5.8 ps. The improvement in pulse amplitude is clearly observed when matched at 1.1 GHz. Some instability of the pulses is observed because the intrinsic laser model based on TLLM is a stochastic model, and the optical pulses are built up from spontaneous emission noise. It should be noted that a purely stochastic model should give different results for different runs but the statistical characteristics should remain the same.

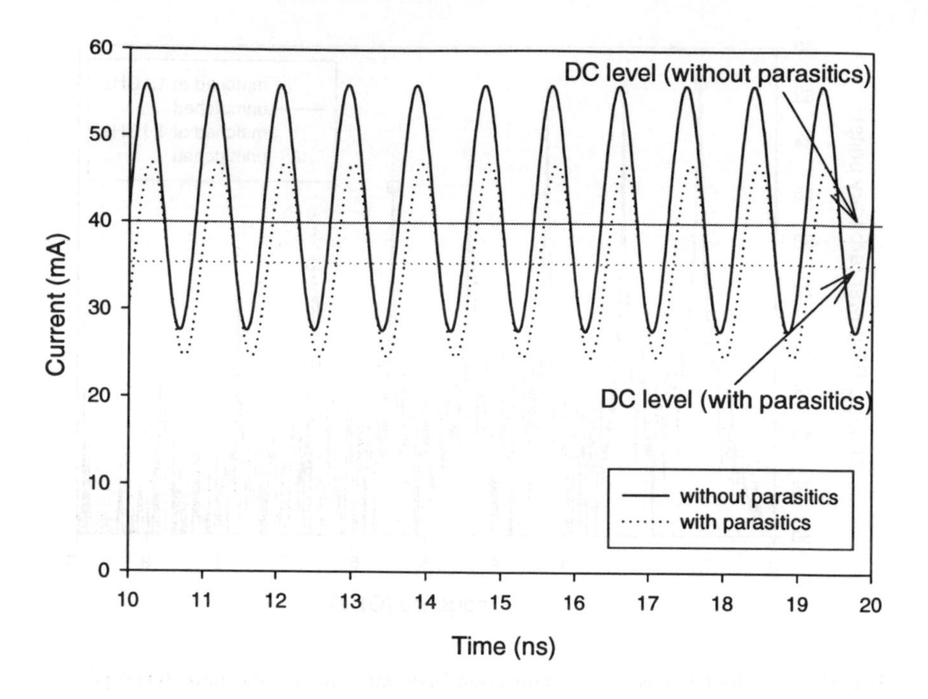

Fig. 15.28. Effect of parasitics on modulation current waveform.

In Fig. 15.29(b), the optical pulses from the same simulation were averaged over a period of 30ns using the stable averaging technique [86], which is equivalent to a digital oscilloscope in *averaging mode*. The inclusion of the matching circuit provided a reduction of 11.6 ps in the full width half maximum (FWHM) of the averaged gain-switched pulse (i.e. from 98.5 ps to 86.9 ps). The peak optical power was also enhanced by matching, from 1.7 mW to 2.1 mW. However, there was a time delay of about 0.2 ns, attributed to the additional reactive elements used for matching.

Next, four other different subharmonic frequencies of 6.6 GHz (i.e. 1.32 GHz, 1.65 GHz, 2.2 GHz, and 3.3 GHz) were used to generate gain-switched optical pulses from the integrated TLLM model. In Fig. 15.30(a), the magnitudes of RF power are plotted against their respective injected subharmonic frequencies. The highest RF power at a single harmonic was generated by 1.32 GHz, due to its proximity to the extrinsic relaxation oscillation frequency (f_r) of 1.5 GHz, given a DC bias level of 1.3 times threshold. The model in [64] also had a relaxation oscillation frequency between 1 and 2 GHz.

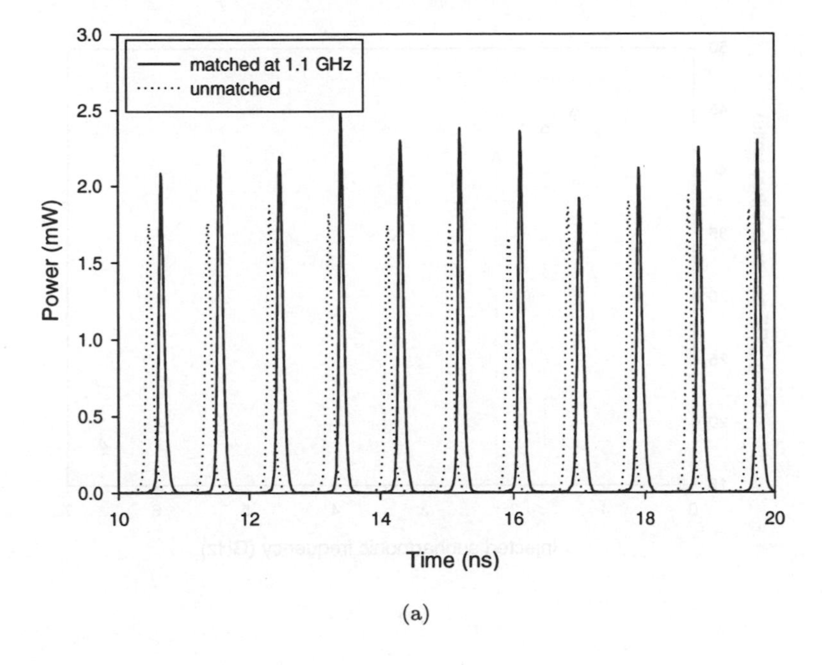

(a)

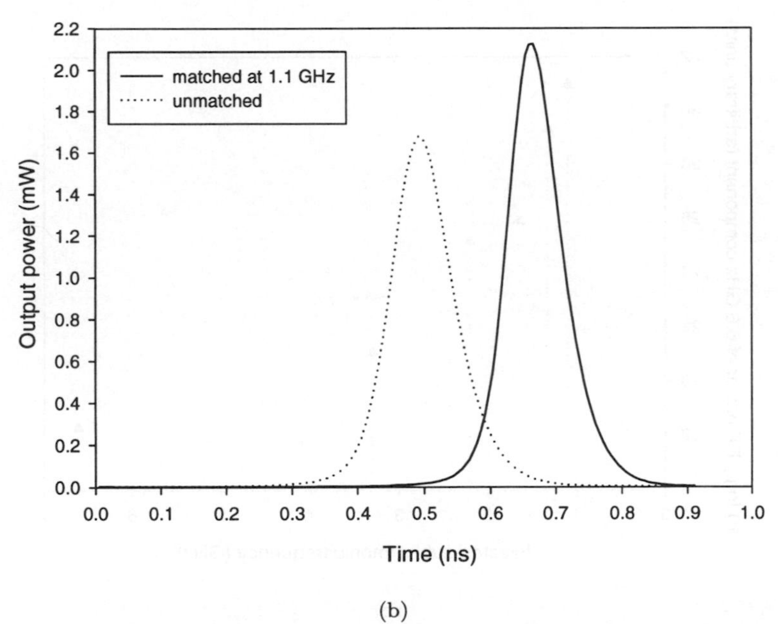

(b)

Fig. 15.29. Comparison between matched and unmatched gain-switched optical pulses: (a) a train of time-averaged pulses (10–20 ns) (b) the stable-averaged pulse over 30 ns.

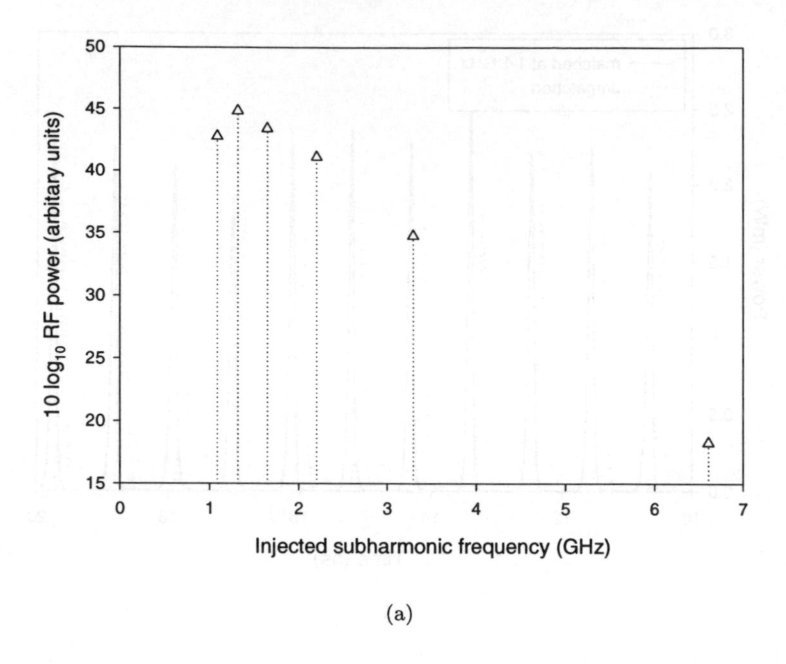

(a)

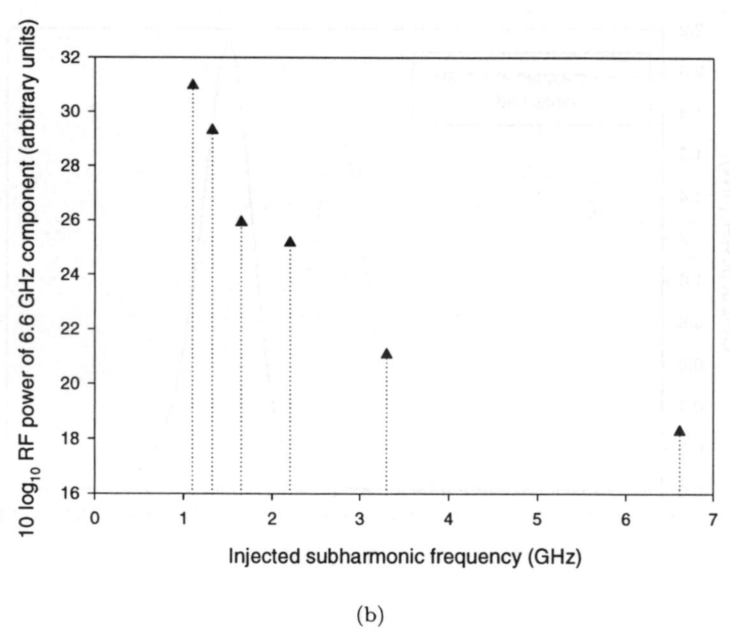

(b)

Fig. 15.30. (a) RF power at the injected subharmonic frequencies (b) RF power of the 6.6 GHz component.

In Fig. 15.30(b), the magnitudes of RF power at 6.6 GHz are plotted against the injected subharmonic frequencies. The RF power at 6.6 GHz was largest when the injected subharmonic frequency was 1.1 GHz. This shows that the relaxation oscillation frequency does not play a major role in generating a large RF power at higher order harmonic frequencies (6.6 GHz in this case). Sum and Gomes [64], using the lumped-element integrated laser model, showed that the highest achievable RF power of the 6.6 GHz component was generated by 1.32 GHz. At first thought, it is predicted that an improvement of RF power at 6.6 GHz can be achieved when the power is shared among fewer harmonic components. However, this improvement is negated by power diffusivity when the laser is gain-switched at high RF frequencies. In our case, no improvement of RF power at 6.6 GHz was observed in the integrated TLLM model when gain-switched was used at higher subharmonic frequencies due to high power diffusivity. This was probably due to the additional damping effect of spontaneous emission noise, which was not included in the deterministic lumped-element integrated laser model of [64].

The power diffusivity effect (i.e. reduction in RF power of the 6.6 GHz component) at lower order subharmonic frequencies (i.e. higher RF frequencies) can further be explained by the time-domain optical pulses in Fig. 15.31. The optical pulses generated by 2.2 GHz and 6.6 GHz were significantly different and distorted from the moderately uniform train of optical pulses generated by 1.1 GHz. At drive frequency of 2.2 GHz, optical pulses could still be observed but they had become severely non-uniform. There was a small delay of the optical pulses and at certain times, the optical pulses were split into two. Their peak amplitudes were also reduced. At a higher frequency of 6.6 GHz, optical pulses could no longer be generated, the output power became more continuous-wave (CW) like instead. This can be explained by the fact that the injected subharmonic frequency exceeded the modulation bandwidth of the laser, and the gain-switching process could not respond fast enough to such a high RF frequency.

15.12 Frequency Chirp

Dynamic frequency chirp is due to carrier-induced refractive index change in the laser [31]. It is significant in gain-switching and therefore should not be neglected. The magnitude of chirp depends on the linewidth enhancement factor (α_{H}), which describes the relationship between refractive index

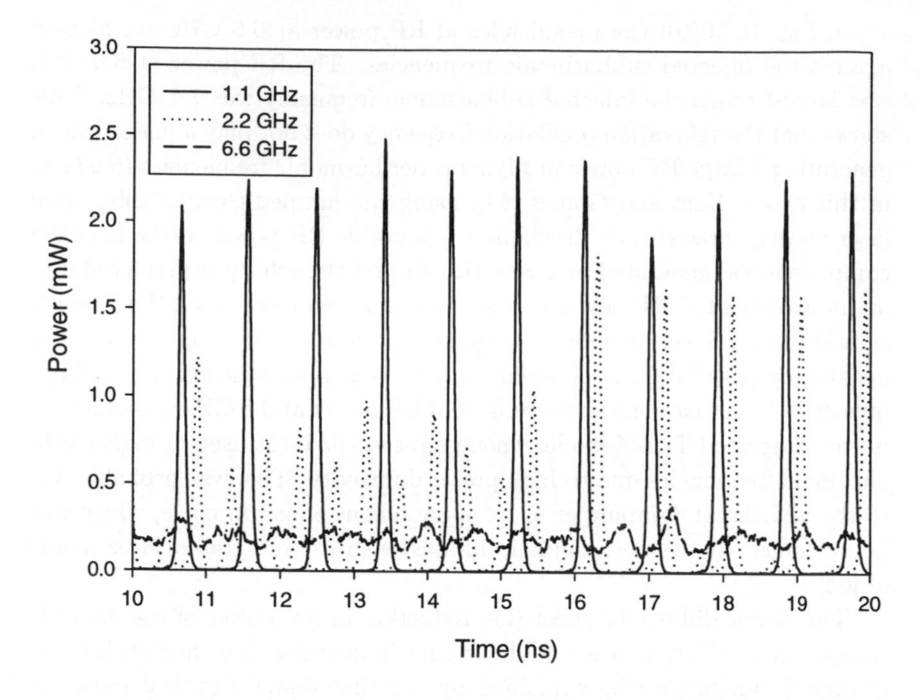

Fig. 15.31. The train of gain-switched optical pulses for different RF drive frequencies.

modulation and gain modulation [31]. The refractive index variation with carrier density has been included in the intrinsic laser model based on the standard TLLM (see Chapter 3). Transmission-line *stubs* of variable stub impedance (equivalent to variable length) depending on the average carrier density and one-way frequency independent *attenuators* are placed at both facets of the intrinsic laser model [87]. These *stub attenuators* model the change in phase length of the optical cavity, which is related to the change in refractive index. The optical spectrum of the laser model at steady-state continuous-wave (CW) operation is shown in Fig. 15.32(a) while the dynamic frequency chirp due to gain-switching is shown in Fig. 15.32(b).

15.13 Carrier-Dependent Laser Diode Impedance

Below threshold, the intrinsic laser diode can be approximated as an effective capacitor (C_t) and a resistor (R_d) in parallel [68, 88, 89]. The effective capacitance, C_t, is a combination of the diffusion capacitance

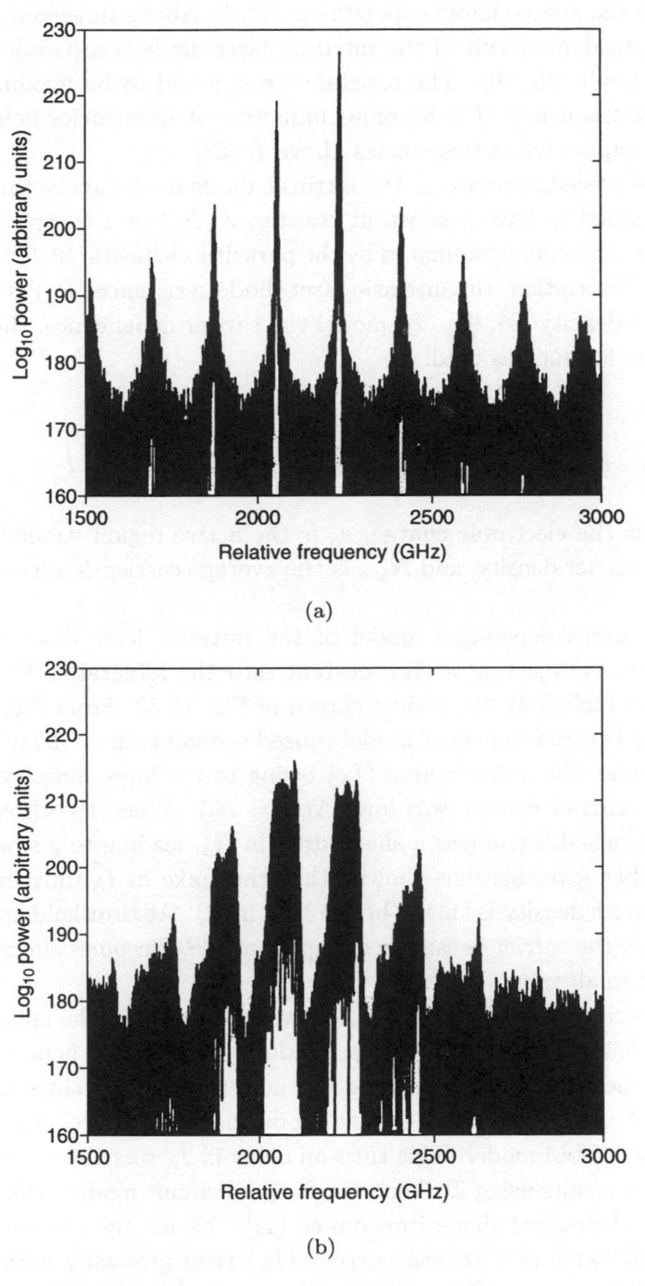

Fig. 15.32. (a) Optical spectrum at CW operation (biased at 1.3 I_{th}) and (b) when gain-switched at 1.1 GHz (with +1 dBm RF power).

(C_d) and the space-charge capacitance (C_{sc}). Above threshold, the small-signal equivalent circuit of the intrinsic laser diode is a parallel RLC resonance circuit [88, 90]. The resistance was found to be maximum at the resonance frequency (f_r), becomes inductive at frequencies below f_r, and becomes capacitive at frequencies above f_r [88].

In the preceding sections, the intrinsic diode model above threshold has been assumed to have a constant resistance (R_d) of 2 Ω, and the overall impedance is usually dominated by the parasitic elements [40, 67]. In a more accurate description, the intrinsic laser diode resistance (R_d) is dependent on carrier density [66, 91]. To model the carrier dependence, the following expression for R_d was used:

$$R_d = \frac{2kT}{q} \cdot \frac{\Delta t}{q v_a (N_{avg} - N_i)} \ln\left(\frac{N_{avg}}{N_i} + 1\right) \tag{15.12}$$

where q is the electronic charge, v_a is the active region volume, N_i is the intrinsic carrier density, and N_{ave} is the average carrier density in the laser cavity.

This carrier-dependent model of the intrinsic laser diode impedance was tested by injecting a step current into the integrated TLLM model (parasitics included), the results shown in Fig. 15.33. From Fig. 15.33, the unmatched carrier-depedent model caused a small turn-on delay of the current reaching the active region (I_A) owing to the large initial value of R_d when the carrier density was low ($N_{init} = N_i$). When the threshold level was approached, there was a sharp drop in R_d, leading to a small spike in I_A. Further investigations showed that the spike in I_A only exists if the initial carrier density is small (below 10^{19} m^{-3}). At threshold current level and above, the carrier density is clamped, and R_d becomes almost constant at a very small value.

The inclusion of the matching network suppressed the spike observed in Fig. 15.33(b) by introducing a gradual drop in R_d, hence the longer delay in the current, I_A, reaching the active region. Besides the turn-on delay of I_A, the carrier dependent model has no major influence on the integrated TLLM model. This turn-on delay in I_A was also observed in the simulation results using Tucker's large-signal circuit model, which included a current-dependent diode impedance [42]. As for the case of R_d set to a constant value of 2 Ω, the current (I_A) rises gradually with negligible turn-on delay.

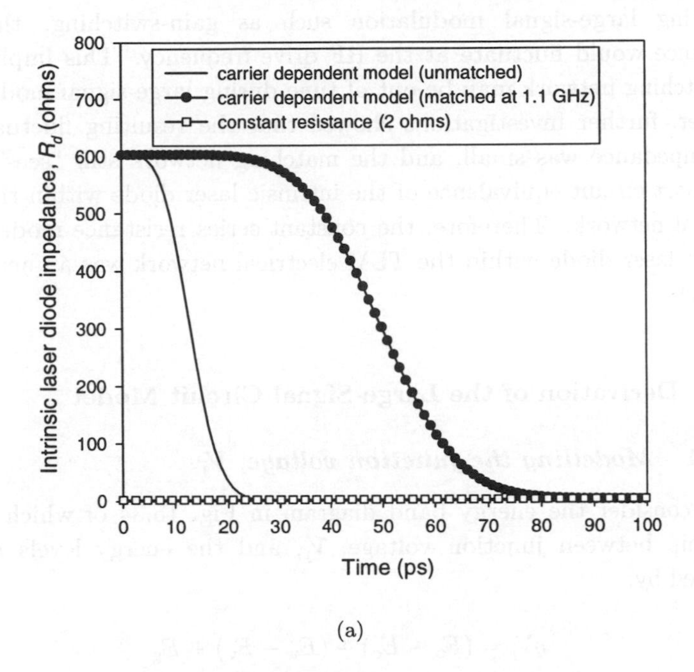

(a)

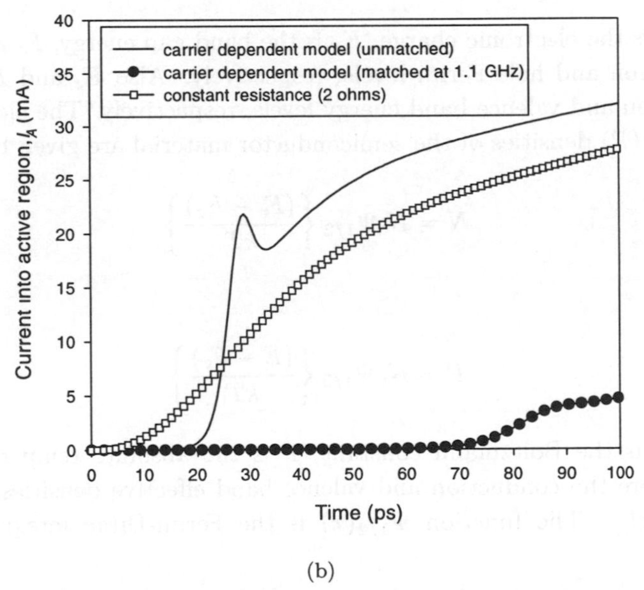

(b)

Fig. 15.33. The effect of carrier dependence on (a) intrinsic laser diode resistance (R_d), and (b) current reaching the active region (I_A).

During large-signal modulation such as gain-switching, the laser impedance would fluctuate at the RF drive frequency. This implies that the matching network may be out of tune during large-signal modulation. However, further investigations showed that the resulting fluctuation of laser impedance was small, and the matching network still "sees" an almost short-circuit equivalence of the intrinsic laser diode within the TLM electrical network. Therefore, the constant series resistance model of the intrinsic laser diode within the TLM electrical network was an acceptable approach.

15.14 Derivation of the Large-Signal Circuit Model

15.14.1 *Modelling the junction voltage, V_j*

Let us consider the energy band diagram in Fig. 15.34 of which the relationship between junction voltage, V_j, and the energy levels may be expressed by:

$$qV_j = (F_c - E_c) + (E_v - F_v) + E_g \tag{15.13}$$

where q is the electronic charge, E_g is the band gap energy, F_c and F_v are the electron and hole Fermi levels, respectively. Also E_c and E_v are the conduction and valence band energy levels, respectively. The electron (N) and hole (P) densities of the semiconductor material are given by:

$$N = N_c \Psi_{1/2} \left\{ \frac{(F_c - E_c)}{kT} \right\} \tag{15.14a}$$

and

$$P = N_v \Psi_{1/2} \left\{ \frac{(E - F_v)}{kT} \right\} \tag{15.14b}$$

where k is the Boltzmann constant, T is the absolute temperature, N_c and N_v are the conduction and valence band effective densities of states, respectively. The function $\Psi_{1/2}(x)$ is the Fermi-Dirac integral defined by [97]:

$$\Psi_{1/2}(x) = \frac{2}{\pi^{1/2}} \int_0^\infty \frac{\varepsilon^{1/2} d\varepsilon}{1 + \exp(\varepsilon - x)} \,. \tag{15.15}$$

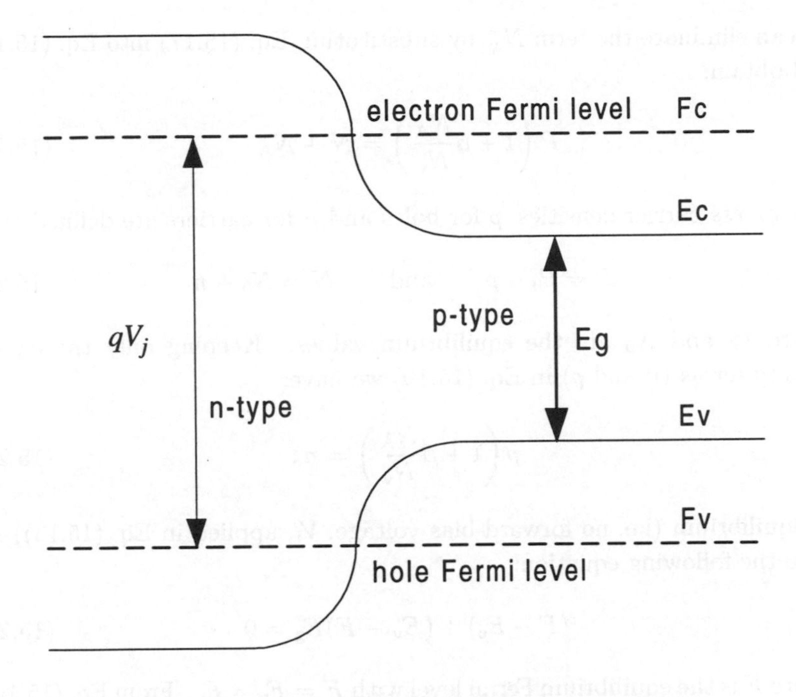

Fig. 15.34. Energy band diagram of n-p heterojunction.

The Fermi-Dirac integral, $\Psi_{1/2}(x)$ could be approximated using a power series truncated after two terms [98]:

$$\frac{(F_c - E_c)}{kT} = \ln\left(\frac{N}{N_c}\right) + \alpha_1 N$$

$$\text{and} \qquad \frac{(E_v - F_v)}{kT} = \ln\left(\frac{P}{N_v}\right) + \alpha_2 P \tag{15.16}$$

where α_1 and α_2 are constants over a carrier density range of interest in a particular semiconductor material. In a doped semiconductor, we can approximate the ratio of N_A (acceptor impurity concentration) and N_A^- (concentration of ionised acceptors) by a linear relationship given as [42]:

$$\frac{N_A^-}{N_A} = 1 - \beta\frac{P}{N_v}\,. \tag{15.17}$$

Since *charge neutrality* states that:

$$P = N + N_A^- \tag{15.18}$$

we can eliminate the term N_A^- by substituting Eq. (15.17) into Eq. (15.18) and obtain:

$$P\left(1+\beta\frac{N_A}{N_v}\right) = N + N_A\,. \qquad (15.19)$$

The *excess* carrier densities, p for holes and n for carriers are defined as,

$$P = P_0 + p \qquad \text{and} \qquad N = N_0 + n \qquad (15.20)$$

where P_0 and N_0 are the equilibrium values. Keeping only the excess density terms (n and p) in Eq. (15.19) we have:

$$p\left(1+\beta\frac{N_A}{N_v}\right) = n\,. \qquad (15.21)$$

In equilibrium (i.e. no forward-bias voltage, V_j, applied in Eq. (15.13)) we have the following equation:

$$(F - E_c) + (E_v - F)E_g = 0 \qquad (15.22)$$

where F is the equilibrium Fermi level with $F = F_c = F_v$. From Eq. (15.16), we have the following equations:

$$\frac{(F - E_c)}{kT} = \ln\left(\frac{N_0}{N_c}\right) + \alpha_1 N_0$$

$$\text{and} \qquad \frac{(E_v - F)}{kT} = \ln\left(\frac{P_0}{N_v}\right) + \alpha_2 P_0 \qquad (15.23)$$

Since the energy band gap is $E_g = E_c - E_v$ (see Fig. 15.34), Eq. (15.22) may be written as:

$$E_g = -kT\ln\left(\frac{P_0}{N_v}\right) - kT\ln\left(\frac{N_0}{N_c}\right) - kT\alpha_1 N_0 - kT\alpha_2 P_0\,. \qquad (15.24)$$

With a forward bias voltage, V_j, applied across the heterojunction we have:

$$qV_j = kT\ln\left(\frac{N}{N_c}\right) + \alpha_1 NkT + kT\ln\left(\frac{P}{N_v}\right) + \alpha_2 PkT - kT\ln\left(\frac{N_0}{N_c}\right)$$

$$- \alpha_1 N_0 kT - kT\ln\left(\frac{P_0}{N_v}\right) - \alpha_2 P_0 kT \qquad (15.25)$$

which can be simplified to:

$$qV_\mathrm{j} = kT\ln\left(\frac{N}{N_0}\right) + \alpha_1 NkT + kT\ln\left(\frac{P}{P_0}\right)$$
$$+\alpha_2 PkT - \alpha_1 N_0 kT - \alpha_2 P_0 kT \tag{15.26}$$

and finally to:

$$V_\mathrm{j} = V_\mathrm{T}\ln\left(1+\frac{n}{N_0}\right) + V_\mathrm{T}\ln\left(1+\frac{p}{P_0}\right) + V_\mathrm{T}(\alpha_1 n + \alpha_2 p) \tag{15.27}$$

where the thermal voltage is given by:

$$V_\mathrm{T} = \frac{kT}{q}\,. \tag{15.28}$$

The junction voltage, V_j can now be represented as three series connected circuit elements with voltage drops V_1, V_2 and V_3 that is:

$$V_\mathrm{j} = V_1 + V_2 + V_3\,. \tag{15.29}$$

The first two voltage drops are across two ideal diodes, which may be expressed as:

$$V_1 = V_\mathrm{T}\ln\left(1+\frac{I_1}{I_{01}}\right) \qquad \text{and} \qquad V_2 = V_\mathrm{T}\ln\left(1+\frac{I_1}{I_{02}}\right). \tag{15.30}$$

These ideal diodes are classical Shockley p.n. junction diodes and have leakage currents (I_{01} and I_{02}) defined as:

$$I_{01} = \frac{qv_\mathrm{a}N_0}{\tau_\mathrm{ns}} \quad \text{and} \quad I_{02} = \frac{qv_\mathrm{a}(N_\mathrm{A}+N_0)}{\tau_\mathrm{ns}} \tag{15.31}$$

and the forward-biased current (I_1) defined as:

$$I_1 = \frac{qv_\mathrm{a}n}{\tau_\mathrm{ns}} \tag{15.32}$$

where v_a is the volume of the active region, and τ_ns is the effective spontaneous recombination lifetime. The last voltage drop, V_3, is across a linear series resistance, R_e, which is related to I_1 by:

$$V_3 = I_1 R_\mathrm{e} \tag{15.33}$$

where series resistance (R_e) is expressed as:

$$R_\mathrm{e} = \left(\alpha_1 + \frac{\alpha_2}{(1+\beta\frac{N_\mathrm{A}}{N_\mathrm{v}})}\right)\frac{N_0 V_\mathrm{T}}{I_{01}}\,. \tag{15.34}$$

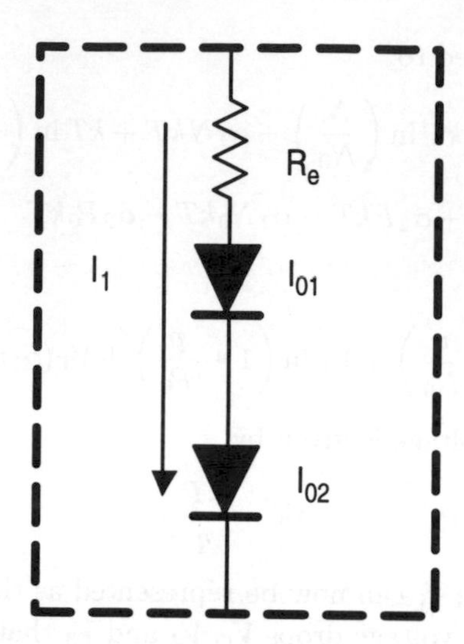

Fig. 15.35. The junction voltage, V_j, at the input port of the large-signal circuit model.

According to Eq. (15.29), the model of the junction voltage drop across the input port of the large-signal circuit model can be schematically drawn out as shown in Fig. 15.35.

15.14.2 *From rate equations to circuit equations*

The first step toward constructing the complete lumped-element circuit model is by mapping the physical parameters of the laser diode (the rate equations) to their equivalent circuit variables. For example, the rate of change of photon density dS/dt can be modelled by a current flowing through a capacitor, which is known to be given as CdV/dt, so that the new circuit variable, V, is proportional to the old variable, S. The single-mode rate equations for the carrier (excess) and photon densities, n and S are, respectively, given as (see Chapter 13):

$$\frac{dn}{dt} = \frac{I}{qv_\mathrm{a}} - \left(\frac{n}{\tau_\mathrm{ns}} + B_1 n^2\right) - gS \tag{15.35}$$

$$\frac{dS}{dt} = gS - \frac{S}{\tau_p} + \beta_\mathrm{sp}\left(\frac{n}{\tau_\mathrm{s}} + B_1 n^2\right) \tag{15.36}$$

where I is the injection current, g is the gain coefficient, τ_p is the photon lifetime, β_sp is the spontaneous emission coupling factor, τ_n is the nonradiative recombination lifetime, τ_s is the low-level injection radiative recombination lifetime, and B_1 is the high-level injection radiative recombination constant. The *excess* radiative and nonradiative spontaneous recombination rates, r_e and r_n are, respectively, expressed as:

$$r_e = \frac{n}{\tau_s} + B_1 n^2 \quad \text{and} \quad r_n = \frac{n}{\tau_n}. \tag{15.37}$$

Therefore, the *total* excess spontaneous recombination rate, r_t is given as:

$$r_t = r_n + r_e \quad \text{or} \quad r_t = \frac{n}{\tau_{ns}} + B_1 n^2 \quad \text{where} \quad \frac{1}{\tau_{ns}} = \frac{1}{\tau_n} + \frac{1}{\tau_s}. \tag{15.38}$$

The total spontaneous recombination can be modelled by a corresponding current element (I_t), which has a non-linear dependence of I_1 (see Eq. (15.32)):

$$I_t = I_1 + bI_1^2 \quad \text{where} \quad b = \frac{B_1 \tau_{ns}^2}{qv_a}. \tag{15.39}$$

By inspection of the carrier rate equation in Eq. (15.35), and using the definition from Eq. (15.39), we can translate the density terms into current terms:

$$I = I_1 + bI_1^2 + \tau_{ns}\frac{dI_1}{dt} + I_g + C_{sc}\frac{dV_j}{dt}. \tag{15.40}$$

Notice that the first two terms on the right hand side (RHS) represent the total recombination terms while the third term on the RHS is proportional to the rate of change of excess carrier density, dn/dt. The fourth term on the RHS is the stimulated emission current (I_g), which is defined as:

$$I_g = qv_a gS. \tag{15.41}$$

The space-charge capacitance [98] was introduced phenomenologically as the fifth term on the RHS of Eq. (15.40), which can easily be included into the circuit model. The space-charge capacitance, C_{sc}, is dependent on junction voltage in the following manner:

$$C_{sc} = C_0 \left(1 - \frac{V_j}{V_d}\right)^{-\frac{1}{2}}. \tag{15.42}$$

From here, it is clear that the carrier rate equation has been transformed into an equivalent circuit model, which is schematically drawn out in Fig. 15.36.

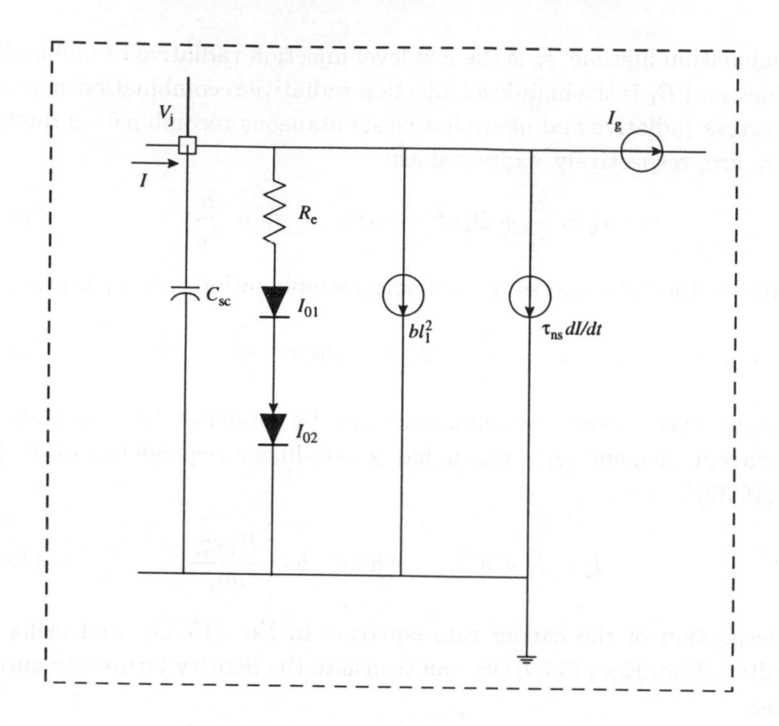

Fig. 15.36. The circuit model of the carrier dynamics.

By careful inspection of Eq. (15.36), and defining the normalised photon density as $S_n = S/S_c$, we can also translate the photon rate equation into the following form:

$$I_g = C_{ph}\frac{dS_n}{dt} + \frac{S_n}{R_{ph}} - \beta_{sp} I_e \tag{15.43}$$

$$I_e = qv_a r_e = aI_1 + bI_1^2 \tag{15.44}$$

where

$$a = \tau_{ns}/\tau_s\,, \qquad C_{ph} = qv_a S_c \qquad \text{and} \qquad R_{ph} = \frac{\tau_p}{C_{ph}}\,. \tag{15.45}$$

The normalisation constant (S_c) is required to avoid working with extremely large numbers which may cause overflow or affect the convergence of the numerical method used to solve the circuit equations. By using Eqs. (15.29), (15.40) and (15.43), the complete large-signal circuit model of the semiconductor laser may be constructed schematically in a time-domain

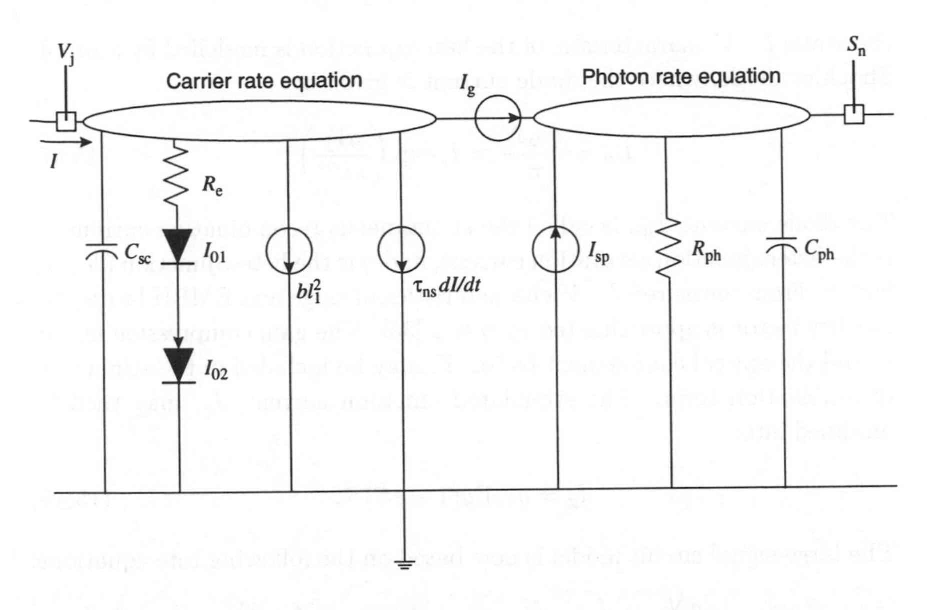

Fig. 15.37. The large-signal circuit model of the stripe geometry DH laser.

circuit simulator such as PSPICE or HP-MDS, as shown in Fig. 15.37. The voltage at the output of the equivalent circuit (S_{n} in Fig. 15.37) is directly proportional to the photon density. Notice that circuit Eqs. (15.40) and (15.43) satisfy Kirchoff's current law (KCL), which account for the two electrical *supernodes* in the complete circuit model of Fig. 15.37. For the stripe-geometry double heterostructure (DH) laser, the optical gain is assumed to have a square law dependence on radiative recombination current per unit volume $J_{\mathrm{nom}} = I_{\mathrm{e}}/v_{\mathrm{a}}$ [42]:

$$g = \frac{D}{qv_{\mathrm{a}}S_{\mathrm{c}}}(J_{\mathrm{nom}} - J_{\mathrm{nom0}})^2 \tag{15.46}$$

where D is the gain constant, and J_{nom0} is the value of J_{nom} at which optical gain goes to zero.

15.14.3 *Simplified large-signal circuit model*

The large-signal circuit model can be simplified by assuming that the laser is biased close to or above threshold, where the carrier density is almost constant. The spontaneous recombination is approximated by N/τ_{n}, and the high-level injection effect (B_1N^2) is neglected. Instead of deriving the junction voltage through the Fermi-Dirac integrals as in Section 15.14.1, the static $I-V$ characteristic of the heterojunction is modelled by a simple Shockley diode, where the diode current is given by:

$$I_{\mathrm{sp}} = \frac{qv_{\mathrm{a}}N}{\tau_{\mathrm{n}}} \approx I_{\mathrm{s}} \exp\left(\frac{qV_{\mathrm{j}}}{\eta kT}\right). \tag{15.47}$$

The diode current, I_{sp}, is called the spontaneous recombination current, I_{s} is the heterojunction saturation current, and η is the heterojunction ideality factor. From measured $I-V$ characteristics of ridge and EMBH lasers, the ideality factor is approximated by $\eta \approx 2$ [35]. The gain compression factor, ε, and the optical confinement factor, Γ, may be included in the stimulated recombination term. The stimulated emission current, I_{g}, may then be modified into:

$$I_{\mathrm{g}} = qv_{\mathrm{a}}\Gamma g(1-\varepsilon S)S\,. \tag{15.48}$$

The large-signal circuit model is now based on the following rate equations:

$$\begin{aligned} \frac{dN}{dt} &= \frac{I}{qv_{\mathrm{a}}} - \frac{N}{\tau_{\mathrm{n}}} - \Gamma g(1-\varepsilon S)S - \frac{C_{\mathrm{sc}}}{qv_{\mathrm{a}}}\frac{dV_{\mathrm{j}}}{dt} \\ \frac{dS}{dt} &= \Gamma g(1-\varepsilon S)S - \frac{S}{\tau_{\mathrm{p}}} + \Gamma\beta_{\mathrm{sp}}\frac{N}{\tau_{\mathrm{n}}} \end{aligned} \tag{15.49}$$

and the simplified large-signal circuit model is shown in Fig. 15.38.

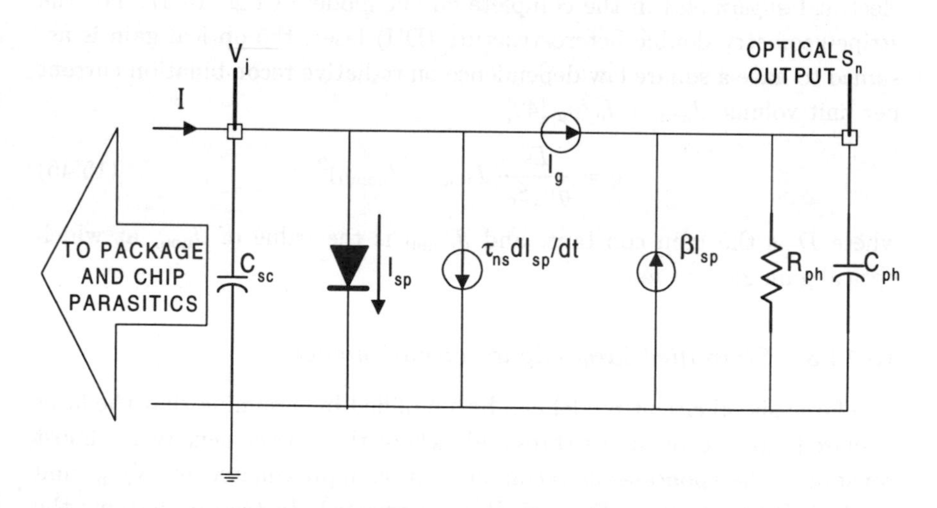

Fig. 15.38. Simplified large-signal circuit model.

15.14.4 *Gain-guided laser — Stripe-geometry laser*

For the GaAs material, the following values were used [89] to find the resistance, R_e, in Eq. (15.33):

$$\alpha_1 = 7.96 \times 10^{-25}\ \text{m}^3$$
$$\alpha_2 = 4.31 \times 10^{-26}\ \text{m}^3 \qquad (15.50)$$
$$\beta = 1.6$$

which were defined for the ranges of electron and hole densities of $N = 0$ to $N = 1.5 \times 10^{24}$ m^3, and $P = 0$ to $P = 1.5 \times 10^{24}$ m^{-3} [89]. The valence band effective density of states, N_v, was assumed to be $8.31 \times 10^{24}/\text{m}^3$ [30, 99]. The equilibrium carrier density, N_0, was taken as $1.5 \times 10^{12}/\text{m}^3$, and the semiconductor was undoped ($N_A = 0$). This gives a value of 0.47 for R_e. By using this value of R_e, the other parameter values used in the large-signal circuit model of the stripe-geometry laser are listed in Table 15.7. Each variable-dependent circuit element in Figs. 15.37 and 15.38 was implemented in HP-MDS as a *symbolically-defined device* (SDD). For example, the SDD used to represent the stimulated emission current,

Table 15.7. Parameter values for the stripe-geometry laser.

Parameter	Value
Resistance	$R_e = 0.47\ \Omega$
Acceptor impurity concentration	$N_A = 4.0 \times 10^{23}\ \text{m}^{-3}$
Zero-bias space-charge capacitance	$C_0 = 10$ pF
Diode built-in potential	$V_d = 1.65$ V
Low-level injection radiative recombination lifetime	$\tau_s = 18$ ns
Constant used in definition of I_{sp}	$b = 6.92\ \text{A}^{-1}$
Effective recombination lifetime	$\tau_{ns} = 2.25$ ns
Leakage current of diode 1	$I_{01} = 2.74 \times 10^{-25}$ A
Leakage current of diode 2	$I_{02} = 18.13 \times 10^{-3}$ A
Normalisation factor for photon desity	$S_c = 10^{21}\ \text{V}^{-1}\text{m}^{-3}$
Photon storage capacitance	$C_{ph} = 0.102$ pF
Electronic charge × active region volume	$qv_a = 1.02 \times 10^{-34}\ \text{Asm}^3$
Photon decay resistance	$R_{ph} = 29.4\ \Omega$
Nominal current density at transparency	$J_{nomo} = 2.0 \times 10^{13}\ \text{Am}^{-3}$
Gain Constant	$D = 1.79 \times 10^{-29}\ \text{V}^{-1}\text{A}^{-1}\text{m}^6$

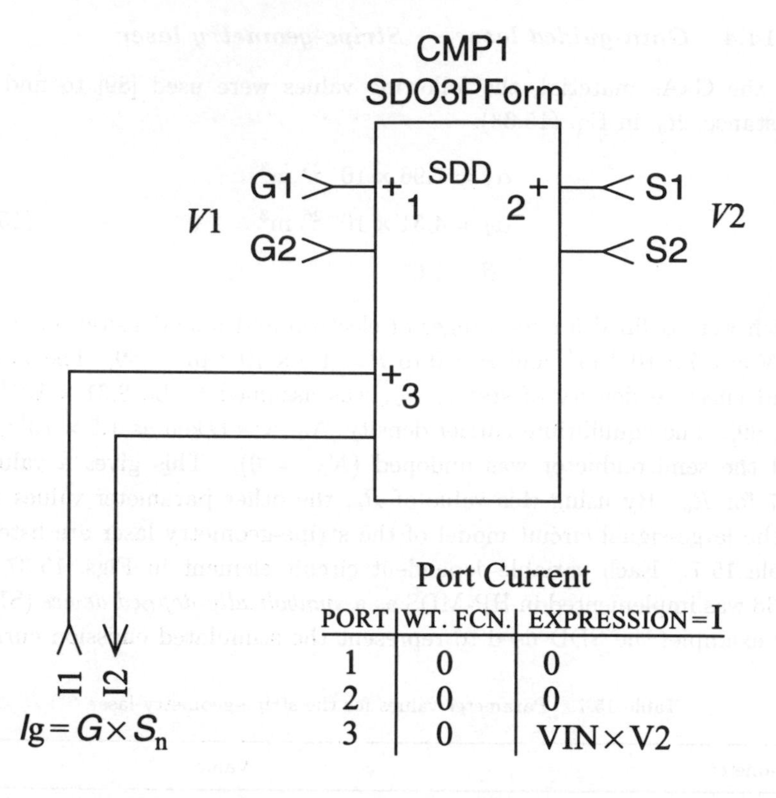

Fig. 15.39. Symbolically defined device (SDD) to model the stimulated recombination current (I_g).

I_g, is shown in Fig. 15.39. The input of port 1 is a voltage drop labelled as **_V1** (following the naming convention of HP-MDS), which is equivalent to the stimulated recombination rate, G. Port 2 takes the voltage **_V2** as input, which is equivalent to the normalised photon density, S_n. The output of the SDD (port 3) is the stimulated recombination current, I_g, which is set to be the multiplication of **_V1** and **_V2**. Note that the current through ports 1 and 2 were set to zero to realise open-circuits.

15.14.5 *Index-guided laser structures — Ridge-waveguide and etched mesa buried heterostructure (EMBH) lasers*

The internal laser parameter values that were used for the ridge waveguide and etched mesa buried heterostructure (EMBH) laser are listed in

Tables 15.8 and 15.10, respectively. Their corresponding parasitics values are also listed in Tables 15.9 and 15.11, respectively.

Table 15.8. Internal parameter values for the circuit model of the ridge waveguide laser.

Parameter	Value
Built-in potential	$V_D = 1.65$ V
Zero bias space-charge capacitance	$C_0 = 2$ pF
Spontaneous emission coupling factor	$\beta_{sp} = 2.0 \times 10^{-4}$
Electronic charge × active region volume	$qv_a = 6.6 \times 10^{-35}$ Cm^3
Differential gain constant × group velocity	$av_g = 3.2 \times 10^{-12}$ m^{3}s^{-1}
Carrier lifetime	$\tau_{ns} = 3$ ns
Saturation current	$I_s = 300$ pA
Transparency carrier density	$N_{tr} = 1.0 \times 10^{24}$ m^{-3}
Optical confinement factor	$\Gamma = 0.3$
Gain compression factor	$\varepsilon = 6.7 \times 10^{-23}$ m^3
Photon lifetime	$\tau_p = 1$ ps
Photon density normalisation constant	$S_c = 1.0 \times 10^{20}$ V^{-1}m^{-3}

Table 15.9. Chip and package parasitic values for the ridge waveguide laser.*

Parameter	Value
Source resistance	$R_{in} = 50\ \Omega$
Standoff capacitance	$C_p = 0.23$ pF
Bondwire inductance	$L_p = 0.63$ nH
Bondwire resistance	$R_p = 1\ \Omega$
Contact (dominates) + substrate resistance	$R_s = 5.5\ \Omega$
MIS† (dominates) + reverse-bias blocking junction capacitance	$C_s = 8$ pF
Substrate resistance	$R_{sub} = 1.5\ \Omega$
Leakage current	$I_L = 0$ mA

*Refer to Chapter 14 for more explanation on parasitics.

†Metal-Insulator-Semiconductor (MIS).

Table 15.10. Internal parameter values for the circuit model of the EMBH laser.

Parameter	Value
Built-in potential	$V_D = 1.65$ V
Zero bias space-charge capacitance	$C_0 = 1$ pF
Spontaneous emission coupling factor	$\beta_{sp} = 1.0 \times 10^{-4}$
Electronic charge × active region volume	$qv_a = 2.4 \times 10^{-35}$ Cm^3
Differential gain constant × group velocity	$av_g = 2.4 \times 10^{-12}$ m^{3}s^{-1}
Carrier lifetime	$\tau_{ns} = 3$ ns
Saturation current	$I_s = 100$ pA
Transparency carrier density	$N_{tr} = 1.0 \times 10^{24}$ m^{-3}
Optical confinement factor	$\Gamma = 0.4$
Gain compression factor	$\varepsilon = 4.5 \times 10^{-23}$ m^3
Photon lifetime	$\tau_p = 1$ ps
Photon density normalisation constant	$S_c = 1.0 \times 20^{20}$ V^{-1}m^{-3}

Table 15.11. Chip and package parasitic values for the circuit model of the EMBH laser.

Parameter	Value
Source resistance	$R_{in} = 50\ \Omega$
Standoff capacitance	$C_p = 0.18$ pF
Bondwire inductance	$L_p = 1.1$ nH
Bondwire resistance	$R_p = 1\ \Omega$
Contact (dominates) + substrate resistance	$R_s = 11\ \Omega$
MIS (dominates) + reverse-bias blocking junction capacitance	$C_s = 12$ pF
Substrate resistance	$R_{sub} = 0.5\ \Omega$
Leakage current	$I_L = 15$ mA

15.15 Small-Signal Circuit Model of Laser Diodes

15.15.1 *Small signal circuit model below threshold*

The carrier rate equation *below threshold* current including a space-charge capacitance term (C_{sc}) may be expressed as:

$$\frac{dN}{dt} = \frac{I}{qv_{\mathrm{a}}} - \frac{N}{\tau_{\mathrm{n}}} - \frac{C_{\mathrm{sc}}}{qv_{\mathrm{a}}}\frac{dV_{\mathrm{j}}}{dt} \tag{15.51}$$

where N is the carrier density, V_{j} is the junction voltage, v_{a} is the active layer volume, q is the electronic charge, and τ_{n} is the carrier lifetime. By assuming that the signal consists of a DC term, I_0 and a small AC signal term, i, we have (see Fig. 15.40):

$$\begin{aligned} I &= I_0 + ie^{j\omega t} && \text{(a)} \\ N &= N_0 + ne^{j\omega t} && \text{(b)} \\ V_{\mathrm{j}} &= V_{j0} + v_j e^{j\omega t} && \text{(c)} \end{aligned} \tag{15.52}$$

where ω is the angular frequency in radians. By substituting the expressions of Eq. (15.52) into Eq. (15.51) and neglecting the products of two small-signal terms, we arrive at the linearised equation given by:

$$i = \frac{n\alpha}{\tau_n} + j\omega(n\alpha + v_j C_{\mathrm{sc}}) \qquad \text{where} \quad \alpha = qv_{\mathrm{a}}\,. \tag{15.53}$$

Using the classical Shockley relationship [58] between electron density and junction voltage we have:

$$n = \frac{v_j q N_0}{2kT}\,. \tag{15.54}$$

Substituting Eq. (15.54) into Eq. (15.53), we then have the following circuit equation:

$$i = v_j\left(\frac{1}{R_{\mathrm{d}}} + j\omega(C_{\mathrm{d}} + C_{\mathrm{sc}})\right) \tag{15.55}$$

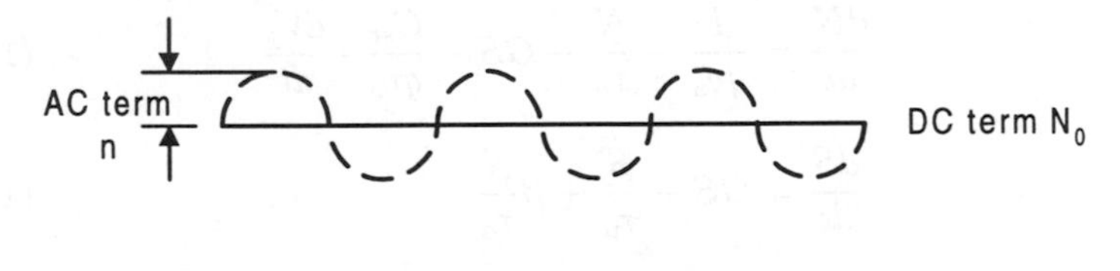

Fig. 15.40. Definition of the DC and AC terms of carrier density (see Eq. (15.52)).

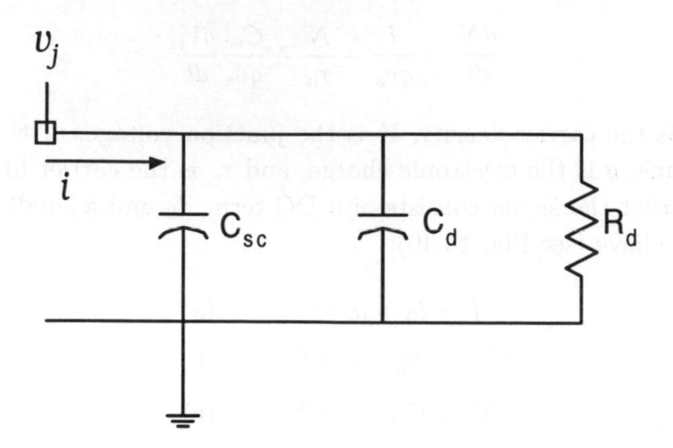

Fig. 15.41. Small-signal circuit model of laser diode below threshold.

where

$$\begin{aligned} R_{\mathrm{d}} &= \frac{2\tau_n kT}{q\alpha N_0} \quad \text{(small-signal diode junction resistance)} \\ C_{\mathrm{d}} &= \frac{\tau_n}{R_{\mathrm{d}}} \quad \text{(active layer diffusion capacitance)} \end{aligned} \tag{15.56}$$

and the circuit representation of Eq. (15.55) is shown in Fig. 15.41.

15.15.2 *Small-signal circuit model above threshold (excluding diffusion)*

The rate equations *above threshold* current including a space-charge capacitance term (C_{sc}) are given as:

$$\frac{dN}{dt} = \frac{I}{qv_{\mathrm{a}}} - \frac{N}{\tau_n} - GS - \frac{C_{\mathrm{sc}}}{qv_{\mathrm{a}}} \cdot \frac{dV_j}{dt} \tag{15.57}$$

$$\frac{dS}{dt} = GS - \frac{S}{\tau_{\mathrm{p}}} + \beta\frac{N}{\tau_n} \tag{15.58}$$

where G is the rate of stimulated emission, S is the photon density, τ_{p} is the photon lifetime, and β is the spontaneous emission coupling factor. As before (see Eq. (15.52), the variables, S and G, in the rate equations are split into DC components and small-signal AC components, which are assumed to have sinusoidal waveforms, that is:

$$S = S_0 + se^{j\omega t}\,, \qquad G = G_0 + ge^{j\omega t} \qquad \text{where} \quad g = \gamma n \quad \text{and} \quad \gamma = a\Gamma v_g \tag{15.59}$$

with a as the differential gain coefficient, Γ is the optical confinement factor, and v_g is the group velocity. At steady state we have:

$$\frac{I_0}{\alpha} - \frac{N_0}{\tau_n} - G_0 S_0 = 0 \qquad \text{and} \qquad S_0 = \left(\frac{I_0\tau_n - \alpha N_0}{G_0\alpha\tau_n}\right). \tag{15.60}$$

Substituting Eqs. (15.52), (15.59) and (15.60) into Eq. (15.57) and neglecting the products of small-signal a.c. terms we have:

$$i = \frac{j\omega\alpha N_0 q v_j}{2kT} + C_{sc} j\omega v_j + \frac{\alpha N_0 q v_j}{2kT\tau_n} + \alpha G_0 s$$

$$+\gamma\frac{\alpha N_0 q v_j}{2kT}\left(\frac{I_0\tau_n - \alpha N_0}{G_0\alpha\tau_n}\right). \tag{15.61}$$

Rearranging Eq. (15.61), we have:

$$i = v_j\left(j\omega\left[\frac{\alpha N_0 q}{2kT} + C_{sc}\right] + \frac{\alpha N_0 q}{2kT\tau_n}\left[1 + \frac{\gamma(I_0\tau_n - \alpha N_0)}{G_0\alpha}\right]\right) + \alpha G_0 s\,. \tag{15.62}$$

Substituting Eq. (15.56) into Eq. (15.62) we have:

$$i = v_j\left(j\omega[C_d + C_{sc}] + \left[\frac{1}{R_d} + \frac{1}{R_4}\right]\right) + i_s \tag{15.63}$$

where the stimulated emission current is given by:

$$i_s = \alpha G_0 s \tag{15.64}$$

and a new resistance, R_4, is defined as:

$$R_4 = R_d\frac{G_0\alpha}{\gamma(I_0\tau_n - \alpha N_0)} = \frac{R_d}{\gamma\tau_n S_0}\,. \tag{15.65}$$

The circuit model that represents Eq. (15.63) is shown in Fig. 15.42. It is worth pointing out that *below threshold*, we have $\tau_n I_0 = N_0\alpha$ and R_4 becomes infinitely large, almost equivalent to an open circuit. The current term, i_s, also vanishes, and the circuit reduces to that of Fig. 15.41. Now, we proceed to derive the complete circuit model of the laser diode under small-signal modulation by including the photon dynamics into it. By using

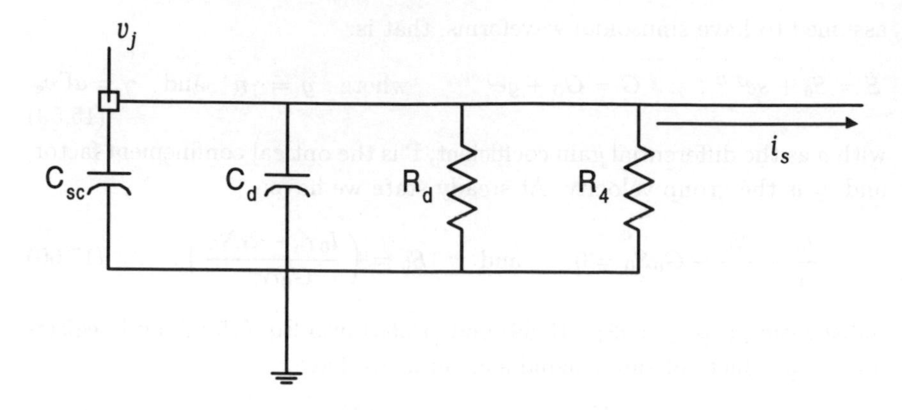

Fig. 15.42. Small signal circuit model above threshold: carrier dynamics only.

Eqs. (15.52) and (15.59) into (15.58), and linearising the result we have another set of equations:

$$j\omega s = \gamma n S_0 + G_0 s - \frac{s}{\tau_{\mathrm{P}}} + \beta\frac{n}{\tau_n}\,. \tag{15.66}$$

Substituting Eqs. (15.54) and (15.60) into Eq. (15.66) we have:

$$s\left(j\omega + \left[\frac{1}{\tau_{\mathrm{p}}} - G_0\right]\right) = \gamma\left(\frac{N_0 q v_j}{2kT}\right)\left(\frac{I_0\tau_n - \alpha N_0}{G_0\alpha\tau_n}\right) + \left(\frac{N_0 q v_j}{2kT}\right)\frac{\beta}{\tau_n}\,. \tag{15.67}$$

From the steady state photon rate equation of Eq. (15.58) we have:

$$\frac{1}{\tau_{\mathrm{p}}} = G_0 + \beta\frac{N_0}{\tau_n S_0}\,. \tag{15.68}$$

Then substituting Eq. (15.68) into Eq. (15.67) we have:

$$\alpha s(G_3 + j\omega) = v_j G_2$$

where

$$G_2 = \left(\frac{\beta}{R_{\mathrm{d}}} + \frac{1}{R_4}\right) = \frac{\beta + \gamma\tau_n S_0}{R_{\mathrm{d}}}$$

and $$G_3 = \frac{\alpha\beta G_0 N_0}{(I_0\tau_n - \alpha N_0)} = \frac{\beta N_0}{\tau_n S_0} \tag{15.69}$$

From Eq. (15.69), the current term, i_s, of Eq. (15.64) may also be expressed as:

$$i_s = \alpha G_0 s = \frac{v_j G_0 G_2}{(G_3 + j\omega)}. \tag{15.70}$$

Therefore, the current i_s flows through circuit elements, which can be identified as a resistor in series with an inductor, that is:

$$i_s = \frac{v_j}{(R_x + j\omega L_x)} \quad \text{where} \quad R_x = \frac{G_3}{G_0 G_2} \quad \text{and} \quad L_x = \frac{1}{G_0 G_2}. \tag{15.71}$$

Finally, the complete small-signal circuit model of the laser diode above threshold is shown in Fig. 15.43. Note the existence of the inductor, L_x, and resistor, R_x, at above threshold, thus forming a resonant circuit.

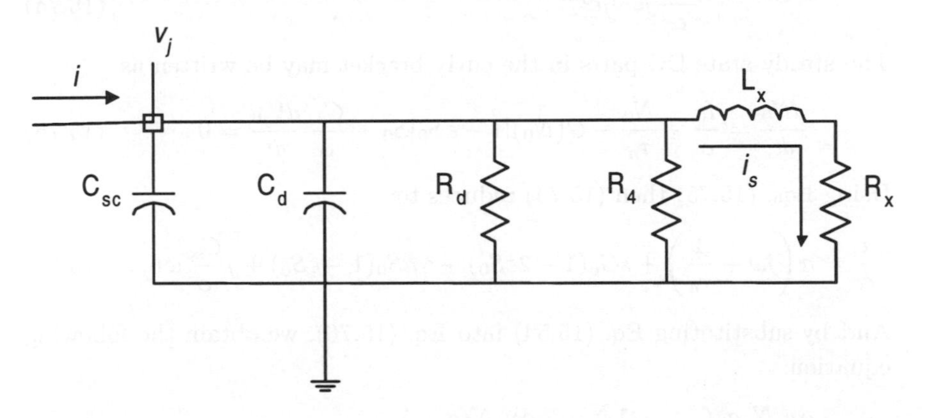

Fig. 15.43. Small signal circuit model of the laser diode above threshold including both carrier and photon dynamics.

15.15.3 *Small-signal model including carrier diffusion effect*

The carrier rate equation including the gain compression term (ε) due to lateral carrier diffusion is expressed as (see derivation in Section 15.16):

$$\frac{dN}{dt} = \frac{I}{\alpha} - \frac{N}{\tau_n} - G(N)[1 - \varepsilon S]S - \frac{C_{sc}}{\alpha}\frac{dV_j}{dt}. \tag{15.72}$$

Substituting the small-signal quantities from Eqs. (15.52) and (15.59) into Eq. (15.72), we have:

$$\begin{aligned}\frac{d}{dt}(N_0 + ne^{j\omega t}) &= \frac{(I_0 + ie^{j\omega t})}{\alpha} - \frac{(N_0 + ne^{j\omega t})}{\tau_n} \\ &\quad - (G_0 + ge^{j\omega t})[1 - \varepsilon(S_0 + se^{j\omega t})](S_0 + se^{j\omega t}) \\ &\quad - \frac{C_{\text{sc}}}{\alpha}\frac{d}{dt}(V_{j0} + v_j e^{j\omega t}) . \end{aligned} \tag{15.73}$$

Rearranging Eq. (15.73) we have:

$$\begin{aligned} j\omega ne^{j\omega t} &= \left\{\frac{I_0}{\alpha} - \frac{N_0}{\tau_n} - \frac{C_{\text{sc}}}{\alpha}\frac{dV_{j0}}{dt} - G_0S_0(1 - \varepsilon S_0)\right\} + \frac{ie^{j\omega t}}{\alpha} - \frac{ne^{j\omega t}}{\tau_n} \\ &\quad - (G_0 se^{j\omega t} - 2\varepsilon S_0 G_0 se^{j\omega t} + gS_0e^{j\omega t} - \varepsilon gS_0^2e^{j\omega t}) \\ &\quad - \frac{C_{\text{sc}}}{\alpha} j\omega v_j e^{j\omega t} \end{aligned} \tag{15.74}$$

The steady-state DC parts in the curly bracket may be written as:

$$\frac{dN_0}{dt} = \frac{I_0}{\alpha} - \frac{N_0}{\tau_n} - G(N_0)[1 - \varepsilon S_0]S_0 - \frac{C_{\text{sc}}}{\alpha}\frac{dV_{j0}}{dt} = 0 . \tag{15.75}$$

Using Eqs. (15.75) then (15.74) reduces to:

$$\frac{i}{\alpha} = n\left(j\omega + \frac{1}{\tau_n}\right) + sG_0(1 - 2\varepsilon S_0) + \gamma nS_0(1 - \varepsilon S_0) + j\frac{C_{\text{sc}}}{\alpha}\omega v_j . \tag{15.76}$$

And by substituting Eq. (15.54) into Eq. (15.76), we obtain the following equation:

$$i = \frac{\alpha v_j N_0 q}{2kT}\left(j\omega + \frac{1}{\tau_n}\right) + \gamma\frac{\alpha v_j N_0 q}{2kT}S_0(1 - \varepsilon S_0) + jC_{\text{sc}}\omega v_j + i_{\text{s}} \tag{15.77}$$

where we have made the approximation that the stimulated emission current term, i_{s}, here (i.e. including gain compression) is almost equal to that defined in Eq. (15.64), that is:

$$i_{\text{s}} = [\alpha G_0 s(1 - 2\varepsilon S_0)] \approx \alpha G_0 s . \tag{15.78}$$

By using Eq. (15.56) (diffusion capacitance and resistance), we can rewrite Eq. (15.77) as:

$$i = v_j\left\{\left[\frac{1 + \tau_n\gamma(1 - \varepsilon S_0)S_0}{R_{\text{d}}}\right] + j\omega[C_{\text{d}} + C_{\text{sc}}]\right\} + i_{\text{s}} . \tag{15.79}$$

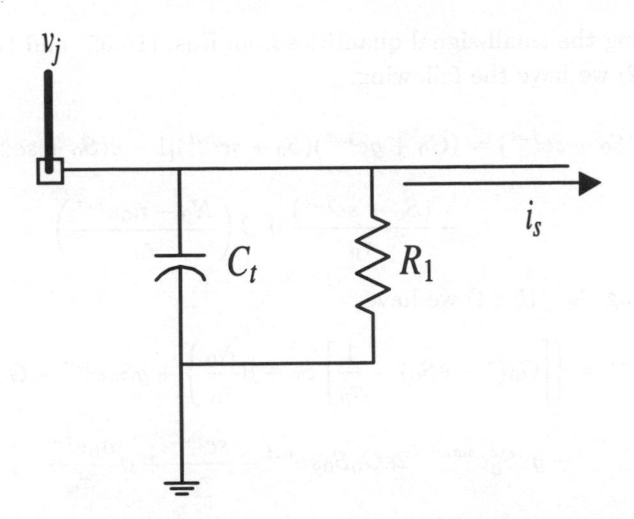

Fig. 15.44. Small-signal circuit model including carrier diffusion: carrier dynamics only.

Now, we may define R_1 as the effective resistance (parallel combination of R_d and R_4) and $C_t = C_\mathrm{d} + C_\mathrm{sc}$ as the effective capactitance (parallel combination of C_d and C_sc):

$$R_1 = \frac{R_\mathrm{d}}{1 + \gamma\tau_n S_0(1 - \varepsilon S_0)} \approx \frac{R_\mathrm{d}}{1 + \gamma\tau_n S_0} \,. \tag{15.80}$$

Therefore, the circuit equation for Eq. (15.72) can finally be written as:

$$i = v_j \left(\frac{1}{R_1} + j\omega C_\mathrm{t} \right) + i_\mathrm{s} \tag{15.81}$$

which can be represented by the circuit shown in Fig. 15.44. At this stage, the circuit is exactly the same as Fig. 15.42, except that the resistors and capacitors have been combined into effective values. Now, we proceed to derive the second part of the circuit that describes the photon dynamics. The photon rate equation including the gain compression term (ε) due to lateral carrier diffusion is expressed as (see derivation in Section 15.16):

$$\frac{dS}{dt} = GS(1 - \varepsilon S) - \frac{S}{\tau_\mathrm{p}} + \beta \frac{N}{\tau_n} \,. \tag{15.82}$$

Substituting the small-signal quantities from Eqs. (15.52) and (15.59) into Eq. (15.82) we have the following:

$$\frac{d}{dt}(S_0 + se^{j\omega t}) = (G_0 + ge^{j\omega t})(S_0 + se^{j\omega t})[1 - \varepsilon(S_0 + se^{j\omega t})] - \frac{(S_0 + se^{j\omega t})}{\tau_p} + \beta\left(\frac{N_0 + n_0 e^{j\omega t}}{\tau_n}\right). \quad (15.83)$$

Rearranging Eq. (15.83) we have:

$$j\omega s e^{j\omega t} = \left\{\left[G_0(1 - \varepsilon S_0) - \frac{1}{\tau_p}\right] S_0 + \beta\frac{N_0}{\tau_n}\right\} + gS_0 e^{j\omega t} + G_0 s e^{j\omega t} - g\varepsilon S_0^2 e^{j\omega t} - 2\varepsilon G_0 S_0 s e^{j\omega t} - \frac{se^{j\omega t}}{\tau_p} + \beta\frac{n_0 e^{j\omega t}}{\tau_n}. \quad (15.84)$$

By recognising that the terms in the curly bracket satisfy the following steady-state rate equation:

$$\frac{dS_0}{dt} = \left\{G_0(1 - \varepsilon S_0) - \frac{1}{\tau_\mathrm{p}}\right\} S_0 + \beta\frac{N_0}{\tau_n} = 0 \quad (15.85)$$

Eq. (15.84) can be simplified as:

$$j\omega s = gS_0 + G_0 s - g\varepsilon S_0^2 - \frac{s}{\tau_\mathrm{p}} + \beta\frac{n}{\tau_n} - 2\varepsilon G_0 S_0 s\,. \quad (15.86)$$

Substituting Eq. (15.54) into Eq.(15.86) we have:

$$v_j\frac{N_0 q}{2\tau_n kT}[\beta + \gamma\tau_n S_0(1 - \varepsilon S_0)] = \left(j\omega + 2\varepsilon S_0 G_0 - G_0 + \frac{1}{\tau_\mathrm{p}}\right) s\,. \quad (15.87)$$

From the steady-state condition of Eq. (15.85), we have the following expression for photon lifetime (τ_p):

$$\frac{1}{\tau_\mathrm{p}} = \frac{\beta N_0}{S_0\tau_n} - G_0\varepsilon S_0 + G_0\,. \quad (15.88)$$

Therefore, by making the substitution of Eq. (15.88) into Eq. (15.87) we obtain:

$$v_j = \frac{2kT\tau_n}{qN_0}\left(j\omega + 2\varepsilon G_0 S_0 - G_0 + \frac{\beta N_0}{S_0\tau_n} - G_0\varepsilon S_0 + G_0\right) \times \frac{s}{[\beta + \gamma\tau_n S_0(1 - \varepsilon S_0)]} \quad (15.89)$$

and finally we have:

$$v_j = \frac{R_d}{G_0(1-2\varepsilon S_0)[\beta+\gamma\tau_n S_0(1-\varepsilon S_0)]}\left(j\omega+\varepsilon G_0 S_0+\frac{\beta N_0}{S_0\tau_n}\right)i_s\,. \tag{15.90}$$

Now, by defining the circuit elements for Eq. (15.90) as:

$$\begin{aligned} L_s &= \frac{R_d}{G_0(1-2\varepsilon S_0)[\beta+\gamma\tau_n S_0(1-\varepsilon S_0)]} \quad \text{(resonant inductor)} \\ R_{s1} &= \varepsilon G_0 S_0 L_s \quad \text{(diffusion damping resistor)} \\ R_{s2} &= \frac{\beta N_0 L_s}{\tau_n S_0} \quad \text{(spontaneous emission damping resistor)} \end{aligned} \tag{15.91}$$

then we can arrive at a circuit equation given by:

$$v_j = (j\omega L_s + R_{s1} + R_{s2})i_s\,. \tag{15.92}$$

Finally, the complete small-signal circuit model including diffusion damping is shown in Fig. 15.45.

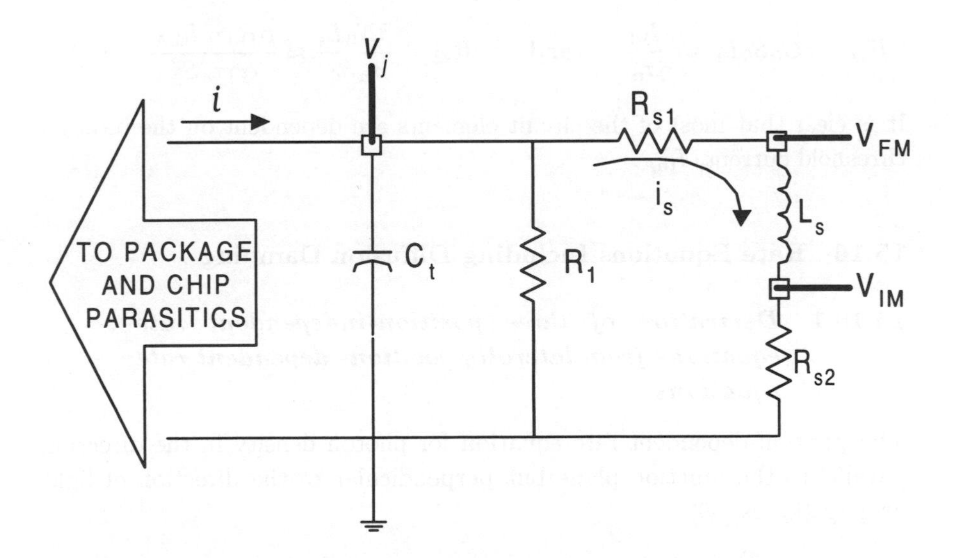

Fig. 15.45. Small-signal circuit model of the laser diode above threshold including diffusion damping.

15.15.4 *Further approximations to the circuit element expressions*

Since above threshold, the carrier density is clamped to its threshold level, the diffusion resistance defined in Eq. (15.56) may be approximated by [35]:

$$R_{\mathrm{d}} \approx \frac{2kT}{qI_{\mathrm{thA}}} \quad \text{where} \quad I_{\mathrm{thA}} = \frac{\alpha N_{\mathrm{th}}}{\tau_n} \approx \frac{\alpha(\frac{1}{\tau_{\mathrm{p}}\gamma} + N_{\mathrm{tr}})}{\tau_n} . \tag{15.93}$$

The term I_{thA} is the *internal* or active layer threshold current (i.e. excluding leakage current). The resonant (lasing) inductor may also be approximated by [35]:

$$L_{\mathrm{s}} = \frac{R_{\mathrm{d}}}{G_0(1-2\varepsilon S_0)[\beta + \gamma\tau_n S_0(1-\varepsilon S_0)]} \approx \frac{R_{\mathrm{d}}\tau_{\mathrm{p}}}{\gamma\tau_n S_0} \tag{15.94}$$

since the following approximation applies (from Eq. (15.88)):

$$\frac{1}{\tau_{\mathrm{p}}} = \frac{\beta N_0}{S_0\tau_n} - G_0\varepsilon S_0 + G_0 \approx G_0 . \tag{15.95}$$

Finally, by using Eq. (15.94), the damping resistors may be approximated by [35]:

$$R_{\mathrm{s1}} = \varepsilon G_0 S_0 L_{\mathrm{s}} \approx \frac{\varepsilon R_{\mathrm{d}}}{\gamma\tau_n} \quad \text{and} \quad R_{s2} = \frac{\beta N_0 L_{\mathrm{s}}}{\tau_n S_0} \approx \frac{\beta R_{\mathrm{d}}\tau_{\mathrm{p}} I_{\mathrm{thA}}}{\alpha\gamma\tau_n S_0^2} . \tag{15.96}$$

It is clear that most of the circuit elements are dependent on the internal threshold current, I_{thA}.

15.16 Rate Equations Including Diffusion Damping

15.16.1 *Derivation of three position-independent rate equations from laterally position- dependent rate equations*

The position-dependent rate equation for photon density in the direction parallel to the junction plane but perpendicular to the direction of light propagation is [59]:

$$\frac{\partial S(x)}{\partial t} = G(x)S(x) - \frac{S(x)}{\tau_{\mathrm{p}}} + \beta\frac{N(x)}{\tau_n} \tag{15.97}$$

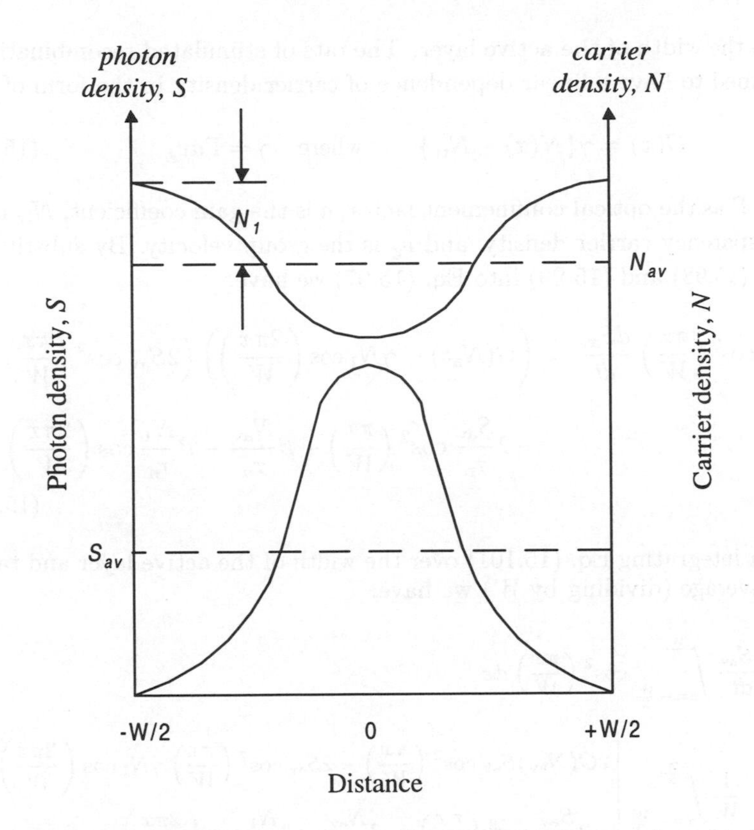

Fig. 15.46. Lateral distribution (x-direction) of photon density and carrier density across the active layer.

where the photon density is given by (see Fig. 15.46):

$$S(x) = 2S_{\mathrm{av}} \cos^2\left(\frac{\pi x}{W}\right) \tag{15.98}$$

and the carrier density is given by (see Fig. 15.46):

$$N(x) = N_{\mathrm{av}} - N_1 \cos\left(\frac{2\pi x}{W}\right) \tag{15.99}$$

where S_{av} is the average photon density, N_{av} is the average carrier density, N_1 is the deviation from the average density, τ_{p} is the photon lifetime, τ_n is the carrier lifetime, β is the spontaneous emission coupling factor, and

W is the width of the active layer. The rate of stimulated recombination is assumed to have a linear dependence of carrier density in the form of [31]:

$$G(x) = \gamma\{N(x) - N_{\mathrm{tr}}\} \qquad \text{where} \quad \gamma = \Gamma a v_{\mathrm{g}} \tag{15.100}$$

with Γ as the optical confinement factor, a is the gain coefficient, N_{tr} is the transparency carrier density, and v_{g} is the group velocity. By substituting Eqs. (15.98) and (15.99) into Eq. (15.97) we have:

$$\begin{aligned} 2\cos^2\left(\frac{\pi x}{W}\right)\frac{dS_{\mathrm{av}}}{dt} &= \left(G(N_{\mathrm{av}}) - \gamma N_1 \cos\left(\frac{2\pi x}{W}\right)\right)\left(2S_{\mathrm{av}}\cos^2\left(\frac{\pi x}{W}\right)\right) \\ &\quad - 2\frac{S_{\mathrm{av}}}{\tau_{\mathrm{p}}}\cos^2\left(\frac{\pi x}{W}\right) + \beta\frac{N_{\mathrm{av}}}{\tau_n} - \beta\frac{N_1}{\tau_n}\cos\left(\frac{2\pi x}{W}\right). \end{aligned} \tag{15.101}$$

Then integrating Eq. (15.101) over the width of the active layer and taking the average (dividing by W) we have:

$$\begin{aligned} &\frac{2}{W}\frac{dS_{\mathrm{av}}}{dt}\int_{x=-\frac{W}{2}}^{\frac{W}{2}} \cos^2\left(\frac{\pi x}{W}\right) dx \\ &= \frac{1}{W}\int_{x=-\frac{W}{2}}^{\frac{W}{2}} \begin{bmatrix} 2G(N_{\mathrm{av}})S_{\mathrm{av}}\cos^2\left(\frac{\pi x}{W}\right) - 2S_{\mathrm{av}}\cos^2\left(\frac{\pi x}{W}\right)\gamma N_1\cos\left(\frac{2\pi x}{W}\right) \\ -2\frac{S_{\mathrm{av}}}{\tau_{\mathrm{p}}}\cos^2\left(\frac{\pi x}{W}\right) + \beta\frac{N_{\mathrm{av}}}{\tau_n} - \beta\frac{N_1}{\tau_n}\cos\left(\frac{2\pi x}{W}\right) \end{bmatrix} dx. \end{aligned} \tag{15.102}$$

By using the trigonometric identity, $\cos^2\theta = (1-\cos 2\theta)/2$ where $\theta = \pi x/W$ and since $\int_{x=-W/2}^{W/2}\cos(m\theta)dx = 0$, Eq. (15.102) reduces to:

$$\frac{dS_{\mathrm{av}}}{dt} = G(N_{\mathrm{av}})S_{\mathrm{av}} - \frac{S_{\mathrm{av}}}{\tau_{\mathrm{p}}} + \beta\frac{N_{\mathrm{av}}}{\tau_n} - \frac{1}{2}\gamma N_1 S_{\mathrm{av}}. \tag{15.103}$$

The above is the position-independent rate equation for the average photon density (S_{av}). Next, the position-dependent rate equation for carrier density in the x-direction [59]:

$$\frac{\partial N(x)}{\partial t} = \frac{I(x)}{\alpha} - \frac{N(x)}{\tau_n} - G(x)S(x) + \frac{L_{\mathrm{eff}}^2}{\tau_n}\cdot\frac{\partial^2 N(x)}{\partial x^2} \qquad \text{where} \quad \alpha = qv_{\mathrm{a}}. \tag{15.104}$$

With q as the electronic charge, v_a is the active layer volume, L_{eff} is the effective carrier diffusion length, and the last term on the right hand side (RHS) describes lateral carrier diffusion. By substituting Eqs. (15.98) and (15.99) into Eq. (15.104) we have:

$$\frac{\partial\{N_{\text{av}} - N_1\cos(2\theta)\}}{\partial t} = \frac{I(x)}{\alpha} - \frac{[N_{\text{av}} - N_1\cos(2\theta)]}{\tau_n}$$
$$- \gamma(N_{\text{av}} - N_1\cos(2\theta) - N_{\text{tr}})(2S_{\text{av}}\cos^2(\theta))$$
$$+ \frac{L_{\text{eff}}^2}{\tau_n} \cdot \frac{\partial^2\{N_{\text{av}} - N_1\cos(2\theta)\}}{\partial x^2}. \tag{15.105}$$

As before, we need to integrate over active layer and take the average of the result. By performing the integration on the third term on the RHS of Eq. (15.105) we have:

$$\frac{1}{W}\int_{x=-\frac{W}{2}}^{\frac{W}{2}} \gamma(N_{\text{av}} - N_1\cos(2\theta) - N_{\text{tr}})(2S_{\text{av}}\cos^2(\theta))dx$$
$$= \frac{1}{W}\int_{x=-\frac{W}{2}}^{\frac{W}{2}} 2S_{\text{av}}\{G(N_{\text{av}}) - \gamma N_1\cos(2\theta)\}\cos^2(\theta)dx$$
$$= \frac{1}{W}\int_{x=-\frac{W}{2}}^{\frac{W}{2}} S_{\text{av}}\{G(N_{\text{av}}) - \gamma N_1\cos(2\theta) + G(N_{\text{av}})\cos(2\theta)$$
$$- \gamma N_1\cos^2(2\theta)\}dx$$
$$= G(N_{\text{av}})S_{\text{av}} - \frac{\gamma N_1 S_{\text{av}}}{2}. \tag{15.106}$$

The carrier diffusion term (last term on the RHS of Eq. (15.105)) will become:

$$\int_{x=\frac{W}{2}}^{\frac{W}{2}} \frac{\partial^2\{N_{\text{av}} - N_1\cos(2\theta)\}}{\partial x^2}dx$$
$$= \int_{x=-\frac{W}{2}}^{\frac{W}{2}} \left\{0 - \left(\frac{2\pi}{W}\right)^2 N_1\cos(2\theta)\right\} dx = \text{NIL}. \tag{15.107}$$

After spatially-integrating all the terms in Eq. (15.105) and taking the average we arrive at the following position-independent rate equation for the average carrier density (N_{av}):

$$\frac{dN_{av}}{dt} = \frac{I}{\alpha} - \frac{N_{av}}{\tau_n} - G(N_{av})S_{av} + \frac{\gamma N_1 S_{av}}{2}. \tag{15.108}$$

In order to obtain another set of rate equation for carrier deviation, N_1, we need to multiply the carrier density rate equation of Eq. (15.105) by $\cos(2\theta)$, integrate over the active layer width and finally take the average. For example, by multiplying the third term on the RHS of Eq. (15.105) by $\cos(2\theta)$, and integrating it across the active layer width we have:

$$\begin{aligned}\frac{1}{W}\int_{x=-\frac{W}{2}}^{\frac{W}{2}} & 2S_{av}\gamma[N_{av}\cos^2(\theta)\cos(2\theta) - N_1\cos^2(2\theta)\cos^2(\theta) \\ & - N_{tr}\cos(2\theta)\cos^2(\theta)]dx \\ &= 2S_{av}\gamma\left[\frac{N_{av}}{4} - \frac{N_1}{4} - \frac{N_{tr}}{4}\right] \\ &= S_{av}\left[\frac{G(N_{av})}{2} - \frac{\gamma N_1}{2}\right].\end{aligned} \tag{15.109}$$

By performing the same spatial-averaging to the last term on the RHS of Eq. (15.105) we get:

$$\begin{aligned}&\frac{L_{eff}^2}{W\tau_n}\int_{-\frac{W}{2}}^{\frac{W}{2}} \cos\left(\frac{2\pi x}{W}\right)\cdot\frac{\partial^2}{\partial x^2}\left\{N_{av} - N_1\cos\left(\frac{2\pi x}{W}\right)\right\}dx \\ &= \frac{L_{eff}^2}{W\tau_n}\int_{-\frac{W}{2}}^{\frac{W}{2}}\left\{N_1\left(\frac{2\pi}{W}\right)^2\cos^2\left(\frac{2\pi x}{W}\right)\right\}dx \\ &= \frac{L_{eff}^2}{\tau_n}\cdot\frac{N_1}{2}\left(\frac{2\pi}{W}\right)^2.\end{aligned} \tag{15.110}$$

Hence, the overall spatial-averaging procedure on Eq. (15.105) yields:

$$\frac{dN_1}{dt} = [G(N_{av}) - \gamma N_1]S_{av} - N_1\left[\frac{1+h}{\tau_n}\right] \qquad \text{where} \quad h = L_{eff}^2\left(\frac{2\pi}{W}\right)^2 \tag{15.111}$$

Notice that the carrier deviation, N_1, has an effective lifetime of $\tau_n/(1 + h)$. Equations (15.103), (15.108) and (15.111) are the three position-independent rate equations derived from the lateral position-dependent rate equations of Eqs. (15.97) and (15.104).

15.16.2 *Reduction of the three position-independent rate equations into two averaged rate equations*

For a narrow-stripe laser, the active layer is assumed to be much smaller than the effective carrier diffusion length, i.e. $(W/L_{\text{eff}}) \ll 1$ or equivalently $h \gg 1$. This means that for τ_n of several nanoseconds, the effective carrier lifetime of N_1 is $\tau_n/(1+h) < 100$ ps. Therefore, a quasi-static condition for N_1 can be assumed and the derivative is set to $(dN_1/dt) = 0$. We can then approximate N_1 as:

$$N_1 \approx \frac{2\varepsilon G(N_{\text{av}})S_{\text{av}}}{(1+2\varepsilon S_{\text{av}})\gamma} \qquad \text{where} \quad \varepsilon = \frac{\gamma\tau_n}{2(1+h)}. \tag{15.112}$$

Since $2\varepsilon S_{\text{av}}$ is negligibly small, we can make a subsequent approximation for N_1 as:

$$N_1 \approx \frac{2\varepsilon G(N_{\text{av}})S_{\text{av}}}{\gamma}. \tag{15.113}$$

By substituting Eq. (15.113) into Eqs. (15.103) and (15.108), the final position-independent rate equations taking into account the damping due to lateral carrier diffusion are expressed by:

$$\begin{aligned} \frac{dS_{\text{av}}}{dt} &= \left\{G(N_{\text{av}})(1-\varepsilon S_{\text{av}}) - \frac{1}{\tau_{\text{p}}}\right\}S_{\text{av}} + \beta\frac{N_{\text{av}}}{\tau_n} \\ \frac{dN_{\text{av}}}{dt} &= \frac{I}{\alpha} - \frac{N_{\text{av}}}{\tau_n} - G(N_{\text{av}})[1-\varepsilon S_{\text{av}}]S_{\text{av}} - \frac{C_{sc}}{\alpha}\frac{dV_j}{dt}. \end{aligned} \tag{15.114}$$

The term ε defined in Eq. (15.112) is often described as the *gain compression factor*, which is a general term used to include gain saturation effects. It is worth pointing out that damping due to lateral carrier diffusion is equivalent to a laser that exhibits gain saturation.

15.17 Summary

Besides the optical characteristics of laser diodes, electrical considerations such as parasitics and matching elements play an important role in determining the high-speed modulation performance [32, 78]. Most numerical models of laser diodes failed to consider the electrical parasitics effect but microwave circuit models of laser diodes allow electrical parasitics to be easily included. The lumped-element circuit models were presented early in this chapter, and their versatility was demonstrated by several application examples. However, these lumped-element circuit models have several shortcomings due to the simplifying assumptions from which these models are derived. In order to overcome these limitations, a distributed-element circuit model, the transmission-line laser model (TLLM) [62], was used. The integrated TLLM model [27] comprising the intrinsic laser diode, electrical parasitics network, and passive matching network was developed to allow a realistic microwave-optoelectronic simulation. The matching network was based on monolithically integrated lumped matching elements [78]. Transmission-line matrix stub-lines and link-lines were used to model both the matching circuit and electrical parasitics network. Simulation results of the distributed-element TLM electrical network were in good agreement with that of the equivalent lumped-element network in a commercial circuit simulator software, HP-MDS.

A microwave application of the integrated TLLM model was demonstrated through harmonic generation by gain-switching [92]. The presence of electrical parasitics tends to broaden the pulsewidth, reduces the peak power of the gain-switched optical pulse, and causes a small time delay in the emitted optical pulses. In addition to the pulse degradation effects, electrical parasitics also affect the extrinsic relaxation oscillation frequency of the laser diode. It was shown that if the RF modulation frequency exceeds the extrinsic modulation bandwidth, stable optical pulse could no longer be generated but the output signal becomes more CW-like instead, leading to low harmonic power content. Simulation results from the integrated TLLM model showed that impedance matching could reduce the pulse width by 10% for a given drive power, and could increase the peak power by over 20% [27] improving the harmonic power content of the gain-switched pulse.

In recent years, it has been noticed that during high-speed modulation, there exist electrical waves travelling on the electrodes of the laser chip, which now behave like transmission lines, and the exact position of the bond-wire should be considered [20, 93–96]. The integrated TLLM model developed in this book, as a distributed-element circuit, may readily include these distributed microwave effects. However, this is beyond the scope of this book.

References

[1] R. J. Helkey, D. J. Derickson, J. G. Wasserbauer, and J. E. Bowers, "Millimeter-wave signal generation using semiconductor diode lasers," *Microw. Opt. Tech. Lett.*, Vol. 6, No. 1, pp. 1–5, 1993.

[2] D. Novak, Z. Ahmed, R. B. Waterhouse, and R. S. Tucker, "Signal generation using pulsed semiconductor lasers for application in millimetre-wave wireless links," *IEEE Trans. Microwave Theory Tech.*, Vol. 43, No. 9, pp. 2257–2261, 1995.

[3] Z. Ahmed, D. Novak, R. B. Waterhouse, and H. F. Liu, "37-GHz Fiber-Wireless System for Distribution of Broad-Band Signals," *IEEE Trans. Microwave Theory Tech.*, Vol. 45, No. 8, pp. 1431–1435, August 1997.

[4] L. Noel *et al.* "Novel Techniques for High-Capacity 60-GHz Fiber-Radio Transmission Systems," *IEEE Trans. Microwave Theory Tech.*, Vol. 45, No. 8, pp. 1416–1423, 1997.

[5] H. Ghafouri-Shiraz, "Integrated Transmit-Receive Circuit Antenna Modules for Radio on Fiber Systems," in *Analysis and design of integrated circuit antenna module*, K. C. Gupta and P. S. Hall, Eds. (NY: John Wiley and Sons, 2000).

[6] J. L. Altman, *Microwave Circuits* (Princeton, New Jersey: D. Van Nostrand Company, Inc., 1964).

[7] R. E. Collins, *Foundations of Microwave Engineering*, (New York: McGraw-Hill, 1966).

[8] S. Iezekiel, C. M. Snowden, and M. J. Howes, "Nonlinear Circuit Analysis of Harmonic and Intermodulation Distortions in Laser Diodes Under Microwave Direct Modulation," *IEEE Trans. Microwave Theory Tech.*, Vol. 38, No. 12, pp. 1906–1915, 1990.

[9] J. Wang, M. K. Haldar, and F. V. C. Mendis, "Equivalent Circuit Model of Injection-Locked Laser Diodes," *Microw. Opt. Tech. Lett.*, Vol. 18, No. 2, pp. 124–126, 1998.

[10] H. Elkadi, J. P. Vilcot, and D. Decoster, "An Equivalent Circuit Model for Multielectrode Lasers: Potential Devices for Millimeter-wave Applications," *Microw. Opt. Tech. Lett.*, Vol. 6, No. 4, pp. 245–249, 1993.

[11] H. A. Tafti, V. S. Sheeba, K. K. Kamath, F. N. Farokhrooz, and P. R. Vaya, "Simulation of gain-switched picosecond pulse generation from quantum well lasers," *Opt. Quant. Elec.*, Vol. 28, pp. 1669–1676, 1996.

[12] N. Bewtra, D. A. Suda, G. L. Tan, F. Chatenoud, and J. M. Xu, "Modeling of Quantum-Well Lasers with Electro-Opto-Thermal Interaction," *IEEE Jour. Select. Top. Quant. Elec.*, Vol. 1, No. 2, pp. 331–340, 1995.

[13] P. J. Probert and J. E. Carroll, "Lumped circuit model for predicting linewidth in Fabry Perot and DFB lasers, including external cavity devices," *IEE Proceedings Pt. J*, Vol. 136, No. 1, pp. 22–32, 1989.

[14] P. J. Probert and J. E. Carroll, "Modelling bistability in injection-laser amplifiers using lumped-circuit models," *IEE Proceedings Pt. J*, Vol. 134, No. 5, pp. 295–302, 1987.

[15] M. F. Lu, J. S. Deng, M. J. Jou, and B. J. Lee, "Equivalent Circuit Model of Quantum-Well Lasers," *IEEE J. Quant. Electron.*, Vol. 31, No. 8, pp. 1419–1422, 1995.

[16] M. Zhang and D. R. Conn, "A dynamic equivalent circuit model for vetical-cavity surface-emitting lasers," *Microw. Opt. Tech. Lett.*, Vol. 20, No. 1, pp. 1–8, 1999.

[17] T. T. Bich-Ha and J. C. Mollier, "Noise Equivalent Circuit of a Two-Mode Semiconductor Laser with the Contribution of Both the Linear and the Nonlinear Gain," *IEEE Jour. Select. Top. Quant. Elec.*, Vol. 3, No. 2, pp. 304–308, 1997.

[18] T. Ikegami and Y. Suematsu, "Resonance-Like Characteristics of the Direct Modulation of a Junction Laser," *Proc. IEEE*, pp. 122–123, Jan 1967.

[19] C. Harder, J. Katz, S. Margalit, J. Shacham, and A. Yariv, "Noise Equivalent Circuit of a Semiconductor Laser Diode," *IEEE J. Quant. Electron.*, Vol. QE-18, No. 3, pp. 333–337, 1982.

[20] K. C. Sum, "Combined Microwave Optoelectronic Modelling for Semiconductor Lasers," Ph.D. Dissertation, University of Kent, 1997.

[21] B. P. C. Tsou and D. L. Pulfrey, "A Versatile SPICE Model for Quantum-Well Lasers," *IEEE J. Quant. Electron.*, Vol. 33, No. 2, pp. 246–254, 1997.

[22] A. S. Daryoush, "Special Issue on Microwave Photonics," *Microw. Opt. Tech. Lett.*, Vol. 6, No. 1, p. 1, 1993.

[23] R. D. Esman and U. Gliese, "Special issue on microwave and millimeter-wave photonics," *IEEE Trans. Microwave Theory Tech.*, Vol. 47, No. 7, pp. 1149–1391, 1999.

[24] A. J. Lowery, "Transmission-line modelling of semiconductor lasers: the transmission-line laser model," *Int. Jour. Num. Model.*, Vol. 2, pp. 249–265, 1989.

[25] A. J. Lowery, "A new time-domain model for spontaneous emission in semiconductor lasers and its use in predicting their transient response," *Int. Jour. Num. Model.*, Vol. 1, pp. 153–164, 1988.

[26] A. J. Lowery, "Amplified spontaneous emission in semiconductor laser amplifiers: validity of the transmission-line laser model," *IEE Proceedings Pt. J*, Vol. 137, No. 4, pp. 241–247, 1990.

[27] W. M. Wong and H. Ghafouri-Shiraz, "Integrated semiconductor laser-transmitter model for microwave-optoelectronic simulation based on transmission-line modelling," *IEE Proceedings Pt. J*, Vol. 146, No. 4, pp. 181–188, 1999.

[28] P. A. Barnes and T. L. Paoli, "Derivative Measurements of the Current-Voltage Characteristics of Double-Heterostructure Injection Lasers," *IEEE J. Quant. Electron.*, Vol. QE-12, No. 10, pp. 633–639, 1976.

[29] T. L. Paoli, "Theoretical Derivatives of the Electrical Characteristic of a Junction Laser Operated in the Vicinity of Threshold," *IEEE J. Quant. Electron.*, Vol. QE-14, No. 1, pp. 62–68, 1978.

[30] W. B. Joyce and R. W. Dixon, "Electrical characterization of heterostructure lasers," *J. Appl. Phys.*, Vol. 49(7), pp. 3719–3728, 1978.

[31] G. P. Agrawal and N. K. Dutta, Semiconductor lasers, 2nd ed. (New York, Van Nostrand Reinhold, 1993).

[32] R. S. Tucker, "High-Speed Modulation of Semiconductor Lasers," *IEEE J. Lightwave Technol.*, Vol. LT-3, No. 6, pp. 1180–1192, 1985.

[33] M. L. Majewski and D. Novak, "Method for Characterization of Intrinsic and Extrinsic Components of Semiconductor Laser Diode Circuit Model," *IEEE Microwave Guided Wave Lett.*, Vol. 1, No. 9, pp. 246–248, 1991.

[34] M. L. Majewski and D. Novak, "Modulation bandwidth limitations in directly modulated semiconductor laser diodes at microwave frequencies," *Microw. Opt. Tech. Lett.*, Vol. 4, No. 13, pp. 581–584, 1991.
[35] R. S. Tucker and I. P. Kaminow, "High-Frequency Characteristics of Directly Modulated InGaAsP Ridge Waveguide and Buried Heterostructure Lasers," *IEEE J. Lightwave Technol.*, Vol. LT-2, No. 4, pp. 385–393, 1984.
[36] M. Maeda, K. Nagano, M. Tanaka, and K. Chiba, "Buried-Heterostructure Laser Packaging for Wideband Optical Transmission Systems," *IEEE Trans. Comm.*, Vol. COM-26, No. 7, pp. 1076–1081, 1978.
[37] R. S. Tucker *et al.* "40 GHz active mode-locking in a 1.5 mum monolithic extended-cavity laser," *Electron. Lett.*, Vol. 25 No. 10, pp. 621–622, 1989.
[38] P. B. Hansen *et al.* "5.5 mm long InGaAsP Monolithic Extended-Cavity Laser with an Integrated Bragg-Reflector for Active Mode Locking," *IEEE Photon. Tech. Lett.*, Vol. 4, No. 3, pp. 215–217, 1992.
[39] W. W. Ng and E. A. Sovero, "An Analytic Model for the Modulation Response of Buried Heterostructure Lasers," *IEEE J. Quant. Electron.*, Vol. QE-20, No. 9, pp. 1008–1015, 1984.
[40] K. Y. Lau and A. Yariv, "Ultra-High Speed Semiconductor Lasers," *IEEE J. Quant. Electron.*, Vol. QE-21, No. 2, pp. 121–138, 1985.
[41] Hewlett-Packard, "MDS Release 7.0 and 7.1", Online Manual.
[42] R. S. Tucker, "Large-signal circuit model for simulation of injection-laser modulation dynamics," *IEE Proceedings Pt. I*, Vol. 128, pp. 180–184, 1981.
[43] M. J. Adams, "Rate equations and transient phenomena in semiconductor lasers," *Opto-electronics*, Vol. 5, pp. 201–215, 1973.
[44] M. Danielson, "A theoretical analysis for Gb/s pulse code modulation of semiconductor lasers," *IEEE J. Quant. Electron.*, Vol. QE-12, pp. 657–659, 1976.
[45] P. M. Boers, M. T. Vlaardingerbroek, and M. Danielsen, "Dynamic Behaviour of Semiconductor Lasers," *Electron. Lett.*, Vol. 11, No. 10, pp. 206–208, 1975.
[46] G. Arnold and P. Russer, "Modulation Behaviour of Semiconductor Injeciton Lasers," *Appl. Phys.*, Vol. 14, pp. 255–268, 1977.
[47] W. I. Way, "Large Signal Nonlinear Distortion Prediction for a Single-Mode Laser Diode Under Microwave Intensity Modulation," *IEEE J. Lightwave Technol.*, Vol. LT-5, No. 3, pp. 305–315, 1987.
[48] K. C. Sum and N. J. Gomes, "Microwave-optoelectronic modelling approaches for semiconductor lasers," *IEE Proceedings Pt. J*, Vol. 145, No. 3, pp. 141–146, 1998.
[49] K. Petermann, *Laser Diode Modulation and Noise* (Dordrecht, Kluwer Academic Publishers, 1991).
[50] I. P. Kaminow and R. S. Tucker, "Mode-controlled semiconductor lasers," in *Guided-Wave Optoelectronics*, 2nd ed., T. Tamir, Ed. (Berlin: Springer-Verlag, 1990), p. Ch. 5.
[51] H. A. Tafti, K. K. Kamath, G. Abraham, F. N. Farokhrooz, and P. R. Vaya, "Circuit modelling of multimode semiconductor laser and study of pulse broadening effect," *Electron. Lett.*, Vol. 29, No. 16, pp. 1443–1445, 1993.

[52] I. Habermayer, "Nonlinear circuit model for semiconductor lasers," *Opt. Quant. Elec.*, Vol. 13, pp. 461–468, 1981.
[53] M. J. Adams and M. Osinski, "Longitudinal mode competition in semiconductor lasers," *IEE Proceedings Pt. I*, Vol. 129, pp. 271–274, 1982.
[54] T. P. Lee, C. A. Burrus, J. A. Copeland, A. G. Dentai, and D. Marcuse, "Short-Cavity InGaAsP Injection Lasers: Dependence of Mode Spectra and Single-Longitudinal-Mode Power on Cavity Length," *IEEE J. Quant. Electron.*, Vol. QE-18, No. 7, pp. 1101–1113, 1982.
[55] K. Y. Lau and A. Yariv, "Intermodulation distortion in a directly modulated semiconductor injection laser," *Appl. Phys. Lett.*, Vol. 45, pp. 1034–1036, 1984.
[56] P. Feldmann, B. Melville, and D. Long, "Efficient frequency domain analysis of large nonlinear analog circuits," *IEEE Custom Integrated Circuits Conference*, pp. 461–464, 1996.
[57] http://www.tec.ufl.edu/~flooxs/cur/floods/ floods_html/node23.html
[58] R. S. Tucker and D. J. Pope, "Microwave Circuit Models of Semiconductor Injection Lasers," *IEEE Trans. Microwave Theory Tech.*, Vol. MTT-31, No. 3, pp. 289–294, 1983.
[59] R. S. Tucker and D. J. Pope, "Circuit Modeling of the Effect of Diffusion on Damping in a Narrow-Stripe Semiconductor Laser," *IEEE J. Quant. Electron.*, Vol. QE-19, No. 7, pp. 1179–1183, 1983.
[60] K. C. Sum and N. J. Gomes, "Harmonic generation limitations in gain-switched semiconductor lasers due to distributed microwave effects," *Appl. Phys. Lett.*, Vol. 71(9), pp. 1154–1155, 1997.
[61] C. F. Tsang, D. D. Marcenac, J. E. Carroll, and L. M. Zhang, "Comparison between 'power matrix mode' and 'time domain model' in modelling large signal response of DFB lasers," *IEE Proceedings Pt. J*, Vol. 141, No. 2, pp. 89–96, 1994.
[62] A. J. Lowery, "New dynamic semiconductor laser model based on the transmission-line modelling method," *IEE Proceedings Pt. J*, Vol. 134, pp. 281–289, 1987.
[63] J. W. Bandler, P. B. Johns, and M. R. M. Rizk, "Transmission-Line Modeling and Sensitivity Evaluation for Lumped Network Simulation and Design in the Time Domain," *Jour. Frank. Inst.*, Vol. 304, pp. 15–32, 1977.
[64] K. C. Sum and N. J. Gomes, "Integrated Microwave-Optoelectronic Simulation of Semiconductor Laser Transmitters," *Microw. Opt. Tech. Lett.*, Vol. 14, No. 6, pp. 313–315, 1997.
[65] M. Sayin and M. S. Ozyazici, "Effect of gain switching frequency on ultrashort pulse generation from laser diodes," *Opt. Quant. Elec.*, Vol. 29, pp. 627–638, 1997.
[66] A. J. Lowery, "A two-port bilateral model for semiconductor lasers," *IEEE J. Quant. Electron.*, Vol. 128, No. 1, pp. 82–92, 1992.
[67] Mitsubishi Electric, "Mitsubishi Optoelectronics — Optical Devices and Optical-Fiber Communication Systems", Data Book.
[68] H. L. Martinez-Reyes, J. A. Reynoso-Hernandez, and F. J. Mendieta, "DC and RF techniques for computing the series resistance of the equivalent

electrical circuit for semiconductor lasers," *Microw. Opt. Tech. Lett.*, Vol. 20, no. 4, pp. 258–261, 1999.

[69] A. Ghiasi and A. Gopinath, "Novel Wide-Bandwidth Matching Technique for Laser Diodes," *IEEE Trans. Microwave Theory Tech.*, Vol. 38, No. 5, pp. 673–675, 1990.

[70] H. Ghafouri-Shiraz and W. M. Wong, "Matching network for microwave applications of semiconductor laser diodes (LDs): Consideration of the effect of electrical parasitics and LD carrier-dependent impedance," *Microw. Opt. Tech. Lett.*, Vol. 25, No. 3, pp. 197–200, 2000.

[71] C. H. von Helmolt, U. Kruger, K. Kruger, and G. Grosskopf, "A Mobile Broad-Band Communication System Based on Mode-Locked Lasers," *IEEE Trans. Microwave Theory Tech.*, Vol. 45, pp. 1424–1429, 1997.

[72] A. J. Lowery and P. C. R. Gurney, "Comparison of Optical Processing Techniques for Optical Microwave Signal Generation," *IEEE Trans. Microwave Theory Tech.*, Vol. 46, No. 2, pp. 142–150, 1998.

[73] P. Young, "Matching techniques optimize high-speed laser interfaces," *Microwaves and RF*, April, pp. 184–189, 1994.

[74] K. W. Ho and C. Shu, "Efficient gain switching of laser diodes with quarter-wave impedance transformer," *Electron. Lett.*, Vol. 29, No. 13, pp. 1194–1195, 1993.

[75] C. L. Goldsmith and B. Kanack, "Broad-Band Reactive Matching of High-Speed Directly Modulated Laser Diodes," *IEEE Microwave Guided Wave Lett.*, Vol. 3, No. 9, pp. 336–338, 1993.

[76] W. E. Stephens and T. R. Joseph, "A 1.3-mum Microwave Fiber-Optic Link Using A Direct-Modulated Laser Transmitter," *IEEE J. Lightwave Technol.*, Vol. LT-3, No. 2, pp. 308–315, 1985.

[77] P. Young, "Matching techniques optimize high-speed laser interfaces," *Microwaves and RF*, Vol. ? April, pp. 184–189, 1994.

[78] S. Maricot *et al.* "Monolithic Integration of Optoelectronic Device with Reactive Matching Networks for Microwave Applications," *IEEE Photon. Tech. Lett.*, Vol. 4, No. 11, pp. 1248–1250, 1992.

[79] D. M. Pozar, *Microwave Engineering.* Reading, Massachusetts: Addison-Wesley, 1990.

[80] S. Maricot, J. P. Vilcot, and D. Decoster, "Improvement of Microwave Signal Optical Transmission by Passive Matching of Optoelectronic Devices," *Microw. Opt. Tech. Lett.*, Vol. 4, No. 13, pp. 591–595, 1991.

[81] J. B. Georges, M. H. Kiang, K. Heppell, M. Sayed, and K. Y. Lau, "Optical Transmission of Narrow-Band Millimter-Wave Signals by Resonant Modulation of Monolithic Semiconductor Lasers," *IEEE Photon. Tech. Lett.*, Vol. 6, No. 4, pp. 568–570, 1994.

[82] P. Callaghan, S. K. Suresh Babu, N. J. Gomes, and A. K. Jastrzebski, "Optical control of MMIC oscillators," *Int. J. Optoelectronics*, Vol. 10, No. 6, pp. 467–471, 1995.

[83] C. R. Lima and P. A. Davies, "Effects of extra low-frequency noise injection on microwave signals generated by a gain-switched semiconductor laser," *Appl. Phys. Lett.*, Vol. 65(8), pp. 950–952, 1994.

[84] C. R. Lima and P. A. Davies, "Degradation of microwave signals generated by a modelocked semiconductor laser due to up-conversion of low-frequency noise," *Electron. Lett.*, Vol. 31, No. 8, pp. 641–642, 1995.
[85] H. F. Liu, M. Fukazawa, Y. Kawai, and T. Kamiya, "Gain-Switched Picosecond Pulse (< 10 ps) Generation from 1.3 μm InGaAsP Laser Diodes," *IEEE J. Quant. Electron.*, Vol. 25, No. 6, pp. 1417–1425, 1989.
[86] J. E. Deardorff and C. R. Trimble, "Calibrated Real-Time Signal Averaging," *HP Journal*, April 1968, pp. 8–13, 1968.
[87] A. J. Lowery, "Model for multimode picosecond dynamic laser chirp based on transmission line laser model," *IEE Proceedings Pt. J*, Vol. 135, No. 2, pp. 126–132, 1988.
[88] M. Morishita, T. Ohmi, and J. I. Nishizawa, "Impedance Characteristics of Double-Heterostructure Laser Diodes," *Sol. Stat. Electron.*, Vol. 22, pp. 951–962, 1979.
[89] R. S. Tucker, "Circuit model of double-heterojunction laser below threshold," *IEE Proceedings Pt. I*, Vol. 128, No. 3, pp. 101–106, June 1981.
[90] J. Katz, S. Margalit, C. Harder, D. Wilt, and A. Yariv, "The Intrinsic Electrical Equivalent Circuit of a Laser Diode," *IEEE J. Quant. Electron.*, Vol. QE-17, No. 1, pp. 4–7, 1981.
[91] G. E. Shtengel, D. A. Ackerman, P. A. Morton, E. J. Flynn, and M. S. Hybertsen, "Impedance-corrected carrier lifetime measurements of semiconductor lasers," IEEE/LEOS'95, San Francisco, USA, p. paper SCL6.3, 1995.
[92] K. Y. Lau, "Gain switching of semiconductor injection lasers," *Appl. Phys. Lett.*, vol. 52(4), pp. 257–259, 1988.
[93] D. A. Tauber, R. Spickermann, R. Nagarajan, T. Reynolds, A. L. Holmes Jr., and J. E. Bowers, "Inherent bandwidth limits in semiconductor lasers due to distributed microwave effects," *Appl. Phys. Lett.*, Vol. 64(13), pp. 1610–1612, 1994.
[94] B. Wu, J. B. Georges, D. M. Cutrer, and K. Y. Lau, "On distributed microwave effects in semiconductor lasers and their practical implications," *Appl. Phys. Lett.*, Vol. 67(4), pp. 467–469, 1995.
[95] R. Vahldieck, S. Chen, H. Jin, and P. Russer, "Field theory analysis of distributed microwave effects in high speed semiconductor lasers and their interconnection with passive microwave transmission lines," *IEEE MTT-S Digest*, WE3F-L3, pp. 861–864, 1995.
[96] Y. C. Chan, A. J. Lowery, and D. Novak, "Semiconductor laser AM response flattening by bond-wire positioning," *IEEE MTT-S Digest*, pp. 125–128, 1994.
[97] H. C. J. Casey and M. B. Panish, Heterostructure Lasers: Part A, Fundamental Principles (New York, Academic Press, 1978).
[98] W. B. Joyce and R. W. Dixon, "Analytic approximations for the Fermi energy of an ideal Fermi gas," *Appl. Phys. Lett.*, Vol. 31, pp. 354–356, 1977.
[99] T. P. Lee, "Effect of junction capacitance on the rise time of LED's and on the turn-on delay in semiconductor lasers," *Bell Syst. Tech. J.*, Vol. 54, pp. 53–68, 1975.

Chapter 16

Transmission-Line Laser Model of Tapered Waveguide Lasers and Amplifiers

16.1 Introduction

In recent years, a new generation of dynamic models for analysis of laser diodes has been developed [1–3]. These models are more accurate and powerful than the laser rate equations [4]. The major drawback of the rate equations is that they cannot handle inhomogeneous longitudinal effects such as carrier density distribution, which are important in distributed feedback (DFB) lasers and travelling-wave amplifiers (TWA). On the other hand, such inhomogeneous spatial effects are considered in this new generation of time-domain laser models. The models are one-dimensional (1-D) in space, as they mostly cater for uniform-width laser structures.

The tapered or flared laser structure has been proposed to overcome some limitations of the conventional uniform-width laser structure. The tapered laser amplifier has been shown to provide better saturation performance [5], and can be used to amplify picosecond pulses with very little distortion [6]. Tapered laser devices have also been used to generate mode-locked [7] and Q-switched [8] short pulses with high peak powers. It has also been shown that the tapered amplifier can be integrated with a DFB laser to produce high contrast ratio and low chirp as an alternative to electro-absorption (EA) modulators [9].

For tapered structures, a more careful consideration must be given to the validity of using 1-D models. Two-dimensional (2-D) models in space by the beam-propagation method (BPM) have been used to model tapered laser structures [10–12]. The BPM is usually used to simulate gain-guided lasers but it does not work well in conditions where the refractive index step is large, e.g. buried heterostructure (BH) lasers; or when reflection and coupling of waves are important, e.g. distributed feedback (DFB) and distributed Bragg reflectors (DBR) laser structures. Most BPM models are static models except those reported in Refs. [11] and [13], which are dynamic models but the perfect travelling-wave condition (i.e. no reflections) was assumed. These 2-D models are not strictly essential unless the spatial beam profile and filamentation effects are to be studied. Light-current characteristics, gain saturation effects, energy conversion efficiencies, relative amplified spontaneous emission (ASE) noise, and output spectra of tapered laser structures can still be studied by using spatially averaged 1-D laser models [5, 14] that take into account inhomogeneous effects.

Computationally intensive 2-D models are unattractive for photonic computer aided design (PCAD) systems, which can consist of many photonic and electronic components. Therefore, efficient 1-D models are more attractive tools for photonic systems designers. In this chapter, we propose a semi-analytical 1-D model of the tapered laser structure based on the transmission-line laser modelling (TLLM) method. The model is semi-analytical to improve computational efficiency, and is in time domain to allow inclusion of nonlinear effects. The TLLM method gives a clear physical interpretation of the laser's internal processes, such as reflections, gain/loss, filtering, noise, and coupling, all in terms of microwave circuit theory. Therefore, this modelling approach is able to provide additional insight into the laser dynamics, in semi-classical terms. This approach is more familiar to those in the field of microwaves who previously had little or no experience in the laser. As for those already in the field of lasers, it opens up a fresh new perspective.

The transmission-line modelling concept is not entirely new and has been proposed by Gordon [15]. However, Gordon's frequency-domain approach assumes linearity, uncoupled modes, uniform saturation along the cavity, and is applicable only in the steady-state condition (i.e. a static model). In contrast, the above restrictions are not required in the TLLM methodology, which is based on purely time-domain scattering matrices. Therefore, laser nonlinear effects, which can be useful or detrimental depending on the application, can be taken into account. Nonlinear effects such as gain saturation and carrier-induced refractive index changes are important in travelling-wave semiconductor laser amplifiers. The novel dynamic model of the tapered waveguide laser in this book may be connected to other existing TLLM component models to form multisection laser structures or to form a complete system simulation. This is because the optical field samples can easily be passed on from one component model to the next. However, the systems simulation is beyond the scope of this book. This chapter concentrates on the construction of the tapered structure TLLM (TS-TLLM).

16.2 Tapered Structure Transmission-Line Laser Model (TS-TLLM)

The schematic diagram of the TS-TLLM is shown in Fig. 16.1. The TS-TLLM is chosen to have an input width of $1\mu m$ in order to maintain single lateral mode condition at the input. The output width is $30\mu m$, and total length of the tapered structure is $1000\mu m$. There is no variation in the y-direction, and thickness (depth) of the bulk active layer is fixed at $0.1\mu m$ to achieve single transverse mode. Active layer width increases in the x-direction in a quasi-adiabatic manner.

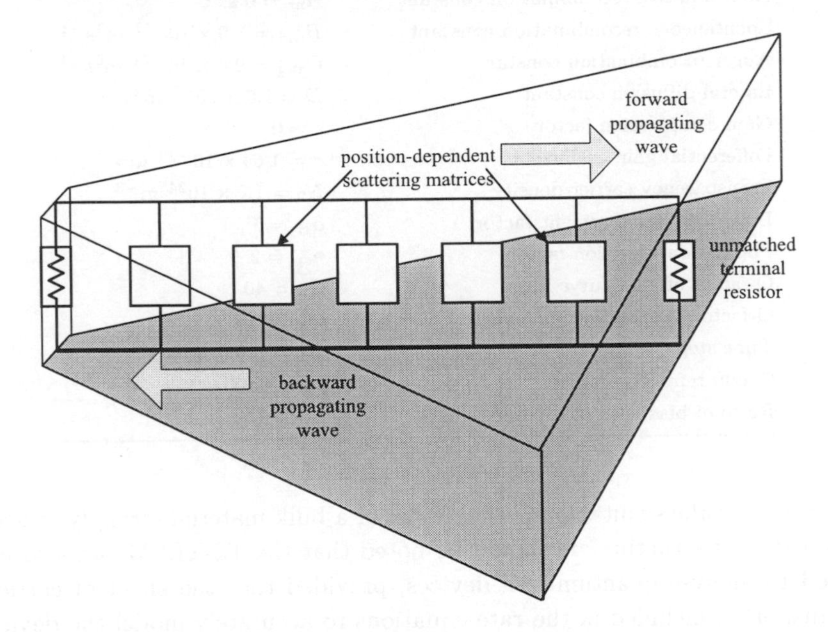

Fig. 16.1. The transmission-line laser model of the tapered structure (TS-TLLM).

The device parameters used in the tapered structure transmission-line laser model (TS-TLLM) are listed in Table 16.1, which were taken from [5] as it contains a comprehensive set of structural and material

Table 16.1. The parameter values used in the tapered structure TLLM.

Physical Quantity	Value
Input width	$W_i = 1\ \mu m$
Output width	$W_o = 30\ \mu m$
Total length	$L = 1000\ \mu m$
Waveguide layer thickness	$d_w = 0.2\ \mu m$
Cladding layer thickness	$d_c = 4\ \mu m$
Active layer thickness	$d = 0.1\ \mu m$
Operating wavelength	$\lambda_o = 1.533\ \mu m$
TE_0 equivalent index of amplifying region	$n_{eq1} = 3.371$
TE_0 equivalent index of cladding region	$n_{eq2} = 3.169$
Optical confinement factor	$\Gamma = 0.2$
Active region loss coefficient	$\alpha_{act} = 1.2 \times 10^4\ m^{-1}$
Cladding region loss coefficient	$\alpha_{clad} = 5.6 \times 10^2\ m^{-1}$
Number of sections in model	$S = 100$
Non-radiative recombination constant	$A_{nr} = 0\ s^{-1}$
Spontaneous recombination constant	$B_{rad} = 8.0 \times 10^{-17}\ m^3s^{-1}$
Auger recombination constant	$C_{aug} = 9.0 \times 10^{-41}\ m^6s^{-1}$
Lateral diffusion constant	$D = 1.0 \times 10^{-3}\ m^2s^{-1}$
Gain compression factor	$\varepsilon = 0$
Differential gain coefficient	$a = 1.64 \times 10^{-20}\ m^2$
Transparency carrier density	$N_0 = 1.0 \times 10^{24}\ m^{-3}$
Linewidth enhancement factor	$\alpha_H = 6$
Population inversion factor	$n_{sp} = 2$
Q-factor of gain curve filter	$Q_g = 40$
Q-factor of spontaneous emission filter	$Q_{sp} = 15$
Time step	$\Delta t = 129.023$ fs
Group refractive index	$n_g = 3.871$
Ratio of bias level of threshold level	$I = 0.7I_{th}$

parameter values suitable for the model of a bulk material strongly index-guided laser structure. It should be noted that the TS-TLLM can also be used to analyse quantum-well devices, provided that the effect of carrier transport is included in the rate equations to accurately model the device dynamics [16]. The quantum well structure improves the energy conversion efficiency, modulation bandwidth, saturation output power, and spectral characteristics of this tapered device [17]. The improvements are one result of enhanced differential gain, reduced linewidth enhancement factor, and smaller confinement factor.

16.3 Effective Index Method (EIM)

In order to reduce the refractive index distribution of the five-layer separate confinement heterostructure (SCH) into a single equivalent index in the transmission-line model, we use the effective index method (EIM) [18, 19] as depicted in Fig. 16.2. The field distributions in a five-layer slab waveguide for *even-order* modes are given by [20]:

$$E_x = \begin{cases} A\cos(py) & 0 \le |y| \le a \\ \dfrac{A\cos(pa)\cos(qb+\phi)}{\cos(qy+\phi)} & a \le |y| \le b \\ \dfrac{A\cos(pa)\cos(qb+\phi)}{\cos(qa+\phi)} e^{r(b-|y|)} & |y| \ge b \end{cases} \tag{16.1}$$

where

$$\begin{aligned} p^2 &= k_0^2 n_1^2 - \beta_{\rm sch}^2 \\ q^2 &= k_0^2 n_2^2 - \beta_{\rm sch}^2 \\ r^2 &= \beta_{\rm sch}^2 - k_0^2 n_3^2 \,. \end{aligned} \tag{16.2}$$

The parameters a and b are shown in Fig. 16.2. The values of a and b are taken as $a = d/2$ and $b = (d/2 + d_{\rm w})$ (see Table 16.1). The refractive indices for different layers (n_1, n_2 and n_3) are also shown. The term k_0 is the free-space wavevector, and ϕ is the phase offset of the electric field distribution.

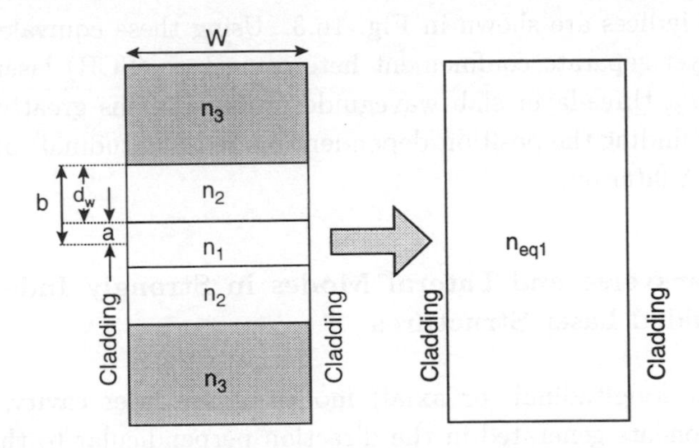

Fig. 16.2. Equivalent index of the five-layer separate confinement heterostructure (SCH) waveguide.

The equivalent refractive index of the tapered amplifying region, n_{eq1}, may be obtained from,

$$n_{eq1} = \frac{\beta_{sch}}{k_0}. \tag{16.3}$$

In the case of $k_0 n_3 \leq \beta_{sch} \leq k_0 n_2$, the value of β_{sch} can be found from the following eigenvalue equation:

$$pa = \tan^{-1}\left\{\eta_{12}\frac{q}{p}\tan\left[\tan^{-1}\left(\eta_{23}\frac{r}{q}\right) - q(b-a)\right]\right\} \tag{16.4}$$

where

$$\eta_{ij} = \begin{cases} 1 & \text{for TE modes} \\ \left(\dfrac{n_i}{n_j}\right)^2 & \text{for TM modes}. \end{cases} \tag{16.5}$$

The equivalent refractive indices, n_{eq1}, of the amplifying region for the TE_0 mode and TM_0 mode are found to be 3.371 and 3.363, respectively. Only the TE modes will be considered in the model. The equivalent refractive index of the InP cladding region, n_{eq2}, will be taken as 3.169. The laser structure is now reduced to a three-layer slab waveguide problem in the lateral x-direction, in contrast to the usual slab waveguide problem across the transverse y-direction [21–22]. The refractive index step, Δn_{eq}, is about 6%, thus the laser structure is strongly index-guided. Strictly speaking, purely TE modes are not achievable in the strongly index-guided laser structure but instead quasi-TE modes exist [23]. The tapered structure and its equivalent indices are shown in Fig. 16.3. Using these equivalent indices, the five-layer separate confinement heterostructure (SCH) laser has been reduced to a three-layer slab waveguide problem. This greatly simplifies the task of finding the position-dependent mode (longitudinal) propagation constant, β, later on.

16.4 Transverse and Lateral Modes in Strongly Index-Guided Laser Structures

Besides the longitudinal (or axial) modes of the laser cavity, there are field components generated in the direction perpendicular to the junction (transverse modes), and field components generated in the direction parallel to the junction (lateral modes). In a real semiconductor laser cavity, the optical modes are neither pure transverse-electromagnetic (TEM) nor pure transverse-electric (TE) (or transverse-magnetic (TM)). The transmission-line laser model (TLLM) is based on the strongly index-guided buried heterostructure semiconductor laser with abrupt discontinuities of the refractive indices leading to so-called "box" modes [24]. The box modes are similar to the resonant modes of a rectangular microwave cavity [25], which are essentially of the TEM type, and are grouped into two sets of hybrid modes known as quasi-TE modes (E^x_{mn}) and quasi-TM modes (E^y_{mn}) [23]. Only the quasi-TE modes, i.e. the electric field predominantly polarised along the x-direction (E_z is also allowed to exist), are used in this book. In the following, although the terms TE and TM modes are used, it is understood that they mean *quasi*-TE and *quasi*-TM modes, respectively.

From Section 16.3, the EIM has reduced the complicated problem of the five-layer SCH (embedded into a rectangular waveguide) into a simplified three-layer slab waveguide problem along the tapered structure (Fig. 16.3). Since the taper is symmetrical, and by assuming that the laser amplifier is excited by an even order TE mode at the input end, odd order TE modes need not be considered. It is important to point out that the slab waveguide problem here consists of three layers *horizontally* placed, instead of the typical problem of vertically placed layers. Therefore, the magnetic field

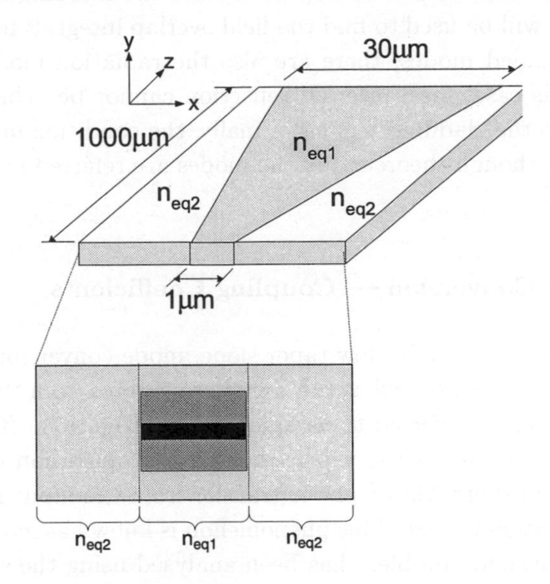

Fig. 16.3. The buried heterostructure (BH) structure reduced into a three-layer slab waveguide problem.

distribution of the TE even order modes (E-field predominantly polarised along x-direction) can be expressed as [26]:

$$\begin{aligned} h_y(x) &= \cos(\kappa x) && \text{for } |x| \le W(z)/2 \\ h_y(x) &= e^{-\gamma(|x|-W(z)/2)}\cos[\kappa W(z)/2] && \text{for } |x| \ge W(z)/2 \end{aligned} \tag{16.6}$$

where

$$\begin{aligned} \kappa &= k_0(n_{\text{eq1}}^2 - n_{\text{eff}}^2)^{1/2} \\ \gamma &= k_0(n_{\text{eff}}^2 - n_{\text{eq2}}^2)^{1/2} \end{aligned} \tag{16.7}$$

κ and γ are the lateral propagation constants, $W(z)$ is the position-dependent width of the middle slab layer, and n_{eff} is the effective index of refraction. The x-polarised electric field e_x can be obtained from the expression:

$$e_x = \frac{j}{\omega\varepsilon}\frac{\partial h_y}{\partial z} = \frac{\beta}{\omega\varepsilon}h_y \tag{16.8}$$

where β is the propagation constant, ε is the dielectric permittivity, and is the angular frequency. The guided modes must satisfy the condition that $n_{\text{eq2}} < \beta/k_0 < n_{\text{eq1}}$ for total internal reflection [24]. The guided modes are completely trapped and do not "leak" into the cladding regions. These guided modes will be used to find the field overlap integrals in Section 16.5. Besides the guided modes, there are also the radiation modes when $\beta < k_0 n_{\text{eq2}}$. In this case, total internal reflection cannot be achieved, and the fields leak into the cladding regions. Finally, the condition of $\beta > k_0 n_{\text{eq1}}$ is not realisable, though theoretically the modes are referred to as evanescent modes [24, 27].

16.5 Mode Conversion — Coupling Coefficients

In a waveguide of infinitely slow taper slope, mode conversion is negligible, and mode power is conserved as the signal propagates from the input to the output end. This is referred to as adiabatic propagation. If small abrupt step transitions occur in the taper, adiabatic propagation can no longer be satisfied, and there will be power transfer from the input mode to other guided or radiation modes. This phenomenon is known as mode conversion. The mode conversion problem has been analysed using the step transition model [5]. We can picture the step transition model as a series of small abrupt transitions from the input to the output of the tapered waveguide.

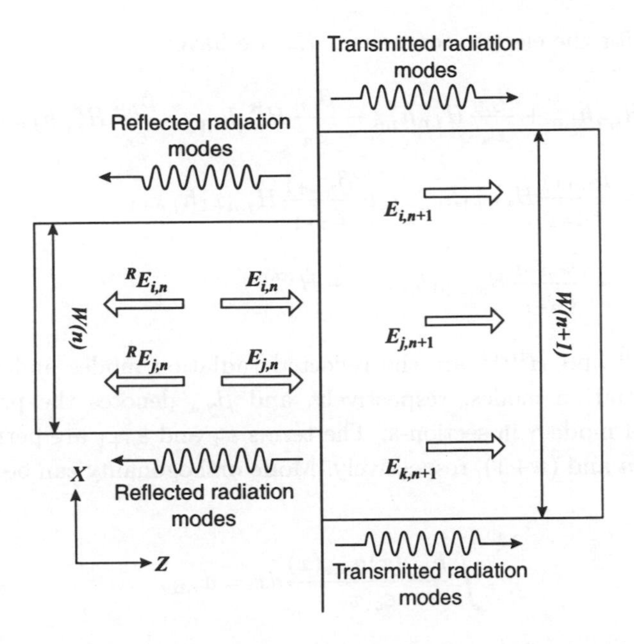

Fig. 16.4. The step transition model (only two adjacent sections shown).

A single abrupt transition is illustrated in Fig. 16.4. Assuming that there are initially two incident modes ($E_{i,n}$ and $E_{j,n}$), this may result in three transmitted modes ($E_{i,n+1}$, $E_{j,n+1}$ and $E_{k,n+1}$) across the step transition, and two reflected modes (${}^{\mathrm{R}}E_{i,n}$ and ${}^{\mathrm{R}}E_{j,n}$) (see Fig. 16.4). Generally, as the input signal (TE_0 mode) propagates along the taper, power will be transferred to higher order modes. In the perfectly symmetric structure, only even guided modes are excited as width increases. The guided modes of the magnetic fields are given by:

$$H_y = H_{i,n} h_{i,n} \tag{16.9}$$

where $H_{i,n}$ is the complex field amplitude of mode-i at section-n, and $h_{i,n}$ is the corresponding x-dependent field distribution, which was discussed in Section 16.4. Boundary conditions for TE modes (electric field polarised along x-direction) require E_x and H_y to be continuous across the step transition [28]. For the magnetic field component, H_y, we have:

$$\begin{aligned} &H_{i,n}h_{i,n} + H_{j,n}h_{j,n} + H_{i,n}^{\mathrm{R}}h_{i,n} + H_{j,n}^{\mathrm{R}}h_{j,n} + H^{\mathrm{refl}} \\ &\quad = H_{i,n+1}h_{i,n+1} + H_{j,n+1}h_{j,n+1} + H_{k,n+1}h_{k,n+1} + H^{\mathrm{trans}}\,. \end{aligned} \tag{16.10}$$

Similarly, for the electric component E_x, we have:

$$\frac{\beta_{i,n}}{\varepsilon_n} H_{i,n} h_{i,n} + \frac{\beta_{j,n}}{\varepsilon_n} H_{j,n} h_{j,n} - \frac{\beta_{i,n}}{\varepsilon_n} H^{\mathrm{R}}_{i,n} h_{i,n} - \frac{\beta_{j,n}}{\varepsilon_n} H^{\mathrm{R}}_{j,n} h_{j,n} + H^{\mathrm{refl}}$$
$$= \frac{\beta_{i,n+1}}{\varepsilon_{n+1}} H_{i,n+1} h_{i,n+1} + \frac{\beta_{j,n+1}}{\varepsilon_{n+1}} H_{j,n+1} h_{j,n+1}$$
$$+ \frac{\beta_{k,n+1}}{\varepsilon_{n+1}} H_{k,n+1} h_{k,n+1} + H^{\mathrm{trans}} \tag{16.11}$$

where H^{refl} and H^{trans} are the reflected radiation modes and the transmitted radiation modes, respectively, and $\beta_{\gamma,\mathrm{s}}$ denotes the propagation constant of mode-γ in section-s. The terms ε_n and ε_{n+1} are permittivities at section n and $(n+1)$, respectively. Mode orthogonality can be expressed as [18]:

$$\int \frac{h_{\gamma,\mathrm{s}}(x) h_{\alpha,\mathrm{s}}(x)}{\varepsilon_{\mathrm{s}}(x)} dx = \delta_{\gamma,\alpha} \,. \tag{16.12}$$

Where $\delta_{\gamma,\alpha}$ is the Kronecker delta. Strictly speaking, when the refractive indices are complex quantities, i.e. with gain or loss, power orthogonality only holds to an approximation [29]. For simplicity, the passive waveguide is assumed for derivation of the mode coupling coefficients. Multiplication of Eq. (16.10) by $\int_{-\infty}^{\infty} (h_{i,n+1}/\varepsilon_{n+1}) dx$, and Eq. (16.11) by $\int_{-\infty}^{\infty} (h_{i,n+1}) dx$, then making use of mode orthogonality gives the approximate equation for $H^{\mathrm{R}}_{i,n}$ as:

$$H^{\mathrm{R}}_{i,n} \cong \frac{\beta_{i,n} I^0_{i,n;i,n+1} - \beta_{i,n+1} I^1_{i,n;i,n+1}}{\beta_{i,n} I^0_{i,n;i,n+1} + \beta_{i,n+1} I^1_{i,n;i,n+1}} H_{i,n} + \vartheta\{H_{j,n} I_{j,n;i,n+1}\} \tag{16.13}$$

where the last term on the right hand side (RHS) is the remainder term proportional to the cross-mode overlap integral, which is assumed to be small. The transmitted radiation modes, H^{trans}, are eliminated because of orthogonality while the reflected radiation modes are small and can be neglected. The overlap integrals are defined as:

$$I^0_{\gamma,\mathrm{s};\delta,\mathrm{q}} = \frac{\int_{-\infty}^{\infty} h_{\gamma,\mathrm{s}}(x) h_{\delta,\mathrm{q}}(x) dx}{\varepsilon_n(x)}$$
$$I^1_{\gamma,\mathrm{s};\delta,\mathrm{q}} = \frac{\int_{-\infty}^{\infty} h_{\gamma,\mathrm{s}}(x) h_{\delta,\mathrm{q}}(x) dx}{\varepsilon_{n+1}(x)} \,. \tag{16.14}$$

Strictly speaking, since the overlap integral extends to infinity at both sides of the axis of propagation, the dielectric permitivitty must be defined for the material (i) within the active layer width, and (ii) outside of it. The definitions are as follows:

$$\varepsilon_n(x) = \begin{cases} n_{\text{eq1}}^2 & \text{if } |x| \le W_n \\ n_{\text{eq2}}^2 & \text{if } |x| > W_n \end{cases}$$
$$\varepsilon_{n+1}(x) = \begin{cases} n_{\text{eq1}}^2 & \text{if } |x| \le W_{n+1} \\ n_{\text{eq2}}^2 & \text{if } |x| > W_{n+1}\,. \end{cases} \tag{16.15}$$

Next, Eqs. (16.10) and (16.11) are multiplied by $\int_{-\infty}^{\infty}(h_{j,n+1}/\varepsilon_{n+1})dx$ and $\int_{-\infty}^{\infty}(h_{j,n+1})dx$, respectively, and the reflected terms $H_{i,n}^{\text{R}}$ and $H_{j,n}^{\text{R}}$ are eliminated upon using Eq. (16.13) to give an expression defining the relationship between the incident and transmitted mode amplitudes only, i.e. $H_{i,n}$, $H_{j,n}$ (incident), and $H_{j,n+1}$ (transmitted). Finally, the relationship between the normalised complex mode amplitudes is found as:

$$A_{j,n+1} = c_{ij}A_{i,n} + c_{jj}A_{j,n} \tag{16.16}$$

where $A_{\gamma,\text{s}}$ are the normalised complex coefficients defined as the ratio of the actual γth mode amplitude, $H_{\gamma,\text{s}}$, to the γth mode amplitude that gives unity power, $H_{\gamma,\text{s}}^n$. That is:

$$A_{\gamma,\text{s}} = \frac{H_{\gamma,\text{s}}}{|H_{\gamma,\text{s}}^n|}\,. \tag{16.17}$$

The term c_{ij} is the coupling coefficient (CC), which couples the ith mode at section-n to the jth mode at section-$(n+1)$. Also the term c_{jj} is the CC which couples the jth mode at section-n to the jth mode at section-$(n+1)$. The coupling coefficient can be found from the following expression [5]:

$$c_{ij} = \frac{2\sqrt{\beta_{i,n}\beta_{j,n+1}}}{\beta_{j,n}I_{j,n;j,n+1}^0 + \beta_{j,n+1}I_{j,n;j,n+1}^1} \times \frac{\beta_{j,n}I_{i,n;i,n+1}^0 I_{j,n;j,n+1}^0 I_{i,n;j,n+1}^1 + \beta_{i,n+1}I_{i,n;i,n+1}^1 I_{i,n;j,n+1}^0 I_{j,n;j,n+1}^1}{\{\beta_{i,n}I_{i,n;i,n+1}^0 + \beta_{i,n+1}I_{i,n;i,n+1}^1\}\sqrt{I_{i,n;i,n}^0 I_{j,n+1;j,n+1}^1}}\,. \tag{16.18}$$

If the total number of guided modes in section-$(n+1)$ exceeds that of section-n, then there is additional coupling from the ith mode to the

kth mode (Fig. 16.4), where $k > M_n$ (total number of guided modes in section-n). In a similar manner, the corresponding coupling coefficient, c_{ik}, can be found by the multiplying Eqs. (16.10) and (16.11) by $\int_{-\infty}^{\infty}(h_{k,n+1}/\varepsilon_{n+1})dx$ and $\int_{-\infty}^{\infty}(h_{k,n+1})dx$, respectively, and by using Eqs. (16.13) to eliminate the reflected guided modes. Hence, the coupling coefficient, c_{ik}, is found to be:

$$c_{ik} = \frac{2\sqrt{\beta_{i,n}\beta_{k,n+1}}}{\beta_{i,n}I^0_{i,n;i,n+1} + \beta_{i,n+1}I^1_{i,n;i,n+1}} \frac{I^0_{i,n;i,n+1}I^1_{i,n;k,n+1}}{\sqrt{I^0_{i,n;i,n}I^1_{k,n+1;k,n+1}}} . \tag{16.19}$$

By applying the duality principle, the coupling coefficients for the TM polarisation can also be obtained from the same procedures as explained above. The necessary substitutions are:

$$H \Rightarrow E$$

$$\varepsilon_n \Rightarrow \mu_n$$

$$\varepsilon_{n+1} \Rightarrow \mu_{n+1} .$$

Before solving the coupling coefficients at each step transition, we first need to find the total number of guided lateral modes, M_n, that exists in each local section. When M_n of section-n is equal to that of section-$(n+1)$, then only the coupling coefficients between the input ith mode $(1 \ldots m_n)$ and the transmitted jth mode $(1 \ldots M_n)$ are used. If the cutoff width of the next higher order transmitted mode has been exceeded, then a new mode (kth) is excited (see Fig. 16.4). In this case, the coupling coefficient c_{jk} must also be used to take into account the coupling between the input ith mode $(1 \ldots M_n)$ and the new transmitted kth mode $(M_n + 1)$. The guided lateral modes in a tapered structure must satisfy the following condition:

$$k_0 n_{\mathrm{eq2}} < \beta_{\gamma,\mathrm{s}} < k_0 n_{\mathrm{eq1}} . \tag{16.20}$$

For TE modes (predominantly polarised normal to the $x \pm W/2$ boundaries), the eigenvalue equation in each section is given by:

$$\tan\left(\frac{\kappa W(n)}{2}\right) = \frac{\sqrt{(n^2_{\mathrm{eq1}} - n^2_{\mathrm{eq2}})k_0^2 - \kappa^2}}{\kappa}\left(\frac{n^2_{\mathrm{eq1}}}{n^2_{\mathrm{eq2}}}\right) \tag{16.21}$$

where $W(n)$ is the active layer width of section-n. The numerical bisection method was applied to find all values of κ, and consequently $\beta_{\gamma,\mathrm{s}} = n_{\mathrm{eff}}k_0$ with error less than 10^{-12}. The total number of local guided modes

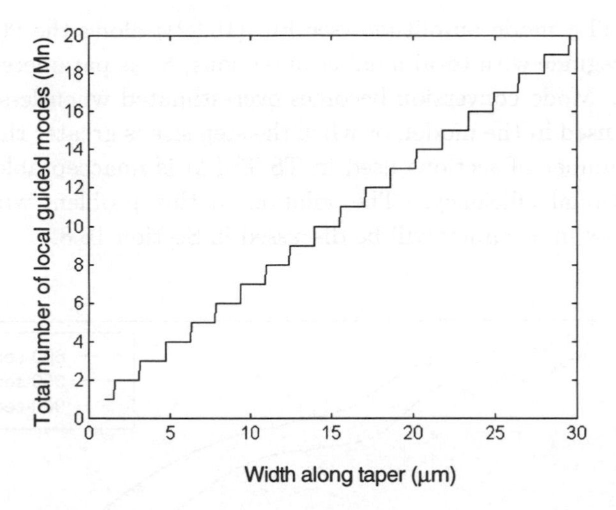

Fig. 16.5. The total number of guided modes in each section of the laser model.

along the taper found by using the procedures described above is shown in Fig. 16.5. Finally, the normalised mode amplitude, $A_{\gamma,\mathrm{s}}$, is calculated by using the following expression:

$$A_{j,n+1} = \sum_i c_{ij} A_{i,n} \exp(-j\{\beta_{i,n} - \beta_{j,n+1}\}\Delta z) \tag{16.22}$$

where Δz is the model's unit section length. The TE_0 mode coupling factor, K_c, is defined as:

$$K_c = \frac{E_{n+1}}{E_n} \tag{16.23}$$

where E_n and E_{n+1} denotes TE_0 field amplitude of the nth and $(n+1)$th sections, respectively. If output facet reflectivity is high, as in laser oscillators or Fabry Perot amplifiers (FPA's), the optical waves will be reflected from the wide output end, and propagate backward toward the narrow input end. In the backward direction of propagation, there is no mode conversion of waves as all higher order modes are radiated away owing to spatial filtering (i.e. reduction of cutoff width), and only the fundamental mode remains when the input end is reached. Hence, the coupling factor, K_c, is only required for the forward quasi-TE_0 propagating waves.

Provided that the tapering profile is adiabatic, optical power will be conserved during propagation. We have assumed quasi-adiabatic propagation, giving total radiation loss of less than 3%. The variation of the normalised TE_0 mode amplitude (see Eq. (16.22)) along the 900μm-long tapered waveguide with total number of sections, S, as parameter is shown in Fig. 16.6. Mode conversion becomes overestimated when less than 600 sections are used in the model, or when the step size is greater than 1.5μm. The large number of sections used in TS-TLLM is unacceptable in terms of computational efficiency. The solution to this problem without any significant loss in accuracy will be discussed in Section 16.8.

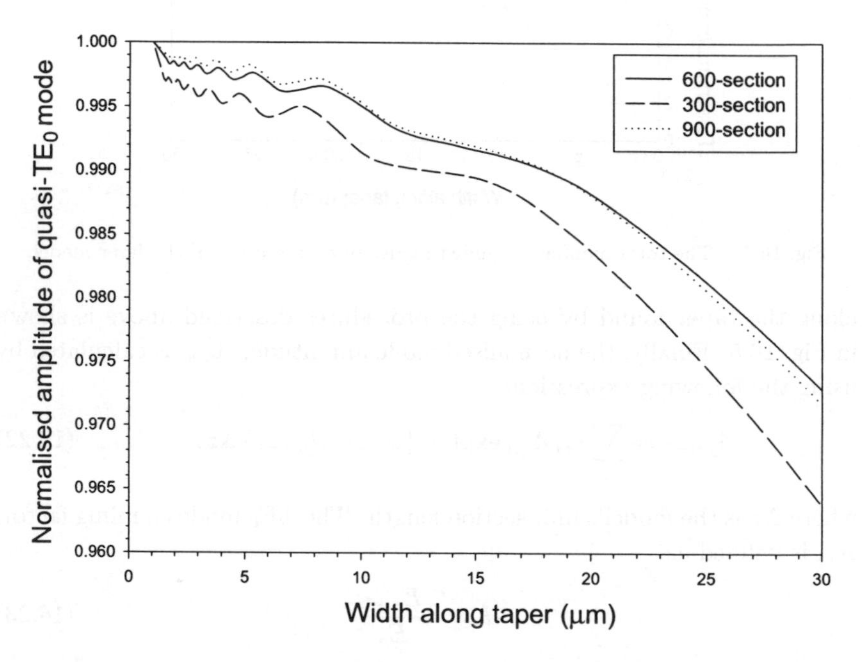

Fig. 16.6. The normalised quasi-TE_0 mode amplitude along the tapered structure.

16.6 Beam-Spreading Factor

By assuming that the tapered waveguide is index-guided, and based upon the principle of conservation of energy, the differential equation that describes the propagation of spatially averaged optical intensity, $I(z,t)$ is given as [5]:

$$\frac{\partial I(z,t)}{\partial z} = \{\Gamma[g(z,t) - a\eta(z)N_1(z,t)] - \alpha_{\text{sc}} \pm \alpha_{\text{tap}}(z)\}I(z,t) \qquad (16.24)$$

where $g(z,t)$ is the intensity gain, Γ is the optical confinement factor, a is the differential gain constant, $N_1(z,t)$ is the carrier density deviation from average value, $\eta(z)$ is the lateral hole-burning term (see Section 16.9 later), and α_{sc} is the conventional loss factor. The additional term α_{tap} for tapered structures takes into account the focusing/spreading of the optical intensity owing to the taper profile. This will be referred to as the *beam-spreading factor*. Notice that $\alpha_{tap}(z)$ can be positive (+) or negative (−) depending on the direction of propagation. According to power conservation, optical intensity increases toward the narrower end but decreases when approaching the wider end in the tapered waveguide. Therefore, α_{tap} acts as a gain-like coefficient (+) toward the narrower end while it becomes a loss-like coefficient (−) toward the wider end. Bendelli *et al.* [5] only considered single-pass propagation toward the wide output end in their model of a perfect TWA. We have generalised α_{tap} to take into account reflections at the facets, and hence also counter-propagating waves in FPAs or laser oscillators. The expression for the beam-spreading factor is given by [5]:

$$\alpha_{tap}(z) = -W(z)\frac{d}{dz}\left(\frac{1}{W(z)}\right) \tag{16.25}$$

where $W(z)$ is the position-dependent width of the active layer. In a single-pass amplification towards to the wider output end, the effect of the beam-spreading factor, α_{tap}, is to reduce the optical intensity along the tapering profile. In this case, the α_{tap} behaves like a loss coefficient (−). In effect, gain saturation can be reduced even as the light is amplified by the active semiconductor material. The intensity dependence of the gain coefficient, g, can be described by [30]:

$$g = \frac{g_o}{1 + \frac{I(z)}{I_{sat}}} \tag{16.26}$$

where g_o is the unsaturated value of the gain coefficient, and I_{sat} is the saturation intensity. Since the beam-spreading effect reduces the optical intensity for increasing width along the taper geometry, the gain becomes saturated at a slower rate. In addition, the effective saturation intensity for the tapered amplifier has been found to be position-dependent as well, increasing toward the wider end, thus allowing higher saturation output power [31, 32]. The impact of the beam-spreading factor is dependent on the tapering profile. The position-dependent widths, $W(z)$, and the corresponding beam-spreading factor, $\alpha_{tap}(z)$, of two simple taper geometries are given in the following [33]:

Linear taper:

$$\alpha_{\rm tap}(z) = \frac{W_{\rm o} - W_{\rm i}}{W(z)L} \tag{16.27}$$

where

$$W(z) = W_{\rm i} + \frac{(W_{\rm o} - W_{\rm i})z}{L}$$

Exponential taper:

$$\alpha_{\rm tap}(z) = \alpha_{\rm e} = \frac{1}{L}\ln\left(\frac{W_{\rm o}}{W_{\rm i}}\right) \tag{16.28}$$

where

$$W(z) = W_i \exp(\alpha_{\rm e} z)\,.$$

In the linear taper, the intensity initially drops rapidly, and then begins to increase as the wider output end is reached. This is because the beam-spreading effect is strongest (i.e. largest value of $\alpha_{\rm tap}$) near to the input, and is weakest at the output. As for the exponential taper, the intensity increases all the way from input to output because the beam spreading effect is the same throughout (i.e. $\alpha_{\rm tap}$ is constant).

16.7 Spectrally-Dependent Gain

To model the gain spectrum, a bandpass filter is placed in each and every scattering node, in both forward and backward propagation directions [34]. The basic construction of the transmission-line matrix (TLM) stub-filter model was explained in Chapter 3. The Q-factor of the filter is determined by the admittances, i.e. capacitive reactance, $Y_{\rm C}$, inductive reactance, $Y_{\rm L}$, and total admittance, Y, of the RLC stub-filter (see Fig. 14.10 in Chapter 14). The RLC stub-filters model the frequency dependence of laser gain, which includes a phase-shift response. These are the Lorentzian lineshape and phase-shift parts of the semiconductor material response assuming a homogeneously broadened two-level system [35].

In reality, there is an asymmetrical spectral dependence of the gain spectrum (which may include spectral hole-burning), which may be modelled by using carefully designed discrete-time digital filters [36]. However, for the sake of computational efficiency, only the simple RLC stub-filter is considered here, which provides a symmetrical spectral dependence. The carrier dependence of refractive index is modelled separately by incorporating localised *phase (adjusting) stubs* (see Section 16.11 later) into the transmission-line model.

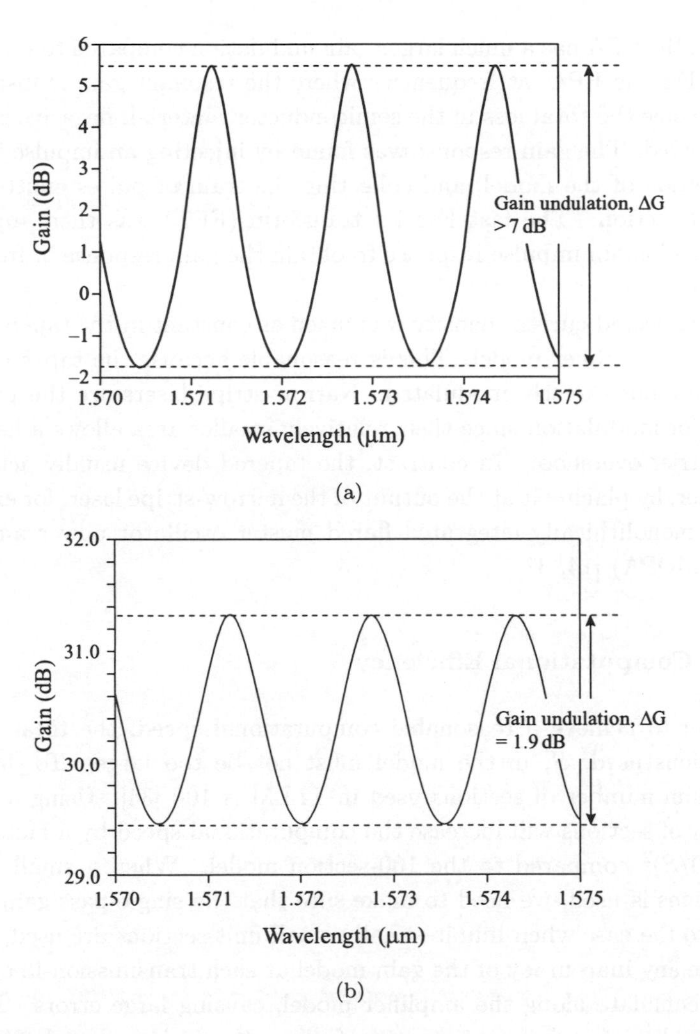

Fig. 16.7. Existence of ripples in the gain spectrum of (a) FPA and (b) TWA owing to finite facet reflectivities.

The gain model was assumed to have a Q-factor of 40, corresponding to a bandwidth of several THz [21]. The expected ripples on the gain spectrum due to residual reflections could be observed in the TLLM simulations because there were voltage pulses recirculating in the bidirectional transmission-line model. The simulation results from TLLM showing the existence of ripples in the gain spectra for the uniform-width FPA ($R = 30\%$ and TWA ($R = 0.01\%$) are shown in Figs. 16.7(a) and (b), respectively.

Clearly, the FPA has a much larger gain undulation compared to that of the TWA. For the FPA, at frequencies where the resonant gain is insufficient to overcome the total loss in the semiconductor material, negative gain can be observed. The gain response was found by injecting an impulse into the first section of the model, and collecting the train of pulses emitted from the last section. The fast Fourier transform (FFT) was then applied to the time-domain impulse response to obtain the gain response in frequency domain.

The injected current density was taken as constant in the tapered semiconductor amplifier model. This is reasonable because the tapered device is usually not directly modulated. Narrow-stripe lasers are the preferred choice for modulation since their relatively smaller area allows a large and fast carrier overshoot. In contrast, the tapered device usually acts as an amplifier, by placing it at the output of the narrow-stripe laser, for example, in the monolithically-integrated flared master oscillator power amplifiers (MFA-MOPA) [14, 37].

16.8 Computational Efficiency

In order to achieve a reasonable computational speed, the total number of sections used, S, in the model must not be too large. To date, the maximum number of sections used in TLLM is 100 [34]. Using a smaller number of sections will increase the computational speed by a factor of up to $(100/S)^2$ compared to the 100-section model. When a small number of sections is used, we need to make sure that the single-pass gain will be equal to the case when infinitesimally small unit-sections are used. This is because any inaccuracy of the gain model at each transmission-line section will accumulate along the amplifier model, causing large errors. To solve this problem, a gain compensating factor, K_{g}, is introduced [38]. The general expression for the Taylor-approximated gain is given as [34]:

$$g(n) = K_{\mathrm{g}}\Gamma\frac{\Delta L}{2}a(N_{\mathrm{av}}(n) - N_{\mathrm{tr}})\left(\frac{1}{1+\varepsilon S(n)}\right) \tag{16.29}$$

where $N_{\mathrm{av}}(n)$ is the average carrier density at the nth-section, N_{tr} is the transparency carrier density, ε is the gain compression factor, and $S(n)$ is the section-dependent photon density. The K_{g} term on the RHS of Eq. (16.29) is the gain compensating factor used to compensate for inaccuracies introduced in the Taylor-approximated gain model owing to

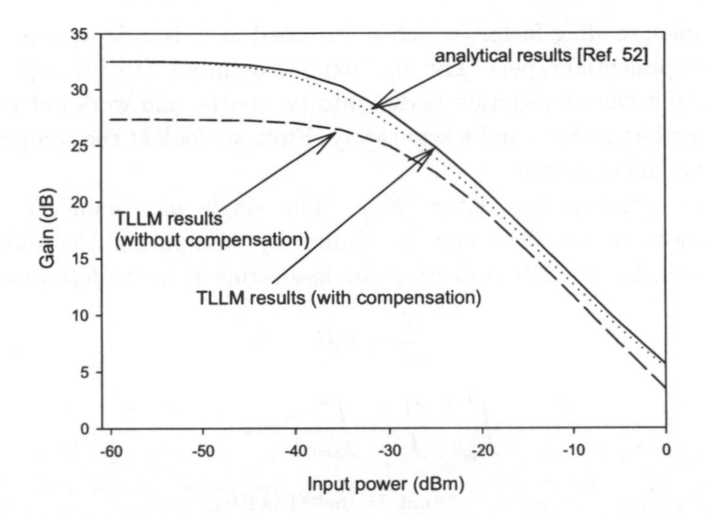

Fig. 16.8. Using the gain compensating factor, K_g, to increase the accuracy of the Taylor-approximated gain model.

a small number of sections used. As an example, the gain characteristics for a 200μm-long uniform width TWA is shown in Fig. 16.8. Eleven sections were used for a transmission-line amplifier model with spatial gain per unit inversion of 5×10^{-20} m^2 (which accounts for the high unsaturated gain). Improvement in the accuracy of the gain saturation characteristics can be observed when the gain compensating factor is used. In the tapered structure model (TS-TLLM), there are several position-dependent variables that complicate the problem. It greatly simplifies matters if we assume no lateral hole-burning but the beam-spreading factor is still considered. The general differential equation that describes the propagation and amplification of optical intensity in the tapered semiconductor amplifier is:

$$\frac{dI}{dz} = \{a - b\}I$$
$$\int \frac{dI}{I} = \int (a - b)dz \tag{16.30}$$
$$I_{n\Delta z} = I_{(n-1)\Delta z} \times \exp\left(\int adz\right) \times \exp\left(-\int bdz\right)$$

where a is the gain term, and b is the loss term, including the beam-spreading effect. By assuming that $N_1 = 0$ during small-signal amplification, a will now be independent of position, z (see Eq. (16.24). In general, the beam-spreading factor, which is included in b, is z-dependent (except for the exponential taper). The final expression in Eq. (16.30) suggests that we can split the propagation kernel into two parts, and work out the compensating factors for a and b separately. First, we look at the compensating factor for the gain term.

Gain compensating factor (K_g): The single pass gain, G_s, of the travelling-wave amplifier can be found by integrating the differential equation in Eq. (16.30) neglecting the loss terms as in the following:

$$\begin{aligned} \frac{dI}{dz} &= \Gamma g I \\ \int_{I_{in}}^{I_{out}} \frac{dI}{I} &= \int_{z=0}^{L} \Gamma g dz \\ I_{out} &= I_{in} \exp(\Gamma g L) \,. \end{aligned} \tag{16.31}$$

In the unsaturated condition, we can equate the analytical expression of single-pass gain, G_s, with the overall power gain of an S-section TLLM model:

$$G_s = \exp(\Gamma g L) = \left(1 + K_g \frac{\Gamma g \Delta L}{2}\right)^{2S} . \tag{16.32}$$

The gain compensating factor, K_g, may then be expressed as:

$$K_g = \frac{2}{\Gamma g \Delta L}\left[\exp\left(\frac{\Gamma g L}{2S}\right) - 1\right] . \tag{16.33}$$

For small-signal amplification, we can make the following approximation for the gain coefficient:

$$g = a\tau_n \left(\frac{J}{qd} - \frac{N_{tr}}{\tau_n}\right) \tag{16.34}$$

where J is the current density, d is the active layer depth, q is the electronic charge, N_{tr} is the transparency carrier density, and τ_n is the carrier lifetime. In general, the carrier lifetime is a function of average carrier density, N_{av}, in the form of [21]:

$$\tau_n(N_{av}) = \frac{1}{A_{nr} + B_{rad} N_{av} + C_{aug} N_{av}^2} \tag{16.35}$$

where A_{nr} is the non-radiative recombination coefficient, B_{rad} is the radiative recombination coefficient, and C_{aug} is the Auger recombination coefficient, which is important in long-wavelength semiconductor lasers

(1.3–1.6μm). Therefore, we first need to find N_{av} by solving the rate equations at steady-state, i.e. $dN_{\text{av}}/dt = 0$. In the small-signal regime, there is negligible lateral hole-burning ($N_1 = 0$). The following cubic equation must then be solved:

$$N_{\text{av}}^3 + \frac{B_{\text{rad}}}{C_{\text{aug}}} N_{\text{av}}^2 + \frac{A_{\text{nr}}}{C_{\text{aug}}} N_{\text{av}} - \frac{J}{qdC_{\text{aug}}} = 0\,. \tag{16.36}$$

The desired root (N_{av}) may be found from [39]:

$$N_{\text{av}} = (s_1 + s_2) - a_2/3 \tag{16.37}$$

where

$$s_1 = \sqrt[3]{r + \sqrt{(q^3 + r^2)}}\,, \tag{16.38a}$$

$$s_2 = \sqrt[3]{r - \sqrt{(q^3 + r^2)}} \tag{16.38b}$$

where

$$q = a_1/3 - a_2^2/9\,, \tag{16.38c}$$

$$r = (a_1 a_2 - 3a_0)/6 - a_2^3/27\,, \tag{16.38d}$$

and

$$a_0 = -J/qdC_{\text{aug}}\,, \quad a_1 = A_{\text{nr}}/C_{\text{aug}}\,, \quad a_2 = B_{\text{rad}}/C_{\text{aug}} \tag{16.38e}$$

From the parameter values given in Table 16.2, the resulting values required to find the gain compensating factor (K_{g}) in Eq. (16.33) are $N_{\text{av}} = 3.541 \times 10^{24}$ m^{-3}, $\tau_n = 0.7$ ns and $g = 4.1 \times 10^4$ m^{-1}. Typical values of the compensating factor, K_{g}, with the total number of sections, S, as parameter are given in Table 16.3. The gain saturation characteristics of a 24-section and 300-section travelling-wave amplifier model (TS-TLLM) are

Table 16.2. Parameter values used in the calculation of K_{g} for TS-TLLM.

Length	$L = 900$ μm
Confinement factor	$\Gamma = 0.2$
Thickness	$d = 0.1$ μm
Nonradiative recombination coefficient	$A_{\text{nr}} = 0$ s^{-1}
Radiative recombination coefficient	$B_{\text{rad}} = 8.0 \times 10^{-17}$ m^{3}s^{-1}
Auger recombination coefficient	$C_{\text{aug}} = 9.0 \times 10^{-41}$ m^{6}s^{-1}
Current density	$J = 8.0 \times 10^7$ Am^{-2}

Table 16.3. Values of S and K_g used in the analysis.

Number of sections, S	Compensating factor, K_g
24	1.081
300	1.006
1000	1.002

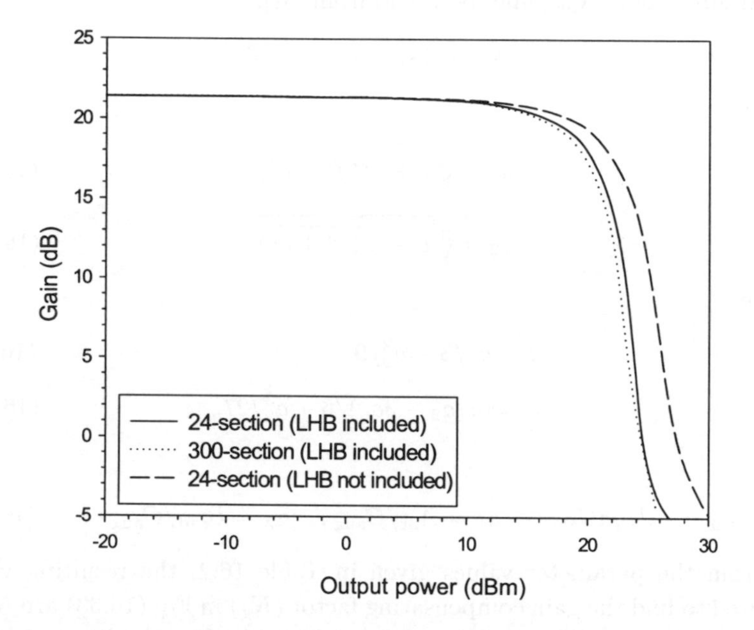

Fig. 16.9. Gain saturation characteristics for (a) 24-section model with and without lateral hole-burning (LHB) and (b) 300-section model with LHB.

then plotted in Fig. 16.9 for comparison. From Fig. 16.9, there is a good agreement between the results for the 24-section and 300-section model by using an appropriate gain compensating factor, K_g, even when the lateral hole-burning (LHB) (see Section 16.9) effect is included. When the LHB effect is neglected, there is an improvement in the saturation output power. This shows that LHB enhances gain saturation at high intensities. The LHB effect has been included in the TS-TLLM, and will be discussed in Section 16.9.

Figure 16.10 shows the output power against input power characteristics of the tapered amplifier models. The amplifier gain is highest during small-signal amplification (at low input signal intensities) with unsaturated

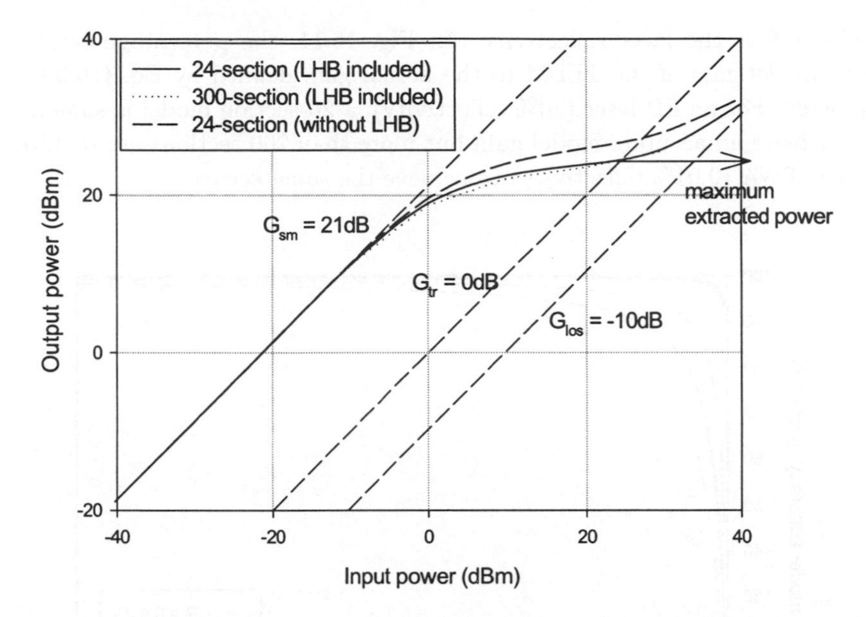

Fig. 16.10. Output power versus input power characteristics for (a) 24-section model with and without lateral hole-burning (LHB) and (b) 300-section model with LHB.

gain, G_{sm}, of 21 dB. At high enough input intensities, the gain saturates down toward the limiting value of unity ($G_{\mathrm{tr}} = 0$ dB) if we neglect loss. This means that the amplifier becomes essentially transparent. In practice, there are losses due to scattering and absorption in the laser material, which reduces the optical intensity along the amplifier, giving an overall loss ($G_{\mathrm{los}} = -10$ dB in the case of Fig. 16.10). The maximum *extracted power* from the amplifier is also shown in Fig. 16.10. The extracted power is defined as the output power minus (in linear scale) the input power, i.e. the net power supplied to the optical radiation by the amplifying medium [35].

Facet reflectivity and gain accuracy: The accuracy of the Taylor-approximated gain Eq. (16.29) also depends on the facet reflectivity used. If high reflectivities are used, for example in an FP laser, the model gain is reasonably accurate even if a small number of sections is used. If a small reflectivity is used, for example in a travelling-wave amplifier, the model gain is very sensitive to the number of sections used. The actual gain required to overcome the mirror loss is given as [21]:

$$g = -\frac{1}{L}\ln(R) \tag{16.39}$$

where R is the facet reflectivity. In Fig. 16.11, the percentage ratio of the model gain of the TLLM to the actual gain defined by Eq. (16.39) is plotted. For an FP laser (30% reflectivity), a 20-section model is sufficient to achieve an accurate model gain but more than 200 sections are required for a TWA (0.01% reflectivity) to achieve the same accuracy.

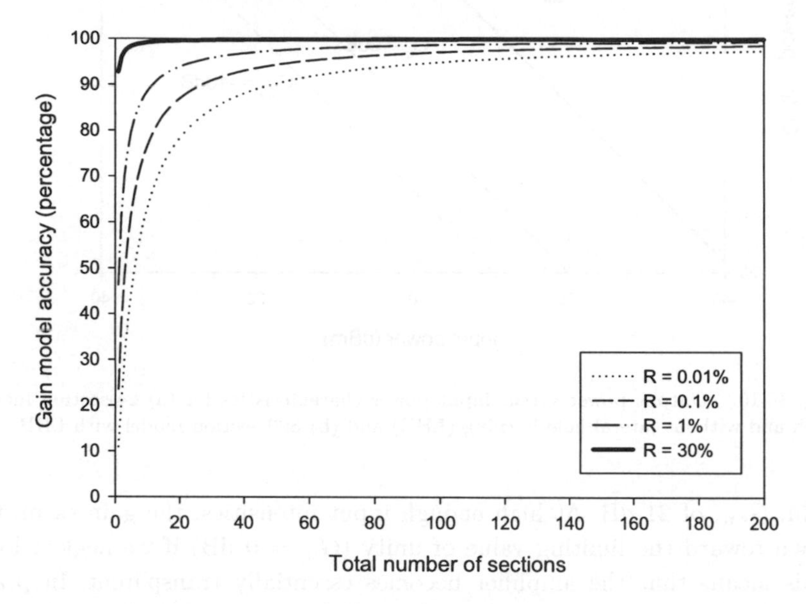

Fig. 16.11. Dependence of gain accuracy on the total number of sections with facet reflectivity as parameter.

Beam-spreading factor (α_{tap}) — *is it a required compensating factor?* We have seen that a compensating factor is necessary for the Taylor-approximated gain model when a small number of sections is used in TS-TLLM. Interestingly, it has been found that no such compensating factor is required for the beam-spreading factor, α_{tap}, regardless of the total number of sections used in TS-TLLM, as we will now prove.

The principle of energy conservation requires that the total power carried by the optical radiation be constant as it propagates along the lossless passive tapered waveguide. This means that the spatially averaged optical intensity drops as width gradually increases. This attenuation effect of optical intensity is described by α_{tap}, which was discussed in Section 16.6. Neglecting material gain and any other loss effects, the propagation of optical intensity may be expressed as:

$$\frac{dI}{dz} = -\alpha_{\text{tap}} I \,(\text{towards wider end})\,. \tag{16.40}$$

To find the relationship between the input intensity and output intensity for a *unit-section*, the following integration is performed,

$$\int_{I_0}^{I_{\Delta z}} \frac{1}{I} dI = -\int_0^{\Delta z} \alpha_{\text{tap}}(z) dz \tag{16.41}$$

where Δz is the step-size of the unit-section, and I_0 is the input intensity. By substituting Eq. (16.27) for the linear taper into Eq. (16.41), the integration can be worked out as the following:

$$\begin{aligned}
\textit{Linear taper}: \quad \ln\left(\frac{I_{n\Delta z}}{I_{(n-1)\Delta z}}\right) x &= -\int \frac{(W_\text{o} - W_\text{i})}{LW_\text{i} + (W_\text{o} - W_\text{i})z} dz \\
&= -\int \frac{1}{\frac{LW_\text{i}}{(W_\text{o} - W_\text{i})} + z} dz \\
&= -\left[\ln\left(z + \frac{LW_\text{i}}{(W_\text{o} - W_\text{i})}\right)\right]_{(n-1)\Delta z}^{n\Delta z} \\
&= \ln \frac{(n-1)\Delta z + \frac{LW_\text{i}}{(W_\text{o} - W_\text{i})}}{n\Delta z + \frac{LW_\text{i}}{(W_\text{o} - W_\text{i})}}\,.
\end{aligned} \tag{16.42}$$

The *space-discretised* relationship of input intensity and output intensity for each unit section from section-1 to section-S is found to be:

$$\begin{aligned}
\textit{Linear taper}: \quad I_{\Delta z} &= I_0 \frac{c}{\Delta z + c} \\
I_{2\Delta z} &= I_{\Delta z} \frac{\Delta z + c}{2\Delta z + c} \qquad \text{where} \\
I_{3\Delta z} &= I_{2\Delta z} \frac{2\Delta z + c}{3\Delta z + c} \\
&\vdots \\
c &= \frac{LW_\text{i}}{(W_\text{o} - W_\text{i})}\,.
\end{aligned} \tag{16.43}$$

By cascading all the individual sections together, the overall output intensity at the amplifier output can be found as the following:

Space-discretised model:

$$I_{\mathrm{L}} = I_0 \cdot \left(\frac{c}{\Delta z + c}\right)\left(\frac{\Delta z + c}{2\Delta z + c}\right)\left(\frac{2\Delta z + c}{3\Delta z + c}\right)\cdots\left(\frac{(S-1)\Delta z + c}{S\Delta z + c}\right)$$

$$= I_0 \cdot \frac{1}{1 + \frac{(W_{\mathrm{o}} - W_{\mathrm{i}})}{W_{\mathrm{i}}}}. \tag{16.44}$$

The analytical expression of output field amplitude may be found by integrating along the total length of the amplifier, L, giving the following result:

$$\textit{Analytical}: \quad I_L = I_0 \cdot \exp\left[-\ln\left(z + \frac{LW_{\mathrm{i}}}{(W_{\mathrm{o}} - W_{\mathrm{i}})}\right)\right]_0^L$$

$$= I_0 \cdot \frac{1}{1 + \frac{(W_{\mathrm{o}} - W_{\mathrm{i}})}{W_{\mathrm{i}}}}. \tag{16.45}$$

It can be seen that Eq. (16.44) and Eq. (16.45) are exactly equal to each other. This proves that the beam-spreading effect along the linear TS-TLLM is exact, regardless of the unit-section step-size (Δz) used in the model. The beam-spreading effect in the exponential TS-TLLM can also be analysed using a similar procedure. Substitution of Eq. (16.29) of the exponential taper into Eq. (16.41) leads to the following results:

Exponential taper:

$$\ln\left(\frac{I_{n\Delta z}}{I_{(n-1)\Delta z}}\right) = -\int \frac{1}{L}\ln\frac{W_{\mathrm{o}}}{W_{\mathrm{i}}}dz = -\frac{1}{L}\ln\frac{W_{\mathrm{o}}}{W_{\mathrm{i}}}\int dz$$

$$= -\frac{1}{L}\ln\frac{W_{\mathrm{o}}}{W_{\mathrm{i}}}[z]_{(n-1)\Delta z}^{n\Delta z}$$

$$= -\frac{\Delta z}{L}\ln\frac{W_{\mathrm{o}}}{W_{\mathrm{i}}}. \tag{16.46}$$

In contrast to the linear taper, the relationship between output and input intensity for each and every individual section from section-1 to section-S is the same, given by:

$$\textit{Exponential taper}: \quad I_{n\Delta z} = I_{(n-1)\Delta z} \cdot \exp\left(-\frac{\Delta z}{L}\ln\frac{W_{\mathrm{o}}}{W_{\mathrm{i}}}\right). \tag{16.47}$$

Again, by cascading all the individual sections together, the output intensity for the exponential taper is found to be:

Space-discretised model:

$$I_{\mathrm{L}} = I_0 \cdot \left\{ \exp\left(-\frac{\Delta z}{L} \ln \frac{W_{\mathrm{o}}}{W_{\mathrm{o}}} \right) \right\}^S = I_0 \cdot \frac{W_{\mathrm{i}}}{W_{\mathrm{o}}} . \tag{16.48}$$

The analytical expression of the output intensity is found as:

$$\textit{Analytical}: \qquad I_{\mathrm{L}} = I_0 \cdot \exp\left(-\frac{1}{L} \ln \frac{W_{\mathrm{o}}}{W_{\mathrm{i}}} [z]_0^L \right) = I_0 \cdot \frac{W_{\mathrm{i}}}{W_{\mathrm{o}}} . \tag{16.49}$$

From Eq. (16.48) and Eq. (16.49), we can conclude that the beam-spreading effect for the exponential TS-TLLM is also exact, regardless of the unit-section length (Δz) used.

16.9 Lateral Hole-Burning (LHB)

The carrier rate equations used in the TS-TLLM model take into account lateral carrier hole-burning. The hole-burning effect was accounted for by keeping the first Fourier term of the lateral carrier distribution as [5, 40]:

$$N(x, z) = N_{\mathrm{av}}(z) - N_1(z) \cos\left(\frac{2\pi x}{W(z)} \right) \tag{16.50}$$

where $N_{\mathrm{av}}(z)$ is the average carrier density, and $N_1(z)$ is the carrier density deviation from the average value. The lateral carrier distribution model is shown in Fig. 16.12.

Therefore, both lateral and longitudinal carrier density distributions are considered in the tapered structure transmission-line laser model (TS-TLLM). Lateral hole-burning is an important consideration in broad-area and tapered laser structures as the amount of lateral carrier diffusion has an impact on the light-current characteristics and the output beam quality [41]. We have also seen that LHB has a significant impact on the gain saturation characteristics of the semiconductor laser amplifier from Figs. 16.10 and 16.12. Longitudinal inhomogeneous effects are important in laser amplifiers [42, 43], especially when the travelling-wave ($R = 0\%$) condition is approached.

The dynamic interplay between the photon density and carrier density is governed by two sets of carrier rate equation models (N_{av} and N_1), placed at each and every section of the TS-TLLM. First, we substitute

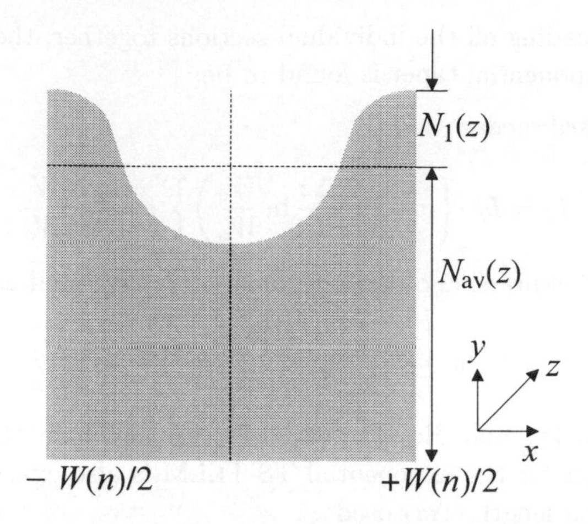

Fig. 16.12. Lateral carrier density distribution exhibiting lateral hole-burning.

Eq. (16.50) into the current continuity equation [21], and eliminate the x and y dependence by spatial averaging (Section 15.16 in Chapter 15). Then by discretising time and space in TS-TLLM the following expressions are obtained:

$$
\begin{aligned}
N_{\rm av}^{k+1} &= N_{\rm av}^{k}(n) + \Delta T\bigg\{\frac{I(n)}{qdW(n)\Delta L} - A_{\rm nr}N_{\rm av}^{k}(n) \\
&\quad - B_{\rm rad}\left[N_{\rm av}^{k}(n)^2 + \frac{(N_1^k(n))^2}{2}\right] \\
&\quad - C_{\rm aug}\left[(N_{\rm av}^{k}(n))^3 + \frac{3}{2}N_{\rm av}^{k}(n)(N_1^k(n))^2\right] \\
&\quad - \Gamma[a(N_{\rm av}^{k}(n) - N_0) - \eta(n)aN_1^k(n)]S(n)\frac{c_0}{n_{\rm g}}\bigg\}
\end{aligned}
\tag{16.51}
$$

$$
\begin{aligned}
N_1^{k+1}(n) &= N_1^k(n) + \Delta T\bigg\{ - D\left(\frac{2\pi}{W(n)}\right)^2 N_1^k(n) - A_{\rm nr}N_1^k(n) \\
&\quad - 2B_{\rm rad}N_{\rm av}^{k}(n)N_1^k(n) - C_{\rm aug}\left[2(N_{\rm av}^{k}(n))^2N_1^k(n) + \frac{3}{4}N_1^k(n)\right] \\
&\quad + 2\frac{c_o}{n_g}\Gamma[\eta(n)a(N_{\rm av}^{k}(n) - N_0) - a\sigma(n)N_1^k(n)]S(n)\bigg\}
\end{aligned}
\tag{16.52}
$$

where c_0 is the speed of light in free space, n_g is the effective group index, d is the thickness of the active layer, $I(n)$ is the total current into section-n, A_{nr} is the non-radiative recombination coefficient, B_{rad} is the radiative recombination coefficient, C_{aug} is the Auger recombination coefficient, and D is the effective lateral carrier diffusion constant. The terms $\eta(n)$ and $\sigma(n)$ are the section-dependent lateral hole-burning terms that resulted from spatial averaging of the carrier rate equations, and are expressed as [5]:

$$\eta(n) = \frac{\int_{-W(n)/2}^{W(n)/2} \cos\left(\frac{2\pi x}{W(n)}\right) e_{x,n}^2(x)dx}{\int_{-W(n)/2}^{W(n)/2} e_{x,n}^2(x)dx} \tag{16.53}$$

$$\sigma(n) = \frac{\int_{-W(n)/2}^{W(n)/2} \cos^2\left(\frac{2\pi x}{W(n)}\right) e_{x,n}^2(x)dx}{\int_{-W(n)/2}^{W(n)/2} e_{x,n}^2(x)dx} \tag{16.54}$$

where $e_{x,n}(x)$ is the lateral field distribution of section-n, which was discussed in Section 16.4. The current density is assumed constant throughout the tapered structure. However, total current into each section along the tapered structure is width-dependent. It is worth pointing out that do not depend on the taper shape but only on local width, given a set of parameters. Therefore, simple analytical expressions for $\eta(n)$ and $\sigma(n)$ can be found and used for arbitrary-shaped structures which assume lateral hole-burning in the form of Eq. (16.50):

$$\eta(n) = \exp\left(-0.6906 + \frac{-0.6164}{W(n)}\right) \tag{16.55}$$

$$\sigma(n) = \frac{1}{1.9997 + \frac{0.2932}{W(n)} + \frac{-0.2164}{W(n)^{1.5}}}\,. \tag{16.56}$$

The accuracy of the fitted curves is shown in Fig. 16.13. In order to increase computational efficiency, the analytical formula of Eqs. (16.55) and (16.56) have been incorporated into the TS-TLLM.

16.10 Spontaneous Emission Spectrum

The spontaneous emission noise model in the TLLM that was based on multiple current sources was first presented in [44]. In laser amplifiers, it is conventional to relate the amplified spontaneous emission (ASE) power in terms of gain. The validity of the amplified spontaneous emission (ASE)

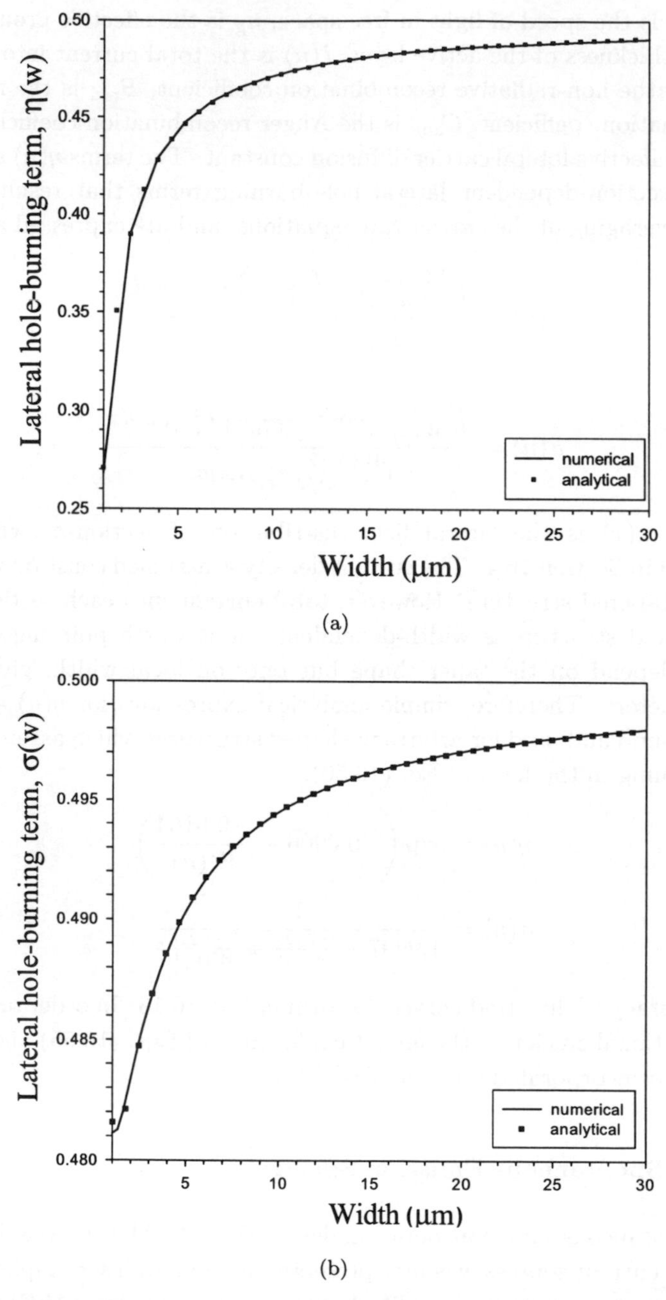

Fig. 16.13. The fitted curves of (a) the lateral hole-burning term, η, and (b) the lateral hole-burning term, σ, as a function of width.

model in TLLM for uniform-width FPAs and TWAs has been proven in [45]. Lowery's ASE model was based on Henry's expression for noise power derived from self-contained classical electromagnetic theory [46]. A similar expression was also derived using microwave circuit theory by Gordon [15]. The spontaneous emission noise power of an ideal TWA (P_{sp}) is expressed as:

$$P_{\mathrm{sp}} = \Delta f \cdot hf_0 \cdot n_{\mathrm{sp}} \cdot (G-1)\left(\frac{g}{g-\alpha_{\mathrm{sc}}}\right) \cdot m_{\mathrm{t}} \tag{16.57}$$

where Δf is the bandwidth of the noise, f_0 is the optical frequency, h is planck's constant, G is the single pass gain, and m_{t} is the total number of transverse modes. The ratio $(g-\alpha_{\mathrm{sc}})/g$ takes into account the stimulated emission not dissipated by material absorption and scattering loss, relative to the total stimulated emission. It is also known as the quantum efficiency, η. The population inversion factor, n_{sp}, is defined as [46]:

$$n_{\mathrm{sp}} = \frac{1}{\left[1-\exp\left(\frac{hf_0-qV}{kT}\right)\right]} \approx \frac{N}{(N-N_{\mathrm{tr}})} \tag{16.58}$$

where qV is the separation between quasi-Fermi levels. The condition $n_{\mathrm{sp}} = 1$ means that there is perfect population inversion, i.e. the ground state is occupied. For good long-wavelength semiconductor lasers, the value of n_{sp} is estimated to be around +2. If the overall gain is negative, n_{sp} changes sign. There is a smooth transition in the value of n_{sp} from below transparency level (loss) to above transparency level (gain) [15, 46].

Similar to other TLLM models, shunt current sources are placed at each and every section in both propagation directions in the tapered structure model (TS-TLLM) to model spontaneous emission. The current sources are modelled by pseudo-random number generators with Gaussian noise statistics. The TS-TLLM is divided into many smaller transmission-line sections, the noise output power of *each* section which satisfies Eq. (16.57). By comparing the noise output power emitted from each transmission-line section with the expression in Eq. (16.57), the mean square value of the noisy current source may be expressed by:

$$\langle i^2 \rangle = m_1 \cdot \frac{2hf_0 \cdot m^2}{W(n) \cdot d \cdot \Delta t \cdot Z_{\mathrm{p}}} \cdot n_{\mathrm{sp}} \cdot \left(\frac{g}{g-\alpha_{\mathrm{sc}}}\right) \cdot [\exp(g\Delta z) - \exp(\alpha_{\mathrm{sc}}\Delta z)] \tag{16.59}$$

where Z_{p} is the wave impedance, and m is a unity constant (dimension of length). In the noise current source, we must also consider the additional higher order transverse modes (in our case, lateral modes, m_1) [47].

White Gaussian noise was used in [45] but this was acceptable since the model bandwidth was only 300 GHz. Here, the model bandwidth is 3.875 THz so the spectral dependence of spontaneous emission must be taken into account. The power spectral response of spontaneous emission is modelled by passing white Gaussian noise with mean square value given by Eq. (16.59) through a TLM stub-filter with Q-factor of 15. The spectrum is approximately Lorentzian, and has a full width at half maximum (FWHM) that extends to over 10 THz. It is worth pointing out that correlated noise sources [1] are needed to accurately model the intensity fluctuation if the unit-section length, Δz, is significantly reduced or when biased well above threshold level.

16.11 Dynamic Wavelength Chirp

The phase-stub model used for the TS-TLLM here is an improved version over the one used in the stub-attenuator model (see Chapter 14) of the integrated TLLM model (see Chapter 15). Instead of placing the phase-stubs only at both ends of the integrated TLLM model, the phase-stubs are now placed in each and every section of the TS-TLLM [48]. This is illustrated in Fig. 16.14.

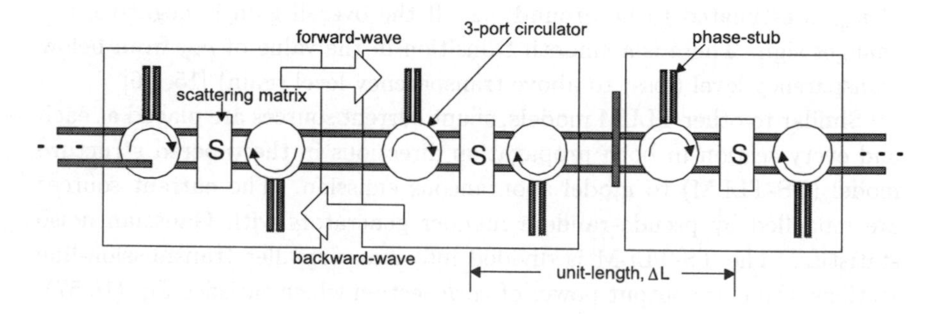

Fig. 16.14. Multiple phase-stub model.

The function of the phase-stubs is to delay the voltage pulses according to the local section's carrier density. Part of the delayed pulses from the current time-iteration (t_k) then continue to propagate in the original direction at the next time-iteration (t_{k+1}). The multiple phase-stubs along the laser cavity allow us to take into account the field amplitude-phase coupling effect described by the linewidth enhancement factor, α_{H}. The localised phase-stub impedance here may be derived in a similar way to that of the stub-attenuator model in Chapter 14 as [48]:

$$Z_S(n) = \begin{cases} -1/\left[\tan(\pi f_{dc}\Delta t)\tan\left(\dfrac{\beta(n)\Delta l_{ph}(n)}{2}\right)\right] \\ \qquad\qquad \text{for } \tan(\beta_z(n)l_{ph}(n)) < 0 \\ \tan(\pi f_{dc}\Delta t)/\tan\left(\dfrac{\beta(n)\Delta l_{ph}(n)}{2}\right) \\ \qquad\qquad \text{for } \tan(\beta_z(n)l_{ph}(n)) > 0 \end{cases} \tag{16.60}$$

where Δt is the time-step, and f_{dc} is the down-converted optical frequency (see Chapter 14 — baseband transformation). The term $\Delta l_{ph}(n)$ is the local section's phase-stub extension as given in Eq. (14.65) of Chapter 14. The modifications required here are to replace the spatially averaged carrier density (N'_{av}) with the local carrier density, $N(n)$, and the cavity length (L) by unit-section length, Δz, that is:

$$\Delta l_{ph} = \frac{\Gamma\Delta z}{n_g}\frac{dn}{dN}(N(n) - N_{ref}) = -\Gamma\left(\frac{\Delta z \alpha_H c_0 a}{4\pi f_0 n_g}\right)(N_{av}(n) - N_{ref}). \tag{16.61}$$

For the tapered structure, $\beta(n)$ in Eq. (16.60) is the propagation constant of the *local* section, n. It should be noted that when modelling chirp, the dynamic spectral shift should not exceed 15% of the model's bandwidth so as to minimise unwanted frequency components [49].

16.12 Scattering Matrix

All the important effects that are included in the TS-TLLM have been outlined from Section 16.4 to 16.11. The flow chart that describes the algorithm of the TS-TLLM is shown in Fig. 16.15. In this section, the scattering matrix that processes the incoming pulses (V^i) into outgoing pulses (V^r) will now be derived for the TS-TLLM. Two sets of transmission-lines are used in the model, one for forward propagating waves, the other for backward propagating waves (see Fig. 16.1). The TS-TLLM thus is able to model laser oscillators, Fabry–Perot amplifiers (FPAs), and take into account residual reflectivity when modelling travelling-wave amplifiers (TWAs).

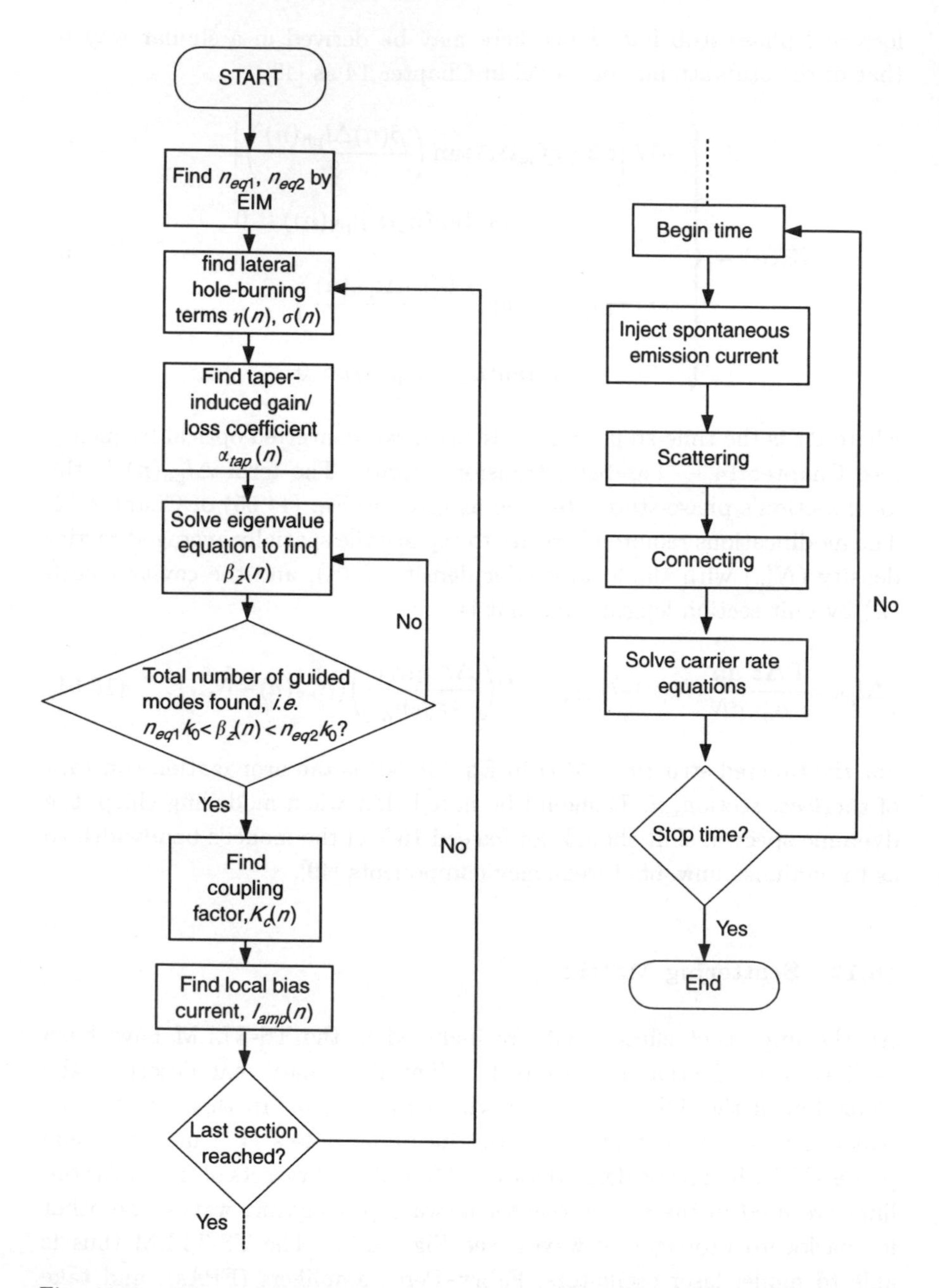

Fig. 16.15. The flowchart of the tapered structure transmission-line laser model (TS-TLLM).

The scattering matrix includes the frequency-dependent gain, attenuation, mode conversion, beam-spreading, and dynamic chirp. The continuous distance, z, is discretised, and replaced by its corresponding section, n. The complete scattering matrix may be derived using the microwave circuit concepts discussed in Chapter 14 as the following [50]:

$$\begin{bmatrix} V_{\mathrm{M}}(n) \\ V_{\mathrm{C}}(n) \\ V_{\mathrm{L}}(n) \\ V_{\mathrm{P}}(n) \end{bmatrix}_k^r = \frac{1}{Y(1+Z_{\mathrm{S}})} \times \begin{bmatrix} K_{\mathrm{C}}\alpha_{\mathrm{t}}(g+Y)(Z_{\mathrm{S}}-1) & 2K_{\mathrm{C}}\alpha_{\mathrm{t}}Y_{\mathrm{C}}(Z_{\mathrm{S}}-1) & 2K_{\mathrm{C}}\alpha_{\mathrm{t}}Y_{\mathrm{C}}(Z_{\mathrm{S}}-1) & 2Y \\ g(Z_{\mathrm{S}}+1) & (2Y_{\mathrm{C}}-Y)(Z_{\mathrm{S}}+1) & 2Y_{\mathrm{L}}(Z_{\mathrm{S}}+1) & 0 \\ g(Z_{\mathrm{S}}+1) & 2Y_{\mathrm{C}}(Z_{\mathrm{S}}+1) & (2Y_{\mathrm{L}}-Y)(Z_{\mathrm{S}}+1) & 0 \\ 2K_{\mathrm{C}}\alpha_{\mathrm{t}}(g+Y)Z_{\mathrm{S}} & 4K_{\mathrm{C}}\alpha_{\mathrm{t}}Y_{\mathrm{C}}Z_{\mathrm{S}} & 4K_{\mathrm{C}}\alpha_{\mathrm{t}}Y_{\mathrm{L}}Z_{\mathrm{S}} & Y(1-Z_{\mathrm{S}}) \end{bmatrix} \times \begin{bmatrix} V_{\mathrm{M}}(n) \\ V_{\mathrm{C}}(n) \\ V_{\mathrm{L}}(n) \\ V_{\mathrm{P}}(n) \end{bmatrix}_k^i . \tag{16.62}$$

In Eq. (16.62), $V_x(n)$ (where $x =$ M, C, L and P) are the voltage pulses in the nth section for the main (M) transmission-line, capacitive (C), inductive (L), and phase (P) stubs, respectively. The gain coefficient is defined by Eq. (16.29) with the admittance terms (Y, Y_{C} and Y_{L}) explained in Section 16.7. The admittance terms are used to include the spectral dependence of the material gain. The term α_{act} is the active region loss coefficient, and α_{act} is the cladding region loss coefficient. The total attenuation factor, α_{t}, is given as:

$$\alpha_{\mathrm{t}}(n) = \exp\left[-\{\alpha_{\mathrm{sc}} + \alpha_{\mathrm{hb}}(n) + \alpha_{\mathrm{tap}}(n)\}\frac{\Delta z}{2}\right] \tag{16.63}$$

where the conventional material loss, α_{sc}, is defined by:

$$\alpha_{\mathrm{sc}} = \Gamma\alpha_{\mathrm{act}} + \{1-\Gamma\}\alpha_{\mathrm{clad}} \tag{16.64}$$

and the position-dependent loss due to the lateral hole-burning effect, $\alpha_{\mathrm{hb}}(n)$, is defined by:

$$\alpha_{\mathrm{hb}}(n) = [\Gamma a\eta(n)N_1(n)] . \tag{16.65}$$

Strictly speaking, Γ in the conventional material loss is position-dependent due to the tapering geometry. However, numerical studies show that the mode confinement factor, Γ, is only weakly dependent on the active layer width above several μm, and hence can be assumed constant in the TS-TLLM [51]. Bendelli *et al.* used an averaged loss factor with α_{sc} being a constant [5]. The term Z_S is the impedance of the phase-stub, which was defined in Section 16.11. The coupling factor, K_c, takes into account power transfer from the quasi-TE_0 mode into higher order modes (i.e. mode conversion), which was defined in Section 16.5.

The scattering matrix in Eq. (16.62) is applicable for the forward travelling-waves only. For the backward-travelling waves, a similar scattering matrix is used but the following modifications are required. The α_{tap} term (+) increases the intensity toward the narrow end. In addition, the term K_c will not be present since the quasi-adiabatic taper filters out the higher-order modes on round trip propagation, thus mitigating filamentation effects. The formal expression of the scattering process in the TS-TLLM from Eq. (16.62) is:

$$V_k^{\mathrm{r}} = S \cdot V_k^{\mathrm{i}} \tag{16.66}$$

where V_k^{r} denotes the reflected waves from the nth scattering node at kth time iteration, V_k^{i} denotes the incident waves into that node, and S is the scattering matrix. In order to include spontaneous emission, Eq. (16.66) can be modified as:

$$V_k^{\mathrm{r}} = S \cdot V_k^{\mathrm{i}} + V_k^{\mathrm{s}} \tag{16.67}$$

where the spontaneous emission voltage is defined by:

$$V_k^{\mathrm{s}} = \left[\frac{I^{\mathrm{r}}(n) Z_{\mathrm{p}}}{2} \exp\left(\frac{-\alpha_{\mathrm{sc}} \Delta z}{2}\right) \; 0 \; 0 \; 0\right]^T . \tag{16.68}$$

The term Z_{p} is the wave impedance (see Eq. (14.37) in Chapter 14), and $I^{\mathrm{r}}(n)$ is the spontaneous emission current of section-n with its mean square value given by Eq. (16.59).

16.13 Connecting Matrix

The connecting matrix, C, describes the propagation of voltage pulses from one scattering node to its adjacent node, that is:

$$V_{k+1}^{\mathrm{i}} = C \cdot V_k^{\mathrm{r}} . \tag{16.69}$$

Equation (16.69) along the main transmission-line and at the facets may be expressed, respectively, as:

$$\begin{bmatrix} V_M^{\rm f}(n) \\ V_M^{\rm b}(n) \end{bmatrix}_{k+1}^{i} = \begin{bmatrix} 1 & 0 \\ 0 & 1 \end{bmatrix} \begin{bmatrix} V_M^{\rm f}(n-1) \\ V_M^{\rm b}(n+1) \end{bmatrix}_{k}^{r} \tag{16.70}$$

and

$$\begin{bmatrix} V_M^{\rm f}(1) \\ V_M^{\rm b}(M) \end{bmatrix}_{k+1}^{i} = \begin{bmatrix} 0 & \sqrt{R_1} \\ \sqrt{R_2} & 0 \end{bmatrix} \begin{bmatrix} V_M^{\rm f}(M) \\ V_M^{\rm b}(1) \end{bmatrix}_{k}^{r}. \tag{16.71}$$

Where $V_M^{\rm f}$ and $V_M^{\rm b}$ denote the forward and backward travelling pulses, respectively. The terms R_1 and R_2 are the power reflectivity of the facets at the input and output ends (section-1 and section-M), respectively. Previous models of the tapered TWA assumed perfect AR coating, i.e. no reflections [5, 6, 11] and [13]. Facet reflectivity plays an important role in determining the amplifier's performance and thus should be considered in simulations [30, 52, 53]. Reflections and feedback effects are easily included in the TS-TLLM.

Besides simulating the propagation of voltage pulses on the main transmission-lines, connecting matrices are also needed for the recirculating voltage pulses on the TLM stub filters, which are used to model the spectral dependence of gain and spontaneous emission. The connecting matrices for the inductive (L) and capacitive (C) stubs are, respectively, given as:

$$\begin{bmatrix} V_{\rm L}^{\rm f}(n) \\ V_{\rm L}^{\rm b}(n) \end{bmatrix}_{k+1}^{i} = \begin{bmatrix} -1 & 0 \\ 0 & -1 \end{bmatrix} \begin{bmatrix} V_{\rm L}^{\rm f}(n) \\ V_{\rm L}^{\rm b}(n) \end{bmatrix}_{k}^{r} \tag{16.72}$$

and

$$\begin{bmatrix} V_{\rm C}^{\rm f}(n) \\ V_{\rm C}^{\rm b}(n) \end{bmatrix}_{k+1}^{i} = \begin{bmatrix} 1 & 0 \\ 0 & 1 \end{bmatrix} \begin{bmatrix} V_{\rm C}^{\rm f}(n) \\ V_{\rm C}^{\rm b}(n) \end{bmatrix}_{k}^{r}. \tag{16.73}$$

The stubs are fixed at half a unit-section length ($\Delta z/2 = v_{\rm g}\Delta t/2$) so that a reflected pulse into it becomes an incident pulse at the scattering node one timestep later. In this way, all the voltage pulses in the transmission-line are synchronised.

The carrier-dependent phase change is simulated by means of voltage pulses being delayed in the adjustable phase-stubs, where they wait until the next iteration before they continue propagating along the main transmission-lines. The amount of delay depends on the local carrier density. The connecting matrices for the phase-stubs can be expressed as:

$$\begin{bmatrix} V_{\mathrm{ph}}^{\mathrm{f}}(n) \\ V_{\mathrm{ph}}^{\mathrm{b}}(n) \end{bmatrix}_{k+1}^{i} = \begin{bmatrix} -1 & 0 \\ 0 & -1 \end{bmatrix} \begin{bmatrix} V_{\mathrm{ph}}^{\mathrm{f}}(n) \\ V_{\mathrm{ph}}^{\mathrm{b}}(n) \end{bmatrix}_{k}^{r} \quad \text{for } \tan[\beta(n)\Delta l_{\mathrm{ph}}(n)] < 0 \tag{16.74}$$

or

$$\begin{bmatrix} V_{\mathrm{ph}}^{\mathrm{f}}(n) \\ V_{\mathrm{ph}}^{\mathrm{b}}(n) \end{bmatrix}_{k+1}^{i} = \begin{bmatrix} 1 & 0 \\ 0 & 1 \end{bmatrix} \begin{bmatrix} V_{\mathrm{ph}}^{\mathrm{f}}(n) \\ V_{\mathrm{ph}}^{\mathrm{b}}(n) \end{bmatrix}_{k}^{r} \quad \text{for } \tan[\beta(n)\Delta l_{\mathrm{ph}}(n)] > 0\,. \tag{16.75}$$

16.14 Results and Discussions

Light-current characteristics: The light-current curves are shown in Fig. 16.16. They are plotted for both the tapered laser amplifiers ($R_1 = R_2 = 0.01\%$) and laser oscillators ($R_1 = 30\%, R_2 = 1\%$). Both linear and exponential tapered structured TLLM were used. Laser amplifiers need to avoid optical feedback and thus require a higher pumping rate than laser oscillators to obtain a reasonable signal gain. The increase in threshold level for the laser amplifiers is clearly shown in Fig. 16.16. The threshold levels of the laser oscillator models are about 600 mA. As for the laser amplifiers, there is a gradual transition from the non-lasing region to the lasing region. The threshold levels for the tapered amplifiers are about 3A. For the same bias level and output width of 30μm, the linear taper provides a higher output power compared to the exponential taper.

The effect of (i) higher order lateral modes and (ii) lateral hole-burning on the light-current characteristics of the linear TS-TLLM amplifier model are shown in Figs. 16.17(a) and (b), respectively. Figure 16.17(a) shows that inclusion of higher order lateral modes in the ASE model causes a smoother transition from the non-lasing to lasing operation. Figure 16.17(b) shows that the lateral hole-burning (LHB) effect reduces the continuous wave (CW) output power of the laser. The LHB effect also contributes to the damping of the laser resonance [54].

Gain saturation performance: In order to achieve travelling-wave amplifiers with wide amplification bandwidth, the gain ripple must be designed to be less than 3 dB (i.e. peak-trough ratio < 2) by using the following expression [52]:

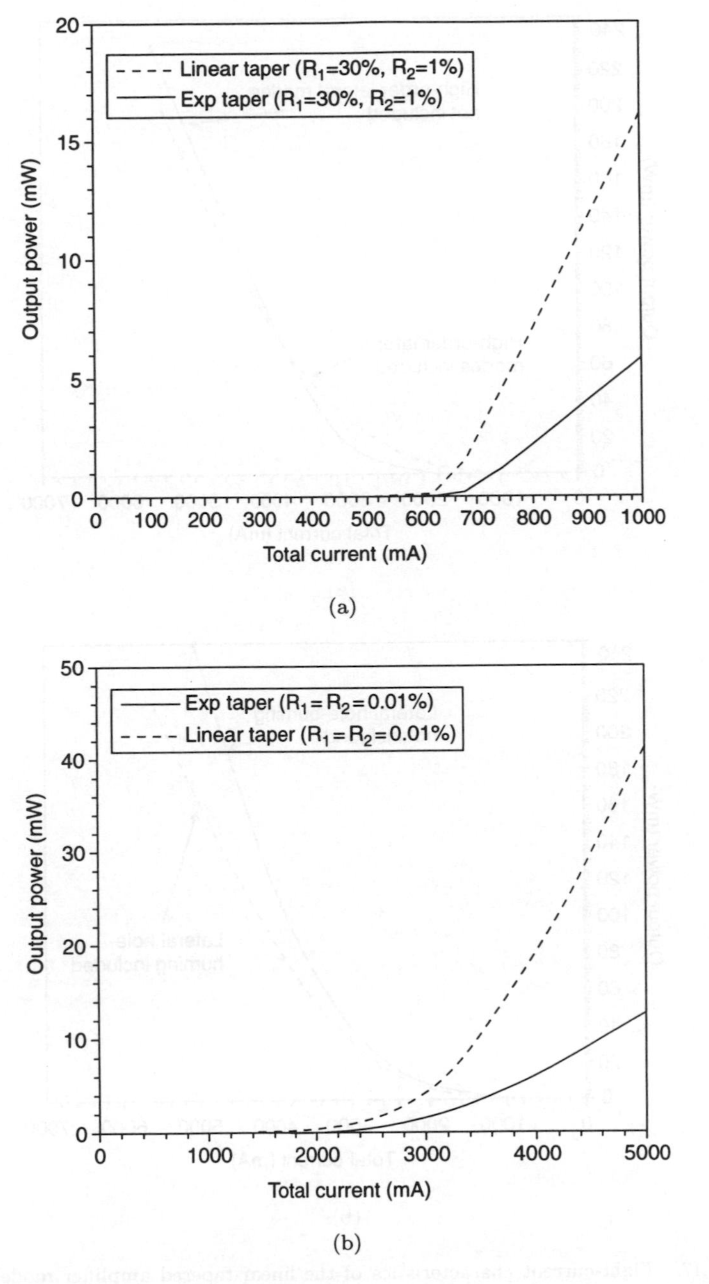

Fig. 16.16. Light-current characteristics of the linear-taper and exponential-taper (a) laser oscillator and (b) amplifier models.

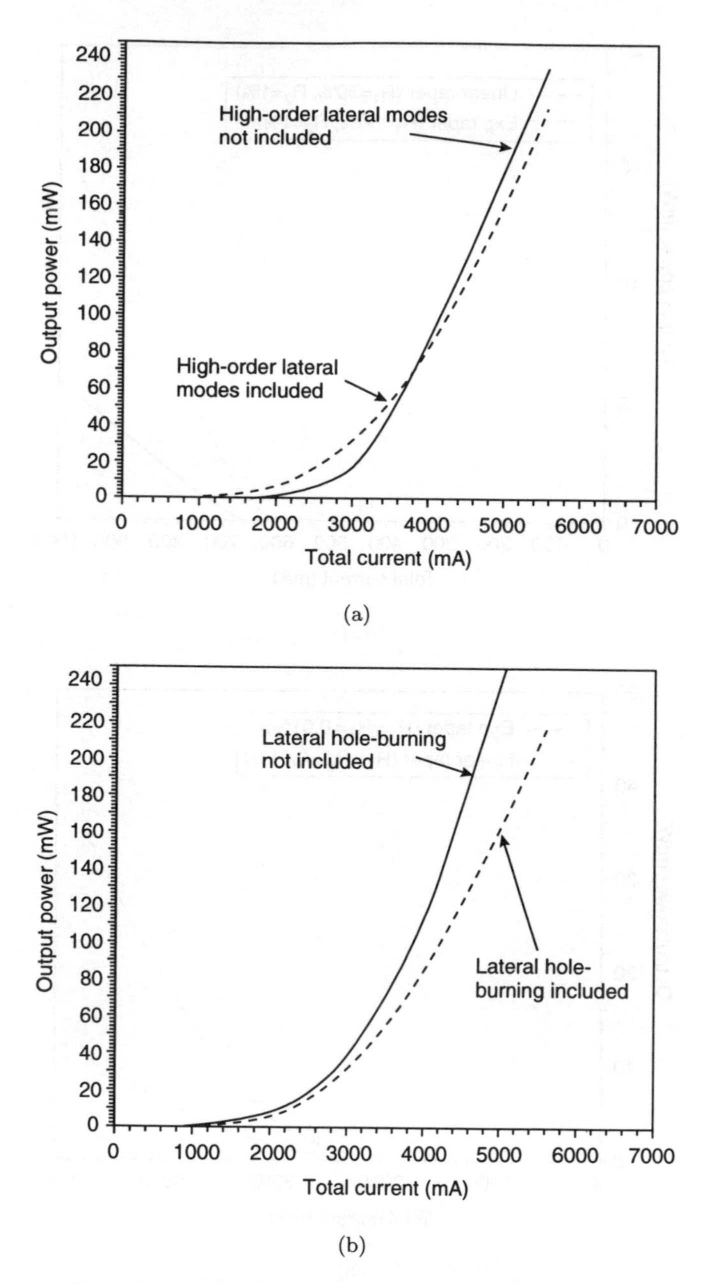

Fig. 16.17. Light-current characteristics of the linear-tapered amplifier model. Effect of (a) higher order lateral modes and (b) lateral hole-burning (LHB).

$$\Delta G = \left(\frac{1 + \sqrt{R_1 R_2} G_s}{1 - \sqrt{R_1 R_2} G_s} \right)^2 \tag{16.76}$$

where $R_1 = R_2$ are the amplifier residual reflectivities, and G_s is the single pass gain. The gain saturation characteristics of the tapered amplifiers when biased at 0.7 times threshold are plotted in Fig. 16.18. They are plotted for the non-tapered, linear tapered, and exponentially tapered structures. The improvement in saturation output power, P_{sat}, for the linear taper is 13 dB whereas it is only 9 dB for the exponential taper, given an output width of $30\mu m$. This agrees well with the results of [5].

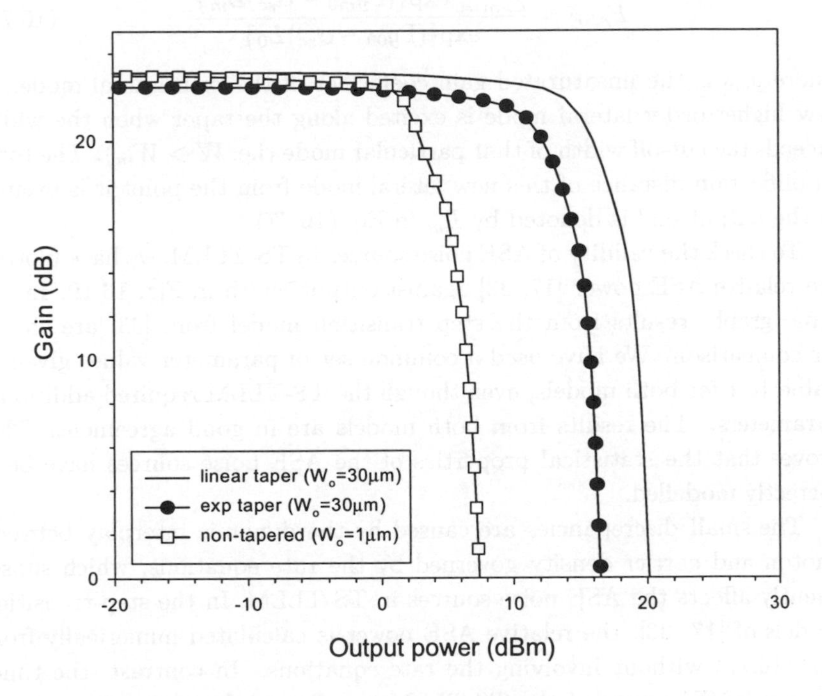

Fig. 16.18. Gain saturation curves of the near travelling-wave TLLM amplifier models.

The results show that the unsaturated gain for the exponential taper is slightly lower than that of the linear taper. This is because the mode conversion is higher in the exponential taper. When the input signal is small (below −40 dBm), the amplified output signal will be dominated by noise, and the signal gain can easily be overestimated. For simplicity, spontaneous emission noise was not included when generating the gain saturation curves. If spontaneous emission noise is be included, there will be ASE noise (including beat noise components) at the output, and a sharp cut-off filter must be then used to obtain the actual amplified output signal. Discrete-time digital filters can be used to implement such a sharp cut-off filter [36]. Alternatively, a simpler but crude approach is by using a TLM stub-filter with a very high Q-factor.

Relative ASE noise: The relative ASE power is a measure of the additional ASE noise in the tapered device compared to the uniform-width conventional amplifier structure with single lateral mode. The relative ASE power is defined as [17]:

$$P_{\text{ASE}} = \frac{\sum_{m=0}^{\text{Mn}} \exp\{(\Gamma g_{m0} - \alpha_{\text{sc}})L_m\}}{\exp\{(\Gamma g_{00} - \alpha_{\text{sc}})L_0\}} \tag{16.77}$$

where g_{m0} is the unsaturated gain coefficient of the mth lateral mode. A new higher order lateral mode is excited along the taper when the width exceeds the cut-off width of that particular mode (i.e. $W > W_m$). The total amplification distance of this new lateral mode from the point it is excited to the output end is denoted by L_m in Eq. (16.77).

To check the validity of ASE noise sources in TS-TLLM, we have plotted the relative ASE power [17, 33] against output width in Fig. 16.19. In the same graph, results from the step transition model from [33] are shown for comparison. We have used a common set of parameter values given in Table 16.1 for both models, even though the TS-TLLM required additional parameters. The results from both models are in good agreement. This proves that the statistical properties of the ASE noise sources have been correctly modelled.

The small discrepancies are caused by the dynamic interplay between photon and carrier density governed by the rate equations, which subsequently affects the ASE noise sources in TS-TLLM. In the step-transition models of [17, 33], the relative ASE power is calculated numerically from Eq. (16.77) without involving the rate equations. In contrast, the time-averaged ASE power of the TS-TLLM is collected for each taper output width and normalised to the ASE power of the single lateral mode device ($W_\text{i} = W_\text{o} = 1\mu$m). All the important effects discussed in Sections 16.4 to 16.11 are included in the TS-TLLM when generating the relative ASE power. In other words, the TS-TLLM is a stochastic model, which is more realistic than the deterministic step transition model. It is assumed that the total ASE noise power is contained in the quasi-TE$_0$ mode. With mode conversion included in TS-TLLM, which is more severe in the exponential taper, total ASE power injected into quasi-TE$_0$ mode begins to fall

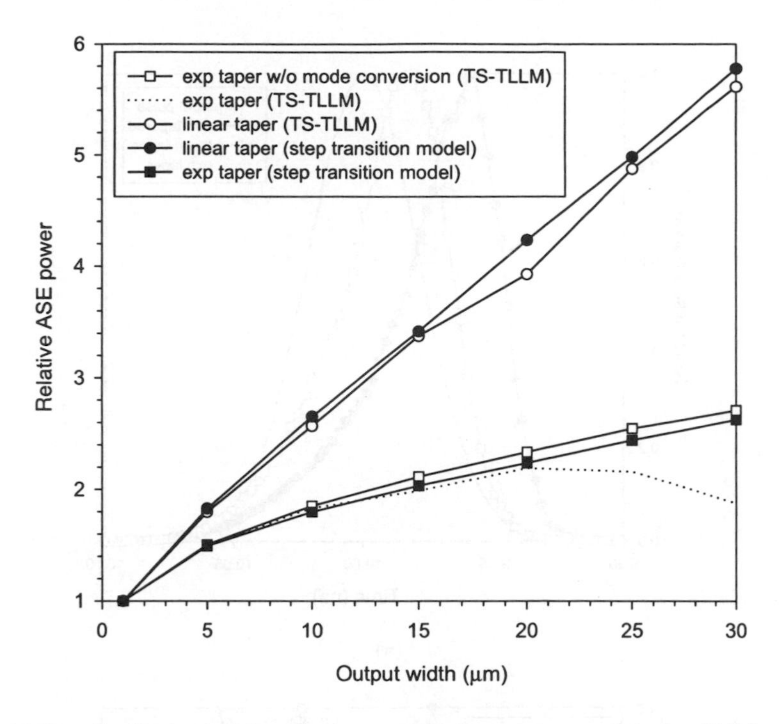

Fig. 16.19. Comparison of the relative amplified spontaneous emission (ASE) noise between two tapered amplifier models.

off at large output widths. This attenuating effect of mode conversion has also been observed for TE_0 gain [32]. If mode conversion is neglected in the exponential TS-TLLM, the results are again in good agreement (see Fig. 16.19).

Picosecond pulse Amplification: An unchirped Gaussian input pulse that has a full width at half maximum (FWHM) of 50 ps was injected to the tapered laser amplifier (input pulse energy, E_{in} = saturation energy, $E_{\text{sat}} = 3.95$ pJ and $R = 0.01\%$). The amplified output pulses (normalised to peak power) are then plotted in Fig. 16.20(a). The output pulse of the non-tapered structure is also shown for comparison. It is clearly seen that the pulse is most distorted for the non-tapered structure — the peak of the pulse has been advanced (i.e. earlier) in time, and the shape is highly non-Gaussian. For the tapered structures, the amplified pulses are less distorted. The linear taper produced the amplified output pulse that most resembles the input pulse. The corresponding normalised optical spectra are plotted in Fig. 16.20(b). The spectrum of output pulse from the linear

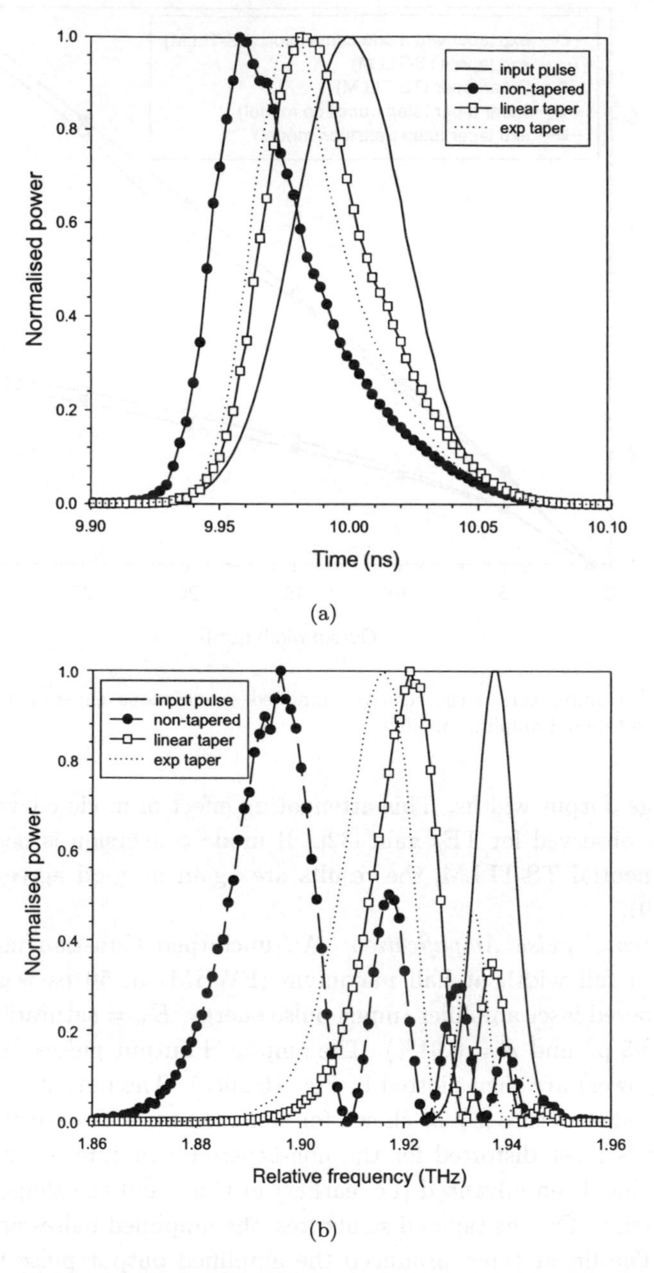

Fig. 16.20. (a) The amplified output pulses of tapered and non-tapered amplifier structures and (b) their corresponding optical spectra.

taper has the smallest frequency shift, only about 20 GHz, whereas it is almost 50 GHz for the non-tapered structure. These results agree well with that of [6, 55].

Next, input Gaussian pulses with different values of energy were injected to the linear tapered amplifier ($R = 0.01\%$). The output pulse shapes are plotted in Fig. 16.21(a). With input pulse energy of $E_{\text{in}} = E_{\text{sat}}$, the normalised output pulse is seen to be asymmetrical due to excessive gain saturation. This is expected since the leading edge is always amplified more than the trailing edge when gain saturation sets in. In this case, the optical spectrum is shifted by more than 20 GHz (see Fig. 16.21(b)). When the input pulse energy is $E_{\text{in}} = 0.01E_{\text{sat}}$, the output pulse resembles a Gaussian pulse except that its peak is delayed slightly in time. This results in an almost unshifted spectrum as seen in Fig. 16.21(b). The results for the exponential taper were found to be very similar.

One of the main advantages of using computer modelling is that ideal cases can be studied. In order to look deeper into the pulse gain saturation mechanism, ideal square pulses of 1ns with the same pulse energies of E_{sat}, $0.1E_{\text{sat}}$ and $0.01E_{\text{sat}}$ are injected into the linear tapered amplifier, and the results are plotted in Fig. 16.22. The highest saturation rate occurred within the first 50 ps of the output pulse, and ceased almost completely after 0.4 ns. With input pulse energy of $E_{\text{in}} = E_{\text{sat}}$ the high amplification rate of the leading edge quickly saturates the gain, and becomes "clipped" for the remaining parts of the square pulse. The interesting feature of the output pulse when $E_{\text{in}} = 0.01E_{\text{sat}}$ is the absence of gain saturation initially, and then we see a sharp drop in gain. Shortly after that, the gain rose a little and then it continued to saturate at a very slow rate. The corresponding output pulse spectra almost coincide with the input pulse spectrum for all cases. There is, however, a noticeable portion of the spectrum "leaking" into the lower frequency side when the input pulse energy is $E_{\text{in}} = E_{\text{sat}}$. The high energy pulse leads to a very quick gain saturation (due to carrier depletion), eventually shifting spectral components of the leading part of the square pulse to lower frequencies.

Residual reflectivity: The amplified output pulses from the linear tapered amplifier with residual reflectivity as parameter (between $R = 0.01\%$ and $R = 1\%$) are shown in Fig. 16.23(a). The corresponding optical spectra are shown in Fig. 16.23(b). An unchirped Gaussian pulse of 3.95 pJ was used as input signal. From Fig. 16.23 we can see that increasing reflectivity caused the peak of the pulse to be delayed in time. With a correct amount of residual reflectivity, and provided that the pulse duration is longer than

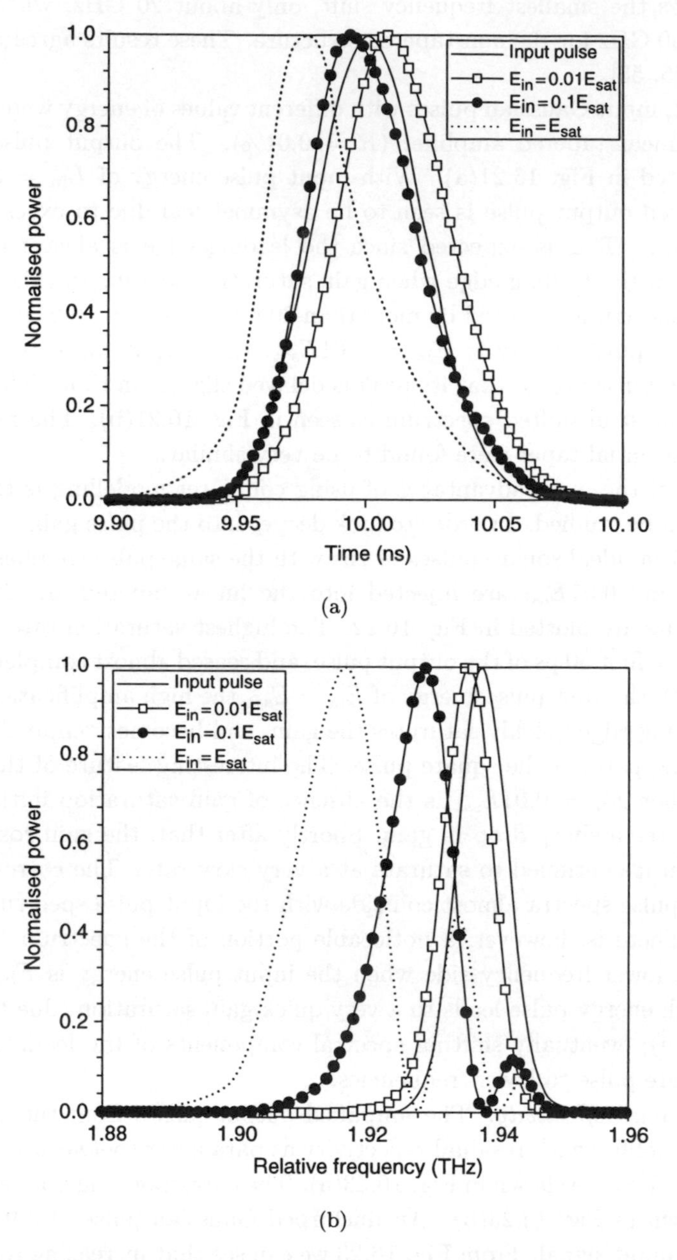

Fig. 16.21. (a) The effect of pulse energy on gain saturation of the amplified output pulses and (b) their corresponding optical spectra for the linear tapered amplifier.

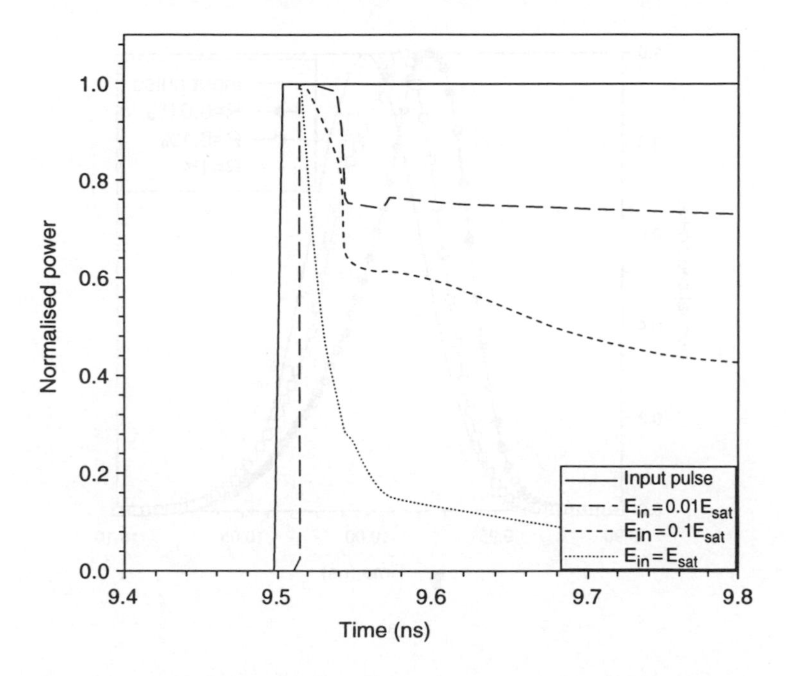

Fig. 16.22. Linear tapered amplifier model: Gain saturation mechanism of an ideal square pulse for various pulse energies.

1.5 times cavity roundtrip (> 38.7 ps), we can achieve an almost undistorted amplified pulse (i.e. Gaussian-like and peak of output pulse coincides with peak of input pulse) from the tapered amplifier. In this case, the dynamic wavelength shift is also smaller. The only drawback is that the output pulse energy is lower than the case with almost zero reflectivity. The pulse energy of the amplified pulse for the case with 0.01% reflectivity is about 165 pJ whereas it is only 32.8 pJ when the reflectivity is 1%. The corresponding pulse energy gains are 16.2 dB and 9.19 dB, respectively. The plots of pulse energy gain against the ratio $E_{\text{in}}/E_{\text{sat}}$ for the linear tapered amplifier model with residual reflectivity as parameter are shown in Fig. 16.24. It is clear that the unsaturated value of pulse energy gain drops with increasing residual reflectivity.

Finally, an ideal square pulse was injected to the amplifier models, each with different values of facet reflectivity. In Fig. 16.25 for the near travelling-wave condition ($R = 0.01\%$), gain saturation is excessively high. With higher facet reflectivity, the leading edge falls off at a slower rate but at the expense of lower pulse energy gain. When the reflectivity is 1% and

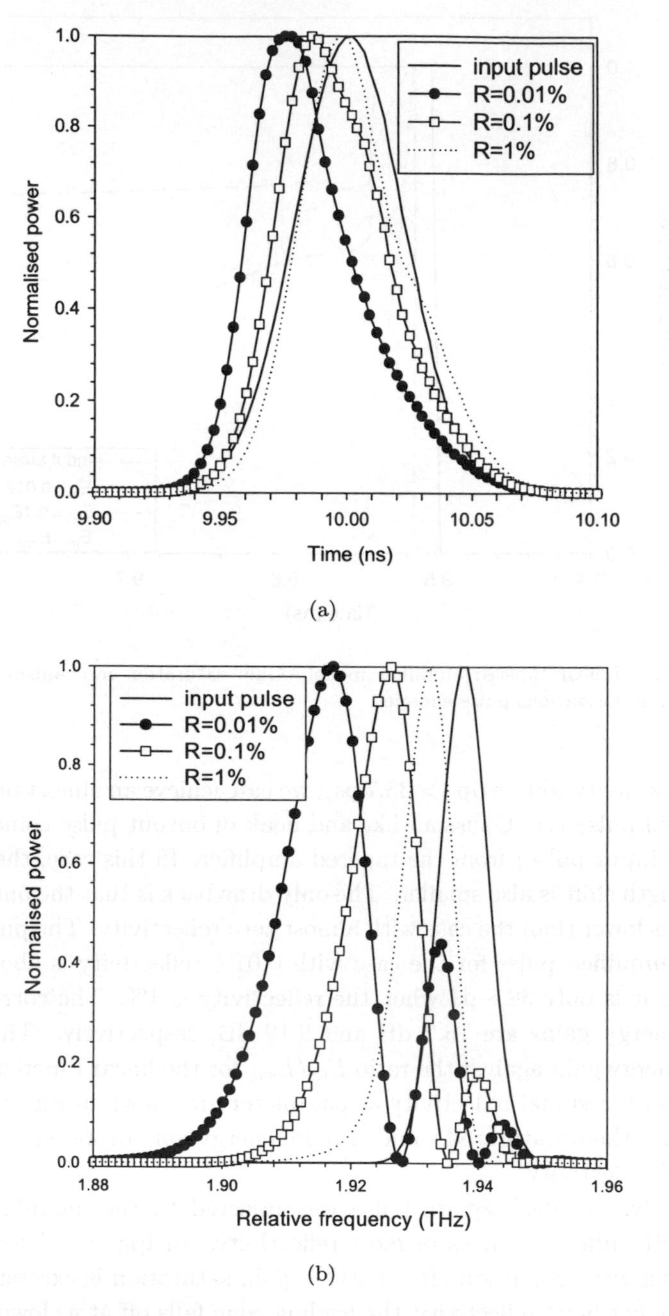

Fig. 16.23. (a) The effect of residual facet reflectivity (R) on pulse saturation and (b) the corresponding optical spectra.

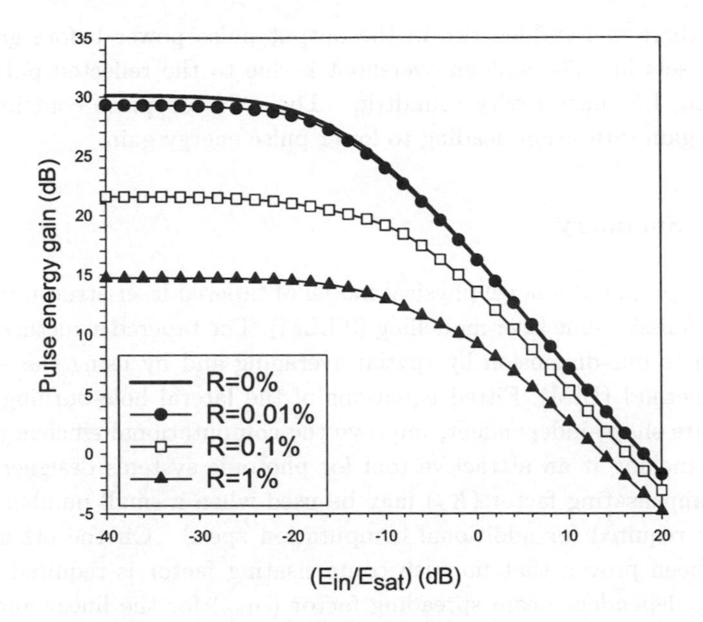

Fig. 16.24. The pulse energy gain against the ratio of input pulse energy to saturation energy for different values of residual reflectivity (R).

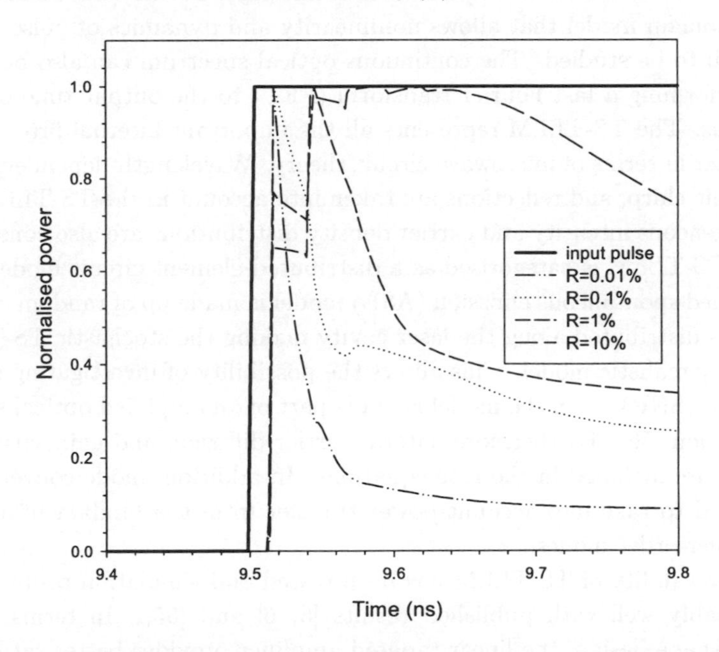

Fig. 16.25. Linear tapered amplifier model: The effect of facet reflectivity on the gain saturation mechanism of an ideal square pulse.

above, there is a sudden rise in the output pulse power before gain saturation sets in. The sudden overshoot is due to the reflected pulse after travelling 1.5 times cavity roundtrip. The reflected pulse contributes to further gain saturation, leading to lower pulse energy gain.

16.15 Summary

We have proposed a novel physical model of tapered laser structures based on transmission-line laser modelling (TLLM). The tapered structure can be reduced to one-dimension by spatial averaging and by using the effective index method (EIM). Fitted equations of the lateral hole-burning terms, which are shape-independent, improve the computational efficiency of the model, making it an attractive tool for photonic systems designers. The gain compensating factor (K_g) may be used when a small number of sections is required for additional computation speed. On the other hand, it has been proven that no such compensating factor is required for the position-dependent beam spreading factor (α_{tap}) for the linear and exponential tapers.

The tapered structure transmission-line laser model (TS-TLLM) is a time-domain model that allows nonlinearity and dynamics of pulse amplification to be studied. The continuous optical spectrum can also be found by performing a fast Fourier transform (FFT) to the output time domain samples. The TS-TLLM represents all the important internal processes of the laser in terms of microwave circuit theory. Wavelength-dependent gain, dynamic chirp, and reflections are taken into account in the TS-TLLM. Inhomogeneous intensity and carrier density distributions are also considered since TS-TLLM is categorised as a distributed-element circuit model. The amplified spontaneous emission (ASE) model is made up of random current sources distributed along the laser cavity making the stochastic TS-TLLM a highly realistic model. This allows the possibility of investigating signal-to-noise (SNR) when the model is made part of an amplified optical system simulation [56]. Furthermore, lateral carrier diffusion and gain saturation effects are included in the rate-equations. In addition, mode conversion is included to take into account power transfer from the fundamental mode to higher-order modes.

The validity of TS-TLLM has been tested and simulation results agree reasonably well with published results [5, 6] and [55]. In terms of the taper shape design, the linear tapered amplifier provides better saturation performance but the exponential tapered amplifier gives lower ASE noise.

The tapered amplifier structures mitigate the detrimental effect of gain saturation on amplified pulse distortion and on its corresponding spectral shift. Pulse energy and residual reflectivity have a significant impact on the dynamics of picosecond pulse amplification in tapered waveguide amplifier structures.

References

[1] D. D. Marcenac and J. E. Carroll, "Quantum-mechanical model for realistics Fabry–Perot lasers," *IEE Proceedings Pt. J*, Vol. 140, No. 3, pp. 157–171, 1993.

[2] L. M. Zhang and J. E. Carroll, "Large-Signal Dynamic Model of the DFB Laser," *IEEE J. Quant. Electron.*, Vol. 28, No. 3, pp. 604–611, 1992.

[3] M. G. Davis and R. F. Dowd, "A Transfer Matrix Method Based Large-Signal Dynamic Model for Multielectrode DFB Lasers," *IEEE J. Quant. Electron.*, Vol. 30, No. 11, pp. 2458–2466, 1994.

[4] M. J. Adams, "Rate equations and transient phenomena in semiconductor lasers," *Opto-electronics*, Vol. 5, pp. 201–215, 1973.

[5] G. Bendelli, K. Komori and S. Arai, "Gain Saturation and Propagation Characteristics of Index-Guided Tapered-Waveguide Traveling-Wave Semiconductor Laser Amplifiers (TTW-SLA's)," *IEEE J. Quant. Electron.*, Vol. 28, No. 2, pp. 447–457, 1992.

[6] H. Ghafouri-Shiraz, P. W. Tan and T. Aruga, "Picosecond Pulse Amplification in Tapered-Waveguide Laser-Diode Amplifiers," *IEEE Jour. Select. Top. Quant. Elec.*, Vol. 3, No. 2, pp. 210–217, 1997.

[7] L. Goldberg, D. Mehuys and D. Welch, "High power mode-locked compound cavity laser using a tapered semiconductor laser amplifier," *IEEE Photon. Tech. Lett.*, Vol. 6, No. 9, pp. 1070–1072, 1994.

[8] B. Zhu, I. H. White, K. A. Williams, F. R. Laughton and R. V. Penty, "High peak power picosecond optical pulse generation from Q-switched bow-tie laser with a tapered traveling wave amplifier," *IEEE Photon. Tech. Lett.*, Vol. 8, No. 4, pp. 503–505, 1996.

[9] Z. Wang, W. Wang, Q. Wang, H. K. Tsang and Z. Jiang, "Amplifier modulation for low-chirp from a monolithic strained-layer MQW InGaAsP/InP distributed-feedback-laser/tapered amplifier," *CLEO/Pacific Rim*, ThD6, pp. 125–126, 1996.

[10] G. P. Agrawal, "Fast fourier transform based beam propagation model for stripe geometry semiconductor lasers: inclusion of axial effects," *J. Appl. Phys.*, Vol. 56, No. 11, pp. 3100–3109, 1984.

[11] S. Balsamo, F. Sartori and I. Monttrosset, "Dynamic beam propagation method for flared semiconductor power amplifiers," *IEEE Jour. Select. Top. Quant. Elec.*, Vol. 2, No. 2, pp. 378–384, 1996.

[12] R. J. Lang, A. H. Hardy, R. Parke, D. Mehuys, S. O'Brien, J. Major and

D. Welch, "Numerical analysis of flared semiconductor laser amplifiers," *IEEE J. Quant. Electron.*, Vol. QE-29, No. 6, pp. 2044–2051, 1993.

[13] B. Dagens, S. Balsamo and I. Monttrosset, "Picosecond pulse amplification in AlGaAs flared amplifiers," *IEEE Jour. Select. Top. Quant. Elec.*, Vol. 3, No. 2, pp. 233–244, 1997.

[14] G. C. Dente and M. L. Tilton, "Modeling multiple longitudinal mode dynamics in semiconductor lasers," *IEEE J. Quant. Electron.*, Vol. QE-34, No. 2, pp. 325–335, 1998.

[15] E. I. Gordon, "Optical maser oscillators and noise," *Bell Syst. Tech. J.*, Vol. 43, pp. 507–539, 1964.

[16] L. V. T. Nguyen, A. J. Lowery, P. C. R. Gurney and D. Novak, "A Time Domain Model for High-Speed Quantum-Well Lasers Including Carrier Transport Effects," *IEEE Jour. Select. Top. Quant. Elec.*, Vol. 1, No. 2, pp. 494–504, 1995.

[17] S. El-Yumin, K. Komori, S. Arai and G. Bendelli, "Taper shape dependence of tapered waveguide traveling wave semiconductor laser amplifier," *Trans. IEICE*, Vol. E77-C, No. 4, pp. 624–632, 1994.

[18] H. Kogelnik, "Theory of optical waveguides," in *Guided-wave optoelectronics*, 2nd ed., T. Tamir, Ed. (Berlin: Springer-Verlag, 1990), p. 419.

[19] J. Buus, "The effective index method and its application to semiconductor lasers," *IEEE J. Quant. Electron.*, Vol. QE-18, No. 7, pp. 1083–1089, 1982.

[20] M. J. Adams, *An introduction to optical waveguide* (Chichester: John Wiley & Sons, 1981).

[21] J. W. Bandler, P. B. Johns and M. R. M. Rizk, "Transmission-Line Modeling and Sensitivity Evaluation for Lumped Network Simulation and Design in the Time Domain," *Jour. Frank. Inst.*, Vol. 304, pp. 15–32, 1977.

[22] H. C. J. Casey and M. B. Panish, *Heterostructure Lasers: Part A, Fundamental Principles* (New York: Academic Press, 1978).

[23] E. A. J. Marcatilli, "Dielectric rectangular waveguide and directional coupler for integrated optics," *Bell Syst. Tech. J.*, Vol. 48, pp. 2071–2102, Sept 1969.

[24] H. Kressel and J. K. Butler, *Semiconductor Lasers and Heterojunction LEDs* (New York: Academic Press, 1977).

[25] R. E. Collins, *Foundations of Microwave Engineering* (New York: McGraw-Hill, 1966).

[26] D. Marcuse, "Radiation losses of tapered dielectric slab waveguides," *Bell Syst. Tech. J.*, Feb, pp. 273–290, 1970.

[27] A. Yariv, *Optical Electronics*, 4th ed. (Fort Worth: Saunders College Publishing, 1991).

[28] W. K. Burns and A. F. Milton, "Mode conversion in planar-dielectric separating waveguides," *IEEE J. Quant. Electron.*, Vol. QE-11, No. 11, pp. 32–39, 1975.

[29] O. Lenzmann, I. Kolchanoc, A. Lowery, D. Breuer and A. Richter, "Photonic Multi-Domain Simulator," *OFC'99*, White Paper, pp. 1–20, 1999.

[30] T. Saitoh and T. Mukai, "1.5 mum GaInAsP Travelling-Wave Semiconductor Laser Amplifier," *IEEE J. Quant. Electron.*, Vol. QE-23, No. 6, pp. 1010–1019, 1987.

[31] J. N. Walpole, "Semiconductor amplifiers and lasers with tapered gain regions," *Opt. Quant. Elec.*, Vol. 28, pp. 623–645, 1996.
[32] H. Ghafouri-Shiraz, P. W. Tan and W. M. Wong, "A Novel Analytical Expression of Saturation Intensity of InGaAsP Tapered Traveling-Wave Semiconductor Laser Amplifier Structures," *IEEE Photon. Tech. Lett.*, Vol. 10, No. 11, pp. 1545–1547, 1998.
[33] H. Ghafouri-Shiraz and P. W. Tan, "Study of a novel laser diode amplifier structure," *Semicon. Sci. Tech.*, Vol. 11, pp. 1443–1449, 1996.
[34] A. J. Lowery, "New dynamic semiconductor laser model based on the transmission-line modelling method," *IEE Proceedings Pt. J*, Vol. 134, pp. 281–289, 1987.
[35] A. E. Siegman, *LASERS* (Mill Valley, California: University Science Books, 1986).
[36] A. V. Oppenheim and R. W. Schafer, *Discrete Time Signal Processing* (NJ: Prentice Hall, 1989).
[37] S. O'Brien *et al.*, "Operating characteristics of a high-power monolithically integrated flared amplifier master oscillator power amplifier," *IEEE J. Quant. Electron.*, Vol. QE-29, No. 6, pp. 2052–2057, 1993.
[38] A. J. Lowery, "New inline wideband dynamic semiconductor laser amplifier model," *IEE Proceedings Pt. J*, Vol. 135, No. 3, pp. 242–250, 1988.
[39] M. Aramowitz and I. A. Stegun, *Handbook of Mathematical Functions*, 1st ed. (NY: Dover Publications, Inc., 1965).
[40] P. C. R. Gurney and R. F. Ormondroyd, "A multimode self-consistent model of thelasing characteristics of buried heterostructure lasers," *IEEE J. Quant. Electron.*, Vol. QE-31, No. 3, pp. 427–437, 1995.
[41] J. W. Lai and C. F. Lin, "Carrier diffusion effect in tapered semiconductor laser amplifier," *IEEE J. Quant. Electron.*, Vol. QE-34, No. 7, pp. 1247–1256, 1998.
[42] D. Marcuse, "Computer model of an injection laser amplifier," *IEEE J. Quant. Electron.*, Vol. QE-19, No. 1, pp. 63–73, 1983.
[43] M. J. Adams, J. V. Collins and I. D. Henning, "Analysis of semiconductor laser optical amplifiers," *IEE Proceedings Pt. J*, Vol. 132, No. 1, pp. 58–63, 1985.
[44] A. J. Lowery, "A new time-domain model for spontaneous emission in semiconductor lasers and its use in predicting their transient response," *Int. Jour. Num. Model.*, Vol. 1, pp. 153–164, 1988.
[45] A. J. Lowery, "Amplified spontaneous emission in semiconductor laser amplifiers: validity of the transmission-line laser model," *IEE Proceedings Pt. J*, Vol. 137, No. 4, pp. 241–247, 1990.
[46] C. H. Henry, "Theory of Spontaneous Emission Noise in Open Resonators and its Application to Lasers and Optical Amplifiers," *IEEE J. Lightwave Technol.*, Vol. LT-4, No. 3, pp. 288–297, 1986.
[47] Y. Yamamoto, "Noise and error rate performance of semiconductor laser amplifiers in PCM-IM optical transmission systems," *IEEE J. Quant. Electron.*, Vol. QE-16, No. 10, pp. 1073–1081, 1980.

[48] A. J. Lowery, "New dynamic model for multimode chirp in DFB semiconductor lasers," *IEE Proceedings Pt. J*, Vol. 137, No. 5, pp. 293–300, 1990.

[49] A. J. Lowery, "Modelling spectral effects of dynamic saturation in semiconductor laser amplifiers using the transmission-line laser model," *IEE Proceedings Pt. J*, Vol. 136, No. 6, pp. 320–324, 1989.

[50] W. M. Wong and H. Ghafouri-Shiraz, "Dynamic model of tapered semiconductor lasers and amplifiers based on transmission line laser modeling," *IEEE Jour. Select. Top. Quant. Elec.*, Vol. 6, No. 4, pp. 585–593, 2000.

[51] P. W. Tan, "Theoretical studies of tapered-waveguide travelling-wave semiconductor laser amplifiers," Ph.D. Dissertation, The University of Birmingham, 1998.

[52] M. J. O'Mahony, "Semiconductor laser optical amplifiers for use in future fiber systems," *IEEE J. Lightwave Technol.*, Vol. LT-6, No. 4, pp. 531–544, 1988.

[53] J. C. Simon, "GaInAsP semiconductor laser amplifiers for single-mode fiber communications," *IEEE J. Lightwave Technol.*, Vol. LT-5, No. 9, pp. 1286–1295, 1987.

[54] R. S. Tucker and D. J. Pope, "Circuit Modeling of the Effect of Diffusion on Damping in a Narrow-Stripe Semiconductor Laser," *IEEE J. Quant. Electron.*, Vol. QE-19, No. 7, pp. 1179–1183, 1983.

[55] G. P. Agrawal and N. A. Olsson, "Self-Phase Modulation and Spectral Broadening of Optical Pulses in Semiconductor Laser Amplifiers," *IEEE J. Quant. Electron.*, Vol. QE-25, No. 11, pp. 2297–2306, 1989.

[56] A. J. Lowery, "Transmission-line laser modelling of semiconductor laser amplified optical communications systems," *IEE Proceedings Pt. J*, Vol. 139, No. 3, pp. 180–188, 1992.

Chapter 17

Novel Integrated Mode-Locked Laser Design

17.1 Introduction

Ultrashort pulse generation using optical techniques is important for research and development in many areas of science and technology. Among its many applications, ultrashort pulse sources are required for realising high-speed optical communication networks [1]. The growing bandwidth demand has led to research into all-optical technology to meet the transmission capacity and speed requirements of today and the future. Although it is foreseen that wavelength division multiplexing (WDM) systems [2] will gradually replace the legacy technology of synchronous optical network/synchronous digital hierarchy (SONET/SDH) [3], another possible solution in the future is optical time division multiplexing (OTDM) systems [4, 5]. In fact, to realise the full potential of the optical fibre, a combination of both WDM and OTDM is envisaged [1, 6, 7]. In such OTDM systems, optical pulses are interleaved before being transmitted over a single optical fibre at aggregate rates in excess of 40 Gb/s [6, 8]. Optical pulse sources are also required for soliton transmission systems, which are designed to overcome the fibre dispersion problem [9, 10]. In addition, optical pulses are used as control signals for ultrafast all-optical signal processing in devices such as the nonlinear optical loop mirror (NOLM) [11–13]. Other applications of optical pulses include measurement of ultrafast relaxation processes for fundamental research in physics and chemistry, laser fusion in manufacturing systems, and optical time domain reflectometry (OTDR) [14]. In millimetre-wave engineering, optical pulse sources can be used for noninvasive circuit probing, offering the advantage of speed over conventional electronic characterisation techniques [15].

Any optical signal in the laser whose time-averaged power envelope contains rapid variation in amplitude or phase compared to the cavity round-trip time, especially during ultrashort pulse generation, will have a frequency spectrum that is wide compared to the cavity intermode spacing. In order to investigate such fast-varying laser signals, a wide bandwidth and multimoded laser model are required. For dynamic simulations of optical picosecond pulses and their transient effects, it is computationally inefficient to use frequency domain laser models such as the transfer matrix model (TMM) [16]. Furthermore, the high peak powers of such short optical pulses induce changes within the semiconductor material, and these nonlinear effects are best modelled by a time-domain laser model. Moreover, we should use travelling-wave rate equations which include spatial variations of carrier and photon densities along the cavity. This is important because of the large amount of amplification and pulse shaping that occurs on each pass through the laser diode. The finite reflectivity of the anti-reflection (AR) coated facet should also be taken into account. The transmission-line laser model (TLLM) is a suitable distributed-element circuit model that meets the criteria as discussed above. The TLLM is well-suited for investigating short pulse generation in laser diodes, such as mode-locking or gain-switching, because it is a time-domain model which takes into account multiple longitudinal modes, inhomogeneous effect along the cavity, and finite facet reflections. The stochastic TLLM also enables us to investigate noise characteristics of the laser, and observe the transient build up of the optical pulses, which are initially seeded by spontaneous emission noise.

In addition, TLLM components models can be combined in any permutation to construct novel compound cavity laser designs. This is possible because TLLM has bidirectional ports allowing passing on and exchange of time-domain waveforms from one component model to another. Using these interconnectable TLLM component models, novel multisegment or compound cavity laser structures can be designed and optimised. This chapter looks into design considerations of a novel actively mode-locked semiconductor laser incorporating Bragg gratings and a tapered amplifier.

17.2 Integrated Mode-Locked Laser Design with Distributed Bragg Gratings and Tapered Waveguide Amplifier

External cavity lasers employing diffraction grating can be used to achieve active mode-locking but they require precision alignment of the bulk optics.

Monolithically integrated devices are more attractive than bulk optics because of reduced costs, compact size, low insertion loss, high reliability, and improved mechanical stability. The integrated mode-locked laser design with a distributed Bragg reflector (DBR) was first proposed by Lowery [17], and later demonstrated by Hansen *et al.* [18]. Hybrid modelocking using a monolithic DBR laser has also been demonstrated by Kim *et al.* [19], where only a low RF power is applied to the saturable absorber. However, the spectrum only consists of two dominant locked modes with large side mode suppression, leading to a beating effect between them, and resulting in sinusoidal-like pulses. In order to achieve sharp temporal pulses, multiple modes are needed.

In this section, a novel integrated mode-locked InGaAsP/InP laser design for generating high-power pulses is proposed. It consists of several sections, which include an active section for RF modulation, a passive waveguide, a bandwidth-limiting element, and an amplifier section. In the monolithic cavity design, a distributed Bragg reflector (DBR) is used as the integrated spectral-filtering element. A narrow-stripe distributed feedback (DFB) laser acts as the modulation section because it has a relatively small area, which allows a large and fast carrier overshoot when excited by high-speed pulsed or RF modulation. The narrow-stripe DFB laser also acts as a spatial filter for single lateral mode selection. The length of the modulated DFB section should be short to allow fast and deep modulation. However, the DFB laser should also be long enough to produce sufficient distributed feedback. A long passive waveguide section is integrated to the device to achieve a reasonable (i.e. low enough) roundtrip frequency for mode-locked operation. A passive waveguide is preferred to an active waveguide as it causes less distortion when the pulse travels through it. Therefore, being able to fabricate a long, low-loss passive waveguide in InP-based systems for devices emitting in the 1.3 μm or 1.55 μm wavelengths is an important design consideration. Size mismatch between the passive waveguide and the thin active region may cause additional coupling losses. In order to generate high power mode-locked pulses, the tapered amplifier is placed at the output section of the device. An etched gap is placed between the tapered amplifier and the DFB laser for electrical isolation. The tapered gain region allows an increase in energy storage capacity and improves energy extraction efficiency due to lower gain saturation. The proposed design is shown in Fig. 17.1.

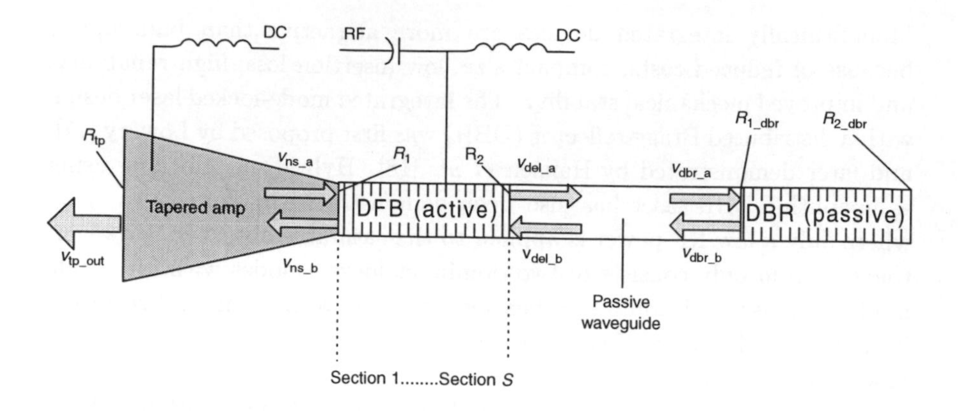

Fig. 17.1. Noval integrated mode-locked laser design: monolithic extended cavity DFB laser with DBR and tapered amplifier (not drawn to scale).

17.3 Detuning Characteristics

The important pulse characteristics in active mode-locking, such as pulsewidth, peak power, spectral width, and pulse stability, are all affected by the detuning from the optimum RF drive frequency. The pulsewidth limiting mechanism due to dynamic detuning was first discovered by Morton *et al.* [20], and was further investigated by Ahmed *et al.* [21]. Owing to the cyclic jitter in active mode-locking, there are stable and unstable regions of mode-locked operation. The total range of frequencies that achieve stable mode-locked operation is known as the "locking bandwidth" of actively mode-locked lasers [21]. The higher the frequency (i.e. the shorter the extended cavity) to achieve mode-locking, the wider the locking bandwidth. It has been shown in [22] that the locking bandwidth is directly proportional to the square of the cavity resonant frequency.

Any significant amount of timing jitter is detrimental to systems employing mode-locked lasers. The timing jitter is partially a result of spontaneous emission noise but is dominated by cyclic instabilities. The cyclic instabilities in active mode-locking have been intensively studied [21–24]. The cyclic jitter was first thought to be due only to the carrier depletion effect [25]. This can be explained as follows. When the first pulse of a sequence of mode-locked pulses has a high peak power, stimulated emission causes the carrier concentration to be depleted, thus suppressing the growth of a subsequent pulse. The subsequent pulse that has a lower peak power causes only a small carrier depletion. Since there is an extra supply of carrier concentration at the next cycle, a high peak-power pulse is again produced. This pattern then repeats itself and the train of mode-locked pulses then exhibits a periodically modulated envelope (large-small-large and so on). Later, it was realised that the dynamic detuning effect, first studied by Morton *et al.* [20], was the dominant factor leading to cyclic instabilities. This lead to further studies of the cyclic timing jitter, and the "pull-in" time is used to relate the locking bandwidth with the cavity resonant frequency [22]. A brief explanation of the how the cyclic jitter arises in active mode-locking is given in Fig. 17.2.

When the recirculating pulse arrives back at the laser chip at exactly the time when the gain modulation reaches its peak value, stable mode-locked pulses will be produced. This is the time-domain picture of active mode-locking. If the drive frequency changes (i.e. detuned away from the optimum frequency), the arriving pulse starts to slip away in time, and the peak of the pulse fails to coincide with the peak of the gain modulation. However, reasonably stable mode-locked pulses can still be achieved provided that the arrival of the pulse peak falls within the pull-in time, T_{pi} (cases A and B). When the drive frequency is too low, large cyclic instability

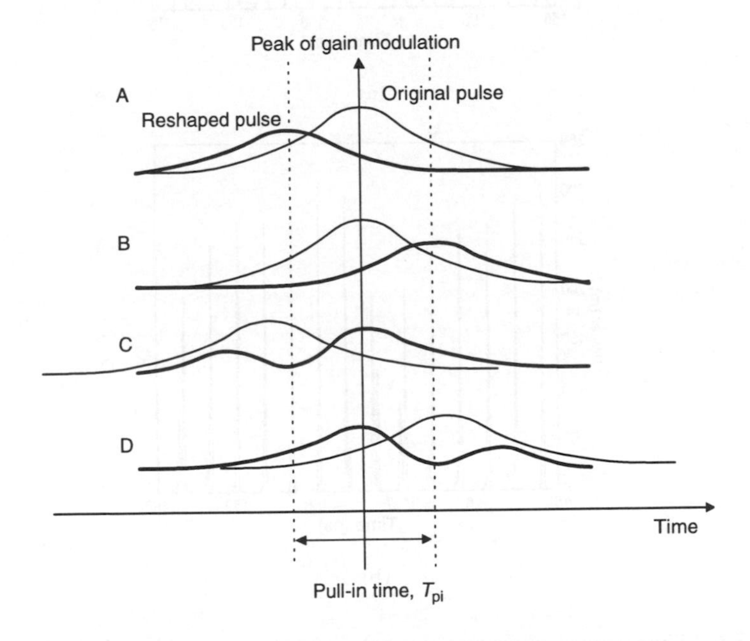

Fig. 17.2. The pulse reshaping mechanism during active mode-locking.

occurs because the pulse arrives back early at the modulated section before the gain has enough time to fully recover to its peak value (case C). Thus, a post-pulse will be preferentially amplified after the returning pulse. On the other hand, if the RF frequency is too high, the pulse arrives back late at the modulated section and the growth of a pre-pulse is favoured (case D). Examples of the time-domain waveforms for cases A and C are shown in Figs. 17.3a and 17.3b, repsectively.

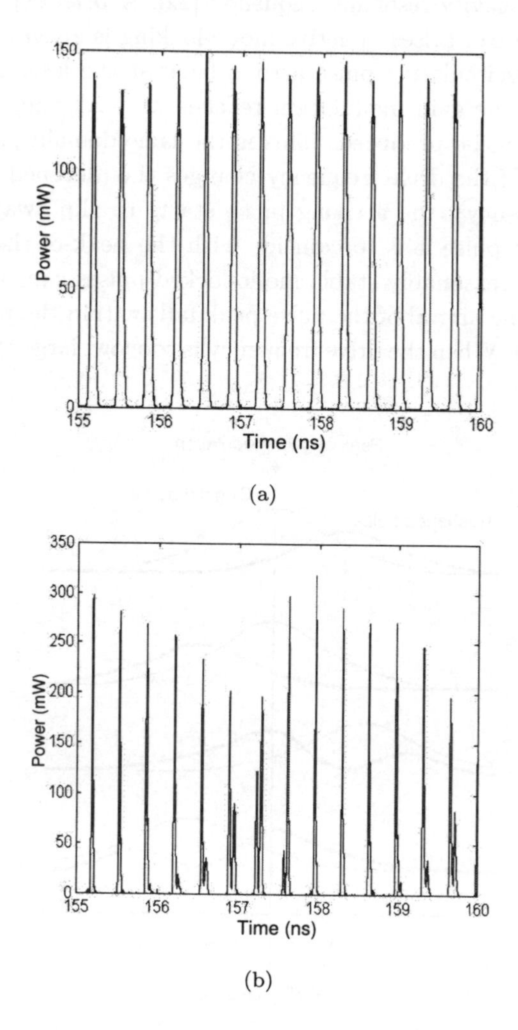

Fig. 17.3. Time-domain waveforms for (a) low jitter (case A) and (b) high jitter (case C).

Although the pulses may show good stability in terms of the RMS timing jitter, it merely indicates that the peaks of a train of pulses are spaced uniformly in time. Zhai *et al.* [22] failed to take into account the pulse-to-pulse peak power variation of the mode-locked pulses, which may be large even with small timing jitter. On the other hand, stable-averaged pulses with multiple peaks may have low peak power variation but large timing jitter exists. The multiple peak situation occurs when there is strong chirp present, an example shown in Fig. 17.4.

In Fig. 17.4, the stable-averaged pulse at the DFB output has multiple peaks but only one dominates, and therefore produces only small timing jitter. In contrast, the amplified pulse has two dominant peaks with almost equal amplitude, and hence large timing jitter. If the peaks dominate alternately between successive pulses, the RMS timing jitter (t_j) is roughly equal to the time spacing between these peaks. In conclusion, a better estimate of the locking bandwidth should include a supporting criterion that the peak power variation remains small, i.e. the ratio of RMS pulse-to-pulse peak power variation (p_{var}) to average peak power (P_{pk}) falling below 10%. Therefore, it is proposed here that p_{var} be used in conjunction with t_j to define the locking range. Upper limits of t_j between 2 and 4 ps

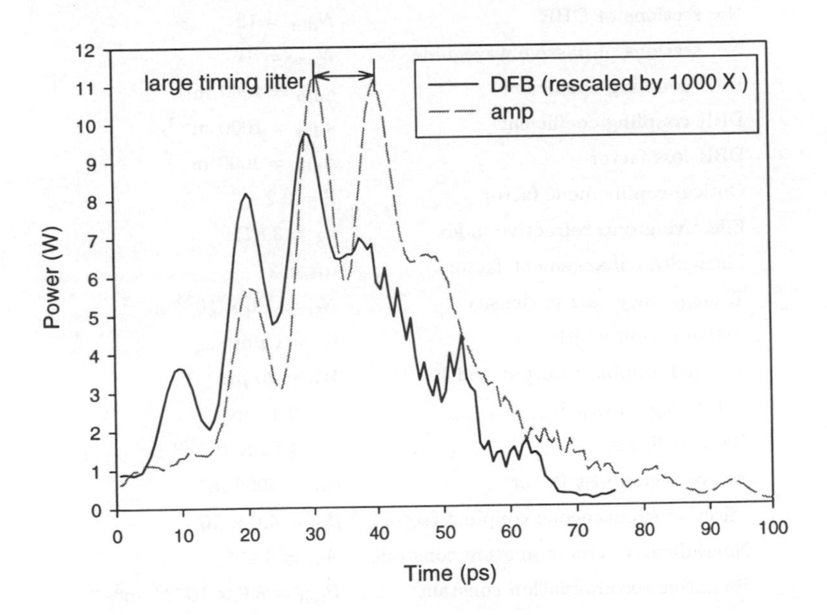

Fig. 17.4. Multiple peaks in the stable-averaged mode-locked pulse.

have been used to define the locking bandwidth [26, 27]. Throughout this book, $t_{j(max)}$ of 2 ps will be used to define the stable region of mode-locked operation (locking range).

The parameter values of the novel integrated mode-locked laser design are given in Table 17.1. A summary of the preliminary simulation results of the laser design will first be presented before a detailed discussion of the design considerations. The detuning characteristics are shown in Fig. 17.5

Table 17.1. Parameters of the novel mode-locked laser design: monolithically integrated DFB laser with DBR and tapered amplifier.

Parameter	Vaslues
Gain peak wavelength	$\lambda_g = 1.497\ \mu m$
DFB laser cavity length	$L_{dfb} = 600\ \mu m$
Passive waveguide length	$L_{pas} = 4200\ \mu m$
DBR length	$L_{dbr} = 600\ \mu m$
Tapered amplifier length	$L_{tap} = 600\ \mu m$
No. sections of DFB laser	$N_{dfb} = 12$
No. sections of tapered amplifier	$N_{tap} = 12$
No. sections of DBR	$N_{dbr} = 12$
No. sections of passive waveguide	$N_{pas} = 84$
DFB coupling coefficient	$\kappa_{dfb} = 4000\ m^{-1}$
DBR coupling coefficient	$\kappa_{dbr} = 1000\ m^{-1}$
DBR loss factor	$\alpha_{dbr} = 4000\ m^{-1}$
Optical confinement factor	$\Gamma = 0.2$
Effective group refractive index	$n_g = 3.871$
Linewidth enhancement factor	$\alpha_H = 2$
Transparency carrier density	$N_{tr} = 1.0 \times 10^{24}\ m^{-3}$
Active region width	$W = 1\ \mu m$
Tapered amplifier output width	$W_o = 20\ \mu m$
Active region depth	$d = 0.1\ \mu m$
Gain coefficient	$a = 1.64 \times 10^{-20}\ m^2$
Active region loss factor	$\alpha_{sc} = 4000\ m^{-1}$
Effective spontaneous coupling factor	$\beta_{sp} = 4.0 \times 10^{-5}$
Nonradioative recombination constant	$A_{nr} = 0\ s^{-1}$
Radiative recombination constant	$B_{rad} = 8.0 \times 10^{-17}\ m^3s^{-1}$
Auger recombination constant	$C_{aug} = 9.0 \times 10^{-41}\ m^6s^{-1}$

and Fig. 17.6. From Fig. 17.5, it is observed that between drive frequencies of 5.5 and 7 GHz, the RMS timing jitter (t_j) is always less than 2 ps. This gives a total locking bandwidth of over 1.5 GHz. Within the stable region of mode-locked operation in Fig. 17.6, the RMS pulse-to-pulse peak power variation (p_{var}) is between 225 mW (0.2 mW) and 500 mW (0.8 mW) for the pulses emitted from the amplifier (DFB laser). Average peak powers (P_{pk}) fall between 8.5 W (7 mW) and 12 W (14 mW). The ratio of p_{var} to P_{pk} satisfies the supporting definition of the locking bandwidth (i.e. $p_{var}/P_{pk} < 10\%$) at all times between 5.5 GHz–7 GHz. Average powers (P_{av}) of the amplifier (DFB) output are between 2.1 W to 2.4 W (1.7 mW to 2.7 mW) across this stable region. The pulsewidths (Δt_{pul}) emitted from the DFB laser *before* passing the amplifier lie between 27.8 ps and 30.4 ps. If a semiconductor device with higher differential gain constant is used, quantum-well devices for example, narrower pulses can be generated. The pulsewidths of the mode-locked pulses collected at the amplifier output fall between 28.4 ps and 31.6 ps. This shows that there is only little pulse broadening owing to the laser amplification. Spectral widths (Δf_{spec}) of the amplifier (DFB) output are between 37 GHz (37 GHz)

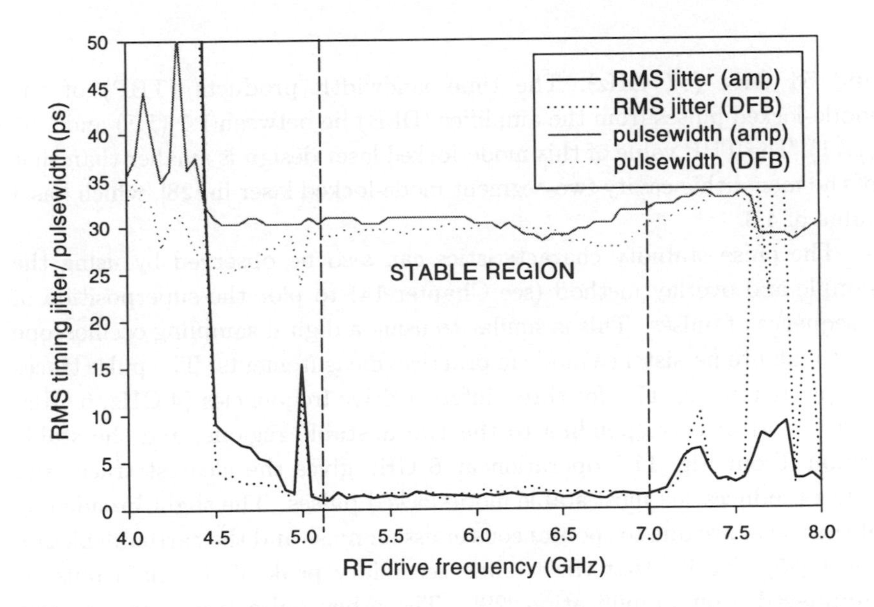

Fig. 17.5. RMS timing jitter and pulsewidth characteristics of amplifier and DFB output.

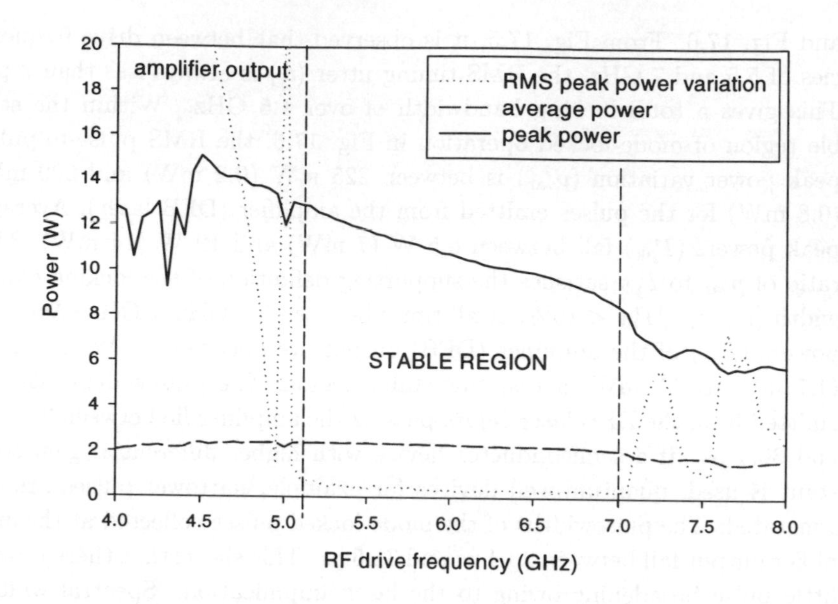

Fig. 17.6. Power vs. frequency characteristics of the amplifier output.

and 54 GHz (57 GHz). The time-bandwidth products (TBP) of the mode-locked pulses from the amplifier (DFB) lie between 1.1 (1.1) and 1.65 (1.71). The TBP value of this mode-locked laser design is smaller than that of the monolithic cavity two-segment mode-locked laser in [28], which has a value of 4.3.

The pulse stability characteristics can also be observed by using the sample and overlay method (see Chapter 14) to plot the superposition of a sequence of pulses. This is similar to using a digital sampling oscilloscope set to infinite persistence mode in practical measurements. The pulse traces are plotted in Fig. 17.7 for three different drive frequencies (4 GHz, 6 GHz, and 7.7 GHz) corresponding to the two unstable regions, and the stable region. From Fig. 17.7 operation at 6 GHz gives the clearest trace, and hence produces the most stable mode-locked pulses. The slight broadening of the traces are due to spontaneous emission noise and the carrier depletion effect [25]. Notice that the trailing secondary peak of the main pulse is suppressed after amplification [29]. The other pulse traces are severely blurred because the consecutive pulses do not lie on top of each other, mainly due to cyclic instabilities.

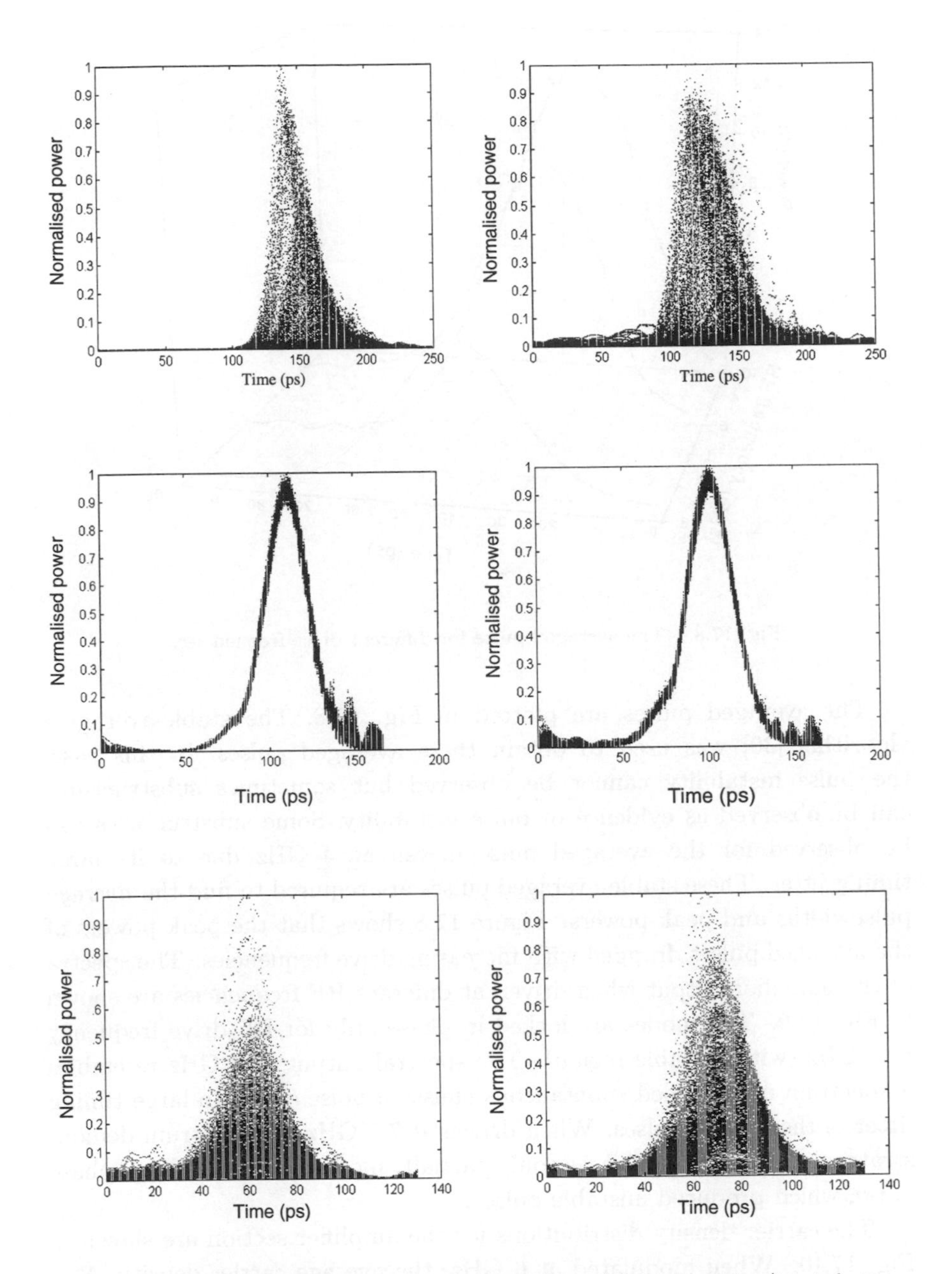

Fig. 17.7. Sample and overlay pulse traces for different drive frequencies (top to bottom: 4 GHz, 6 GHz, 7.7 GHz): DFB output (left), amplifier output (right).

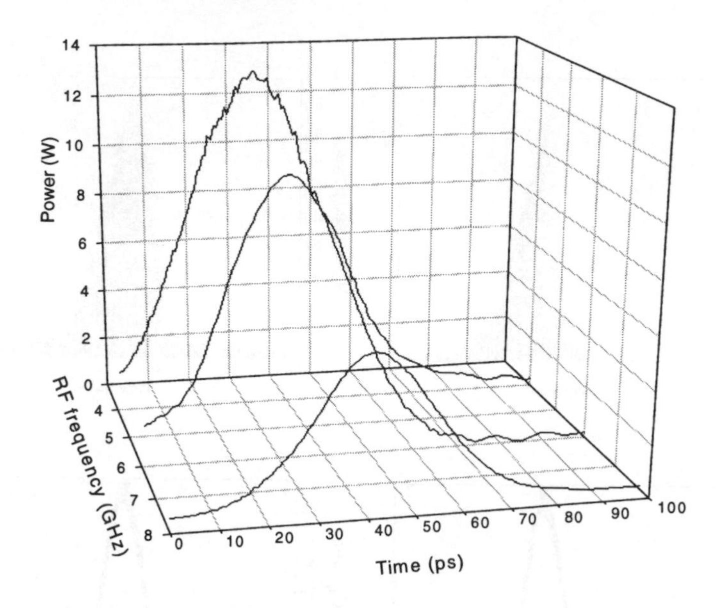

Fig. 17.8. The averaged pulse for different drive frequencies.

The averaged pulses are plotted in Fig. 17.8. The stable-averaging algorithm [30] was used to obtain these averaged pulses. In this case, the pulse instability cannot be observed but sometimes substructures can be observed as evidence of pulse instability. Some substructures can be observed for the averaged pulse driven at 4 GHz due to its large timing jitter. These stable-averaged pulses are required to find the *average* pulsewidths and peak powers. Figure 17.8 shows that the peak powers of the averaged pulses dropped with increasing drive frequencies. The spectra of the amplifier output when driven at different RF frequencies are shown in Fig. 17.9. The modes are locked in phase only for RF drive frequency of 6 GHz (within stable region). The spectral output at 4 GHz resembles a spectrum of amplified spontaneous emission noise owing to large timing jitter of the output pulses. When driven at 7.7 GHz the spectrum demonstrates that the laser device was only partially mode-locked with large phase noise, which produced unstable pulses.

The carrier density distributions for the amplifier section are shown in Fig. 17.10. When modulated at 6 GHz, the average carrier density, N_{av}, was lowest falling from 3.4×10^{24} m^{-3} at the input (adjacent to DFB section) to 1.4×10^{24} m^{-3} at the output. This was because the carriers

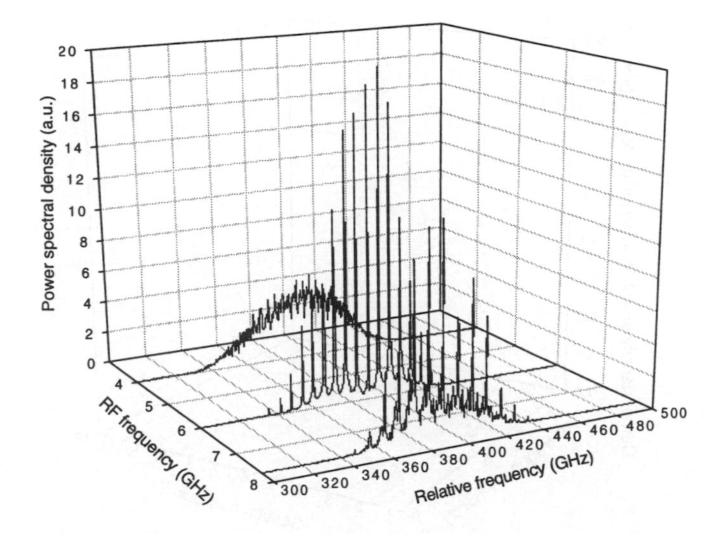

Fig. 17.9. The spectral characteristics of the mode-locked pulses for different drive frequencies.

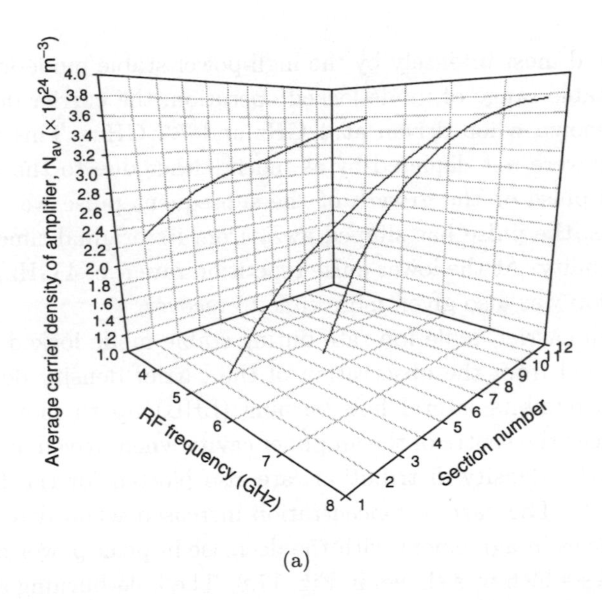

(a)

Fig. 17.10. The carrier density distributions for the tapered amplifier section (a) average carrier density, N_{av}, and (b) carrier density deviation, N_1, for drive frequencies of 4, 6, and 7.7 GHz corresponding to different regions of pulse stability.

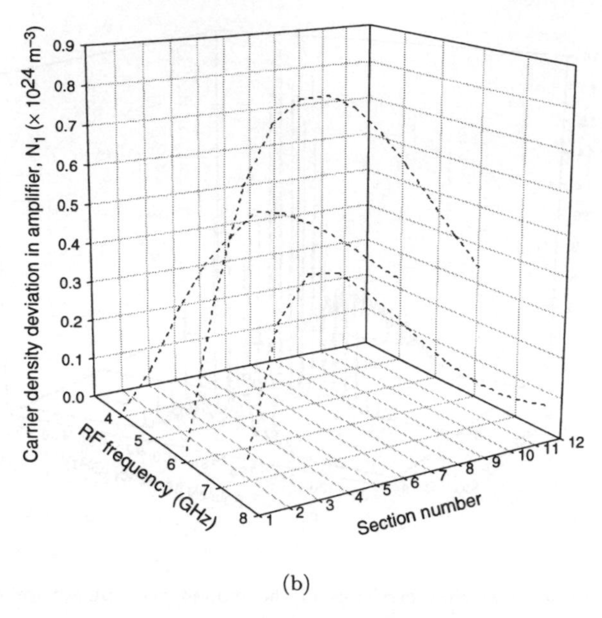

(b)

Fig. 17.10. (*Continued*)

were depleted most intensely by the high-power stable mode-locked pulses. In the unstable range of mode-locked operation, the carrier densities were higher as shown when driven at 4 GHz and 7.7 GHz. This was because the carriers were not depleted by as much, either due to the absence of a subsequent pulse or the growth of the subsequent pulse was delayed. In other words, the pulse has slipped away from its original time slot due to cyclic instability. At the lower pulse repetition rate (i.e. 4 GHz), the carrier concentration was also given more time to recover.

The high carrier depletion rate during stable mode-locked operation is also confirmed from the observation of the carrier density deviation, N_1, plot, which exhibits lateral hole-burning (LHB) up to more than $0.8 \times 10^{24}\ \mathrm{m}^{-3}$ near the centre of the amplifier cavity when driven at 6 GHz. The average carrier density distributions are also plotted for the DFB section in Fig. 17.11. The carrier concentration increased when driven at higher RF frequencies in agreement with the decrease in peak power of the mode-locked pulses which was shown in Fig. 17.8. The hole-burning at the centre of the cavity is clearly seen as expected for the quarter-wave-shifted (qws) DFB laser with a high coupling factor ($\kappa L = 2.4$). Again, the hole-burning was most severe for the case of stable mode-locking, i.e. driven at 6 GHz.

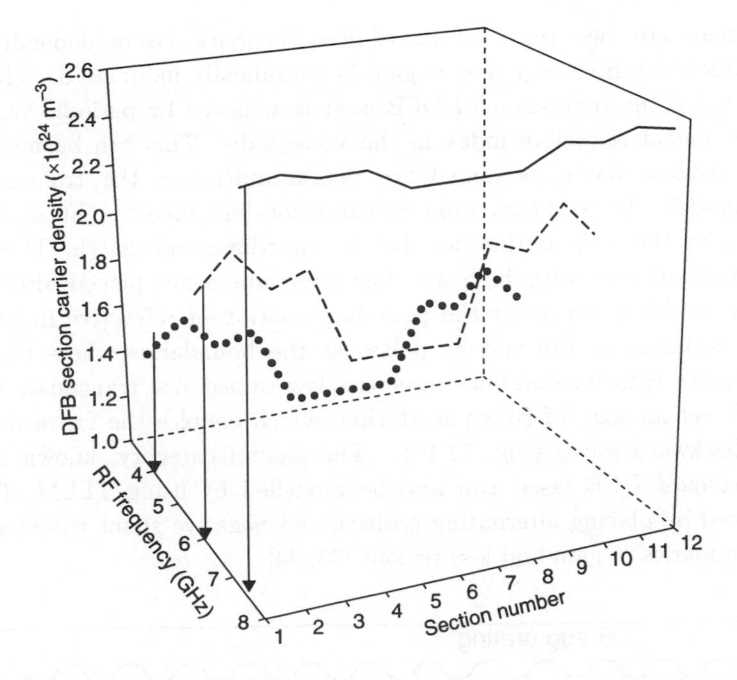

Fig. 17.11. The carrier density distributions for the DFB section for drive frequencies of 4, 6, and 7.7 GHz corresponding to different regions of pulses stability.

17.4 TLM Model of Corrugated Grating Structures

Before moving on to study the design considerations of the DFB and DBR sections, the transmission-line matrix (TLM) model of corrugated gratings is first introduced. Corrugated gratings are wavelength selective structures found in distributed Bragg reflectors (DBR), distributed feedback lasers (DFB), and fibre Bragg grating (FBG) lasers. Kogelnik and Shank [31] were the first to observe lasing action in a periodic structure that utilised the distributed feedback mechanism. This means that feedback for the lasing action is not localised at the cavity facets but is distributed throughout the cavity length. When an etched corrugated grating is introduced into the waveguide, the perturbation of the layer causes periodic modulation of the refractive index or gain resulting in distributed feedback by means of Bragg scattering, which couples the counter-propagating waves. In practice, there are uncontrollable phase shifts associated with incomplete periods of the grating near the edges. For this study, we assume complete periods for the grating.

There are two types of distributed feedback lasers depending on whether the refractive index or gain is periodically modulated. The first type called the *index-coupled* DFB laser is achieved by periodic variation of the modal refractive index in the waveguide. This can be modelled, in equivalent basis, as impedance discontinuities in the transmission-line model. It is known from transmission-line theory that scattering occurs at these discontinuities due to impedance mismatch. Therefore, transmission-lines with high and low impedances are placed alternately in the model to represent the periodic variation of refractive index [32]. The scattering of the voltage pulses at the boundaries where the high impedance transmission-lines meet the low impedance transmission-lines is a direct analogy of Bragg scattering, which couples the forward-waves and backward-waves (Fig. 17.12). The second category, known as the *gain-coupled* DFB laser, can also be modelled by using TLLM. This is achieved by placing alternating positive and negative shunt conductances at boundaries of gain and loss regions [33, 34].

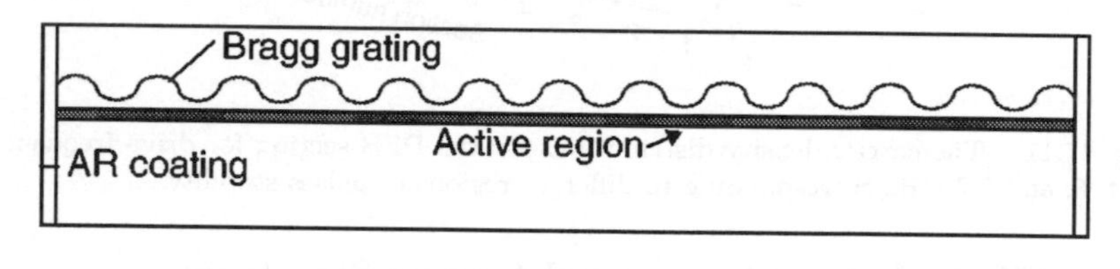

(a)

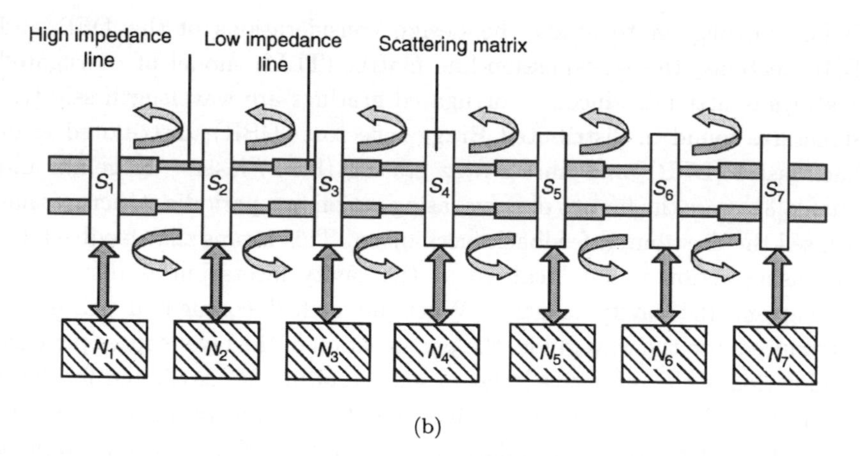

(b)

Fig. 17.12. (a) The uniform index-coupled DFB laser and its (b) transmission-line model.

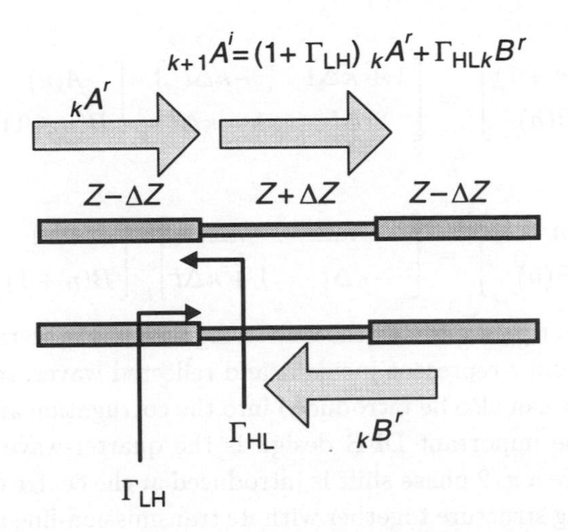

Fig. 17.13. Alternating high and low impedance transmission-lines representing the periodic modulation of refractive index.

In the index-coupled DFB laser model, the line impedances are modulated by $\pm\Delta Z$ from their mean impedance value, Z (Fig. 17.13). From transmission-line theory, the reflection coefficients can be found as [36]:

low-high (LH) impedance boundary:
$$\Gamma_{LH} = \frac{(Z+\Delta Z)-(Z-\Delta Z)}{(Z+\Delta Z)+(Z-\Delta Z)} = \frac{\Delta Z}{Z} \tag{17.1}$$

high-low (HL) impedance boundary:
$$\Gamma_{HL} = \frac{(Z-\Delta Z)-(Z+\Delta Z)}{(Z-\Delta Z)+(Z+\Delta Z)} = -\frac{\Delta Z}{Z}. \tag{17.2}$$

This reflection coefficient is related to the coupling factor per unit section, $\kappa\Delta l$ by [32]:

$$\Gamma_{LH} = \frac{\Delta Z}{Z} = \kappa\Delta l \tag{17.3}$$

$$\Gamma_{HL} = -\frac{\Delta Z}{Z} = -\kappa\Delta l \tag{17.4}$$

where Δl is the unit section length. By denoting the forward-waves at section-n as $A(n)$, and the backward-waves as $B(n)$, it is straightforward to find the scattering matrices as the following. The first section is selected to be a low-impedance line.

if n is odd:

$$
{}_{k+1}\begin{bmatrix} A(n+1) \\ B(n) \end{bmatrix}^{\mathrm{i}} = \begin{bmatrix} 1+\kappa\Delta 1 & -\kappa\Delta l \\ \kappa\Delta l & 1-\kappa\Delta l \end{bmatrix}_k \begin{bmatrix} A(n) \\ B(n+1) \end{bmatrix}^{\mathrm{r}} \tag{17.5}
$$

if n is even:

$$
{}_{k+1}\begin{bmatrix} A(n+1) \\ B(n) \end{bmatrix}^{\mathrm{i}} = \begin{bmatrix} 1-\kappa\Delta l & \kappa\Delta l \\ -\kappa\Delta l & 1+\kappa\Delta l \end{bmatrix}_k \begin{bmatrix} A(n) \\ B(n+1) \end{bmatrix}^{\mathrm{r}} \tag{17.6}
$$

where the subscript k denotes the scattering at the kth iteration, and the superscripts i and r represent incident and reflected waves, respectively.

Phase-shifts can also be introduced into the corrugation anywhere along its length. One important DFB design is the quarter-wave-shifted (qws) structure, where a $\pi/2$ phase shift is introduced at the centre of the grating. The qws grating structure together with its transmission-line representation is shown in Fig. 17.14. This will give a lowest-threshold mode at exactly the Bragg wavelength (centre of the stop band). The qws DFB laser is attractive because it is able to achieve single-mode operation with large MSR of above 30 dB even when directly modulated at high speeds in the

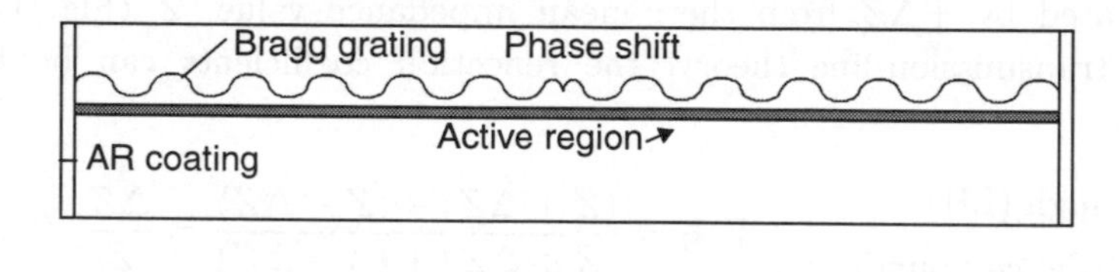

(a)

High impedance line
Low impedance line
Scattering matrix
S_1 S_2 S_3 S_4 S_5 S_6

(b)

Fig. 17.14. (a) The $\lambda/4$-shifted DFB laser and its (b) transmission-line laser model.

GHz range. In order to mitigate severe spatial hole-burning, the coupling factor should not be too high. Injection current can be increased at the location of the phase-shift to replenish the 'burned' carriers. Alternatively, multiple phase-shifts can be placed at various locations along the grating to achieve a more spatially uniform carrier density distribution [39]. For the novel mode-locked laser design in this book, the single phase-shift qws DFB laser is chosen as the RF-modulated section for its unique spectral characteristics. The model can be implemented in TLLM by placing a zero reflection interface at the centre of the cavity. This is simply represented by a 2×2 identity matrix given as:

$$_{k+1}\begin{bmatrix} A(n+1) \\ B(n) \end{bmatrix}^{i} = \begin{bmatrix} 1 & 0 \\ 0 & 1 \end{bmatrix}_{k} \begin{bmatrix} A(n) \\ B(n+1) \end{bmatrix}^{r} \tag{17.7}$$

where n is S/2, and S is an even number denoting the number of sections used in the model.

17.5 Design of Grating (DFB and DBR) Sections

In active mode-locking, a bandwidth-limiting element such as a dispersive grating filter or DBR is required to ensure narrow spectral output for stable mode-locked pulses. In bulk optics, the diffraction grating is typically used as the bandwidth-limiting element which provides both filtering and feedback. The wavelength-selective feedback can also be provided by a fibre Bragg grating, which can be fabricated by using a two-beam interference ultraviolet irradiation technique. For the monolithic device used in this book, the passive DBR section has been chosen as the integrated bandwidth-limiting element. The coupling coefficient, length, and the attenuation coefficient of the Bragg region can be altered to achieve different feedback efficiencies (reflectivity) and bandwidths. For simple geometries such as rectangular-shaped grooves with duty-cycle of 1:2, the coupling coefficient can be approximated by [37]:

$$\kappa \cong k_0 \Delta n \Gamma \frac{\sin\left(\frac{m\pi}{2}\right)}{m\pi} \tag{17.8}$$

where k_0 is the free space wavenumber, Δn is the refractive index difference on both sides of the grating, and m is the order of the grating.

In order to achieve a narrow spectral filtering for stable mode-locking, a weak coupling coefficient is required but there is a trade-off in terms of feedback efficiency. To achieve low values of coupling coefficient, the corrugation depth should be shallow [37]. The weak coupling factor will produce a spectral response with a "rounded-off" reflection band with very small ripples at the sidemodes, i.e. the DBR spectral response resembles that of a dispersive grating element used in conventional active mode-locking. This is in contrast to the "flattened" reflection band with large ripples at the sidemodes observed in the spectral response for strongly coupled DBRs. The reflection peak broadens and flattens when $\kappa L \gg 1$ but such high coupling factors give better feedback efficiencies. The evolution of the spectral characteristics for different coupling strengths is shown in Fig. 17.15.

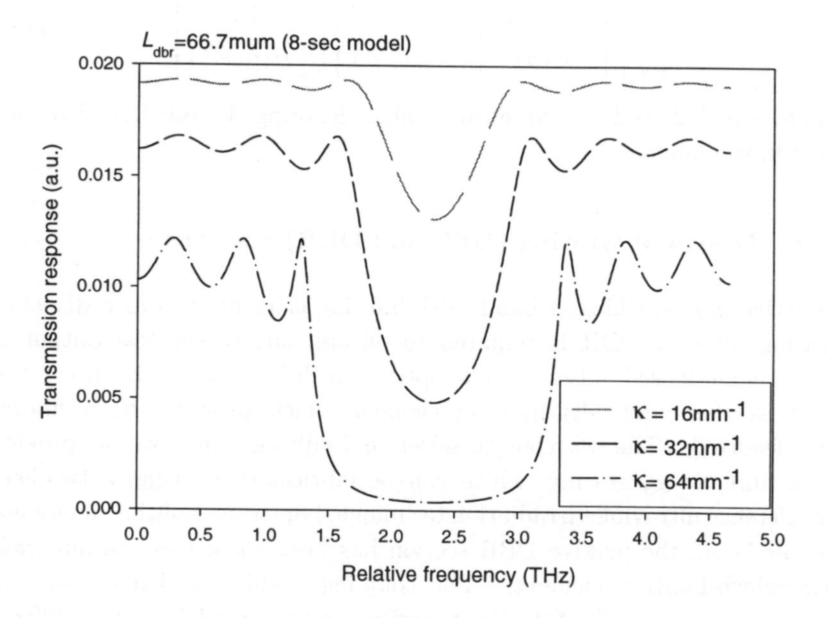

Fig. 17.15. The spectral characteristics of the uniform DBR for various coupling strengths.

Since there are two sets of distributed Bragg gratings used in the integrated device (one of them is DC-biased and RF-modulated while the other remains unbiased and unmodulated), it is important to ensure that any interactions (i.e. spectral feedback and filtering) between them do not cause any unwanted effects during the mode-locked operation. Priority of choice is given to the modulated DFB section as this will eventually determine the output signal to be emitted into the tapered amplifier.

If the uniform (uni) DFB laser is used in combination with the DBR, the CW spectral output from the DFB section (refer to Fig. 17.1) will consist of two widely-separated (in frequency) lasing modes. If the DFB section is modulated, there would be "clusters" of extended cavity modes around each of these two lasing modes. Consequently, the optical pulses will appear to have many temporal substructures due to the intense beating effects between these two widely-separated clusters of extended cavity modes. Furthermore, the two different clusters of extended cavity modes are fed by different portions of the spontaneous emission spectrum, which is weakly correlated causing even more beating noise and large timing jitter between pulses. In addition, the uniform grating laser structure has been found to give self-pulsations when biased well above threshold [38]. These self-pulsations are due to mode instabilities induced by spatial hole-burning. Therefore, the uniform grating design will not be considered. An alternative, the quarter-wave-shifted DFB laser, was chosen as the RF-modulated section because it has only one transmission peak (i.e. the lowest threshold mode) at the centre of the stop band, in contrast to the two widely-separated transmission peaks for the uniform grating structure. Therefore, the spectral output would not have any unwanted beating effects, which may cause temporal substructures to appear in the mode-locked pulses (neglecting chirp for the moment).

The steady-state spectra at CW operation comparing the two different DFB-DBR combinations are shown in Fig. 17.16. A few extended cavity modes (intermode spacing of several GHz) are observed for the qws DFB — uni DBR device owing to some self-pulsations because the device was biased well above threshold. The qws DFB — uni DBR device has a lower threshold current than its uni DFB — uni DBR counterpart. On the other hand, the lasing modes of the uni DFB — uni DBR combination are separated by more than 150 GHz. The clusters of extended cavity modes around each of these lasing modes are not observed because the DFB section was not modulated.

The FP-DBR mode-locked laser has been demonstrated by Hansen *et al.* [18]. However, it is not possible to include a travelling-wave amplifier (TWA) section next to a FP laser because this means the elimination of the FP cleaved facet, which provides the feedback required for mode-locked operation. Hence, the DFB-DBR mode-locked laser design was chosen since it allows the inclusion of a TWA section at the output end to provide high-power output pulses. Simulation studies showed that the most stable mode-locked pulses free from substructures could only be achieved by using

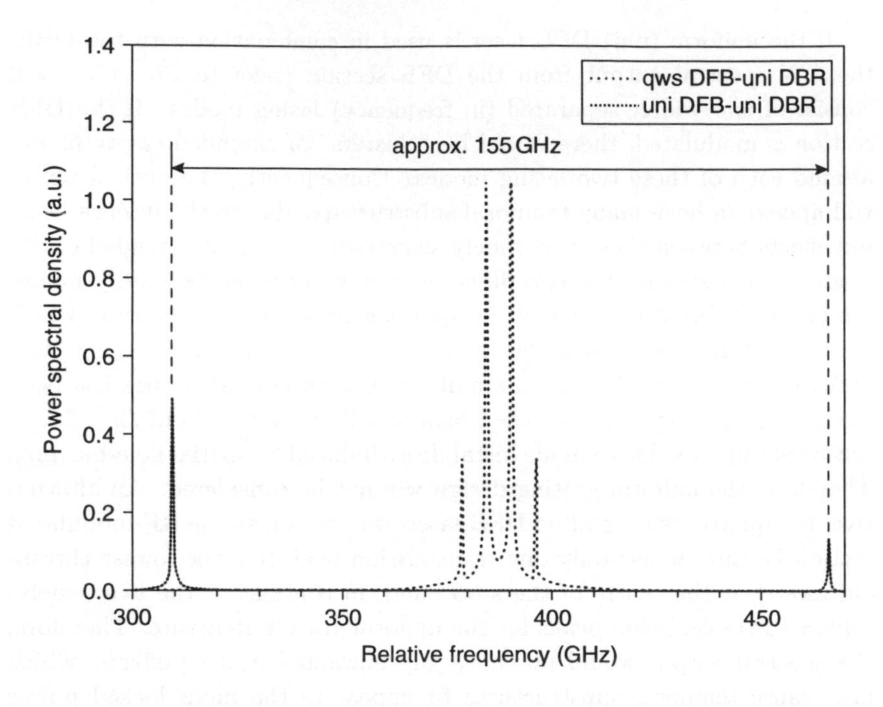

Fig. 17.16. Steady-state spectra at CW operation ($I_{dc} = 40$ mA) for two combinations (i) uni DFB — uni DBR and (ii) qws DFB — uni DBR (qws — quarter-wave-shifted grating, uni — uniform grating).

either the qws DFB — uni DBR or the qws DFB — qws DBR laser design. The CW spectral characteristics of the integrated DFB-DBR and FP-DBR mode-locked laser devices are compared in Fig. 17.17. It was found that the intermode spacings (corresponding to the cavity resonant frequencies) are 6 GHz and 5.5 GHz for the qws DFB — uni DBR and qws DFB — qws DBR combinations, respectively. As for the FP — uni DBR combination, the resonant frequency is higher, at around 8 GHz. The discrepancies are due to the complex spectral filtering and feedback provided by the integrated DFB-DBR devices.

The spectral widths (FWHM) of the uniform DBR and qws DBR grating for a coupling coefficient of 10 cm^{-1} are found to be 130 GHz and 230 GHz, respectively (as shown in Fig. 17.18). Strictly speaking, there are two passbands for the qws DBR but here we assume that they are both merged into one and consider the effective passband only. Therefore, a qws DBR with the same coupling coefficient will have a wider equivalent band-

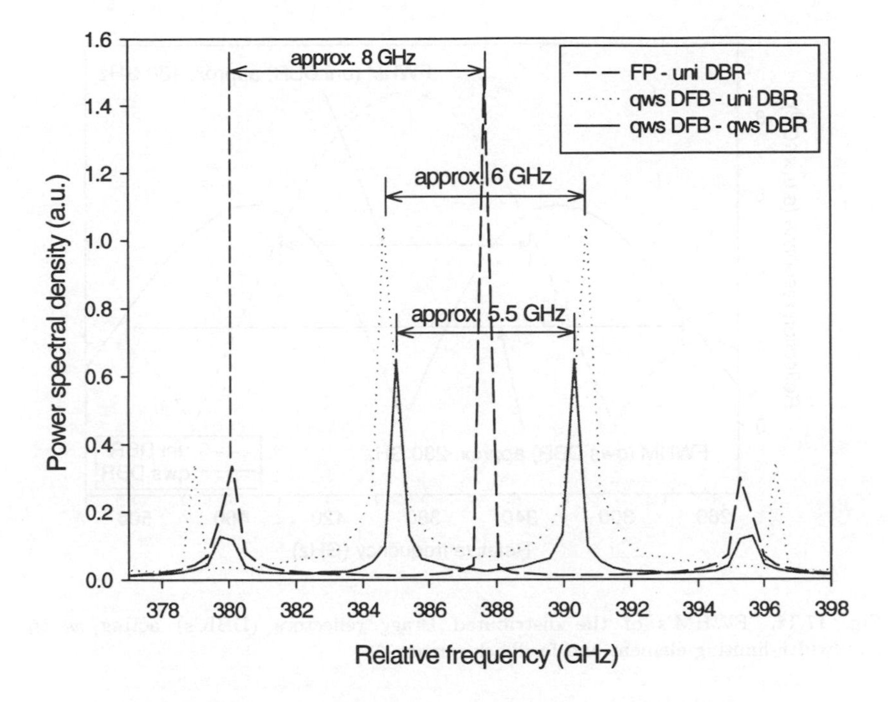

Fig. 17.17. Steady-state spectra at CW operation for bias level of $I/I_{\text{th}} \approx 1.2$: (i) FP-DBR was biased at 80 mA (ii) DFB-DBR combinations were biased at 40 mA.

width. Notice that there is a sharp transmission peak (thus poor feedback efficiency) at the Bragg wavelength for the qws grating, hence there can be no cavity resonance for signal frequencies that lie within that region. If no chirp is present, the spectra for the qws DFB combinations are symmetrical about the Bragg wavelength (e.g. Fig. 17.17).

Simulations showed that the optical pulses generated from the qws DFB — qws DBR combination produced mode-locked pulses with narrower pulsewidths and higher peak powers than the qws DFB — uni DBR combination. The absence of a reflection peak at the Bragg wavelength in the qws DFB and qws DBR leads to a weaker suppression ratio between the two dominant modes nearest to the Bragg wavelength and their sidemodes. This produced a wider bandwidth for the same number of locked extended cavity modes in the spectrum or possibly a greater number of locked modes together, and hence narrower pulse width in the time domain. Thus, the novel integrated mode-locked laser was based on the qws DFB — qws DBR combination. The final design of the monolithically integrated device has

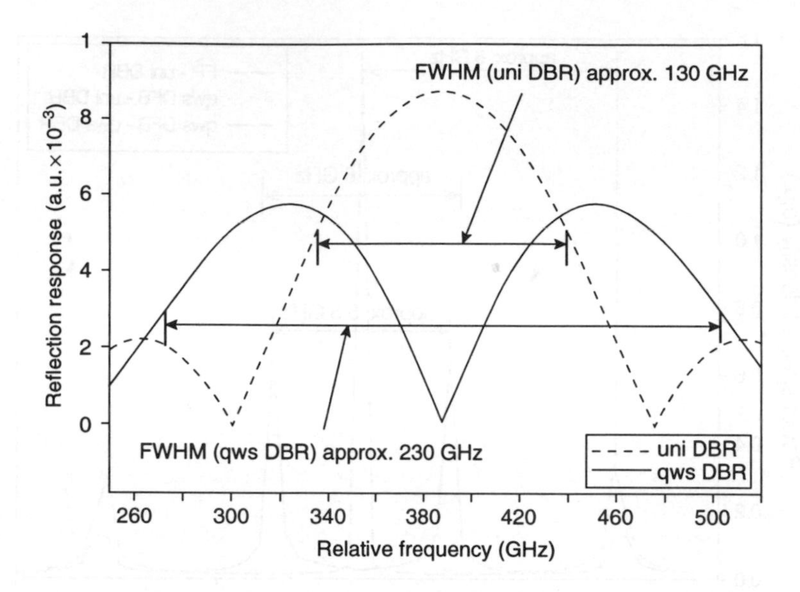

Fig. 17.18. FWHM's of the distributed Bragg reflectors (DBR's) acting as the bandwidth-limiting element and feedback mirror.

an overall threshold current of around 25 mA, given amplifier bias current, I_{amp}, of 630 mA. The design has coupling factors of $\kappa L = 2.4$ for the DFB section, and $\kappa L = 0.6$ for the DBR section.

In order to induce mode-locking, a bias current of 40 mA and a RF modulation current (peak-zero amplitude) of 300 mA were used. The model was run for 20 ns of real laser time to allow the mode-locked operation to stabilise. Another 10 ns was required to perform time-averaging on the output signal to filter out noise and the high frequency components. Simulation studies showed that all transients would have settled within 10 ns. For computational efficiency during the early design stage, a small number of sections (S) could be used. This is because the computation time is proportional to S^2. The DFB and DBR models were each divided into 12 sections with model timestep, Δt, of 645 fs. The coupling per section must be increased accordingly to compensate for the small number of sections. Nevertheless, the spectral response of the reduced-section model closely mimics that of the model with infinitesimally small sections, but only within a limited bandwidth as shown in Fig. 17.19.

Strictly speaking, the effective length of the DBR is a function of its coupling coefficient. The effective length of the Bragg region is given in

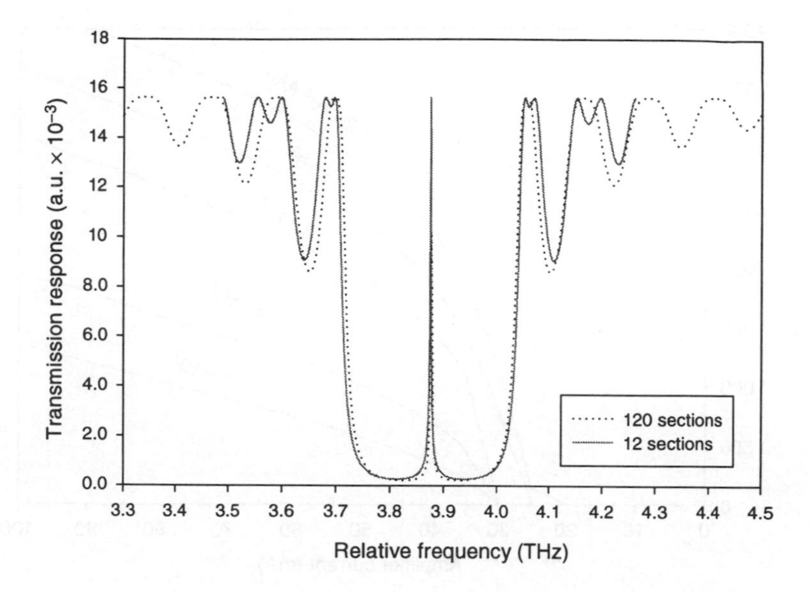

Fig. 17.19. Spectral response of the passive distributed Bragg grating for coupling coefficient of $\kappa = 100\ \text{cm}^{-1}$ and $L = 600\ \mu\text{m}$ for $S = 12$ and $S = 120$.

[39–40]. However, the coupling factor dependence of the effective length of the grating was not taken into account in TLLM because dispersionless transmission-lines were used to satisfy the synchronisation criterion (i.e. all pulses must arrive at the scattering nodes at the same time). Hence, the design assumed the exact length of the grating.

17.6 Amplifier Design

Figure 17.20 shows the light output emitted from the amplifier section (output width, W_0 equals to 20 μm) as a function of injection current into the DFB laser, with amplifier current density, J_{amp}, as parameter. At lower values of J_{amp} ($0.8 \times 10^8\ \text{A/cm}^2$ and below), the threshold current of the DFB laser section is around 30 mA. At higher amplifier current densities however, the threshold current begins to shift to lower values. This is due to the higher contribution of amplified spontaneous emission from the amplifier section into the DFB section. Also observed in the figure is the presence of saturation when the DFB laser current is increased for a fixed value of amplifier current density.

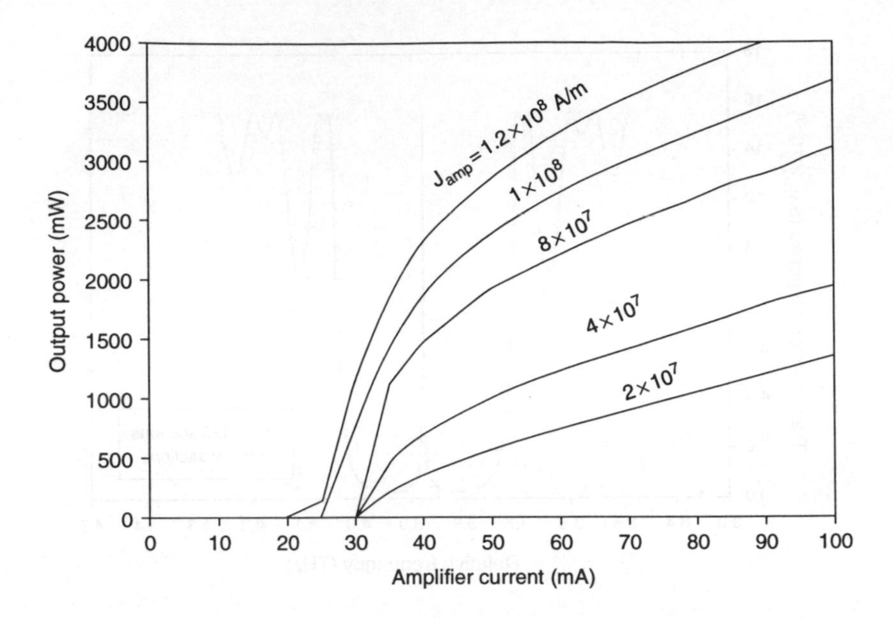

Fig. 17.20. Light-current characteristics of the integrated mode-locked laser device as a function of DFB laser current and with amplifier current density, J_{amp} as parameter.

Figure 17.21 shows the light output from the amplifier section as a function of amplifier current density, J_{amp}, with DFB laser current, I_{dfb} as parameter. When the DFB laser is biased below 40 mA, threshold currents exist for the amplifier section. In this case, the threshold level is only reached when the contribution of ASE from the amplifier section into the DFB laser is high enough to cause lasing behaviour in the integrated device. When the DFB laser is biased at 40 mA and above, lasing action is achieved even without contribution from the ASE. In this case, the output from the DFB section is merely amplified by the tapered amplifier. The output power emitted is a linear function of the amplifier current. There is no sign of saturation in the light-current characteristics of Fig. 17.21 in contrast to Fig. 17.20 because now the amplifier current is increased, supplying more carriers to avoid gain saturation.

The gain spectrum of a travelling-wave amplifier can have a wide bandwidth of up to 10 THz and above, provided the gain ripple is less than 3 dB. This is satisfied since the reflectivity of the output end of the amplifier is only 1×10^{-4}. This value is achievable with a single-layer AR coating and with appropriate combination of refractive index and film thickness [41].

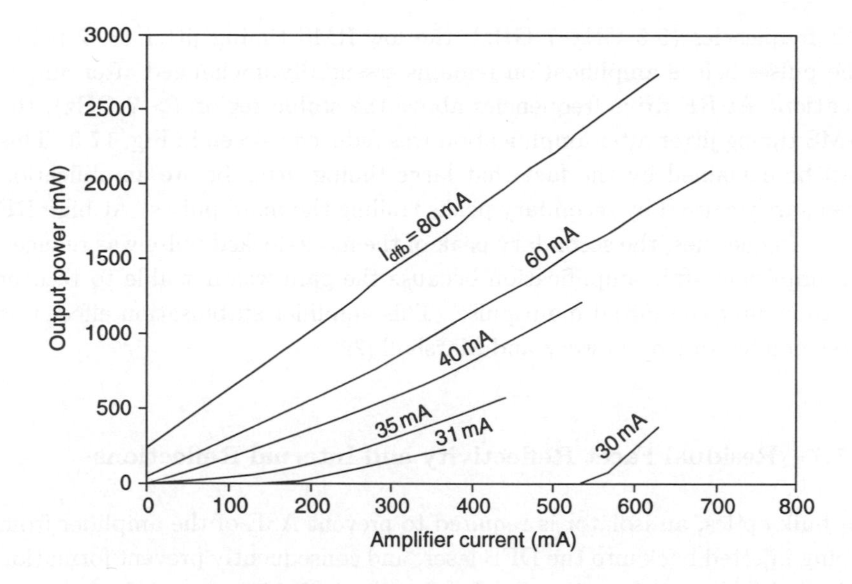

Fig. 17.21. Light-current characteristics of the integrated mode-locked laser device as a function of amplifier current and with DFB laser current I_{dfb}, as parameter.

The exact gain spectrum bandwidth is not important in the mode-locked laser design since the bandwidth of the output signal will eventually be limited by the Bragg reflector.

Using the parameter values in Table 17.2, the unsaturated single pass gain was around 22 dB when spontaneous emission noise was deliberately switched off in the model. When spontaneous emission noise is present in the model, the amplified output power includes ASE, and therefore the signal gain must be filtered to avoid overestimation of the amplifier gain. An accurate filter with sharp cut-off can only be achieved using higher order digital filters [42]. In a crude manner, however, the output signal can be filtered by using a TLM stub filter with a very high Q-factor. Simulations showed that when spontaneous emission noise was included in the amplifier model, the pulse-to-pulse peak power variation increased, reducing pulse stability. The inclusion of ASE in the amplifier model gives us a more realistic description of the output pulse waveform in the study of picosecond pulse generation.

Simulations showed that for drive frequencies below the stable region (< 5.5 GHz) there was an increase in the RMS timing jitter after amplification of the pulses from the DFB section. Within the stable range of

RF frequencies (5.5 GHz–7 GHz), the low RMS timing jitter (< 2 ps) of the pulses before amplification remains essentially unchanged after amplification. At RF drive frequencies above the stable region (> 7 GHz), the RMS timing jitter after amplification was reduced as seen in Fig. 17.5. This can be explained by the fact that large timing jitter before amplification was partly caused by secondary peaks trailing the main pulses. At high RF drive frequencies, the secondary peak of the mode-locked pulse was reduced in amplitude after amplification because the gain was not able to recover in time after the initial main pulse. This amplifier stabilisation effect was first pointed out by Lowery and Marshall [29].

17.7 Residual Facet Reflectivity and Internal Reflections

In bulk optics, an isolator is required to prevent ASE of the amplifier from being injected back into the DFB laser, and consequently prevent formation of another resonant cavity. In the monolithically integrated device, such an isolator could not be used. However, the output facet is AR-coated to ensure that the feedback from the amplifier section does not severely affect the mode-locked operation in the DFB-DBR sections. The near-zero reflectivity also suppresses laser oscillation within the amplifier, which may otherwise cause excessive gain saturation. The gradual increase in the width of the active region allows for higher saturation output power but at the expense of higher ASE, which is inevitably emitted back into the DFB section.

In the standard design, the amplifier section is AR-coated to a residual reflectivity of 1×10^{-4}, which makes it qualify as a travelling wave amplifier (TWA) to amplify the picosecond pulses generated by the mode-locked DFB-DBR section. In order to investigate the influence of finite reflections at the output facet on the mode-locked operation, the pulse characteristics are plotted in Figs. 17.22, 17.23 and 17.24 with output reflectivity (R_{tp}) as parameter. The spike at 5 GHz (see later discussion on internal reflections) that shows large amplitude and phase jitter is deliberately left out in these figures for better clarity. Increasing the value of R_{tp} caused the locking bandwidth to shrink from over 1.5 GHz to less than 0.4 GHz (Fig. 17.22). This is because the higher reflectivity caused stronger feedback into the DFB section, thus destabilising the mode-locked operation.

At higher output reflectivities, the peak power of mode-locked pulses emitted from the amplifier section drops (Fig. 17.23). From Fig. 17.24,

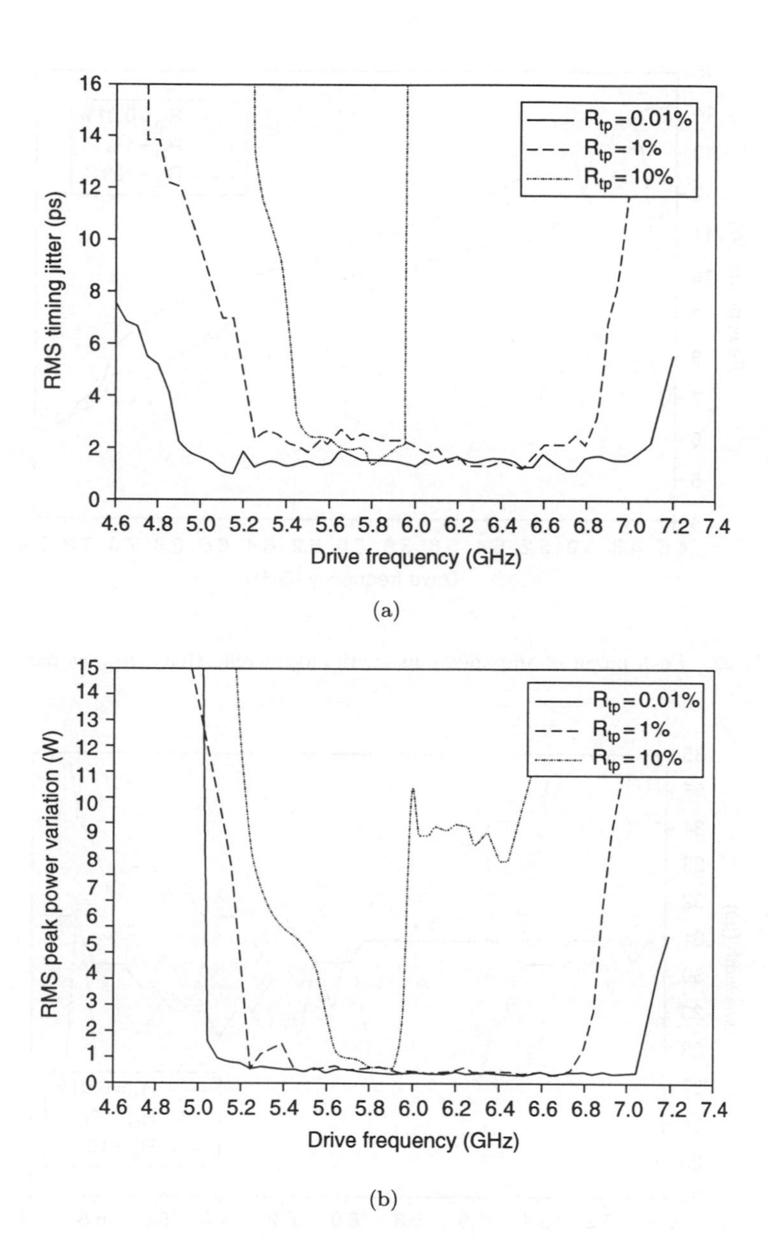

Fig. 17.22. Amplified pulse stability characteristics of the integrated mode-locked laser design with output reflectivity, R_{tp} as parameter. (a) RMS timing jitter and (b) peak power variation.

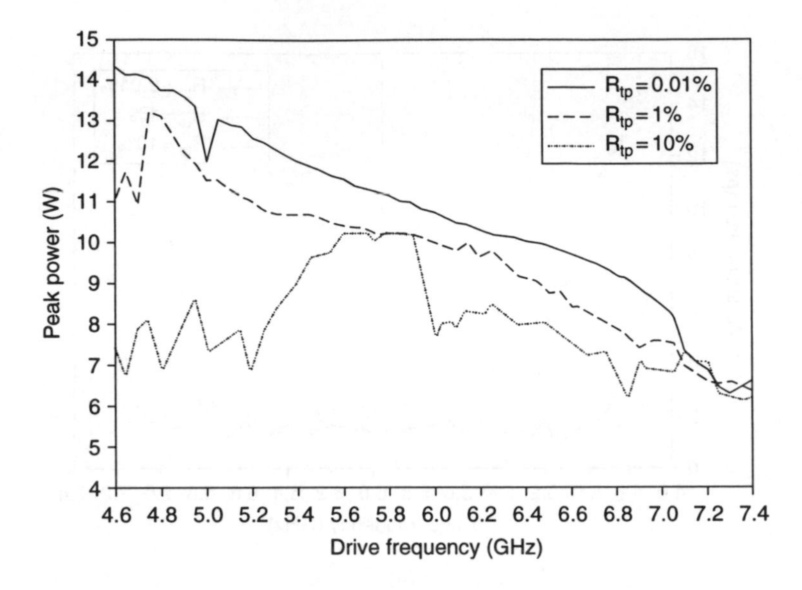

Fig. 17.23. Peak power of amplified pulse with output reflectivity, $R_{\rm tp}$, as parameter.

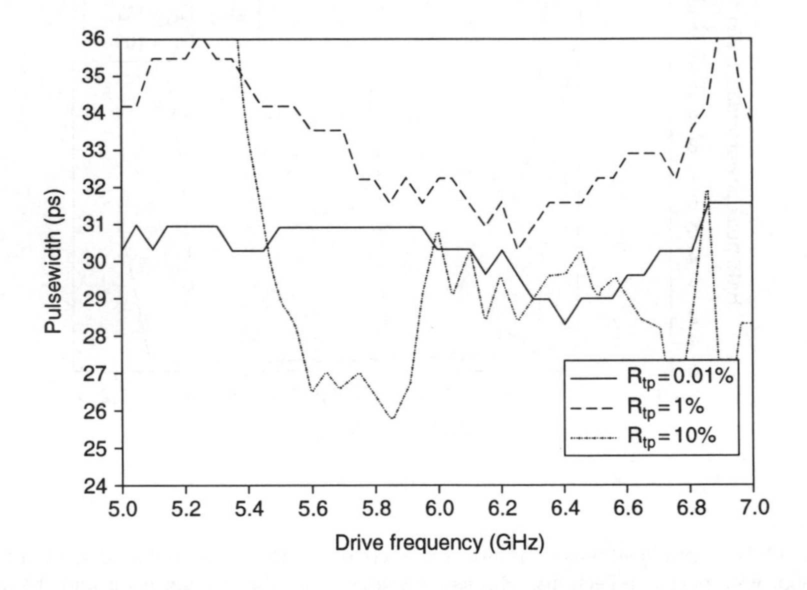

Fig. 17.24. Pulsewidth of amplified pulse with output reflectivity, $R_{\rm tp}$, as parameter.

the pulsewidth increased with increasing reflectivity at low values of $R_{\rm tp}$ (0.01%–1%). Interestingly, when a moderately high value of $R_{\rm tp}$ (10%) was used, the narrowest pulsewidths could be achieved within a small range of drive frequencies (5.6–5.9 GHz). These frequencies coincide with the new stable mode-locked region, which also gives the highest peak powers of the mode-locked pulse. The peak power rises to a pedestal within the stable region similar to the behaviour of conventional mode-locked lasers using only one wavelength selective element [17, 26] and [43]. This is probably because the mode-locked operation is now dominated by a new resonant cavity, formed between the amplifier output facet ($R_{\rm tp}$) and the DBR section, that resembles the extended cavity laser used in conventional active mode-locking.

In the monolithically integrated device, any imperfect active-passive waveguide transition and slightly different refractive index between them may cause finite intercavity reflections, which then leads to multiple pulses recirculating in the compound cavity device. Simulation of these undesirable reflections can easily be incorporated into the transmission-line laser model of the integrated mode-locked laser design. In order to see the effect of finite internal reflections on the mode-locked operation, the following three different cases were investigated: (1) $R_1 = 1\%$, $R_2 = 0\%$ (2) $R_1 = 0\%$, $R_2 = 1\%$ and (3) $R_1 = 1\%$, $R_2 = 1\%$ where the positions of R_1 and R_2 within the device have been defined in Fig. 17.1. These internal reflections are shown in Fig. 17.25 and the corresponding scattering matrices in transmission-line model of the novel mode-locked laser design are:

$$_{k+1}\begin{bmatrix} V_{\rm a}(1) \\ V_{\rm ns\_b} \end{bmatrix}^{\rm i} = \begin{bmatrix} \sqrt{R_1} & (1-\sqrt{R_1}) \\ (1-\sqrt{R_1}) & \sqrt{R_1} \end{bmatrix}_{\rm k} \begin{bmatrix} V_{\rm b}(1) \\ V_{\rm ns\_a} \end{bmatrix}^{\rm r} \tag{17.9}$$

$$_{k+1}\begin{bmatrix} V_{\rm b}(S) \\ V_{\rm del\_a} \end{bmatrix}^{\rm i} = \begin{bmatrix} \sqrt{R_2} & (1-\sqrt{R_2}) \\ (1-\sqrt{R_2}) & \sqrt{R_2} \end{bmatrix}_{\rm k} \begin{bmatrix} V_{\rm a}(S) \\ V_{\rm del\_b} \end{bmatrix}^{\rm r} \tag{17.10}$$

where $V_{\rm a}$ are $V_{\rm b}$ the forward and backward travelling waves in the DFB section, and the section number, n, is labelled from left to right as in Fig. 17.1.

The simulation results are plotted in Fig. 17.26. There is little change in the locking range for all three different cases of internal reflections. When the reflectivity at the front of the DFB laser (R_1) is absent, a spike around 5 GHz in the pulse stability characteristics (RMS peak power vari-

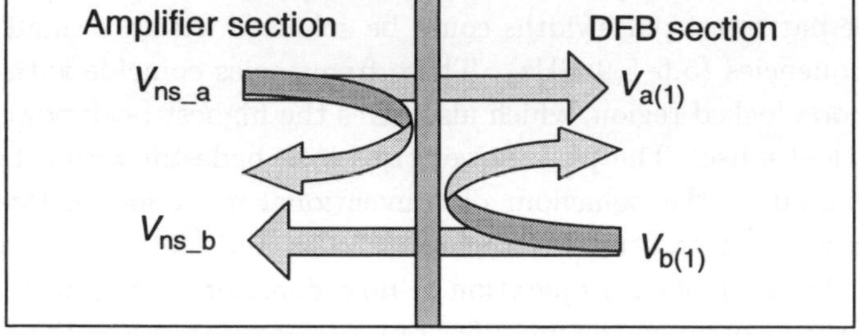

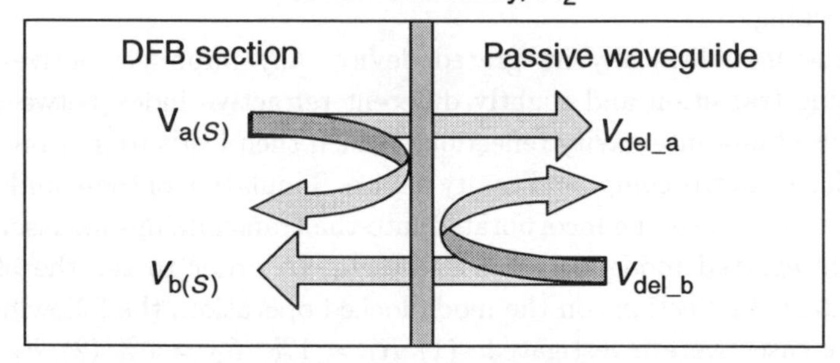

Fig. 17.25. Internal reflections in the monolithic cavity.

ation) is observed. This spike has been observed in the simulation results of the standard design (Fig. 17.5), where no internal reflections have been assumed. Interestingly, with only 1% reflectivity in the DFB front facet, this instability can be suppressed.

The time-domain pulses and their respective spectra with R_1 as parameter are shown in Fig. 17.27. The bifurcation of peak power into three different amplitudes is clearly observed. These pulses would still give relatively low RMS timing jitter, and could be deceptive if the RMS peak power variation was not considered (as discussed in Section 17.3). The effect of a finite value of R_1 is to filter out the periodic noise-like signal (depicted by arrows in Fig. 17.27) in between the resonant cavity modes. This filtering effect is probably due to the formation of separate cavities within the integrated device, similar to spectral filtering using a Fabry Perot interferometer. However, when internal reflections are present in both the front facet (R_1) and back facet (R_2) of the DFB laser, the pulsewidth broadens within the stable region of mode-locked operation (Fig. 17.26(b)).

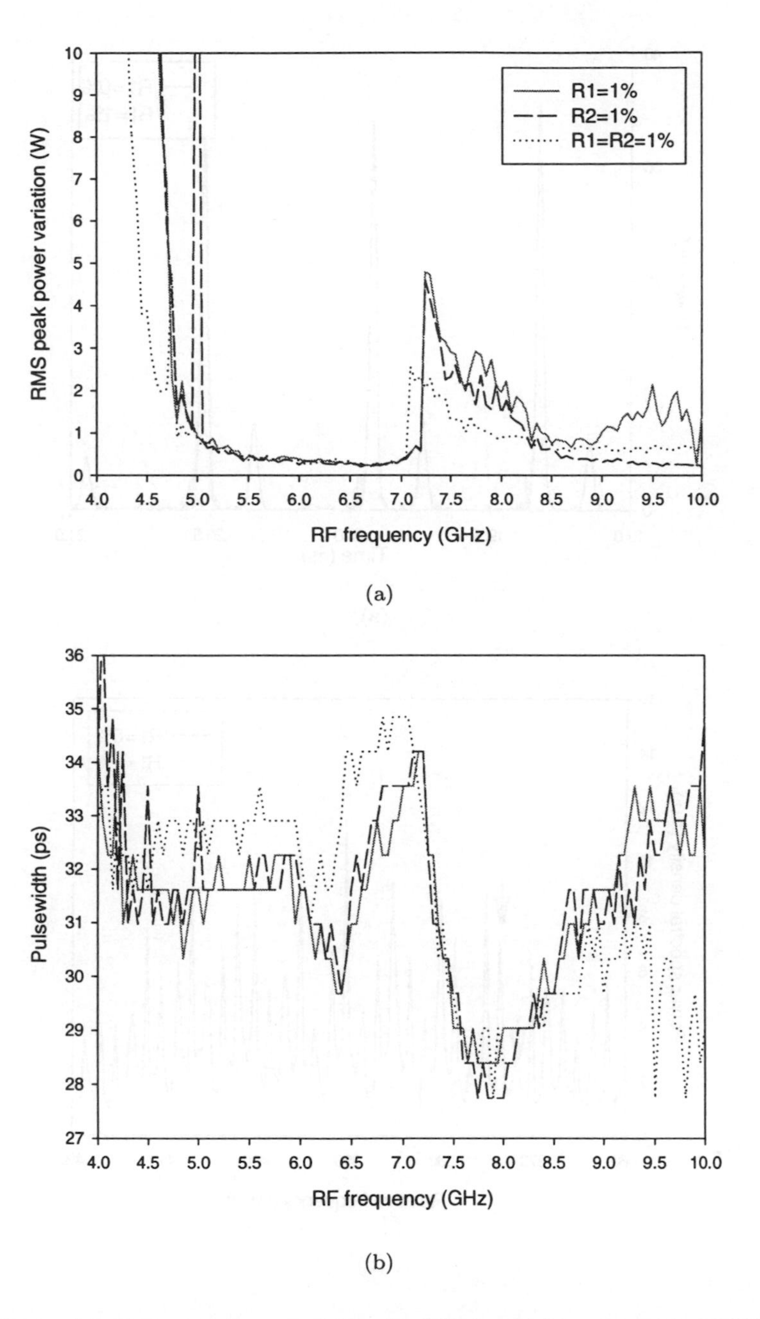

Fig. 17.26. (a) RMS peak power variation (stability), and (b) pulsewidth (FWHM) of the mode-locked pulses when internal reflections are present.

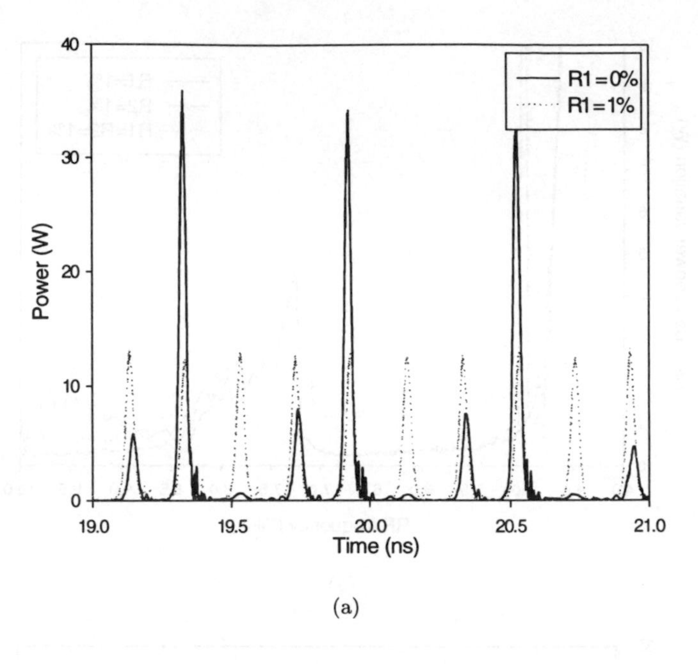

(a)

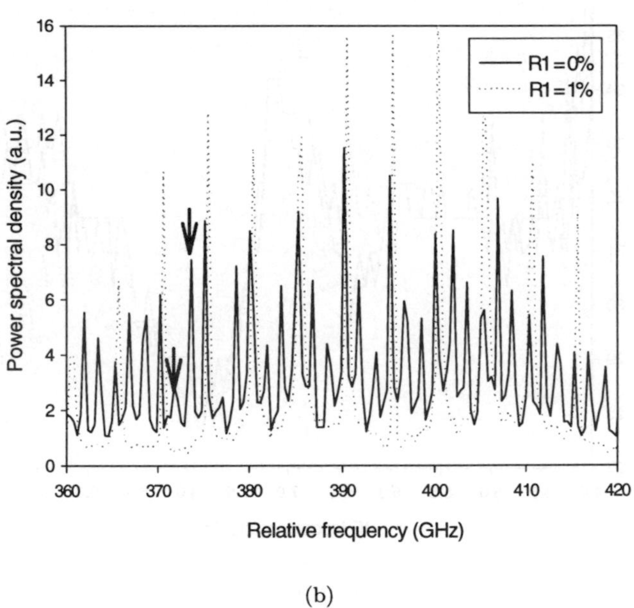

(b)

Fig. 17.27. The effect of front facet reflectivity of DFB section, R_1, on the mode-locked pulses (a) in the domain, and (b) the respective spectra.

Simulation studies showed that the power (average and peak) emitted at the amplifier output dropped when these internal reflections were present. This is related to the higher Q-factor of the integrated device.

Finally, the reflectivity at the open end of the DBR section, R_{2_dbr}, may also affect the mode-locked operation. For the standard design, the DBR is assumed to have a perfect AR-coating ($R_{2\_\mathrm{dbr}} = 0$). The stability and pulsewidth characteristics when higher reflectivities are present at the open end of the DBR section have been investigated, and plotted in Fig. 17.28. The figure shows that the locking range is not affected by the DBR facet reflectivity. However, the pulsewidth seems to widen with increasing values of R_{2_dbr}. Further investigation shows that the peak power emitted from the amplifier output remains unaffected by R_{2_dbr}, which means that the average power emitted increases with R_{2_dbr}, since the pulse has widened.

17.8 Passive Waveguide Loss

The passive waveguide is usually used for reducing the drive frequency required for active mode-locking to reasonable values ($<$ 20 GHz). The design here included a passive waveguide section with a length of 4.2 mm, seven times longer than the other individual sections. Although high attenuation is not desirable, controlled loss can be introduced in the passive waveguide (or the DBR section) to eliminate any large recirculating pulse due to energy storage within the DBR, which may otherwise cause large timing jitter. The presence of leading and trailing secondary pulses are detrimental to OTDM systems. However, if the loss is too high then the mode-locked pulses broadens [17], and the locking bandwidth shrinks [27]. The average output power of the mode-locked pulses will also fall if the loss becomes excessive. Therefore, the value of loss in the passive waveguide and the DBR section must be chosen carefully to improve both pulseshape and stability. Passive waveguide loss as low as 3–4 dB/cm has been achieved in monolithic devices [18] and [44].

The passive waveguide loss contributes to the total loss in the monolithic device. The overall loss affects the threshold carrier concentration in the active DFB section. The threshold carrier concentration required for stimulated emission increases with total loss. This will subsequently affect the characteristics of the mode-locked pulses since they are dependent on DC bias level relative to threshold level, as well as RF modulation depth (see Section 17.10). Excessive loss will produce noisier output signal and lower extinction ratio. For a fixed bias level, excessive total loss may

even cause the carrier concentration to fail to reach threshold level and subsequently inhibit any mode-locked operation.

The timing jitter characteristics with the passive waveguide loss as parameter are shown in Fig. 17.28. As observed earlier in Section 17.6, the timing jitter for the output pulse emitted from the tapered amplifier is smaller than that of the pulse entering the amplifier from the DFB section at RF frequencies above the stable region (> 7 GHz). There is little change to the RMS timing jitter within the stable region even as the waveguide loss is increased up to 8 dB/cm. However, there is a formation of an unstable region between 6.5 GHz and 7 GHz when the loss is excessive (at 16 dB/cm). Furthermore, at higher waveguide losses (8 dB/cm and 16 dB/cm), the large timing jitters have been significantly reduced in the unstable region above 7 GHz, leaving a moderately stable region around 8.5 GHz. It has been shown that waveguide loss can be introduced so that multiple regions of stability can "merge" into one stable region but at expense of smaller locking bandwidth [27]. Although low timing jitter is achieved in this new "stable" region at around 8.5 GHz, it will not be considered for mode-locking because the RMS pulse-to-pulse peak power variation is large relative to the average peak power (i.e. ratio > 10%). Further simulations confirmed that the waveguide loss should be below 12 dB/cm to prevent formation of unstable regions within the original stable region of mode-locked operation (i.e. 5.5 GHz–7 GHz).

The influence of the waveguide loss to the output powers of the amplifier and DFB output is shown in Figs. 17.29 and 17.30. It is clear that increasing the loss will result in lower peak power as well as average power from both the amplifier and DFB sections. The smallest influence of waveguide loss is seen for peak power of the *amplified* mode-locked pulses at RF drive frequencies between 5.5 GHz to 6 GHz as shown in Fig. 17.29(a). For all cases, both the peak power and average power drop as drive frequency is increased. The steepest fall in average power occurs near 7 GHz, where excessive pulse instability begins to set in (Fig. 17.30). In conventional mode-locking using a single wavelength selective element such a fibre Bragg grating [43], the peak power of the mode-locked pulse reaches a peak value near the middle of the locking range. This behaviour has also been found in the conventional method of active mode-locking using diffraction gratings [27]. This is in contrast to the novel mode-locked laser design incorporating two wavelength selective structures (DFB laser and DBR filter sections), where the peak power falls almost linearly by increasing drive frequency within the stable region, as shown in Fig. 17.29.

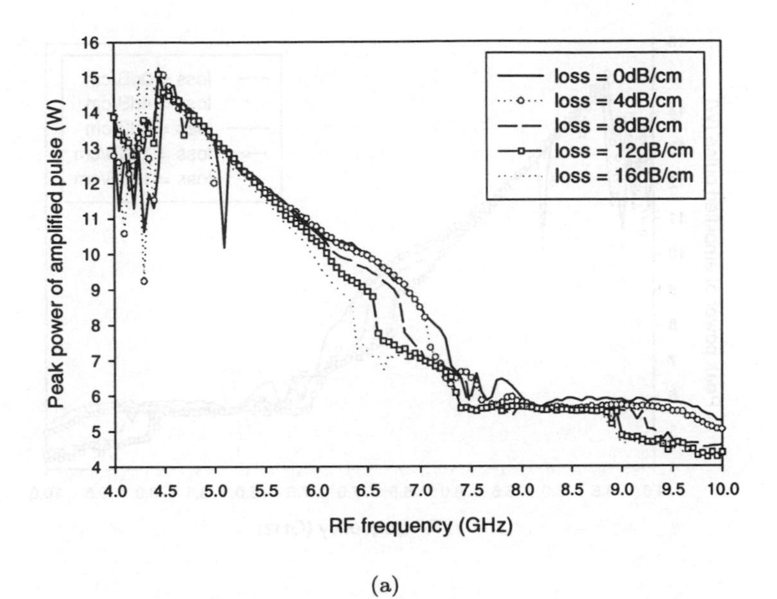

(a)

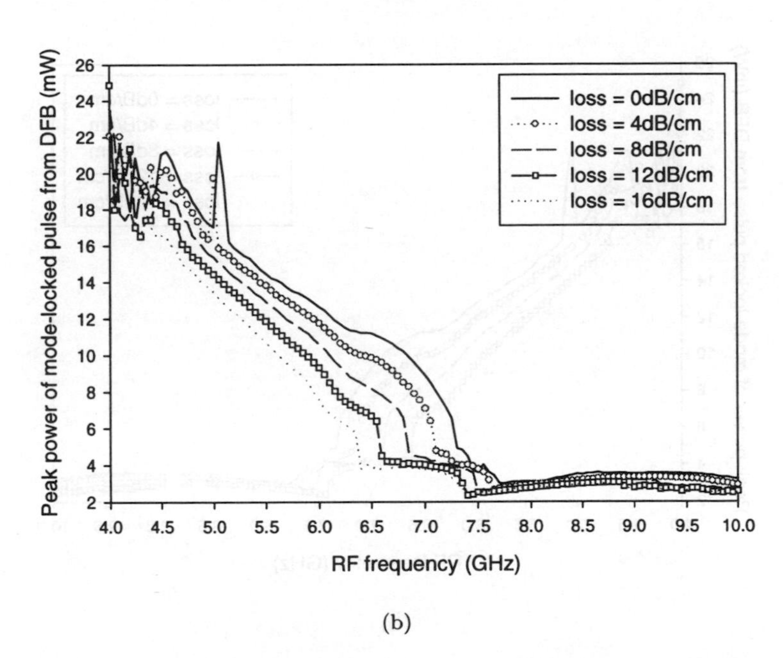

(b)

Fig. 17.29. Peak power against RF frequency for (a) amplifier output, and (b) DFB output with passive waveguide loss as parameter.

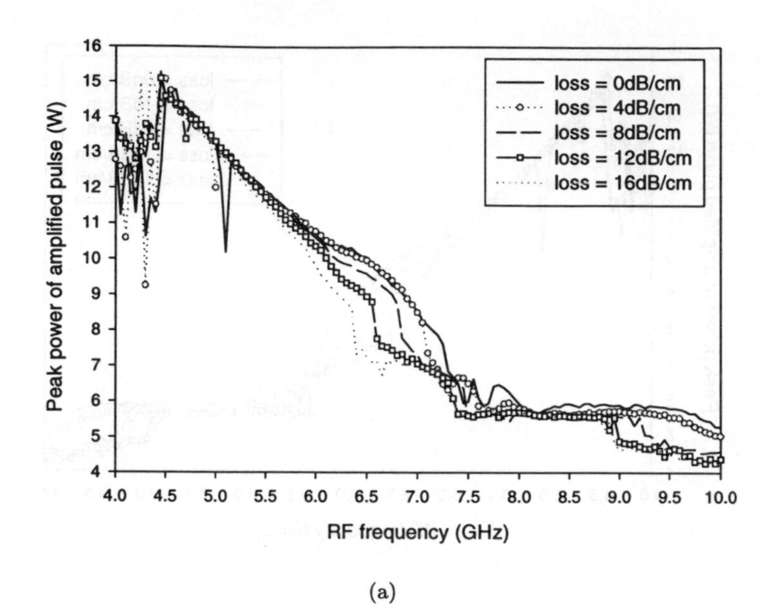

(a)

(b)

Fig. 17.30. Average power against RF frequency for (a) amplifier output, and (b) DFB output with passive waveguide loss as parameter.

The pulsewidth characteristics for various values of passive waveguide loss are shown in Fig. 17.31. There are two regions of shortest pulsewidth, one situated near 6.5 GHz, and the other lies near 7.7 GHz. The deep "valley" near the higher RF frequency of 7.7 GHz gives the shortest pulses down to 26.5 ps. However, this region falls outside the stable region and will not be considered. This shows that the range of frequencies that produces the shortest pulsewidth does not necessary give stable pulses, which is why the pulse stability characteristics (RMS timing jitter and pulse-to-pulse peak power variation) should always be plotted to check whether they fall within the stable region of mode-locked operation. When the passive waveguide loss is first increased from 0 to 8 dB/cm, the pulse narrows within the stable region of mode-locked operation. As the loss is increased further to 16 dB/cm, the pulse begins to widen within the locking range. In this case, there is also a flattening of the "valley" within the stable region (see Fig. 17.31). The pulsewidth of the mode-locked pulses (within the locking range) is broadened only a little after being amplified. The broadening is due to the gain saturation of the amplifier, which amplifies the leading edge more than the trailing edge of the pulse [45]. The tapered amplifier provides some improvement in the saturation characteristics, hence only a small amount of broadening in the pulsewidth [46].

17.9 Effect of Dynamic Chirp

The amplitude-phase coupling governed by the Kramers-Kronig relationship leads to dynamic chirping during short pulse generation. The carriers are depleted quickly owing to the high stimulated emission rate at the onset of the optical pulse, and recovers to above threshold before the next subsequent pulse. This carrier-photon interaction causes frequency shifts toward the blue (shorter wavelength) and red sides during the leading and trailing edges of the pulse, respectively. The different parts of the pulse are now at different wavelengths, and hence travel at different speeds due to group velocity dispersion (GVD). This may result in pulse broadening after propagation in the optical fibre. In an optical fibre with optimised length for normal GVD (positive value), the red-shifted components travel faster than blue-shifted components. Thus, the trailing edge is able to catch up with the leading edge during propagation in the optical fibre, and the output pulse can be compressed by a factor of 4–5 [37]. Besides its effect on the time-domain characteristics of the optical pulse, the dynamic chirp produces new frequency components leading to a wider spectrum. The role

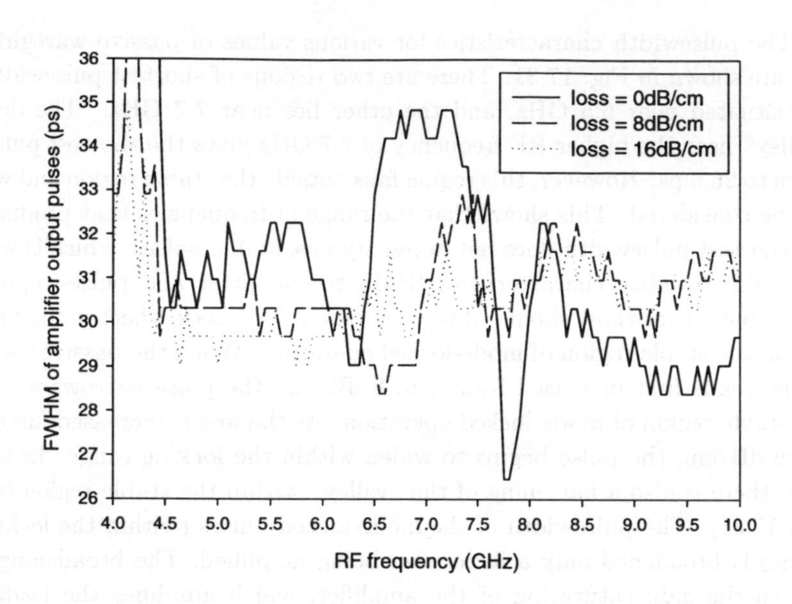

(a)

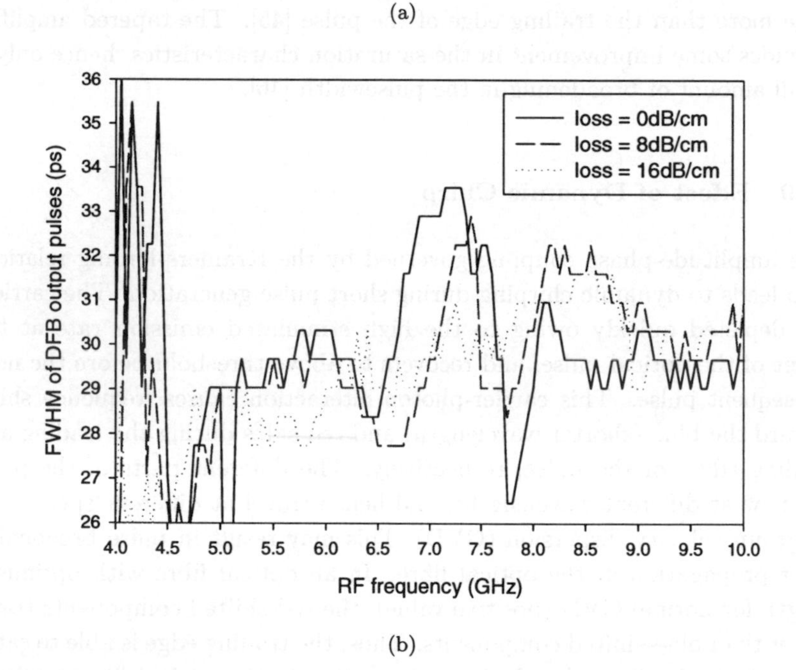

(b)

Fig. 17.31. Pulsewidth characteristics for (a) amplifier output, and (b) DFB output with passive waveguide loss as parameter.

of dynamic chirp in mode-locking has been included in the transmission-line model (see Chapters 14 and 16). It has been found that the value of Henry's linewidth enhancement factor (α_H) is smaller in tapered waveguide structures [47]. Hence, a value of $\alpha_H = 2$ was chosen as the standard parameter for all simulations.

The comparison of the time-domain mode-locked pulse train for $\alpha_H = 0$ and $\alpha_H = 2$ is shown in Fig. 17.32. The RMS timing jitter of the mode-locked pulses when chirp is not included ($\alpha_H = 0$) is 1.13 ps and the percentage ratio of RMS peak power variation to average peak power is only 1.7%. The corresponding average pulsewidth is 31 ps. On the other hand, when dynamic chirp is included, the mode-locked pulses have small bursts of trailing pulses, evidence of the onset of pulse instability. The RMS timing jitter is now increased to 1.57 ps and the percentage ratio of RMS peak power variation to average peak power is 2.6%. However, the average pulsewidth has reduced to 30.3 ps. This confirms that the presence of chirp may lead to narrowing of mode-locked pulses [48]. With chirp included, temporal substructures also become visible in the optical pulse as shown at the inset of Fig. 17.32(a). This is probably due to the mixing effect between the new spectral components generated by dynamic chirp.

In Fig. 17.32(b), the spectral-widening effect of dynamic chirp during mode-locked operation is shown. When chirp is present, more resonant cavity modes are present inside the spectral envelope. The central dominant modes are reduced in magnitude as the sidemodes grow larger, causing the FWHM of the time-averaged spectrum to increase from 18.6 GHz to 50.7 GHz when chirp is present. This corresponds to an increase in the time-bandwidth product (TBP) from 0.577 to 1.54. Therefore, it is more difficult to achieve transform-limited mode-locked pulses when a semiconductor device with a large value of linewidth enhancement factor (α_H) is used.

In a more general analysis, the influence of dynamic chirp on the spectral width (FWHM) and its time-bandwidth product (TBP) during mode-locked pulse generation is shown in Fig. 17.33. Linear regression lines of order 10 are fitted to the fluctuating curves for better clarity. It is clear from the figures that stronger chirp will lead to widening of the spectral width, and subsequently larger values of TBP. There is a dip within the locking range (5.5 GHz–7 GHz) giving a minimum value of TBP. These values are approximately 0.6, 0.8, and 1.3 for the cases of $\alpha_H = 0$ 1 and 2, respectively. At the lower RF frequencies (< 2 GHz), where dips in the TBP curves are also observed near 1 GHz, optical pulses as narrow as

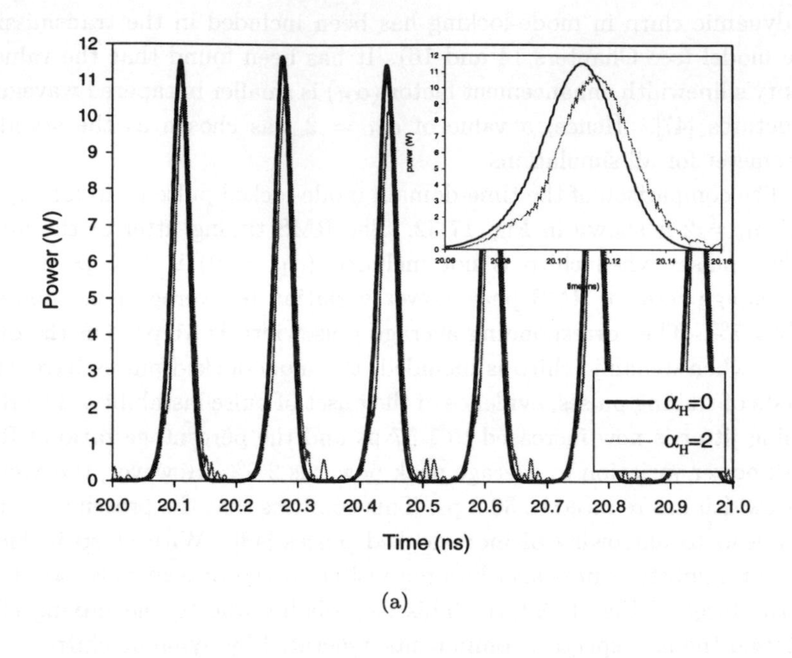

(a)

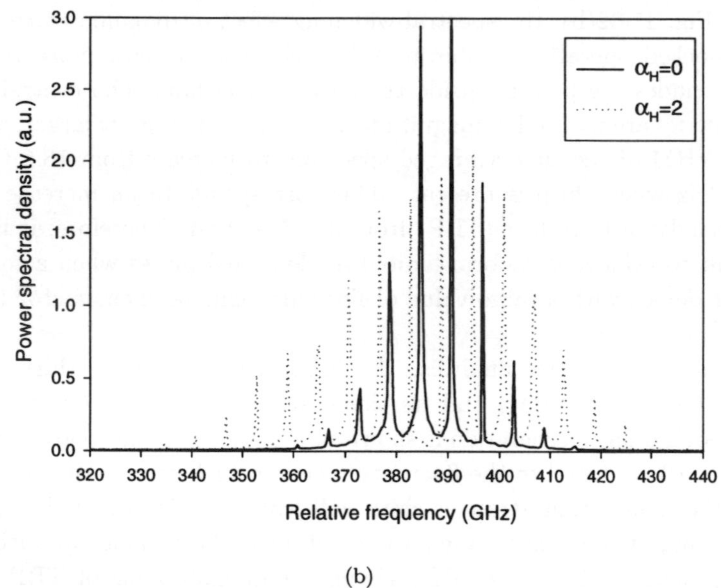

(b)

Fig. 17.32. (a) The mode-locked pulses (inset: single pulse), and (b) their respective spectra with linewidth enhancement factor (α_H) as parameter.

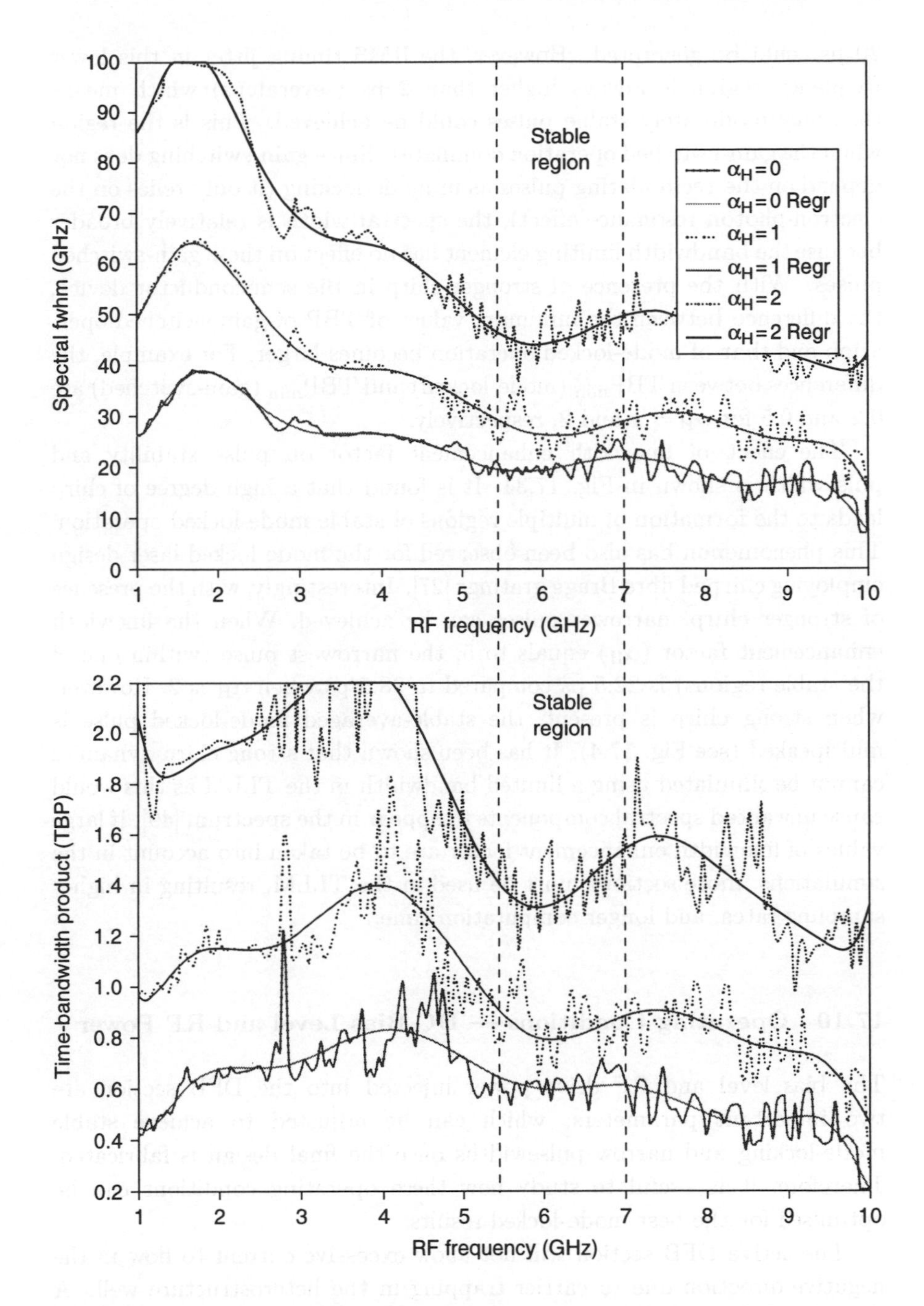

Fig. 17.33. The spectral width and time-bandwidth product of the mode-locked pulses with linewidth enhancement factor (α_H) as parameter.

20 ps could be generated. However, the RMS timing jitter in this lower frequency region is always higher than 2 ps (several ps) which means that only moderately stable pulses could be achieved. This is the region when the gain-switched operation dominates. Since gain-switching does not depend on the recirculating pulses as in mode-locking (it only relies on the electron-photon resonance effect), the spectral width is relatively broader because the bandwidth limiting element has no effect on these gain-switched pulses. With the presence of stronger chirp in the semiconductor device, the difference between the minimum values of TBP of gain-switched operation and that of mode-locked operation becomes larger. For example, the differences between TBP_{min} (mode-locked) and TBP_{min} (gain-switched) are 0.2 and 0.5 for $\alpha_H = 1$ and 2, respectively.

The effect of linewidth enhancement factor on pulse stability and pulsewidth is shown in Fig. 17.34. It is found that a high degree of chirp leads to the formation of multiple regions of stable mode-locked operation. This phenomenon has also been observed for the mode-locked laser design employing chirped fibre Bragg gratings [27]. Interestingly, with the presence of stronger chirp, narrower pulses can be achieved. When the linewidth enhancement factor (α_H) equals to 5, the narrowest pulse (within one of the stable regions) is 22.5 ps compared to 28.5 ps when α_H is 2. However, when strong chirp is present, the stable-averaged mode-locked pulse is multipeaked (see Fig. 17.4). It has been shown that strong chirp dynamics cannot be simulated using a limited bandwidth in the TLLM as this would cause unwanted spectral components to appear in the spectrum [49]. If large values of linewidth enhancement factor are to be taken into account in the simulations, more sections must be used in the TLLM, resulting in higher sampling rates, and longer computation time.

17.10 Operating Conditions — DC Bias Level and RF Power

The bias level and RF drive power injected into the DFB section are two important parameters, which can be adjusted to achieve stable mode-locking and narrow pulsewidths once the final design is fabricated. Therefore, it is useful to study how these operating conditions can be optimised for the best mode-locked results.

The active DFB section will not allow excessive current to flow in the negative direction due to carrier trapping in the heterostructure well. A high drive power level is not desirable due to destructive heating and

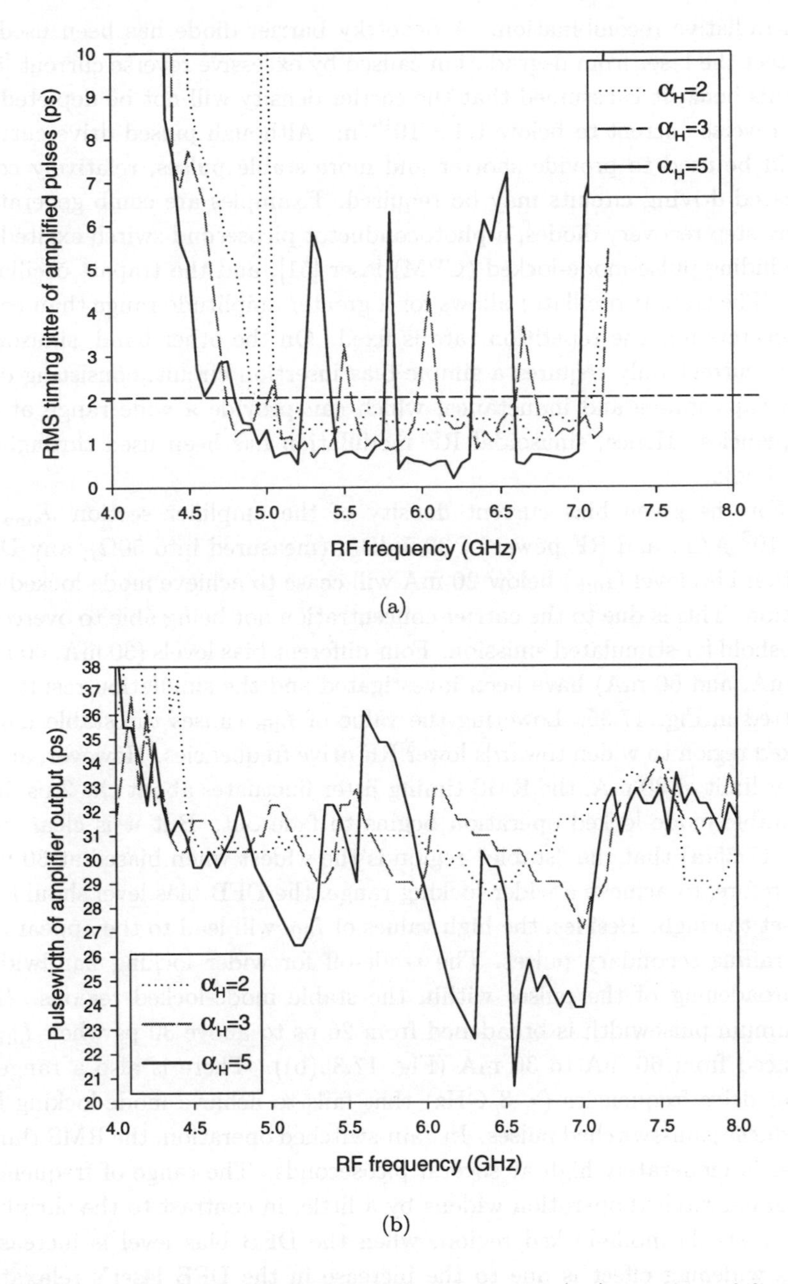

Fig. 17.34. The influence of linewidth enhancement factor (α_H) on pulse stability and pulsewidth.

non-radiative recombination. A Schottky barrier diode has been used to protect the laser from degradation caused by excessive reverse current [50]. In this book, it is assumed that the carrier density will not be depleted by the reverse current to below 1.4×10^{16}/m. Although pulsed drive current could be used to provide shorter and more stable pulses, relatively complicated driving circuits may be required. Examples are comb generators using step recovery diodes, a photoconductor picosecond switch excited by a colliding-pulse-mode-locked (CPM) laser [51], and the trapatt oscillator [52]. The trapatt oscillator allows for a greater amplitude range than comb generators but the repetition rate is fixed. On the other hand, sinusoidal drive current only requires a simple bias insertion circuit, consisting only of a capacitance and inductance, which can provide a wide range of RF frequencies. Hence, sinusoidal RF modulation has been used throughout this book.

For the given bias current density of the amplifier section J_{amp} at 1×10^8 A/m, and RF power of 30.5 dBm (measured into 50Ω), any DFB section bias level (I_{dfb}) below 20 mA will cease to achieve mode-locked operation. This is due to the carrier concentration not being able to overcome threshold for stimulated emission. Four different bias levels (30 mA, 40 mA, 50 mA, and 60 mA) have been investigated and the simulation results are plotted in Fig. 17.35. Lowering the value of I_{dfb} causes the stable mode-locked region to widen towards lower RF drive frequencies. However, at the lower limit of 30 mA, the RMS timing jitter fluctuates about the 2 ps limit as stable mode-locked operation begins to fade out. But it is clear from Fig. 17.35(a) that the "stable" region is the widest when biased at 30 mA. Therefore, to achieve a wider locking range, the DFB bias level should not be set too high. Besides, the high values of I_{dfb} will lead to the appearance of trailing secondary pulses. The trade-off for wider locking bandwidths is broadening of the pulses within the stable mode-locked regions. The minimum pulsewidth is broadened from 26 ps to above 30 ps when I_{dfb} is reduced from 60 mA to 30 mA (Fig. 17.35(b)). There is also a range of lower drive frequencies (< 3 GHz) that fails to achieve mode-locking but produces gain-switched pulses. In gain-switched operation, the RMS timing jitter is moderately high at several picoseconds. The range of frequencies for gain-switched operation widens by a little, in contrast to the shrinking of the stable mode-locked region, when the DFB bias level is increased. This widening effect is due to the increase in the DFB laser's relaxation oscillation frequency with higher bias levels.

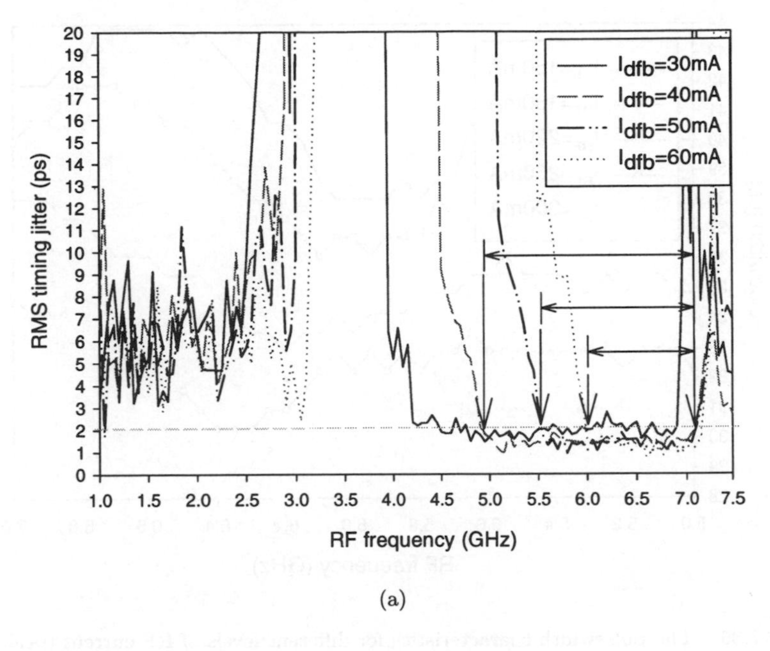

(a)

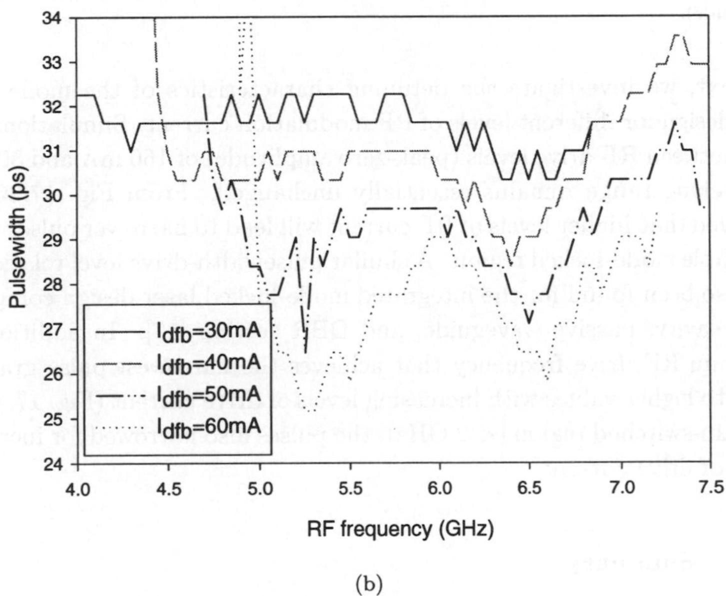

(b)

Fig. 17.35. The effect of bias level of DFB section on (a) pulse stability, and (b) pulsewidth.

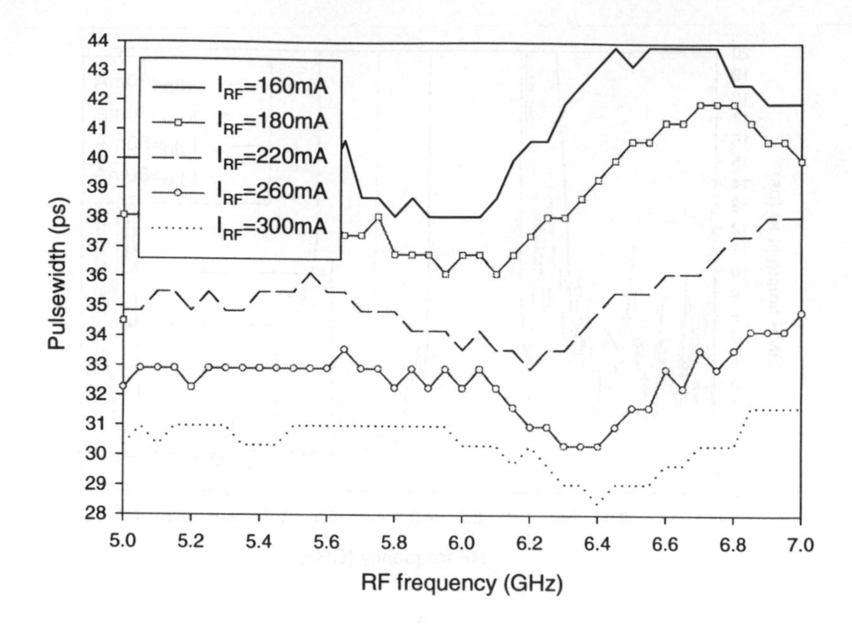

Fig. 17.36. The pulsewidth characteristics for different levels of RF current (peak-zero amplitude).

Next, we investigate the detuning characteristics of the mode-locked laser design for different levels of RF modulation current. Simulations show that between RF drive levels (peak-zero amplitude) of 160 mA and 300 mA, the locking range remains essentially unchanged. From Fig. 17.36, it is observed that higher levels of RF current will lead to narrower pulses within the stable mode-locked region. A similar pulsewidth-drive level relationship has also been found for the integrated mode-locked laser design comprising a FP cavity, passive waveguide, and DBR section [17]. In addition, the optimum RF drive frequency that achieves the narrowest pulse gradually shifts to higher values with increasing levels of drive current (Fig. 17.36). In the gain-switched region (< 2 GHz), the pulses also narrowed for increasing levels of drive current.

17.11 Summary

The transmission-line laser model (TLLM) is a wide-bandwidth time-domain laser model that covers multiple resonant cavity modes. This makes it a suitable choice for modelling ultrashort pulse generation in laser diodes.

The transient build-up of optical pulses from spontaneous emission as well as nonlinearity due to dynamic chirp and gain saturation are all included in TLLM simulations. The bidirectional interfaces of TLLM component models make it straightforward to combine several devices together in any permutation to form larger multisegment or compound cavity devices. A novel mode-locked laser design based on a multisegment monolithic cavity device has been proposed in this chapter, and the design considerations are thoroughly investigated using TLLM. The mode-locked laser design was simulated by combining transmission-line models of the tapered amplifier (TS-TLLM), DFB laser, passive waveguide, and Bragg reflector.

Pulse stability and its spectral characteristics of the mode-locked pulses are among the important design considerations that have been investigated. Low timing jitters are essential for stable mode-locked operation. To achieve stable mode-locked operation, we must prevent the formation of separate clusters of extended cavity modes in the optical spectrum. The DBR section acts as the bandwidth-limiting element to achieve only one cluster of locked modes. Simulations showed that the quarter-wave-shifted grating structures gave narrower as well as "cleaner" pulses (free from temporal substructures), and hence were chosen for the DFB and DBR sections. In addition, low coupling factor of the DBR reflector is required to achieve narrow spectral filtering. At the same time, the coupling must also be strong enough to provide sufficient feedback to induce active mode-locking.

The amplifier section should be AR-coated to prevent formation of a composite cavity between the amplifier output facet and the DFB section. The amplifier bias level should not be set too high as this will cause ASE to be injected back into the DFB section, which is detrimental to stable mode-locked operation. The dynamic gain saturation mechanism in the amplifier section provides some stabilisation effect to the mode-locked output pulses.

Internal reflections between the different sections in the monolithic cavity device can also affect the overall mode-locked operation. Moderate values of reflectivity ($< 10\%$) at the amplifier output facet will cause the locking range to shrink, and the mode-locked pulses to broaden. A small reflectivity at the front facet of the DFB section, however, is able to suppress large pulse-to-pulse peak power variations without any significant change in the locking bandwidth. On the other hand, high values of reflectivity at the DBR section end do not affect the locking bandwidth but cause broadening of the mode-locked pulses.

Passive waveguide loss is another important design factor used to optimise the mode-locked operation. Moderate values of loss in the passive waveguide can be used to attenuate any secondary pulses recirculating in the extended cavity as well as to obtain narrower pulses. However, when the loss becomes excessive, new unstable regions start to appear in the original stable mode-locked region, and the mode-locked pulses begin to broaden. For all cases, higher waveguide loss means lower output power emitted from the device.

In the monolithic multisegment device, the quarter-wave-shifted DFB laser section is modulated with deep sinusoidal current at RF frequencies. There will be an optimum drive frequency, at which the narrowest stable mode-locked pulses can be achieved. The detuning characteristics show how the pulse characteristics change as the drive frequency deviates from the optimum value. The stable mode-locked region (or locking range) and its locking bandwidth can be found from the detuning characteristics. Lower DC bias levels of the DFB section have been found to provide a wider locking bandwidth for a given RF drive level. However, to achieve narrower pulses, higher DC bias levels are required. By increasing the amplitude of the RF current, narrower mode-locked pulses could also be obtained, and the optimum drive frequency (for narrowest pulsewidth) would gradually shift to higher values.

This chapter presented a detailed investigation of a novel mode-locked laser design based on a monolithic compound cavity device, which incorporated two distributed feedback structures. Since the mode-locked laser design comprised several sections, the design considerations were more complicated than single cavity devices. The quality of the mode-locked pulses have been studied mainly in conjunction with their detuning characteristics. It has been found that there is no single best design solution but the final design depends on the specific requirements of the novel mode-locked laser device.

References

[1] P. E. Green Jr., "Optical networking update," *IEEE J. Select. Areas Comm.*, Vol. 14, No. 5, pp. 764–779, 1996.

[2] C. A. Brackett, "Dense wavelength division multiplexing networks: principles and applications," *IEEE J. Select. Areas Comm.*, Vol. 8, No. 6, pp. 948–964, 1990.

[3] S. P. Ferguson, "Implications of SONET and SDH," *IEE Electron. & Comm. Engr. Journal*, June 1994, pp. 133–142, 1994.

[4] R. S. Tucker, G. Eisenstein and S. K. Korotky, "Optical Time-Division

Multiplexing For Very High Bit-Rate Transmission," *IEEE J. Lightwave Technol.*, Vol. LT-6, No. 11, pp. 1737–1749, 1988.
[5] D. M. Spirit, A. D. Ellis and P. E. Barnsley, "Optical time division multiplexing: systems and networks," *IEEE Comm. Mag.*, Vol. 32, No. 12, pp. 56–62, 1994.
[6] M. J. O'Mahoney, "Optical multiplexing in fiber networks: progress in WDM and OTDM," *IEEE Comm. Mag.*, Dec, pp. 82–88, 1995.
[7] B. K. Mathason *et al.*, "All-optical multiwavelength semiconductor switching for hybrid WDM/TDM demultiplexing and signal processing," *CLEO'99, Summaries*, Vol. ?, pp. 347–348, 1999.
[8] D. M. Spirit, A. D. Ellis and P. E. Barnsley, "Optical time division multiplexing: systems and networks," *IEEE Comm. Mag.*, Vol. 32, No. 12, pp. 56–62, 1994.
[9] G. P. Agrawal, *Nonlinear Fiber Optics*, (Boston: Academic Press Inc., 1989).
[10] G. P. Agrawal, *Fiber-Optic Communication Systems*, (New York: John Wiley & Sons, Inc., 1992).
[11] R. J. Manning, A. D. Ellis, A. J. Poustie and K. J. Blow, "Semiconductor laser amplifiers for ultrafast all-optical signal processing," *J. Opt. Soc. Am. B*, Vol. 14, No. 17, pp. 3204-3216, 1997.
[12] R. J. Manning, A. D. Ellis, A. J. Poustie and K. J. Blow, "Semiconductor laser amplifiers for ultrafast all-optical signal processing," *J. Opt. Soc. Am. B*, Vol. 14, No. 17, pp. 3204-3216, 1997.
[13] J. P. Solokoff, P. R. Prucnal and M. Kane, "A Terahertz Optical Asymmetric Demultiplexer (TOAD)," *IEEE Photon. Tech. Lett.*, Vol. 5, No. 7, pp. 787–790, 1993.
[14] M. Nakawaza, "Special Issue on Ultrashort Optical Pulse Technologies and their Applications," *IEICE Trans.*, Vol. E81-C, No. 2, pp. 93–276, 1998.
[15] W. M. Robertson, *Optoelectronics Techniques for Microwave and Millimeter-Wave Engineering*, (Boston: Artech House, 1995).
[16] M. G. Davis and R. F. Dowd, "A Transfer Matrix Method Based Large-Signal Dynamic Model for Multielectrode DFB Lasers," *IEEE J. Quant. Electron.*, Vol. 30, No. 11, pp. 2458–2466, 1994.
[17] A. J. Lowery, "Integrated mode-locked laser design with a distributed Bragg reflector," *IEE Proceedings Pt. J*, Vol. 138, No. 1, pp. 39–46, 1991.
[18] P. B. Hansen *et al.*, "5.5 mum long InGaAsP Monolithic Extended-Cavity Laser with an Integrated Bragg-Reflector for Active Mode Locking," *IEEE Photon. Tech. Lett.*, Vol. 4, No. 3, pp. 215–217, 1992.
[19] D. Y. Kim *et al.*, "Ultrastable millimetre-wave signal generation using hybrid modelocking of a monolithic DBR laser," *Electron. Lett.*, Vol. 31, No. 9, pp. 733–734, 1995.
[20] P. A. Morton, R. J. Helkey and J. E. Bowers, "Dynamic Detuning in Actively Mode-Locked Semiconductor Lasers," *IEEE J. Quant. Electron.*, Vol. 25, No. 12, pp. 2621–2633, 1989.
[21] Z. Ahmed, L. Zhai, A. J. Lowery, N. Onodera and R. S. Tucker, "Locking Bandwidth of Actively Mode-Locked Semiconductor Lasers," *IEEE J. Quant. Electron.*, Vol. 29, No. 6, pp. 1714–1721, 1993.

[22] L. Zhai, A. J. Lowery, Z. Ahmed, N. Onodera and R. S. Tucker, "Locking bandwidth of mode-locked semiconductor lasers," *Electron. Lett.*, Vol. 28, No. 6, pp. 545–546, 1992.

[23] N. Onodera, A. J. Lowery and R. S. Tucker, "Cyclic wavelength jitter in actively mode-locked semiconductor lasers," *Electron. Lett.*, Vol. 27, No. 3, pp. 220–222, 1991.

[24] A. J. Lowery, N. Onodera and R. S. Tucker, "Stability and Spectral Behavior of Grating-Controlled Actively Mode-Locked Lasers," *IEEE J. Quant. Electron.*, Vol. 27, pp. 2422–2430, 1991.

[25] A. J. Lowery, "Cyclic three-phase jitter in mode-locked semiconductor lasers," *Electron. Lett.*, Vol. 25, No. 12, pp. 799–800, 1989.

[26] L. Zhai, A. J. Lowery and Z. Ahmed, "Diffraction grating model for transmission-line laser model of actively mode-locked semiconductor lasers," *IEE Proceedings Pt. J*, Vol. 141, No. 1, pp. 21–26, 1994.

[27] L. Zhai, A. J. Lowery and Z. Ahmed, "Locking bandwidth of actively mode-locked semiconductor lasers using fiber-grating external cavities," *IEEE J. Quant. Electron.*, Vol. 31, No. 11, pp. 1998–2005, 1995.

[28] D. J. Derickson *et al.*, "Short Pulse Generation using Multisegment Mode-Locked Semiconductor Lasers," *IEEE J. Quant. Electron.*, Vol. QE-28, No. 10, pp. 2186–2202, 1992.

[29] A. J. Lowery and I. W. Marshall, "Stabilisation of mode-locked pulses using travelling-wave semiconductor laser amplifier," *Electron. Lett.*, Vol. 26, No. 2, pp. 104–105, 1989.

[30] J. E. Deardorff and C. R. Trimble, "Calibrated Real-Time Signal Averaging," *HP Journal*, April 1968, pp. 8–13, 1968.

[31] H. Kogelnik and C. V. Shank, "Coupled-Wave Theory of Distributed Feedback Lasers," *J. Appl. Phys.*, Vol. 43, No. 5, pp. 2327–2335, 1972.

[32] A. J. Lowery, "Dynamic Modelling of Distributed-Feedback Lasers Using Scattering Matrices," *Electron. Lett.*, Vol. 25, No. 19, pp. 1307–1308, 1989.

[33] A. J. Lowery and D. F. Hewitt, "Large-signal dynamic model for gain-coupled DFB lasers based on the transmission-line model," *Electron. Lett.*, Vol. 28, No. 21, pp. 1959–1960, 1992.

[34] A. J. Lowery and D. Novak, "Performance comparison of gain-coupled and index-coupled DFB semiconductor lasers," *IEEE J. Quant. Electron.*, Vol. QE-30, No. 9, pp. 2051–2063, 1994.

[35] G. H. B. Thompson, *Physics of semiconductor laser devices*, (Chichester: John Wiley & Sons, 1980).

[36] L. V. T. Nguyen, A. J. Lowery, P. C. R. Gurney and D. Novak, "A Time Domain Model for High-Speed Quantum-Well Lasers Including Carrier Transport Effects," *IEEE Jour. Select. Top. Quant. Elec.*, Vol. 1, No. 2, pp. 494–504, 1995.

[37] G. P. Agrawal and N. K. Dutta, *Semiconductor lasers*, 2nd ed. (New York: Van Nostrand Reinhold, 1993).

[38] A. J. Lowery, "Dynamics of SHB-induced mode instabilities in uniform DFB semiconductor lasers," *Electron. Lett.*, Vol. 29, No. 21, pp. 1852–1854, 1993.

[39] T. L. Koch and U. Koren, "Semiconductor Lasers for Coherent Optical Fiber Communications," *IEEE J. Lightwave Technol.*, Vol. LT-8, No. 3, pp. 274–293, 1990.

[40] F. Koyama, Y. Suematsu, S. Arai and T. Tanbun-Ek, "1.5–1.6 mum GaInAsP/InP Dynamic Single Mode (DSM) Lasers with Distributed Bragg Reflectors," *IEEE J. Quant. Electron.*, Vol. QE-19, No. 6, pp. 1042–1051, 1983.

[41] Y. Yamamoto, Ed., *Coherence, Amplification, and Quantum Effects in Semiconductor Lasers* (New York: John-Wiley and Sons, Inc., Wiley Series in Pure and Applied Optics, 1991).

[42] A. V. Oppenheim and R. W. Schafer, *Discrete Time Signal Processing* (NJ: Prentice Hall, 1989).

[43] P. A. Morton *et al.*, "Mode-locked hybrid soliton pulse source with extremely wide operating frequency range," *IEEE Photon. Tech. Lett.*, Vol. 5, pp. 28–31, 1993.

[44] R. S. Tucker *et al.*, "40 GHz active mode-locking in a 1.5 mum monolithic extended-cavity laser," *Electron. Lett.*, Vol. 25, No. 10, pp. 621–622, 1989.

[45] G. P. Agrawal and N. A. Olsson, "Self-Phase Modulation and Spectral Broadening of Optical Pulses in Semiconductor Laser Amplifiers," *IEEE J. Quant. Electron.*, Vol. QE-25, No. 11, pp. 2297–2306, 1989.

[46] H. Ghafouri-Shiraz, P. W. Tan and T. Aruga, "Picosecond Pulse Amplification in Tapered-Waveguide Laser-Diode Amplifiers," *IEEE Jour. Select. Top. Quant. Elec.*, Vol. 3, No. 2, pp. 210–217, 1997.

[47] M. Heinrich and M. Claasen, "Chirp and linewidth enhancement factor of current modulated Ga(Al)As-GRINSCH-SQW-MCRW flared waveguide lasers," *Electron. Lett.*, Vol. 26, No. 19, pp. 1598–1600, 1990.

[48] A. J. Lowery and I. W. Marshall, "Numerical Simulations of 1.5 mum Actively Mode-Locked Semiconductor Lasers Including Dispersive Elements and Chirp," *IEEE J. Quant. Electron.*, Vol. 27, pp. 1981–1989, 1991.

[49] A. J. Lowery, "Modelling spectral effect of dynamic saturation in semiconductor laser amplifiers using the transmission-line laser model," *IEE Proceedings Pt. J*, Vol. 136, No. 6, pp. 320–324, 1989.

[50] N. Onodera, H. Ito and H. Inaba, "Generation and Control of Bandwidth-Limited, Single-Mode Picosecond Optical Pulses by Strong RF Modulation of a Distributed Feedback InGaAsP Diode Laser," *IEEE J. Quant. Electron.*, Vol. QE-21, No. 6, pp. 568–575, 1985.

[51] P. M. Downey, J. E. Bowers, R. S. Tucker and E. Agyekum, "Picosecond Dynamics of a Gain-Switched InGaAsP Laser," *IEEE J. Quant. Electron.*, Vol. QE-23, No. 6, pp. 1039–1047, 1987.

[52] G. J. Aspin and J. E. Carroll, "Gain-switched pulse generation with semiconductor lasers," *IEE Proceedings Pt. I*, Vol. 129, No. 6, pp. 283–290, 1982.

Chapter 18

Summary, Conclusion and Suggestions

18.1 Summary of Part I (Chapters 1 to 7)

In Part I of this book, the performance characteristics of semiconductor laser amplifiers (SLAs) were studied both theoretically and experimentally. As discussed in the introductory Chapter 1, these laser amplifiers can be used in optical fibre communication systems as in-line repeaters or can be integrated with the optical receiver chip as an optical pre-amplifier. The basic principles of optical amplification were explored in Chapter 2 by considering the interaction of electromagnetic radiation with a two-level quantum mechanical system. This was shown to give a qualitatively accurate picture of the principles and performance characteristics of all types of optical amplifier, including both SLAs and fibre amplifiers. The discussion was narrowed down to SLAs in Chapter 3 by considering in further detail how semiconductor lasers have been used as optical amplifiers, both by examining their physical principle of operation as well as reviewing the historical development of using semiconductor lasers as optical amplifiers. In this discussion, SLAs were classified by their structural differences (i.e. the difference between Fabry–Perot amplifiers (FPAs) and travelling-wave amplifiers (TWAs)) and the subsequent differences in their performance characteristics were examined. Typical applications of SLAs as in-line repeaters, pre-amplifiers, and simple routing switches were also explored.

A systematic theoretical study of the performance characteristics of SLAs was the major objective in Chapters 4 to 6. The SLAs were treated as an active dielectric optical waveguide in Chapter 4, in which the propagation characteristics of optical signals were analysed using electromagnetic theory. The resulting modal gain coefficient of the SLA, which determines the single pass gain and hence overall gain of the amplifier for a particular input signal wavelength, were analysed using a perturbation technique proposed by Ghafouri-Shiraz and Chu [1]. This technique allows the polarisation sensitivity of SLAs with a simple buried heterostructure (BH), the most common SLA structure, as well as more complex structures like stripe loaded structures and buried channel (BC) structures, to be analysed quickly and accurately. Furthermore, the modal gain coefficients for SLAs with more complicated transverse modal field distributions (e.g. rib structures) can also be calculated. For these structures, a single optical confinement factor, as used commonly in index-guiding BH SLAs, cannot be used to calculate the modal gain coefficients directly. Often, several similar "optical confinement factors" have to be defined in each region in the transverse plane before the modal gain coefficients can be calculated [1]. Hence the proposed technique provides an alternative and yet effective way to calculate the modal gain coefficient directly without involving conventional formulae with optical confinement factors. The structural design for polarisation insensitive SLAs has been explored, showing that a thick active layer will minimise the polarisation sensitivity of amplifiers with BH. It can be minimised further by using alternative structures such as BC structures. Incidentally, BC structures become more and more common for integration of electronic and optical devices. The importance of maintaining single-transverse mode (STM) operation in SLAs was also discussed in this chapter.

The gain characteristics and the corresponding saturation behaviour of SLAs were analysed in Chapter 5. The analysis is a complex one as there are continuous interactions between the photons and injected carriers due to the bias current, both in a temporal and a spatial sense. The temporal relation between the photons in the SLA cavity and the injected carrier density is governed by a set of rate equations, which form the basis of the analysis of SLAs. The spatial relation between the photons and the injected carrier density can be included in the analysis by introducing travelling-wave equations to analyse the variation of signal field and the intensity of spontaneous emissions inside the SLA. The solution procedures can be very tedious, even for simple structures such as BH FPAs and TWAs. An approximation based on averaging the photon density along the SLA has been examined, which will simplify the solution procedures. However, both the exact and the approximate technique requires the field distribution in the SLA cavity to be solved analytically before numerical iterations are applied to solve the equations. For more complex structures (e.g. two-section amplifiers, SLAs with a tapered active region, etc.), this may not be possible. A robust and efficient technique based on transfer matrix analysis, as proposed by Chu and Ghafouri-Shiraz [2–3], has been described in Chapter 5. This technique does not require any averaging approximation and is able to account for the spatial relationship between photon density and carrier density along the amplifier, and has the advantage of ease of implementation on computers and greater flexibility compared with the approximation technique. The application of the proposed technique for analysing the effects of stray reflections from fibre ends on the gain characteristics were demonstrated in a later chapter. The effects of the structural parameters on the gain characteristics of BH SLAs were also studied, and the implications for system design were briefly discussed.

The effects of spontaneous emissions were neglected in the discussion of Chapter 5. In Chapter 6, spontaneous emissions first examined by using the photon statistics master equations approach, the rate equation approach and the travelling-wave equation approach. From these basic investigations, it was found that the spontaneous emissions from SLAs generate additional beat noises when their output is detected by photodiodes. Then the technique explained in Chapter 5 was extended to construct a full equivalent circuit model for analysing SLAs. It was seen that any SLA can be modelled by a cascaded network of negative active resistors and lossless transmission lines. By terminating the network with an appropriate load impedance, and by changing the characteristic impedance of the transmission lines, any refractive index steps occurring along the amplifier can be modelled. This allows complex structures to be analysed. The resulting noise due to quantum mechanical fluctuations in the negative resistors is found to be both qualitatively and quantitatively accurate for modelling the effects of spontaneous emissions in actual SLAs. This has been justified by comparing the deduced spontaneous emission power emitted in a SLA without input signal, using the equivalent circuit model, with that calculated by Green's function. Therefore, simple circuit analysis techniques, supplemented by some knowledge of signal analysis, can be used to compute the noise generated in SLAs, rather than involving the more complicated concepts of Langevin forces and statistical physics. It can also take into account the spatial dependence of the photon density and carrier density along the

SLA cavity. The power of the equivalent circuit model was illustrated by analysing the effects of stray reflections from fibre ends on the noise characteristics of SLAs. Finally, the structural design of low noise BH SLAs was discussed, and the effects of noise on the design of optical communication systems with SLAs as in-line repeaters and pre-amplifiers was examined.

Any theoretical study on SLAs needs to be supported by a complementary experimental study. This was described in Chapter 7, where the experimental setup, and the principles of SLA measurements and experimental procedures were described in detail. Three particular aspects of SLAs were studied experimentally. First, the recombination mechanisms of injected carriers in SLAs were investigated by measuring the amplified spontaneous emission power of the SLA, and it has been found that non-radiative recombinations, such as Auger recombinations, in the 1.5 μm wavelength region are extremely important. In fact, these recombination mechanisms allow the structure of the active region to be tailored to achieve STM operation, with a relatively thick active layer to reduce polarisation sensitivity. In SLAs with materials for optical amplification of shorter wavelengths, where non-radiative recombinations are not important, such dimensions of the active region will result in a lower threshold current for oscillation, hence limiting the range of injection current which is allowed to flow through the SLA, as illustrated in Chapter 5. Secondly, the gain characteristics were measured for both TE and TM polarisations of the input optical signal. The low polarisation sensitivity claimed by the supplier were observed. The effects of the output power level and the injection current on the overall gain and output saturation power of the SLA were examined. These results were compared with the theoretical predictions made by the equivalent circuit model [2–3] and the approximation method, and these illustrate the importance of the spatial dependence of photon density and carrier density in the SLA. Finally, the noise current generated in an InGaAsP pin photodiode by a SLA with different input optical power levels was measured.

18.1.1 *Limitations of the research study*

The strength of any scientific or engineering model cannot be fully utilised without being aware of their limitations. Similarly, the significance of any research cannot be fully appreciated without examining its limitations. In this section, we will examine some of the limitations in this SLA study. Three particular aspects will be examined: the limitations of the analytical models used in this book, the limitations of the experimental studies and the limitations in the scope of the overall research work presented here.

18.1.2 *Limitations on theoretical studies*

There are mainly two techniques used in the first part of this book. In Chapter 4, the analysis of modal gain coefficients in SLAs by perturbation techniques was examined. In Chapters 5 and 6, an equivalent circuit model was developed for analysing SLAs and this can be implemented on computers using transfer matrix methods. To analyse an SLA, whatever its structure, both of these techniques should be combined to form a systematic framework for this purpose.

The discussion in Chapters 4 to 6 may lead one to conclude that the proposed model may only work for index guided buried heterostructures (BH), because, as we have seen in Chapters 5 and 6 the optical confinement factor was used to calculate the modal gain coefficient in the SLA, which is an essential parameter in solving for the coefficients of the transfer matrices. Indeed, the analysis and discussion have been limited to BH SLAs because they are the most common structures for SLAs at present. However, there is no reason why we cannot extend the framework to analyse other structures for which a single optical confinement factor cannot be used to calculate the modal gain coefficient (e.g. buried channel structures, rib structures, DFB structures, etc.). The only modification required is to replace the modal gain coefficients in the TMM analysis with the perturbation expression discussed in Chapter 4, which does not require the use of an optical confinement factor in the calculation.

The real limitation of the model is that, because of the way the equivalent circuit has been constructed, it will only work for index guided structures. For gain guided structures, the longitudinal propagation constant β_z is not independent of z, nor the transverse positions x and y. In principle, however, it may be possible to modify the framework of analysis following the arguments of Buus [4]. Nevertheless, most of the SLAs and laser diodes manufactured now are index guided. This is because the coupling efficiency to optical fibres is better for the index guided laser structure [5]. There are recent trends, especially for DFB lasers, in using gain guided structures to stabilise the single longitudinal mode (SLM) operation under direct modulation [6–8]. However, for the purpose and scope of the present book, concentrating on index guided structures is sufficient.

18.1.3 *Limitations on experimental studies*

To be in line with the theoretical studies, the experimental measurements discussed in Chapter 7 concentrate only on gain, gain saturation and noise characteristics of SLAs. The complexity involved in the experimental arrangement required to perform tests in simple optical communication systems, or to measure characteristics such as cross-talk, harmonic distortion, gain saturation in amplification of short optical pulses, etc., is significantly higher. The major problem encountered in the measurements, as discussed in Chapter 7, is the need to have very good alignment between optical components in order to maintain a stable and optimum coupling. Compared with fibre amplifiers, SLAs indeed suffer from this disadvantage of relative difficulty in coupling. This problem was greater in this case because the SLA used in the experiment was an unpacked type. It is necessary, in more complex experiments, to have SLAs which are packaged, if reliable and accurate measurements are to be made.

18.2 Summary of Part II (Chapters 8 to 12)

Part II consists of the following five chapters

In Chapter 8 we introduced a new semiconductor laser diode amplifier structure, which consists of a semi-exponential-linear tapered waveguide active layer to improve the saturation output power. Various characteristic performances of this device were analyzed based on the step-transition approach, taking into account the effects of the spatial distributions of the carrier density and the optical field. The effects of various tapered structures on the gain saturation, intensity distribution and amplified spontaneous emission noise were been investigated.

In Chapter 9 the amplification characteristics of picosecond Gaussian pulses in conventional non-tapered and both linear and exponential tapered-waveguides (TWs) laser diode amplifier (LDA) structures were studied [9–10]. The analysis was based on numerical simulation of the rate equation which also takes into account the effect of lateral carrier density distribution. The amount of pulse distortion experienced within the amplifier for input pulses having energies $E_{\text{in}} = 0.1E_{\text{sat(in)}} = 0.475$ pj (where $E_{\text{sat(in)}}$ is the input saturation energy of the amplifier) and $E_{\text{in}} = E_{\text{sat(in)}} = 4.75$ pj were analysed for each structure which has a length of 900 μm and an input width of 1 μm. It was found that the TW-LDA provided higher gain saturation and hence imposed less distortion on the amplified pulse as compared with a conventional non-tapered LDA. The amplified 10*ps* pulse used in this study experiences almost no broadening in the tapered-waveguide LDA, whereas it suffers from broadening in the conventional non-tapered LDA. The carrier density distribution and the dependence of the amplifier gain on the input pulse energy were also studied for both non-tapered and tapered amplifier structures. For example, in a TW-LDA with an output width of 20 μm and a length of 900 μm, the exponential structure provides 9 dB improvement in saturation energy as compared with the conventional amplifier. This improvement is about 10.5 dB in linear TW-LDA.

In Chapter 10, the sub-picosecond gain dynamic in a highly index-guided tapered-waveguide laser diode amplifier was studied in this chapter. The analysis is based on a set of rate equations for the carrier density N and carrier temperature T, where the effects of lateral distributions of both N and T are considered. It was found that both N and T have significant effects on the gain dynamic of the tapered-waveguide laser diode amplifier. The gain saturation of the tapered-waveguide amplifier for both 20 ps and 200 fs pulses was also calculated.

In Chapter 11 we introduced a novel approximate analytical expression for saturation intensity for tapered travelling-wave semiconductor laser amplifier (TW-SLA) structures. The application of this analytical expression of saturation intensity was demonstrated by considering the effect of gain saturation on polarisation sensitivity of two tapered amplifier structures, linear and exponential tapered amplifier structures. It was found that polarisation sensitivity of the tapered amplifier structure was several dB higher than that of passive tapered waveguides in unsaturated condition. Polarisation sensitivity of the two tapered amplifier structures was also investigated in a highly saturated condition.

In Chapter 12 wavelength conversion using cross-gain modulation (XGM) in linear tapered-waveguide semiconductor laser amplifiers (SLAs) was studied and compared with the non-tapered semiconductor laser amplifier. For example, we have found that for the linear tapered-waveguide SLA with a 450 μm cavity length, a higher signal power is required to achieve a similar extinction ratio as obtained by the conventional SLA with the same cavity length. However, a similar extinction ratio is observed for both conventional and linear tapered-waveguide with the cavity length of 1200 μm. The difference between the extinction ratio for up and down conversion is less severe in a linear tapered-waveguide SLA as compared to that for a conventional SLA. The rise time for linear tapered-waveguide is observed to be higher than that for conventional SLA. It is possible to attain extinction ratio conservation using XGM for both up and down conversions in a SLA with a very long cavity length of 2400 μm.

18.3 Summary of Part III (Chapters 13 to 17)

Part III of this book is microwave circuit principles applied to semiconductor laser modelling. The advantages and additional insight provided by circuit models that have been used for *analytical* analysis of laser diodes have long been acknowledged [11–14]. We have mostly concentrated on the derivation, implementation, and application of *numerical* circuit-based models of semiconductor laser devices.

In Chapter 13, a brief historical background and the relevant physics behind the semiconductor laser were presented. Chapter 14 introduced the transmission-line matrix (TLM) method that provides the basic microwave circuit concepts used to construct the time-domain semiconductor laser model known as the transmission-line laser model (TLLM). We then proceeded to compare two categories of equivalent circuit models, i.e. lumped-element and distributed-element, of the semiconductor laser in Chapter 15. In the same chapter, a comprehensive laser diode transmitter model was developed for microwave optoelectronic simulation. The microwave optoelectronic model is based on the transmission-line modelling technique, which allows propagation of optical waves as well as lumped electrical circuit elements to be simulated. In Chapter 16, the transmission-line modelling technique was applied to a new time-domain model of the tapered waveguide semiconductor laser amplifier, useful for investigating short pulse generation and amplification when finite internal reflectivity is present. The new dynamic model is based on the strongly index-guided laser structure, and quasi-adiabatic propagation is assumed. Finally, Chapter 17 demonstrated the usefulness of the microwave circuit modelling techniques that have been presented in this book through a design study of a novel mode-locked laser device. The novel device is a multisegment monolithically integrated laser employing distributed Bragg gratings and a tapered waveguide amplifier for high power ultrashort pulse generation.

Photonics computer aided design (PCAD) for innovative design and optimisation of novel laser devices is an invaluable asset in today's fast-paced photonics industry. The deregulation of the telecommunications market is creating an urgent need for novel and cost-effective photonic devices that can perform new functions, especially for WDM systems that are currently being deployed. The PCAD approach allows designs to be virtually constructed, carefully analysed, and optimised, thereby minimising the risk involved in fabricating costly prototypes without first gaining a proper understanding of the devices. Among several PCAD modelling approaches available, microwave circuit techniques applied to semiconductor laser modelling are attractive because the microwave and RF aspects of the laser diode can be conveniently included.

Hence, the main theme of this book is microwave circuit techniques applied to semiconductor laser modelling. This approach provides a simple and clear physical insight into the laser dynamics, especially for microwave design engineers who have little experience in the laser but would like to embark on the emerging research area of microwave photonics. The feasibility of using lumped-element and distributed-element circuits for modelling the dynamics and modulation response of the laser diode has been demonstrated in this book, and the difference between the two types of circuit models has been compared. The transmission-line laser model (TLLM) is categorised as a distributed-element circuit, which meets most of the criteria of an ideal laser model. Besides its many advantages over the lumped-element circuit, the flexibility of TLLM [15] also makes itself favourable over other types of time-domain laser models, such as the power matrix model (PMM) [16], transfer matrix model (TMM) [17], time domain model (TDM) [18], and beam propagation model (BPM) [19].

18.4 Summary of New Contributions

18.4.1 *Microwave optoelectronic models of the laser diode transmitter* (*Chapter* 15)

First, the construction of the large-signal and small-signal *lumped-element* circuit models of the semiconductor laser diode was explained in detail, and their usefulness was demonstrated through several applications such as the light-current characteristics, transient response, and modulation response, with the effect of electrical parasitics included. The nonlinear circuit models were implemented as symbolically defined devices (SDD) in the commercial microwave design software, HP-MDS. A harmonic balance simulation was applied to Tucker's large-signal circuit model to analyse the nonlinear distortion that occurs during two-tone modulation of the laser diode. The harmonic balance method is more efficient than the time-domain simulation approach that was proposed by Way [20].

Comparisons between the lumped-element and distributed-element circuit models of the semiconductor laser were presented in Table 4.4.1. Although the lumped-element circuit models are simple to implement, their usefulness in semiconductor laser modelling is limited by many simplifying assumptions that have been used in the model construction [21–23]. This is because the lumped-element models are based on low-frequency concepts, where the operating wavelength is much larger than the dimensions of the device, thereby only allowing us to simulate the optical intensity or power envelope in which case, phase information of the optical carrier is lost. Moreover, the lumped-element laser models use average values of photon and carrier densities, thereby neglecting any inhomogeneous distribution along the laser cavity. In other words, there is no spatial correspondence between the circuit model and the real physical device. All these limitations can be overcome using a distributed-element circuit approach — transmission-line laser modelling/model (TLLM) [24].

For this reason, a realistic microwave optoelectronic model based on TLLM has been developed (i.e. the *integrated TLLM model* [25]). The integrated TLLM model includes the electrical parasitics, matching network, and the intrinsic laser diode. Most numerical models of laser diodes have failed to consider the inclusion of a matching circuit and the electrical parasitics effect [16–19] but since TLLM is based on microwave circuit principles, it allows matching elements and electrical parasitics to be easily included in the overall laser model. The matching network has been, for the first time, included in the standard TLLM model, and is based on monolithically integrated lumped elements [26]. The transmission-line matrix (TLM) stub-lines and link-lines [27] have been used to model both the electrical parasitics and matching networks. The integrated TLLM was used to demonstrate improvement in harmonic content of gain-switched pulses when the matching network was included.

18.4.2 *Transmission-line laser model of tapered waveguide semiconductor laser amplifier structures* (*Chapter* 16)

Computational intensive two dimensional (2-D) models of laser devices are not strictly essential for simulation of laser performance characteristics such as the light-current characteristics, modulation response, transient response, energy conversion efficiency, gain saturation effects, and spectral characteristics. Spatially-averaged one-dimensional (1-D) models make a more attractive simulation tool for photonic system designers because they offer faster computation speed. However, such 1-D models require careful consideration for tapered waveguide laser devices [28], which have been proposed to overcome some limitations of the conventional uniform-width laser structure, e.g. to provide better saturation performance [29], to amplify picosecond pulses with minimal distortion [30], and to generate high power optical pulses [31]. Furthermore, tapered waveguide laser amplifier structures have been found to give lower ASE noise compared to broad-area laser structures [32].

A novel 1-D time-domain model of the tapered laser structure, based on transmission-line laser modelling (TLLM), has been proposed and developed. This dynamic laser model, referred to as the tapered structure transmission-line laser model (TS-TLLM), is able to provide a simple and clear physical interpretation of all the laser's internal processes in terms of microwave circuit theory [33].

In TS-TLLM, the tapered waveguide structure has been reduced to one dimension in space by taking the spatial average of the optical field across the transverse plane, and by applying the effective index method (EIM). In addition, position-dependent lateral hole-burning terms are implemented in empirical form to improve the computational efficiency of the model. In a separate study, it has been shown that the steady-state characteristics of the tapered semiconductor amplifier can be computed in a highly efficient manner by using a closed form expression of the width-dependent saturation intensity [34]. For additional computation speed, a smaller number of sections may be used in the TS-TLLM but a compensating factor (K_g) is essential in such cases. Although the compensating factor is required for the Taylor approximated gain model, no such compensating factor is required for the position-dependent beam spreading factor (α_{tap}), regardless of the unit section length (Δz) used in the linear and exponential tapered structures.

The time domain nature of TS-TLLM allows nonlinearity and dynamics of pulse amplification to be studied, and the continuous optical spectrum can be found by performing a fast Fourier transform (FFT) to the output time-domain samples. Previous models of tapered semiconductor amplifier structures failed to consider residual reflectivity [29–35] but in this book reflections have been taken into account in all TS-TLLM simulations. Furthermore, the stochastic nature of TS-TLLM is favoured over previous deterministic models [29–35] because it allows the influence of noise to be studied, making it a highly realistic dynamic laser model.

18.4.3 *Design and optimisation of a novel multisegment mode-locked laser device* (*Chapter* 17)

The TLLM is a realistic, flexible, and yet computationally efficient time-domain semiconductor laser model that is useful for investigating ultrashort pulse generation in laser diodes because it takes into account multiple longitudinal modes, inhomogeneous effects along the cavity, and finite facet reflections. The intense ultrashort pulses generated by the laser diode inevitably lead to nonlinear effects such as gain saturation and dynamic chirp, which is best modelled by a time-domain model. The random nature of spontaneous emission is also included in the stochastic TLLM, which enables us to investigate laser noise characteristics, and observe the initial transient build up of the ultrashort optical pulses from spontaneous emission.

The TLLM component models have bidirectional ports allowing the passing on and exchange of time-domain waveforms from one component model to another [36]. Hence, these interconnnectable component models may be combined in any permutation to form novel multisegment or compound cavity laser structures. This feature of the TLLM approach has been utilised by performing a design and optimisation study of a novel multisegment mode-locked semiconductor laser incorporating Bragg gratings and a tapered semiconductor amplifier. Important design considerations and operating conditions of the novel multisegment mode-locked laser device have been identified, and these can be used as guidelines for the device designer.

The principle behind the novel design is to employ a monolithically integrated DBR section as the spectral filtering and feedback element, a DFB section for synchronously modulating the signal, and an AR-coated tapered semiconductor amplifier section for amplifying the mode-locked pulses. Hence, the multisegment laser model that was used for the design study was simply a concatenation of the newly developed transmission-line model of the tapered semiconductor amplifier, i.e. the TS-TLLM [33], and other existing TLLM component models, i.e. DFB laser, passive waveguide, and distributed Bragg reflector (DBR) [37, 38].

The novel multisegment mode-locked laser design was more complicated than the standard two-section cavity device (i.e. laser chip and external passive cavity) used for conventional active mode-locking, thereby requiring a more careful consideration of the interactions between the multiple sections. Moreover, the monolithic device contained two Bragg grating sections (i.e. DFB and DBR), which made the interactions too complex for a simple analytical description, hence the numerical TLLM approach was used.

In the analysis of the multisegment mode-locked laser design, the RF detuning characteristics were plotted to find the stable region of mode-locked operation. In all previous work, the RMS timing jitter has been used as the sole parameter in defining the locking range [39–40] but in this book, it has been pointed out that low RMS timing jitter could be misleading when a bifurcation of the pulses' peak powers exists. Therefore, an additional criterion for stable mode-locking is to have a small variation of peak power between successive pulses, by using a new parameter defined as the ratio of RMS pulse-to-pulse peak power variation (P_{var}) to average peak power (P_{pk}). This ratio should be less than 10% for stable mode-locked operation.

18.5 Suggestions for Future Work

18.5.1 *Parts I and II of the book*

As discussed above, the purpose of parts I and II of this book is to present a new method of analysing the performance characteristics of SLAs. The power of this method has been illustrated by analysing the basic characteristics of simple index guided BH SLAs, and this is supported by experimental measurements. Two new research directions are possible, based on the work presented in this book. The first new direction is to extend it to analyse index guided SLAs with more complex structures. This will include rib structures [41–42], DFB laser amplifiers [43], and two or multi-section laser amplifiers [44]. There is also potential to apply the model to quantum well structures [45–48]. The major differences between quantum well laser amplifiers and bulk laser amplifiers which we have been examining so far, are on the recombination mechanisms [49], dependence of material gain coefficients, refractive index and optical confinement factor on the band structure and injected carriers to the quantum wells [50–51]. One can replace the models used for bulk structures by those appropriate for quantum wells, and the entire analysis for quantum well laser amplifiers will follow an identical framework to that used for the bulk structureds described in this book. In addition, since transfer matrix methods have been readily applied to analyse semiconductor lasers [52–54], the equivalent circuit model may also be extended to model semiconductor lasers as well. Together with other models used to analyse passive optical devices [55] and active electronic components [56], this may form the basis of a general computer-aided design package to design integrated opto-electronic devices [57].

The second new direction is to use the equivalent circuit model to investigate other performance characteristics of SLAs. Three specific areas can be pursued further using the present model without significant modifications. The first area is to extend the analysis of noise to consider the effect of spontaneous emissions of SLAs on the linewidth of the signal [58–60] and the resulting phase noise introduced into the system [61]. This can be used to consider the bit-error-rate of the system in more detail than has been considered in this book [62–63]. The second area is in the analysis of analogue intensity modulated systems in the radio frequency (RF) and microwave frequency regions. In such applications, the harmonic distortion introduced by the non-linearity in the SLAs becomes important. Finally, the amplification of short optical pulses can be examined by this method. Such analysis is useful for predicting the performance limits of SLAs used for high speed digital transmission in optical fibre communication systems.

18.5.2 *Part III of the book*

18.5.2.1 *Wavelength conversion and all-optical regeneration in semiconductor optical amplifiers*

In wavelength-routed WDM optical networks, wavelength conversion is useful to set up links that involve more than one wavelength, thereby increasing the flexibility of the network [41]. Furthermore, wavelength conversion techniques such as cross gain modulation (XGM) and cross phase modulation (XPM) can be used to implement 2R (Re-amplifying, and Re-shaping) and 3R (Re-amplifying, Re-shaping, and Re-timing) optical signal regeneration [42–45]. These XGM and XPM regeneration methods rely on semiconductor optical amplifier (SOA)-based Mach Zehnder interferometeric (MZI) devices (see Fig. 18.1).

We can extend the work of the transmission-line model of the semiconductor laser amplifier to model wavelength conversion by injecting both pump and probe signals into the device model. The multimode feature of the transmission-line laser model (TLLM) means that multiple signals of different wavelengths can be included in the simulation, making it suitable for modelling WDM applications. Furthermore, the nonlinearity involved in wavelength conversion techniques is readily modelled by the

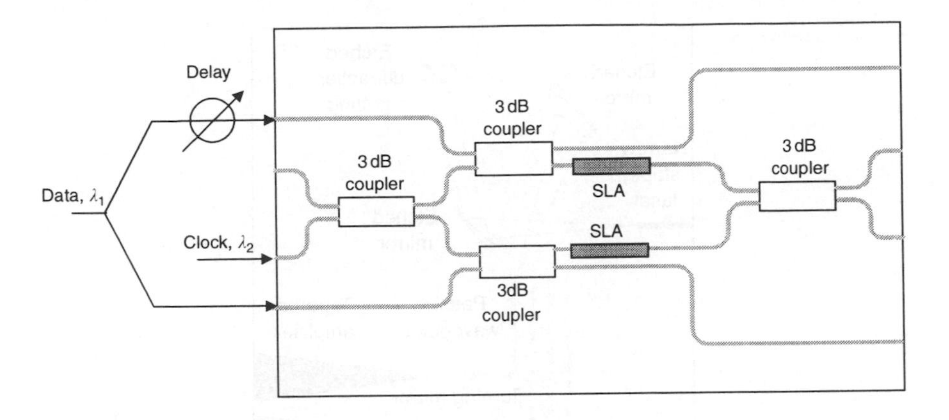

Fig. 18.1. Semiconductor laser diode amplifier (SLA)-based interferometer for 3R regeneration [42–45].

time-domain nature of the TLLM. The 3 dB couplers and MZI arms may simply be modelled by including coupling losses and relative phase delays in the TLLM. Both the uniform-width and tapered semiconductor amplifier structures may be investigated and compared in terms of their performance in wavelength conversion and optical signal regeneration techniques.

18.5.2.2 *Mode-locked laser design based on the multichannel grating cavity (MGC) laser*

The explosive growth of bandwidth demand owing to broadband distributive and interactive services can be met by combining wavelength division multiplexing (WDM) and optical time-division multiplexing (OTDM) to provide an even larger transmission capacity [41]. Therefore, multiwavelength ultrashort optical pulse sources have been proposed by using the intracavity spectral shaping technique [46, 47]. Alternatively, the multiwavelength grating cavity (MGC) laser is a potential source of multiwavelength picosecond pulses for such WDM/OTDM transmission systems [48]. Furthermore, the MGC laser can be used to perform novel network functions including space-switching, wavelength routing, simultaneous wavelength conversion, and demultiplexing [49–53]. The integrated MGC laser has been demonstrated [48] but there is very little reported on design considerations of this device.

As further work, it is proposed that the MGC laser device be modelled by combining a set of interconnectable bidirectional TLLM component

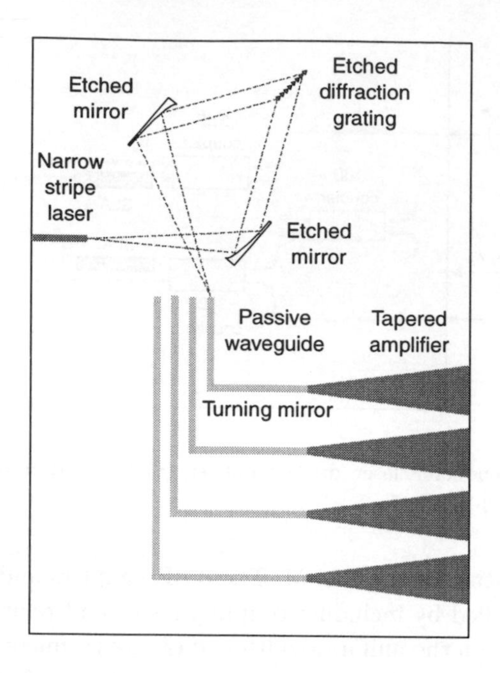

Fig. 18.2. Improved design of the multiwavelength grating cavity (MGC) laser.

models of the master stripe laser, slave stripe lasers, and the transmission grating. From the simulations we can identify the important design considerations, such as structural parameters, internal laser physical quantities, and optimum operating conditions. An improved design of the MGC laser is proposed here by using tapered waveguide semiconductor amplifiers as the slave (output) waveguides (see Fig. 18.2). By using the tapered amplifiers, we can generate picosecond pulses of higher peak powers, thus resulting in higher extinction ratios. The tapered amplifier also provides higher saturation power, thus minimising distortion during pulse amplification [30].

18.5.2.3 *Chirped fibre Bragg gratings for novel applications*

The TLM model of the distributed Bragg reflector (DBR) can be used in future work to simulate fibre Bragg gratings for applications in WDM networks and microwave photonics. The fibre Bragg grating laser has been demonstrated to exhibit low chirp and good temperature stability when modulated at speeds of up to 10 Gb/s [54]. By using the optical heterodyne technique, millimetre-wave signals with a tunable frequency span of up to

40GHz could also be generated using the fibre Bragg grating laser [55]. A novel mode-locked laser with a linearly chirped fibre Bragg reflector has also been demonstrated as a potential soliton pulse source with a wide operating frequency range because of its self-tuning mechanism [56].

The linearly *chirped* grating may be modelled as a set of transmission-lines of linearly increasing (or decreasing) length along the standard TLM model of the uniform grating [37]. The schematic diagram of the TLM model of the linearly chirped fibre Bragg grating is shown in Fig. 18.3. To simulate the linearly chirped Bragg grating, phase-stubs are again to be used to alter the phase length of each individual section of the uni-

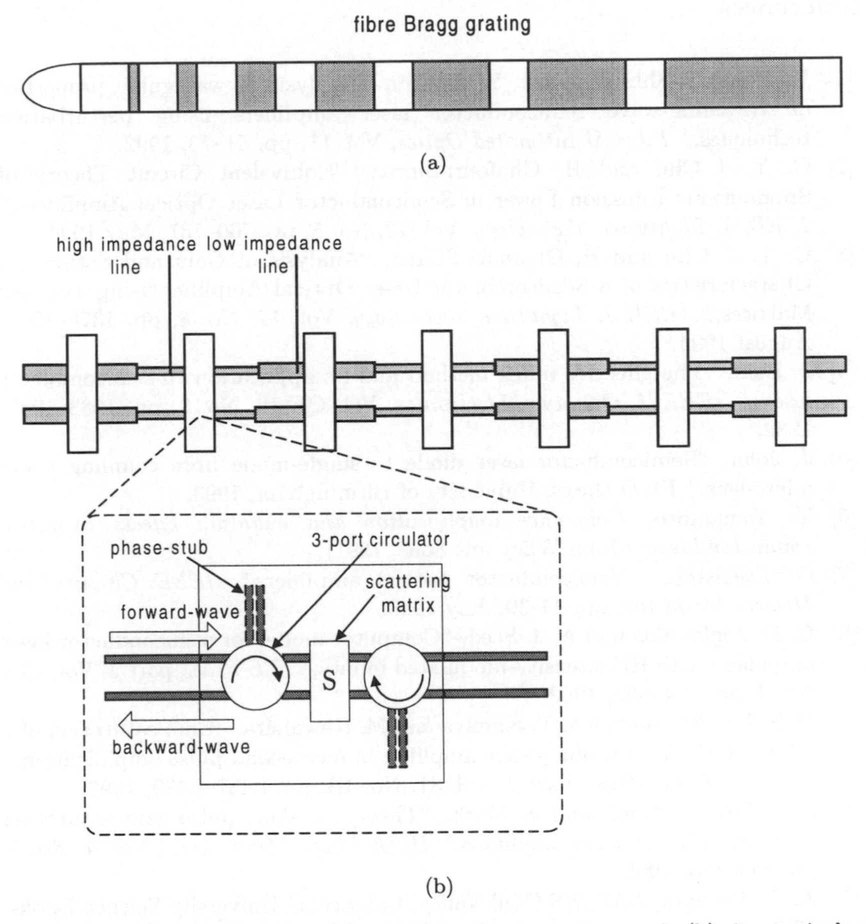

Fig. 18.3. (a) The linearly chirped fibre Bragg grating and (b) its equivalent transmission-line model.

form grating model. When a total chirp of $\Delta f(\mathrm{H_z})$ is required across the fibre grating of K sections, the phase length may be adjusted by $\Delta 1_{\mathrm{ph}}$ at section n, equivalent to a phase-stub impedance (Z_s) of [57]:

$$Z_s = Z_0 \tan\left(\frac{\pi n \Delta f}{K[\frac{1}{4\Delta t} - \Delta f]}\right) \tag{18.1}$$

where Z_0 is the characteristic impedance of the main transmission-line, and Δt is the model time step.

References

[1] H. Ghafouri-Shiraz and C. Y. J. Chu, "Analysis of waveguide properties of travelling-wave semiconductor laser amplifiers using perturbation techniques," *Fiber & Integrated Optics*, Vol. 11, pp. 51–70, 1992.

[2] C. Y. J Chu and H. Ghafouri-Shiraz, "Equivalent Circuit Theory of Spontaneous Emission Power in Semiconductor Laser Optical Amplifiers," *IEEE, J. Lightwave Technology*, Vol. 12, No. 5, pp. 760–767, May 1994.

[3] C. Y. J. Chu and H. Ghafouri Shiraz, "Analysis of Gain and Saturation Characteristics of a Semiconductor Laser Optical Amplifier using Transfer Matrices," *IEEE J. Lightwave Technology*, Vol. 12, No. 8, pp. 1378–1386, August 1994.

[4] J. Buss, "The effective index method and its application to semiconductor lasers," *IEEE J. Quantum Electronics*, Vol. QE-18, No. 7, pp. 1083–1089, 1982.

[5] J. John, "Semiconductor laser diode to single-mode fibre coupling using micro-lens," Ph.D thesis, University of Birmingham, 1993.

[6] Y. Yamamoto, *Coherence amplification and quantum effects in semiconductor lasers* (John Wiley ans Sons, 1991).

[7] G. Eisenstein, "Semiconductor optical amplifiers," *IEEE Circuits and Devices Magazine*, pp. 25–30, July 1989.

[8] C. D. Zaglanakis and A. J. Seed, "Computer model for semiconductor laser amplifiers with RF intensity-modulated inputs," *IEE Proc.*, part J, Vol. 139, No. 4, pp. 254–262, 1992.

[9] H. F. Liu, M. Tohyama, T. Kamiya and M. Kawahara, "Gain saturation of a 1.3 μm InGaAs travelling-wave amplifier in picosecond pulse amplification," *Japan J. Appl. Phys.*, Part 2, Vol. 31, No. 4B, pp. L487–L489, 1992.

[10] A. Uskov, J. Mϕrk and J. Mark, "Theory of short-pulse gain saturation in semiconductro laser amplifiers," *IEEE Photo. Tech. Lett.*, Vol. 4, No. 5, pp. 443–446, 1992.

[11] A. E. Siegman, *LASERS* (Mill Valley, California: University Science Books, 1986).

[12] M. Morishita, T. Ohmi and J. I. Nishizawa, "Impedance Characteristics

of Double-Heterostructure Laser Diodes," *Sol. Stat. Electron.*, Vol. 22, pp. 951–962, 1979.

[13] J. Katz, S. Margalit, C. Harder, D. Wilt and A. Yariv, "The Intrinsic Electrical Equivalent Circuit of a Laser Diode," *IEEE J. Quant. Electron.*, Vol. QE-17, No. 1, pp. 4–7, 1981.

[14] H. A. Haus, *Waves and fields in optoelectronics* (New Jersey: Prentice-Hall, Inc., 1984).

[15] A. J. Lowery, "Transmission-line modelling of semiconductor lasers: the transmission-line laser model," *Int. Jour. Num. Model.*, Vol. 2, pp. 249–265, 1989.

[16] L. M. Zhang and J. E. Carroll, "Large-Signal Dynamic Model of the DFB Laser," *IEEE J. Quant. Electron.*, Vol. 28, No. 3, pp. 604–611, 1992.

[17] M. G. Davis and R. F. Dowd, "A Transfer Matrix Method Based Large-Signal Dynamic Model for Multielectrode DFB Lasers," *IEEE J. Quant. Electron.*, Vol. 30, No. 11, pp. 2458–2466, 1994.

[18] C. F. Tsang, D. D. Marcenac, J. E. Carroll and L. M. Zhang, "Comparison between "power matrix mode" and "time domain model" in modelling large signal response of DFB lasers," *IEE Proceedings Pt. J*, Vol. 141, No. 2, pp. 89–96, 1994.

[19] S. Balsamo, F. Sartori and I. Monttrosset, "Dynamic beam propagation method for flared semiconductor power amplifiers," *IEEE Jour. Select. Top. Quant. Elec.*, Vol. 2, No. 2, pp. 378–384, 1996.

[20] W. I. Way, "Large Signal Nonlinear Distortion Prediction for a Single-Mode Laser Diode Under Microwave Intensity Modulation," *IEEE J. Lightwave Technol.*, Vol. LT-5, No. 3, pp. 305–315, 1987.

[21] R. S. Tucker and I. P. Kaminow, "High-Frequency Characteristics of Directly Modulated InGaAsP Ridge Waveguide and Buried Heterostructure Lasers," *IEEE J. Lightwave Technol.*, Vol. LT-2, No. 4, pp. 385–393, 1984

[22] R. S. Tucker, "Large-signal circuit model for simulation of injection-laser modulation dynamics," *IEE Proceedings Pt. I*, Vol. 128, pp. 180–184, 1981.

[23] R. S. Tucker and D. J. Pope, "Microwave Circuit Models of Semiconductor Injection Lasers," *IEEE Trans. Microwave Theory Tech.*, Vol. MTT-31, No. 3, pp. 289–294, 1983.

[24] A. J. Lowery, "New dynamic semiconductor laser model based on the transmission-line modelling method," *IEE Proceedings Pt. J*, Vol. 134, pp. 281–289, 1987.

[25] W. M. Wong and H. Ghafouri-Shiraz, "Integrated semiconductor laser-transmitter model for microwave-optoelectronic simulation based on transmission-line modelling," *IEE Proceedings Pt. J*, Vol. 146, No. 4, pp. 181–188, 1999.

[26] S. Maricot *et al.*, "Monolithic Integration of Optoelectronic Device with Reactive Matching Networks for Microwave Applications," *IEEE Photon. Tech. Lett.*, Vol. 4, No. 11, pp. 1248–1250, 1992.

[27] J. W. Bandler, P. B. Johns and M. R. M. Rizk, "Transmission-Line Modeling and Sensitivity Evaluation for Lumped Network Simulation and Design in the Time Domain," *Jour. Frank. Inst.*, Vol. 304, pp. 15–32, 1977.

[28] J. N. Walpole, "Semiconductor amplifiers and lasers with tapered gain regions," *Opt. Quant. Elec.*, Vol. 28, pp. 623–645, 1996.
[29] G. Bendelli, K. Komori and S. Arai, "Gain Saturation and Propagation Characteristics of Index-Guided Tapered-Waveguide Traveling-Wave Semiconductor Laser Amplifiers (TTW-SLA's)," *IEEE J. Quant. Electron.*, Vol. 28, No. 2, pp. 447–457, 1992.
[30] H. Ghafouri-Shiraz, P. W. Tan and T. Aruga, "Picosecond Pulse Amplification in Tapered-Waveguide Laser-Diode Amplifiers," *IEEE Jour. Select. Top. Quant. Elec.*, Vol. 3, No. 2, pp. 210–217, 1997.
[31] B. Zhu, I. H. White, K. A. Williams, F. R. Laughton and R. V. Penty, "High peak power picosecond optical pulse generation from Q-switched bow-tie laser with a tapered traveling wave amplifier," *IEEE Photon. Tech. Lett.*, Vol. 8, No. 4, pp. 503–505, 1996.
[32] S. El-Yumin, K. Komori, S. Arai and G. Bendelli, "Taper shape dependence of tapered waveguide traveling wave semiconductor laser amplifier," *Trans. IEICE*, Vol. E77-C, No. 4, pp. 624–632, 1994.
[33] W. M. Wong and H. Ghafouri-Shiraz, "Dynamic model of tapered semiconductor lasers and amplifiers based on transmission line laser modeling," *IEEE Jour. Select. Top. Quant. Elec.*, Vol. 6, No. 4, pp. 585–593, 2000.
[34] H. Ghafouri-Shiraz, P. W. Tan and W. M. Wong, "A Novel Analytical Expression of Saturation Intensity of InGaAsP Tapered Traveling-Wave Semiconductor Laser Amplifier Structures," *IEEE Photon. Tech. Lett.*, Vol. 10, No. 11, pp. 1545–1547, 1998.
[35] B. Dagens, S. Balsamo and I. Monttrosset, "Picosecond pulse amplification in AlGaAs flared amplifiers," *IEEE Jour. Select. Top. Quant. Elec.*, Vol. 3, No. 2, pp. 233–244, 1997.
[36] Author to provide
[37] A. J. Lowery, "Dynamic Modelling of Distributed-Feedback Lasers Using Scattering Matrices," *Electron. Lett.*, Vol. 25, No. 19, pp. 1307–1308, 1989.
[38] A. J. Lowery, "Integrated mode-locked laser design with a distributed Bragg reflector," *IEE Proceedings Pt. J*, Vol. 138, No. 1, pp. 39–46, 1991.
[39] Z. Ahmed, L. Zhai, A. J. Lowery, N. Onodera and R. S. Tucker, "Locking Bandwidth of Actively Mode-Locked Semiconductor Lasers," *IEEE J. Quant. Electron.*, Vol. 29, No. 6, pp. 1714–1721, 1993.
[40] L. Zhai, A. J. Lowery, Z. Ahmed, N. Onodera and R. S. Tucker, "Locking bandwidth of mode-locked semiconductor lasers," *Electron. Lett.*, Vol. 28, No. 6, pp. 545–546, 1992.
[41] I Maio, "Gain saturation in travelling-wave ridge waveguide semiconductor laser amplifiers," *IEEE Photo. Tech. Lett.*, Vol. 3, No. 7, pp. 629–631.
[42] N. Storkfelt, B. Mikkelsen, D. S. Olesen, M. Yamaguchi and K. E. Stubkjaer, "Measurement of carrier lifetime and line width enhance factor for 1.5 μm ridge-waveguide laser amplifier," *IEEE Photo. Tech. Lett.* Vol. 3, No. 7, pp. 632–634, 1991.
[43] T. Makino and T. Glinski, "Transfer matrix analysis of the amplified

spontaneous emission of DFB semiconductor laser amplifiers," *IEEE J. Quantum Electronics*, Vol. QE-24, No. 8,pp. 1507–1518, 1988.

[44] R. M. Fortenberry, A. J. Lowery and R. S. Tucker, "Up to 16 dB improvement in detected voltage using two section semiconductor optical amplifier detector," *Electronics Letters*, Vol. 28, No. 5, pp. 474–476, 1992.

[45] P. J. Stevens and T. Mukai, "Predicted performance of quantum-well GaAs-(GaA1) As optical amplifiers," *IEEE J. Quantum Electronics*, Vol. 26, No. 11, pp. 1910–1917, 1990.

[46] K. S. Jepsen B. Mikkelsen, J. H. Povlsen, M. Yamaguchi and K. E. Stubkjaer, "Wavelength dependence of noise figure in InGaAs/InGaAsP multiple-quantum-well laser amplifier," *IEEE Photo. Tec. Lett.*, Vol. 4, No. 6, pp. 550–553, 1992.

[47] K. Komori, S. Arai and Y. Suematsu, "Noise study of low-dimension quantum-well semiconductor laser amplifiers," *IEEE J. Quantum Electron.*, Vol. 28, No. 9, pp. 1894–1990, 1992.

[48] D. Tauber, R. Nagar, A. Livne, G. Eisenstein, U. Koren and G. Raybon, "A low-noise-figure 1.5m multiple-quantum-well optical amplifier," *IEEE Photo. Tech. Lett.*, Vol. 4, No. 3, pp. 238–240, 1992

[49] G. P. Agrawal and N. K. Dutta, *Long-wavelength semiconductor lasers* (New York, Van-Nostrad Reinhold, 1986).

[50] A Yariv, *Quantum Electronics*, 3rd Edition (John Wiley and Sons, 1989).

[51] C. Aversa and K. Iizuka, "Gain of TE-TM modes in quantum-well lasers," *IEEE I. Quantum Electron.*, Vol. 28, No. 9, pp. 1864–1873, 1992.

[52] G. Bjork and O. Nilsson, "A new exact and efficient numerical matrix theory of complicated laser structures: properties of asymmetric phase-shifted DFB lasers," *J. Lightwave Technology.*, Vol. LT-5, No. 1, pp. 140–146, 1987.

[53] G. Bjork and O. Nilsson, "A tool to calculate the linewidth of complicated semiconductor lasers," *IEEE J. Quantum Electronics*, Vol. QE-23, No. 8, pp. 1303–1313, 1987.

[54] M. G. Davis and R. F. O'Dowd, "A new large-signal dynamic model for multielectrode DFB lasers based on the transfer matrix method," *IEEE Photo. Tech. Lett.*, Vol. 4, No. 8, pp. 838–840, 1992.

[55] C. H. Henry and Y. Shani, "Analysis of mode propagation in optical waveguide devices by Fourier expansion," *IEEE J. Quantum Electron.*, Vol. QE-27, No. 3, pp. 523–530, 1991.

[56] Y. Imai, E. Sano and K. Asai, "Design and performance of wideband GaAs MMICs for high-speed optical communication systems," *IEEE Trans. Microwave Theory Tech.*, Vol. 40, No. 2, pp. 185–189, 1992.

[57] U. Koren *et al.*, "High power laser-amplifier photonic intergrated circuit for a 1.48 μm wavelength operation," *Appl. Phys. Lett.*, Vol. 59, No. 19, pp. 2351–2353, 1991.

[58] M. J. Connelly and R. F. O'Dowd, "Designing optically repeated links using a new semiconductor laser amplifier noise model," *Optical Amplifiers and their Applications Topical Meeting of OSA*, Vol. 13, pp. 184–187, 1990.

[59] M. J. Connely and R. F. O'Dowd, "Theory of signal degradation in semiconductor laser amplifiers with finite facet reflectivities," *IEEE J. Quantum Electron.*, Vol. 27, No. 11, pp. 2397–2403, 1991.

[60] K. Hinton, "Optical carrier linewidth broadening in a travelling wave semiconductro laser amplifier," *IEEE J. Quantum Electronics*, Vol. QE-26, No. 7 pp. 1176–1182, 1990.

[61] K. Kikuchi, C.E. Zah and T. P. Lee, "Measurement and analysis of phase noise generated from semiconductor optical amplifiers," *IEEE J. Quantum Electonics*, Vol. QE-27, No. 3, pp. 416–422, 1991.

[62] P. A. Humblet and M. Azizoglu, "On the bit error rate of lightwave systems with optical amplifiers," *IEEE J. Lightwave Technology*, Vo. LT-9, No. 11, pp. 1576–1582, 1991.

[63] D. Marcuse, "Calculation of bit-error probability for a lightwave system with optical amplifiers," *IEEE J. Lightwave Technology*, Vol. 9, No. 4, pp. 505–513, 1991.

Index